D1523140

STP 1245

Zirconium in the Nuclear Industry: Tenth International Symposium

Anand M. Garde and E. Ross Bradley, Editors

ASTM Publication Code Number (PCN):
04-012450-04

ASTM
1916 Race Street
Philadelphia, PA 19103
Printed in the U.S.A.

ISBN: 0-8031-2011-7
PCN: 04-012450-04
ISSN: 1050-7558

Copyright © 1994 AMERICAN SOCIETY FOR TESTING AND MATERIALS, Philadelphia, PA. All rights reserved. This material may not be reproduced or copied, in whole or in part, in any printed, mechanical, electronic, film, or other distribution and storage media, without the written consent of the publisher.

Photocopy Rights

Authorization to photocopy items for internal or personal use, or the internal or personal use of specific clients, is granted by the AMERICAN SOCIETY FOR TESTING AND MATERIALS for users registered with the Copyright Clearance Center (CCC) Transactional Reporting Service, provided that the base fee of $2.50 per copy, plus $0.50 per page is paid directly to CCC, 222 Rosewood Dr., Danvers, MA 01923; Phone: (508)750-8400; Fax: (508)750-4744. For those organizations that have been granted a photocopy license by CCC, a separate system of payment has been arranged. The fee code for users of the Transactional Reporting Service is 0-8031-2011/7-94 $2.50 + .50.

Peer Review Policy

Each paper published in this volume was evaluated by three peer reviewers. The authors addressed all of the reviewers' comments to the satisfaction of both the technical editor(s) and the ASTM Committee on Publications.

The quality of the papers in this publication reflects not only the obvious efforts of the authors and the technical editors, but also the work of these peer reviewers. The ASTM Committee on Publications acknowledges with appreciation their dedication and contribution to time and effort on behalf of ASTM.

Printed in Ann Arbor, MI
December 1994

Foreword

The Tenth International Symposium on Zirconium in the Nuclear Industry was held in Baltimore, MD on 21–24 June 1993. The sponsor of the event was ASTM Committee B-10 on Reactive and Refractory Metals and Alloys in cooperation with the Minerals, Metals and Materials Society.

The Symposium Chairman was A. M. Garde, ABB Combustion Engineering Nuclear Operations, and the Editorial Chairman was E. R. Bradley, Sandvik Special Metals. Serving as Editors of this publication were A. M. Garde and E. R. Bradley.

Contents

Overview—A. M. GARDE AND E. R. BRADLEY — xi

KROLL AWARD PAPERS

Development of Zirconium-Barrier Fuel Cladding—J. S. ARMIJO, L. F. COFFIN, AND H. S. ROSENBAUM — 3

Zirconium Alloy Performance in Light Water Reactors: A Review of UK and Scandanavian Experience—D. O. PICKMAN — 19

HYDROGEN EFFECTS

The Effect of Fluence and Irradiation Temperature on Delayed Hydride Cracking in Zr-2.5Nb—S. SAGAT, C. E. COLEMAN, M. GRIFFITHS, AND B. J. S. WILKINS — 35
Discussion — 60

A Study of the Hydrogen Uptake Mechanism in Zirconium Alloys—M. B. ELMOSELHI, B. D. WARR, AND S. McINTYRE — 62
Discussion — 78

Hydrogen Absorption Kinetics During Zircaloy Oxidation in Steam—D. CHARQUET, P. RUDLING, M. MIKES-LINDBÄCK, AND P. BARBERIS — 80
Discussion — 96

Scanning Electron Microscope Techniques for Studying Zircaloy Corrosion and Hydriding—D. I. SCHRIRE AND J. H. PEARCE — 98
Discussion — 115

Growth and Characterization of Oxide Films on Zirconium-Niobium Alloys—V. F. URBANIC, P. K. CHAN, D. J. KHATAMIAN, AND O.-T. T. WOO — 116
Discussion — 131

PRESSURE TUBES

Correlation Between Irradiated and Unirradiated Fracture Toughness of Zr-2.5Nb Pressure Tubes—P. H. DAVIES, R. R. HOSBONS, M. GRIFFITHS, AND C. K. CHOW — 135
Discussion — 167

Variability of Irradiation Growth in Zr-2.5Nb Pressure Tubes—R. G. FLECK,
    J. E. ELDER, A. R. CAUSEY, AND R. A. HOLT   168
    Discussion   182

Change of Mechanical Properties by Irradiation and Evaluation of the Heat-Treated Zr-2.5Nb Pressure Tube—M. H. KOIKE, T. AKIYAMA,
    K. NAGAMATSU, AND I. SHIBAHARA   183
    Discussion   201

On the Anisotropy of In-Reactor Creep of Zr-2.5Nb Tubes—A. R. CAUSEY,
    J. E. ELDER, R. A. HOLT, AND R. G. FLECK   202
    Discussion   220

Fabrication of Zr-2.5Nb Pressure Tubes to Minimize the Harmful Effects of Trace Elements—J. R. THEAKER, R. CHOUBEY, G. D. MOAN, S. A. ALDRIDGE,
    L. DAVIS, R. A. GRAHAM, AND C. E. COLEMAN   221
    Discussion   242

FABRICATION

Modeling of Damage in Cold Pilgering—J. L. AUBIN, E. GIRARD, AND
    P. MONTMITONNET   245
    Discussion   263

Mitigation of Harmful Effects of Welds in Zirconium Alloy Components—
    C. E. COLEMAN, G. L. DOUBT, R. W. L. FONG, J. H. ROOT, J. W. BOWDEN,
    S. SAGAT, AND R. T. WEBSTER   264
    Discussion   283

Fabrication Process of High Nodular Corrosion-Resistant Zircaloy-2 Tubing—H. ABE, K. MATSUDA, T. HAMA, T. KONISHI, AND M. FURUGEN   285
    Discussion   305

Corrosion Behavior of Zircaloy-4 Sheets Produced Under Various Hot-Rolling and Annealing Conditions—H. ANADA, K. NOMOTO, AND Y. SHIDA   307
    Discussion   326

Optimization of PWR Behavior of Stress-Relieved Zircaloy-4 Cladding Tubes by Improving the Manufacturing and Inspection Process—
    J.-P. MARDON, D. CHARQUET, AND J. SENEVAT   328
    Discussion   347

LITHIUM EFFECTS AND SECOND-PHASE PARTICLES

Experimental and Theoretical Studies of Parameters that Influence Corrosion of Zircaloy-4—Ph. BILLOT, J.-C. C. ROBIN, A. GIORDANO, J. PEYBERNÈS,
    J. THOMAZET, AND H. AMANRICH   351
    Discussion   376

Aqueous Chemistry of Lithium Hydroxide and Boric Acid and Corrosion of
  Zircaloy-4 and Zr-2.5Nb Alloys—N. RAMASUBRAMANIAN AND
    P. V. BALAKRISHNAN   378
  Discussion   398

Corrosion Behavior of Irradiated Zircaloy—B.-C. CHENG, R. M. KRUGER, AND
    R. B. ADAMSON   400
  Discussion   417

Correlation of Transmission Electron Microscopy (TEM) Microstructure
  Analysis and Texture with Nodular Corrosion Behavior for
  Zircaloy-2—B. J. HERB, J. M. McCARTHY, C. T. WANG, AND H. RUHMANN   419
  Discussion   436

Micro-Characterization of Corrosion-Resistant Zirconium-Based Alloys—
    T. ISOBE, Y. MATSUO, AND Y. MAE   437
  Discussion   449

An Experimental Investigation into the Oxidation of Zircaloy-4 at Elevated
  Pressures in the 750 to 1000°C Temperature Range—I. L. BRAMWELL,
    T. J. HASTE, D. WORSWICK, AND P. D. PARSONS   450
  Discussion   466

MECHANICAL PROPERTIES

Prediction of Creep Anisotropy in Zircaloy Cladding—R. A. PERKINS AND
    S.-H. SHANN   469
  Discussion   481

Microstructure and Properties of Corrosion-Resistant Zirconium-Tin-Iron-
  Chromium-Nickel Alloys—E. R. BRADLEY AND A.-L. NYSTRÖM   483
  Discussion   497

Fatigue Behavior of Irradiated and Unirradiated Zircaloy and Zirconium—
    S. B. WISNER, M. B. REYNOLDS, AND R. B. ADAMSON   499
  Discussion   520

Effect of Irradiation on the Microstructure of Zircaloy-4—D. GILBON AND
    C. SIMONOT   521
  Discussion   548

Irradiation Effect on Fatigue Behavior of Zircaloy-4 Cladding Tubes—
    A. SONIAK, S. LANSIART, J. ROYER, J.-P. MARDON, AND N. WAECKEL   549

Grain-by Grain Study of the Mechanisms of Crack Propagation During Iodine
  Stress Corrosion Cracking of Zircaloy-4—R. E. HADDAD AND
    A. O. DORADO   559
  Discussion   575

Oxide Characterization

Microstructure of Oxide Layers Formed During Autoclave Testing of Zirconium Alloys—B. WADMAN, Z. LAI, H.-O. ANDRÉN, A.-L. NYSTRÖM, P. RUDLING, AND H. PETTERSSON   579
Discussion   597

Corrosion Performance of New Zircaloy-2 Based Alloys—P. RUDLING, M. MIKES-LINBÄCK, B. LETHINEN, H.-O. ANDRÉN, AND K. STILLER   599
Discussion   613

Examinations of the Corrosion Mechanism of Zirconium Alloys—H. J. BEIE, A. MITWALSKY, F. GARZAROLLI, H. RUHMANN, AND H.-J. SELL   615
Discussion   642

On the Initial Corrosion Mechanism of Zirconium Alloy: Interaction of Oxygen and Water with Zircaloy at Room Temperature and 450°C Evaluated by X-Ray Absorption Spectroscopy and Photoelectron Spectroscopy—U. DÖBLER, A. KNOP, H. RUHMANN, AND H.-J. BEIE   644

How the Tetragonal Zirconia is Stabilized in the Oxide Scale that is Formed on a Zirconium Alloy Corroded at 400°C in Steam—J. GODLEWSKI   663
Discussion   683

John Schemel Award Paper

Oxidation of Intermetallic Precipitates in Zircaloy-4: Impact of Radiation—D. PÊCHEUR, F. LEFEBVRE, A. T. MOTTA, C. LEMAIGNAN, AND D. CHARQUET   687

In-Reactor Corrosion

Corrosion Optimized Zircaloy for Boiling Water Reactor (BWR) Fuel Elements—F. GARZAROLLI, R. SCHUMANN, AND E. STEINBERG   709
Discussion   722

In-Reactor Corrosion Performance of ZIRLO and Zircaloy-4—G. P. SABOL, R. J. COMSTOCK, R. A. WEINER, P. LAROUERE, AND R. N. STANUTZ   724
Discussion   743

Phenomenological Study of In-Reactor Corrosion of Zircaloy-4 in Pressurized Water Reactors—Y. S. KIM, K. S. RHEEM, AND D. K. MIN   745

Corrosion Behavior of Zircaloy-4 Cladding with Varying Tin Content in High-Temperature Pressurized Water Reactors—A. M. GARDE, S. P. PATI, M. A. KRAMMEN, G. P. SMITH, AND R. K. ENDTER   760
Discussion   778

Effects of Pressurized Water Reactor (PWR) Coolant Chemistry on Zircaloy Corrosion Behavior—T. KARLSEN AND C. VITANZA 779

Author Index 791

Subject Index 793

Overview

This volume contains papers presented at the Tenth International ASTM Symposium on Zirconium in the Nuclear Industry, held in Baltimore, MD in June of 1993. The symposium was attended by 200 zirconium experts from 16 countries. The symposium consisted of seven platform sessions and a poster session. The platform session titles, and cochairmen for each session were as follows:

- Hydrogen Effects—D. G. Franklin and F. Garzarolli
- Pressure Tubes—B. A. Cheadle and D. Pickman
- Fabrication—E. Tenckhoff and J. B. Narayan
- Lithium Effects and Second-Phase Particles—G. P. Sabol and C. Lemaignan
- Mechanical Properties—R. B. Adamson and G. D. Moan
- Oxide Characterization—B. Cox and C. M. Eucken
- In-reactor Corrosion—L. van Swam and P. Rudling

Without the thorough technical review of each manuscript performed by the 34 reviewers, publication of these proceedings would have been impossible. The best paper presented at the platform session of the symposium was selected for the John Schemel award on the basis of technical merit, originality, and general presentation. The paper selected for the John Schemel award for the Tenth Zirconium Symposium is entitled "Oxidation of Intermetallic Precipitates in Zircaloy-4: Impact of Irradiation," by D. Pêcheur, F. Lefebvre, A. T. Motta, C. Lemaignan and D. Charquet. The award will be presented at the Eleventh Zirconium Symposium, Sept. 1995, in Germany.

This volume also contains papers written by the recipients of two Kroll awards presented at the Symposium. The Kroll papers are "Development of Zirconium-Barrier Fuel Cladding," by J. S. Armijo, L. F. Coffin, and H. S. Rosenbaum and "Zirconium Alloy Performance in Light Water Reactors: A Review of UK and Scandinavian Experience," by D. O. Pickman. The editors acknowledge the cooperation of the Kroll Award Committee Chairman, G. P. Sabol, in facilitating publication of the Kroll papers. Twenty papers were presented as posters in the poster session. The titles and authors of the poster presentations are listed in the Appendix following this section.

The recent trends in the nuclear industry towards higher fuel discharge burnups for better fuel use and higher heat ratings and coolant temperatures for better thermal efficiency have significantly increased the technical interest on corrosion of zirconium alloys. Accordingly, papers relating to corrosion had a dominant position at the symposium. It is clear that the corrosion resistance of the current zircaloys will not be adequate for these more aggressive service conditions. Consequently, considerable effort has been devoted to evaluating the effects of processing history and alloying elements on the corrosion resistance of zircaloys and also to new alloys outside the composition range of conventional materials for nuclear reactor applications.

Four separate sessions were devoted to various aspects of zirconium alloy corrosion: Hydrogen Effects, Lithium Effects and Second-Phase Particles, Oxide Characterization, and In-Reactor Corrosion. The remaining symposium sessions were: Pressure Tubes, Fabrication, and

Mechanical Properties. The papers in this volume are organized by the topical sessions in which they were presented.

The technical information contained in this book is valuable to zirconium alloy producers, nuclear fuel fabricators, reactor materials designers and development engineers, utility plant operators, and regulators. The data are useful in achieving safe, economic, and efficient power generation from the nuclear energy.

Hydrogen Effects

Hydrogen pickup and its effect on the mechanical and corrosion behavior of the zirconium alloy components used in the nuclear industry is becoming an important issue. The low hydrogen uptake of Zr-2.5Nb is associated with the β-Zr phase that contains 20% Nb. This observation supports the relationship between the electronic conductivity of the oxide and hydrogen pickup by the underlying metal. The delayed hydrogen cracking velocity of Zr-2.5Nb cold-worked material increases due to irradiation and K_{ID} is reduced by radiation hardening. For Zircaloy, the hydriding rate is proportional to the corrosion rate and hydrogen absorption depends on the final heat treatment, integrated annealing parameter, tin level, Ni level, Cr Level, and Fe/Cr ratio. The hydrogen pickup fraction is not constant and appears to be inversely proportional to the extent of corrosion. A scanning electron microscopic technique (SEM) applicable to both irradiated and unirradiated zirconium alloys was described to study the distribution and morphology of hydrides in zirconium alloys. It was proposed that hydroxyl groups are probably picked up by the oxide layer, mainly at the grain boundaries of the outer porous layer, and that protons are the species permeating the dense barrier oxide layer to charge hydrogen in the underlying metal layer.

Pressure Tubes

Eighty-seven of the world's reactors that either have been built or are under construction use pressure tubes rather than pressure vessels to contain the hot pressurized coolant. Zircaloy-2 was used for the pressure tubes in the early reactors. Currently, Zr-2.5Nb, an alloy that is stronger than Zircaloy-2, is used for the pressure tubes. The metallurgical condition of the Zr-2.5Nb tubes varies in different countries: cold worked and annealed (Russia); cold worked and stress relieved (Canada); and water quenched, cold worked, and aged (India and Japan).

The main concern for pressure tubes is their service life. The hot pressurized water and neutron flux produce an environment of stress, temperature, corrosion, and lattice damage. These result in changes in dimensions (creep and growth); changes in mechanical properties (increases in strength and DHC velocity, decreases in ductility and fracture toughness); and changes in chemical composition (hydrogen ingress).

The papers in this book that are associated with the pressure tube materials addressed all of these concerns. In CANDU pressure tubes, there is an initial decrease in fracture toughness with fluence and then little further change; this correlates well with the dislocation density. There is also a direct relationship between unirradiated and irradiated fracture toughness. Hence, from tests on new pressure tubes, their behavior during service can be predicted. In addition, it has been found that chlorine, which is an impurity element, reduces the fracture toughness if there is more than 1 ppm present in the material. The Fugen heat-treated pressure tubes had similar increases in strength and reductions in ductility and fracture toughness with fluence as the CANDU tubes. This is surprising as their microstructures are very different. However, the magnitude of the changes may be different and since different fracture toughness specimens were used, the results are difficult to compare directly. The irradiation growth behavior of small specimens from CANDU pressure tubes is quite variable. A key parameter has been found to be the iron concentration; increasing iron from 500 to 2000 ppm reduces the growth by about half. It is suggested that the volume expansion due to hydride precipitation can contribute to the irradiation growth strain. Biaxial creep behavior was reported for Zr-2.5Nb tubes up to fluences of $7 \cdot 10^{25}$ n·m^{-2}, showing the effect of texture on diametral strain rates. Advances have been made in the modeling of in-reactor creep, and the latest model takes into account grain interaction effects and shows how the texture can be optimized for the best in-reactor deformation behavior.

The big advance in pressure tube technology in the last few years has been the work showing the importance of 'impurity' elements and texture on the properties of the tubes. Reducing chlorine, carbon, and phosphorous increases fracture toughness, and increasing the iron decreases growth. In addition, careful attention to fabrication can considerably reduce the hydrogen concentration and control the distribution of oxygen between the α and β phases.

Fabrication

Fabrication process optimization for zirconium alloy components is important for two reasons. Firstly, the in-reactor performance (specifically the in-reactor corrosion resistance) of zirconium alloys strongly depends on the manufacturing history. Secondly, with the objective of 'zero defect' in-reactor performance, there is a continuing drive to eliminate the low-frequency manufacturing defects that contribute to the in-reactor failures. Important fabrication parameters enhancing the uniform corrosion resistance of Zircaloy-4 include higher annealing temperature immediately after the hot rolling step, low (<30%) second phase particle density in finished tube microstructure for fine (<100 nm) particles, Fe/Cr ratio of ~1.9, Si level of ~70 ppm, high integrating annealing parameter, and low β quenching rate. Lowering of the tin content and increase of stress relief annealing temperature are beneficial to improve the in-PWR corrosion resistance. The stress corrosion cracking resistance is improved by flush pickling of the cladding tube inside surface. Irradiation growth is controlled by limiting the carbon level below 200 ppm. The nodular corrosion resistance of Zircaloy-2 is enhanced by introduction of an additional quench step from the ($\alpha + \beta$) phase region followed by only two tube reduction steps with higher value of Q. The severity and frequency of surface defects occasionally observed in cold pilgered Zircaloy-4 tubes are correlated to a damage function based on stresses and strains generated in the material during the pilgering operation.

Welding of zirconium alloy tubes introduces changes in texture within the weld region and also residual tensile stresses. These factors decrease the burst ductility, burst strength of the material, and promote delayed hydrogen cracking. To mitigate these harmful effects of weld, the tube wall thickness at the weld location is increased by 25% and the welded tube is subjected to a stress-relief annealing treatment.

Lithium Effects and Second-Phase Particles

Lithium is added to the PWR coolant to decrease the possibility of crud deposition on the fuel rods, and thereby reducing the transfer of activated products to the rest of the plant. While higher levels of lithium in the primary coolant are expected to lead to lower radiation levels in the secondary circuit components, possible concerns exist regarding (a) higher fuel cladding corrosion rate with higher lithium levels and (b) stress corrosion cracking of steam generator tubing. There are also indications that boron added to the primary coolant to control the reactivity would negate the lithium effect of corrosion acceleration of zirconium alloys. Two papers dealing with the lithium effect on zirconium alloy corrosion are included in this section of the book. (A third paper related to the lithium effect is part of the in-reactor corrosion section of the book.) The corrosion model in one paper predicts up to 30% corrosion enhancement at

50 GWd/MTU burnup for Zircaloy-4 irradiated in a high-temperature PWR when the lithium concentration in the coolant is increased from 2.2 to 3.5 ppm. The model considers the lithium-boron coordinated chemistry with a pH of 7.2 and 7.4 for these lithium levels. In another paper, the acceleration of corrosion in lithiated water is correlated with undisassociated LiOH forming OLi groups on the oxide surface. In the presence of boron, the OLi groups are removed and thereby the corrosion rate decreases. The effect of lithium on corrosion of Zr-2.5Nb appears to be stronger than that on corrosion of Zircaloy-4.

The second phase particles in zirconium alloys have a strong effect on their uniform and nodular corrosion resistance. The important parameters in this regard are the particle size, particle composition, particle crystallanity, inter-particle distance, and so forth. Dissolution of second-phase particles in Zircaloys during irradiation is proposed to accelerate uniform corrosion at high burnups while eliminating nodular corrosion. The size of the second-phase particles depends on the processing history of the alloy. A higher amount of deformation during the tube reduction of Zircaloy-2 produces higher radial textured tubing with superior nodular corrosion resistance. A small precipitate size and short inter-particle distance also leads to superior nodular corrosion resistance. Analysis of second-phase particles in a new promising Zr-Sn-Fe-Cr-Nb alloy indicates a significantly lower solubility for niobium in this alloy compared to the Zr-Nb binary alloy. The new alloy has superior corrosion resistance and significantly lower hydrogen uptake than Zircaloy-4, while its mechanical properties are comparable to those of Zircaloy-4.

Mechanical Properties

Mechanical properties of irradiated zirconium alloy components strongly influence their in-reactor performance. Some of the mechanical properties of interest are creep, fatigue, stress corrosion cracking, embrittlement, delayed hydrogen cracking, and so forth. These properties depend on the evolution of the material microstructure due to the neutron bombardment. The phases present in the microstructure also depend on the material chemical composition and the fabrication history. The irradiation growth of the component that is an indication of the microstructural stability also depends on the above mentioned factors. Two important areas of technical interest are the correlation of in-reactor behavior to the laboratory behavior and modeling of mechanical properties to satisfy design requirements and mitigate operational concerns. The papers included in this proceeding address some of the points listed above.

Investigations related to the fatigue behavior of Zircaloys and zirconium show that irradiation does not affect the low-cycle fatigue behavior but decreases fatigue life in the high-cycle regime. This decrease in the fatigue life is related to strain localization and saturates with the extent of radiation damage. Fatigue crack growth rate is insensitive to the radiation damage but increases in the presence of water. Moreover, the increase is greater at higher oxygen content of the water. Microstructural evaluation of irradiated zircaloy shows that the breakaway growth in recrystallized and β-quenched Zircaloy at high fluences is associated with the generation of basal plane c-component dislocations. The irradiation induced changes in the second-phase particles (crystalline to amorphous transformation, dissolution, new phase particle precipitation) are shown to be temperature dependent. An elegant technique to study the stress corrosion crack propagation is described. Results confirm that the pseudo-cleavage of zirconium alloys produced during iodine SCC occurs along the basal planes, and cracking starts in those grains with maximum tensile stress on the basal plane. In another paper, the mechanical and corrosion properties of a new promising alloy are investigated as a function of the alloy composition. Possible correlations between anisotrophy parameters of the irradiated and unirradiated states of Zircaloy-2 materials with different fabrication history are also discussed.

Oxide Characterization

Several aspects of the oxide layer morphology affect the corrosion resistance of the substrate zirconium alloy. These features include the phases present and their crystallanity, different sublayers formed in the oxide layer (including the layer next to the metal oxide interface called the "barrier layer"), the grain structure and the grain boundary orientations in the different sublayers, stresses developed in different sublayers during oxidation, porosity in different sublayers, chemical concentration of coolant additives within the oxide pores (for example, lithium hideout), and incorporation of the second-phase particles present in the substrate metal into the oxide layer as the oxidation proceeds. Recent results associated with most of the phenomena listed above are presented in papers included in this publication.

New information presented at the symposium is summarized below. The initial reaction rate of clean zirconium surface with H_2O molecules appears to be about one thousand times slower than the reaction rate with oxygen molecules. The laser Raman microprobe results indicate that at the metal-oxide interface, up to 40% of the oxide may be in the tetragonal phase. However, transmission electron microscopy (TEM) of the metal-oxide interface does not show the tetragonal phase, possibly due to a phase transformation associated with the relaxation of compressive stress during the specimen preparation. The stresses generated in the oxide and the proportion of tetragonal oxide near the metal-oxide interface vary cyclically in step with the cycles of oxidation kinetics. The magnitude of stresses developed in the oxide films may be related to the corrosion resistance of the substrate alloy. The oxidation of the intermetallic precipitates (as they are incorporated in the oxide layer) appear to modify the local stresses due to the volume change associated with such oxidation. The oxidation process may leave behind part of the iron content of the precipitate as iron particles in the oxide. This evolution of iron may have a major impact on the long-term in-reactor corrosion of zirconium alloys.

Morphological studies of oxide films continue to show that columnar ZrO_2 crystallites are associated with good oxidation resistance, and equiaxed microcrystallites are associated with nodular corrosion behavior and with more extensive cracking in post-transition uniform oxides. TEM results clearly show that the "barrier" layer at the metal-oxide interface is crystalline rather than amorphous. Interconnected porosity is observed at the crystallite boundaries in oxides formed in LiOH solutions accompanying the higher oxidation kinetics. The effects of the minor elements in the Zircaloys are only just becoming evident in terms of optimum Fe:Cr ratios in the intermetallics and the incorporation of Si in $Zr_2(Fe/Ni)$ precipitates, that may provide a mechanism for the effect of Si on hydrogen uptake properties. If this incorporation of Si can modify the hydrogen absorption or recombination properties of the intermetallic, then it would probably also affect hydrogen uptake. In a similar manner, the precise stoichiometry of the $Zr(Fe/Cr)_{2+x}$ phase may also have an impact on hydrogen absorption.

In-Reactor Corrosion

Corrosion resistance of zirconium alloys irradiated in both BWRs and PWRs to extended burnups is being investigated. The extended burnup data presented in this publication show new trends not apparent from the low-burnup data previously available. For optimum cladding corrosion resistance for BWR applications at both the low- and high-burnups, an intermediate range of second-phase particle size and beta quench rates are suggested. Also, for high-burnup BWR cladding applications, higher levels of Sn and Cr in Zircaloy-2 appear to be beneficial. For the high-burnup PWR application, reducing the tin level in Zircaloy-4 is shown to be beneficial. A superior in-PWR performance data (lower corrosion and lower irradiation growth compared to conventional Zircaloy-4) are presented for a Zr-Sn-Nb-Fe alloy.

Regarding the ex-reactor autoclave tests with good correlation to the in-reactor corrosion behavior, the following information is presented. For an indication of Zircaloy-2 corrosion performance in BWRs, a long-term water test at 300°C appears to be more appropriate than the 520°C steam test. For an indication of corrosion performance of zirconium alloys in PWR, a 360°C water test (pure water or with 70 ppm lithium addition) gives better correlation than the 400°C steam test.

The in-pile tests in the PR loop at Halden indicate no acceleration of corrosion rate of Zircaloy-4 when the lithium level in coolant was increased at 4.5 ppm. This result contradicts the results of loop testing in France. In the past few years, conflicting results are reported in the literature regarding the effect of lithium in the coolant on the in-PWR corrosion resistance of Zircaloy-4. The results vary from no effect to a 30% corrosion acceleration. It is clear that more work is needed to determine conditions under which lithium accelerates the in-reactor corrosion rate of zirconium alloys.

A comparison of the corrosion behavior of a defective rod and adjacent nondefective rod irradiated for two cycles in a PWR indicates a significantly higher corrosion rate in the defective rod. This corrosion rate acceleration appears to be due to the hydride precipitates at the metal oxide interface in the defective rod.

Anand M. Garde

ABB Combustion Engineering Nuclear Operations;
Windsor, CT 06095;
Symposium Chairman and STP Editor.

E. Ross Bradley

Sandvik Special Metals;
Kennewick, WA 99336;
Symposium Editorial Chairman and STP Editor.

APPENDIX

LIST OF POSTER PRESENTATIONS
10TH ASTM ZIRCONIUM SYMPOSIUM
BALTIMORE, MD; JUNE 1993

Hydride Characterization in Irradiated Zirconium Alloys

J. Thomazet – FRAGEMA, Lyon, France
F. Barcelo – Commissariat a l'Energie Atomique, Gif Sur Yvette, France
M. Trotabas
D. Gilbon
J. Y. Blank

Evaluation of Oxide Layer Thickness on Zircaloy-4 Tube by Electrochemical Impedance Spectroscopy

A. Frichet – PECHINEY, C. R. Voreppe, France
J. P. Mardon – FRAGEMA, Lyon, France
J. Senevat – ZIROTUBE, Paimboeuf, France

Correlation Study Among Corrosion Properties, Microstructure, and Additional Heat Input History of Zircaloy-2 Tubing

Y. Suzuki – Japan Nuclear Fuel Co., Kanagawa, Japan
T. Iwasaki
Y. Ito

X-Ray Small Angle Scattering of the Small Precipitates in Zircaloy

C. Miyake – Osaka University, Osaka, Japan
M. Uno

T. Kamiyama – Tohok University, Miyagi, Japan
K. Suzuki

Electron Spin Resonance Study of Oxide Films on Zircaloy Cladding

C. Miyake – Osaka University, Osaka, Japan
M. Uno
G. Abe

Magnetic Observation on Corrosion Susceptibility of Zircaloy Cladding

C. Miyake – Osaka, University, Osaka, Japan
G. Abe

M. Shimamoto – Nuclear Fuel Industry, Osaka, Japan
T. Kimura

Study on the Zirconium-Hydrogen System

S. Yamanaka – Osaka University, Osaka, Japan
K. Higuchi
M. Miyake

Observation on Corrosion Oxide Film Formed on Zirconium Alloy

T. Kimura – Nuclear Fuel Industry, Osaka, Japan
K. Kawanishi
M. Shimamoto
T. Okada

Interaction Between Factors Acting on the Uniform Corrosion Behavior of Zircaloy-4

D. Charquet – CEZUS, Ugine, France

Effect of Hydriding on Mechanical Behavior of Zircaloy-4 Structural Components of the Fuel Assembly Under Various Conditions

F. Prat – FRAGEMA, Lyon, France
J. Bard

E. Andrieu – Ecole Des Mines, Paris, France

Characterization of Post-Transition Corrosion Kinetics of Stress-Relieved, Recrystallized and Beta Quenched Zircaloy-4 Tubes

G. Brun – Commissariat a l'Energie Atomique, Gif Sur Yvette, France
J. Blanchet
P. Julia
J. P. Martinetti
C. Blain

Effects of Hydriding and Oxidizing on the Mechanical Properties of Unirradiated Zircaloy-4 Cladding Tubes

R. Limon – Commissariat a l'Energie Atomique, Gif Sur Yvette, France
J. Pelchat
R. Maury

J. P. Mardon – FRAGEMA, Lyon, France

Coralline Facility Study on Irradiation Behavior Under PWR Conditions of Zirconium-Based Alloys to be Used in the Fuel Assembly Structure

F. Lefebvre – Commissariat a l'Energie Atomique, Gif Sur Yvette, France
C. Millet
R. Limon

J. Bard – FRAGEMA, Lyon, France
V. Rebeyrolle

Long-Term Out-of-Pile Corrosion of Zirconium Alloys

R. Bordoni – Comision Nacional de Energia Atomica, Buenos Aires, Argentina
M. A. Blesa
A. M. Iglesias
A. J. G. Maroto
A. M. Olmedo

G. Rigotti – Universidad Nacional de La Plata, La Plata, Argentina
M. Villegas

Contribution of the Beta Flux to Corrosion Enhancement of Zr Alloys Under Irradiation

R. Salot – Commissariat a l'Energie Atomique, Grenoble, France
I. Schuster
F. Lefevre
C. Lemaignan

Superimposed Effects of Grain Shape Anistropy and Crystallographic Texture on Anisotropic Biaxial Creep of Zircaloy Cladding

K. L. Murty – North Carolina State University, Raleigh, North Carolina
J. C. Britt

Y. S. Kim – Korean Atomic Energy Research Institute, Taejon, South Korea
Y. H. Jung

Mechanism of Oxidation of Zircaloy Fuel Cladding in Pressurized Water Reactors

K. Forsberg – ABB Atom AB, Vasteras, Sweden
M. Limback
A. Massih

Evaluation of Non-Destructive Hydrogen Detection Methods in Zirconium Alloys

L. Goldstein – S.M. Stoller Corp., Pleasantville, New York
R. Klein
A. A. Strasser

The Ductile-Brittle Transition of Zircaloy-4 Due to Hydrogen

J-H. Huang – National Tsing Hua University, Hsinchu, Taiwan
S. P. Huang
C. S. Ho

Influence of Irradiation on Iodine Stress Corrosion Cracking Behavior of Zircaloy-4

I. Schuster – Commissariat a l'Energie Atomique, Grenoble, France
C. Lemaignan

J. Joseph – FRAMATOME, Lyon, France

Kroll Award Papers

Joseph Sam Armijo,[1] *Louis F. Coffin,*[2] *and Herman S. Rosenbaum*[3]

Development of Zirconium-Barrier Fuel Cladding

REFERENCE: Armijo, J. S., Coffin, L. F., and Rosenbaum, H. S., "**Development of Zirconium-Barrier Fuel Cladding,**" *Zirconium in the Nuclear Industry: Tenth International Symposium, ASTM STP 1245,* A. M. Garde and E. R. Bradley, Eds., American Society for Testing and Materials, Philadelphia, 1994, pp. 3–18.

ABSTRACT: This paper was prepared for the 1991 Kroll Award. A review is presented of the development of barrier fuel. It includes the recognition of the pellet-cladding interaction (PCI) fuel failure mode and of a coordinated program to develop understanding, mitigating strategies, and a fuel that is resistant to this failure mode. The efforts to understand PCI led to the conclusion that the dominant mechanism is stress-corrosion cracking of the Zircaloy. The invention and development of zirconium-barrier fuel was intended to provide a materials solution to this fuel failure mode. This review includes the work to understand the failure mechanism as well as the program to develop PCI-resistant fuel designs. Ultimately, the zirconium-barrier fuel was tested in power ramps to ascertain and to quantify the resistance to PCI under expected service conditions in commercial boiling water reactors (BWRs). The program that led to a large-scale demonstration in a commercial power plant (Quad Cities-2) is described briefly. Subsequent to that, program work continued with in-reactor load following and experiments in a test reactor on power cycling of barrier fuel. Finally, the performance of failed fuel is discussed briefly.

KEY WORDS: nuclear fuel cladding, zirconium, pellet-cladding interaction, large-scale demonstrations, in-reactor fuel tests, in-reactor fuel performance, zirconium alloys, nuclear materials, nuclear applications, radiation effects

On this occasion, marking both the ASTM Tenth International Symposium on Zirconium in the Nuclear Industry and our receipt of the 1991 William J. Kroll Medal, we accept this high honor with a deep sense of appreciation. The list of previous Kroll Medal recipients is truly impressive; we consider ourselves very fortunate to be among them. Furthermore, this series of symposia represents a very high standard of materials research and technology.

In this paper, we review the development of the zirconium-barrier fuel as a science and technology effort and ultimately as a widely accepted commercial product. As of September 1992, GE has supplied six initial cores and 163 reloads of barrier fuel. This corresponds to >1.7 million fuel rods, of which >1.4 million had completed more than one cycle of operation. The lead exposure bundle was at 43 MWd/kgU (bundle average). Other vendors in the industry have also offered barrier fuel and have sold both lead use assemblies and full reloads. Recently, product variations have come into the market featuring low-alloy versions of the zirconium-barrier fuel design.

---

[1] General manager, Nuclear Fuel, GE Nuclear Energy, Wilmington, NC 28402.

[2] GE Corporate Research and Development (retired); now, distinguished research professor, Department of Mechanical Engineering, Aeronautical Engineering and Mechanics, Rensselaer Polytechnic Institute, Troy, NY 12181.

[3] Senior program manager, Fuel Technology, GE Nuclear Energy, San Jose, CA 95125.

The development of barrier fuel came from effective cooperation between the commercial nuclear fuel business of GE and the GE Corporate Research and Development Laboratories (GE R&D); subsequently, there developed a cooperative program among the U.S. Department of Energy, the Commonwealth Edison Company, and GE. In addition, supporting tests and analyses were done with cooperation of the U.S. Nuclear Regulatory Commission, the Japan Atomic Research Institute, Toshiba, and Hitachi. We are indeed fortunate to have been able to garner this technological and financial support.

Our review will start with the discovery of the problem, pellet-cladding interaction (PCI). Then we will proceed to the recognition that it is intrinsic to the fuel rod designs comprising urania fuel and Zircaloy cladding. We will then review the invention of the zirconium-barrier fuel design, the program which qualified that design for commercial use in boiling water reactors (BWRs), and finally, we will discuss some recent information about its performance and use, including recent concern about the degradation of failed fuel.

The PCI Phenomenon and Efforts to Cope

In 1964, while engaged in tests to explore the practical performance limits of light water reactor fuel with respect to fuel rod power and heat flux, failures occurred in two test fuel rods operated at extremely high power, $\approx$180 kW/m ($\approx$55 kW/ft) [1]. Gamma scans showed the presence of fission product iodine (and tellurium) near the axial location of these "pinhole" cladding breaches. From this observation, the question of whether fission product iodine might not react with the cladding to form a zirconium iodide was suggested. Even the possibility that fission-product iodine might be recycled to the hotter central region of the fuel column to decompose as in the well-known van Arkel-de Boer process used for zirconium purification was also considered. By such a mechanism even a small amount of iodine might adversely affect fuel cladding perfomance. This thought stimulated some simple experiments of iodine with Zircaloy in which it was discovered that in the combined presence of iodine vapor and a stress at 300 to 400°C, Zircaloy cracked [2,3]. Cracks occurred before there was significant evidence of the expected general or pitting corrosion, or both.

At this point, we acknowledge the recent literature reviews by our fellow Kroll Medal recipient, Professor Brian Cox, on both the PCI fuel failure phenomenon [4] and the stress-corrosion cracking of zirconium alloys [5]. Within GE, testing in 1966 of some early commercial BWR fuel at the GE Test Reactor (GETR) at the Vallecitos Nuclear Center in California led Rowland [6] to correlate fuel failures with power increase. This fuel was from Dresden-1 at burnup 10 to 15 MWd/kgU. Rowland also cited the circumstantial evidence linking these unexpected fuel failures in commercial BWRs with a possible stress-corrosion cracking (SCC) mechanism [7]. By 1971, it was recognized within GE that the PCI failure of BWR fuel was a potentially serious operational problem, and a coordinated effort was started involving the scientific and engineering capabilities of both the GE Nuclear Energy Business in San Jose and in Pleasanton, California, and the GE Corporate Research and Development Center in Schenectady, New York. See the chronology in Table 1.

By 1972, we had fractographic evidence linking the PCI failures with SCC. This was based on laboratory testing and on correlation of fracture morphologies of failed fuel rod cladding from commercial reactors with those in laboratory tests where the conditions were well controlled. In analyzing the fracture morphologies, painstaking stereological fractographic work was done to assure that the fracture surface had nucleated at and was interconnected with the cladding inner surface, as is expected in SCC.

During this period, PCI fuel failures occurred in commercial reactors and were documented in test reactors. Initial hypotheses that failures were due to minor variations in cladding

TABLE 1—*Chronology of GE experience with PCI and development of zirconium-barrier fuel.*

First evidence that I_2 causes stress corrosion cracking of Zircaloy	1965
Power ramp tests on fuel rods from Dresden-1 documenting that failures occur on power increase	1966
GE interpretive review linking power history to fuel failures	1969
PCI fuel failures in commercial reactor	1971
Intensive PCI study and development effort started	1971
Fractographic evidence linking PCI with stress corrosion cracking	1972
Zirconium-barrier fuel invented (substantially pure)	1973
Plant operating rules implemented to avoid PCI fuel failures	1973
Unequivocal fractographic evidence for stress corrosion cracking (incipient crack in unfailed fuel rod)	1974
Fission product species tested; I_2 and Cd cause stress corrosion cracking	1974
Segmented rod bundles to test potential remedies placed in commercial reactors: Quad Cities-1, Monticello, Millstone	1974
Potential solutions evaluated	1975
Accelerated PCI in-reactor ramp tests indicate zirconium-barrier resistant to PCI	1976
Moderate purity zirconium-barrier invention	1976
Joint GE/CECo/DOE demonstration program	1977–1984
Lead test assemblies irradiation began at Quad Cities-1	1979
Defected rod test (in-reactor)	1979
Loss of coolant accident tests (laboratory)	1980
Multiple ramp tests	1980–1983
Start of demonstration irradiation in Quad Cities-2 (144 bundles)	1981
First demonstration ramp at Quad Cities-2, Cycle 6	1983
Second demonstration ramp at Quad Cities-2, Cycle 7	1985
Operational limit recommendation removed from zirconium-barrier fuel	1985
Unrestricted operations at Quad Cities-2, Cycle 8	1986

properties or to lapses in as-fabricated quality proved unable to solve the problem. Although the failure mechanism was not fully understood, effective operational procedures were developed and implemented. These were based on the recognition of the importance of power history, in which GE used results of tests in the GETR (Table 1). (Similar technology as derived from the experience in Canada was published by Penn et al. [8].) Nonetheless, the resulting loss in plant capacity factor provided ample incentive to develop fuel rods that are not susceptible to failure by PCI.

The invention of zirconium-barrier fuel in 1973 (based on a crystal bar zirconium liner metallurgically bonded to the Zircaloy) occurred in the midst of this intensive effort to gain understanding of this phenomenon [9]. Armijo did early fabrication experiments with several barrier concepts, including the coextrusion of Zircaloy tubeshells with crystal bar zirconium liners. The tubing made of these tubeshells provided material for the early tests.

By 1974, we had found an incipient PCI crack in an unfailed fuel rod that provided unequivocal evidence of SCC. Because Zircaloy is notch-sensitive, such events are difficult to find; most PCI cracks progress quickly through the cladding wall. Before this finding, it could not be said with certainty that the apparent SCC features in the fracture morphology had not been affected by the ingress of steam.

GE Program for Development of PCI-Resistant Fuel

In 1974, GE placed into three BWRs (Quad Cities-1, Monticello, and Millstone-1) segmented rod bundles in which potential remedy designs could acquire exposure for subsequent power ramp testing in the GETR. The crystal bar zirconium-barrier was among these candidate remedy designs. In the meantime, Grubb, while exploring which fission products could reasonably be the agent for cracking, found that in addition to iodine, cadmium (especially in the presence of cesium) could also give an SCC (or liquid metal embrittlement) effect [10]. The mechanistic studies at GE have been discussed by Davies et al. [11] along with the laboratory methods that had been developed at GE R&D [12,13]. These included localized ductility and expanding mandrel tests that were designed to simulate the loading conditions of the cladding during PCI in the chemical environments: iodine, cadmium or cesium-cadmium. In both tests, the loading was noncompliant and in plane-strain.

While the remedies were receiving their pre-test irradiation at relatively low linear power, accelerated tests were started wherein the test fuel rods were designed to have strong mechanical interaction (collapsed clad design) and the fuel was enriched to acquire burnup and fission product inventory quickly. By this time, two basic barrier types had been selected for comprehensive evaluation: the zirconium-barrier and the copper-barrier. Unlike the zirconium-barrier, which was metallurgically bonded and is ≈ 75 μm thick, the copper was a thin layer (≈ 10 μm) electrodeposited on the inner surface of the Zircaloy tube. In these early, low-burnup power ramp tests, both barrier types were successful. In 1976, the zirconium-barrier based on use of sponge zirconium was disclosed for patent [14], and test fuel of this type was also prepared for subsequent in-reactor tests.

An expanding mandrel test configuration was used to test irradiated cladding tubes [15]. As mentioned, this test applies a noncompliant loading of short tubing segments in the presence of the chemical agent; the circular symmetry of the tubing is preserved in this test. In this way, Adamson and coworkers further evaluated the PCI performance of both the zirconium-barrier and the copper-barrier cladding types. These tests were done with either unfueled tubes or with the unfueled plenum regions of fuel rods; tests included both the iodine and the cadmium environments. In such tests, both barrier types were clearly superior to the conventional Zircaloy cladding.

Large-Scale Demonstration Program

Encouraged by our early power ramp test data with both the copper-barrier and the zirconium-barrier fuel types, we began to consider how a PCI-resistant fuel might be introduced. The possibility of a demonstration in a power reactor took shape, and we began to speak in terms of a "100-bundle demonstration." Such a venture required, of course, substantial funding and a close cooperation with a host utility in whose reactor the demonstration would occur. The Commonwealth Edison Company of Chicago expressed keen interest in our work and actively encouraged our efforts. Simultaneously, we, with encouragement from Commonwealth Edison, submitted proposals to the U. S. Department of Energy for a large-scale demonstration program at one of the reactors at the Quad Cities Nuclear Power Station. These reactors are of the BWR/3 type with cores of low enough power density that a design to deliberately ramp certain bundles (those that are PCI-resistant) while safeguarding the conventional fuel seemed possible. Thus, with substantial financial support from the Department of Energy, we embarked on a program that would lead to the selection of the zirconium-barrier fuel and ultimately to a demonstration with 144 barrier fuel bundles. The program was begun in 1977 [16] and was completed in 1984 [17].

At the start of this program, the term "barrier fuel" applied equally to the copper-barrier and the zirconium-barrier types, both having encouraging ramp test results to the burnups tested, ≈ 9 MWd/kgU. One of the objectives of the program was to choose the better candidate. But, as was mentioned earlier, these two candidates had already been distilled from a much larger group, including graphite coatings, foil barriers, and others.

The program had an initial Phase 1 (1977 to 1978), which included:

1. a generic nuclear engineering study to show that the demonstration was feasible in a reactor of the BWR/3 type,
2. laboratory and reactor tests to verify the PCI resistance of the copper-barrier and the zirconium-barrier fuel types,
3. laboratory test of barrier cladding under simulated loss-of-coolant accident (LOCA) conditions, and
4. design, licensing documentation, fabrication, and preirradiation characterization of four lead test assemblies (LTAs) for irradiation in Quad Cities Unit 1.

The four LTAs represented the two barrier types with a variation of each. The copper-barrier could be plated directly onto the inner surface of a Zircaloy tube. Alternatively, the copper could be plated onto an oxidized (in an autoclave) Zircaloy surface. The purpose of the oxide layer was to inhibit interdiffusion of the copper with the Zircaloy. The zirconium-barrier could be made with a liner material made either from crystal bar zirconium or from sponge zirconium. Thus, the set of four LTAs represented each of the four barrier types in this program.

Phase 2 (1979 to 1984) included:

1. Selection of the fuel design for the demonstration.
2. Nuclear design and core management of the demonstration, expanding from the generic feasibility study in Phase 1 to a specific reactor and cycle and bundle designs (Quad Cities Unit 2 was chosen, beginning in Cycle 6 in December 1981).
3. Design, licensing documentation, and manufacturing of the demonstration fuel.
4. The demonstration per se, that is, the irradiation (including specially designed power ramps) to demonstrate PCI resistance.
5. Laboratory and test reactor evaluations of the behavior of zirconium-barrier fuel after water ingress.
6. In-reactor tests for reactivity-initiated accident.
7. Continued irradiation and evaluation of the four LTAs.
8. Continued testing of the barrier fuel in a test reactor to assure PCI resistance at burnup levels relevant to the demonstration. (This included multiple ramps with intermediate "deconditioning" irradiations at low power to prove that the barrier fuel did not lose its ability to resist PCI by having sustained a severe power ramp.)

This program was extensively reported as can be seen from the references in the Final Report as compiled and edited by Rowland [17]. Here, we shall review some of the highlights.

Although the program had many facets, all of which were essential, the most salient parts of the program were: (*a*) the power ramp tests of individual irradiated test fuel rods (called either "segments" or "rodlets") in test reactors (mainly at Studsvik), and (*b*) the deliberate power ramps on the barrier fuel bundles in the actual demonstration at the commercial power reactor, Quad Cities-2.

Davies et al. [18,19] described the ramp tests in test reactors; these tested the resistance of the barrier fuel to PCI. Care had to be taken to provide an "incubation irradiation" at a low linear power level (<20 kW/m) to the desired burnup. These were done in the segmented rod bundle in either the Monticello or the Millstone-1 reactors. (The segmented rod bundle in Quad Cities-1 was of the older 7 × 7 bundle design. This program concentrated on the segments from the reactors with the 8 × 8 design.) The initial tests were done at the GETR at the Vallecitos site, but when the reactor was closed in 1977, the testing was moved to the Studsvik R2 Reactor. The Studsvik staff contributed significantly in devising ways to meet our test specifications. Some six different power paths were used (Fig. 1). The A-ramp was important in defining the power level at which failure occurred. That power path was a staircase form having a hold time of 1 h at each level. The faster ramp rates (B', C, and C') were attempts to achieve the most severe power ramps to provide a thorough test. The fastest rate, ≥328 kW/m · min in the C and C' ramps, simulated the rate when a BWR control blade is moved; it corresponds to infinite power ramp rate, limited only by the thermal time constant of the fuel rod.

In these tests, it was seen that the PCI-resistance of the copper-barrier fuel did not persist beyond burnup ≈15 MWd/kgU. Thus, the copper-barrier was rejected as a candidate remedy [18].

In cooperation with Hitachi and Toshiba, multiple ramp tests of zirconium-barrier fuel were done [20]. After a successful ramp test at burnup, 7 to 23 MWd/kgU, the fuel was "deconditioned" by reirradiation at low power for an additional ≈5 MWd/kgU and then again power ramped. This was done to assure that the PCI resistance persists even after the fuel rod has survived an initial power shock.

The ramp test data for the zirconium-barrier are shown in Fig. 2 [21]. The power/burnup data for the tests on the barrier fuel is superposed on the statistical representation for similar data (mainly A-Ramps) for conventional nonbarrier reference rodlets. The statistical analysis was described by Rowland et al. [22]. The advantage of the zirconium-barrier fuel in comparison with conventional fuel is apparent.

The demonstration per se involved 144 barrier bundles in Quad Cities-2. Licensing approval was received in November 1980, and irradiation was begun at the start of Cycle 6 in December 1981. Sixty-four of the 144 had special enrichment distributions to increase local power peaking in the "wide-wide" corner rods. These were called "special barrier bundles"; half were of the crystal bar type, and half of the sponge type. The remaining 80 were of standard nuclear design. The first demonstration ramp was done at the end of Cycle 6 in March 1983. The step-wise withdrawals of the control blades resulted in local instantaneous power increases of up to 32.8 kW/m, to power levels up to 41.3 kW/m as each location along the length of the fuel was uncovered by its control blade. Although the burnups were quite modest (≈6 MWd/kgU bundle average), the test was clearly successful with no failures. This gave us confidence that the test was sound, providing a significant test of the barrier fuel while safeguarding the nonbarrier fuel in the core [21].

The second demonstration ramp [22] at the end of Cycle 7 (March 1985) had more significant exposure, ≈14 MWd/kgU. Now there was much more barrier fuel in the core, reducing the danger of inadvertently failing a nonbarrier fuel rod during the end-of-cycle blade pulls. The dramatic power ramps achieved on the wide-wide corner rods of the special barrier bundles are shown in Fig. 3 [23]. Each region uncovered by the step-wise movement of the control blade resulted in sudden power increments of >30 kW/m to power levels >44 kW/m. Had nonbarrier fuel been subjected to such a test, numerous failures would have been expected.

Before ending this discussion of the large-scale demonstration program, the mechanism underlying the PCI resistance should be explained as best we can. Tested by itself, zirconium resists SCC far better than does Zircaloy. Finite element code circulation has been done to explore the effect of the metallurgical bond, and expanding mandrel experiments were done to compare the PCI protection offered by the zirconium-barrier compared with an unbonded foil barrier. It was shown [24] that three attributes of the zirconium-barrier are essential to provide PCI resistance: (a) the metallurgical bond, (b) the thickness of the liner, and (c) the purity or softness of the liner material. The bond is quite remarkable in that recystallization occurs across the zirconium-Zircaloy interface. Thus, the interface is transgranular with no physical boundary that might impede the flow of heat, not even a grain boundary. The

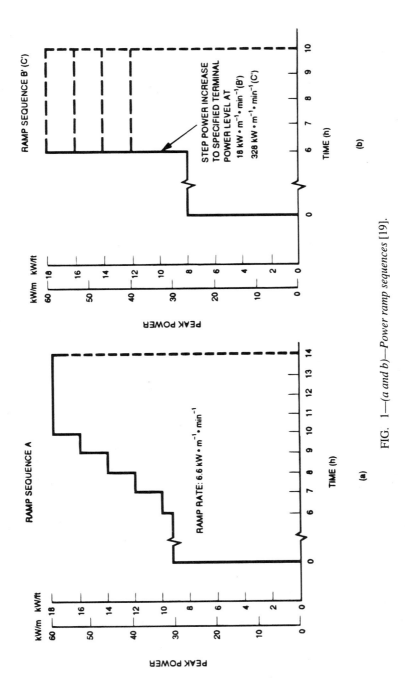

FIG. 1—(a and b)—Power ramp sequences [19].

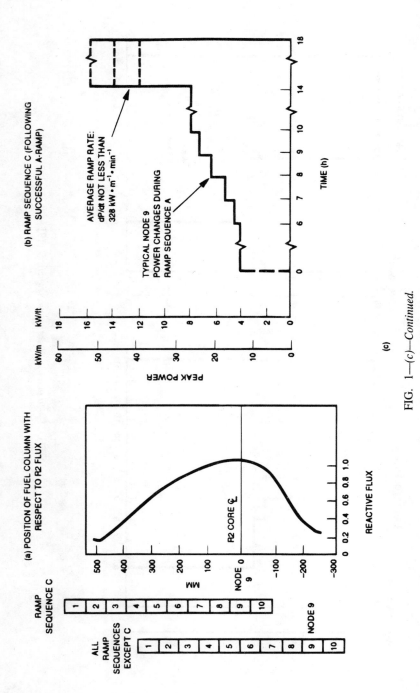

FIG. 1—(c)—Continued.

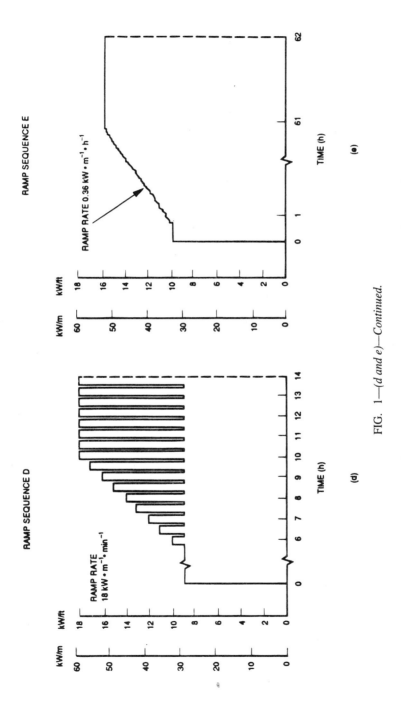

FIG. 1—(d and e)—Continued.

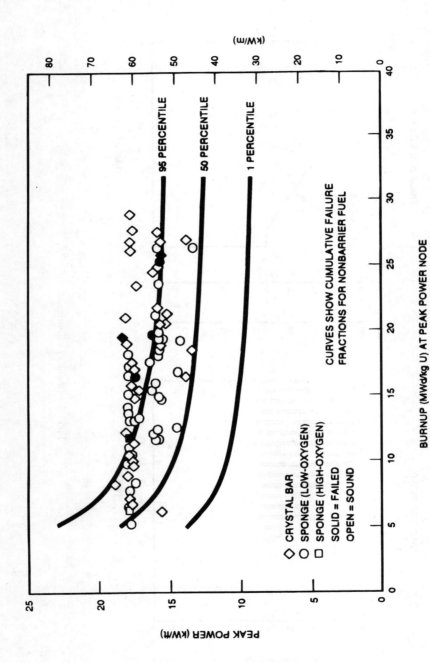

FIG. 2—*Power ramp test results through 1984 for zirconium-barrier fuel* [21].

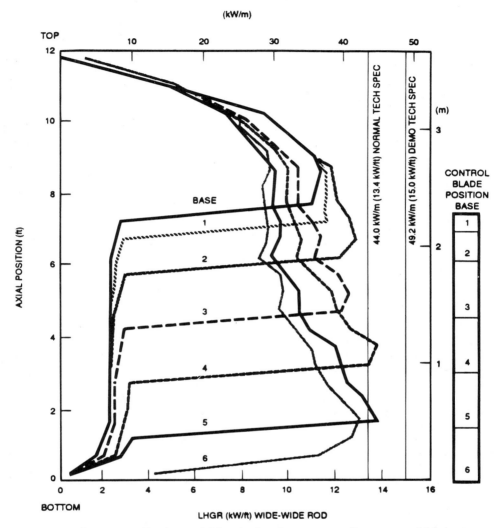

FIG. 3—*Axial power profiles in special barrier bundles corresponding to control blade steps during end of Cycle 7 power ramps in Quad Cities-2 (second demo ramp)* [23].

softness implies high localized ductility of the liner material and ability to dissipate localized stress concentrations.

Beyond the Formal Demonstration Program

GE was not alone in testing barrier fuel. (We already cited cooperation with Hitachi and Toshiba [20] in connection with multiple ramp tests.) In a separate program involving a large-scale demonstration of the copper-barrier, ABB also had a test program in cooperation with Tokyo Electric Power Company, Hitachi, and Toshiba. Results from that program have been reported with corroborating ramp test data on both the copper-barrier and the zirconium-barrier fuel types [25]. They too found that the PCI resistance of the copper-barrier diminished with burnup, while that of the zirconium-barrier persisted. Successful power ramp tests for zirconium-barrier fuel to burnup up to 54 MWd/kgU and to power levels to 57.5 kW/m have been reported [26].

As the Quad Cities-2 reactor acquired additional zirconium-barrier fuel, tests were done subjecting the fuel to load follow and to economic operation control (automatic frequency control) [23]. A recent set of tests in the Halden reactor of re-instrumented sponge zirconium-barrier fuel rod segments from the barrier lead test assembly (Quad Cities-1) subjected the fuel to various power cycle schemes [27]. Both the fission gas release and the cladding diametral deformation were measured on-line. The ability of barrier fuel to sustain unrestricted operation was further confirmed. That zirconium-barrier fuel is in fact a PCI-resistant fuel product has been thoroughly corroborated and accepted [26,28].

Failed Fuel Degradation

Recently, there have been some non-PCI failures of barrier fuel that have resulted in very high releases of radioactive fission products and off-gas [29,30]. Most barrier fuel failures have resulted in low releases of activity to the coolant. While the variations in post-defect behavior are not yet fully understood, it appears that there is a correlation with the practice of operating even failed barrier fuel at high linear power levels with unrestricted rates of power increase.

Davies [31] has considered the degradation of Zircaloy-clad fuel rods in operation and developed the concept that under oxygen-starvation conditions the cladding absorbs hydrogen and degrades by secondary hydriding. During the development of the zirconium-barrier fuel, qualification of zirconium-liner tubing to oxidation and hydriding in simulated water ingress conditions were done [24]. These tests showed that although the zirconium-liner oxidizes rapidly (zirconium does not form a protective oxide film), the oxidation rate returns to that expected of Zircaloy when the oxide-metal interface reaches the underlying Zircaloy.

Davies et al. [32] then did an in-reactor defected fuel test in the Studsvik R2 Reactor comparing the fuel degradation of zirconium-barrier fuel to that of nonbarrier. They reported that although the barrier fuel cladding absorbed more hydrogen, as was expected, the overall post-defect behavior of the barrier fuel with regard to the release of fission products was similar to the nonbarrier fuel. The fuel was tested within the so-called Locke curve and no adversely large release was seen. It appears that the Locke curve applies to barrier fuel reasonably well, although barrier fuel was not in Locke's data base, see Fig. 4. In a subsequent paper, Davies [33] further analyzed that experiment, showing that fission gas release is greatly increased due to the higher fuel temperature when the helium gas in the fuel rod is replaced by steam from the ingress of coolant.

Davies and Potts [30] have shown that most of the failed barrier fuel discussed in their study rods released modest amounts of fission products to the coolant and that nonbarrier fuel is also subject to large off-gas incidents. Nonetheless, this important topic is being addressed currently in the nuclear industry by GE and by others, with definitive results and increased understanding of post-defect fuel behavior expected within the next few years.

Summary

In this Kroll Award paper for the development of barrier fuel, the recognition of the PCI fuel failure problem was described. The work to gain a mechanistic understanding of the PCI phenomenon was reviewed as was the evaluation of various potential remedies. The cooperative

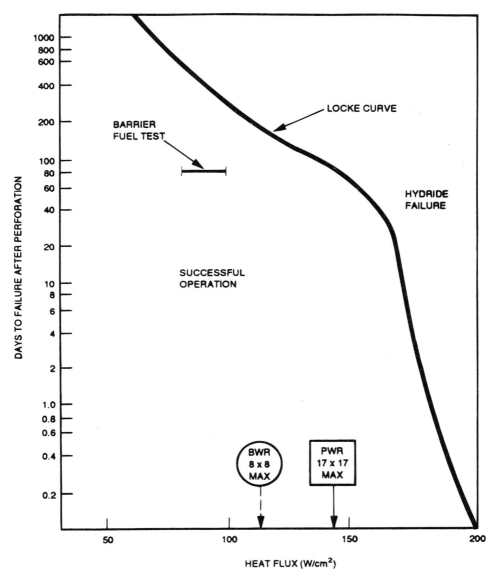

FIG. 4—*Behavior of defective LWR fuel showing barrier fuel test specification* [33].

program with Commonwealth Edison Co. and the U.S. Department of Energy that led to the lead test assemblies in Quad Cities-1 and the 144 bundle demonstration in Quad Cities-2 was described. That program included support tests of the materials behavior under SCC conditions, simulated LOCA situation, RIA, and operation after water ingress. The main support tests were the power ramp tests of fuel rod segments to ascertain and to quantify the PCI resistance of barrier fuel.

The events subsequent to that formal program were also reviewed. These include tests of the behavior of barrier fuel under severe power cycling. This paper also addressed the recent observations of some non-PCI fuel rod failures that have resulted in high off-gas.

Acknowledgments

We gratefully acknowledge the many who have contributed to this project. Dr. J. H. Davies was responsible for the segmented rod program for the PCI evaluation of the various PCI remedies, including the barrier fuel concepts; he also planned and executed the in-reactor defected fuel rod test. T. C. Rowland did the statistical analyses of the ramp test data; he planned the RIA safety tests; and he took over as the program manager for this effort from 1983 to 1984. Dr. R. B. Adamson directed work on the expanding mandrel tests of irradiated cladding, the mechanical properties tests, and the LOCA tests. He was ably assisted by Dr. R. P. Tucker (acknowledged here posthumously and very much missed by all of us), S. B. Wisner, and G. H. Henderson. Dr. R. A. Rand performed finite element analyses in support of the program. Dr. B. Cheng did critical laboratory testing for behavior after water ingress. R. E. Blood assisted in this corrosion work, and he did the LOCA tests. G. R. Lundeen did the metallography. J. A. Baumgartner headed the lead test assembly work. E. Rosicky played a key role in the execution of the ramp tests at both GETR and at Studsvik; E. L. Esch and E. V. Hoshi did the PIE. H. H. Paustian and D. T. Furtado were the principal contributors of the core design. C. B. Patterson provided help and guidance in the fuel testing and evaluations. Dr. R. A. Proebstle is acknowledged for encouragement and support of the early segmented rod program, and both he and Dr. H. H. Klepfer are acknowledged for strongly supporting this effort throughout.

From the GE Corporate R&D Center in Schenectady, we thank Dr. R. P. Gangloff and Dr. D. S. Tomalin for contributions in the mechanical testing and the expanding mandrel tests. Dr. W. T. Grubb explored the ability of various simulated fission product species to cause cracking. Drs. A. M. Bueche, H. W. Schadler, and M. G. Benz helped to focus the people and the resources of the Center to this area of technology. D. C. Lord did mechanical tests, and J. R. Hughes did the early fabrication studies and provided test materials. At our factory in Wilmington, North Carolina, Dr. C. D. Williams, F. Eddens, D. E. Allen, and R. J. Mack and many of their colleagues are thanked for their work with the lead test assemblies and the demonstration bundles.

Dr. G. Ronnberg and H. Tomani from the Studsvik staff were very helpful in devising means to meet the demanding test specifications on the ramp tests. We are grateful to Hitachi, Ltd., and to the Toshiba Corp. for their participation in the multiple ramp tests; special thanks go to Drs. H. Maki of Hitachi and N. Oi of Toshiba. Y. Wakashima and Dr. K. Ito of Nippon Nuclear Fuel Development, Ltd., actively helped to execute these tests.

As was mentioned earlier, we are very grateful for the encouragement by the Commonwealth Edison Co. (particularly W. M. Kiefer and Dr. D. O'Boyle) and for both encouragement and financial support from the U.S. Department of Energy, Dr. P. M. Lang and A. S. Mehner. The staff of the Quad Cities Nuclear Power Station is thanked for their cooperation and help with irradiation of the lead test assemblies and with the demonstration. The cooperation of Drs. W. V. Johnston and R. van Houton of the U.S. Nuclear Regulatory Commission (NRC) and that of Dr. M. Ishikawa of the Japan Atomic Research Institute are acknowledged in connection with the in-reactor safety tests. Drs. R. O. Meyers and M. Tokar of the NRC provided helpful discussions and encouragement in connection with the demonstration.

References

[1] Lyons, M. F., et al., "High Performance UO_2 Program First Pinhole Fuel Rod Failure during Center-Melt Operation—Assembly EPT-12-C-A," General Electric Company Report GEAP-5100K, 1966.

[2] Rosenbaum, H. S., "The Interaction of Iodine with Zircaloy-2," *Electrochemical Technology*, Vol. 4, 1966, pp. 153–156.
[3] Rosenbaum, H. S., Davies, J. H., and Pon, J. Q., "Interaction of Iodine with Zircaloy-2," General Electric Company Report GEAP-5100-5, 1966.
[4] Cox, B., "Pellet-Clad Interaction (PCI) Failures of Zirconium-Alloy Fuel Cladding—A Review," *Journal of Nuclear Materials*, Vol. 172, 1990, pp. 249–292.
[5] Cox, B., "Environmentally-Induced Cracking of Zirconium Alloys—A Review," *Journal of Nuclear Materials*, Vol. 170, 1990, pp. 1–23.
[6] Rowland, T. C. and Mathay, P. W., "Continued Irradiation of Dresden Type-I Segments in the GETR," unpublished work, 1967.
[7] Rowland, T. C., "Internal Impurity Failure of Zircaloy Clad Fuel—An Interpretive Review," unpublished work, 1969.
[8] Penn, W. J., Lo, R. K., and Wood, J. C., "CANDU Fuel—Power Ramp Performance Criteria," *Nuclear Technology*, Vol. 34, 1977, pp. 249–268.
[9] Armijo, J. S. and Coffin, L. F., U.S. Patent 4, 372, 817, 1983.
[10] Grubb, W. T. and Morgan, M. H., "Cadmium Embrittlement of Zircaloy-2," *Water Reactor Fuel Performance*, American Nuclear Society (ANS) Topical Meeting, St. Charles, IL, 1977, pp. 295–304.
[11] Davies, J. H., Rosenbaum, H. S., Armijo, J. S., Proebstle, R. A., Rowland, T. C., Thompson, J. R., Esch, E. L., Romeo, G., and Rutkin, D. R., "Irradiation Tests to Characterize the Fuel Failure Mechanism," *Water Reactor Fuel Performance*, American Nuclear Society (ANS) Topical Meeting, St. Charles, IL, 1977, pp. 230–242.
[12] Coffin, L. F. and Gangloff, R. P., "Integrated Laboratory Methods for Evaluating PCI in Zircaloy-2 Fuel Cladding," *Water Reactor Fuel Performance*, American Nuclear Society (ANS) Topical Meeting, St. Charles, IL, 1977, pp. 346–355.
[13] Tomalin, D. S., "Localized Ductility of Irradiated Zircaloy-2 Cladding in Air and Iodine Environments," *Zirconium in the Nuclear Industry (3rd International Conference), ASTM STP 633*, A. L. Lowe, Jr., and G. W. Parry, Eds., American Society for Testing and Materials, Philadelphia, 1977, pp. 557–572.
[14] Armijo, J. S. and Coffin, L. F., U.S. Patent 4,200,492, 1980.
[15] Tomalin, D. S., Adamson, R. B., and Gangloff, R. P., "Performance of Irradiated Copper and Zirconium Barrier-Modified Zircaloy Cladding under Simulated Pellet-Cladding Interaction Conditions," *Zirconium in the Nuclear Industry (4th International Conference), ASTM STP 681*, American Society for Testing and Materials, Philadelphia, 1979, pp. 122–144.
[16] "Demonstration of Fuel Resistant to Pellet-Cladding Interaction," First Semiannual Report, July-Dec. 1977, compiled by H. S. Rosenbaum, General Electric Company Report, GEAP-23773, Feb. 1978.
[17] Demonstration of Fuel Resistant to Pellet-Cladding Interaction, Phase 2—Final Report, General Electric Company Report GEAP 25163-10, compiled by T. C. Rowland, Nov. 1984.
[18] Davies, J. H., Rosenbaum, H. S., Armijo, J. S., Rosicky, E., Esch, E. L., and Wisner, S. B., "Irradiation Tests of Barrier Fuel in Support of a Large-Scale Demonstration," *LWR Extended Burnup—Fuel Performance and Utilization*, American Nuclear Society (ANS) International Topical Meeting, Williamsburg, April 1982, pp. 5–51–5–67.
[19] Davies, J. H., Rosicky, E., Esch, E. L., and Rowland, T. C., "Fuel Ramp Tests in Support of a Barrier Fuel Demonstration," General Electric Company Report GEAP-22076, July 1984.
[20] Wakashima, Y., Ito, K., Rosicky, E., Oi, N., and Maki, H., "Multiple Ramp Tests of Zirconium Barrier Fuel," *Light Water Reactor Fuel Performance*, American Nuclear Society (ANS) Topical Meeting, Orlando, April 1985, pp. 7–79–7–91.
[21] Rosenbaum, H. S., Davies, J. H., Adamson, R. B., Tucker, R. P., Rowland, T. C., Paustian, H. H., Thompson, J. R., and O'Boyle, D. R., "Large-Scale Demonstration of Barrier Fuel," *Light Water Reactor Fuel Performance*, American Nuclear Society (ANS) Topical Meeting, Orlando, April 1985, pp. 7–63–7–78.
[22] Rowland, T. C., Davies, J. H., Thompson, J. R., and Rosenbaum, H. S., "Statistical Analysis of Power Ramp PCI Data," *Proceedings*, IAEA Specialists' Meeting on PCI, International Atomic Energy Agency, Vienna, 1983, pp. 61–68.
[23] Rosenbaum, H. S., Rowland, T. C., Davies, J. H., Furtado, D. T., and Paustian, H. H., "Large-Scale Demonstration of Barrier Fuel," *Proceedings*, International Symposium on Improvements in Water Reactor Fuel Technology and Utilization, International Atomic Energy Agency, Vienna, 1987, pp. 117–134.
[24] Rosenbaum, H. S., Rand, R. A., Tucker, R. P., Cheng, B., Adamson, R. B., Davies, J. H., Armijo, J. S., and Wisner, S. B., "Zirconium Barrier Cladding Attributes," *Zirconium in the Nuclear Industry*

(7th International Conference), ASTM STP 939, R. B. Adamson and L. F. P. Van Swam, Eds., American Society for Testing and Materials, Philadelphia, 1987, pp. 675–699.

[25] Inoue, K., Suzuki, K., Maki, H., Yasuda, T., Oi, N., Hayashi, Y., Wakashima, Y., Ogata, K., Junkrans, S., Vesterlund, G., Lysell, G., and Ronnberg, G., *Light Water Reactor Fuel Performance,* American Nuclear Society (ANS) Topical Meeting, Orlando, April 1985, pp. 6–1–6–16.

[26] Hayashi, Y., Koyama, T., Koizumi, S., Yasuda, T., and Nomura, M., "BWR Fuel Performance and Recent R&D Activities in Japan," *Fuel for the 90's,* American Nuclear Society–European Nuclear Society (ANS-ENS) International Topical Meeting on LWR Fuel Performance, Avignon, France, April 1991, pp. 36–46.

[27] Rowland, T. C., Rosenbaum, H. S., Iwazaki, R., Nakatsuka, N., Iwano, Y., Maru, A., Matsumoto, T., and Yoshizawa, A., "Power Cycling Operation Tests of Zr-Barrier Fuel," *Fuel for the 90's,* American Nuclear Society–European Nuclear Society (ANS-ENS) International Topical Meeting on LWR Fuel Performance, Avignon, France, April 1991, p. 808–817.

[28] Holzer, R. and Knaab, H., "Recent Fuel Performance Experience and Implementation of Improved Products," *Proceedings,* International Topical Meeting on LWR Fuel Performance, Williamsburg, 1988, pp. 69–80.

[29] Jonsson, A., Hallstadius, L., Grapengiesser, B., and Lysell, G., "Failure of a Barrier Rod in Oskarshamn-3," *Fuel for the 90's,* American Nuclear Society–European Nuclear Society (ANS-ENS) International Topical Meeting on LWR, Fuel Performance, Avignon, France, April 1991, pp. 371–377.

[30] Davies, J. H., and Potts, G. A., "Post-Defect Behavior of Barrier Fuel," *Fuel for the 90's,* American Nuclear Society–European Nuclear Society (ANS-ENS) International Topical Meeting on LWR Fuel Performance, Avignon, France, April 1991, pp. 272–284.

[31] Davies, J. H., "Secondary Damage in LWR Fuel Following PCI Defection: Characteristics and Mechanisms," *Proceedings,* IAEA Specialists' Meeting, Chalk River, Ontario, Canada, International Atomic Energy Agency, Vienna, 1979, pp. 135–140.

[32] Davies, J. H., Berry, C. M., Esch, E. L., and Wisner, S. B., "Irradiation Performance of Zr-Barrier Fuel Operating with Simulated Cladding Defects," General Electric Company GEAP-25481, July 1982.

[33] Davies, J. H., "Fission Product Release from Artificially Perforated Fuel Rods Operating in a Test Reactor Loop," *Fission Product Behavior and Source Term Research,* American Nuclear Society (ANS) Topical Meeting, Snowbird, UT, July 1984, pp. 12–1–12–13.

David O. Pickman[1]

Zirconium Alloy Performance in Light Water Reactors: A Review of UK and Scandinavian Experience

REFERENCE: Pickman, D. O., "**Zirconium Alloy Performance in Light Water Reactors: A Review of UK and Scandinavian Experience,**" *Zirconium in the Nuclear Industry: Tenth International Symposium, ASTM STP 1245,* A. M. Garde and E. R. Bradley, Eds., American Society for Testing and Materials, Philadelphia, 1994, pp. 19–32.

ABSTRACT: Various aspects of zirconium alloy development for light water reactors in the UK and Scandinavia are reviewed, including the contribution made by some unique nuclear testing facilities. Among the problems encountered were the irradiation enhancement of corrosion and hydrogen pickup, crud deposition, iodine-induced stress-corrosion cracking on power ramping, and severe cladding deformation in loss-of-coolant accident conditions. The causes and behavior of defects, including hydride defects and fretting corrosion, are discussed.

KEY WORDS: zirconium alloys, nuclear fuel, corrosion, hydriding, stress corrosion, fretting corrosion, pellet-clad mechanical interaction, loss-of-coolant accidents, fuel rod defects, neutron irradiation, zirconium, nuclear materials, nuclear applications, radiation effects

The combination of low neutron capture cross section and good corrosion resistance in water, at least up to 400°C, together with adequate strength and ductility, led to the early selection of zirconium for use in the U.S. nuclear submarines by Admiral Rickover [1]. Other candidate materials were aluminum, magnesium, beryllium, stainless steels, and nickel-base alloys, but most had serious limitations, although stainless steel was used as fuel cladding in early boiling water reactors (BWRs) and pressurized water reactors (PWRs).

With the advent of the prototype LWR power reactors, the need to understand the behavior of zirconium alloy cladding on the now familiar light water reactor (LWR) type fuel rods became urgent, to ensure that costly mistakes were not made on the large power reactors.

This paper is concerned mainly with experience with this fascinating metal over the past 30 years, with some of the problems and phenomena encountered, and with the particular contributions of the UK and the Scandinavian countries.

The European Scene

The mid to late 1950s saw the first significant studies of zirconium alloys for nuclear applications in the UK. The first use was as cladding on the fuel elements for nuclear submarines, with minor applications of high-strength zirconium-copper-iron (ZrCuFe) and zirconium-copper-molybdenum (ZrCuMo) alloys as support beams in Magnox reactors. Consideration of zirconium alloy cladding in various conceptual reactor studies followed on the successful

---

[1] Former head of UKAEA Springfields Nuclear Laboratory, UK Atomic Energy Authority, Salwick, Preston, Lancashire, UK PR4 ORR.

development of a hafnium separation process. Corrosion remained a major concern as behavior was erratic and early work by Cox [2] at AERE Harwell was aimed at understanding the corrosion behavior. Eventually, in 1956, the UK selected a heavy-water-moderated, light-water-cooled, pressure tube BWR, the steam-generating heavy-water reactor (SGHWR), to succeed the gas-cooled advanced gas-cooled reactors (AGRs). A 100-MWe prototype was built and commenced operation in late 1967. Plans to build large commercial power plants were later shelved and PWRs were introduced.

In Sweden, the Ågesta pressurized heavy water reactor started operation with Zircaloy-2 fuel cladding and shroud tubes in 1964 and operated successfully for many years after early fretting problems. Research reactors at the Studsvik site participated in the materials development for Ågesta and for the Swedish designed BWRs that followed.

Both Denmark and Norway established nuclear research facilities in the 1950s, the Danish research (DR) reactors at Riø and the Halden BWR, an Organization for Economic Cooperation and Development (OECD) project. Neither country has seriously contemplated a nuclear power program, but they have contributed to zirconium clad fuel development.

Finland has four nuclear power plants in operation, two Soviet-designed PWRs and two Swedish-designed BWRs.

It is a strange commentary on the energy scene that although at 31 December 1991 power reactors in Western Europe had accumulated 2300 reactor years of safe operation, compared with 1800 reactor years in the United States [3], there is now no reactor operating with zirconium alloy clad fuel in the UK. Moreover, the power reactors in Sweden are under threat of closure by 2010, although recent opinion polls suggest a change of public opinion in favor of nuclear power.

The European scene as a whole is very different, however, with 11 countries (out of a world total of 14) generating more than 25% of their electricity from LWRs using zirconium alloy clad fuel. Two countries, Belgium and Sweden generated over 50% and France was the clear leader with over 75%. For the UK, generation was 21%, but all from stainless steel or magnesium alloy clad fuel in gas-cooled reactors. The 1200-MWe Sizewell B PWR will commence operation early in 1994.

Nuclear Fuel Development Facilities

From the beginning of the 1960s, the need to prove designs of nuclear fuel rods required irradiation facilities to accommodate fuel rods of substantial length and with appropriate coolant and flow conditions. At this time, the UK and Canada entered into a collaboration agreement that gave the UKAEA access to the excellent boiling water loop, X-6, in the NRX reactor. A superheated, steam-cooled loop, X-4, was also used to test stainless steel clad superheat fuel rods that behaved impeccably. It was in these X-6 loop experiments that the type and quality of cladding tubes and the basic design parameters were first tested. Fuel rods that were 1.8 m long could be irradiated, and eventually a trefoil of rods that allowed different types of spacer and rod support could be tested.

Behind this irradiation testing program lay a lot of alloy and tubing development, from its manufacturing route (three UK and one U.S. supplier were used), texture, tensile strength and ductility, creep strength, corrosion and hydrogen pickup, hydrogen solubility limits, and hydride orientation [4–9]. The irradiation testing in NRX, backed up by specimen testing in MTRs at Harwell and Dounreay, led to the identification of the important irradiation-enhanced and induced effects.

In 1962, the OECD Halden reactor in Norway had completed its initial reactor physics program and the terms of reference were re-written to embrace a fuel test program. Very soon, a series of UK irradiation experiments were loaded, involving many design and materials variants including Zr-1Nb cladding, 0.38-mm wall cladding, hollow pellets, spacer grid variants, uranium metal fuel, and a number of instrumented fuel rods to measure stack length change and fission gas release. All of these experiments were comprised of five rod clusters 1.8-m long and, although the environmental temperature and pressure were low (510 K and 2.9 MPa), useful information was obtained, especially on certain irradiation processes for which the enhancement factor was high at the low temperature; for example, the corrosion rate enhancement was about 100! The Halden BWR was later fitted with loops providing representative BWR and PWR coolant conditions as well as special facilities for power ramp and loss-of-coolant testing. A major feature of the Halden Project has always been its capability to perform sophisticated instrumented fuel rod experiments.

Some experimental work was also carried out in other European facilities, notably the HFR at Petten in the Netherlands, BR3 in Belgium, R2 at Studsvik in Sweden, DR3 in Denmark, ESSOR at Ispra in Italy, SILOE and PHEBUS in France and FR2 at Karlsruhe in the FRG. Much of this work involved fuel testing rather than cladding development, but almost all these facilities became involved in loss-of-coolant accident (LOCA) testing and the R2 reactor has been pre-eminent in the world in conducting a series of power ramp test projects. Some interesting work was carried out in the DR3 reactor at Risø in Denmark on dispersion-hardened zirconium alloys, but was eventually unsuccessful in producing a usable cladding material.

Like the UK, the Swedes made good use of the Halden Reactor and their domestic facilities at Studsvik in the development of their own version of BWR. Finland, unlike the UK and Sweden, built its reactors under licence, but did participate fully in the Halden Project and some other joint European programs.

Some dynamic engineering facilities were needed to test fuel assembly stability, control rod stability, and fuel charge/discharge operations. Such facilities were built at the UKAEA Springfields laboratory, at the then ASEA-ATOM laboratory at Västerås in Sweden and at Studsvik. Fuel assemblies for the SGHWR were exhaustively tested in the Springfields facility and the occasional incidence of fretting damage eventually led to the use of spring-type spacer grids as used in most other LWRs.

When the SGHWR in the UK was commissioned in late 1967, a large whole assembly irradiation test program was initiated covering design, material, and performance variables [10]. Within a few weeks, this program was hit by a severe incidence of crud overheating defects, but after the copper removal from the feedwater had been improved, the program was resumed and the fuel rod and assembly designs were fine-tuned to achieve reliable high burnup within specified safe operating limits. This reactor proved to be an excellent test bed and was widely used by other customers.

Performance Experience with Zirconium Alloy Cladding

The cladding on LWR fuel rods operates in a hostile environment with radiolytically produced radicals of H_2 and O_2, such as HO_2 and H_2O_2, as well as corrosion products, in the coolant. There is a high heat flux across the cladding and aggressive fission products being released inside it. The cladding is subjected to continuous fast neutron bombardment, causing frequent atomic displacement, typically about 10 to 15 displacements per atom in the lifetime of LWR cladding. There is a resultant population of point defects, interstitials, vacancies, and their aggregates that influence all mechanical properties. The cladding is also subjected to both steady and variable stresses from the pressure differential across it and from interactions with the UO_2 fuel. Finally, cladding may be subjected to high-temperature excursions, resulting from power/coolant mismatch or loss-of-coolant, and to severe mechanical and thermal shocks from reactivity excursions.

Corrosion and Hydrogen Pickup

World-wide experience from corrosion testing and from fuel cladding examination has confirmed that irradiation causes a significant enhancement of corrosion rate [11,12]. For BWR and PWR clad surface temperatures (about 560 and 630 K), the enhancement factors are of the order of 10 and 3, respectively, although they may vary widely with coolant composition and purity, flow velocity and coolant pressure, Zr/ZrO_2 interface temperature, fast neutron flux and, indirectly, on exposure time. The irradiation enhancement becomes very small at temperatures of 700 K and above, but thermal oxidation rates increase rapidly. Oxidation in the early stages has a parabolic/cubic rate curve, followed by a transition to an effectively linear rate. Irradiation reduces the time to transition and increases the weight gain at transition. The irradiation effects are probably caused partially by the concentration and types of oxidizing radicals in the water in contact with the cladding surface and partially by fast neutron damage to the ZrO_2 layer. The exact mechanism for the transfer of oxygen ions through the oxide layer remains uncertain, it may be assisted by point defects, or result from a higher concentration gradient resulting from the oxidizing radicals on the surface.

The difference between the irradiation enhancement rate in BWR and PWR is partly an effect of the H_2 overpressure in reducing the concentration of oxidizing radicals and partly a result of the more rapid annealing out of neutron bombardment damage in the oxide film in the PWR case.

A form of corrosion described, aptly, as nodular corrosion has a more rapid rate of growth, although not covering the whole surface in its early stages. This type of corrosion is more common in BWRs, although it can occur under PWR conditions [11,12]. This type of corrosion is clearly associated with the size and distribution of intermetallic precipitates that may act as sites for local cathodic depolarization and may be influenced by stress-induced recrystallization in the oxide. A mechanism involving H_2 precipitation and pressure buildup at the Zr/ZrO_2 interface has also been proposed that is dependent on intermetallic precipitate size. A further powerful influence on nodular corrosion, is the electrical coupling to a stainless steel spacer or other component close to the clad surface. Both flow and steam quality also influence the formation of nodular corrosion. Some alloys, notably single-phase alloys, are resistant to nodular corrosion, and in recent years improved heat-treatment schedules and minor modifications to composition have had beneficial effects.

In the UK development program, a range of alloys was tested, including Zircaloy-2 and Zircaloy-4, both cold-worked and stress-relieved (CWSR) and recrystallized, Zr-1Nb, Zr-2½Nb, Zr-3Nb-1Sn, ZrCrFe, and Ozhennite 0.5. From the results, it was concluded that none was significantly better than CWSR Zircaloy-2, although the niobium alloys showed promise, despite heavy corrosion in weld regions. At this time, the value of a β-quench (from 1150 K) was recognized as a means of reducing intermetallic precipitate size, so minimizing nodular corrosion.

Additions of H_2 or NH_3 to the coolant were studied in both NRX and SGHWR irradiations and, while showing some benefit, the extra cost involved could not be justified.

Corrosion in zirconium alloys is accompanied by pickup of H_2, ranging from about 10% of the so-called corrosion hydrogen in BWRs to 20 to 30% in PWRs. Zirconium hydride precipitating under stress, or forming in tubing with an adverse fabrication history (sinking greater than axial extension), can have very adverse effects on ductility, especially at temperatures below 420 K. The terminal solubility limits at BWR and PWR cladding temperatures are about 80 and 200 ppm, respectively, and hydrogen concentrations exceeding these values have been reported in cladding as well as in spacer grids and pressure tubes. Such levels may not be harmful, and in UK experience there was no problem with SGHWR, as solubility limits were not exceeded.

Hydrogen pickup is a more serious problem in pressure tubes, as used in SGHWR, CANDU, and RBMK reactors, especially in any cool regions or if the reactor is cooled down under pressure. One further H_2 pickup problem is the segregation to cooler surfaces, unlikely to be a problem with fuel rods, except possibly in end-cap regions, but which proved to be a major problem in CANDU PHW reactors. The mechanism is that H_2 diffusing in a component will precipitate out first in a cooler region that then acts as a sink for further diffusion, thus leading to a major and damaging buildup of zirconium hydride.

Hydrogen pickup during nodular corrosion is very low and this, together with the good thermal conductance through this porous type of oxide, minimizes the effect of relatively thick oxidation.

SCANUK Alloy Development—In the early 1970s, there was a view in some quarters that improved zirconium alloys would be needed, especially to show reduced corrosion at temperatures of 675 K and above. Following discussions between the UK, Sweden, Denmark, and Norway, a joint enterprise to develop improved alloys was established. Finland joined in this program at a later date. A group of six alloys was selected, based on theoretical considerations and particularly representing the views of Tyzack (UK) and Lunde (Norway). The selected alloys were manufactured in Scandinavia and fabricated to BWR tube size. Alloys 1 and 2 were 1% niobium, Alloys 3 and 4 were 1 and 0.5%, respectively, of niobium with 0.5% chromium, and Alloys 5 and 6 were 0.5% niobium with 0.3% molybdenum and 0.3% chromium plus 0.2% molybdenum [*13*].

After a very extensive development program covering out and in-reactor corrosion and H_2 pickup, heat treatment, mechanical testing and microstructural studies, and high temperature burst testing, the work was terminated and none of the alloys was selected for commercial reactor use. The alloys were compared with Zr-2^1/$_2$Nb and Zircaloy-2. The SCANUK alloys were superior at high temperatures and showed less H_2 absorption than Zircaloy-2. The molybdenum bearing Alloys 5 and 6, chosen for high temperature strength, were disappointing in their corrosion resistance; and Alloys 3 and 4, the NbCr alloys had the best all-round performance and certainly compared favorably with Zircaloy-2. With the present increased interest in new alloy development, this work may represent a good starting point to develop an improved BWR cladding alloy.

Pellet-Clad Mechanical Interaction

In the mid-1960s, some defects in zirconium-alloy-clad fuel rods were reported that showed brittle, partly intergranular, radially penetrating cracks. The first reports came from experiments by General Electric at Vallecitos but were soon followed by a remarkable fuel rod that defected in the Halden BWR. This was a rod in the UK experiment, IFA22, which on post-irradiation examination showed 47 cracks on one transverse section, some fully penetrating. The now familiar phenomenon of pellet/clad mechanical interaction (PCMI), sometimes known as power ramp or power bump defects, had appeared. This has turned out to be a most important defect mechanism that is still the subject of investigation today. Rosenbaum [*14*] reported the stress-corrosion cracking of zirconium alloys in iodine vapor, and Garlick and co-workers at UKAEA Windscale Laboratory undertook an extensive investigation over several years to study the phenomenon and the limiting conditions of stress and iodine partial pressure [*15*]. There is now a worldwide consensus that zirconium alloys are susceptible to iodine stress-corrosion cracking, as demonstrated in many laboratory experiments, but still a minority view favors liquid metal embrittlement by caesium or cadmium as the cause of PCMI defects. The typical fluted fracture surfaces seen on stress-corrosion induced fractures are believed to represent tearing of ligaments between cleavage cracks [*16*]. The mechanism probably involves a nucleation time, during which corrosion pitting penetrates sufficiently deeply to allow the crack-tip stress intensity factor for iodine stress corrosion conditions to be reached. Stress corrosion crack propagation then takes place until a critical crack length is reached and a crack propagates through the cladding. The extent of axial propagation depends on axial variations in the hoop stress and can be substantial.

Zirconium-alloy-clad fuel rods are particularly susceptible to high localized tensile stresses on the tube bore surface if local fuel rod rating is increased above a previously prevailing level. The cause is basically that the short UO_2 fuel pellets have radial cracks and "wheatsheaf" on thermal expansion, the diametral expansion being greatest at pellet ends. This effect is compounded by a radial crack opening displacement caused by the increase in pellet center temperature. A flux tilt across a fuel rod will magnify the stress concentration and can influence the position of major stress. The stress gradient close to a radial crack is also influenced by the friction between fuel and cladding. The end result is that stresses high enough to initiate cracking, usually close to a pellet end opposite a radial crack, can readily occur. The stress state at these pellet end positions can be close to equi-biaxial, and the peak stress can exceed twice the mean stress calculated for the power increase.

The clad stress level at which cracking will occur is important to know, as is the influence of type, condition, and irradiation dose of the cladding. A threshold of 500 MPa was established in UKAEA gas pressurization tests, for cold-worked stress-relieved Zircaloy-2, and this was used as a criterion for assessing failure probabilities for a series of fuel rods power ramped in the SGHWR [17]. The UKAEA fuel performance code, SLEUTH-SEER, was used for these calculations, but an important assumption that had to be made was how the crack opening displacement was distributed between the radical cracks, typically five to ten in number. It was assumed that half the displacement occurred at one crack and this led to cracking threshold values in the range 400 to 450 MPa. These values are probably high because of the crack opening displacement assumption, and other recent work suggests that the threshold stress reduces with irradiation, probably to a minimum of about 200 MPa.

The understanding of the cause of PCMI defects has led to a steady stream of remedies being proposed. The conditions for these defects arise from a variety of causes, from control rod movements, peaking factors caused by partial reloads, following power demands, sometimes rapidly, or from flux tilt caused by spatial instabilities. The various remedies mainly depend on reducing the clad bore surface stress by:

1. reducing fuel rod diameter;
2. using hollow, duplex, or shorter pellet fuel;
3. use of a bore surface lubricant; and
4. use of multiple crack initiators on the fuel pellet surface;

or by reducing the availability of iodine by:

1. reducing fuel rod diameter (less iodine release),
2. use of increased helium pre-pressurization, and
3. use of rifled bore cladding or axially serrated UO_2 pellets.

PCMI defects can in practice always be avoided by a slow power rise, such that the clad creeps to accommodate the power change without reaching the threshold stress. This same principle is embodied in the remedy developed by General Electric and now used in most BWR fuel. In this design, a duplex cladding tube is used with a thin bore surface layer of pure soft zirconium. This bore surface layer yields at a stress below the threshold stress and so does not crack at low strain values. The role of this bore surface layer in influencing the type of secondary hydride defects has been questioned but remains uncertain.

The other successful remedy in regular use was developed by AECL for use in CANDU-PHW fuel that is axially shuffled on power in the horizontal channels, therefore experiencing a recurring power ramp. This remedy comprises a graphite bore surface lubricant, known as CANLUB, to reduce the friction and hence the peak stress. This remedy may not be so effective for reactors in which the fuel has a much higher burn-up, but no limitation has yet been found.

In early UK experiments, it was shown that Zircaloy-2 tubing with a predominantly radial basal pole texture, which can be produced by an appropriate tube-reduction schedule, was optimum for resisting iodine stress-corrosion cracking and tubing with such a texture is generally used by fuel manufacturers.

International Power Ramp Projects—The need to know PCMI performance limits has led to a series of experiments being performed in BWR and PWR loops in the R2 reactor at Studsvik, Sweden. Power is altered by changing He3 pressure in a coil surrounding the test section.

These projects, INTER-RAMP [*18*], OVER-RAMP [*19*], DEMO-RAMP I and II, SUPER-RAMP [*20*], SUPER-RAMP EXTENSION, and TRANS-RAMP I and II [*21,22*], covered all current BWR and PWR designs of interest with certain design variations, including type, condition, and thickness of cladding, pellet design (and UO_2 manufacturing route), pellet density and grain size, filling gas pressure and pre-ramp rating history and burn-up (10 to 40 GWd/Tu).

An over-simplified view of the merit of a particular fuel rod design that has been in common use in recent years is the so-called ramp terminal level (RTL), being the limiting rating to which rods can be ramped without failure. Used strictly comparatively, such a measure can be useful, but so many factors are involved, such as type and condition of cladding, including texture, fuel-to-clad gap, pre-ramp rating and burn up, ramp rate and power increment in the ramp, that no single parameter can define the merit of a design. The most important parameters are undoubtedly the pre-ramp irradiation that will have caused irradiation damage and creep-down of the cladding and some fission gas release, the power increment after hard fuel-to-clad contact occurs and the ramp rate following this hard contact. The actual rating levels at hard contact and at the ramp terminal level are also important as they influence the clad stress level because of their influence on UO_2 creep (temperature dependence) and cladding creep (fast flux dependence).

There were some surprising findings in the Studsvik ramp projects. Some rods failed at a RTL between 30 and 35 kW/m and with a power increment of only 10 to 15 kW/m, while others survived a ramp to over 60 kW/m with a power increment of over 30 kW/m. Overall, surprising features were the relative unimportance of burnup level and ramp rate, the ineffectiveness of a large fuel-to-clad gap, the contradictory evidence on the merits of hollow pellet fuel, and the lack of any correlation between failures and fission gas release in the ramps, some rods surviving with almost 50% release. A most surprising feature was the number of cases in which fuel rods failed when no fuel-to-clad interaction was calculated to have occurred. For some reason, some UO_2 pellets undergo a large diametral expansion that cannot be accounted for by purely thermal expansion.

The contributions on this important aspect of fuel rod behavior made by facilities in the UK and Scandinavia have been recognized by the worldwide interest and participation in the European projects.

Load Following—The need for daily load following, to match diurnal variations in demand, arises when nuclear power forms a large part of total electricity supply, as in France. The large power reduction needed, followed by a fast return to full power represents a special type of power ramp, particularly as it may be repeated up to 1000 times in the life of a fuel assembly.

Experiments carried out in the SGHWR, and in France, have shown that some pessimistic code predictions of progressive clad thinning were not confirmed, and the time at reduced power was not sufficient to generate a high clad stress, the creepdown being minimal. A further concern, that load following might generate additional fission gas release, was also found to be incorrect, so there is no fear of fuel damage from a daily load cycle.

Performance in a Loss-of-Coolant Accident (LOCA)

The possible consequences of a major LOCA had been a persistent worry for LWR operators for many years and led to a major experimental effort to establish the probable behavior starting in the mid-1970s. The accident scenario is a break in a main coolant supply line, for example, a double guillotine break in a cold leg inlet header in a BWR, which results in a rapid loss of the coolant inventory. This is followed by actuation of emergency core cooling systems that refill and reflood the core and are designed to keep it covered for as long as necessary. A major pressure vessel break is deemed impossible based on periodic inspections and crack growth rate data.

The important aspects of a postulated LOCA are the time-temperature transient experienced by the fuel cladding before it is quenched and the differential pressure across the cladding during the transient.

The typical transient of this type has a blowdown phase lasting about 20 s followed by refill and reflood phases with full core covering taking up to about 300 s. The clad temperature rise starts early in the blowdown phase and reaches a peak dependent on any residual stored energy and fission product decay heat, typically after 50 to 100 s.

The safety concerns in a LOCA are twofold. First, that the zirconium alloy cladding will oxidize so heavily that on quenching it will shatter, leading to an uncontrolled spread of the fuel rod contents. The second concern is that the fuel clad swelling may cause a flow blockage sufficient to prevent effective cooling during reflood. Such a situation would lead to a core melt-out.

Work on the high-temperature oxidation and embrittlement, mainly in the United States, but supported by UKAEA studies [23] and work in the FRG, led to a definition of safe limits in terms of maximum clad temperature, 1477 K, and maximum local cladding oxidation of 17%. Alternative embrittlement criteria were based on minimum β-phase thickness with a maximum oxygen content [24]. The USNRC also restricted total H_2 generated from oxidation to 1% of the total core inventory of zirconium.

The retention of a coolable geometry was a more difficult problem to solve, and the early UKAEA experiments which showed that extensive axial lengths of cladding could strain by large amounts [25] spawned a huge international effort involving major laboratory rigs in many countries and in-reactor experiments in the United States, UK, FRG, Japan, France, Canada and Norway. The phenomenon of axially extended straining, widely referred to as "ballooning," was at one time believed to involve superplasticity, but was later shown to be controlled by convective cooling conditions during the experiments. As the clad diameter increased, the clad temperature reduced to a level at which little further strain took place. This was, therefore, a temperature-controlled stabilization that occurred both with direct-resistance-heated or internally heated cladding specimens. Following local stabilization, straining could continue in upstream and downstream regions, so extending the length of the balloon.

Apart from crucial in-reactor experiments, many sophisticated out-reactor test series were conducted, especially to examine multi-rod behavior. As expected, interactions between neighboring fuel rods aided the axial extension of clad strain. Maximum clad strains were found at about 1125 K, just below the α to β transus, and could exceed 100% diameter increase. The amount of blockage produced by a given strain in neighboring rods was influenced by the boundary conditions, since, without an immovable boundary, clad swelling interaction resulted in radial movement of rod centers that would be inhibited in a large array, such as a PWR fuel assembly.

Many factors were found to influence swelling and blockage, especially circumferential temperature variations in the cladding that may result from pellet fragment movement, flux tilt or the presence of spacer grids, control rod guide tubes, control rods, and fuel boxes in BWRs. The consensus that eventually emerged was that the maximum local blockage would be of the order of 90%. Many experiments confirmed this as an upper bound, with a low probability of blockage exceeding 60%.

Two things were needed to satisfy safety authorities in the UK and elsewhere. The first was a definitive in-reactor experiment to study clad swelling and the second was a heat-transfer experiment to show that blockages of the maximum expected extent were coolable.

To satisfy the first requirement, experiments were performed on five-rod assemblies in the Halden BWR (IFA543-547) and compared with results on similar electrically heated assemblies. Clad temperature transients achieved the required levels of about 1100 K and all rods were ruptured in the in-reactor experiments, but problems of premature quenching by a steam/water froth at lower levels were difficult to overcome. As a result of this series of experiments, no large co-planar swellings were produced, almost certainly because of large circumferential temperature variations. A more ambitious series of in-reactor experiments, jointly sponsored by USNRC and UKAEA, were carried out in the NRU reactor using a full-length 32-rod section of a 17 × 17 PWR assembly. These assemblies known as MT-1 to MT-4 had 12 test rods surrounded by 20 guard rods [26]. The largest strains were seen in Test MT-3, reaching about 95% and significant strains occurred over about 0.75 m, with co-planar strains up to about 50% existing over 0.1 m. A major influence of spacer grids was seen in these experiments, with maximum strains at the downstream end of an inter-grid span. The spacers promoted improved mixing in the rising coolant that caused de-superheating in the upstream end of the inter-grid spans with resultant improved heat transfer and a tendency to form carrot-shaped balloons. Clad strains were less than would be expected with greater external restraint. Modeling to assess the expected effect of increased restraint showed it to be relatively small. None of these experiments would suggest that blockages of as great as 90% could arise in a worst accident. Good support for these conclusions came from the 5 × 5 rod tests in PHEBUS, France [27].

Blockage Coolability—Having established a worst scenario for a LOCA in terms of clad swelling and blockage, the coolability has to be established. To an extent, this had been examined in the limited number of in-reactor experiments with reflood, but more definitive experiments were needed. A multi-rod heat transfer rig, THETIS, was built at UKAEA Winfrith for this purpose and tests were performed on a 7 × 7 array of full length SGHWR-type rods with internal heaters [28]. An off-center array of 4 × 4 rods were fitted with simulated co-axial balloons based on results from the KfK REBEKA experiments [29]. A 200-mm-long parallel section had a 50% clad strain (blockage factor 90%) with upstream and downstream conical tapers of 200 and 50 mm, respectively. The results of tests in this rig showed the blockage to be well cooled with heat-transfer coefficients in the blockage and by-pass not very different. A key factor was whether sufficient water, as droplets, passed through the blockage to prevent excessive steam superheating. Similar tests in a heat transfer rig, FEBA, at KfK Karlsruhe resulted in similar conclusions [30], and other tests, FLECHT/SEASET in the United States and SCTF in Japan, gave generally similar results. In worst cases, there could be up to a 10-s delay in the cooling of blocked regions compared with by-pass regions, and the maximum peak temperature difference was 50 to 100 K, although the norm was zero or negative. The effectiveness of the heat transfer is largely due to the disruption of the parallel flow with impingement of water droplets and a mixture of droplets and steam resulting in some de-superheating. Spacer grids were found in all these experiments to have similar effects in promoting de-superheating. The refill stage following a LOCA was studied separately in THETIS rig experiments. As soon as the water level rose to start covering the fuel rods, a low-density froth rose rapidly and started quenching rods some distance in front of the single-phase water front. This effect is known as "level swell."

Other Transients—There are many events that can cause short-term disturbances in LWRs and have an influence on reactivity and hence power output. Such events as a turbine trip, a loss of feedwater, pressure increase, decrease in coolant temperature and increase or decrease in coolant inventory are regarded as anticipated transients. Most such events will trigger a reactor scram, but some may occur so rapidly that some damage may be caused to the fuel, for example, from a sudden power increase, or an excursion into dryout.

Little experimental work has been performed in Europe on such events, but much has been done in the United States and Japan on reactivity insertion accidents (RIAs). Safe limits for such accidents have been established (200 to 250 cal/g UO_2 energy deposition), and this has fed back into reactor design. It is perhaps ironic that the most serious reactor accident at Chernobyl was a RIA, while the only important LOCA, at Three Mile Island in 1979, was caused by the malfunction of a small valve, not by any break in circuitry.

A flow reduction or power excursion may result in cladding dryout and the need to prevent it happening inadvertently, possibly without any indication to the operator, has led to the use of large margins to dryout. Most dryout testing has been performed in electrically heated rigs to establish safe operating power levels. Some such tests were done in a 9-MW rig at UKAEA Winfrith. They were later supplemented by in-reactor tests using full size SGHWR fuel assemblies [*31*]. The heat-transfer post-dryout was more effective than expected, because of the high flow velocity of the high steam quality fluid, and peak clad temperature did not exceed 875 K. Excursions into dryout lasted up to 2.5 min and fuel assemblies showed no abnormal damage or deterioration, although it would be expected that some additional fission gas will have been released. Fuel assemblies having experienced such a transient would be capable of continued operation under normal conditions.

Fuel Defects—Causes and Behavior

Not surprisingly, in view of the very large number of zirconium alloy clad fuel rods that have been manufactured and irradiated, there have been some defects that have released fission products with undesirable effects. Generally, the release has been of gaseous fission products only, but on occasions there has been more severe disintegration leading to spillage of UO_2 into the coolant.

Apart from PCMI defects, which have been described earlier, there are other defect mechanisms that are related specifically to the properties of zirconium alloys, some of which are discussed in this section. Other defects have been caused by the use of defective components or faulty manufacture, but generally the quality of zirconium alloy components has been high.

Crud Deposition

The deposition of a generally iron-oxide-rich layer on fuel rod surfaces has been found in all LWRs. These deposits result from trace amounts of corrosion products in the circulating coolant, they generally reach an equilibrium thickness, and because they are very porous they have a negligible effect on clad surface temperature. The amount of deposit, commonly called crud, depends on circuit materials, coolant chemistry, and clean-up plant.

In the early years of BWR operation in the United States, large crud deposits were found in inlet nozzles that reduced flow, and there were also substantial deposits on fuel rods, especially in the lower part of the core. Some deposits contained copper and were responsible for failures in Big Rock Point. This problem was overcome by full flow filtration and ion-exchange purification. A similar phenomenon was encountered within a few weeks of the start-up of the SGHWR in the UK. Several fuel rods developed leaks and, on examination, it was found that the cladding had oxidized through from the outside and structural changes indicated that the cladding temperature had exceeded 875 K. On further examination, it was seen that the iron oxide crud layer, up to about 100 μm thick, was infilled with copper and was no longer porous. More importantly, the crud had become hard, like an eggshell, and in places had separated from the cladding surface. This left a small gap that filled with steam and acted as an excellent insulator, giving a temperature drop of about 600 K at the fuel rod ratings in use [32]. The problem was solved by an improved ion-exchange clean-up plant, and later by replacement of certain copper alloy components in the condensate and feed trains.

Hydriding

Hydrogen gets into zirconium alloys during corrosion in water or steam. The concentration builds up, in solution, until the solubility limit is reached, when precipitation as small platelets of zirconium hydride starts, initially in cooler areas where the solubility limit is less. The orientation of hydrides so formed is important and depends in part on the fabrication history of the component and partly on the stress level [5]. It was shown many years ago by Marshall and Louthan [33] that hydride precipitates oriented normal to an applied stress could result in brittle fractures at low temperature. Hydride precipitation has not been a problem for fuel cladding in the UK because terminal solubility is not reached at operating temperature and cladding is not stressed in tension during cooldown. This is potentially more of a problem for pressure tubes, but with current and proposed increases in burnup, solubility limits may be exceeded in fuel cladding. However, early UK work [34] showed only a small reduction in elongation in closed-end burst tests on samples with 340 ppm hydrogen at up to 625 K.

The most common cause of defects in zirconium alloy clad fuel rods has been internal hydriding, which is unique to zirconium among possible cladding materials. The mechanism is that H_2, in the form of moisture or other H_2 containing materials, is sealed into the fuel rod during manufacture and reacts by a rather special process to produce a zirconium hydride inclusion on the bore surface. The volume change involved produces local tensile stress in the cladding such that radial arms of hydride spread from the initial inclusion and penetrate the clad wall. Most of the H_2 originates from H_2 contained in the UO_2 and from moisture picked up by the UO_2 after sintering and prior to fuel rod manufacture. These defects generally form during the first few weeks of operation, during which the H_2O contained in the rod is progressively reduced by oxidation of the UO_2 and clad bore surface. This results in an increase in the H_2 partial pressure, and when the H_2O/H_2 ratio is reduced to less than 10^{-2}, or thereabouts, the H_2 is able to attack the clad bore surface, forming the local inclusion. These defects can occur at any position, but there has been a tendency to occur at bottom end welds. The solution to this type of defect is to control the H_2 content of the UO_2 pellets by careful drying. A maximum level of 1.5 mg H_2 per fuel rod was set by UKAEA and has proved completely satisfactory. Meeting such a limit in high humidity conditions, or with low density UO_2 would be difficult and require controlled atmosphere assembly. The use of a pickled and autoclaved tube bore surface proved to lead to an increased risk of defects, but H_2 getters, which are used in General Electric BWR fuel, have been successful.

Fretting Corrosion

Fretting corrosion involves a local loss of metal where contact with another component occurs intermittently, generally as a result of vibration. The loss of metal is accentuated if impingement is followed by some sliding before contact is lost. The mechanism is believed to be in part oxidation, resulting from the generation of a high local temperature at the contact position, and in part the welding and tearing of small surface asperities. Zirconium alloys are particularly susceptible to fretting corrosion, and the damage is typified by a very smooth surface and a precise indentation from the other contacting component.

Fretting problems normally arise at contacts between fuel rods and spacers, sometimes due to spring relaxation or breakage. It was a frequent occurrence with early designs and extended burnup requirements have necessitated a review of spring relaxation. Other causes of clad fretting have been the trapping of debris at spacer grids and cross flow caused by leaks in pressure vessel baffles.

Irradiation Growth

While not directly a cause of fuel rod defects, unexpectedly high increases in fuel rod length and inadequate end clearances were a potential cause of fuel rod bow that could lead to local overheating. Some additional fuel rod length increase was caused by a type of ratcheting interaction between the UO_2 and the cladding. Irradiation growth is a phenomenon peculiar to anisotropic metals, caused by a combination of interstitial absorption, enabling dislocations to climb over obstacles, and absorption or nucleation of interstitial loops on prism planes. This process is a volume-conserving type of stress-free creep. Some intergranular stresses are generated at grain boundaries, because of differences in orientation and the anisotropy of thermal expansion, but they are not sufficient to cause significant creep strain. Some additional mechanism has been reported to operate at high fluences and high temperatures [35]. Adamson [36] has attributed this to the formation of larger dislocation arrays.

Defect Behavior

It may be neither convenient nor necessary to remove leaking fuel rods as soon as they are detected, providing the release does not exceed any permitted levels. Defects have been shown to give an increasing release with time, probably due to a minor jacking effect as UO_2 local to the primary leak site is oxidized, but a more important factor is the formation of secondary hydride defects and this has been shown to be influenced by rating (surface heat flux), typically with very long times to formation at low ratings, and about 100 to 200 days at typical power reactor ratings [37].

The mechanism of secondary hydriding is interesting. Following the formation of a primary leak by any mechanism, water/steam enters and fills the clad-to-fuel gap at a pressure close to coolant pressure. Immediately, reaction between H_2O and UO_2 and zirconium starts and some of the O_2 is consumed, helping to reduce the H_2O/H_2 ratio towards the critical level (less than about 10^{-2}) at which rapid H_2 attack on the zirconium alloy can take place. This process will take place more rapidly the higher the rating, but another important factor is that more H_2O is entering the rod at the leak site to compensate for the O_2 that has reacted. This sets up a gradient in the H_2O/H_2 ratio, so that the critical level for rapid hydriding will form at some axial distance away from the primary leak. As a partial remedy for secondary hydriding, any device to assist rapid diffusion of fresh H_2O away from the leak site might be effective. Designs aimed at facilitating axial flow of water/steam entering the primary leak, such as the Swedish rifled bore cladding, show promise of at least delaying secondary hydriding.

Conclusions

This review has covered most of the important aspects of the behavior of zirconium alloys in nuclear reactors experienced by the author over the past 30 years. There are some notable omissions, such as the irradiation enhancement of creep rate and the very high creep strains before rupture, also, the irradiation hardening effects on strength and ductility. Although important in some contexts, these effects have not had an important influence on fuel design or performance. As far as pressure tubes are concerned, creep could have proved a limiting factor in reactors of the SGHW type.

This 30-year journey has always been interesting and stimulating, never dull, and the award of the W. J. Kroll Zirconium Medal gives me great satisfaction at this late stage of my career.

References

[1] Rickover, H. G., "History of the Development of Zirconium Alloys for Use in Nuclear Reactors," Denver, 1975.
[2] Cox, B., et al., "Oxidation and Corrosion of Zirconium and Alloys," *Journal,* Electrochemical Society, Vol. 108, 1961, pp. 24 and 129; also Vol. 109, 1962, p. 6.
[3] Lewiner, C., "50 Years of Nuclear Energy—a Glance at the Past, a Look into the Future," *Nuclear Energy,* Vol. 32, No. 1, 1993, p. 9.
[4] Slattery, G. F., "The Prediction of Collapse Pressures for Anisotropic Zircaloy-2 Tubing Using Tensile Stress—Strain Data," UKAEA Report TRG 1476(S), UK Atomic Energy Authority, Preston, 1966.
[5] Hindle, E. D. and Slattery, G. F., "The Influence of Processing Variables on the Grain Structure and Hydride Orientation in Zircaloy-2 Tubing," *Journal,* Institute of Metals, Vol. 94, 1966, p. 245.
[6] Hindle, E. D. and Slattery, G. F., "Preferred Orientation in Zircaloy-2 Tubing Manufactured by Various Routes," *Journal,* Institute of Metals, Vol. 93, 1965, p. 565.
[7] Slattery, G. F., "The Terminal Solubility of Hydrogen in Zirconium Alloys Between 30°C and 400°C," *Journal,* Institute of Metals, Vol. 95, 1967, p. 43.
[8] Allen, P. L., Moore, D. A., and Trowse, F. W., "The Relation of Proof Testing to Long Term Corrosion Behaviour of Zirconium Alloys," UKAEA Report TRG 1134(S), UK Atomic Energy Authority, Preston, 1966.
[9] Nichols, R. W. and Watkins, B., "Pressure Tube Materials for SGHWR's," *Proceedings,* British Nuclear Energy Society Conference on Steam Generating and Other Heavy Water Reactors, 1968, p. 115.
[10] Pickman, D. O., "SGHWR Fuel Design and Materials," *Proceedings,* British Nuclear Energy Society Conference on Steam Generating and Other Heavy Water Reactors, 1968, p. 29.
[11] Trowse, F. W., Sumerling, R., and Garlick, A., "Nodular Corrosion of Zircaloy-2 and Some Other Zirconium Alloys in Steam Generating Heavy Water Reactors and Related Environments," *Zirconium in the Nuclear Industry: Third International Symposium, ASTM STP 633,* A. L. Lowe, Jr., and G. W. Parry, Eds., American Society for Testing and Materials, Philadelphia, 1977, p. 236.
[12] Sumerling, R., Garlick, A., Stuttard, A., Hartog, J. M., Trowse, F. W., and Sims, P., "Further Evidence of Zircaloy Corrosion in Fuel Elements Irradiated in a Steam Generating Heavy Water Reactor," *Zirconium in the Nuclear Industry: Fourth International Conference, ASTM STP 681,* American Society for Testing and Materials, Philadelphia, 1979, p. 107.
[13] Tyzack, C., Hurst, P., Slattery, G. F., Trowse, F. W., Garlick, A., Sumerling, R., Stuttard, A., Videm, K., Lunde, L., Warren, M., Tolksdorf, E., Tarkpea, P., and Forsten, J., "SCANUK: A Collaborative Programme to Develop New Zirconium Cladding Alloys," *Journal of Nuclear Materials,* Vol. 66, 1977, pp. 163–186.
[14] Rosenbaum, H. S., *Electrochemical Technology,* Vol. 4, March/April 1966.
[15] Garlick, A. and Wolfenden, P. D., "Fracture of Zirconium Alloys in Iodine Vapour," *Journal of Nuclear Materials,* Vol. 41, 1971, p. 274.
[16] Leach, N. A. and Garlick, A., "The Influence of Environment on the Formation of Fluting Microstructures During Fracture of Zircaloy," *Journal of Nuclear Materials,* Vol. 125, 1984, p. 19.
[17] Bond, G. G., Cordall, D., Cornell, R. M., Fox, W. N., Garlick, A., and Howl, D. A., "SGHWR Fuel Performance Under Power Ramp Conditions," *Journal,* British Nuclear Energy Society, Vol. 16, 1977, pp. 225–235.

[18] Mogard, H., et al, "The Studsvik INTER-RAMP Project," *An International Power Ramp Experimental Programme,* American Nuclear Society Topical Meeting on LWR Fuel Performance, Portland, OR, 1979, pp. 284–294.
[19] Hollowell, T. E., Knudsen, P., and Mogard, H., "The International OVER-RAMP Project at Studsvik," *Proceedings,* American Nuclear Society Topical Meeting on LWR Extended Burn-up Fuel Performance and Utilisation, Williamsburg, Vol. 1, 1982, pp. 4–18.
[20] Djurle, S., "Final Report on the SUPER-RAMP Project," U.S. Department of Energy, Washington, DOE/ET/34032-1, 1985.
[21] Mogard, H., et al, "The International TRANS-RAMP I Fuel Project. Fuel Rod Internal Chemistry and Fission Product Behaviour," IAEA IWGFPT/25, International Atomic Energy Agency, Vienna, 1985, pp. 157–167.
[22] Mogard, H., Howl, D. A., and Grounes, M., "The International TRANS–RAMP II Fuel Project—A Study of the Effects of Rapid Power Ramping on the PCI Resistance of PWR Fuel," *Proceedings,* American Nuclear Society Topical Meeting on LWR Fuel Performance, Williamsburg, 1988, pp. 232–244.
[23] Parsons, P. D., "Zircaloy Cladding Embrittlement in a Loss-of-Coolant Accident," *Proceedings,* 6th Water Reactor Safety Information Meeting, Gaithersburg, MD, 1978.
[24] Chung, H. M. and Kassner, T. F., "Embrittlement Criteria for Zircaloy Fuel Cladding Applicable to Accident Situations in Light Water Reactors," NUREG/CR-1344, U.S. Nuclear Regulatory Commission, 1980.
[25] Rose, K. M., Mann, C. A., and Hindle, E. D., "The Axial Distribution of Deformation in the Cladding of PWR Fuel Rods in a Loss-of-Coolant Accident," *Nuclear Technology,* Vol. 46, 1979, p. 220.
[26] Parsons, P. D., Hindle, E. D., and Mann, C. A., "The Deformation, Oxidation and Embrittlement of PWR Cladding in a Loss-of-Coolant Accident," UKAEA Report ND-R-1351(S), UK Atomic Energy Authority, Preston, 1982, (prepared as a state-of-the-art report for OECD/CSNI).
[27] Adroguer, B., Hueber, C., and Trotobas, M., "Behaviour of PWR Fuel in LOCA Conditions—PHEBUS Test 215," *Proceedings,* CSNI/IAEA Specialists Meeting on Water Reactor Fuel Safety and Fission Product Release in Off-Normal and Accident Conditions," Risø, Denmark, May 1983.
[28] Pearson, K. G., Cooper, C. A., Jowitt, D., and Kinnear, J. H., "Reflooding Experiments on a 49-Rod Cluster Containing a Long 90% Blockage," UKAEA Report AEEW-R1591, UK Atomic Energy Authority, Preston, Jan. 1983.
[29] Erbacher, F. J., "LWR Fuel Cladding Deformation in a LOCA and its Interaction with the Emergency Core Cooling," *Proceedings,* American Nuclear Society/European Nuclear Society Topical Meeting on Safety Aspects of Fuel Behaviour, Sun Valley, ID, Aug. 1981.
[30] Erbacher, F. J., Neitzel, H. J., and Wiehr, K., "Effects of Thermohydraulics on Clad Ballooning, Flow Blockage and Coolability in a LOCA," *Proceedings,* CSNI/IAEA Specialists Meeting on Water Reactor Fuel Safety and Fission Product Release in Off-Normal and Accident Conditions, IAEA, Risø, Denmark, May 1983.
[31] Redpath, W., "Winfrith SGHWR In-Reactor Dryout Tests," *Journal,* British Nuclear Energy Society, Vol. 13, 1974, pp. 87–97.
[32] Pickman, D. O., "Fuel Performance in the Prototype SGHWR Power Station," UKAEA Report TRG 1943 (S), UK Atomic Energy Authority, Preston, 1969.
[33] Marshall, R. P. and Louthan, M. R., "Tensile Properties of Zircaloy with Oriented Hydrides," *Transactions,* American Nuclear Society, Vol. 56, 1963, p. 693.
[34] Pickman, D. O., "Properties of Zircaloy Cladding," *Nuclear Engineering and Design,* Vol. 21, No. 2, 1972, pp. 212–236.
[35] Rogerson, A. and Murgatroyd, R. A., *Journal of Nuclear Materials,* Vol. 113, 1983, p. 256.
[36] Adamson, R. B., Tucker, R. P., and Fidleris, V., "High Temperature Irradiation Growth in Zircaloy," *Zirconium in the Nuclear Industry, Fifth Symposium, ASTM STP 754,* D. Franklin, Ed., American Society for Testing and Materials, Philadelphia, 1981, pp. 208–234.
[37] Locke, D. H., "Defective Fuel Behaviour in Water Reactors," *Proceedings,* Topical Meeting on Water Reactor Fuel Performance, American Nuclear Society, St. Charles, IL, May 1977.

Hydrogen Effects

Stefan Sagat,[1] *Christopher E. Coleman,*[2] *Malcolm Griffiths,*[3] *and Brian J. S. Wilkins*[4]

The Effect of Fluence and Irradiation Temperature on Delayed Hydride Cracking in Zr-2.5Nb

REFERENCE: Sagat, S., Coleman, C. E., Griffiths, M., and Wilkins, B. J. S., **"The Effect of Fluence and Irradiation Temperature on Delayed Hydride Cracking in Zr-2.5Nb,"** *Zirconium in the Nuclear Industry: Tenth International Symposium, ASTM STP 1245,* A. M. Garde and E. R. Bradley, Eds., American Society for Testing and Materials, Philadelphia, 1994, pp. 35–61.

ABSTRACT: Zirconium alloys are susceptible to a stable cracking process called delayed hydride cracking (DHC). DHC has two stages: (*a*) crack initiation that requires a minimum crack driving force (the threshold stress intensity factor, K_{IH}) and (*b*) stable crack growth that is weakly dependent on K_I. The value of K_{IH} is an important element in determining the tolerance of components to sharp flaws. The rate of cracking is used in estimating the action time for detecting propagating cracks before they become unstable. Hence, it is important for reactor operators to know how these properties change during service in reactors where the components are exposed to neutron irradiation at elevated temperatures. DHC properties were measured on a number of components, made from the two-phase alloy Zr-2.5Nb, irradiated at temperatures in the range of 250 to 290°C in fast neutron fluxes ($E \geq 1$ MeV) between 1.6×10^{17} and 1.8×10^{18} n/m$^2 \cdot$ s to fluences between 0.01×10^{25} and 9.8×10^{25} n/m^2. The neutron irradiation reduced K_{IH} by about 20% and increased the velocity of cracking by a factor of about five. The increase in crack velocity was greatest with the lowest irradiation temperature. These changes in the crack velocity by neutron irradiation are explained in terms of the combined effects of irradiation hardening associated with increased <a>-type dislocation density, and β-phase decomposition. While the former process increases crack velocity, the latter process decreases it. The combined contribution is controlled by the irradiation temperature. X-ray diffraction analyses showed that the degree of β-phase decomposition was highest with an irradiation temperature of 290°C while <a>-type dislocation densities were highest with an irradiation temperature of 250°C.

KEY WORDS: delayed hydride cracking, crack velocity, threshold stress intensity factor, irradiation fluence, irradiation temperature, zirconium, zirconium alloys, nuclear materials, nuclear applications, radiation effects

---

[1] Research Metallurgist, AECL Research, Materials and Mechanics Branch, Chalk River Laboratories, Chalk River, Ontario KOJ 1JO, Canada.
[2] Branch manager, AECL Research, Fuel Channel Components Branch, Chalk River Laboratories, Chalk River, Ontario KOJ 1JO, Canada.
[3] Research metallurgist, AECL Research, Reactor Materials Research Branch, Chalk River Laboratories, Chalk River, Ontario KOJ 1JO, Canada.
[4] Senior research metallurgist, AECL Research, Materials and Mechanics Branch, Whiteshell Laboratories, Pinawa, Manitoba ROE 1LO, Canada.

Delayed hydride cracking (DHC) is a stable crack growth mechanism in zirconium alloys. Hydrogen accumulates at a stress raiser. If sufficient hydrogen is present, hydrides form and, if the stress is high enough, the hydrides fracture and the crack advances. The process is then repeated until the crack becomes unstable. The two main characteristic parameters of DHC are the crack velocity, V, and the threshold loading below which cracks do not grow; with sharp cracks at moderate loads, linear elastic fracture mechanics is used, and the threshold stress intensity factor is called K_{IH}. Neutron irradiation at some temperatures increases V and reduces K_{IH} [1,2]. At least two methods exist to evaluate the effect of irradiation on DHC: (1) prepare appropriate specimens, then irradiate them in a reactor, or (2) machine specimens from components removed from reactors for surveillance. In this paper, we describe the results of experiments using both methods. Many cold-worked Zr-2.5Nb pressure tubes were removed from CANDU power reactors as part of a planned large-scale retubing program and thus a comprehensive evaluation program was possible. The objectives were to measure the effects of irradiation fluence and irradiation temperature on V and K_{IH} and to relate the results to the microstructural changes produced by irradiation.

Experimental Procedure

Material

Pressure tubes in CANDU reactors are made from Zr-2.5Nb alloy by hot extrusion of hollow billets followed by 25% cold work. The extrusion process results in a crystallographic texture with the following resolved basal pole fractions: 0.32 (radial), 0.61 (transverse), and 0.07 (axial). The pressure tubes are about 6 m long and have an inside diameter of 103 mm and a wall thickness 4.1 mm. They are joined to stainless steel end fittings by internal rolling. The material used in these experiments came from 45 pressure tubes that were removed from power reactors and from offcuts. The offcuts are small pieces of archived material cut off from each end of a pressure tube prior to operation. The tubes were removed from different lattice positions and sampled at different axial positions along the tube to provide a range of irradiation fluences and temperatures. A schematic diagram of a pressure tube and related terminology used throughout this paper are shown in Fig. 1. The rolled joint (R/J) regions are

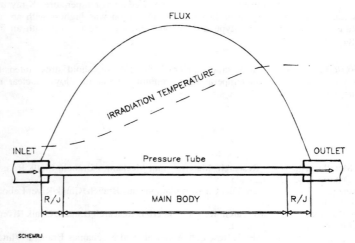

FIG. 1—*A schematic diagram of a fuel channel in CANDU reactor.*

defined as the pieces that are within about 200 mm from each end of the tube, and the remainder of the tube is referred as the main body.

Zr-2.5Nb material was also irradiated in a high flux reactor (OSIRIS) at Centre d'Etudes Nucleaires de Saclay. The purpose of this irradiation was to achieve end-of-life fluences substantially ahead of the life of the tubes in the reactors. Two materials were used for this irradiation: standard, cold-worked Zr-2.5Nb (tube H737) and Zr-2.5Nb material produced by a modified route, TG3R1 (Task Group 3 Route 1 material developed at Ontario Hydro). The TG3R1 pressure tubes were produced using a lower extrusion ratio and higher percent coldwork than standard pressure tubes, and had an extra stress relieve at 500°C for 6 h after cold drawing. Details about production routes for the fabrication of cold-worked Zr-2.5Nb tubes are given in Ref 3. About 1 atomic percent of hydrogen was added gaseously to these materials prior to irradiation. To achieve a uniform distribution of hydrogen, the materials were homogenized at 400°C for 72 h. The irradiation of H737 material was called ERABLE and the irradiation of TG3R1 material was called TRILLIUM. Both irradiations have been carried out at a temperature of 250°C.

Specimens

Two types of specimens were prepared from the tubes: (1) cantilever beam (CB) and (2) curved compact toughness (CCT) specimens, Fig. 2. The CB specimens were used for measuring

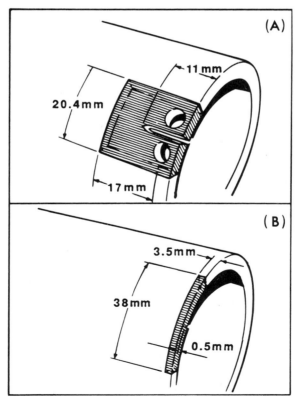

FIG. 2—*Curved compact toughness (A) and cantilever beam (B) specimens used in DHC tests.*

V and K_{IH} in the radial direction, and the CCT specimens for measuring V in the axial direction. Cracks were started in CB specimens from 0.5-mm-deep notches machined on the inside surface. The CCT specimens were fatigue pre-cracked. Because the initial hydrogen concentration in the specimens was low, hydrogen was added to some specimens electrolytically to allow testing up to 290°C. In this technique a layer of solid zirconium hydride is deposited on the surface of the specimen. The specimen is then homogenized at a solvus temperature corresponding to the required hydrogen concentration. This treatment results in uniform distribution of fine hydrides through the bulk of the specimen.

Experimental Techniques

Crack velocity of DHC is sensitive to the temperature history; the maximum value of V is attained by cooling (without undercooling) to the test temperature from above the solvus temperature of hydrogen in zirconium [4]. Thus, a standard procedure for measuring V was developed as shown in Fig. 3. To minimize annealing of irradiation damage, the maximum peak test temperature in irradiated material should not be greater than the irradiation temperature plus 10°C. The load can be applied either at the end of the high temperature soak, or after attaining the test temperature. The latter avoids premature cracking during cooling and allows an accurate evaluation of the time to crack initiation. The loads applied should be within ASTM criteria for linear elastic fracture mechanics, that is, ASTM Test Method for Plane-Strain Fracture Toughness of Metallic Materials (E 399–90). CB specimens were loaded in

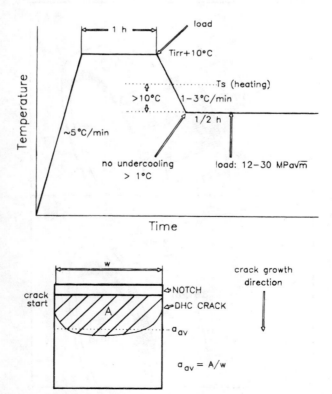

FIG. 3—*Standard temperature and loading history for DHC testing.*

bending and cracking was detected by acoustic emission [5]. CCT specimens were dead weight loaded in tension in a creep frame and cracking was monitored by a d-c potential drop method [6]. The cracks were grown from about 0.5 to 2.0 mm. Crack velocity was derived from the average crack depth divided by the time over which steady cracking occurred. Crack depth was defined as the area of DHC, delineated by heat-tinting, divided by the specimen thickness.

In the literature, two methods of determining K_{IH}, have been described as follows:

1. A number of specimens are loaded to different K_I and incubation times for cracking are measured; K_{IH} is estimated by a projection to infinite time [7]. Measuring K_{IH} using this technique takes several years, but provides realistic values.
2. One specimen only is used and K_{IH} is measured over a relatively short period of time by either increasing the load until cracking starts or decreasing the load until cracking ceases [5,8].

The method we have standardized on for quick comparisons is based on decreasing the load. After the standard temperature cycle, the specimen is loaded to 17 MPa$\sqrt{m}$. Once cracking is initiated, the K_I is reduced 2% per 5 μm of crack growth, derived from the calibration of the acoustic emission through an automatic feedback system [5]. The load is reduced until acoustic emission stopped and no further indications are obtained for at least 24 h. K_{IH} is derived from this final load and the average crack depth.

Experimental Errors

The main sources of errors in the measurement of DHC velocity are:

1. uncertainty in the determination of the time for steady cracking, and
2. uncertainty in the determination of the crack depth.

It takes a certain amount of time, called incubation time, before cracking starts. Before steady crack growth is attained, the cracking may go through a transient period, during which the cracking may be erratic or intermittent. Neither the incubation time nor the transient time should be included in the crack velocity calculation. The uncertainty in determining the time of steady cracking depends on the sensitivity of the acoustic emission or potential drop techniques and on estimating the onset of steady cracking, which is not always clear-cut.

Crack propagation is not always uniform across the specimen section; hence, the average crack depth, derived by dividing the cracked area by the specimen thickness, is sensitive to the crack shape. Crack velocity in nonuniform cracks and cracks that tunnel excessively, may be different from those in uniform cracks, even if crack areas are the same.

The errors arising from temperature and load measurements are small in comparison with the errors in determining crack depth and time of cracking.

Based on data, we estimate that the error in measurement of DHC velocity is ±25% and ±50% from the mean, in unirradiated and irradiated material, respectively.

The error in the measurement of K_{IH}, using the standard load reducing method, depends on the following:

1. uniformity of DHC crack,
2. accuracy of acoustic emission equipment,
3. accuracy of load measurement, and
4. the required "wait time" of no cracking.

As far as the crack shape is concerned, similar arguments apply to K_{IH} tests as those discussed for crack velocity tests. In a K_{IH} test, it is absolutely mandatory that the acoustic emission equipment is not noisy, because each time false counts are recorded, the load is reduced and this leads to underestimating K_{IH}.

An accurate load measurement is important because the loads are very small when K_{IH} is approached.

We estimate the error in measuring K_{IH}, using the load reducing method, to be ± 1 MPa$\sqrt{m}$ ($\pm 15\%$).

Microstructural Examinations

The microstructure of cold-worked Zr-2.5Nb consists of flat, α-zirconium grains surrounded by thin layers of continuous β-zirconium, Fig. 4. To study the evolution of the microstructure, notably the decomposition of the β-phase during the irradiation, and the evolution of the dislocation substructure, X-ray diffraction (XRD) was used; transmission electron microscopy was not able to resolve the changes occurring in the β-phase or evaluate the dislocation substructure quantitatively. Specimens for analysis by XRD were prepared by cutting thin slices from the tube perpendicular to each of the three principal tube axes, that is, longitudinal normal (LN), transverse normal (TN), radial normal (RN). The LN($10\bar{1}0$)α, TN(0002)α, LN(110)β, and RN(200)β diffraction lines were measured using a Rigaku rotating anode diffractometer with CuK$_\alpha$ radiation. The RN($11\bar{2}0$)α diffraction lines were measured using a Siemens diffractometer also operating with CuK$_\alpha$ radiation. The line shapes of the X-ray profiles for the α-phase were analyzed using the Fourier method and these were interpretted in terms of dislocation density for <a>-type and <c>-component dislocations [9,10].

Results

The factors that affect DHC were expected to be irradiation fluence, irradiation temperature, test temperature, and direction of testing on the plane normal to the transverse direction.

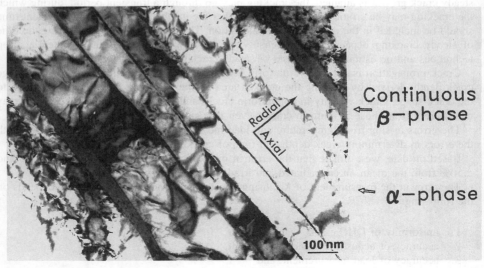

FIG. 4—*Microstructure of cold-worked Zr-2.5Nb pressure tube.*

Effect of Irradiation Fluence

The effect of irradiation fluence was studied on specimens prepared from the R/J and main body regions of pressure tubes removed from power reactors, and on specimens from the end-of-life irradiations in France. DHC velocities in both the radial and axial directions were determined as a function of fluence. Tensile properties in the transverse direction were determined as a function of fluence.

In the R/J region, the crack velocity increased rapidly with fluence, Figs. 5 and 6. The results from the end-of-life irradiations (TRILLIUM and ERABLE) showed a weak dependence of the crack velocity on fluence, Fig. 6. In specimens from the main bodies of pressure tubes, the crack velocity was three to five times higher than that of unirradiated material and was found to be independent of fluence, Figs. 5 and 6. These observations suggest that the effect of irradiation saturates at about 1×10^{25} n/m². The yield stress follows a similar behavior, Fig. 7 [11]; it saturates at about 0.3×10^{25} n/m².

Effect of Irradiation Temperature

In CANDU reactors, the temperature along the channel increases from about 250°C at the inlet end to about 290 to 300°C at the outlet end. As shown in Figs. 5 and 6, the crack velocities are higher at the inlet end than at the outlet end. Figures 8 and 9 show the crack velocity as a function of position along the fuel channel. The velocity drops off sharply towards the R/Js and decreases gently from the inlet end towards the outlet end. These results are obtained on 45 pressure tubes and the scatter includes tube-to-tube variability.

Direction of Testing

Shapes of cracks in early Pickering and Bruce reactors indicated that the rate of crack growth in the axial direction of the tube is about twice that in the radial direction. This

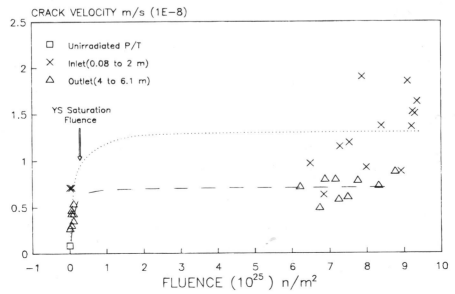

FIG. 5—*Dependence of axial DHC velocity on irradiation fluence (test temperature: 130°C).*

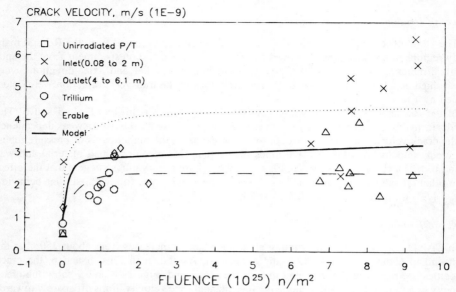

FIG. 6—*Dependence of radial DHC velocity on irradiation fluence (test temperature: 130°C).*

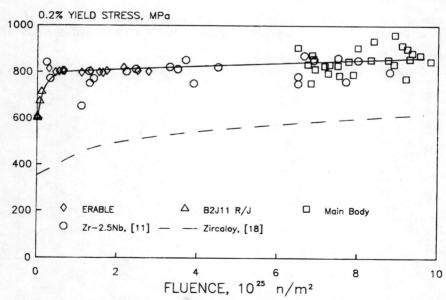

FIG. 7—*Dependence of yield stress on irradiation fluence (test temperature: 250°C).*

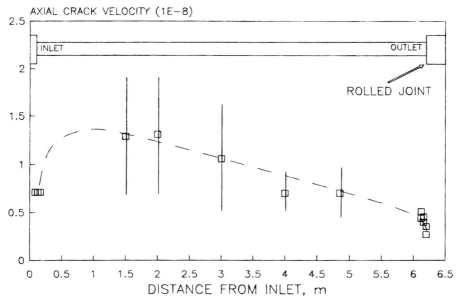

FIG. 8—*Variation of axial DHC velocity along pressure tube (test temperature: 130°C).*

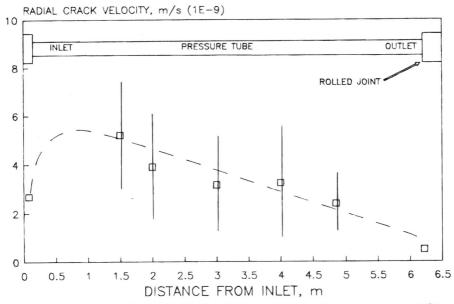

FIG. 9—*Variation of radial DHC velocity along pressure tube (test temperature: 130°C).*

observation was confirmed in a round-robin test performed at Chalk River Laboratories (CRL), Whiteshell Laboratories (WL), and Ontario Hydro Research Division (OHRD). Standard, unirradiated Zr-2.5Nb pressure tube material with two hydrogen concentrations, 0.36 and 0.50 atomic percent, was used in this round robin. WL and OHR used CCT specimens and measured crack velocity in the axial direction of the pressure tube, and CRL used CB specimens and measured crack velocity in both the radial and axial directions. The objective of this round robin was to determine the variability in crack velocity between CCT and CB specimens and to determine the effect of crack growth direction on DHC velocity. In general, there was good agreement between all three laboratories in the axial crack velocity measurements, which confirmed that the specimen type was not a factor in crack velocity determination. The experimental uncertainty in the results from all three laboratories was less than 25% of the mean at a given hydrogen concentration and crack growth direction. The crack velocity in the axial direction was found to be 1.7 to 1.9 times higher than that in the radial direction, Table 1.

Crack velocities were also measured, in both the radial and axial directions, on the 45 pressure tubes removed from different lattice positions in four power reactors. The majority of the tubes from the power reactors had low hydrogen isotope concentrations (0.05 to 0.1 atomic percent) and had to be tested at temperatures between 120 and 140°C. Some tubes had high hydrogen isotope concentrations at the outlet end, and some specimens were hydrided electrolytically to allow testing up to 300°C.

To measure the radial velocity, cantilever beam specimens were prepared from different axial positions of these tubes. The temperature dependence of the crack velocity followed an Arrhenius relationship as shown in Fig. 10. A least-squares regression produced the following expression for V in m/s

$$V_R = 0.0863 \exp(-58/RT) \qquad (1)$$

where

$R = 8.314 \times 10^{-3}$ kJ/mol,
T = the absolute temperature, and 95% confidence limits are a factor of ± 2.6 from the mean.

Crack velocities, measured on the offcut material from two tubes, are grouped around the lower bound line of the irradiated material, Fig. 10. In these tests, many specimens were prepared from sections near the inlet and outlet ends of pressure tubes. Comparing crack velocities in these sections, it was observed that those coming from the inlet sections had consistently higher crack velocities than those from the outlet sections, as shown in Fig. 11. The average crack velocity at the inlet end is about twice that at the outlet end. The least-square regression analysis gave the following expression for the mean V, in m/s, for inlet and outlet, respectively

TABLE 1—*Comparison of axial (V_A) and Radial (A_R) velocities determined at CRL.*

H concentration, atomic %	Test Temperature, °C	Axial V, m/s	Radial V, m/s	V_A/V_R
0.5	250	$11.8 \times 10^{-8} \pm 3.1$	$6.1 \times 10^{-8} \pm 1.3$	1.9
0.5	200	$2.3 \times 10^{-8} \pm 0.15$	$1.2 \times 10^{-8} \pm 0.16$	1.9
0.36	250	$5.9 \times 10^{-8} \pm 0.87$	$3.5 \times 10^{-8} \pm 0.46$	1.7
0.36	200	$2.2 \times 10^{-8} \pm 0.23$	$1.3 \times 10^{-8} \pm 0.17$	1.7

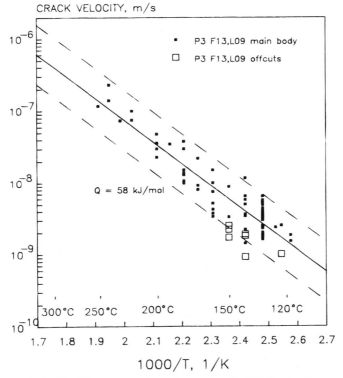

FIG. 10—*Temperature dependency of radial DHC velocity.*

$$V_{Rinl} = 0.169 \exp(-59/RT) \tag{2}$$

with 95% confidence limits equal to a factor of ±1.7 from the mean, and

$$V_{Rout} = 0.068 \exp(-58/RT) \tag{3}$$

with 95% confidence limits equal to a factor of ±2.1 from the mean.

Figure 12 shows the temperature dependence of the axial crack velocity. The least-square analysis gave the following expression for the mean V, in m/s

$$V_A = 4.02 \times 10^{-3} \exp(-43/RT) \tag{4}$$

with 95% confidence limits equal to a factor of ±2.3 from the mean.

These results show that the temperature dependency of V in the axial direction is smaller than that in the radial direction, Fig. 13. At low temperature, $V_A > V_R$, but at 300°C, both velocities have about the same value.

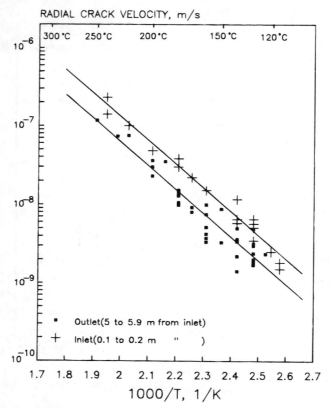

FIG. 11—*Difference between radial DHC velocity at the outlet and inlet ends of pressure tube.*

Threshold Stress Intensity Factor

The threshold stress intensity factor, K_{IH}, was measured using the load-reducing method on several tubes removed from reactors and on their offcuts. The results showed a small decrease in K_{IH} as the fluence increased to about 1×10^{25} n/m², but no decrease in K_{IH} was observed with further increase in fluence, Fig. 14. The average K_{IH} in the offcuts was 7.5 ± 1.3 MPa$\sqrt{m}$ and, in the irradiated material, it was 6.2 ± 0.9 MPa$\sqrt{m}$ at the 95% confidence level. Within the temperature range of 140 to 250°C, K_{IH} remained constant, Fig. 15.

Microstructural Examinations

XRD analysis shows that enrichment of the β-phase by niobium occurs during service and increases significantly from the inlet toward the outlet, Fig. 16. The enrichment is the result of a gradual decomposition of metastable 20 to 40 atomic percent niobium-enriched β-Zr towards a stable β-Nb phase about 88 atomic percent niobium at 300°C. X-ray diffraction indicates a variation in <a>-type dislocation density along the length of the tubes, being highest at the inlet end and decreasing by about 20% toward the outlet end, Fig. 17. The data from samples taken from within 0.5 m of the centers of the tubes show that the <a>-type dislocation density has saturated by fluences of about 2×10^{25} n/m², Fig. 18.

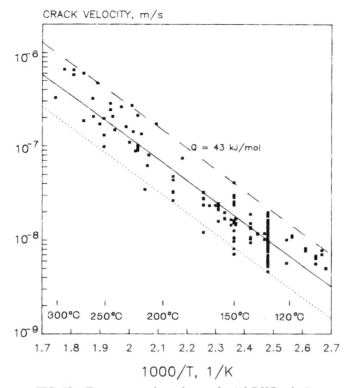

FIG. 12—*Temperature dependency of axial DHC velocity.*

Discussion

The results are summarized as follows. The crack velocity, V, of DHC in cold-worked Zr-2.5Nb is increased by a factor of up to five times, and K_{IH} is slightly reduced by neutron irradiation. The effects saturate after about 1×10^{25} n/m². V decreases with increase in irradiation temperature. In unirradiated material, V in the axial direction is twice that in the radial direction, but in irradiated material, the dependence of V on direction changes with testing temperature. Neutron irradiation produces damage in the crystal lattice of Zr-2.5Nb in the form of predominantly type-<a> dislocation loops that harden the material. The irradiation hardening saturates at a fluence of about 2×10^{25} n/m² and so does the yield stress. The β-phase decomposition is controlled by the irradiation temperature—the higher the temperature, the greater is the decomposition. Both the yield stress and the state of the β-phase decomposition have been shown to affect DHC velocity [12,13]. The results will be discussed in the context of these observations.

Effect of Irradiation Fluence

The effect of irradiation fluence on DHC can be explained by the theory of delayed hydride cracking developed by Dutton and Puls [14], and later improved by Puls [8,15] and Ambler [16]. According to this theory, when there is a hydride present at the crack tip, its rate of growth is determined by the rate of diffusion of hydrogen into the crack-tip region. The driving

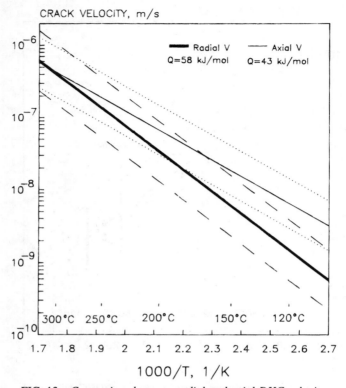

FIG. 13—*Comparison between radial and axial DHC velocity.*

forces for diffusion are: hydrogen concentration gradient, stress gradient, and thermal gradient. The theory assumes that the crack velocity is equal to the rate of growth of the hydride at the crack tip. The rate of diffusion of hydrogen atoms, *dN/dt*, into the cylinder of unit length and radius, *r*, is

$$\frac{dN}{dt} = -2\pi r J_H \tag{5}$$

where J_H is the hydrogen flux (atoms · m^{-2} · s^{-1}).

Assuming that the crack velocity is equal to the rate of growth of the hydride at the crack tip, the crack velocity, *da/dt*, can be written as

$$\frac{da}{dt} = \frac{-2\pi r J_H}{\alpha a_c N_H} \tag{6}$$

where

α = the thickness to length ratio of the hydride,
αa_c = the thickness of the hydride at fracture, and
N_H = the atomic density of the hydride.

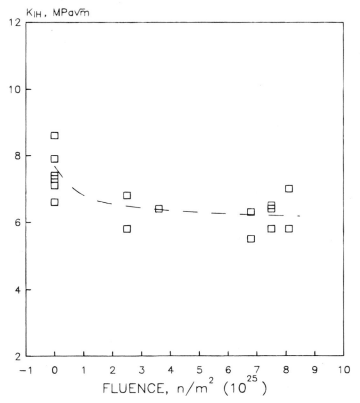

FIG. 14—*Dependence of K_{IH} on irradiation fluence.*

Substituting for hydrogen flux

$$J_H = \frac{D_0 C_0}{r\Omega_{Zr}\phi} (E_L - E_1) \qquad (7)$$

the crack velocity, *da/dt*, can be written as

$$\frac{da}{dt} = \frac{2\pi D_0 C_0}{\Omega_{Zr}\alpha a_c N_H \phi} (E_L - E_1) \qquad (8)$$

where

$$E_{L,1} = \exp\left\{\frac{(w_t^{inc})_{L,1} + (w_t^a)_{L,1}}{xR(T)_{L,1}}\right\} \qquad (9)$$

$$\phi = \int_1^L \frac{1}{r} \exp\frac{(F - w_H^a + Q_D)}{RT} dr \qquad (10)$$

The remaining parameters in Eqs 5 to 10 are given in Appendix I.

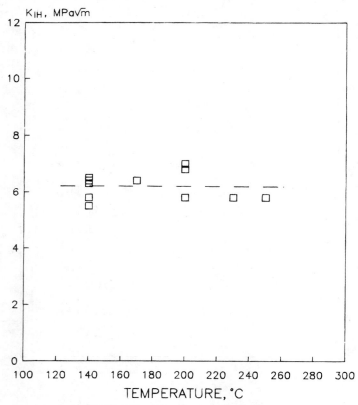

FIG. 15—*Temperature dependency of* K_{IH}.

In this model, the effect of irradiation fluence can be accounted for through the value of the yield stress of the material as shown by Puls [15]. There are two terms that are sensitive to yield stress: the position of the maximum stress, a_c, and the interaction energy of the crack-tip hydride with the locally applied stress, $(w_t^a)_1$. a_c is related to the plastic zone size at the crack tip that depends on the yield stress. $(w_t^a) = -6.66\sigma_y$ for δ-hydrides. Using the average yield stress from Fig. 7, the model predicts an increase in the mean DHC velocity due to the irradiation hardening to be a factor of about 3, Fig. 19. The current model does not account for either the direction of the crack growth or for the decomposition of the β-phase. However, as diffusion data become available for the anisotropic microstructures of pressure tubes, the model can be expanded to include the crack growth in two-phase material in both the radial and axial directions.

Crack velocity in unirradiated cold-worked Zircaloy-2 is much lower than that in Zr-2.5Nb [2,17]. Results of experiments show that irradiation to a fluence of 7.7×10^{25} n/m² increases the crack velocity in Zircaloy-2 by a factor of 50, making it similar to that of irradiated Zr-2.5Nb, Fig. 19 (for Harvey tubes in Ref 2 that had a similar texture to that of standard Zr-2.5Nb pressure tubes). Using an increase in yield stress from 350 to 600 MPa from irradiation [18], and the diffusion coefficient for α-Zr, the preceding model predicts an increase in the crack velocity, but only by an order of magnitude. Clearly, the increase in yield stress alone cannot fully account for the observed behavior in Zircaloy-2.

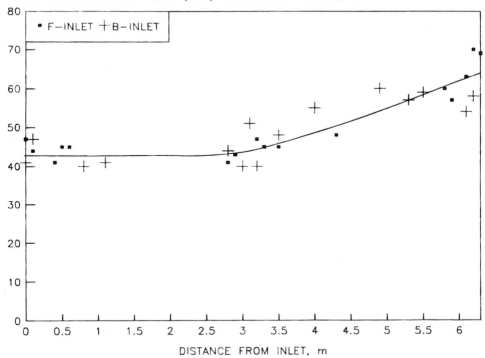

FIG. 16—*Percentage of niobium in the β-phase of Zr-2.5Nb pressure tubes as a function of position relative to the inlet, for back (B) and front (F) ends relative to the extrusion process.*

A conceptual model for K_{IH} was recently developed [*19*]. The theoretical expression for K_{IH} is given in terms of the elastic constants, the yield stress of the material, the hydride fracture stress (σ_f^h), and the thickness of the crack-tip hydride (t) as follows

$$(K_{IH})^2 = \frac{E \cdot t}{1 - \nu^2} \left\{ \frac{0.043 E \cdot \epsilon_{yy}}{(1 - \nu^2)\left[\dfrac{1}{1 - 2\nu} - \dfrac{\sigma_f^h}{\sigma_y}\right]} - \gamma \cdot \sigma_y \right\} \qquad (11)$$

where

σ_y = the yield stress of the material, and
γ = a constant (~0.025).

Other parameters in this equation are defined in Appendix I. The values of K_{IH} calculated by this model are smaller than those measured experimentally, Table 2. The model assumes that the crack-tip hydride covers the entire fracture surface. The higher values measured in experi-

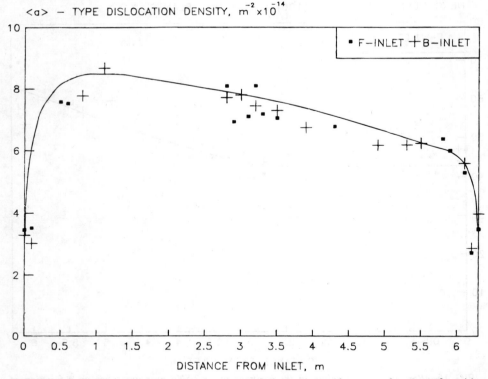

FIG. 17—*<a>*-type dislocation density in Zr-2.5Nb pressure tubes as a function of position relative to the inlet, for back (B) and front (F) ends relative to the extrusion process.

ments were assumed to be caused by the fact that the hydrides may not cover the whole front of the crack and the experimental K_{IH}^{exp} was related to the theoretical K_{IH} as follows

$$K_{IH}^{exp} = fK_{IH} + (1 - f)K_i \qquad (12)$$

where

f = the fraction of hydride coverage at the crack tip, and
K_i = the initiation fracture toughness of the matrix (zirconium alloy) material.

Substituting $K_{IH} = 3.7$ MPa$\sqrt{m}$, $K_{IH}^{exp} = 6.2$ MPa$\sqrt{m}$, and $K_i = 30$ MPa$\sqrt{m}$ into Eq 12, f is 0.9, that is, hydrides should cover nearly all of the area ahead of the crack tip. Our preliminary results based on metallographic examinations of hydrides at crack tips show that, at low values of K_I (near K_{IH}), the hydrides are well developed and continuous. However, further examinations need to be done to quantify the state of the crack-tip hydrides more accurately.

The model predicts a slight reduction in K_{IH} with irradiation that agrees with experimental results. This reduction is mainly due to an increased yield stress by neutron irradiation in the irradiated material. Our preliminary results on unirradiated Zr-2.5Nb indicate that a material with a lower yield stress has a higher value of K_{IH}, but most of such material also has a

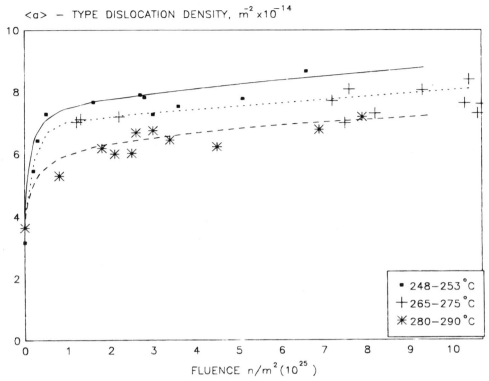

FIG. 18—*Variation of dislocation density as a function of fluence for samples of Zr-2.5Nb pressure tubes taken from within 0.5 m of the center line of the reactors.*

different crystallographic texture, hence, the effect of yield stress alone cannot be easily isolated. Clearly, the relationship between K_{IH} and yield stress in unirradiated Zr-2.5Nb material needs to be established.

Effect of Irradiation Temperature

It has been shown that the diffusion coefficient in the pure β-phase is about two orders of magnitude above the diffusion coefficient in α-phase [20,21]. In two-phase, cold-worked Zr-2.5Nb material, the presence of the β-phase enhances (up to an order of magnitude) the hydrogen diffusivity, compared with that in the α-phase. Experiments have shown that the crack velocity in Zr-2.5Nb alloy is greater in the material with the continuous β-phase than that in materials with decomposed β-phase [13,21]. These observations can be used in explaining the different crack velocities at the inlet and outlet ends of irradiated pressure tubes. The DHC experiments showed higher crack velocities at inlet ends of irradiated pressure tubes than at the outlet ends, while X-ray diffraction showed a smaller degree of β-phase decomposition at inlet ends than at outlet ends, Figs. 5, 6, 11, and 16. The profile of the crack velocity in an irradiated pressure tube, Figs. 8 and 9, is mainly the result of differences in the β-phase decomposition along the tube, Fig. 16, and to a lesser degree because of decreasing irradiation hardening, Fig. 17.

TABLE 2—*Predicted and measured K_{IH} values in unirradiated and irradiated Zr-2.5Nb materials.*

Temperature, °C	Predicted K_{IH}, MPa$\sqrt{m}$	Measured K_{IH}^{exp}, MPa$\sqrt{m}$
	UNIRRADIATED	
150	4.1	7.5 ± 1.3
250	5.3	7.5 ± 1.3
	IRRADIATED	
150	3.7	6.2 ± 0.9
250	4.6	6.2 ± 0.9

Temperature Dependence of DHC Velocity

The temperature dependence in the DHC model derives primarily from the sum of the activation energies for diffusion and solubility contained in the $D_H C_H$ product, which is 70 kJ/mol, Appendix I. A secondary effect comes from the temperature dependence of the yield stress, which reduces the total activation energy for DHC to about 60 kJ/mol, Fig. 19. DHC experiments have shown that the crack velocity follows an Arrhenius type of relationship with temperature, for Zr-2.5Nb alloy

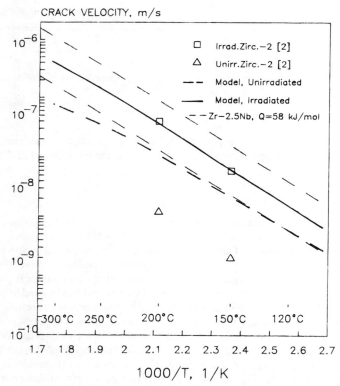

FIG. 19—*Comparison between DHC velocities in Zr-2.5Nb and Zircaloy-2.*

$$V = A \exp\left(-\frac{Q}{RT}\right) \tag{13}$$

with the coefficients listed in Table 3. All unirradiated materials, but one [7], have temperature dependencies in the range from 58 to 72 kJ/mol, which is reasonably close to the theoretical value. The results in Ref 7 were not obtained using the standard temperature cycling procedure. Most of the unirradiated materials have the β-phase decomposed from homogenization treatment at 400°C for several days after gaseous hydriding, hence, the cracking in both directions is governed mainly by the diffusivity and solubility in the α-phase. In the radial direction, the temperature dependence of cracking of the irradiated material is similar to that of unirradiated material, suggesting that the factors that dominate the temperature dependence, solubility and diffusivity in the α-phase, are not much affected by radiation. Lower temperature dependence of the axial crack velocity in the irradiated material can be explained in terms of the contribution of the β-phase (which is only partially decomposed in the irradiated material) to the diffusion of hydrogen in this direction (the activation energy for diffusion in the β-phase is lower than that in the α-phase [20,21]). These ideas are being explored in our current research programs.

Conclusions

1. Neutron irradiation increases the velocity of DHC in cold-worked Zr-2.5Nb by a factor of 3 to 5 times, and reduces K_{IH} by a small amount, about 20%. The effect of irradiation saturates at a fluence of about 1×10^{25} n/m^2.

2. Temperature dependence of the crack velocity in both the radial and axial directions in the unirradiated material, and in the radial direction in the irradiated material, follows the theoretical predictions based on a diffusional model for the α-phase and the change in yield stress with temperature and irradiation. In the axial direction, the dependence on test temperature does not follow the theory based on diffusion in the α-phase; the theory needs to incorporate a two-phase (α + β) diffusion model.

3. Crack velocity decreases with irradiation temperature primarily through the change in the configuration of the β-phase and, secondarily, through changes in dislocation density.

TABLE 3—*Temperature dependence of DHCV velocity for unirradiated and irradiated Zr-2.5Nb material.*

Direction	A	Q, kJ/mol	Ref
	UNIRRADIATED		
Radial[a]	6.9×10^{-1}	72	[4]
Radial[a]	2.1×10^{-2}	59	this paper
Radial[b]	4.0×10^{-2}	58	this paper
Radial[a]	1.4×10^{-4}	42	[7]
Axial[a]	1.5×10^{-1}	66	[8]
Axial[a]	5.3×10^{-2}	60	this paper
	IRRADIATED		
Radial[c]	8.6×01^{-2}	58	this paper
Axial[c]	4.0×01^{-3}	43	this paper

[a]Hydrogen was added gaseously at 400°C followed by homogenization at 400°C for 72 h.
[b]The material was in the autoclaved condition, that is, it was heated at 400°C for 24 h.
[c]Some irradiated specimens had hydrogen added electrolytically at 90°C followed by solution treatment at 290°C for 14 days.

Acknowledgments

We would like to thank the following people for assistance with experiments: G. Brady, P. Dhanjal, R. W. Gilbert, J. Kelm, W. G. Newell, D. Sage, J. M. Smeltzer, K. D. Weinsenberg, J. E. Winegar, and the staff of the General Chemistry Branch and the Universal Cells at CRL. We enjoyed useful discussions with Drs. M. P. Puls and S. Q. Shi. We thank OHRD for permission to use the TG3R1 material information. The work was funded through Working Party 31 and 33 of the CANDU Owners Group.

APPENDIX I

Parameters used in DHC Model

D_0 2.17×10^{-7} is a pre-exponential term in the expression for the diffusion coefficient of hydrogen in zirconium: $D_H = 2.17 \times 10^{-7} \exp(-35\,100/RT)$ m²/s

C_0 10.2 is a pre-exponential term in the expression for the concentration of hydrogen in solid solution when there is no external stresses: $C_H^S = 10.2 \exp(-35\,000/RT)$ atomic %

Ω_{Zr} 2.3×10^{-29} m³/atom is the atomic volume of zirconium

α The thickness to length ratio of the hydride

a_c†   The distance from the crack tip to the position of the maximum stress = $(0.144\, K_I/\sigma_y)^2$

N_H 6.13×10^{28} atom/m³ is the atomic density of the hydride

w_H^{at} $0.577(K_I/\sqrt{r})$ J/mol is the molal interaction energy of the stresses with the hydrogen in solid solution

Q_D 35 100 J/mol is the activation energy for diffusion of hydrogen in zirconium from the expression for D_H

$(w_t^{inc})_{L,1}$ 4912 J/mol is the molal elastic strain energy of the matrix and fully constrained δ-hydrides with plate normals parallel to direction [0001]

$(w_t^a)_L^{\ddagger}$ $-0.9447(K_I/\sqrt{L})$ J/mol is the interaction energy of matrix δ-hydrides with the applied stress

$(w_t^a)_1^{\dagger}$ $-6.6563\,\sigma_y$ J/mol is the interaction energy of crack-tip δ-hydrides with the applied stress

x 1.66 is the mole fraction of hydrogen in ZrH$_{1.66}$ hydride

Q_H 35 kJ/mol is the activation energy of the equilibrium hydrogen concentration given in the expression for C_H^S

K_I The Mode I stress intensity factor in MPa$\sqrt{m}$

ν The Poisson's ratio in zirconium = 0.3

V_H 16.7×10^{-7} m³/mol is the partial molar volume of hydrogen in zirconium

V_{hydr} 16.3×10^{-6} m³/mol is the molal volume of hydride with a composition ZrH$_x$

Q^* 25 100 J/mol is the heat of transport of hydrogen in zirconium

ϵ_{yy} The stress-free strain of the hydride in the y direction = 0.054

E Young's modulus in the transverse direction = $102.47 - 0.011\,743\,T - 8.068\,69 \times 10^{-5}\,T^2$, where E is in GPa and T is in °C

σ_f^h The hydride fracture stress = $0.007\,357\,E$

The expressions marked by a dagger (†) are derived in Appendix II.

APPENDIX II

Derivation of formulas marked by a dagger (†) in Appendix I.

Position of the Maximum Stress, a_c

The maximum stress, a_c, is assumed to be equal to the position of the maximum stress in the plastic zone that is derived by equating the hydrostatic stress at the crack tip, p_h, to that in the matrix and then solving for r (r is defined in Fig. 20)

$$p_h(r) = -(\sigma_{11} + \sigma_{22} + \sigma_{33})/3 = -\tfrac{2}{3}(K_I/\sqrt{2\pi r})(1 + \nu) = -0.346\, K_I/\sqrt{r}$$

According to Rice and Johnson [22], the maximum hydrostatic stress is

$$p_h(1) = -2.4\,\sigma_y$$

hence,

$$2.4\,\sigma_y = 0.346\, K_I/\sqrt{r}$$

solving for r

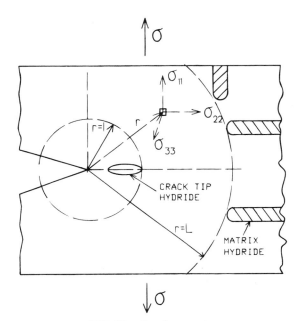

FIG. 20—*Crack geometry.*

$$r(1) = (0.144 \, K_I/\sigma_y)^2$$

where σ_y is in Mpa and K_I in MPa$\sqrt{m}$.

Molal Interaction Energy of the Stresses

The molal interaction energy of the stresses with the hydrogen in solid solution is expressed by

$$w_H^a = p_h V_H$$

where $V_H = 16.7 \times 10^{-7}$ m^3/mol and $p_h = (\sigma_{11} + \sigma_{22} + \sigma_{33})/3$ for plane strain conditions

$$\sigma_{11} = \sigma_{22} = K_I/(\sqrt{2\pi L})$$

$$\sigma_{33} = \nu(\sigma_{11} + \sigma_{22}) = 2\nu K_I/\sqrt{2\pi L}$$

At L (matrix)

$$w_H^a = 2/3(K_I/\sqrt{2\pi L})(1 + \nu)V_H = 0.577 \, K_I/\sqrt{L}$$

where K_I is in MPa$\sqrt{m}$.

Interaction Energy of Matrix δ-Hydrides with the Applied Stress

At L (matrix)

$$(w_t^a)_L = -V_{\text{hydr}} \Sigma \sigma_{ij} e_{ij} = -V_{\text{hydr}} \, K_I/\sqrt{2\pi L}(e_{11} + e_{22} + e_{33})$$

for δ-hydrides $e_{11} = 0.72$, $e_{22} = 0.0458$, $e_{33} = 0.0458$

$$(w_t^a)_L = -0.9447 \, K_I/\sqrt{L}$$

where K_I is in MPa$\sqrt{m}$ and $V_{\text{hydr}} = 16.3 \times 10^{-6}$ m^3/mol.

At l (crack tip)

$$(w_t^a)_l = -V_{\text{hydr}}(3\sigma_y e_{11} + 1.8\sigma_y e_{22} + 2.4\sigma_y e_{33})$$

$$(w_t^a)_l = -6.6563 \sigma_y$$

for δ-hydride.

References

[1] Moan, G. D., Coleman, C. E., Price, E. G., Rodgers, D. K., and Sagat, S., "Leak-Before-Break in the Pressure Tubes of CANDU Reactors," *International Journal of Pressure Vessels & Piping*, Vol. 43, 1990, pp. 1–21.

[2] Huang, F. H. and Mills, W. J., "Delayed Hydride Cracking Behavior for Zircaloy-2 Tubing," *Metallurgical Transactions*, Vol. 22A, 1991, pp. 2049–2060.

[3] Fleck, R. G., Price, E. G., and Cheadle, B. A., "Pressure Tube Development for CANDU Reactors," *Zirconium in the Nuclear Industry: Sixth International Symposium, ASTM STP 824*, D. G. Franklin

and R. B. Adamson, Eds., American Society for Testing and Materials, Philadelphia, 1984, pp. 88–105.
[4] Ambler, J. F. R., "Effect of Direction of Approach to Temperature on the Delayed Hydrogen Cracking Behaviour of Cold-Worked Zr-2.5Nb," *Zirconium in the Nuclear Industry, ASTM STP 824*, D. G. Franklin and R. B. Adamson, Eds., American Society for Testing and Materials, Philadelphia, 1984, pp. 653–674.
[5] Sagat, S., Ambler, J. F. R., and Coleman, C. E., "Application of Acoustic Emission to Hydride Cracking," 29th Acoustic Emission Working Group Meeting, The Royal Military College of Canada, Kingston, Ontario, 1986.
[6] Simpson, L. A. and Clarke, C. F., "Application of the Potential-Drop Method to Measurement of Hydrogen-Induced Sub-Critical Crack Growth in Zirconium-2.5Nb," AECL Research Report, AECL-5815, Whiteshell Laboratories, Pinawa, Manitoba, 1977.
[7] Coleman, C. E. and Ambler, J. F. R., "Susceptibility of Zirconium Alloys to Delayed Hydrogen Cracking," *Zirconium in the Nuclear Industry (3rd Conference), ASTM STP 633*, A. L. Lowe, Jr., and G. W. Parry, Eds., American Society for Testing and Materials, Philadelphia, 1977, pp. 589–607.
[8] Simpson, L. A. and Puls, M. P., "The Effect of Stress, Temperature and Hydrogen Content on Hydride-Induced Crack Growth in Zr-2.5Nb," *Metallurgical Transactions*, Vol. 10A, 1979, pp. 1093–1105.
[9] Holt, R. A., "Recovery of Cold-Work in Extruded Zr-2.5Nb," *Journal of Nuclear Materials*, Vol. 59, 1976, pp. 234–242.
[10] Griffiths, M., Winegar, J. E., Mecke, J. F., and Holt, R. A., "Determination of Dislocation Densities in HCP Metals Using X-ray Diffraction and Transmission Electron Microscopy," *Advances in X-ray Analysis*, Vol. 35, 1992, pp. 593–599.
[11] Chow, C. K, Coleman, C. E., Hosbons, R. R., Davies, P. H., Griffiths, M., and Choubey, R., "Fracture Toughness of Irradiated Zr-2.5Nb Pressure Tubes from CANDU Reactors," *Zirconium in the Nuclear Industry: Ninth International Symposium, ASTM STP 1132*, C. M. Eucken and A. M. Garde, Eds., American Society for Testing and Materials, Philadelphia, 1991, pp. 246–275.
[12] Shek, G. K. and Graham, D. B., "Effects of Loading and Thermal Manoeuvres on Delayed Hydride Cracking in Zr-2.5Nb Alloys," *Zirconium in the Nuclear Industry: Eighth International Symposium, ASTM STP 1023*, L. F. P. Von Swam and C. M. Eucken, Eds., American Society for Testing and Materials, Philadelphia, 1989.
[13] Simpson, L. A. and Cann, C. D., "The Effect of Microstructure on Rates of Delayed Hydride Cracking in Zr-2.5Nb Alloy," *Journal of Nuclear Materials*, Vol. 126, 1984, pp. 70–73.
[14] Dutton, R. and Puls, M. P., "A Theoretical Model for Hydrogen Induced Sub-Critical Crack Growth," *Proceedings*, Conference on Effect of Hydrogen on Behavior of Materials, American Institute of Mining, Metallurgical, and Petroleum Engineers, 7–11 Sept. 1975, pp. 516–525.
[15] Puls, M. P., "Effect of Crack Tip Stress States and Hydride-Matrix Interaction Stresses on Delayed Hydride Cracking," *Metallurgical Transactions*, Vol. 21A, 1990, pp. 2905–2917.
[16] Ambler, J. F. R., "The Effect of Thermal Gradients on the Delayed Hydride Cracking Behaviour of Zirconium Alloys—Mathematical Model," AECL Research, unpublished work, Chalk River Laboratories, Chalk River, Ontario, 1985.
[17] Puls, M. P., Simpson, L. A., and Dutton, R., "Hydride-Induced Crack Growth in Zirconium Alloys," AECL Research Report, AECL-7392, Whiteshell Laboratories, Pinawa, Manitoba, 1982.
[18] Coleman, C. E., Cheadle, B. A., Causey, A., Chow, C. K., Davies, P. H., McManus, M. D., Rodger, D. K., Sagat, S., and van Drunen, G., "Evaluation of Zircaloy-2 Pressure Tubes from NPD," *Zirconium in the Nuclear Industry: Eighth International Symposium, ASTM STP 1023*, L. F. P. Van Swam and C. M. Eucken, Eds., American Society for Testing and Materials, Philadelphia, 1989, pp. 35–49.
[19] Shi, S. Q. and Puls, M. P., "Criteria of Fracture Initiation at Hydrides in Zirconium," to be published.
[20] Sawatzky, G. A., Ledoux, R. L., Tough, R. L, and Cann, C. D., "Hydrogen Diffusion in Zirconium-Niobium Alloys," *Metal-Hydrogen Systems, Proceedings*, Miami International Symposium, 1981, Pergamon Press, Oxford, 1982, pp. 109–120.
[21] Skinner, B. C. and Dutton, R., "Hydrogen Diffusivity in α-β Zirconium Alloys and its Role in Delayed Hydride Cracking," *Hydrogen Effects on Material Behaviour*, N. R. Moody and A. W. Thomson, Eds., The Minerals, Metals & Materials Society, Jackson Lake Lodge, 1989.
[22] Rice, J. R. and Johnson, M. A. in *Inelastic Behavior of Solids*, M. F. Kanninen, Ed., McGraw-Hill, New York, 1970, p. 641.

DISCUSSION

Brian Cox[1] (written discussion)—One of the questions that D. Franklin hoped to answer in his introduction was whether DHC cracks could be initiated from smooth surfaces, given a high enough stress maintained for a long enough time. Can you comment on whether or not DHC cracks can be initiated at smooth surfaces, and the times and stresses needed to do this.

S. Sagat et al. (authors' closure)—Early experiments by Cheadle and Ells[2] showed that DHC cracks can be initiated at smooth surfaces. In these experiments, smooth (smooth refers to the original surface quality of the as-received pressure tube) cantilever beam specimens were machined from cold-worked Zr-2.5Nb pressure tubes. Some of the specimens were gaseously hydrided at 400°C to a hydrogen concentration of 40 to 120 ppm, the remainder of specimens were tested with the as-received hydrogen concentration that was between 10 and 15 ppm. The specimens were tested in pure bending. After loading, the specimens were heated to 300°C and then cooled to the test temperature to reorient some of the hydrides under the applied stress. The as-received specimens were tested at 80°C and the hydrided specimens at either 150 or 250°C.

In specimens that contained 10 to 15 ppm hydrogen, cracks initiated at outerfibre stresses >550 MPa, either at surface nicks or angular SiO_2 particles that were embedded to the surface of the tubes by sand blasting operations used to clean the tubes during their fabrication. The range of failure times in these tests was between 53 and 6600 h.

In specimens with hydrogen concentration between 40 and 120 ppm, the crack initiation was not associated with any surface flaws or embedded particles. In these specimens, the cracks initiated at reoriented radial hydrides, near the surface of the specimen, at outerfiber stresses ≥413 MPa at a temperature of 150°C, and at stresses ≥585 MPa at 250°C. The range of failure times in these tests was between 2 and 1920 h.

In conclusion, crack initiation by DHC at smooth (as-fabricated) surfaces in cold-worked Zr-2.5Nb pressure tubes requires a large tensile stress at a surface asperity or hydrides close to the surface and perpendicular to the stress direction.

C. Lemaignan[3] (written discussion)—Is there any tendency for the hydrides to precipitate in α, β phase or at the interface between the two?

S. Sagat et al. (authors' closure)—The microstructure of cold-worked Zr-2.5Nb pressure tubes consists of lath-shaped α-grains (hexagonal close packed) highly elongated in the axial direction. The α-grains (with a thickness of about 0.5 μm) are surrounded by a continuous network of body centered cubic β-phase (about 0.05 μm thick and comprising approximately 7% volume fraction). The original hydrides in this material are oriented parallel to the circumferential-axial plane along the α/β phase boundaries. Very few β-grain boundaries are in the radial direction. The precipitation of hydrides in the hexagonal α-phase is known to occur on habit planes that are very close to the basal planes. The precipitation of hydrides in the β-phase is to my knowledge not very well known. If there is a crack present in the material, the hydride will precipitate at the crack tip that in most cases is the α-phase. No systematic study has been done to see what happens if the β-phase is at the crack tip. However, the DHC crack does not seem to have any problems growing through both, the α and β-phase, as it grows through the pressure tube wall.

---

[1] University of Toronto, Toronto, Ontario, Canada.
[2] Cheadle, B. A., "Crack Initiation in Cold-Worked Zr-2.5Nb by Delayed Hydrogen Cracking," *Proceedings*, Second International Congress on Hydrogen in Metals, Paris, 6-11/VI/1977; also as AECL-5799.
[3] CEA, Grenoble, France.

Suresh K. Yagnik[4] *(written discussion)*—What is the typical hydrogen concentration at the crack-tip for DHC mechanism? Is there a threshold value for this?

S. Sagat et al. (authors' closure)—The hydrogen concentration at the crack tip that is required for DHC initiation has to be greater than the terminal solid solubility (TSS) limit of hydrogen in zirconium for a given temperature. The Zr-2.5Nb material exhibits hysteresis between the heatup and cooldown TSS. TSS is also affected by the applied stress. To derive the required hydrogen concentration for DHC to occur, one has to chose the appropriate TSS. A theoretical treatment of hydrogen solubility in zirconium alloys can be found in two publications written by Puls.[5,6]

---

[4] Electric Power Research Institute, Palo Alto, CA.
[5] Puls, M. P., "The Effect of Misfit and External Stress on Terminal Solid Solubility in Hydride-Forming Metals," *Acta Metallurgica,* Vol. 29, 1981, pp. 1961–1968.
[6] Puls, M. P., "On the Consequences of Hydrogen Supersaturation Effects in Zr Alloys to Hydrogen Ingress and Delayed Hydride Cracking," *Journal of Nuclear Materials,* Vol. 165, 1989, pp. 128–141.

Mohammad B. Elmoselhi,[1] *Brian D. Warr*,[1] *and Stewart McIntyre*[2]

A Study of the Hydrogen Uptake Mechanism in Zirconium Alloys

REFERENCE: Elmoselhi, M. B., Warr, B. D., and McIntyre, S., "**A Study of the Hydrogen Uptake Mechanism in Zirconium Alloys**," *Zirconium in the Nuclear Industry: Tenth International Symposium, ASTM STP 1245,* A. M. Garde and E. R. Bradley, Eds., American Society for Testing and Materials, Philadelphia, 1994, pp. 62–79.

ABSTRACT: Hydrogen uptake in zirconium alloy CANDU (CANada Deuterium Uranium) pressure tubes and other core components is controlled by the rate of transport of atomic/ionic species across the oxide film. The importance of understanding the mechanism of transport stems from the need to predict and control the rate of uptake. Samples of Zr-2.5Nb and Zircaloy-2 were prefilmed in steam (H_2O, 400°C at ~2 MPa) and subsequently exposed to D_2O (10^{-3} Pa to ~18 MPa) and D_2 (~10^{-3} Pa) at a temperature range of 250 to 380°C in the laboratory. Samples from Zr-2.5Nb pressure tubes removed from CANDU power reactors were also examined. Hydrogen mobility in oxides was investigated by secondary ion mass spectroscopy (SIMS) following these exposures. Diffusional-type through-oxide-thickness deuterium profiles have been observed adjacent to the oxide-metal interface for samples exposed to environments containing D_2O for 4 h out-reactor and up to ~10 years in-reactor. These profiles probably represent the density of accessible sites on surfaces of intergranular porosity through-thickness. Although, in small regions observed by transmission electron microscopy (TEM) such porosity has not been found. Nevertheless, from observations of grain size, sufficient sites would be available to produce deuterium concentration observed near oxide surfaces. The observed deuterium concentration profiles appear to result predominantly from deuteroxyl groups bonded to such sites. Deuterium content at the oxide-metal interface provides an indication of the extent of interfacial intergranular porosity. High deuterium contents at the interface may imply local regions with absent oxide barrier at the interface. In the presence of sufficient D_2O, the oxide is continually healed, and deuterium uptake is relatively low where short-circuit routes such as intermetallics in Zircaloy-2 are not present. In environments with relatively high $D_2:D_2O$ ratios, deuterium atoms may diffuse through the oxide to the interface and react directly with the metal resulting in high deuterium uptake rates. It is proposed that observed deuterium profiles may be the sum of mainly two components. The predominant component is due to deuteroxyl groups residing on accessible sites on surfaces of intergranular porosity with no direct link to hydrogen uptake by the bulk alloy. The second masked component would be due to another mobile hydrogen species (for example, H) that is diffusing to the bulk alloy. Further work is needed to substantiate the proposed hypothesis that would include exposures with varying $D_2:D_2O$ ratios and further TEM examination.

KEY WORDS: zirconium, zirconium alloys, nuclear materials, nuclear applications, hydrogen ingress, hydrogen diffusion, zirconium oxides, secondary ion mass spectrometry, hydrogen mobility, Zr-2.5Nb, radiation effects

---

[1] Engineer-scientist and unit head, respectively, Zr Alloy Corrosion Mechanisms Unit, Ontario Hydro Research Division, Toronto, Ontario, Canada M8Z 5S4.

[2] Director, Surface Science Western, University of Western Ontario, London, Ontario, Canada N6A 5B7.

The importance of understanding the mechanisms of hydrogen[3] uptake stems from the need to predict and control its rate. Two principal routes have been identified for hydrogen ingress into the body of operating CANDU pressure tubes. The primary heat transport system (PHTS) coolant, in contact with tube inside surface, is heavy water (D_2O) at ~10 MPa with a pH of ~10.3 maintained by LiOH, and with 3 to 10 cm^3 H_2/kg coolant added to suppress radiolytically produced oxygen. The outside surface of the tubes is exposed to a dry (D_2O partial pressure $<$~250 Pa) CO_2 annulus gas (AG) at approximately atmospheric pressure with up to ~130 Pa deuterium.

In the operating temperature range between ~250° and 310°C, hydrogen is absorbed by the alloy as the corrosion reaction (see Eq 1) is taking place. The amount absorbed by the bulk alloy is usually expressed as a percentage of the hydrogen liberated in the reaction

$$Zr + 2D_2O \rightarrow ZrO_2 + 2D_2 \tag{1}$$

Since uniform corrosion rates are generally low (<2 and ~0.1 μm/year at the inside and outside surfaces of the tubes, respectively), hydrogen ingress from corrosion is expected to be low. However, hydrogen gas present in the AG may also permeate slowly through the oxide into the bulk alloy resulting in relatively high overall rates of ingress. The objectives of research programs in this area are to identify rate-controlling processes, in order to predict and minimize ingress from either of the above-mentioned sources of hydrogen. The scope of the present work includes examination of samples removed from pressure tubes for power reactors as well as out-reactor exposures to D_2O and gases containing D_2O and D_2 at a range of temperatures and partial pressures.

Several attempts [1–9] have been made to elucidate the mechanism of transport of hydrogen through the oxide and into the bulk alloy during corrosion and due to exposure to hydrogen-containing gas. The subject has been reviewed by Cox [10]. It has been shown, by corrosion experiments using T_2:H_2O mixtures, that the hydrogen that diffuses into the bulk alloy during corrosion, does so as part of the reaction of zirconium with the water molecules, and not with dissolved hydrogen gas in the water [6]. Proposed mechanisms for hydrogen gas absorption usually include molecular or dissociative adsorption of the hydrogen molecule at the oxide surface followed by permeation through the oxide. In all cases, the diffusing species as well as permeation routes through the oxide are unclear. In the case of hydrogen gas exposure, tritium auto-radiography experiments at a relatively high temperature of 800°C supported permeation through pores and cracks in the oxide, since tritium was picked up by the base alloy without being observed in the oxide [4]. However, when tritium was injected into Zircaloy-2 specimens and its concentration profile and release from the surface monitored, a diffusion coefficient of 7.17×10^{-21} m^2/s at 300°C was inferred for a region of ~5 μm containing the oxide lending support to solid-state lattice diffusion [7]. Although the presence of hydrogen in the oxide film has been established, its mobility within the oxide has not been adequately investigated.

Experimental Procedure

Specimen Preparation and Exposures

Specimens were cut (typically ~10 by 10 by 1 mm) from cold-worked Zr-2.5Nb and Zircaloy-2 CANDU pressure tubes and offcuts. The samples were machined and oxides were

---

[3] Hydrogen may refer to hydrogen or its isotope deuterium.

grown by steam autoclaving in light water (H_2O) at ~2 MPa and 400°C for 24 h to reproduce the as-received pressure tubes condition. Controlled gaseous exposures were performed in an ultra-high vacuum (uhv) chamber (~10^{-8} Pa) equipped with sample heaters capable of maintaining temperature within ±1°C. Residual gas analysis was performed with a mass spectrometer. Gas dosing was controlled by a pressure/flow controller. Aqueous exposures were conducted with D_2O in static Inconel 600 autoclaves. The pH was maintained at 10.5 by addition of LiOH from a standard 1-M solution. Hydrogen and oxygen concentrations of 700 to 1000 μL/L and <25 ppb, respectively, were measured in samples of autoclave water following purging with H_2 gas prior to exposures.

Specimen Characterization

Secondary ion mass spectrometry (SIMS) provides a sensitive technique for obtaining the hydrogen concentration distribution through the oxide thickness at a depth resolution of ~10 nm and lateral resolution in the order of 300 nm. The technique uses a beam of energetic (~10 KeV Cs^+) primary ions rastered over an area 250 by 250 μm square to sputter (remove) several atomic layers of the sample surface. The positive and negative secondary ions from a region 60 μm diameter are extracted by electric fields and energy and mass analyzed using a mass spectrometer.

In through-oxide thickness depth profiling, the intensities of the various sputtered ions measured as a function of sputtering time produce a depth-composition profile of principal atomic species within the oxide. The intensities of the ions $^1H^-$, $^2D^-$, and $^{106}ZrO^-$ are typically measured through the oxide and into the base alloy. The $^2D^-$ profile provides an indication of the penetration of this isotope into the oxide during exposures. The $^1H^-$ content is mostly associated with hydrogen present in the oxide formed by steam prefilming (pressure tube as-received condition). By using ion-implanted standards of D and an absolute technique such as nuclear reaction analysis (NRA) for measuring concentration, quantitative SIMS depth profiles for deuterium and hydrogen have been determined. The $^{106}ZrO^-$ profile provides information on the location and width of the metal-oxide interface. The width of the interface inferred from SIMS can be quite variable and probably arises from differences in interfacial roughness.

Among the available techniques to determine the average level of hydrogen in metals in general and in zirconium alloys, in particular, the two main and most accurate practical methods used in this investigation are; (1) hot vacuum extraction mass spectrometry (HVEMS) also called isotope dilution mass spectrometry (IDMS) that measures both deuterium and hydrogen [11], and (2) differential scanning calorimetry (DSC) that measures the total hydrogen equivalent content of the sample [12]. DSC is an appropriate technique for measuring accumulated hydrogen uptake during permeation experiments, since it is insensitive to hydrogen in the oxide.

Experimental Results and Analysis

Parameters and conditions of the experiments performed in this work are given in Tables 1 and 2. The experiments are explained and the results are analyzed in the following subsections.

Deuterium Content Near Oxide Surface

Exposing zirconium oxide to a deuterium-containing species usually results in a through-oxide-thickness deuterium concentration distribution. Since the profile starts with the deuterium-concentration near the oxide surface, a series of experiments were first performed to examine the effect of some key parameters on such concentrations. The parameters examined included exposures to gases containing D_2 and D_2O at various ratios, partial pressures, and

TABLE 1—*Experiments to determine influence of gas composition and partial pressure on deuterium concentration near the oxide surface using SIMS depth profiling.*

Experiment Number	Exposure Environment Composition	Partial Pressure, Pa	Temperature, °C	Exposure Duration, min	Near Surface Concentration, ppm
1	D_2 (gas) (filament on)	1.33	330	60	32
2	D_2 (gas)	1.33	330	60	2
3	D_2 (gas)	7.98	330	60	9
4	D_2 (gas)	1064	330	60	19
5	D_2O (vapor)	798	350	34	350

TABLE 2—*Experiments to study the mobility of deuterium within the oxide and bulk alloy deuterium uptake.*

Experiment Number	Exposure Environment Composition	Partial Pressure, Pa	Temperature, °C	Exposure Duration	Results Presented in
1	D_2O (vapor)	10^{-3}	300	1, 2, 4, 48, 165 h	Fig. 1
2	D_2O (vapor)/vacuum	$10^{-3}/10^{-5}$	300	2 h/12 h	Fig. 2
3	D_2 gas ($D_2/D_2O \sim 250$)	10^{-3}	300	48 h	Fig. 1
4	D_2O (argon stream bubbled through water)	~3200	300	168 h	Fig. 3
5	D_2O (aqueous)	up to ~18 MPa	250 to 360	up to 600 days	Fig. 4
6	CANDU-PHTS D_2O	~10 MPa	250 to 310	up to 12 years	Fig. 5a
7	CANDU-AGS	~100 KPa CO_2 ~1 to 250 Pa D_2O ~10 to 100 Pa D_2	250 to 310	up to 13 years	Fig. 5b
8	D_2O (vapor)	10^{-3}	380	360 h	Fig. 8
9	D_2 (gas)	10^{-3}	380	330 h	Fig. 8

dissociation of D_2. The range of exposures used and results are given in Table 1. Changing D_2 partial pressure from ~1 to ~1000 Pa resulted in a change in the near surface concentration of deuterium from 2 to 19 ppm. Using a hot tungsten filament with a maximum temperature of 1400°C (at which most of D_2 near the filament will be dissociated) to assist in dissociating D_2 in the gas, resulted in an increase from 2 to 32 ppm of deuterium at the surface. However, exposure to D_2O (at ~800 Pa) resulted in a near surface concentration of 350 ppm (Table 1: Row 5), which was almost 20 times higher than the approximately equivalent exposure to D_2 gas (Table 1: Row 4). Near-surface concentration was considered as the concentration at the outer layer of the oxide (up to 20 nm from the oxide surface) measured by SIMS depth profiling. In further experiments concerning deuterium mobility within the oxide, D_2O was initially chosen over D_2 due to its significant interaction with the oxide surface (Table 2).

Through-Oxide-Thickness Deuterium Concentration Profiles

A set of exposures was designed to examine how the deuterium concentration profile in the oxide evolves with time (Table 2: Row 1). Oxides on Zr-2.5Nb were exposed to ~10^{-3}

Pa of D_2O for 1, 2, 4, 48, and 164 h, respectively. The through-oxide-thickness deuterium concentration profiles obtained by SIMS for the exposed samples are shown in Fig. 1. Deuterium concentration near the oxide surface increases and deuterium penetrates deeper into the oxide with increasing exposure time up to 4 h. Beyond 4 h and up to 164 h of exposure, the profile remains practically unchanged (within observed statistical scatter).

The sample exposed to D_2O for 2 h was subsequently exposed to vacuum ($\sim 1 \times 10^{-5}$ Pa) at 300°C for 12 h (Table 2: Row 2). The deuterium profile in the oxide was practically unchanged (within observed scatter) due to such exposure indicating a relatively strong bonding of the deuterium-containing species with the oxide, compared to weak bonding expected if they were physisorbed on the surfaces of cracks and pores. The results are shown in Fig. 2. However, when the oxide was re-exposed to D_2O at 300°C for more than 2 h, the profile moved to the previously observed equilibrium profile produced after 4 h exposure shown in Fig. 1.

The transport process in the first 4 h of exposure may be modeled effectively as Fickian diffusion, that is, concentration gradient driven and possibly temperature dependent. The parabolic differential equation for one-dimensional diffusion

$$\partial^2 C/\partial x^2 = 1/D_e \partial C/\partial t \qquad (2)$$

where C is deuterium concentration, x is inward distance from oxide surface, D_e is effective diffusion coefficient, and t is the exposure time, was solved numerically with the following initial and boundary conditions:

1. Deuterium concentration near the oxide surface is initially zero and increases with time according to a curve fitted to the observed values of Fig. 1. The near-surface concentration

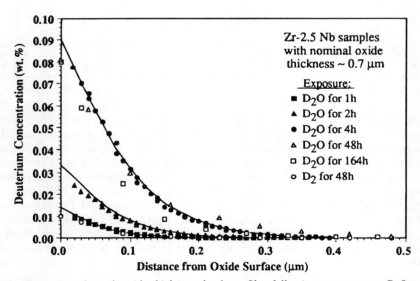

FIG. 1—*Deuterium through-oxide-thickness depth profiles following exposures to D_2O and D_2 at 300°C. A Fickian diffusion model fit to the 1, 2, and 4 h profiles resulted in an inferred average effective diffusion coefficient of $\sim 1 \times 10^{-18}$ m²/s. The solid lines represent calculations using the inferred coefficient.*

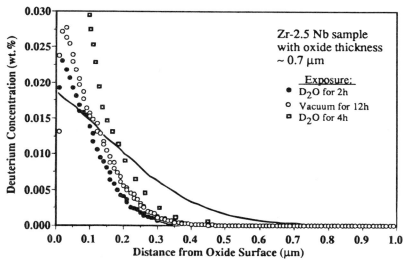

FIG. 2—*Deuterium profiles for 2-h exposure to D_2O followed by exposure to vacuum at 300°C for 12 h. The solid line represents the expected profile after vacuum exposure using inferred diffusion coefficient.*

values are taken from the D_2 content at ~0.02 μm from outer oxide surface. The values are 0.0097, 0.026, and 0.076% by weight corresponding to exposures for 1, 2, and 4 h, respectively.
2. Initial deuterium concentration throughout the oxide is assumed to be zero.

The oxide was assumed to be a homogeneous medium from a diffusion point of view with a constant value of diffusion coefficient throughout. The profiles of Fig. 1 were used to infer effective diffusion coefficients by least squares minimization fits to the observed data points. The fits are shown as the solid lines in Fig. 1. An inferred average value of 1.1×10^{-18} m²/s for D_e is in the same order of magnitude of the reported value of 3.8×10^{-18} m²/s for the diffusion of deuterium implanted into oxides on Zr-2.5Nb at 300°C [13]. This latter value was further explained as a result of diffusion through interconnected micropores at grain boundaries as opposed to solid-state lattice diffusion [14]. Although the observed agreement in D_e values may suggest similar mechanism or routes of transport, the absence of further movement after 4 h and longer exposures requires further explanation that will be addressed in the discussion section.

In order to further investigate the mobility of hydrogen in the oxide, the effect of isotope exchange on deuterium profiles in the oxide was examined by exposing steam-prefilmed samples (in H_2O at 400°C for 48 h) to a stream of argon bubbled through D_2O at 300°C for 168 h (Table 2: Row 4). The $^1H^-$ and $^2D^-$ through-oxide-thickness profiles were obtained and are shown in Fig. 3. The profiles are showing complete replacement of the hydrogen profile by deuterium indicating that isotope exchange has taken place throughout the oxide thickness. Another possible explanation for the results of Fig. 3 is that the flow over the sample surface might have depleted the hydrogen near the surface resulting in a gradient driving the hydrogen out of the oxide while deuterium is building up at the surface and diffusing into the oxide. However, since the flow was relatively slow (~0.5 cm/s), the aerodynamic effects may not

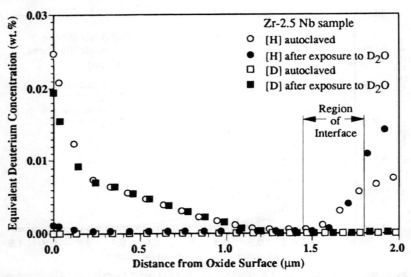

FIG. 3—*Deuterium and hydrogen through-oxide-thickness profiles after exposure to a stream of argon bubbled through D_2O at 300°C for 168 h. The samples were prefilmed by steam autoclaving at 400°C for 48 h.*

be significant and isotope exchange is more likely to have occurred. The profiles of Fig. 3 are slightly different than those of Fig. 1, which may be due to the different exposure procedure (minor aerodynamic effects or higher pressure or both) or different thicknesses. However, the influence of the isotope exchange reaction on development of deuterium profiles is demonstrated and it appears that hydrogen and hydrogen-containing species are mobile throughout the oxide thickness when the surface is exposed to water vapor.

Further evolution of the through-oxide-thickness $^2D^-$ profiles shown in Fig. 1 during long-term out-reactor aqueous exposures has been examined by autoclave exposures of samples of prefilmed Zircaloy-2 and Zr-2.5Nb to aqueous D_2O at 250 to 360°C for periods up to 600 days (Table 2: Row 5). The resulting profiles are shown in Figs. 4a and c for Zr-2.5Nb and Zircaloy-2, respectively. Figure 4b shows a typical example of observed variability in profiles obtained for the 250°C samples of Zr-2.5Nb. The profiles appear to follow the general expected trend of developing porosity in the outer region of the oxide with relatively high deuterium content (up to 0.1% by weight) while the region of decreasing deuterium content (similar to that observed after the 4 h exposure in Fig. 1) remains next to the oxide-metal interface. In thin oxides (<3 μm), the deuterium content decreases from the outside surface to the metal-oxide interface comparable to short-term exposure profiles of Fig. 1. Both materials show similar overall shape of deuterium profiles in these oxides, suggesting that the number and distribution of pores in the outer oxide and the mechanism of deuterium permeation may be similar.

Typical deuterium profiles in inside surface oxides of Zr-2.5Nb pressure tubes removed from operating reactors after up to 12 years operation (Table 2: Row 6) are shown in Fig. 5a. These exhibit similar trends to those observed following out-reactor exposures, although higher deuterium contents (up to 0.4% by weight) are found in the outer regions of these irradiated oxides. In thick oxides, for both in- and out-reactor exposures, the region near the oxide-alloy interface containing decreasing deuterium concentration is greater than in thin oxides that

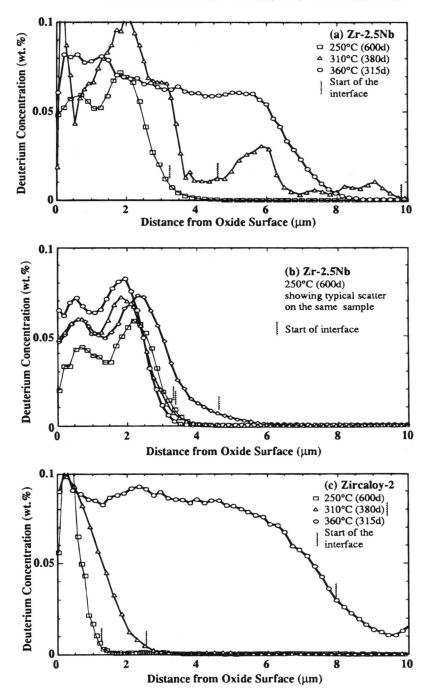

FIG. 4—*Deuterium through-oxide-thickness depth profiles resulting from out-reactor aqueous exposures to D_2O in autoclaves at 250° to 360°C for up to 600 days.*

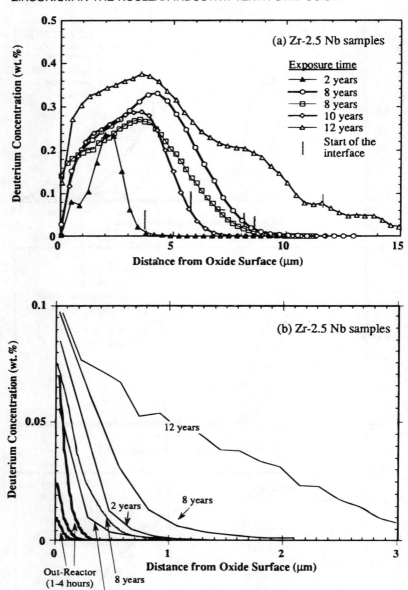

FIG. 5—*Deuterium through-oxide-thickness depth profiles of samples from (a) inside surfaces and (b) outside surfaces of removed outlet regions (~300°C) of pressure tubes from CANDU power reactors. The end of each profile represents the start of the interface.*

probably results from a difference in the distribution of accessible sites. Differences in the deuterium concentration profiles in oxide outer regions in Figs. 4 and 5 (for example, peaks and surface depletion) are probably due to loss of deuterium during storage.

Outside surface oxides in removed pressure tubes were also examined (Table 2: Row 7). These had been exposed to reactor gas annuli (N_2 or CO_2) at ~300°C with D_2O partial pressures ranging from ~0.25 to ~200 Pa and D_2 partial pressures between ~10 and 100 Pa for up to 13 years. The resulting profiles are shown in Fig. 5b. The profiles, where in-reactor oxidation was minimal (<~1 μm in 10 years), show trends similar to out-reactor D_2O vapor exposures except for a tube with 12 years in the reactor and high bulk-alloy uptake showing higher deuterium concentration at the start of the oxide-alloy interface. This tube will be discussed later in the text.

Figure 6 shows cross-sectional micrographs obtained by transmission electron microscopy (TEM) for typical thin oxides (~1 μm) grown on Zr-2.5Nb used in the present study, that is, steam prefilmed at 400°C for 24 h. The columnar grains (~20 to 50 nm wide and ~200 to 500 nm long) of the oxide may be seen in Fig. 6a extending to the oxide surface with no apparent intergranular porosity. Few pores were observed throughout the oxide and these appeared not to be interconnected. A band of niobium-rich region corresponding to beta-niobium phase in the bulk alloy and apparently containing porosity is shown in Fig. 6b crossing the columnar grains. Although TEM has not yet been performed on thick oxides with high deuterium content, TEM micrographs of thin porous oxides on pure zirconium (with uniform high deuterium content) have been reported [9]. The pores were seen to form an interconnected network of up to 10-nm-wide following equiaxed grain boundaries. It would thus be expected that porous regions in thick oxides on zirconium alloys, as observed in Figs. 4 and 5a, would probably follow the same pattern.

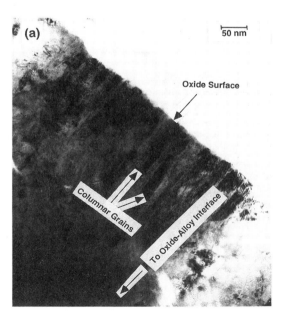

FIG. 6—*TEM micrographs of a typical thin oxide formed on Zr-2.5Nb in steam showing* (a) *columnar grains extending to the outer surface of the oxide (low magnification), and* (b) *niobium-rich band crossing the columnar grains and no apparent interconnected pores.*

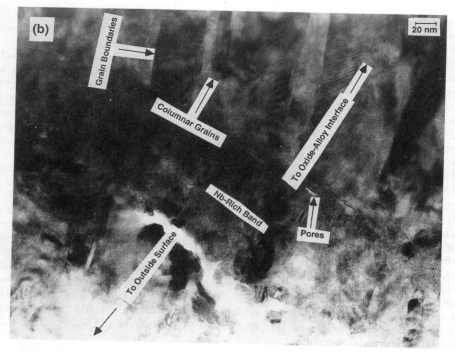

FIG. 6—Continued.

Near oxide surface deuterium concentration resulting from out- and in-reactor exposure to a gaseous environment with minimal oxide growth was generally <0.1% by weight (see Fig. 5b). In view of Fig. 6, the available adsorption or bonding sites on grain boundary surfaces (defined as grain surface atoms) are estimated to be $\sim 10^{15}$ sites/cm^2 assuming an average columnar grain size of $\sim$30 nm wide and 500 nm long. This results in a site density of $\sim 1.4 \times 10^{21}$ sites/cm^3. If each surface site were occupied by deuterium atoms, this would represent $\sim$0.07% by weight. This is in reasonable agreement with the previously-mentioned measured concentration of 0.1% by weight suggesting that the measured profiles may represent deuterium species residing on the accessible sites at grain boundaries. Greater concentrations observed could be accommodated as multiple layers on surfaces of intergranular porosity or finer grain size resulting in higher site density. From the obtained profiles for thin oxides, the density of accessible sites on grain boundaries seems to be highest near the outer surface and gradually decreases through-thickness to reach its minimum value near the interface. This may be partly explained by the through-thickness variation in compressive stress in the oxide being highest at the metal-oxide interface [15].

Bulk Alloy Hydrogen Uptake

Bulk alloy hydrogen analyses using HVEMS of samples from CANDU reactors and laboratory long-term aqueous exposures (Figs. 4 and 5a) showed that generally less than 5% of the deuterium evolved from the corrosion reaction (Eq 1) was picked up by the samples (Fig. 7) regardless of SIMS profile.

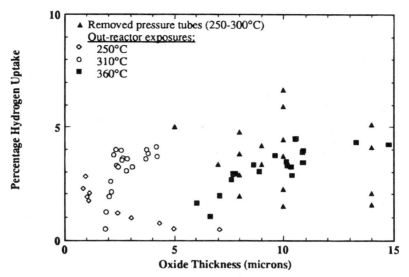

FIG. 7—*Inferred percentage hydrogen uptake in Zr-2.5Nb samples exposed out-reactor to aqueous D_2O at 250, 310, and 360°C, and from removed pressure tubes.*

A thin sample of Zr-2.5Nb was exposed to 1×10^{-3} Pa of D_2O vapor at 380°C for 360 h to determine the rate of permeation into the bulk alloy (Table 2: Row 8). Practically no deuterium in the bulk alloy was detected by DSC following this exposure. It was also confirmed by the deuterium concentration distribution in Fig. 8 showing low concentration near the interface and in the bulk alloy.

In order to investigate the mechanism of high bulk-alloy hydrogen uptake, samples were exposed to pure deuterium gas at different temperatures. Samples of Zr-2.5Nb exposed to a gaseous environment consisting of predominantly deuterium gas (10^{-3} Pa) with only 10^{-5} Pa D_2O at 300°C for 48 h (Table 2: Row 3) exhibited lower deuterium concentration at the surface and within the oxide as shown in Fig. 1. The resulting bulk alloy deuterium uptake was still below detectability. However, when samples were exposed to the same gaseous environment at 380°C (to accelerate uptake) for 330 h, a significant uptake by the bulk alloy was observed (Table 2: Row 9). Typical residual gas partial pressures before admitting D_2 gas and during the exposure are given in Table 3 showing the $D_2:D_2O$ partial pressure ratio during exposure to be ~250. Following sample exposure, bulk alloy deuterium uptake was ~12 $\mu g/cm^2$ determined by DSC. The resulting deuterium profile in Fig. 8 is showing significant uptake by the bulk alloy. The thickness of the oxide remained practically unchanged.

Since deuterium gas molecules had a collision frequency ~500 times higher than that of D_2O during the previously mentioned exposure, deuterium molecules had a better chance of interacting with the oxide surface. Oxygen diffusion at the oxide-alloy interface due to the relatively high temperature and without enough oxygen diffusing from the outer surface, might have created extra diffusion pathways for the permeating hydrogen species such as high vacancy concentration or line defects due to the resulting substoichiometry. The relatively high deuterium concentration observed near the interface in Fig. 8 may be an indication of the dissolution processes occurring at the interface. However, the significant peaks observed in the alloy may represent hydrides formed upon cooling the sample with its high deuterium-content to room temperature.

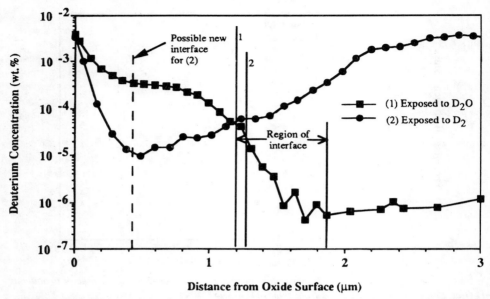

FIG. 8—*Deuterium depth profiles for prefilmed Zr-2.5Nb samples exposed to D_2O and D_2 at 380°C for up to 360°C.*

Discussion

The self-limiting feature of the observed mobility after 4 h exposure depicted in Fig. 1 suggests that the observed profile may reflect the density of accessible sites through the oxide thickness. Diffusion pathways implied in the time-dependent profiles of Fig. 1 appear to be intergranular porosity according to the measured near-surface concentrations. Although such intergranular porosity was not found by TEM in the regions of oxide observed, sufficient grain boundary surface area exists to provide the near-surface deuterium concentration observed (that is, ~0.1% by weight). The practically unchanged 2-h profile after being exposed to vacuum at 300°C (Fig. 2) appears to indicate the presence of relatively strong bonding of deuterium along diffusion pathways. The mobility of deuterium, however, appears to be

TABLE 3—*Lists of partial pressures in the test chamber produced by mass spectrometer for an exposure to a predominantly D_2 environment (that is, experiment of Fig. 8).*

Mass Number	Species	Partial Pressures before Admitting D_2, Pa	Partial Pressures after Admitting D_2, Pa
2	H_2	8.3×10^{-8}	1.5×10^{-5}
4	D_2	1.1×10^{-7}	1.5×10^{-3}
18	H_2O	1.0×10^{-7}	1.7×10^{-6}
20	D_2O	1.7×10^{-9}	6.0×10^{-6}
28	N_2	2.3×10^{-8}	9.3×10^{-6}
32	O_2	1.7×10^{-9}	2.4×10^{-7}
44	CO_2	8.0×10^{-10}	1.3×10^{-10}

triggered by re-exposure to water vapor. This is also confirmed by the result of the isotope exchange experiment shown in Fig. 3; this result implies that for aqueous exposure, there is continued interchange of deuterium in the environment and bulk oxide.

The nature of the deuterium-containing species in the oxide is a key in investigating the transport mechanism. For oxides grown by steam autoclaving, hydroxyl groups have been detected in their SIMS sputtered ions by high precision mass spectrometry [16]. Also, infrared examination of relatively thick films has provided evidence of hydroxyl groups being present in oxides grown in water or steam [17]. It is thus reasonable to assume that hydroxyl groups and hydrogen atoms from dissociated water and hydrogen molecules present at the surface are the main constituents of the diffusing species through the oxide. Hydroxyl groups, as a result of their charge are more reactive and therefore more likely to be bonded on accessible sites along grain boundaries. On the other hand, neutral hydrogen atoms present from $D_2:D_2O$ ratio in the environment are more likely to permeate through narrow pathways to the bulk alloy.

From a hydrogen ingress point of view, the results generally indicate that exposing the oxide to a predominantly D_2O environment results in much lower bulk alloy uptake than exposure to a high temperature (>300°C) environment where deuterium gas is present. At lower temperatures (~300°C) and following ~330 h of exposure to D_2O, the bulk alloy uptake is undetectable by DSC. This suggests that the environment ($D_2:D_2O$ ratio) as well as temperature are key factors in the mechanism leading to bulk uptake; that is increasing $D_2:D_2O$ ratio leads to increasing bulk alloy uptake.

The obtained through-oxide-thickness deuterium concentration profiles may be the sum of two main components. The predominant component appears to be due to hydroxyl groups residing on grain boundaries. The second component may be due to the mobile hydrogen species in case of $D_2:D_2O$ ratios greater than a certain value, for example deuterium atoms, diffusing through narrow grain boundaries, or extra diffusion pathways such as line defects. Line defects may be created by oxide dissolution at the oxide-alloy interface resulting in localized substoichiometry. Oxide dissolution may occur due to the lack of oxidizing species, for example, when oxides are exposed to a predominantly deuterium gas environment [18].

Thus, in all cases shown here, the through-oxide-thickness deuterium concentration profile probably results predominantly from OD groups bonded to the surfaces of intergranular porosity. The deuterium content at the interface may provide an indication of the extent of intergranular porosity in the oxide at this location. High deuterium contents at the interface in Figs. 4 and 5 imply that local regions with extremely thin oxides are present with the aqueous environment having access to regions close to the interface via intergranular porosity. In the presence of sufficient D_2O, the oxide may be continually healed, and in the absence of short-circuit routes (such as intermetallics in zircaloys), deuterium uptake is relatively low (<5% shown in Fig. 7 for the zirconium-niobium alloy exposed here). Where $D_2:D_2O$ ratios are relatively high, however (~250 out-reactor and ~0.5 in-reactor) in the case of Fig. 5b for the 12 years exposure in-reactor, deuterium may permeate through the oxide to the interface and may react directly with the metal at local regions providing high uptake rates.

In the case of out-reactor exposures to predominantly deuterium gas at high temperatures ($D_2:D_2O$ ratio of 250 and 380°C), the oxide-metal interface shifts towards the original oxide surface as a result of oxide dissolution as shown in Fig. 8. If this oxide were to be exposed to D_2O, then the deuterium content at the location of the new interface (arising from hydroxyl groups) would be similar to Curve 1 in Fig. 8; that is, the extent of intergranular porosity would be relatively high. As shown in Curve 2, the deuterium content is low at the new interface (following exposure to deuterium gas) although bulk alloy deuterium content is high. Thus, Curve 2 may represent the species that are responsible for deuterium uptake, for example, deuterium atoms. It is also possible, however, that part of the observed profile may be due to hydroxyl groups arising from the reaction with the relatively low D_2O partial pressure in the environment.

In regard to the hydrogen uptake mechanism, this study shows that the mobility of the hydrogen species permeating through the oxide into the bulk alloy depends on the through-oxide-thickness distribution of intergranular porosity as well as the extent to which they are filled with bonded hydroxyl groups.

This may suggest that the hydroxyl groups at the grain boundaries, although resulting in the observed hydrogen profiles, may not be the hydrogen-containing species diffusing into the bulk alloy at interfacial regions where intergranular porosity exists. Also, despite the observed high hydrogen content in the outer region of the oxides exposed to water, the oxides are protective from a hydrogen bulk alloy uptake point of view, as long as short circuit routes such as intermetallics are not present. It follows that the profile representing the driving gradient for the mobile hydrogen diffusing through the oxide is usually masked by the profile due to the residing hydroxyl groups at the accessible sites on the grain boundaries.

With the two-component deuterium-profile model in mind, more experiments are being planned for exposures with varying $D_2:D_2O$ ratios in a monitored gaseous environment. Also, further TEM examinations are planned to further elucidate the relationship between the oxide microstructure and deuterium distribution and ingress into the bulk alloy.

Conclusions

1. Samples exposed to environments containing D_2O at $\sim 300°C$ show similar diffusional-type through-oxide-thickness deuterium profiles adjacent to the oxide-metal interface following 4 h out-reactor and up to ~ 10 years in-reactor. The shape of these profiles probably represents the density of accessible sites through-thickness.

2. Accessible sites are probably on surfaces of intergranular porosity present in the oxide. Deuterium concentration profiles appear to result predominantly from deuteroxyl groups bonded to such sites on the surfaces of intergranular porosity. In small regions observed by TEM such porosity has not been found, although from observations of grain size, sufficient sites would be available to provide deuterium content observed near oxide surfaces.

3. Interfacial deuterium content at the oxide-metal interface provides an indication of the extent of interfacial intergranular porosity. High deuterium contents at the interface imply local regions with thin or absent oxide barriers at the interface.

4. In the presence of sufficient H_2O as in aqueous exposures, the oxide is continually healed, most of the accessible sites on intergranular porosity are occupied, and deuterium uptake is relatively low where short-circuit routes (for example, intermetallics in Zircaloy-2) are not present.

5. Where $D_2:D_2O$ ratios are relatively high, deuterium atoms may permeate through the oxide to the interface and react directly with the metal (zirconium is known to be a getter for hydrogen) providing high deuterium uptake rates.

6. Deuterium profiles may be the sum of mainly two components. The predominant component is due to deuteroxyl groups residing on accessible sites on surfaces of intergranular porosity. The second component that is usually minor would represent the driving gradient of the mobile hydrogen species, for example, H, that is permeating to the bulk alloy.

7. Further work is needed to substantiate the proposed hypothesis that would include exposures with varying $D_2:D_2O$ ratios and further TEM examination especially for thick oxides.

Acknowledgments

Funding for this work was provided by Ontario Hydro and the CANDUOwners Group (COG). Ontario Hydro Research Division staff, J. Mummenhoff, L. Grant, and J. DeLuca, carried out much of the experimental program. Valuable discussions with Dr. N. Ramasubramanian are gratefully acknowledged. Dr. Y. P. Lin provided the TEM micrographs. SIMS analysis was performed by G. Mount and C. Weisener at the University of Western Ontario. Discussions with staff at the Atomic Energy of Canada Limited-Chalk River Nuclear Laboratories are acknowledged.

References

[1] Gulbransen, E. A. and Andrew, K. F., "Reaction of Hydrogen with Preoxidized Zircaloy-2 at 300° to 400°C," *Journal*, Electrochemical Society, Vol. 104, No. 12, Dec. 1957.

[2] Aronson, S., "Some Experiments on the Permeation of Hydrogen through Oxide Films on Zirconium," U.S. Report, *Bettis Technical Review*, WAPD-BT-19, June 1960, pp. 75–81.

[3] Shannon, D. W., "Effect of Oxidation Rates on Hydriding of Zirconium Alloys in Gas Mixtures Containing Hydrogen," *Corrosion*, Vol. 19, Dec. 1963.

[4] Roy, C., "Hydrogen Distribution in Oxidized Zirconium Alloys by Auto-radiography," Atomic Energy of Canada Ltd. (AECL), Report No. AECL-2085, Sept. 1964.

[5] Smith, T., "Diffusion Coefficients and Anion Vacancy Concentration for the Zirconium-Zirconium Dioxide System," *Journal*, Electrochemical Society, June 1965, pp. 560–567.

[6] Cox, B. and Roy, C., "The Use of Tritium as a Tracer in Studies of Hydrogen Uptake by Zirconium Alloys," Atomic Energy of Canada Ltd., Report AECL-2519, Dec. 1965.

[7] Austin, J. H., Elleman, T. S., and Verghese, K., "Tritium Diffusion in Zircaloy-2 in the Temperature Range −78 to 204C," *Journal of Nuclear Materials*, Vol. 51, 1974, p. 321.

[8] Woolsey, I. S. and Morris, J. R., "A Study of Zircaloy-2 Corrosion in High Temperature Water Using Ion Beam Methods." *Corrosion*, Vol. 37, 1981 pp. 575–585.

[9] Warr, B. D., Elmoselhi, M. B., McIntyre, N. S., Brennenstuhl, A. B., Lichtenberger, P. C., and Newcomb, S. B., "Oxide Characteristics and Their Relationship to Hydrogen Uptake in Zirconium Alloys," *Zirconium in the Nuclear Industry: Ninth International Symposium, ASTM STP 1132*, C. M. Eucken and A. M. Garde, Eds., American Society for Testing and Materials, Philadelphia, 1991, pp. 740–757.

[10] Cox, B., "Mechanisms of Hydrogen Absorption by Zirconium Alloys," Atomic Energy of Canada Ltd., Report No. AECL-8702 1985.

[11] Green, L. W., James, M. W. D., Leeson, P. K., and Lamarche, T. G., "The Determination of Hydrogen in Zirconium Using the First CRL Hot Vacuum Extraction Mass Spectrometry System," Atomic Energy of Canada Ltd. (AECL), Report No. 91-252, RC-730, Oct. 1991.

[12] Tashiro, K., "Calorimetric Study of Hydrogen in Zr-2.5Nb Preliminary Report," Ontario Hydro Research Division, Report No. 87-41-K, March 1987.

[13] Khatamian, D. and Manchester, D., "An Ion Beam Study of Hydrogen Diffusion in Oxides of Zr and Zr-Nb," *Journal of Nuclear Materials*, Vol. 166, Jan. 1989, pp. 300–306.

[14] Khatamian, D., "Diffusion of Hydrogen in the Oxides of Annealed Zr-1 Nb, Zr-2.5 Nb and Zr-20 Nb Alloys," Atomic Energy of Canada Ltd. (AECL), Report No.RC-843-92-138, March 1992.

[15] Godlewski, J., "How the Tetragonal Zirconia is Stabilized in Oxide Scale Formed on Zirconium Alloy Corroded at 400°C in Steam Water?" in this volume.

[16] Godlewski, J., Gros, J. P., Lambertin, M., Wadier, J. F., and Weidinger, H., "Raman Spectroscopy Study of the Tetragonal-to-Monoclinic Transition in Zirconium Oxide Scales and Determination of Overall Oxygen Diffusion by Nuclear Microanalysis of O^{18}," *Zirconium in the Nuclear Industry: Ninth International Symposium, ASTM STP 1132*, C. M. Eucken and A. M. Garde, Eds., American Society for Testing and Materials, Philadelphia, 1991, pp. 416–436.

[17] Ramasubramanian, N. and Balakrishnan, P. V., "Aqueous Chemistry of Lithium Hydroxide and Boric Acid and Corrosion of Zircaloy-4 and Zr-2.5 Nb Alloy," in this volume.

[18] Newcomb, S. B., Warr, B. D., and Stobbs, W. M, "The TEM Characterization of Oxidation-Reduction Processes in ZrNb Alloys," *Proceeding*, EMAG, Bristol, UK, Sept. 1991, F. J. Humphreys, Ed., Institute of Physics Conference, Serial No. 119, p. 221.

DISCUSSION

A. Hermann[1] *(written discussion)*—Your paper is concerned with the mechanism of hydrogen uptake in zirconium alloys but you mainly showed us results with Zr-2.5Nb. Do you mean your proposed mechanism is also valid in the case of other zirconium alloys and, if so, I would like to make the following comment: It is known, for instance from impedance spectroscopy measurements that the electolyte in the case of aqueous corrosion penetrates far into the oxide scale. Therefore, high concentrations of hydrogen in the outside regions of the oxide scale may not be essential for the mechanism of hydrogen in the non-penetrated layer at the metal-oxide interface. In order to make proper conclusions, the measured hydrogen profiles should be seen in connection with the amount and distribution of pores and cracks (among their microcracks) as well as with the dimensions of the protective non-penetrated oxide layer. Would you agree with this?

M. B. Elmoselhi et al. (authors' closure)—I totally agree with your comment. The paper actually revolves mainly around thin oxides and the mechanism of hydrogen penetrating through their microstructure. For thin oxides, we are proposing diffusion along surfaces of intergranular porosity. The outer region of thick oxides that was shown to be filled with bonded hydrogen species and that is known to be porous was only mentioned for completeness rather than as a direct part of the uptake mechanism.

Brian Cox[2] *(written discussion)*—The experiments in which you were able to completely exchange hydrogen for deuterium (and vice-versa) in water atmospheres, but not change the deuterium profiles in vacuum were very interesting. Have you done similiar exchanges of hydrogen and deuterium for specimens exposed in gaseous hydrogen atmospheres?

M. B. Elmoselhi et al. (authors' closure)—No, the reason being that exposure to gaseous hydrogen environment results in almost an order of magnitude less hydrogen in the oxide than exposure to water vapor, that is, reduced sensitivity for the measuring technique (SIMS). However, such experiments are currently being planned.

R. A. Ploc[3] *(written discussion)*—(1) Have you measured deuterium uptake as a function of D_2O partial pressure?

(2) Can you give us an idea of the "roughness" of the oxide/metal interface compared to the "thin layer" region shown in your figures?

M. B. Elmoselhi et al. (authors' closure)—(1) No, we have not measured deuterium uptake as a function of D_2O partial pressure directly. However, our experiment of exposure to a predominantly D_2O environment as compared to an exposure to a predominantly D_2 environment would represent the two extremes of high and low D_2O partial pressures.

(2) The roughness of the oxide/metal interface is approximately 0.3 μm, while the thin layer region in most of the graphs is between 1 and 2 μm.

Bo Cheng[4] *(written discussion)*—Your comments that Zr-2.5Nb and Zircoly-2 have about the same characteristics in the H_2 distribution or uptake characteristics surprised me. As you know, Zircaloy-2 and Zr-2.5Nb have very significant differences in hydrogen pickup fractions in reactors. Please comment whether this is true. If so, please explain why Zircaloy-2 and Zr-2.5Nb differ.

M. B. Elmoselhi et al. (authors' closure)—My comment that there may be some similarities in the mechanism of hydrogen transport through the oxide of Zircaloy-2 and Zr-2.5Nb is in terms of the transporting species and the mechanistic modeling of the transport process, for

---

[1] Paul Scherrer Institute, Wurenlingen, Switzerland.
[2] University of Toronto, Ontario, Canada.
[3] AECL Research, Chalk River, Ontario, Canada.
[4] Electric Power Research Institute (EPRI), Palo Alto, CA.

example, steps such as dissociation, adsorption, diffusion, etc. may be common for the two systems. However, the differences in terms of the microstructure such as the presence of intermetalics (acting as short-circuit routes) in Zircaloy-2 oxides and the effect of the two-phase structure of Zr-2.5Nb on its oxide properties are certainly recognized, and these could be reflected in the values of the coefficients associated with the involved mechanistic steps controlling the quantity of hydrogen transport that is known to be significantly different.

D. Charquet,[1] P. Rudling,[2] M. Mikes-Lindback,[2] and P. Barberis[1]

Hydrogen Absorption Kinetics During Zircaloy Oxidation in Steam

REFERENCE: Charquet, D., Rudling, P., Mikes-Lindback, M., and Barberis, P., **"Hydrogen Absorption Kinetics During Zircaloy Oxidation in Steam,"** *Zirconium in the Nuclear Industry: Tenth International Symposium, ASTM STP 1245,* A. M. Garde and E. R. Bradley, Eds., American Society for Testing and Materials, Philadelphia, 1994, pp. 80–97.

ABSTRACT: The oxidation and hydriding kinetics in 400 and 500°C steam have been determined for various grades of Zircaloy. Over a wide range of corrosion rates, the hydriding kinetics essentially depend on the oxidation rate.

Two types of parameters can be distinguished: those that modify only the corrosion rate, with a concomitant effect on the hydriding kinetics, and those that also have an intrinsic influence on the latter. Among factors of the first type are the final heat treatment, the "cumulative annealing parameter," the tin content, and the iron/chromium (Fe/Cr) ratio. The nickel content falls in the second category. When the oxidation rate increases, the hydriding rate also rises, but the fraction of the available hydrogen that is absorbed decreases. These kinetic observations explain the sometimes opposite variations of the integrated parameters that are usually measured, for instance, the change in hydrogen absorption as a function of weight gain. The interpretation of conventional integrated data requires consideration of the successive modifications in the kinetics during corrosion.

KEY WORDS: zirconium, zirconium alloys, steam corrosion, hydriding, hydrogen analysis, chemical composition, processing route, nuclear materials, nuclear applications, radiation effects

Development work performed over the last few years has led to an improvement in the corrosion resistance of Zircaloy-type alloys, both in autoclave tests and in reactor exposures. In addition to slower growth of the oxide scale, an important consequence is a reduction in hydrogen uptake. In effect, when the Zircaloys are corroded by water or steam, hydrogen is liberated and is distributed among the corrosive medium, the scale, and the metal [1,2]. It is the hydrogen picked up by the metal that is troublesome, since it leads to embrittlement at high concentrations. Preliminary results on the effects of heat treatment and alloying additions (tin, iron, chromium, and nickel) have already been published [3,4]. However, these parameters also affect the oxidation kinetics, and it is difficult to determine whether their influence on hydrogen absorption is a direct consequence of the oxidation rate or whether it is the result of a supplementary intrinsic effect.

Materials and Experimental Procedures

Various batches of Zircaloy-4 sheets with standard composition corresponding to ASTM Specification for Zirconium and Zirconium Alloy Strip, Sheet, and Plate (B 551-92), produced

---

[1] Senior scientist and research engineer, respectively, Centre de Recherches CEZUS, 73403 Ugine Cedex, France.
[2] Specialist, nuclear fuel materials and manager, materials fuel division, respectively, ABB ATOM AB, S-72163 Västerås, Sweden.

TABLE 1—*Zircaloy sheet tested at 400 and 500°C.*

Sample	Metallurgical Condition	Cumulative Annealing Parameter	Corrosion Resistance	
			400°C Test	500°C Test
1a	recrystallized	$1.4 \cdot 10^{-19}$	high corrosion rate	low corrosion rate
1b	recrystallized	$1.4 \cdot 10^{-19}$	high corrosion rate	low corrosion rate
2a	beta heat treated low cooling rate	...	intermediate corrosion rate	not tested
2b	beta heat treated high cooling rate	...	not tested	low corrosion rate
3a	recrystallized	$1.3 \cdot 10^{-17}$	low corrosion rate	high corrosion rate
3b	recrystallized	$1.3 \cdot 10^{-17}$	low corrosion rate	high corrosion rate

either industrially or on a laboratory scale, were selected. Their compositions were tin = 1.4 to 1.5%, iron = 0.20 to 0.21%, chromium = 0.10 to 0.11%, and carbon = 0.013 to 0.014%. To obtain a wide range of oxidation kinetics, the processing sequence was varied by, in particular, the "cumulative annealing parameter" and the final heat treatment. These materials were principally used to study the oxidation and hydriding kinetics in steam at 400 and 500°C.

The strips were manufactured by the conventional sequence, involving quenching from the beta field at 20 mm thickness, followed by hot rolling in the alpha phase range down to 6 mm, then are two or three cold-rolling and annealing cycles down to 1, 2, and 3 mm, and final heat treatment. Table 1 summarizes the main parameters that were varied to obtain different corrosion behaviors. The cumulative annealing parameter was calculated from the relationship [5]

$$\Sigma A = \Sigma t \exp(-40\,000/T)$$

where the time, t, is in hours and the temperature, T, in Kelvin.

Zircaloy-4 sheets with variable tin contents and iron/chromium (Fe/Cr) ratios were also produced on a laboratory scale to assess the influence of these two parameters on hydrogen pickup.

Buttons containing 1.4% tin, with a Fe/Cr ratio ranging from 0.2 to 5.8 for a constant sum Fe + Cr = 0.33%, were melted and processed to sheet using a conventional procedure: treatment in the beta field followed by cooling in argon, hot rolling, several cold-rolling/annealing cycles, and final recrystallization. The value of the cumulative annealing parameter for these sheets was $3 \cdot 10^{-18}$ h.

All tests were performed in a static autoclave on samples pickled before oxidation. The oxidation kinetics were monitored by weighing, assuming that the weight gains were due essentially to oxygen.

Samples were taken at regular intervals for hydrogen analysis, which were performed in a standard manner, by extraction from a molten metal bath under an argon carrier gas, using a LECO RH1 Model 260–300 apparatus.

Hydrogen Distribution

In the present study, only the hydrogen contained in the metal is of interest, since this causes embrittlement.

Hydrogen analyses were performed on oxidized sheet samples or sandblasted and pickled samples. A large difference is observed between oxidized samples and specimens descaled and pickled in this way, as shown in Fig. 1. This indicates that a large amount of hydrogen is contained in the oxide. However, part of this hydrogen could be contained in water absorbed in the oxide pores. In the hydrogen analysis, the total amount of hydrogen in the oxide and the metal is measured and divided by the sample weight. This ratio gives the amount of hydrogen in ppm. However, if the amount of hydrogen in the oxide is large compared to that in the metal, the calculated weight ppm of hydrogen in the metal will be too high.

One way to obtain the correct hydrogen analyses in the Zircaloy is to mechanically remove the oxide prior to the analyses. However, if there exists a hydride gradient in the Zircaloy material with the highest hydride concentration at the metal/oxide interface [6], this hydride layer at the interface may be removed together with the oxide. This will result in an analyzed hydrogen content in the metal that is too low.

The micrographs in Fig. 2 show no evidence of a significant hydrogen gradient in the metal. Therefore, it may be considered as negligible.

So, all the following hydrogen contents were measured after removing of oxide scale.

Oxidation and Hydriding Kinetics of Zircaloy-4 at 400°C

The oxidation kinetics at 400°C of the five materials tested (Table 1) are shown in Fig. 3. Marked differences in behavior are observed. Extremely poor corrosion resistance for Materials 1a and 1b is due principally to the very low value of the cumulative annealing parameter. Excellent resistance for Materials 3a and 3b are the result of their fully recrystallized structure and high values of the cumulative annealing parameter. Finally, an intermediate performance is observed for the material treated in the beta field, that is, Material 2a.

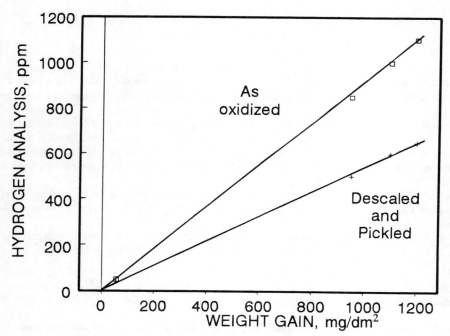

FIG. 1—*Influence of sample conditioning on the hydrogen analysis.*

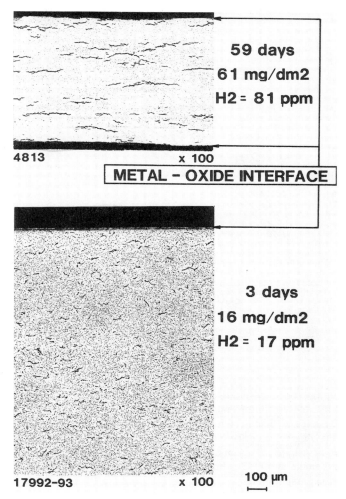

FIG. 2—*Hydride distribution after corrosion testing.*

Figure 4 shows the hydriding kinetics for these same materials. The hydrogen uptake increases continuously with time, and the materials can be ranked in the same order as for the oxidation rates.

Figure 5 shows the hydrogen pickup ratio as a function of exposure time. It is defined as the ratio between the quantity of hydrogen absorbed by the metal (calculated from the hydrogen analyses) and the amount theoretically liberated by the reaction (calculated from the weight gains, assuming them to be due solely to oxygen). It can be seen in Fig. 5 that the ranking of the materials is the opposite of the previous ones, the relative pickup being lower for higher corrosion rates. Moreover, this ratio is not stable, varying with exposure time; when the oxide gets thicker, the absorbed fraction increases.

To explain these results, we could make the two following assumptions:

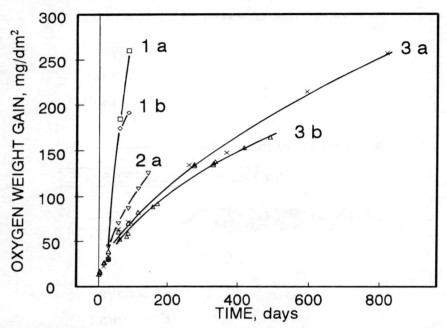

FIG. 3—*Weight gains of Zircaloy-4 samples as a function of test duration in 400°C, 10.3 MPa steam.*

1. The limiting process of hydrogen ingress in the metal is the diffusion through a barrier oxide scale, close to the metal/oxide interface.
2. The diffusion coefficient of hydrogen in this scale is smaller than that for oxygen. Elmoselhi [7] gives an effective diffusion coefficient of 10^{-18} m²/s at 300°C for deuterium. Godlewski [8] gives an effective diffusion coefficient for oxygen in the 10^{-12} to 10^{-10} m²/s range at 400°C. Although these data are not really comparable, it supports the previous assumption.

For the high corrosion rate, the quantity of hydrogen produced is larger than can diffuse through the barrier oxide film. This leads therefore to a low absorbed fraction.

For the low corrosion rate, a greater fraction diffuses through the barrier and leads to high pickup fraction.

Hydrogen in excess can escape out of the sample by recombining with the electrons [9] or to be trapped in the oxide film (in compliance with Fig. 1).

This kinetic approach explains both the influence of the test duration and of the oxidation kinetics (Figs. 3 and 5). Obviously, the properties of the oxide film close to the metal oxide interface change with these two parameters [10,11]. It also probably changes the possible maximum hydrogen flux through the barrier oxide film and explains a part of the scatter of the results.

Figure 6 plots the hydrogen pickup ratio as a function of the weight gain and confirms the existence of three families of materials:

1. those that corrode slowly, but with a high pickup ratio (3a and 3b)

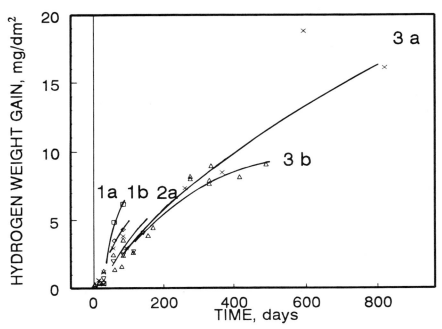

FIG. 4—*Hydrogen weight gains of Zircaloy-4 samples as a function of test duration in 400°C, 10.3 MPa steam.*

2. those that corrode rapidly, but with a low pickup ratio (1a and 1b); and
3. an intermediate case (2a).

These curves provide cumulative data, which integrate successive variations in kinetics during the corrosion, and are therefore difficult to interpret. For this reason, it is preferable [12] to consider instantaneous oxidation and hydriding rates that are deduced from Figs. 3 and 4.

Figure 7 shows that the hydriding rate depends essentially on the corrosion kinetics, the same curve being obtained over a wide range of corrosion rates, whatever the material tested. The hydriding rate increases with corrosion rate. As regards the relative hydrogen absorption, by comparison with the curve for a pickup ratio of 50%, it can be seen to be higher than this level for the slowest corrosion rates and lower than 50% when corrosion increases. The variations of corrosion rate are associated to the variations of the properties of the part of the oxide film close to the metal/oxide interface. As mentioned previously, this probably changes the hydrogen diffusion rate, but this influence is less marked than that of the oxidation rate.

It should be noted that, although the pickup ratio decreases, the overall rate of hydrogen uptake increases continuously. This curve suggests that the manufacturing parameters that were varied for these materials affect only the oxidation kinetics, which in turn determine the rate of hydriding, and that they have no supplementary intrinsic effect on the latter. Figure 7 can be considered to be a characteristic curve for Zircaloy-4 relating hydriding and oxidation rates. In the present operating conditions, and for oxidation rates between 0.1 and 10 mg/dm^2/day, this relationship can be described as

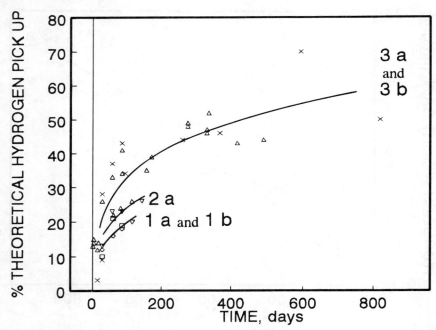

FIG. 5—*Calculated percent hydrogen uptake in Zircaloy-4 tested in 400°C, 10.3 MPa steam as a function of test duration.*

$$\ln V_{hyd} = 0.533 \ln V_{ox} - 3.163$$

where V_{hyd} and V_{OX} are respectively the hydriding and oxidation rates; the rates considered are only the post-transition rates.

Influence of Tin at 400°C

The beneficial influence of lowering the tin content on corrosion at 400°C is well known [13] and is shown in Fig. 8 for recrystallized sheets. For these two materials, the hydrogen content was analyzed after exposures of 195 and 335 days. The results summarized in Table 2 show that less absorption is obtained for the lower tin content for a given time. We can also consider the hydrogen absorption for a given weight gain: 102 mg/dm^2 after 195 days for high-tin content and 103 mg/dm^2 but after 335 days for low tin content. The hydrogen absorption is higher for the low tin content at a given weight gain. Assuming the hydriding and oxidation rates to be constant between 195 and 335 days, the following values were obtained for the alloy with 0.54% tin:

1. an oxidation rate of 0.26 mg/dm^2/day, and
2. a hydriding rate of 0.023 mg/dm^2/day.

As shown in Fig. 9, these results fall close to the typical curve for Zircaloy-4 giving the hydriding rate as a function of the oxidation rate. This suggests that the favorable effect on

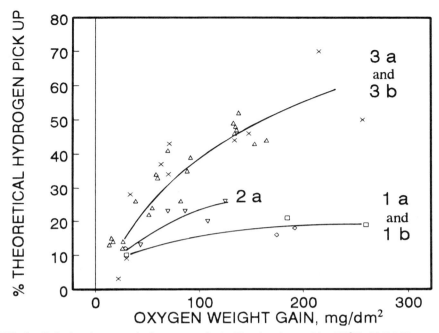

FIG. 6—*Calculated percent hydrogen uptake in Zircaloy-4 tested in 400°C, 10.3 MPa steam as a function of weight gain.*

hydriding of decreasing the tin content is due principally to the reduction in corrosion rate, rather than to any direct intrinsic influence on hydrogen pickup.

Influence of Fe/Cr Ratio at 400°C

The oxidation kinetics at 400°C are given in Fig. 10. They show that the influence of the Fe/Cr ratio is small over the 2 and 5.8 range, whereas for a ratio of 0.2, an extremely high and virtually linear corrosion rate, is observed. Table 3 shows that, for the Fe/Cr ratios from 2 to 5.8, for which very similar corrosion kinetics are obtained (Fig. 10), all the relative hydrogen pickup levels are also close to 50%, indicating the absence of an intrinsic effect of Fe/Cr ratio on hydriding. The pickup ratio is extremely small for the grade, with Fe/Cr = 0.2. For this alloy, assuming a linear corrosion rate between 0 and 333 days (Fig. 10), together with a linear hydriding rate, the following values are obtained:

1. an oxidation rate of 8.9 mg/dm^2/day, and
2. a hydriding rate of 0.2 mg/dm^2/day.

When this result is plotted on Fig. 9, it also falls close to the characteristic curve for Zircaloy-4. Thus, the value of 19% obtained for the pickup ratio is not the consequence of an intrinsic effect of the Fe/Cr ratio, but is simply the reflection of the influence of the latter on the corrosion rate.

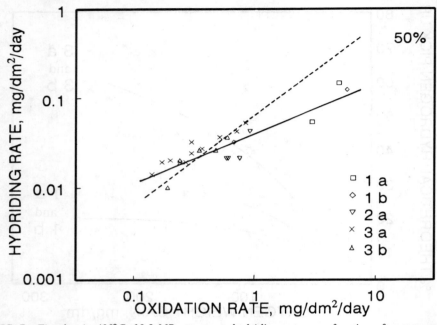

FIG. 7—*Zircaloy in 400°C, 10.3 MPa steam = hydriding rate as a function of post transition oxidation rate; dashed line corresponds to 50% hydrogen uptake.*

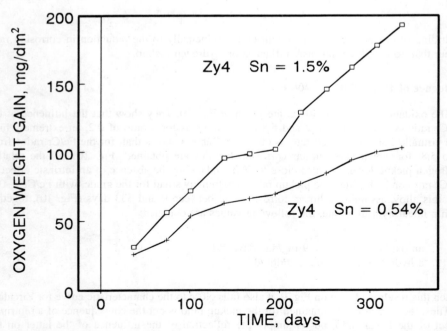

FIG. 8—*Influence of tin on the corrosion resistance of zirconium alloy in 400°C, 10.3 MPa steam.*

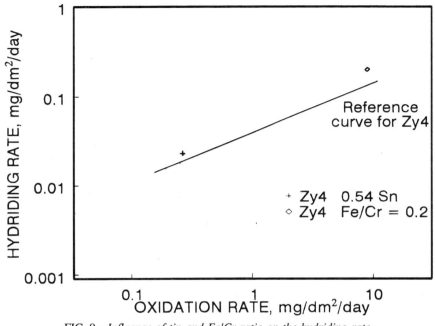

FIG. 9—*Influence of tin and Fe/Cr ratio on the hydriding rate.*

TABLE 2—*Influence of tin in the zirconium alloy on the corrosion resistance at 400°C.*

Tin Content, %	195 Days		335 Days	
	Oxygen Weight Gain, mg/dm^2	Hydrogen Weight Gain, mg/dm^2	Oxygen Weight Gain, mg/dm^2	Hydrogen Weight Gain, mg/dm^2
0.54	68	3.45	103	6.7
1.5	102	5.80	193	8.5

Comparison of Zircaloy-2 and Zircaloy-4 at 400°C

A sheet of each grade was tested at 400°C. These were recrystallized sheets with a cumulative annealing parameter of 5.10^{-18} h. For this value, the uniform corrosion behavior was identical for Zircaloy-2 and Zircaloy-4, as shown in Fig. 11. Figure 11 also gives the hydriding kinetic curves for Zircaloy-2 and Zircaloy-4; they are within the range of variance of Zircaloy-4 Samples 3a and 3b in Fig. 4. Nevertheless, we consider that the difference between Zircaloy-2 and Zircaloy-4 shown in Fig. 11 is a significant example. First, because the samples have been tested together in the same autoclave, and the hydrogen analysis was performed at the same time. Second, this example is in compliance with all our observations concerning the difference between Zircaloy-2 and Zircaloy-4. So for an equivalent corrosion rate, the greater hydrogen absorption attributed to the presence of nickel [14] is confirmed. This indicates that nickel has a specific intrinsic effect on hydrogen uptake, which may add to its influence on the corrosion resistance.

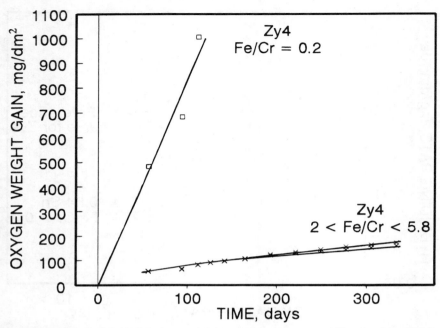

FIG. 10—*Influence of the Fe/Cr ratio on the corrosion resistance of Zircaloy-4 in 400°C, 10.3 MPa steam.*

TABLE 3—*Influence of the Fe/Cr ratio in Zircaloy on the fraction of hydrogen absorbed at 400°C.*

Fe/Cr Ratio	Exposure Time, days	Absorbed Fraction, %
0.2	112	19
2	333	52
2.7	333	49
3.9	333	56
5.8	333	52

Oxidation and Hydriding Kinetics of Zircaloy-4 at 500°C

The products tested at 500°C are given in Table 1. Most of these materials were also evaluated at 400°C. The test durations were varied from 3 to 72 h. Figure 12 summarizes the corrosion results obtained. Marked differences in behavior are observed for the various materials and confirm the interest of a low value of the cumulative annealing parameter, together with rapid cooling after treatment in the beta field. Figure 13 shows the variation of hydrogen pickup ratio as a function of the extent of corrosion. It decreases with weight gain at 500°C, whereas the opposite behavior is observed at 400°C (Fig. 6). The same mechanism as mentioned earlier can be invoked here. If at 400°C, kinetics are the "parabolic" type, then at 500°C, kinetics are in our case "exponential." So, there is more and more hydrogen in excess that cannot diffuse through the metal-oxide interface, giving a lower absorbed fraction. Of course,

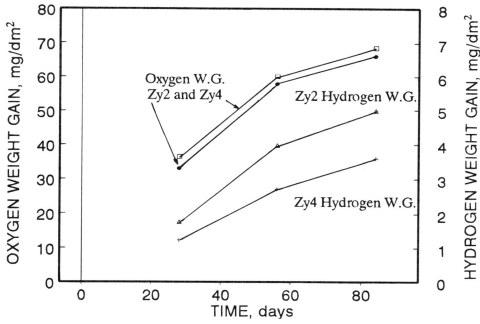

FIG. 11—*Comparison of Zircaloy-2/Zircaloy-4 in 400°C, 10.3 MPa steam.*

this explanation still supposes that the oxygen diffuses faster at 500°C than the hydrogen. We have no data to support this assumption.

If it is assumed that the oxidation and hydriding rates are linear between two measurement points, as shown in Fig. 12, it is possible to estimate the influence of oxidation rate on the hydriding kinetics. The results are reported in Fig. 14. The same type of curve is obtained as for 400°C, an increase in corrosion rate being accompanied by a rise in the quantity of hydrogen absorbed, but with a decrease in the pickup ratio, which is greater than 25% for slow oxidation rates and less than 25% when the kinetics are rapid. In the case of Fig. 14, the hydriding rate is described by the relationship $\ln V_{hyd} = 0.856 \ln V_{ox} - 2.838$.

Conclusions

All of the results obtained on Zircaloy-4, both at 400 and 500°C, are summarized in Fig. 15, which shows that the hydriding kinetics increase with the corrosion rate at both temperatures. The shift in the abscissa and the ordinate of the two curves reflects the higher rates at 500°C, probably related to the effect of temperature on the diffusion phenomena. Figure 15 also shows that when the corrosion rates and, therefore, the hydriding rates increase, the relative absorption ratio decreases, indicating the existence of a mechanism that limits the penetration of hydrogen.

We assume this mechanism to be the diffusion of hydrogen through the barrier oxide layer that could be slower than the oxygen diffusion. These data explain the integrated results that are obtained experimentally, when the oxidation rate is considered a function of time. Thus, when the oxidation rate decreases with time (the case often encountered at 400°C), the pickup ratio is observed to increase; that is why the pickup ratio increases as a function of the test duration (Fig. 5). Conversely, when the oxidation accelerates (as frequently occurs at 500°C), the relative absorption diminishes.

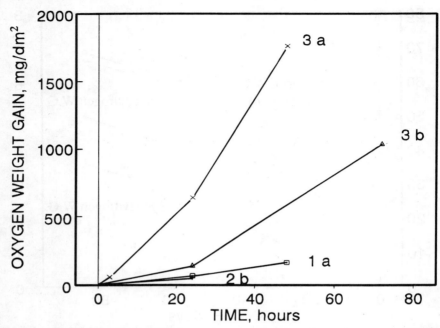

FIG. 12—*Weight gains of Zircaloy-4 samples as a function of test duration in 500°C, 10.3 MPa steam.*

Cases occur where the corrosion initially shows parabolic behavior and then accelerates. In terms of pickup ratio, this should correspond to an initial rise, followed by a decrease.

As regards the hydrogen, the important service parameter is the amount absorbed by the metal and not the pickup ratio, which is a phenomenological characteristic. To decrease the hydriding rate, the most important factor to consider is the reduction in the oxidation rate [*15*].

The available parameters can be divided into two types:

1. those that act principally on the corrosion rate, with little additional intrinsic effect on the hydrogen pickup ratio. For the exposure times employed here, the results show that this is the case for the final heat treatment, the cumulative annealing parameter, the tin content, and the Fe/Cr ratio; and
2. those with a specific supplementary influence on relative hydrogen absorption, as was found to be the case for nickel additions.

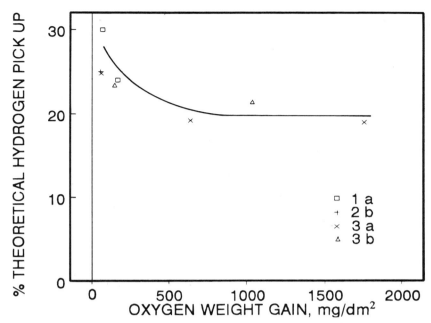

FIG. 13—*Calculated percent hydrogen uptake in Zircaloy-4 tested in 500°C, 10.3 MPa steam as a function of weight gain.*

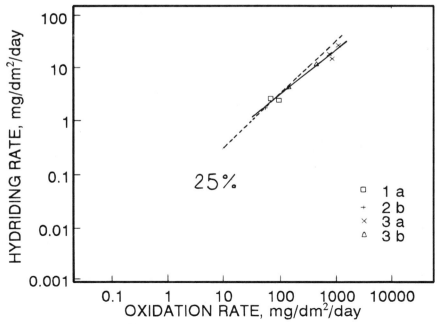

FIG. 14—*Zircaloy-4 in 500°C, 10.3 MPa steam: hydriding rate as a function of oxidation rate; dashed line corresponds to 25% hydrogen uptake.*

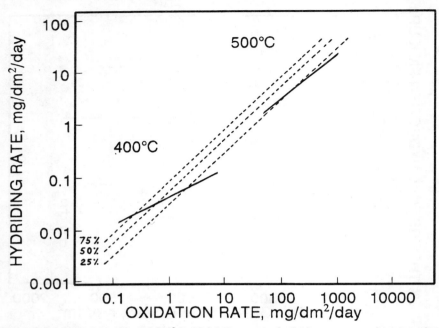

FIG. 15—*Zircaloy-4 in 400 and 500°C, 10.3 MPa steam: hydriding rate as a function of oxidation rate.*

References

[1] Laursen, T., Leslie, J. R., and Tapping, R. L., "Deuterium Depth Distributions in Oxidized Zr-2.5 wt% Nb Measured by Nuclear Reaction Analysis," *Journal of the Less-Common Metals*, Vols. 172–174, 1991, pp. 1306–1312.
[2] McIntyre, N. S., Weisener, C. G., Davidson, R. D., Brennenstuhl, A., and Warr, B., "Analysis of Zr-Nb Fuel Channel Surfaces for Hydrogen and Other Elements Using Secondary Ion Mass Spectrometry (SIMS)," *Journal of Nuclear Materials*, Vol. 178, 1991, pp. 80–92.
[3] Berry, W. E., Vaughan, D. A., and White, E. L., "Hydrogen Pickup During Corrosion of Zirconium Alloys," *Corrosion*, Vol. 17, March 1961, pp. 81–89.
[4] Kass, S. and Kirk, W. W., "Corrosion and Hydrogen Absorption Properties of Nickel-Free Zircaloy-2 or Zircaloy-4," *Transactions*, American Society for Metals, Vol. 55, 1962, pp. 77–100.
[5] Steinberg, E., Weidinger, H. G., and Schaa, A., "Analytical Approaches and Experimental Verification to Describe The Influence of Cold Work and Heat Treatment on the Mechanical Properties of Zircaloy Cladding Tubes," *Zirconium in the Nuclear Industry, Sixth International Symposium, ASTM STP 824*, D. G. Franklin and R. B. Adamson, Eds., American Society for Testing and Materials, Philadelphia, 1984, pp. 106–122.
[6] Garde, A. M., "Enhancement of Aqueous Corrosion of Zircaloy-4 Due to Hydride Precipitation at the Metal/Oxide Interface," *Zirconium in the Nuclear Industry, Ninth International Symposium, ASTM STP 1132*, American Society for Testing and Materials, Philadelphia, C. M. Eucken and A. M. Garde, Eds., 1991, pp. 566–594.
[7] Elmoselhi, M. B., Warr, B. D., and McIntyre, N. S., "A Study of the Hydrogen Uptake Mechanism in Zirconium Alloys," in this volume.
[8] Godlewski, J., thesis, Universite de Technologie de Compiegne, France, 1990.
[9] Klepper, H. H., "Hydrogen Uptake of Zirconium Alloys During Water and Steam Corrosion," *Corrosion*, Vol. 19, Aug. 1963, pp. 225–291.
[10] Godlewski, J., Gros, J. P., Lambertin, M., Wadier, J. F., and Weidinger, H., "Raman Spectroscopy Study of the Tetragonal–to–Monoclinic Transition in Zirconium Oxide Scales and Determination of Overall Oxygen Diffusion by Nuclear Microanalysis of O^{18}," *Zirconium in the Nuclear Industry,*

Ninth International Symposium, ASTM STP 1132, C. M. Eucken and A. M. Garde, Eds., American Society for Testing and Materials, Philadelphia, 1991, pp. 416–436.

[11] Wadman, B., Lai, Z., Andren, H.-D., Nystrom, A.-L., Rudling, P., and Pettersson, H., "Microstructure of Oxide Layers Formed during Autoclave Testing of Zirconium Alloys," in this volume.

[12] Cox, B., *Progress in Nuclear Energy,* Series IV, Chapter 3-3, Pergamon Press, New York, 1961.

[13] Mishima, Y., "Forty-One Years with Zirconium," *Journal of Nuclear Science and Technology,* Vol. 27, No. 3, March 1990, pp. 285–294.

[14] Kirk, W. W., *Zirconium Highlights,* WAPD-ZH-21, Contract AT-11-1-GEN-14, Operated for the U.S. Atomic Energy Commission by Bettis Atomic Power Division, Westinghouse Electric Corporation, 1959.

[15] Graham, R. A. and Eucken, C. M., "Controlled Composition Zircaloy-2 Uniform Corrosion Resistance," *Zirconium in the Nuclear Industry: Ninth International Symposium, ASTM STP 1132,* C. M. Eucken and A. M. Garde, Eds., American Society for Testing and Materials, Philadelphia, 1991, pp. 279–303.

DISCUSSION

J. B. Bai[1] (written discussion)—Have you noticed any modifications in the nature and the morphology of the oxide and hydride formed at 400 and 500°C?

D. Charquet et al. (authors' closure)—We have no clear data about the influence of temperature on the nature of the oxide and hydrides. Concerning the morphology of the oxide film, we have observed, only on poor corrosion resistance materials, some spots of white oxide at 400 and 500°C. The morphology of these spots of white oxide depends on the testing temperature: that is, cracks are parallel to the metal-oxide interface at 500°C, and cracks are perpendicular to the metal-oxide interface at 400°C.

The morphology of the hydrides depends mainly on the structure of the metal, on the hydrogen content, and on the autoclave cooling rate but does not seem to depend on the testing temperature.

B. Warr[2] (written discussion)—Your observations of decreasing percentage of hydrogen pickup with increasing corrosion rate and decreasing tin content are empirical. Do you have any information on the oxide structure and composition to explain the reasons for this behavior?

D. Charquet et al. (authors' closure)—The characteristics of the oxide film depend on the corrosion rate. If we consider the oxide close to the metal-oxide interface, we find more tetragonal phase where we have the lower corrosion rate.

Nobuaki Yamashita[3] (written discussion)—(1) How did the surface after autoclave look in the first slide that shows low ΣA_i material, poor corrosion?

(2) What is the atmosphere for 300°C, 5-min heat treatment?

D. Charquet et al. (authors' closure)—(1) For low ΣA material, that is, poor corrosion resistance material, in 400°C steam, the oxide scale is not regular and there are some areas of white oxide.

(2) The samples were subjected to flowing argon gas during the heat treatment to reduce the tendency of sample oxidation. This worked fine for 300 and 450°C material. However, at 600°C, the samples started to oxidize.

Brian Cox[4] (written discussion)—(1) The ZrO_2 catalyst studies show that heating at temperatures well above 300°C (up to 600°C) is needed to completely eliminate absorbed hydrogen from ZrO_2 powders. What other temperatures did you use to pretreat specimens for analysis and did these give lower analyses than the 300°C treatment?

(2) You did not show any plots of ΔH versus ΔO for your samples, so it was difficult to tell whether the percent hydrogen uptake changed as the oxides got thicker. Can you supply curves plotted in such a manner?

D. Charquet et al. (authors' closure)—(1) The temperatures are given in the text; that is, the second paragraph on page 6. The temperatures were 300, 450, and 600°C. The hydrogen analysis results obtained at 300 and 450°C were about the same. The results obtained at 600°C were misleading since the samples started to oxidize during the heat treatment at 600°C.

(2) Figure 6 shows that the absorbed fraction changes as the oxide gets thicker.

Suresh K. Yagnik[5] (written discussion)—How did you ensure that all of the moisture from the specimen was driven off prior to making hydrogen analyses? It appears that a 5-min heating at 300°C may not be adequate for this purpose.

---

[1] Lab. MSS/MAT, CNRS URA 850, Ecole Centrale Paris, 92295 Chatennay Malabry, France.
[2] Ontario Hydro, Toronto, Ontario, Canada.
[3] GE Nuclear Energy, Wilmington, NC.
[4] University of Toronto, Toronto, Ontario, Canada.
[5] Electric Power Research Institute (EPRI), Palo Alto, CA.

D. Charquet et al. (authors' closure)—We believe that most of the hydrogen in the oxide was driven off because the heat-treated specimen (300°C for 5 min) with the oxide showed about the same hydrogen content as the non-heat-treated reference specimen (from the same sample) where the oxide had been removed. If not, most of the hydrogen in the oxide had been driven off, the hydrogen content in the heat-treated specimen with the oxide would be much higher than that in the non-heat-treated specimen without oxide. This, since the hydrogen content in the oxide (up to ~5 atom percent) may be much higher than that in the metal.

N. Ramasubramanian[6] (written discussion)—(1) What is the steam pressure used in the experiments?

(2) Was a correction applied for the initial hydrogen impurity in the alloy from the surface preparation pickling?

D. Charquet et al. (authors' closure)—(1) The steam pressure is 10.3 MPa.

(2) The samples are pickled before testing, and we have verified that the hydrogen absorption after pickling was lower than 2 ppm.

Bo Cheng[7] (written discussion)—You concluded that nickel can reduce corrosion, but significantly increase hydrogen pickup fraction. My question is whether your steam test results are applicable to the in-reactor performance of Zircaloy-2 in boiling water reactors (BWRs). There are large databases in the literature of irradiated Zircaloy-2 indicating that the pickup fraction is in the neighborhood of 10% or less; the pickup fraction of Zircaloy-2 has little difference from that of Zircaloy-4. Samples of the <10% pickup data were taken from very thin oxide (1 to 3 μm). So, please comment on the relevance of the steam test results for the in-reactor application.

D. Charquet et al. (authors' closure)—Your point is well taken. Today there exists no out-of-pile test that predicts in-pile hydriding characteristics. We need to establish an autoclave test that will predict BWR hydriding performance of Zircaloy materials.

[6] Ontario Hydro, Toronto, Ontario, Canada.
[7] Electric Power Research Institute (EPRI), Palo Alto, CA.

David I. Schrire[1] *and John H. Pearce*[2]

Scanning Electron Microscope Techniques for Studying Zircaloy Corrosion and Hydriding

REFERENCE: Schrire, D. I. and Pearce, J. H., "**Scanning Electron Microscope Techniques for Studying Zircaloy Corrosion and Hydriding**," *Zirconium in the Nuclear Industry: Tenth International Symposium, ASTM STP 1245,* A. M. Garde and E. R. Bradley, Eds., American Society for Testing and Materials, Philadelphia, 1994, pp. 98–115.

ABSTRACT: A procedure has been developed for preparing scanning electron microscope (SEM) samples of irradiated or unirradiated Zircaloy, suitable for oxide layer imaging, hydride concentration and morphology determination, and X-ray microanalysis (EPMA).

The area fraction of the hydride phase is determined by image analysis of backscattered electron images (BEI). Measurements performed on unirradiated laboratory-hydrided samples, as well as cladding samples from pressurized water reactor (PWR) fuel irradiated to a burnup of about 40 MWd/kg U, gave good agreement with hot extraction hydrogen analysis over a wide range of hydrogen concentrations, based on the assumption that all the hydrogen is present as the δ-phase hydride.

The local hydrogen concentration can be determined quantitatively with a spatial resolution of less than 100 μm. This capability was used to determine the radial hydrogen concentration profiles across the cladding wall for PWR samples with different total hydrogen contents, surface oxide thicknesses, and local heat rating. The results indicated that the hydrogen concentration profile was essentially flat (uniform) across the wall thickness for the samples with a low total hydrogen content ($\approx$200 ppm) or a negligible radial heat flux (plenum), while the samples from fueled sections with >200 ppm H had a steep increase in the hydrogen concentration close to the outer surface. Analysis of a longitudinal section showed peak hydrogen concentrations opposite pellet interfaces a factor of two higher than in the mid-pellet region.

The hydride morphology can be readily studied by image analysis in connection with the area fraction measurements. Surface oxide layers are easily imaged and measured in the SEM, for determining the local hydrogen pickup fraction. Electron channeling grain contrast can be seen in unirradiated Zircaloy. Local tin concentrations in irradiated PWR cladding were determined by EPMA with an accuracy better than 0.05% by weight.

KEY WORDS: zirconium, zirconium alloys, hydrogen concentration, hydride morphology, scanning electron microscopy, X-ray microanalysis, nuclear materials, nuclear applications, radiation effects

Corrosion and hydriding often limit the in-reactor lifetime of zirconium alloy components. It is frequently important to study the oxide films, the hydride distribution, and the microstructure and composition of the metal after irradiation or autoclave exposure. When characterizing in-pile corrosion, it is desirable to assess the nature of the oxide layer (nodular, uniform, spalling, porous, etc.), its extent, and its relation to other features (local composition and microstructure of the underlying metal, hydrides, etc.). Both the nature and the extent of the

---

[1] Fuel performance, ABB Atom AB, S-721 63 Västerås, Sweden.
[2] Fuel examination, AEA Technology, Technical Services, Windscale, Seascale, Cumbria CA20-1PF, UK.

corrosion can affect the service performance of the component; for instance, a porous or blistered oxide layer may provide a greater thermal resistance than a dense, uniform one.

The local hydrogen content in Zircaloy components may vary greatly as a result of differences in local hydrogen pickup or redistribution. Extreme variations in local hydrogen absorption in fuel cladding give rise to "sunburst" hydrides, which may be a primary cause of fuel rod failure if the hydrogen originates from inside the fuel rod, or a cause of severe secondary failures following a breach of cladding integrity by some other mechanism [1]. Concentration, temperature, and stress gradients lead to hydrogen redistribution in zirconium alloys [2]. Hydrogen can severely reduce the mechanical properties of zirconium and its alloys, resulting in failure by embrittlement, delayed hydride cracking, or sunburst-type defects. The common feature is that the hydrides, and not the hydrogen in solid solution, cause the mechanical failure.

In order to assess the effect of hydriding on the mechanical properties of a zirconium alloy component, it is necessary to determine the spatial distribution of the hydriding, as well as the hydride morphology. The hydride distribution as determined after irradiation is in practice equivalent to the total hydrogen distribution (hydrogen in solid solution as well as the hydride phase) at the service temperature, due to the low hydrogen solubility at room temperature. The spatial distribution data is also invaluable for verifying models for the redistribution of hydrogen. Hydride morphological parameters that have been shown to influence embrittlement include the hydride orientation relative to the applied load [3–6], the hydride platelet length [7], the hydride continuity coefficient (HCC) [8,9], the general continuity of the hydride network [10], the shape of the hydrides [11], the presence of a massive surface hydride layer [11–13], and the interhydride spacing [4].

While the local corrosion and hydriding are important for the mechanical properties, the average oxide and hydrogen levels are needed to determine the overall hydrogen pickup fraction. However, even this relationship is complicated by the interaction between oxidation and hydriding; the hydride phase is suspected of oxidizing faster than the metal [14], and the pickup fraction may possibly vary with the oxide layer thickness. In order to elucidate these effects, it is necessary to determine the oxide thickness, average hydrogen level, and the hydride distribution locally.

The chemical composition is known to influence the corrosion and hydriding behavior of zirconium alloys; for instance, a reduced tin content has been shown to improve the corrosion resistance of zirconium alloys in an autoclave environment [15]. However, a certain variation in the chemical composition may occur within a specific lot of material. It is natural to expect that the local alloy composition may influence the local behavior; it is thus useful to be able to analyze the composition locally in connection with postirradiation measurements of the corrosion in order to more precisely define possible composition effects.

Current methods suffer from a number of shortcomings. Hot vacuum extraction (HVE) techniques for determining the hydrogen content are nonrepeatable (the sample cannot be remeasured), and only provide the average hydrogen concentration in the sample. In addition, instruments that depend on standards for calibration are often limited in their range (or sample size at high hydrogen levels), as well as their inherent accuracy. There may also be some uncertainty in the analysis of samples with a significant oxide layer, both as regards the effective sample weight and the hydrogen evolved from the oxide. Traditionally, optical microscopy of etched metallographic samples has been employed to assess the local hydride fraction [16]. However, this method has severe limitations: the etching conditions are not always exactly identical (as regards etchant concentration, temperature, and application) that leads to poor etching reproducibility; and the hydride appearance is not directly proportional to the hydrogen/hydride concentration up to high hydrogen levels (overetching at high hydrogen concentrations leaves the surface completely "black" at less than solid hydride fractions).

Similarly, traditional techniques for characterizing the oxide layer after irradiation are not always satisfactory. Eddy current testing (ECT) using the lift-off principle is frequently used for determining the oxide layer thickness. However, this only measures the total distance (lift-off) from the underlying metal, and cannot distinguish between oxide and the gap in the event of blistering or cracking. This technique is also sensitive to variations in the underlying metal composition, hydriding below the oxide layer, and is difficult to apply to multilayer components such as "duplex" tubing. Metallographic techniques are recognized as being more accurate, and they also reveal the nature of the oxide (nodular, spalling, etc.). However, optical microscopy provides very little focal depth, so that it is in practice difficult to image the surface oxide at the same time as the hydrides, since chemical etching for hydrides produces significant relief. Sample "rounding" during polishing can also sometimes cause difficulty in maintaining focus across a thick oxide layer.

A method has now been developed for determining the local hydride volume fraction, and thereby the hydrogen concentration, in irradiated or unirradiated Zircaloy using a backscattered electron imaging technique in a scanning electron microscope (SEM). This method offers a number of advantages compared to optical microscopy:

1. greater reproducibility, since the analysis is performed on unetched samples;
2. better correlation between apparent hydride area fraction and true hydrogen concentration;
3. a larger range of magnification, including far higher magnifications than are possible in the optical microscope;
4. the possibility of combining hydride analysis with X-ray analysis (EPMA), for instance, of alloying elements, oxygen, etc; and
5. one-step sample preparation that enables both oxide layer and hydride imaging (and electron channeling grain contrast in unirradiated material).

This paper presents the results of a series of measurements performed on irradiated pressurized water reactor (PWR) cladding samples, as well as unirradiated, gaseously hydrided samples. These results illustrate the type of information that may be obtained by this technique and are not intended as a full study of Zircaloy corrosion.

Experimental Procedure

Sample Preparation

Nine irradiated PWR cladding samples, including both Zircaloy-2 and Zircaloy-4, were analyzed. Eight of the samples were taken from fuel rods irradiated in Assembly 5F1 (which had reached an assembly average burnup of 40 MWd/kg U after four cycles) and one from Assembly 5F2 (47 MWd/kg U after five cycles). The sample identifications, locations, and types of analysis performed are listed in Table 1.

Unirradiated laboratory-hydrided samples were prepared by gaseous hydriding of a number of Zircaloy-2 tubing sections simultaneously at 400°C. The system was filled with hydrogen at a pressure corresponding to the nominal target hydrogen concentration for the particular sample batch weight. No particular attention was paid to homogenization heat treatment after the hydrogen absorption: a single constant temperature cycle was used in all cases. Essentially, complete hydrogen absorption was ensured by noting that the system pressure had dropped to a low, steady value prior to the cooling.

Both the irradiated and laboratory-hydrided samples consisted of segments (arcs) of rings cut from cladding tubing. In the case of the irradiated material, 2-mm-high rings were cut

TABLE 1—*Irradiated Zircaloy samples for SEM/EPMA analysis.*

Sample Identity	Assembly Number	Rod Number	Distance from Bottom of Rod, mm	Type of Analysis
01A	5F1	096-40001	3094	SEM (H)
13E	5F1	096-40013	3755 (plenum)	SEM (H)
13G	5F1	096-40013	2033	SEM (H)
13I	5F1	096-40013	870	SEM (H)
65A	5F1	906-40065	3141	SEM (H)
1536	5F2	096-40015	3110	SEM (H)
Sn1	5F1	096-40013	3790 (plenum)	EPMA (Sn)
Sn2	5F1	096-40012	3790 (plenum)	EPMA (Sn)
Sn3	5F1	096-40001	3790 (plenum)	EPMA (Sn)

from the fuel rods, and the fuel was removed. The rings were sectioned, with one segment being analyzed by HVE. All the cutting and defueling operations were performed in a hot cell. Similar or slightly larger ring segments were prepared from the laboratory-hydrided material, and the hot extraction analyses were performed on adjacent ring samples instead. The irradiated cladding samples were cut to different arc lengths, to facilitate identification. One axial sample (1536) was also prepared from a defueled cladding half extending over the length of almost two fuel pellets. A single HVE analysis was performed on each irradiated clad sample. Single or multiple hot extraction analyses were performed on the laboratory hydrided tubing.

The irradiated cladding samples were generally mounted together with unirradiated samples in a 10- or 25-mm diameter ring. A graphite-loaded two-component epoxy, a standard epoxy followed by sputtered conductive coating, and a copper-loaded phenolic were all successfully tested for embedding the samples. The mounted sample was then ground and polished according to standard hot cell Zircaloy metallographic preparation procedures, using a colloidal silica suspension or attack polish as the final step.

Image Acquisition and Analysis

The hydrogen analysis was performed on a JEOL 8600 analytical SEM (AEA Technology) or a JEOL 840 analytical SEM (Studsvik Nuclear and Royal Institute of Technology). Images were digitized and analyzed by a LINK X-ray microanalysis system at a resolution of 512 by 512 pixels. "Hard" contrast (high gain) was used for imaging the hydrides. Reduced gain (low contrast) was used for imaging the surface oxide. The images typically had areas of about 70 by 80 μm, with seven or eight such areas just covering the entire cladding wall thickness in a "radial scan." Six or more radial scans were analyzed for each of the irradiated samples, although less were usually performed on the laboratory-hydrided samples.

The LINK image analysis software could not perform shading correction or sophisticated feature detection compared to the local background, so a single threshold (grey level) was used for the hydride detection. The thresholds were set by the operator by comparing the binary image (detected hydrides) with the grey level backscattered image. The hard contrast and good resolution of the backscattered images ensured satisfactory reproducibility in the threshold settings, although some operator dependence cannot be ruled out. The total area fraction of detected hydrides was measured and recorded for each field. Very small features were rejected as noise.

SEM images have also been used for characterizing the hydride morphology using a more sophisticated image analyzer (Quantimet 570). Examples of the morphological parameters determined in this way include the Fn factor, the orientation distribution of all hydrides or hydride segments longer than a given length, and the length distribution of all hydrides or hydride segments within a specific orientation interval. The hydride area fraction was also determined at the same time. Shading correction (for intensity variations over the image) and automatic feature detection were also utilized.

Hydrogen Determination

This technique is based on backscattered electron imaging, exploiting the compositional (atomic number) contrast available in a SEM with symmetrically placed backscattered electron detectors (thereby avoiding any topological contrast effects). The hydride phase has a much lower mean atomic number than the metallic (zirconium) phase, and thus provides good contrast. An automatic image analysis system is used to measure the area fraction of the hydride phase in the plane of polish. Provided that the analyzed area is representative of the material or region (irrespective of the orientation of the plane of polish), the area fraction of the intercepted hydrides is equivalent to their volume fraction [17].

The following assumptions form the basis of the determination of the hydrogen content from the volume fraction of hydrides:

1. the hydrided material has been cooled slowly (in-reactor or furnace-cooled), and the resulting hydrogen in solid solution is negligible compared to the hydrogen present in the hydrides; and
2. all the hydrides are of the δ-phase (62 atomic % H). This phase is favored under the conditions prevailing in most irradiated and laboratory-hydrided Zircaloy.[3]

From the preceding assumptions, the following relationship between the hydrogen concentration and the measured hydride area (and thus volume) fraction, F, is obtained

$$Wt_H = Wt_\delta \cdot F \left(\frac{\rho_\delta}{\rho_{Zr}(1 - F) + \rho_\delta \cdot F} \right)$$

where

Wt_H = wt ppm[4] H,
Wt_δ = wt ppm H in δ-phase hydride (17 570 ppm),
F = measured hydride area fraction (equivalent to volume fraction),
ρ_δ = density of δ-phase hydride (5.65 g/cm^3 at room temperature), and
ρ_{Zr} = density of α-phase metal (6.54 g/cm^3 at room temperature).

This formula has been used to calculate the hydrogen concentration from the measured hydride fraction.

---

[3] The following conditions are known to favor the formation or growth of δ-phase hydrides [18]:
   (a) oxygen (>1000 ppm) that is usual in both irradiated and laboratory-hydrided Zircaloy. It is thought that a high oxygen content tends to favor the δ-phase instead of the γ-phase due to the increase in the matrix yield stress; irradiation hardening would therefore be expected to further strengthen this trend;
   (b) slow cooling (furnace or in-reactor); and
   (c) high hydrogen contents (>100 ppm), that is, the area for which this method is intended.
[4] In this paper, hydrogen concentrations are reported as wt ppm H, that is, mg H/kg sample.

Electron Probe Microanalysis (EPMA)

The irradiated samples, embedded with a sample of fresh Zircaloy of known composition, were mounted in a SEM sample holder with zirconium and tin X-ray standards. The EPMA measurements were performed on a JEOL 840 analytical SEM, with partly shielded wavelength dispersive X-ray spectrometers, controlled by a LINK AN10000 X-ray microanalysis system. After calibration and standardization, the samples were analyzed with beam current and ZAF correction. Ten micro-areas (approximately 5 µm by 4 µm each) were analyzed in each run. A number of such runs were performed close to the outer surface, midwall, and bore of each sample. These measurements took place over a few days, so that it was possible to check the reproducibility of the measurements. In addition, a series of measurements was performed on the zirconium standard each day. The "detected tin concentration" in the zirconium standard (with a true tin content of only a few ppm) was subtracted from the detected concentration in the samples. This procedure ensured that any bias due to the sample radioactivity, or incorrect background levels, was accounted for.

Results

Imaging

"Hydrides" could be discerned in both the optical microscope (OM) under normal illumination and SEM (backscattered electron imaging) after the polishing procedure previously outlined. However, the features seen in the OM were considerably larger (broader) than the same features seen in the SEM (Fig. 1), and the area fraction of the OM "hydrides" greatly overestimated the hydrogen concentration in the sample tested. The laboratory hydrided samples with the highest hydrogen concentrations had relatively thick surface hydride layers (Fig. 2a). Although such layers usually appear to be solid hydride in optical microscopy, the SEM images clearly reveal the fine structure in the hydride layer. By comparison, even the cold outer surface of the irradiated samples usually had a much less dense concentration of hydrides (Fig. 2b).

Grain contrast (presumably electron channeling contrast) was clearly visible in the laboratory-hydrided samples. However, the grain contrast was far less pronounced in the irradiated samples under the same conditions (Fig. 3). This reduction in the grain contrast in the irradiated Zircaloy is presumably due to irradiation damage affecting the electron channeling effect. The reduced grain contrast in the irradiated samples facilitates the hydride imaging and analysis.

The outer surface oxide layer in an irradiated cladding sample (Rod 096-40001 peak oxide elevation) is shown in Fig. 4 (low contrast imaging). Under these imaging conditions, the oxide layer is readily detected by the image analyzer, enabling the average amount of oxide (area fraction) excluding porosity or internal cracks to be determined. The total oxide thickness including cracks and porosity can also be determined using suitable image processing routines to fill the internal voids.

Hydrogen Concentration

The results of the hydride fraction radial profile measurements for three samples from Rod 096-40013 are plotted in Fig. 5. The radial profile for Sample 1536 averaged over all the longitudinal positions measured is plotted in Fig. 6a, and in Fig. 6b, the hydrogen concentration averaged over the entire wall thickness is plotted as an axial profile. The hydrogen concentration data is normalized for presentation, and is plotted as the mean $\pm$ the sample standard error $(s/\sqrt{n})$ for the six to ten measurements at each radial position.

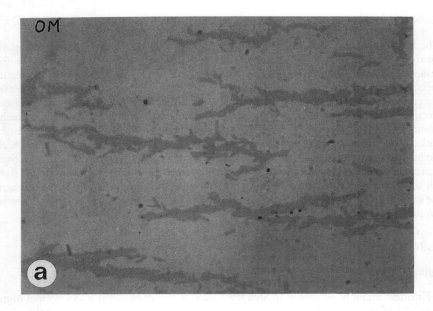

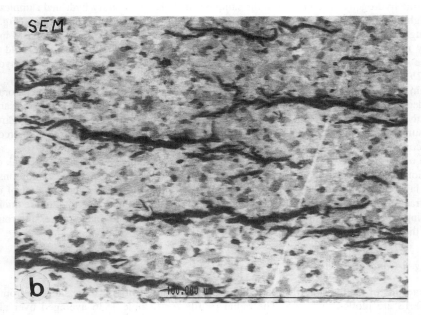

FIG. 1—(a) *OM image of hydrides (as polished), and (b) SEM image of hydrides. Note the overestimate of the hydride platelet thickness in the OM image compared to the SEM image.*

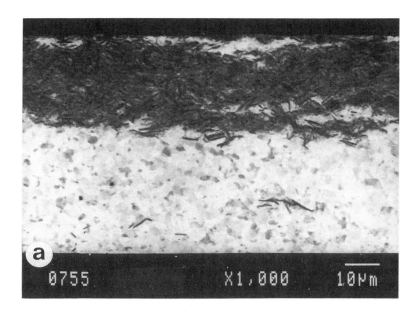

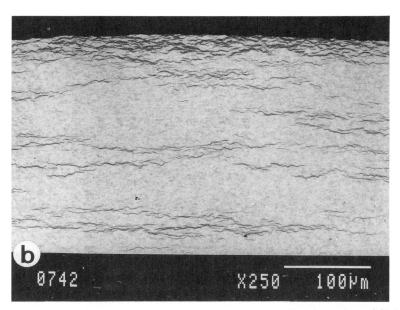

FIG. 2—*Examples of surface hydride layers in:* (a) *laboratory-hydrided sample, and* (b) *irradiated sample, outer surface.*

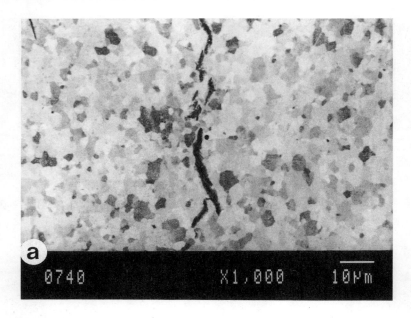

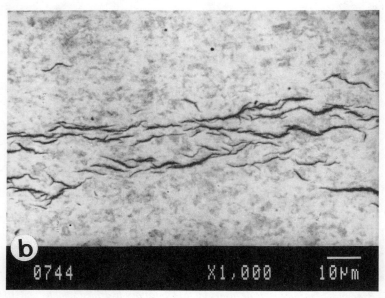

FIG. 3—(a) Grain contrast in unirradiated recrystallized Zircaloy-2 and (b) grain contrast in irradiated stress relief annealed (SRA) Zircaloy-4.

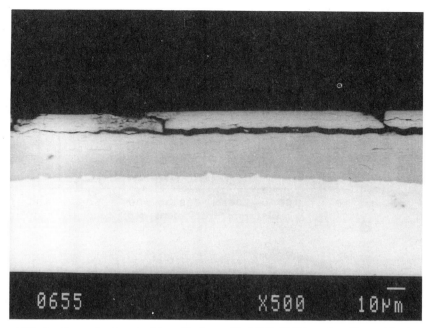

FIG. 4—*Example of waterside oxide layer (Rod 096-40001 peak oxide position).*

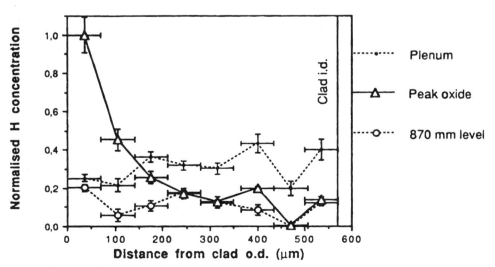

FIG. 5—*Cladding radial hydrogen profiles for Rod 096-40013 at three elevations.*

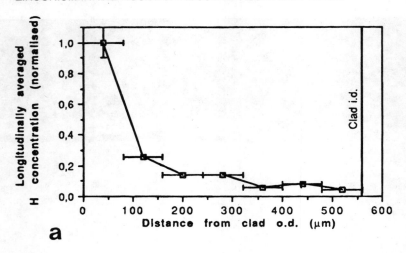

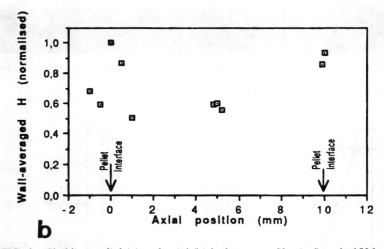

FIG. 6—*Cladding radial (a) and axial (b) hydrogen profiles in Sample 1536.*

The hydrogen concentration determined from the hydride fraction measurements was averaged over the entire wall thickness and compared with hot extraction data. These results are shown in Fig. 7. The hydrogen pickup fraction for the irradiated cladding samples from Assembly 5F1, based on both SEM analysis and HVE measurements, is plotted against the relative oxide thickness determined from the SEM images in Fig. 8.

Tin Content

The mean and standard deviation of the mean of the tin measurements on the three-irradiated samples as well as the reference sample are given in Table 2 for each run of ten measurements. The results are grouped by date, together with the "zirconium standard bias" determined for each day. The results are corrected for the zirconium standard bias.

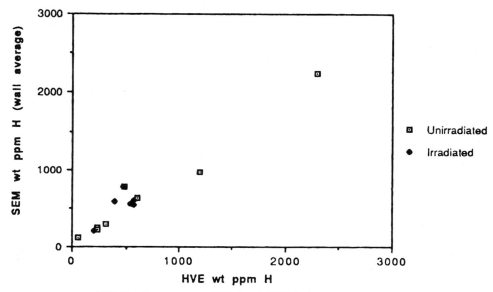

FIG. 7—*Comparison of SEM and HVE hydrogen results.*

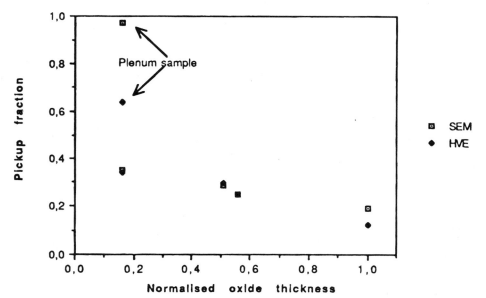

FIG. 8—*Hydrogen pickup fraction as a function of relative oxide thickness, for all samples from Assembly 5F1.*

TABLE 2—*EPMA tin analysis results. Percent by weight tin (mean ± standard deviation of the mean, ten measurements). Corrected for zirconium standard bias.*

Day	Zr Standard Bias	Sample[a]	Outside Diameter	Mid-Wall	Bore
1	0.113	Reference	1.29 ± 0.01	1.53 ± 0.01	1.49 ± 0.02
		Sn1	1.55 ± 0.02	1.50 ± 0.02	1.47 ± 0.01
		Sn2	1.55 ± 0.01	1.54 ± 0.02	1.51 ± 0.01
		Sn3	1.50 ± 0.01	1.50 ± 0.02	1.48 ± 0.01
2	0.113	Reference	1.43 ± 0.01	1.41 ± 0.01	1.41 ± 0.01
		Sn1	1.44 ± 0.01	1.41 ± 0.01	1.41 ± 0.01
3	0.123	Reference	1.43 ± 0.02	1.48 ± 0.02	1.46 ± 0.01
		Sn1	1.55 ± 0.01	1.50 ± 0.01	1.51 ± 0.01
		Sn2	1.46 ± 0.01	1.45 ± 0.02	1.44 ± 0.02
		Sn3	1.44 ± 0.02	1.47 ± 0.01	1.42 ± 0.02
4	0.110	Reference	1.38 ± 0.02	1.40 ± 0.01	1.45 ± 0.01
		Sn1	1.42 ± 0.02	1.43 ± 0.02	1.39 ± 0.02
		Sn2	1.45 ± 0.01	1.41 ± 0.01	1.44 ± 0.01
		Sn3	1.42 ± 0.02	1.40 ± 0.02	1.40 ± 0.02

[a]Summary of results for each sample, based on the mean values for each run:

Sample	Reference	Sn1	Sn2	Sn3
Mean	1.43	1.47	1.47	1.45
Standard deviation of mean	0.02	0.02	0.02	0.01

Discussion

Hydrogen Determination Accuracy

The wall-averaged hydrogen content as determined by this method gave results in good general agreement with the hot extraction results for both the irradiated and the laboratory-hydrided samples (Fig. 7). The two methods agree over a wide range of hydrogen concentrations. The discrepancies between the two methods could be due to imprecision or bias in the SEM analysis and the hot extraction measurements, as well as real variations in the local hydrogen content between the samples. Potential sources of error in the current SEM analysis method are discussed later.

The precision of the hot extraction method is commonly about ±10% of the measured value for closely spaced samples of unirradiated, oxidized Zircaloy tubing [19]. Measurement bias in the hot extraction method is more difficult to assess. Hydrogen present in the oxide layer is known to be a (potentially large) source of error. This particular problem is avoided in the SEM measurements, since only the hydrides in the metal are measured. Large local differences in the hydrogen content occur in irradiated cladding, and significant variation may also occur in laboratory-hydrided material. The sample variability is probably the dominant source of variation in the SEM results (shown as error bars in the figures).

There are a number of sources of experimental error that affect the overall bias and precision of the SEM hydride image analysis method. Some of the most important error sources are discussed in the following paragraphs with respect to their cause, effect, and ways of minimizing the error.

Artefacts—Artefacts (scratches gouges, and dirt) can cause both bias (tending to increase the apparent hydride fraction) and imprecision. This error is minimized by good sample preparation. It can also be reduced by judicious selection of imaging fields (while taking care to maintain the requirement for "representative sampling areas"). Artefacts were not thought to be a large source of error in the current study.

Incorrect Threshold Setting—Incorrect threshold setting results in a bias in the hydride area fraction. It can also reduce the precision, particularly the reproducibility with different operators. This problem is also instrument related, since some image analyzers can automatically detect the second phase irrespective of variations in the local image intensity. This may have been a significant error source in this study.

Instrument Drift and Sample Shading—Instrument drift and sample shading are related to the previous problem, since the threshold setting is no longer correct in relation to all parts of every image. Software or instrumentation are available that can virtually eliminate these errors. Errors due to instrument drift were noticed in early measurements during this study. Careful SEM operation (ensuring constant beam current) was found to significantly reduce this problem.

Depth of Information—Depth of information causes a systematic bias (overestimate) of the hydride fraction, since hydrides below the surface are also detected, thus invalidating the stereological equivalence between observed area fraction and bulk volume fraction. This error increases with increasing hydride fraction, and with decreasing feature (hydride platelet) size. The depth of information for backscattered electrons is related to the Kanaya-Okayama range and is proportional to $V^{1.6}$ (where V is the accelerating voltage). Reducing the accelerating voltage is thus an efficient way to minimize this error; however, the image resolution deteriorates with decreasing voltage, so that a balance must be found in practice. No attempt was made to evaluate this error. This error could be quantified in practice by measuring the apparent hydride area fraction in a particular image field at different accelerating voltages.

Hydride Phase Composition and Density—Hydride phase composition and density affect the calculated hydrogen concentration, but not the measurement of the hydride fraction. Characterization of the hydride phases formed under different conditions is an important area for future work. The results of this study do not give cause for any revision of the assumptions used to determine the hydrogen concentration from the measured hydride fractions.

Hydride Heterogeneity—Hydride heterogeneity, on the micro-scale, gives rise to a spread in the sampled data. This effect is particularly significant in the case of a low hydride fraction, where the individual hydride strings are relatively large, so that only a few separate hydride "features" are present in each field. It is clear that a larger total image area has to be sampled in order to attain the same sampling precision for a low hydrogen level as for a higher level, if the hydrides have a similar size in both cases.

Hydrogen Profiles and Pickup Fractions

Hydride concentration gradients have frequently been observed qualitatively in optical microscopy of etched cladding samples. The results of these quantitative measurements on high burnup PWR cladding confirm previously observed qualitative trends:

1. high bulk hydrogen levels together with a heat flux (Sample 13G in Fig. 5, Fig. 6a) cause steep increases in the local hydrogen content towards the outer surface;
2. hydrogen is axially redistributed to the cooler part of the cladding opposite pellet interfaces (Fig. 6b);
3. in the absence of a radial heat flux (plenum region), the hydrogen is more or less uniformly distributed across the cladding wall (Sample 13E in Fig. 5); and
4. for bulk hydrogen levels close to or below the solubility limit, there is no clear radial hydrogen concentration profile (Sample 13I in Fig. 5).

The laboratory-hydrided samples were also found to have nonuniform radial hydrogen distributions. However, in this case, a solid hydride layer formed on the surfaces of the tubing, while the hydrogen concentration in the interior never rose above 600 to 800 ppm.

The hydrogen pickup fraction in the samples from Assembly 5F1 appeared to decrease with increasing oxide thickness. The plenum sample had an apparent pickup fraction approaching 1; hydrogen apparently migrated from the top of the fueled cladding to the cooler plenum zone that caused the high local hydrogen concentration.

Tin Content

The tin analyses were performed on micro-areas larger than the Zircaloy grain size, in order to avoid any possible problems due to microstructural heterogeneity in the tin concentration. Such heterogeneity could be expected if an area contained intergranular secondary phase particles, which could be either enriched or depleted in tin.

The zirconium standard bias was found to be fairly constant from day to day, resulting in an overestimate of the apparent tin concentration of approximately 0.11% by weight. The variation in the results for ten analyses at a similar location was similar to the variation in the mean values of different runs performed on the same sample (compare the standard errors of the individual runs in the upper part of Table 2 with the standard errors of the means in the lower part of the table). This implies that the analysis reproducibility from day to day was no worse than the measurement precision on any particular occasion.

The mean tin concentration determined for the reference sample was $1.43 \pm 0.02\%$ by weight, which agrees fairly well with the values of 1.46% and 1.48% by weight for this material obtained by two different bulk analysis methods. There was no significant difference in the measured tin concentration among the three irradiated samples, which led to the conclusion that the difference in the oxide thickness between these rods (of over 30%) was not attributable to differences in the tin content.

Conclusions

A procedure has been developed for preparing SEM samples of irradiated or unirradiated Zircaloy that is straightforward and easily applied to irradiated samples under hot cell conditions.

SEM hydride fractions measured on unirradiated laboratory-hydrided samples, as well as highly irradiated PWR fuel cladding samples, verified that the wall-average hydrogen concentrations determined by this method gave good agreement with hot extraction hydrogen analysis over a wide range of concentrations, based on the assumption that all the hydrogen was present as the δ-phase hydride. The technique was used to determine both radial and axial hydrogen concentration profiles in PWR cladding samples representing different conditions as regards total hydrogen content, surface oxide thickness, and local heat rating, with a spatial resolution of less than 100 μm.

It was found that in addition to excellent imaging of the hydrides, both the surface oxide layer and (in the case of the unirradiated samples) the grain structure could be clearly seen in the SEM using backscattered electron imaging. X-ray microanalysis was also used to determine the local tin content in irradiated cladding samples to detect possible variations that might affect the in-pile corrosion behavior. This is clearly a powerful method for the characterization of Zircaloy (and presumably other zirconium-based alloys), where the advantages of a simple, single-step preparation procedure can be combined with SEM imaging and X-ray microanalysis.

Acknowledgments

The SEM hydrogen analysis on samples from Assembly 5F1 was performed by AEA Technology, Windscale. The sample preparation and HVE analyses on the irradiated samples, the EPMA (tin) measurements, and the SEM hydrogen analysis of Sample 1536, were performed by Studsvik Nuclear. The laboratory hydriding and the hot extraction measurements on the unirradiated samples were performed by ABB. Additional SEM image analysis on unirradiated material was performed by Sofia Bank (Royal Institute of Technology, Stockholm). The Quantimet 570 image analyzer was provided by courtesy of Leica AB. The work was financed by the Swedish State Power Board and ABB Atom.

References

[1] Clayton, J. C., "Internal Hydriding in Irradiated Defected Zircaloy Fuel Rods," *Zirconium in the Nuclear Industry: Eighth International Symposium, ASTM STP 1023,* L. F. P. Van Swam and C. M. Eucken, Eds., American Society for Testing and Materials, Philadelphia, 1989, pp. 266–288.

[2] Northwood, D. O. and Kosasih, U., "Hydrides and Delayed Hydrogen Cracking in Zirconium and Its Alloys," *International Metals Reviews,* Vol. 28, No. 2, 1983, pp. 92–121.

[3] Marshall, R. P. and Louthan, M. R., "Tensile Properties of Zircaloy with Oriented Hydrides," *Transactions, American Society for Metals,* Vol. 56, 1963, pp. 693–700.

[4] Coleman, C. E. and Hardie, D., "The Hydrogen Embrittlement of α-Zirconium," *Journal of the Less-Common Metals,* Vol. 11, 1966, pp. 168–185.

[5] Aitchison, I., "The Effect of Orientation of Hydride Precipitates on the Fracture Toughness of Cold-Rolled Zircaloy-2 and 2.5 Nb Zirconium," *Applications-Related Phenomena for Zirconium and Its Alloys, ASTM STP 458,* American Society for Testing and Materials, Philadelphia, 1969, pp. 160–178.

[6] Bai, J., Prioul, C., Pelchat, J., and Barcelo, F., "Effect of Hydrides on the Ductile-Brittle Transition in Stress-Relieved, Recrystallized and β-Treated Zircaloy-4," *Proceedings,* International Topical Meeting on LWR Fuel Performance. Avignon, 21–24 April 1991, pp. 233–241.

[7] Puls, M. P., "The Influence of Hydride Size and Matrix Strength on Fracture Initiation at Hydrides in Zirconium Alloys," *Metallurgical Transactions A,* Vol. 19A, 1988, pp. 1507–1522.

[8] Bell, L. G. and Duncan, R. G., "Hydride Orientation in Zr-2.5%Nb; How it is Affected by Stress, Temperature and Heat Treatment," Report AECL-5110, Atomic Energy of Canada Limited, 1975.

[9] Davies, P. H. and Stearns, C. P., "Fracture Toughness Testing of Zircaloy-2 Pressure Tube Material with Radial Hydrides Using Direct-Current Potential Drop," *Fracture Mechanics: Seventeenth Volume, ASTM STP 905,* Underwood et al., Eds., American Society for Testing and Materials, Philadelphia, 1986, pp. 379–400.

[10] Bai, J. B., Prioul, C., Lansiart, S., and François, D., "Brittle Fracture Induced by Hydrides in Zircaloy-4," *Scripta Metallurgica et Materialia,* Vol. 25, 1991, pp. 2559–2563.

[11] Lin, S.- C., Hamasaki, M., and Chuang, Y. -D., "The Effect of Dispersion and Spheroidization Treatment of δ-Zirconium Hydrides on the Mechanical Properties of Zircaloy," *Nuclear Science and Engineering,* Vol. 71, 1979, pp. 251–266.

[12] Slattery, G. F., "The Mechanical Properties of Zircaloy-2 Tubing Containing Circumferentially Aligned Hydride," *Applications-Related Phenomena for Zirconium and Its Alloys, ASTM STP 458,* American Society for Testing and Materials, Philadelphia, 1969, pp. 95–110.

[13] Price, E. G., "Hydride Orientation and Tensile Properties of Zr-2.5 wt%Nb Pressure Tubing Hydrided while Internally Pressurized," *Canadian Metallurgical Quarterly,* Vol. 11, 1973, pp. 129–138.

[14] Garde, A. M., "Enhancement of Aqueous Corrosion of Zircaloy-4 due to Hydride Precipitation at the Metal-Oxide Interface," *Zirconium in the Nuclear Industry: Ninth International Symposium, ASTM STP 1132*, C. M. Eucken and A. M. Garde, Eds., American Society for Testing and Materials, Philadelphia, 1991, pp. 566–594.

[15] Eucken, C. M., Finden, P. T., Trapp-Pritsching, S., and Weidinger, H. G., "Influence of Chemical Composition on Uniform Corrosion of Zirconium-Base Alloys in Autoclave Tests," *Zirconium in the Nuclear Industry: Eighth International Symposium, ASTM STP 1023*, L. F. P. Van Swam and C. M. Eucken, Eds., American Society for Testing and Materials, Philadelphia, 1989, pp. 113–127.

[16] Hyatt, B. Z., "Metallographic Standards for Estimating Hydrogen Content of Zircaloy-4 Tubing," WAPD-TM-1431, Westinghouse Electric Corporation, Feb. 1982.

[17] Hilliard, J. E., "Measurement of Volume in Volume" *"Quantitative Microscopy,"* R. T. DeHoff and F. N. Rhines, Eds., McGraw-Hill, New York, 1968.

[18] Cann, C. D., Puls, M. P., Sexton, E. E., and Hutchings, W. G., "The Effect of Metallurgical Factors on Hydride Phases in Zirconium," *Journal of Nuclear Materials,* Vol. 126, 1984, pp. 197–205.

[19] Rudling, P., ABB Atom, Västerås, Sweden, private communication.

DISCUSSION

Brian Cox[1] (written discussion)—(1) In order to convert your area fraction of hydride to a volume fraction, you need to know that the aspect ratio, continuity, and randomness of the hydrides in your fuel cladding and your standards are the same. If they differ too much you would need to do a number of sections to assess this. Have you usually done this?

(2) You mentioned "long distance diffusion" at fuel pin ends. For hydrogen diffusion, what do you consider to be a long distance?

D. I. Schrire and J. H. Pearce (authors' closure)—(1) No standards were used in determining the hydride area fraction for either the irradiated or unirradiated samples. The stereological equivalence between the area fraction and volume fraction of a phase is valid irrespective of orientation or shape. The validity of the assumption of hydride randomness at a particular location is tested in practice by repeating the analyses at a number of equivalent locations, and determining the variation in the measured area fraction. This was done for all the irradiated cladding samples, and the variation is represented in the figures by the standard deviation of the mean.

(2) "Long distance diffusion" in this context refers to hydrogen diffusion over distances much greater than the pellet length, as apparently occurred at the top of Rod 096-40013, where an unexpectedly high level of hydrogen was found in the plenum approximately 6 cm above the top of the fuel stack.

J. B. Bai[2] (written discussion)—When one uses image processing to determine the hydrogen content, one should give (or assume) a thickness of the hydride. According to our experience, this given thickness will be multiplied if several hydrides are stocked together or the hydride thickens for high hydrogen content. How have you chosen this parameter?

D. I. Schrire and J. H. Pearce (authors' closure)—It is only necessary to assume a thickness for the hydrides if one is trying to determine the apparent hydride area indirectly, by multiplying the (measured) hydride length by the thickness. In this study, the area fraction was determined directly by image analysis, so it was unnecessary to make any assumptions regarding the thickness of the hydride platelets.

---

[1] University of Toronto, Toronto, Ontario, Canada.
[2] Lab. MSS/MAT, CNRS URA 850, Ecole Centrale, Paris, France.

Vincent F. Urbanic,[1] Paul K. Chan,[1] Djamshid Khatamian,[1] and On-Ting T. Woo[1]

Growth and Characterization of Oxide Films on Zirconium-Niobium Alloys

REFERENCE: Urbanic, V. F., Chan, P. K., Khatamian, D., and Woo, O.-T. T., **"Growth and Characterization of Oxide Films on Zirconium-Niobium Alloys,"** *Zirconium in the Nuclear Industry: Tenth International Symposium, ASTM STP 1245,* A. M. Garde and E. R. Bradley, Eds., American Society for Testing and Materials, Philadelphia, 1994, pp. 116–132.

ABSTRACT: Pressure tubes for CANDU reactors are made from extruded and cold-drawn Zr-2.5Nb alloy. Their microstructure consists of elongated α-Zr grains (0.3 to 0.5 μm thick), containing about 1 atom percent niobium, surrounded by a thin (30 to 50 nm) network of metastable β-Zr phase, containing about 20 atom percent niobium. Alloys of Zr-1Nb and Zr-20Nb were prepared, heat treated, and oxidized in 573 K water to produce bulk microstructures and oxides that would simulate those normally found on a much finer scale in pressure tubes. These were subsequently characterized by chemical analyses, scanning electron microscopy (SEM), X-ray diffraction (XRD), analytical electron microscopy (AEM), secondary ion mass spectroscopy (SIMS), X-ray photoelectron spectroscopy (XPS), and nuclear reaction analyses (NRA).

Oxidation of Zr-20Nb (β-Zr phase) was more rapid than that for the Zr-1Nb (predominantly α-Zr phase) but, despite this, the hydrogen absorption was considerably lower. During corrosion testing, the metastable β-Zr undergoes partial decomposition to omega phase. The oxides show contrasting morphologies in terms of crystallite size (20 to 60 nm for oxides on α-Zr versus about 15 nm for oxides on β-Zr). In addition to monoclinic ZrO_2, there is evidence for either tetragonal ZrO_2 or the mixed oxide, $6ZrO_2 \cdot Nb_2O_5$ in the β-Zr oxide. Scanning transmission electron microscopy (STEM) imaging shows niobium associated with the oxide formed over the β-Zr phase in oxidized pressure tube material.

Deuterium distributions obtained by SIMS depth profiling through the oxides are radically different for each oxide type. The concentration of deuterium in the β-Zr oxide was substantially less than that in the α-Zr oxide, which was consistent with the observed lower deuterium uptake measured for the Zr-20Nb alloy. Complementary XPS results also suggest that some unoxidized niobium is present in these water-formed oxides.

Hydrogen depth profiling by ^{15}N nuclear reaction analyses has been used to investigate the diffusion of hydrogen in these oxides. The oxide films were implanted with hydrogen and the progressive dispersion of the implanted hydrogen, as a result of annealing, was used to investigate hydrogen diffusion as a function of temperature. The nondispersive nature of the implanted hydrogen peaks in the Zr-1Nb oxide after annealing was suggestive of the presence of interconnected porosity in those oxides. However, the broadened peaks in the Zr-20Nb oxide after annealing are indicative of a normal diffusion process in a nonporous medium.

The implications of these observations will be discussed in terms of corrosion and hydrogen uptake in Zr-2.5Nb pressure tube material.

KEY WORDS: zirconium-niobium alloys, zirconium, corrosion, hydrogen uptake, oxides, microscopy, spectroscopy, diffusion, zirconium alloys, nuclear materials, nuclear applications, radiation effects

---

[1] Senior staff scientist and research scientists, respectively, Atomic Energy of Canada Ltd., Chalk River Laboratories, Chalk River, Ontario, Canada.

Pressure tubes in CANDU reactors are made from Zr-2.5Nb alloy. The tubes are extruded at about 1100 K, cold drawn 20 to 30%, and stress-relieved by autoclaving in 673 K steam for 24 h. The resulting microstructure consists of elongated α-Zr platelets, containing up to 1 atom percent niobium in solution, with an aspect ratio of about 1:5:50 in the radial, transverse, and longitudinal directions, respectively. They are surrounded by a thin network of metastable β-Zr phase, containing about 20 atom percent niobium in solution. The thickness of the α-Zr phase in the radial direction varies from 0.3 to 0.5 μm, whereas that of the β-Zr is about ten times less. The corrosion and hydrogen ingress processes for this two-phase system are complex and not well understood. Earlier work has shown that there is preferential oxidation occurring in the vicinity of the β-Zr regions of the two-phase structure [1] and that there is deuterium present in oxide films grown in heavy water [2]. However, the fine α-Zr + β-Zr structure typical of CANDU pressure tubes presents some limitations with respect to studying the relative importance of each phase to corrosion and hydrogen ingress behavior. In this paper, we investigate the corrosion and deuterium pickup in predominantly α-Zr and pure β-Zr separately. The oxide films were characterized using analytical electron microscopy (AEM), secondary ion mass spectroscopy (SIMS), X-ray photoelectron spectroscopy (XPS), and nuclear reaction analysis (NRA) techniques. The oxide characterization results are correlated with the corrosion and deuterium pickup data in an attempt to increase our understanding of corrosion and deuterium ingress into pressure tube material.

Experimental Procedure

Alloys of Zr-1Nb and Zr-20Nb were obtained from Teledyne Wah Chang, Albany. Samples about 3 by 1.5 by 0.1 cm were vacuum annealed at 1123 K for 1 h and air cooled. Prior to corrosion testing, the sample surfaces were abraded to a 600 grit finish and subsequently pickled in an acid mixture comprised of 10% HF + 15% HNO_3 + 30% H_2SO_4 + 45% H_2O. Samples were exposed in autoclaves to 573 K lithiated heavy water, (4×10^{-4} molar LiOD in D_2O), for up to 163 days with periodic removal for weight gain measurements. After the final exposure, samples were selected for scanning electron microscope (SEM) examination and deuterium analyses by hot vacuum extraction mass spectrometry.

Oxide examination by AEM was carried out by preparing samples for examination in plan view. Thin slices cut from the coupon with oxide on one side were masked and then electropolished from the metal side to produce a web of oxide. The web was ion milled to transparency from the oxide side using 4 keV Ar^+ ions in a Gatan Duomill with a cold stage. This procedure would expose oxide from a region close to the metal-oxide interface. Characterization was carried out using a Philips CM30 AEM equipped with an ultrathin window link X-ray detector.

The water-formed oxides were characterized by XPS and SIMS and compared with oxides formed by anodization and to Nb_2O_5, NbO_2, and NbO oxide standards in the form of powders obtained from Alpha Products. The anodized oxides, ~0.13 μm thick, were grown in saturated sodium nitrite solution at a current density of 0.3 mA/cm^2, corresponding to a final forming voltage of 65 V.

XPS spectra were recorded using an ESCALAB II spectrometer (VG Scientific Ltd.). In these experiments, the Mg-Kα anode was operated at 12 kV and 20 mA. The hemispherical analyzer was set to a pass energy of 20 eV in a constant analyzer energy mode. The total photoelectron current was detected with a triple-channeltron detector. Ion bombardment for depth profiling was carried out using a hot cathode duoplasmatron Ar^+ ion source (VG DP-51), using a current of 5 μA at 7 kV, rastered over an area of ~2.2 cm^2 with the analyzer chamber back-filled with O_2 (to 10^{-7} torr) during sputtering. This method has been carefully evaluated, and it was determined, using standard samples, that no reduction of zirconium oxide occurred during sputtering.

SIMS depth profiles were generated using the Cameca IMS-3f ion microscope. A Cs^+ ion primary beam current of 0.7 to 0.8 μA at 10 kV in a 50 μm diameter spot was rastered over a sample area of 250 by 250 μm. The ion current density was ~50 mA/cm^2. This current density was sufficient to prevent memory effects resulting from deuterium redeposition. Several depth profiles were recorded for each specimen to determine the reproducibility of the particular reaction on the surface. The position of the alloy/oxide interface was determined from the intensity change of $^{106}ZrO^-$ as the beam profiled through the interface. All SIMS spectra were collected with an energy filter of 400 V. This high-energy filter ensured that only elemental and very strong molecular signals were detected. The quantitative yield of D^- in ZrO_2 was calibrated against a standard oxide film determined by nuclear reaction analysis to contain 6.5×10^{15} atm/cm^2 of deuterium.

Hydrogen diffusion in the water-formed oxides on both alloys was studied using combined ion implantation and the ^{15}N profiling technique on 1 by 1.5 by 0.05 cm samples oxidized in H_2O at 633 K to form a 2-μm-thick film. The oxide films were implanted with hydrogen at an energy of 125 keV. This energy was chosen, based on TRIM[2] calculations, to place the hydrogen peak close to the mid-depth of the oxide layers. The samples were first profiled using the ^{15}N hydrogen profiling technique [3], then diffusion annealed at various temperatures in air for selected time intervals and profiled again for diffusion analysis to determine the hydrogen diffusivity as a function of temperature.

The 6.405 MeV resonance of the $^1H(^{15}N, \alpha\gamma)^{12}C$ nuclear reaction allows one to measure the hydrogen concentration as a function of depth through the oxide. At this resonant energy, the intensity of gamma rays measured is proportional to the hydrogen concentration. If the ^{15}N beam energy is varied upwards from 6.405 MeV, resonance will occur at various depths below the surface, and the amount of hydrogen as a function of depth can then be determined based on the measured gamma spectrum and the stopping power of the oxide. For these samples, the depth resolution is about 3 nm at the surface and about 50 nm at an oxide depth of 1 μm. The experimental procedure and standards used to verify the quantitative reliability of the data have been described previously [4].

Results and Discussion

Microstructure

The starting microstructures are shown in Figs. 1a and b. The Zr-1Nb alloy had a much finer, equiaxed structure than the Zr-20Nb alloy. X-ray diffraction confirmed the presence of single-phase β-Zr in the Zr-20Nb alloy and identified the presence of both α-Zr and β-Zr phases in the Zr-1Nb alloy. The majority of its structure was α-Zr with the niobium-rich β-phase present both at the grain boundaries and within the α-Zr grains of this alloy (Fig. 1c); however, the volume fraction of α-Zr compared to β-Zr was deemed sufficiently high that the relative difference in response of both the α-Zr and β-Zr phases could be evaluated. A single phase α-Zr structure was not achieved with the 1123 K heat treatment. Following corrosion testing at 573 K, ω-phase was observed in both the Zr-1Nb alloy (Fig. 1d) and the Zr-20Nb alloy (Fig. 1e) due to thermal decomposition of the β-Zr phase to ω-phase and $β_{enr}$ (β-Zr enriched in niobium).

Corrosion and Deuterium Pickup

The surface appearance of samples after testing for 163 days in 573 K water is shown in Figs. 2a and b. The outer oxide replicates the starting surface in both cases; however, there

---

[2] The TRIM (TRansport of Ions in Matter) program has been revised several times and the current version is available from J. F. Ziegler, IBM Research, Yorktown, NY.

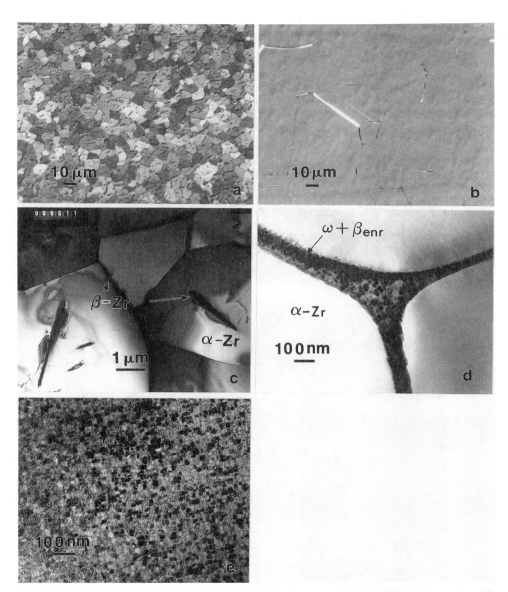

FIG. 1—*Alloy microstructures: (a) and (c) Zr-1Nb before testing, (b) Zr-20Nb before testing, (d) Zr-1Nb after testing, and (e) Zr-20Nb after testing.*

is some evidence of microspalling occurring in the Zr-20Nb alloy. The corrosion results are given in Fig. 3, and analysis shows the kinetics are near cubic for the Zr-20Nb alloy but closer to parabolic for the Zr-1Nb alloy. Although the kinetic data for the Zr-20Nb contain some uncertainty due to the microspalling, of greater significance is the fact that mechanistic informa-

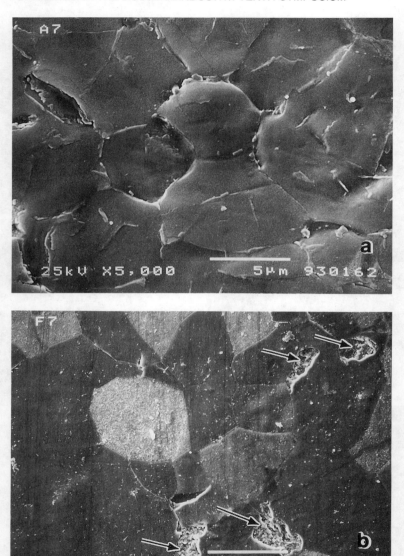

FIG. 2—*SEM examination of oxide surfaces on* (a) *Zr-1Nb and* (b) *Zr-20Nb after corrosion testing for 163 days in 573 K water (arrows indicate microspalling).*

tion cannot be obtained from the kinetic behavior due to a changing microstructure during the test.

Zr-20Nb displays higher weight gains than Zr-1Nb throughout the duration of the tests, although at longer exposures the instantaneous rate for Zr-20Nb is approaching that of the Zr-1Nb alloy. Thick oxide ridges are observed in pressure tube oxides when the metal-oxide

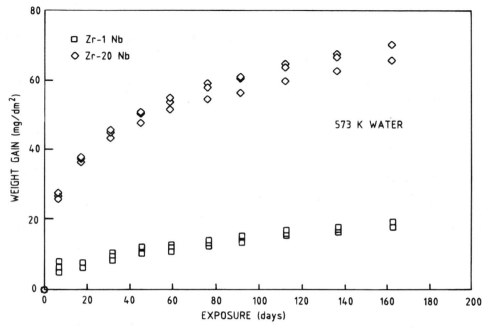

FIG. 3—*Oxidation kinetics for Zr-1Nb and Zr-20Nb in 573 K lithiated (4×10^{-4} molar) water.*

interface is revealed by etching away the metal substrate [1,5]. These ridges are coincident with the β-Zr phase of the pressure tube structure. The kinetic results of this work suggest that the oxide ridges are established early in life and explain why the ridges do not propagate deeper into the underlying metal substrate with time.

Selected samples were analyzed for deuterium concentration by hot vacuum extraction mass spectrometry (Table 1). The deuterium pickup in the Zr-20Nb alloy was as much as 15 times less than that for Zr-1Nb alloy after 137 days of exposure despite the oxide thickness being about four times greater. A second sample analyzed after 163 days exposure confirmed the low pickup for the Zr-20Nb alloy. Based on these values, the percent theoretical uptake for the Zr-20Nb alloy is 20 to 60 times less than that for Zr-1Nb in these tests.

Oxide Structure—Typical results of oxide examination by transmission electron microscopy (TEM) are shown in the micrographs in Fig. 4. TEM examination showed that the oxides

TABLE 1—*Corrosion and deuterium pickup in lithiated D_2O at 573 K.*

Structure	Oxide Thickness, μm	D Pickup, atomic %	Theoretical Uptake, %
Zr-1Nb (α-Zr)	1.1[a]	0.041	5.9
Zr-20Nb (β-Zr)	4.5[a]	0.003	0.1
Zr-20Nb (β-Zr)	4.8[b]	0.007	0.3

[a]137 days exposure.
[b]163 days exposure.

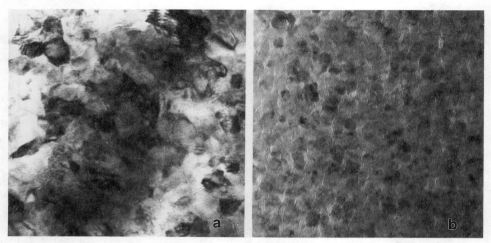

FIG. 4—*TEM micrographs comparing oxides formed on* (a) *Zr-1Nb and* (b) *Zr-20Nb alloys (plan view close to the metal-oxide interface).*

grown on the Zr-1Nb had oxide crystallites ranging in size from 20 to 60 nm in diameter. The oxide grown on the Zr-20Nb was quite different in appearance having much finer crystallites about 15 nm in diameter. The crystallites that were uniformly distributed in the Zr-20Nb oxide appeared to be associated with the cuboidal morphology of the ω-phase that developed in the underlying substrate during corrosion testing (Fig. 1e). Both X-ray diffraction and electron diffraction indicate that the oxide formed on the Zr-1Nb alloy is predominantly monoclinic ZrO_2. The oxide formed on the Zr-20Nb alloy, in addition to monoclinic ZrO_2, contains another component associated with a d-spacing of about 0.295 nm, which cannot be attributed to monoclinic ZrO_2. The nature of this component could be tetragonal ZrO_2, as suggested by Ding and Northwood [6], or the mixed oxide, $6ZrO_2 \cdot Nb_2O_5$ [7], and is the subject of additional investigation.

Examinations of water-formed oxide grown on pressure tube material showed differences between the oxides formed over the β-Zr and α-Zr regions, Fig. 5a, consistent with those observed in oxides grown on the Zr-1Nb and Zr-20Nb alloys (Figs. 4a and b). The crystallite size in the oxide over the β-Zr filaments was typically in the range 5 to 15 nm in diameter, much smaller than the oxide crystallites grown over the α-Zr grains, which were typically 20 to 60 nm in diameter. Scanning transmission electron microscope (STEM) X-ray imaging showed niobium to be associated with the β-Zr oxide filaments while the α-Zr oxide did not contain appreciable amounts of niobium (Fig. 5b). X-ray spectra, not shown, also showed that the β-Zr oxide contained small amounts of iron, which is known to partition to the β-Zr phase in the metal. Thus, the finer oxide structure in the pressure tube oxide is associated with the niobium-rich β-Zr phase of the metal substrate.

SIMS/XPS Analyses

Based on SIMS composition versus depth profiles, the deuterium profiles through the Zr-20Nb oxide were significantly different from those through the Zr-1Nb oxide. For the Zr-1Nb oxide, the deuterium concentration decreased uniformly across the oxide film to a base level typical of that in the metal substrate, while for the Zr-20Nb oxide, the deuterium

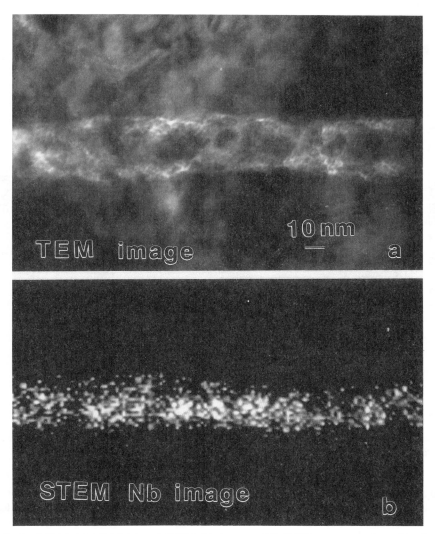

FIG. 5—*TEM bright field (a) and STEM (b) images of Zr-2.5Nb pressure tube oxide showing fine crystallites and niobium enrichment associated with an oxidized β-Zr region.*

concentration dropped rapidly to the detection limit of the instrument, remained low in the oxide, then increased through the interface region to a constant concentration in the metal substrate (Fig. 6). The vertical dashed lines define the region over which the intensity of the $^{106}ZrO^-$ signal is seen to drop from an upper plateau in the oxide to the background level in the metal. From these observations, the roles of the α-Zr and β-Zr phases in determining the corrosion and hydrogen ingress behavior are different.

To provide information on the role of niobium with respect to corrosion and hydrogen ingress in CANDU pressure tube material, the chemical form of niobium in the oxides formed on Zr-20Nb (β-Zr) has been characterized using XPS. The XPS binding energies of Zr3d and

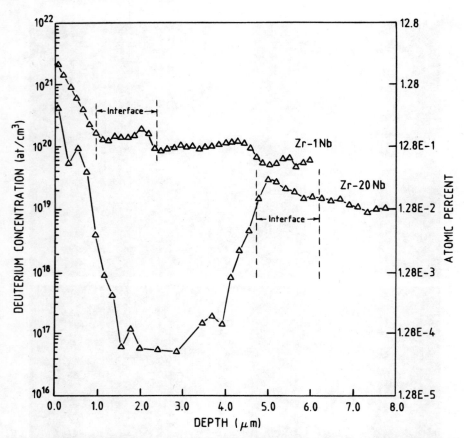

FIG. 6—*Deuterium distribution through Zr-1Nb and Zr-20Nb oxide films formed by corrosion in water at 573 K for 163 days.*

Nb3d electronic levels were monitored by depth profiling using Ar^+ ion sputtering. Although the chemical state of niobium in the oxide films cannot be unambiguously determined due to sputter-induced reduction, the reduction can be minimized by backfilling the analyzer chamber with O_2 (from 10^{-9} to 10^{-7} torr) during sputtering. This has been demonstrated by the XPS analysis of the Nb_2O_5 standard, which indicated that although a change in the oxidation state of niobium from 5^+ to 2^+ occurred with oxygen addition during sputtering, there was no evidence for a zero valent niobium peak near 202 eV, nor was there any change in the spectrum despite considerable sputtering. Therefore, during sputtering, a low partial pressure of oxygen was used in subsequent measurements.

The XPS spectra for niobium in the anodic oxide (Fig. 7) show that the surface oxide is Nb_2O_5 (Spectrum a). After ~3 h of sputtering (Spectrum b), corresponding to removal of about one-third of the film, the main component of the peak is still Nb_2O_5, but a shoulder near 204 eV, corresponding to NbO, is observed. Once the metal substrate is reached after ~8 h of sputtering, a peak attributable to unoxidized niobium is observed, as expected (Spectrum c). The XPS spectra for zirconium (not shown) in the water-formed oxide are identical to those obtained with anodic oxide, that is, the oxidation state is 4^+, with no evidence of a

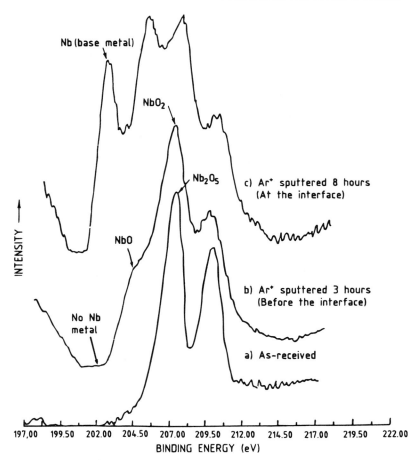

FIG. 7—*XPS Nb3d$_{5/2,3/2}$ spectra of anodic oxide grown on Zr-20Nb alloy in saturated sodium nitrate solution.*

change observed in the oxide during sputtering. However, unlike the anodic oxide, the XPS spectra for niobium in the water-formed oxide (Fig. 8), show a peak at 202 eV indicative of zero valent niobium, and a significant amount of NbO, more than can be accounted for by sputter reduction alone after 86 h of sputtering (Spectrum c). The XPS spectra for zirconium at this point still displayed the doublet characteristic of ZrO_2, indicating that all signals were still being generated from within the oxide film. Therefore, in addition to the presence of niobium oxides, niobium also exists in a valence state lower than 2^+ in the thermally-grown oxide.

Hydrogen Diffusion in Oxides—A typical example of the hydrogen profiles in Zr-20Nb oxide before and after the diffusion anneal is shown in Fig. 9a. These profiles show that the implanted hydrogen peak has broadened after diffusion annealing, which is an indication of a normal diffusion process in a homogeneous medium. The profiles were analyzed using a model in which the diffusion equation

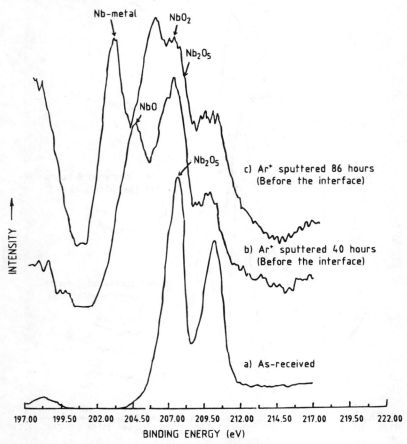

FIG. 8—*XPS Nb3d$_{5/2,3/2}$ spectra of oxide on Zr-20Nb alloy formed by corrosion in water at 573 K for 163 days.*

$$\partial C(x, t)/\partial t = D\partial^2 C(x, t)/\partial x^2 \tag{1}$$

was solved with a Gaussian distribution, approximately representing the implanted peak, as the initial condition, and $C(0, t) = C(\infty, t) = 0$ as the boundary conditions [8]. The solid curves in Fig. 9a are obtained as a result of using this model in the analysis. The diffusion constants, D, obtained at different temperatures are presented in Fig. 10. The solid line represents a least squares fit to the data yielding the expression

$$D(m^2/s) = 1.64 \times 10^{-8} \exp(-118\,700/RT)$$

where T is in K and R is the universal gas constant, 8.314 J/mol · K.

Comparing these results with those obtained from oxide films grown in 673 K air [9] shows that hydrogen diffusion in water-formed and air-formed oxides is similar (Fig. 10), particularly at reactor operating temperatures (520 to 580 K).

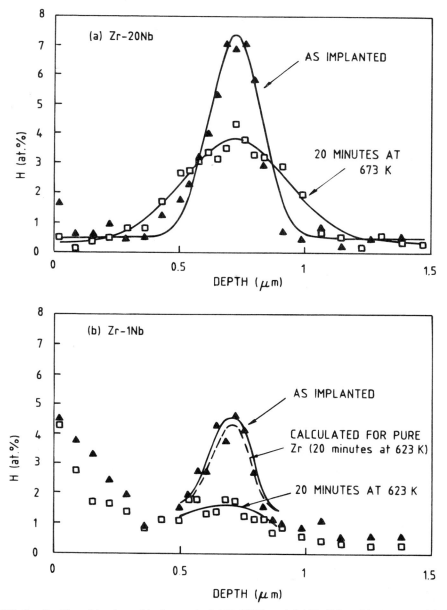

FIG. 9—*Profiles of implanted hydrogen in (a) Zr-20Nb and (b) Zr-1Nb oxides measured by ^{15}N profiling before and after a diffusional anneal (calculated profile for pure zirconium oxide shown for comparison).*

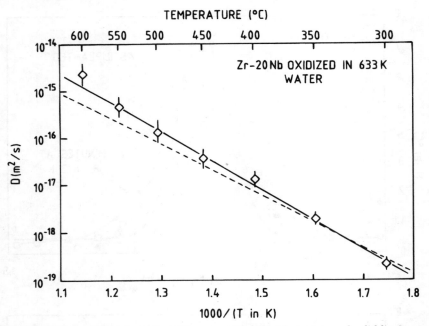

FIG. 10—*Hydrogen diffusivity in Zr-20Nb water-formed oxide (points and solid line) compared with previous measurements for Zr-20Nb air-formed oxide (dashed line).*

The profiles measured in the Zr-1Nb oxide (Fig. 9b) display several different features compared with those measured in the Zr-20Nb oxides. Although both oxides were implanted under identical conditions, the height of the as-implanted peak is lower for the Zr-1Nb oxide compared to the Zr-20Nb oxide. This indicates that some of the implanted hydrogen has escaped from the Zr-1Nb oxides during implantation. Figure 9b also shows that during the diffusion anneal, the peak height has decreased with no change in the peak width, again indicating some loss of hydrogen from the oxide sample. Another difference between the two is the background hydrogen level near the outer oxide surface. Profiles taken before hydrogen implantation have shown that this hydrogen in the Zr-1Nb oxides is deposited in the oxide during corrosion in water. This supports the differences observed in the SIMS profiles for deuterium in those oxides described earlier.

These differences displayed by the Zr-1Nb oxides suggest they are more heterogeneous than those formed on Zr-20Nb, and possibly comprised of homogeneous, nonporous cells surrounded by a network of fast diffusion paths for hydrogen. With the exception of the high background hydrogen levels near the oxide surface, the features of the Zr-1Nb oxide profiles described earlier are also typical of those seen with air-formed oxides on Zr-1Nb and Zr-2.5Nb [9], and on pure polycrystalline zirconium [10] in similar experiments. The dashed curve in Fig. 10 represents the hydrogen profile expected in oxides on pure zirconium tested under identical conditions based on these earlier experiments. By comparison, hydrogen diffusion in Zr-1Nb oxides is significantly faster than that in pure zirconium oxides. In the past, such profiles were analyzed using a model in which the diffusion equation was solved with the width of the implanted peak kept constant [10]. This boundary condition imposes the restriction that the size of the nonporous oxide cell is equivalent to the width of the peak. Since the derived values for D and the cell size are interdependent, diffusion constants derived in this manner are not absolute and can only be compared with others if similar diffusional behaviors are observed. To obtain absolute values for material displaying this behavior, some knowledge of the cell sizes and a modified solution to the diffusion equation are required.

Conclusions

Further insight into the understanding of corrosion and hydrogen ingress into Zr-2.5Nb pressure tube material has been achieved by studying the behavior of Zr-20Nb and Zr-1Nb separately.

The observation that Zr-20Nb (β-Zr) corrodes faster than Zr-1Nb (predominantly α-Zr) in high-temperature water, explains the formation of oxide ridges that penetrate into the underlying metal substrate at the β-Zr regions of Zr-2.5Nb pressure tube structures. As the β-Zr phase decomposes with time, its corrosion rate approaches that of the α-Zr phase explaining why the ridges observed in pressure tube oxides eventually reach some limiting depth of penetration. The growth of finer, niobium-rich oxide crystallites on the Zr-20Nb oxide compared to those on the Zr-1Nb oxide is similar to observations on oxides formed over the α-Zr and β-Zr regions of pressure tube material. The fine structure characteristic of the Zr-20Nb oxide appears to be associated with the cuboidal morphology of the ω-phase formed by thermal decomposition of the β-Zr phase during the corrosion test.

The deuterium uptake in Zr-20Nb is considerably lower than that for Zr-1Nb despite the higher corrosion rate displayed by the former. Low deuterium concentration in the oxide seems to be associated with the low uptake. The higher deuterium concentrations in the Zr-1Nb oxide could be a reflection of increased porosity in this oxide, with deuterium-bearing species adsorbed on the pore surfaces. This is certainly consistent with the diffusion measurements, which show diffusion profiles representative of a homogeneous structure for the Zr-20Nb oxide but a heterogeneous structure for the Zr-1Nb oxide. The heterogeneity could be associated with oxide porosity.

The implications of finding metallic niobium in the Zr-20Nb oxide are significant and require further investigation. This finding implies that electronic conduction and hence proton discharge would be easiest at the oxide formed over the β-Zr phase of pressure tube material. Despite the production of cathodic deuterium that would occur at these preferred regions, the resistance of this phase to deuterium ingress may be a factor in the low deuterium uptake traditionally seen in Zr-2.5Nb pressure tubes.

Acknowledgments

The authors gratefully acknowledge the technical assistance of A. Audet, K. Irving, V. C. Ling, L. G. Laurin, and R. W. Gilbert of Chalk River Laboratories, and staff at the University of Western Ontario (SSW), Cornell University, and McMaster University for assistance with SIMS analyses, hydrogen implantation, and profiling, respectively. The work presented here was funded by the Candu Owner's Group (COG).

References

[1] Urbanic, V. F. and Gilbert, R. W., "Effect of Microstructure on the Corrosion of Zr-2.5Nb Alloy," *Proceedings,* Technical Committee Meeting on Fundamental Aspects of Corrosion on Zirconium Base Alloys in Water Reactor Environments, International Atomic Energy Agency, Vienna, IWGFPT/34, 1990, pp. 262–272.

[2] Warr, B. D., Elmoselhi, M. B., Newcomb, S. B., McIntyre, N. S., Brennenstuhl, A. M., and Lichtenberger, P. C., "Oxide Characteristics and Their Relationship to Hydrogen Uptake in Zirconium Alloys," *Zirconium in the Nuclear Industry: Ninth International Symposium, ASTM STP*

1132, C. M. Eucken and A. M. Garde, Eds., American Society for Testing and Materials, Philadelphia, 1991, pp. 740–757.
[3] Lanford, W. A., "Use of Nuclear Reaction Analysis to Characterize the Elemental Composition and Density of Thin Film Amorphous Silicon," *Solar Cells,* Vol. 2, 1980, p. 351.
[4] Stern, A., Khatamian, D., Laursen, T., Weatherly, G. C., and Perz, J. M., "Hydrogen and Deuterium Profiling at the Surface of Zirconium Alloys; I. The Effects of Preparation," *Journal of Nuclear Materials,* Vol. 144, 1987, p. 35.
[5] Warr, B. D., Rasile, E. M., and Brennenstuhl, A. M., "Electron Microscopical Analyses of Oxides in Zr-2.5wt%Nb," *Proceedings,* Technical Committee Meeting on Fundamental Aspects of Corrosion on Zirconium Base Alloys in Water Reactor Environments, International Atomic Energy Agency, Vienna, IWGFPT/34, 1990, pp. 124–134.
[6] Ding, Y. and Northwood, D. O., "The Formation of a Barrier Oxide Layer on a Zr-2.5wt.%Nb Alloy During Corrosion in High Temperature Pressurized Water," *Journal of Alloys and Compounds,* Vol. 187, 1992, p. 317.
[7] Roth, R. S. and Coughanour, L. W., *Journal of Research,* National Bureau of Standards, Vol. 54–55, 1955.
[8] Schimco, R., Schwarz, G., and Rogge, K., "Diffusion from Implanted Layers with Consideration of Out-Diffusion on the Surface," *Physica Status Solidi,* Vol. A28, 1975, p. K163.
[9] Khatamian, D., "Diffusion of Hydrogen in the Oxides of Zr-1Nb, Zr-2.5Nb and Zr-20Nb Alloys," *Zeitschrift für Physikalische Chemie,* Bd. 181, 1993, pp. 435–440.
[10] Khatamian, D. and Manchester, F. D., "An Ion Beam Study of Hydrogen Diffusion in Oxides of Zr and Zr-Nb (2.5 wt%), I. Diffusion Parameters for Dense Oxides," *Journal of Nuclear Materials,* Vol. 166, 1989, p. 300.

DISCUSSION

N. Ramasubramanian[1] (written discussion)—(1) In the CANDU pressure tube material, the β phase with 20% niobium and α phase with 1% niobium are galvanically coupled. Have you done experiments coupling the two materials in your work?

(2) The oxide thicknesses on 20% niobium and 1% niobium alloy are different. SIMS concentration profiles show different near-surface concentrations of deuterium? How is hydrogen diffusivity that is calculated from nuclear reaction analysis (NRA) related to hydrogen pick-up by metal?

V. F. Urbanic et al. (authors' closure)—(1) We have considered the possibility that galvanically coupling the two phases, as in the actual microstructure, could affect the responses of the individual phases. Therefore, tests are in progress to evaluate the significance of this.

(2) The results obtained from the hydrogen-implanted samples indicate that the Zr-1Nb oxide may have a more porous structure than the Zr-20Nb oxide. We think this may have an effect on the hydrogen pickup rate of these two alloys. Also, the porous nature of the Zr-1Nb oxide may be a factor contributing to the more deuterium-bearing species found in the Zr-1Nb oxide compared to that in the Zr-20Nb oxide leading to the differences in the SIMS profiles.

Brian Cox[2] (written discussion)—Since your Zr-20Nb had decomposed into a β-enriched phase with ≥40% niobium at the end of the oxidation experiments, have you done any experiments on single-phase Zr-40Nb to look at the oxide morphology, in this case, and compare it with the Zr-20Nb material?

V. F. Urbanic et al. (authors' closure)—In our tests, the maximum niobium content investigated was with the Zr-20Nb alloy. We have not yet done any experiments with higher niobium levels, but in view of the aging processes that occur both thermally and under neutron irradiation, we plan to investigate the responses of higher niobium alloys for this reason.

A. Strasses[3] (written discussion)—Please comment on the unique characteristics of oxide films on zirconium-niobium alloys that make them different from those on Zircaloys, and how these differences might affect the hydrogen diffusion and distribution in the Zircaloy oxide films in comparison to zirconium-niobium oxide films?

V. F. Urbanic et al. (authors' closure)—The major difference between the Zircaloys and Zr-2.5Nb alloy is the presence of $Zr(Fe,Cr)_2$ and $Zr_2(Fe,Ni)$ type intermetallics in the Zircaloys and their virtual absence in Zr-2.5Nb. These intermetallics are considered to play a vital role in the hydrogen uptake mechanism by providing easy pathways for electron transfer from the metal substrate, through the barrier oxide layer to the solution phase. Following proton discharge at the solution–oxide interface, atomic hydrogen is then available for entry into the metal, probably via diffusion at the boundaries between the oxide crystallites and the intermetallic particles. In the case of Zr-Nb oxide films, electron transport through the barrier film is uniform and more difficult due to the relative absence of these intermetallics, most likely contributing to the lower pickups commonly observed in zirconium-niobium systems.

R. A. Ploc[4] (written discussion)—In response to the presentation and a question from the floor on why niobium in zirconium could lead to lower hydrogen pickup, it was indicated that several years ago Kofstads group in Norway showed that niobium in its highest valency state within ZrO_2 could decrease the solubility of protons in the oxide. If such regions exist (Nb_2O_5) would it limit hydrogen?

---

[1] Ontario Hydro, Toronto, Ontario, Canada.
[2] University of Toronto, Toronto, Ontario, Canada.
[3] S. M. Stoller Corporation, Pleasantville, NY.
[4] ACEL Research, Chalk River Laboratories, Chalk River, Ontario, Canada.

V. F. Urbanic et al. (authors' closure)—XPS results certainly indicate the presence of Nb_2O_5 in these films. Thus, a reduction in the solubility of protons in these oxides, as a result of this certainly, must be considered as a possible explanation for the lower hydrogen pickup commonly observed with zirconium-niobium alloys. However, this line of thinking is based on solid-state diffusion arguments and the role of line diffusion in the hydrogen ingress process would have to be factored into any such arguments on the mechanism of hydrogen uptake.

Pressure Tubes

Pauline H. Davies,[1] Robert R. Hosbons,[1] Malcolm Griffiths,[1] and C. K. (Peter) Chow[2]

Correlation Between Irradiated and Unirradiated Fracture Toughness of Zr-2.5Nb Pressure Tubes

REFERENCE: Davies, P. H., Hosbons, R. R., Griffiths, M., and Chow, C. K., "**Correlation Between Irradiated and Unirradiated Fracture Toughness of Zr-2.5Nb Pressure Tubes,**" *Zirconium in the Nuclear Industry: Tenth International Symposium, ASTM STP 1245*, A. M. Garde and E. R. Bradley, Eds., American Society for Testing and Materials, Philadelphia, 1994, pp. 135–167.

ABSTRACT: Three coordinated research programs were undertaken on the fracture toughness of Zr-2.5Nb pressure tubes to determine relationships between irradiated and unirradiated values, the effect of long-term irradiation, and the causes of the variation in toughness. The present paper describes results from these programs and their implications.

It is shown that a correlation exists between the toughness of the pressure tube material before and after irradiation to a fast-neutron fluence of 1×10^{26} n · m^{-2} ($E > 1$ MeV). Further, for the majority of tubes, degradation occurs early in the lifetime of a CANDU reactor; that is, at fluences $<2 \times 10^{24}$ n · m^{-2}. Limited results on relatively low-toughness materials indicate that the flux does not have a major effect on the irradiated toughness and that saturation is complete by about 1.6×10^{25} n · m^{-2}.

The results are discussed in the light of detailed fractographic and microstructural (X-ray diffraction, transmission electron microscopy) studies carried out on the irradiated and unirradiated pressure tube material. It is shown that in the presence of void nucleating sites, such as microsegregated species (chlorine and carbide) and particles (phosphides and carbides), very large reductions in toughness occur with irradiation. In the absence of such species, there is little deterioration in toughness, and the high crack-tip strains (toughness) achieved with unirradiated material can be maintained after irradiation.

KEY WORDS: fracture toughness, zirconium alloys, irradiation, pressure tubes, crack growth resistance, *J-R* curves, zirconium, nuclear materials, nuclear applications, radiation effects

The thin-walled pressure tubes in a CANDU reactor are fabricated from cold-worked Zr-2.5Nb and have a length of 6.3 m, inside diameter of 103 mm, and wall thickness of 4.2 mm. During service, irradiation and deuterium pickup from the pressurized heavy water, which acts as the primary heat transport coolant, can reduce the toughness; periodic assessments of surveillance tubes removed from the reactor are conducted to ensure that the tubes remain "fit-for-service" [1]. Recently, some high values have been observed in the toughness of surveillance tubes [2], raising the possibility of improving the toughness of future CANDU reactor pressure tubes and increasing their service life.

---

[1] Research scientist, assistant branch manager, and research scientist, respectively, AECL Research, Chalk River Laboratories, Chalk River, Ontario, Canada K0J 1J0.
[2] Research scientist, AECL Research, Whiteshell Laboratories, Pinawa, Manitoba, Canada R0E 1L0.

Variations in the toughness of the unirradiated tube material had been observed in 1987. Subsequently, two major programs were launched to determine whether a correlation existed between the toughness of irradiated and unirradiated material that might indicate some variation in the intrinsic microstructure. For low-neutron fluences ($<1 \times 10^{25}$ n · m^{-2}), a program was initiated to evaluate the response of preselected material irradiated to different fast-neutron fluences in the NRU reactor, Chalk River, Ontario. For high-neutron fluences, a test matrix was established for Zr-2.5Nb pressure tube material removed from Pickering NGS A Unit 3 before re-tubing in 1990. Matched irradiated rings and unirradiated offcuts from 40 different tubes were tested to investigate any correlation between the irradiated and as-installed toughness for fast-neutron fluences up to 1×10^{26} n · m^{-2}. The microstructural factors responsible for the variability in toughness were also investigated. A third program was initiated to determine the long-term effects of irradiation and flux on the mechanical properties of a single material. Mechanical test specimens were irradiated in the high-flux OSIRIS test reactor at Saclay, France, and in the lower flux NRU reactor. This paper summarizes the results from these three programs.

Material

The tubes used in the test programs were all fabricated as standard cold-worked (about 26%) Zr-2.5Nb pressure tube material, with a nominal hydrogen concentration <25 ppm by weight (0.23 atom percent) (Table 1). Billets are produced from double-vacuum-arc remelted ingots that are cut and forged, machined into hollow billets, extruded at about 815°C at an extrusion ratio of 10.5:1 into tubes and cold drawn [3]. The tubes are finally autoclaved at 400°C for 24 h, to relieve residual stresses and produce an adherent protective surface oxide film [3]. Some refinements in the manufacturing process have been made over the years to improve the homogeneity of the final product. For example, billets were later rotary forged after press forging, and later still an homogenization treatment was introduced by heating the billets into the β-phase and water quenching before extrusion. The aim was to produce a more uniform grain structure in the final extruded tube by producing a more uniform grain structure in the billet [4].

The final dual-phase microstructure of a typical Zr-2.5Nb pressure tube consists of a network of elongated α-phase (hexagonal-close-packed), containing about 1% by weight niobium in

TABLE 1—*Chemical specification for CANDU Zr-2.5Nb pressure tubes used in present study (tubes fabricated before 1987).*

	Before 1987	Current
Niobium	2.4 to 2.8% by weight	2.5 to 2.8% by weight
Oxygen	900 to 1300 ppm[a]	1000 to 1300 ppm
Carbon	<270 ppm	<125 ppm
Chromium	<200 ppm	<100 ppm
Hydrogen	<25 ppm[b]	<5 ppm
Iron	<1500 ppm	<650 ppm
Nickel	<70 ppm	<35 ppm
Nitrogen	<65 ppm	<65 ppm
Silicon	<120 ppm	<100 ppm
Tantalum	<200 ppm	<100 ppm
Zirconium and other permitted impurities	balance	balance

[a] ppm by weight.
[b] 20 ppm hydrogen for ingot, 25 ppm hydrogen for final tube.

solution, surrounded by a thin film of β-phase (body-centered-cubic), with about 20% by weight niobium. The α grains are about 0.2 to 0.5 μm wide in the radial direction, with an aspect ratio of 1:5:50 in the radial, transverse, and axial directions of the tube, respectively. The extrusion process produces a strong texture in the final tube, with the α grains being preferentially oriented such that the basal plane normals are distributed in the radial-transverse plane with the maximum in the transverse direction. Rings of material, that is, offcuts, are removed from each end of the tube before installation. Front and back end offcuts refer to material cut from the front and back ends of the pressure tubes. The front end is the end that emerged first in the extrusion operation.

Test Programs

Low-Fluence Irradiation—Preselected Material Irradiated in the NRU Reactor

For this program, material was preselected from the centers of eight different Zr-2.5Nb pressure tubes fabricated between 1969 and 1986. Previous testing had shown this material to exhibit the required wide distribution in transverse tensile strength and fracture toughness [5]. The tubes were identified as; 622, 1037, B439, C70, G1175, H190M, H737, and H850. Only Tube 622 had been autoclaved before being received by AECL. Also, the billets for Tubes G1175, H190M, H737, and H850 had been heated into the β-phase and water-quenched before extrusion at the hollow billet stage for Tube H190M and as solid logs for the remaining three tubes.

The specimen configurations and preparation were identical to those used previously for selection [5]. Transverse tensile specimens (3.6 mm nominal diameter) were machined from flattened rings cut from the centers of each tube. For fracture toughness testing, standard curved compact toughness specimens (17 mm wide) [6] were machined directly from the tube centers, the material for the irradiation being axially adjacent to that used in the selection process [5]. The toughness specimens were oriented for crack growth in the axial direction on the radial-axial plane. To produce a stress-relieved condition for all material similar to that produced by the autoclave treatment, the tensile and fracture toughness specimens (except those from Tube 622) were stress-relieved at 400°C for 24 h in vacuum; the flattened transverse tension specimens from Tube 622 were given a minimum stress relief at 400°C for 1 h to avoid further significant microstructural changes compared with the remaining specimens.

Three canisters of specimens were prepared, each containing 16 transverse tension (two per tube) and 24 curved compact specimens (three per tube). Each canister was irradiated in the NRU reactor to a different target fast-neutron fluence of 2, 5, and 10×10^{24} n · m^{-2} ($E > 1$ MeV), at a flux of 1.6×10^{17} n · m^{-2} · s^{-1} and an irradiation temperature of 255°C ± 5°C. Light water is the cooling medium pressurized at 10.4 MPa. Mechanical testing and post-irradiation examinations of the first specimen set irradiated to 2.1×10^{24} n · m^{-2} have been completed, and the results are reported here.

High-Fluence Irradiation—Pickering NGS Unit 3 Test Matrix

The Zr-2.5Nb pressure tubes in Pickering NGS (PNGS) A Unit 3 operated between 1972 and 1989 (117 330 effective full power hours) before being shut-down for re-tubing. Reactor operation resulted in a fast-neutron flux profile that was a flattened cosine along the tube length with a maximum value of about 2.5×10^{17} n · m^{-2} · s^{-1} ($E > 1$ MeV) at the center. The tubes were cooled by pressurized heavy water passing over the UO$_2$-fuel at an internal pressure of about 10 MPa, with the operating temperature of the tubes varying from about 250°C (inlet) to about 290°C (outlet).

All the tubes in Pickering NGS (PNGS) A Unit 3 were fabricated in the late 1960s by the earliest production techniques for Zr-2.5Nb pressure tubes. This included press forging of the billets, but not rotary forging nor homogenization by β-quenching before extrusion.

Irradiated rings (75 mm wide) cut from various axial locations of the 390 reactor tubes were available for testing. A matched matrix of irradiated rings and unirradiated offcuts was established for 40 different tubes (about 10% of the total, to be statistically representative) based on the following criteria:

1. availability of sufficiently large back-end offcut (>24 mm wide ring),
2. single autoclave treatment of the tube before installation, and
3. highest possible fluence for irradiated ring.

The selection of back-end offcuts, rather than front, was based on previous tests that indicated minor variations in toughness along the length of an unirradiated tube, with the toughness generally being lower at the back end. It was also known that up to 15% of the tubes in PNGS A Unit 3 had received multiple autoclave treatments, which could influence their toughness, especially for the unirradiated material. Such tubes were excluded from the final matrix. Imposition of these criteria produced a test matrix with a range of fluence varying from 6.5 to 9.8×10^{25} n · m^{-2}. Finally, no consideration was given to the irradiation temperature (axial location) of each irradiated ring that varied from 254 to 287°C.

Both transverse tension and curved compact toughness specimens were made from the offcuts and irradiated rings. The transverse tension specimens were machined from flat (2-mm-thick) plates produced by grinding or milling curved (4-mm-thick) pressure tube blanks. The final machined tension specimen had a section width (axial direction) of 4 mm and thickness (radial direction) of 2 mm. Standard curved compact toughness specimens (17 mm wide) were machined directly from the tube sections. For the irradiated material, both specimen configurations were produced by spark-machining blanks. Further details are given elsewhere [2,6,7].

High-Flux Irradiation—Preselected Material Irradiated in the OSIRIS Reactor

This high-flux irradiation test was designed to determine the mechanical properties of pressure tubes at the end of their design life of 40 years. The specimens are being irradiated in an insert in the OSIRIS reactor in Saclay, France. The flux profile along the length of the insert is a cosine with a maximum flux of 1.89×10^{18} n · m^{-2} · s^{-1} at the center and a gradual drop off such that the end specimens receive 71% of the maximum. The coolant is liquid NaK, and the irradiation temperature for the specimens is controlled by heaters to nominally 250°C ± 5°C.

Material was selected from pressure Tube H737 for this irradiation. Transverse tension specimens (3.6 mm nominal diameter) were machined from flattened rings cut from the center of the tube and stress relieved for 24 h at 400°C after flattening. Curved compact specimens (17 mm wide) were machined directly from the tube center. The standard height of a 17-mm-wide curved compact specimen is 20.4 mm, or 1.2 × width [6]; that is, the in-plane dimensions are in the same proportions as those for a standard compact specimen in the ASTM Test Method for Determining *J-R* Curves (E 1152-87). However, due to space limitations, it was necessary to reduce the height of the specimens for this irradiation to 18.4 mm; that is, 1.08 × width.

After machining, all specimens were heat-treated at 400°C in vacuum to produce a similar stress-relieved condition; the tension specimens were heat-treated for an additional 48 h (total of 72 h), and the fracture toughness specimens for 72 h. This treatment was 48 h longer than that used for the preselected material from eight different tubes. The heat treatment matched that of hydrided cantilever beams, hydrided gaseously and homogenized at 400°C for 72 h that were included in this program for delayed hydrogen cracking tests.

The insert contains 16 transverse tension and 16 reduced-height curved compact specimens and is designed for specimens in the bottom half to be replaced periodically. The reactor operates in cycles of about 20 days; the bottom half of the insert was replaced after 2, 7, and 16 cycles, providing three sets of eight transverse tension and eight curved compact specimens that had been irradiated for 2, 5, and 9 cycles. These specimens had received fast-neutron fluences in the range of 4.5×10^{24} to 2.75×10^{25} n · m^{-2} ($E > 1$ MeV).

The results from this experiment are required to predict the performance of pressure tubes in a CANDU power reactor for which the flux is typically 2.5 to 3.7×10^{17} n · m^{-2} · s^{-1}. Therefore, a complementary irradiation was undertaken in the NRU reactor, which has a flux similar to a CANDU power reactor, to determine whether there is a flux effect. Additional sets of transverse tension and curved compact specimens are being irradiated in the NRU reactor at 255°C to fast-neutron fluences of 3, 6, and 9×10^{24} n · m^{-2} ($E > 1$ MeV) at a flux of 1.6×10^{17} n · m^{-2} · s^{-1}. The first specimen set irradiated to 2.9×10^{24} n · m^{-2} has been tested, and the results are reported here for comparison with the high-flux results.

Conversion to Displacements per Atom

Although there is a factor of ten difference in the flux of the OSIRIS and the CANDU and NRU reactors, the number of displacements per atom (dpa) (thermal and fast neutrons) per unit fluence is similar in each case. For a threshold energy of 25 eV for zirconium [8], the conversion of fluence to dpa for the reactors is about 1 dpa = 4.1×10^{24} n · m^{-2} [9]. For example, the specimens from eight preselected tube materials irradiated in the NRU to a fast-neutron fluence of 2.1×10^{24} n · m^{-2} received 0.51 dpa. In comparison, the irradiated specimens machined from PNGS A Unit 3 reactor tubes received from 15.8 to 23.9 dpa. Finally, for the high-flux irradiation in OSIRIS, the specimens from Tube H737 received from 1.1 to 6.7 dpa compared with 0.71 dpa for the NRU control specimens.

Transverse Tensile Strength (UTS)

The transverse tension specimens were tested at a nominal strain rate of 10^{-3} s^{-1} at 240°C (low-fluence or high-flux irradiation programs) or 250°C (high-fluence irradiation program) in an air furnace. A test temperature of 250°C was selected for the PNGS A Unit 3 reactor material as corresponding to the lowest (operating) irradiation temperature of the reactor tubes to avoid annealing out any irradiation damage. Duplicate specimens were tested for each tube irradiated in NRU (low-fluence irradiation), each fluence achieved in OSIRIS (high-flux irradiation), and for the unirradiated and irradiated (low-flux irradiation in the NRU) controls for Tube H737. For the Pickering material, a single specimen was available from each irradiated ring, with duplicates available from the unirradiated back-end offcuts.

The UTS generally occurred at very low plastic strains close to the 0.2% offset, indicating the low work-hardening capacity of the irradiated material up to maximum load. Strain localization or discontinuous yield was observed in most cases, as indicated by a series of serrations in the load-displacement output beyond the load maximum. This was also shown by the presence of intense deformation bands on the surface of the specimens at an angle of 45°C to the tensile axis. For the majority of specimens, the final fracture surface consisted of a central flat fracture zone, with shear lips following these intense bands. One round tension specimen from Tube H850 was unusual in producing a full shear fracture with no central flat fracture region.

Low-Fluence Irradiation—Preselected Material Irradiated in the NRU Reactor

For this test matrix, the yield stress of the unirradiated material from the centers of the eight different tubes ranged from 553 MPa (Tube B439) to 634 MPa (Tube H190M); that is, 81 MPa. After irradiation in NRU to a fluence of 2.1×10^{24} n · m^{-2}, the range in yield stress increased to 802 MPa (Tube B439) to 884 MPa (Tube H190M); that is, an increase in yield stress of about 250 MPa. In comparison, the UTS ranged from 596 MPa (Tube B439) to 679 MPa (Tube H190M) before irradiation; that is, 83 MPa. After irradiation, the range in UTS had increased to 803 MPa to 889 MPa, respectively, an increase in UTS of about 210 MPa.

The correlation between the strength of the irradiated and unirradiated material tested at 240°C is shown clearly in Figs. 1a and b for the yield stress and UTS, respectively, where the average result from duplicate tests is given. Linear regression analyses produced the following relationships for material from the tube centers

$$\sigma_{yi} = 1.05 * \sigma_{yu} + 224 \text{ MPa}, R = 0.962$$

$$\sigma_{ui} = 1.05 * \sigma_{uu} + 180 \text{ MPa}, R = 0.971$$

where the subscripts i and u refer to the strength of the irradiated and unirradiated material, respectively. The results for Tube 622 were excluded from the regression analysis for the yield strength as the yield stress of the unirradiated material was unusually low. This may have been because the specimens, machined from flattened strips, from this tube received a shorter stress-relief treatment (1 h at 400°C) than specimens from the other tubes (24 h at 400°C), because Tube 622 had previously been autoclaved. Residual tensile stresses would tend to lower the applied or apparent stress necessary to initiate yield.

High-Flux Irradiation—Preselected Material Irradiated in the OSIRIS Reactor

The transverse tensile strength results for Tube H737 from the high-flux irradiation program are summarized in Figs. 2a and b, where the yield stress and UTS are shown as a function of the fast-neutron fluence. Each data point is the average of two test results. For the irradiated material, the difference between the yield and UTS ranged from 0 to 6 MPa, confirming the low work hardening of irradiated cold-worked Zr-2.5Nb pressure tube material up to the UTS. Comparison of results obtained from specimens irradiated in the high-flux (OSIRIS) and lower-flux (NRU) facilities showed little evidence of a flux effect on strength within the standard deviation of results of about ±7 MPa for the yield stress and about ±8 MPa for the UTS. In addition, comparison with the unirradiated control specimens suggested that the transverse strength increased rapidly at low fluences, with saturation occurring at fluences $\leq 2.9 \times 10^{24}$ n · m^{-2}. The increase in the 0.2% offset yield stress was about 200 MPa and for the UTS about 190 MPa, in good agreement with the previous results for the eight different tube materials irradiated in NRU to 2.1×10^{24} n · m^{-2}.

High-Fluence Irradiation—Pickering NGS Unit 3 Test Matrix

The range in yield stress and UTS for the back-end offcuts from the 40 different Pickering tubes was 518 MPa (Tube 609) to 581 MPa (Tube 120) and 633 MPa (Tube 64) to 703 MPa (Tube 469), respectively; that is, 63 MPa and 70 MPa, respectively. However, although the previous results suggest that the increase in the transverse tensile strength should have saturated, no correlation was obtained between the strength of the tube centers irradiated to fluences between 6.5 and 9.8×10^{25} n · m^{-2} and the unirradiated back-end offcuts. In fact, the range in yield stress and UTS measured from the 40 irradiated rings was 754 MPa (Tube 120, Fuel

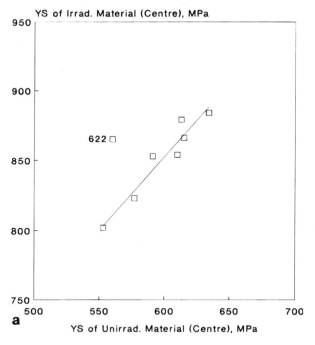

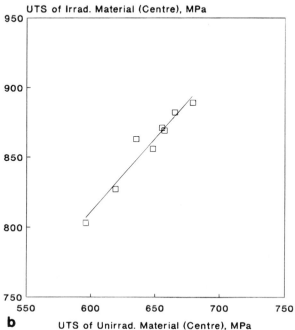

FIG. 1—*Correlation between transverse tensile strength of irradiated and unirradiated material at 240°C (preselected material irradiated in NRU at 255°C to a fast-neutron fluence of 2.1 × 10^{24} n · m^{-2})*: (a) *0.2% offset yield stress and* (b) *ultimate tensile strength.*

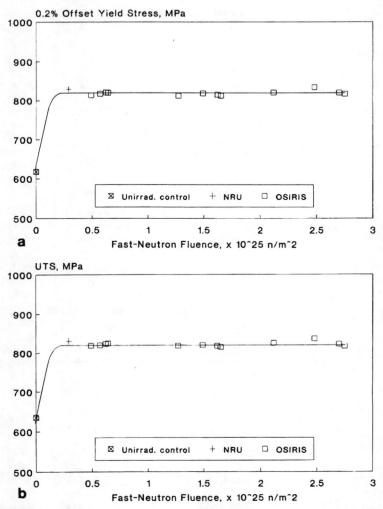

FIG. 2—*Effect of neutron fluence on transverse tensile strength of tube H737 at 240°C (preselected material irradiated in OSIRIS and NRU at 250/255°C):* (a) *0.2% offset yield stress and* (b) *ultimate tensile strength.*

Channel 020, fluence of 6.8×10^{25} n · m^{-2}) to 961 MPa (Tube 329, Fuel Channel G14, fluence 8.9×10^{25} n · m^{-2}) and 782 MPa (Tube 120) to 965 MPa (Tube 329), respectively; that is, a range of 207 MPa and 183 MPa, respectively, much larger than that for the offcuts. Any differences in irradiation temperature could not account for the range in irradiated strength.

It is known that the strength of as-installed Zr-2.5Nb pressure tubes varies along the length, generally being higher at the back end of the tube. The current results suggest that the range in transverse tensile strength for the back-end offcuts may not be characteristic of the range for the main body of the as-installed tube. This may be due to a smaller variation in grain size at the back, as indicated by previous studies on grain size variations between front- and back-end offcuts in early Pickering 3 tubes [10].

Fracture Toughness

The curved compact specimens were tested following the standard method for fracture toughness testing of CANDU reactor pressure tubes [6]. Specimens were fatigue precracked approximately 2 mm at room temperature to produce a relative crack length, a/W of 0.5, where a is the crack length and W is the specimen width. The tests were carried out at either 240°C (low-fluence and high-flux irradiation programs) or 250°C (high-fluence irradiation program) in an air furnace. However, a 10°C difference in test temperature should have little effect on the test results in the upper shelf (operating temperature) regime, since in all cases the test temperature was below the irradiation temperature.

Testing was carried out at constant displacement at a rate corresponding to an initial rate of increase of the stress intensity factor of about 1 MPa$\sqrt{m}$ · s^{-1}. The d-c potential drop method was used to monitor the crack growth, and tests were terminated after about a 3- to 4-mm crack extension. For the reduced-height curved compact specimens used in the high-flux irradiation program, a new potential drop calibration was obtained using the analog method. At the end of a test, the final crack front position was marked either by fatigue (unirradiated specimens) or by heat tinting (irradiated specimens) before breaking open the specimens to determine the initial and final crack lengths.

Three curved compact specimens were tested for each different tube for the low-fluence irradiation in NRU. For the high-flux irradiation program, a single irradiated specimen was available for each different fluence, with duplicate tests performed on the unirradiated and irradiated (NRU controls) specimens. Finally, for the high-fluence irradiation program, a single specimen was available from each irradiated ring, with duplicates available from the back-end offcuts.

A crack growth resistance or J-R curve was produced for each test following the procedures in Ref 6. For the reduced-height curved compact specimens, appropriate corrections were made to the standard equations for the elastic compliance and elastic stress intensity factor. Each J-R curve was characterized by means of the following four parameters:

1. the crack initiation toughness, $J_{0.2}$, J-integral value at the intersection of the 0.2 mm offset line and J-R curve;
2. the maximum load toughness, J_{ml}, J-integral value corresponding to the maximum load point;
3. the $J_{1.5}$, J-integral value at the intersection of the 1.5-mm offset line and J-R curve; and
4. the initial crack growth toughness, dJ/da, linear regression slope of the J-R curve between the 0.15- and 1.5-mm offset lines.

Low-Fluence Irradiation—Preselected Material Irradiated in the NRU Reactor

The different tubes that exhibited a wide variation in crack growth resistance before irradiation also showed a wide variation in toughness after irradiation. However, the ranking was similar, with Tubes 622 and H850 having the lowest and highest crack growth resistance (J-R) curves, respectively. Figure 3 shows the distribution of J-R curves at 240°C for the irradiated material; the results for the unirradiated material were presented earlier [5]. The majority of tubes had J-R curves between those for Tubes 622 and H190M.

The J-R curves for the irradiated material showed a stepped appearance, with an initial sharp reduction in slope close to crack initiation (Stage 1) followed by a sharp increase in slope (Stage 2) and steadily decreasing slope beyond a load maximum (Stage 3). This three-stage crack growth behavior for irradiated Zr-2.5Nb pressure tube material has been reported previously [11]. The exceptions to this were J-R curves for material from the highest toughness

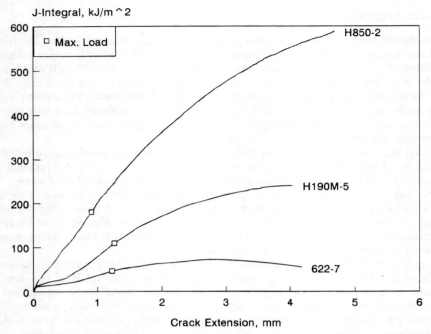

FIG. 3—*Distribution of* J-R *curves for irradiated material tested at 240°C (preselected material irradiated in NRU at 255°C to a fast-neutron fluence of* 2.1×10^{24} $n \cdot m^{-2}$).

Tube H850 that were similar to those for the unirradiated material; that is, generally exhibiting a steep rise followed by a steadily decreasing slope.

Discrete increments of crack growth or "crack jumps" were noticeable for some specimens. Crack jumps of up to 0.1 mm were observed at crack extensions less than the 1.5-mm offset. At larger crack extensions, crack jumps between 0.2 and 0.5 mm were occasionally observed for lower toughness specimens. The decrease in J-R curve slope at larger crack extensions is known to be associated with significant breakdown of the J-controlled crack growth regime for plane stress crack growth [12]. The increased incidence and length of crack jumps at longer crack extensions suggested that such behavior was exacerbated by the increasing crack-tip constraint as the remaining ligament became much less than the wall thickness. This implied that the use of J-R curve data at crack extensions less than 2 mm was reasonable and should be more characteristic of thin-walled pressure tube material deforming under plane stress crack-tip conditions (ligament/wall thickness of about 1).

The toughness of the irradiated material correlated well with that of the unirradiated material, irrespective of the toughness parameter used. This is demonstrated in Fig. 4 for the initial crack growth toughness, dJ/da, where for the irradiated toughness each data point is the average of two or three test results and, for the unirradiated controls, it is the average of three test results. A linear regression analysis produced a correlation coefficient of 0.967 (excluding results from the highest toughness Tube H850, as explained in the fractographic section). In comparison, linear regression analyses of results for the other toughness parameters, $J_{0.2}$, J_{m1}, and $J_{1.5}$, produced correlation coefficients of 0.824, 0.917, and 0.963, respectively.

The good correlation obtained for the crack growth toughness may seem surprising in view of the stepped nature of the J-R curve. However, for the J-R curves at 240°C, the linear

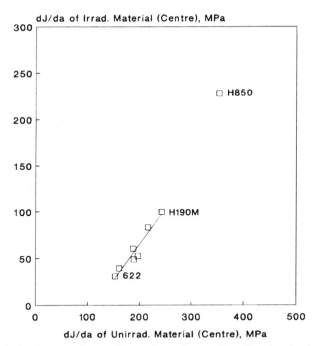

FIG. 4—*Correlation between the crack growth toughness of irradiated and unirradiated material at 240°C (preselected material irradiated in NRU at 255°C to a fast-neutron fluence of 2.1×10^{24} $n \cdot m^{-2}$).*

regression slope between the 0.15 and 1.5 mm offsets was mainly characteristic of Stage 2 or the rising portion of the curve. This region has been shown to be associated with the transition in fracture mode from flat to shear as the surface yield zones develop [11].

High-Fluence Irradiation—PNGS Unit 3 Test Matrix

The *J-R* curves for specimens from the 40 Pickering tubes also exhibited a wide variation in toughness both before and after irradiation. This is shown in Fig. 5 by the *J-R* curves for irradiated specimens from a high-, intermediate-, and low-toughness tube. In particular, Tube 230, removed from Fuel Channel M21, had the highest toughness both on installation and after irradiation to a fast-neutron fluence of 7.3×10^{25} $n \cdot m^{-2}$. The remaining tubes had a toughness between that of the low-toughness Tube 394 (Fuel Channel D18, fluence of 7.2×10^{25} $n \cdot m^{-2}$) and Tube 331 (Fuel Channel C16, 7.9×10^{25} $n \cdot m^{-2}$).

The *J-R* curves for the irradiated material generally exhibited the three-stage crack growth behavior, with the extent of Stage 1 or plateau region increasing with decreasing toughness. There was also an increase in the extent of discontinuous crack growth with decreasing toughness, with crack jumps of up to 0.23 mm at less than 2 mm crack growth noted for the lowest toughness material. The *J-R* curve for material from the highest toughness tube showed a steadily decreasing slope, similar to that of the unirradiated material. However, for the current test matrix, unirradiated material from the lowest toughness tubes also showed the three-stage crack growth behavior.

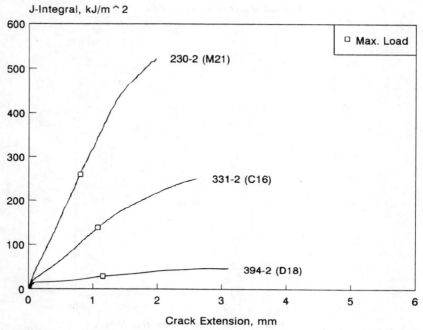

FIG. 5—*Distribution of J-R curves for irradiated material tested at 250°C (PNGS A Unit 3 material irradiated at 254 to 287°C to a fast-neutron fluence of 6.5 to 9.8 × 10^{25} n · m^{-2}).*

A correlation was also found between the toughness of the irradiated and unirradiated material for this test matrix. This is demonstrated in Fig. 6, where the toughness parameter, *dJ/da*, of material from the irradiated ring (single test result for each ring, except for an average of two tests for Tube 230) has been plotted versus that of the unirradiated back-end offcut (average of two tests). For the lower toughness material in this test matrix, the crack growth toughness may be considered an average slope characteristic of both Stages 1 and 2, with the proportion of Stage 1 decreasing with increasing toughness. Examination of the different J-R curves as well as the three remaining toughness parameters, $J_{0.2}$, J_{m1}, and $J_{1.5}$, confirmed the trend with the latter, revealing similar scatter to Fig. 6. Therefore, there was little benefit in selecting a different toughness parameter to *dJ/da*.

The results in Fig. 6 confirm the high toughness of Tube 230, and also indicate little change in the toughness of this tube as a result of irradiation for 17 years in-reactor. In comparison, most tubes showed a significant reduction in toughness with irradiation, having a crack growth toughness on installation <150 MPa at 250°C, which decreased to <50 MPa after 17 years of service. Examination of the irradiated test results did not reveal any effect of fast-neutron fluence on the toughness for fluences between 6.5 and 9.8 × 10^{25} n · m^{-2}. However, the results suggested a possible effect of irradiation temperature for the lower toughness tubes (*dJ/da* < 150 MPa on installation); that is, less reduction in toughness with increasing irradiation temperature. More test results are required from higher toughness material showing less variation to confirm any trend.

For comparison, Fig. 6 includes results from Tubes 622, B439, H190M, and H850, where the data obtained for the unirradiated back ends have been plotted versus those from the tube centers irradiated to a fluence of 2.1 × 10^{24} n · m^{-2} in NRU. The latter were in good agreement

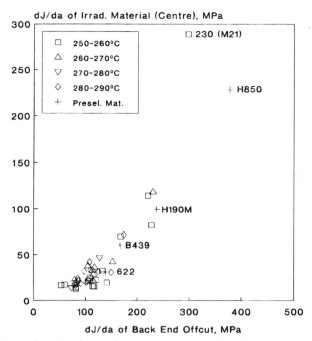

FIG. 6—*Correlation between the crack growth toughness of irradiated and unirradiated material at 250°C (PNGS A Unit 3 material irradiated at 254 to 287°C to a fast-neutron fluence of 6.5 to 9.8×10^{25} n · m^{-2}).*

with the current results, suggesting that the large reduction in crack growth toughness with irradiation occurs at fluences $<2.1 \times 10^{24}$ n · m^{-2}, with little further reduction to a fast-neutron fluence of 1×10^{26} n · m^{-2}.

High-Flux Irradiation—Preselected Material Irradiated in the OSIRIS Reactor

The crack growth toughness of Tube H737 before and after irradiation in the OSIRIS and NRU facilities is shown in Fig. 7. Although there is some scatter, the results from the NRU reactor were within the scatterband of those for the OSIRIS reactor. This indicates that the high-flux facility can provide results equivalent to those of a lower-flux facility; that is, there is no significant effect of neutron flux on toughness. Furthermore, the results confirm that the major degradation in toughness of Zr-2.5Nb pressure tube material occurs at relatively low fluences $<2.9 \times 10^{24}$ n · m^{-2}, with little further change with increasing irradiation. Saturation appears to be complete by a fast-neutron fluence of about 1.6×10^{25} n · m^{-2}.

The variation in toughness of OSIRIS specimens may be partly due to the effects of γ-heating associated with a high-flux facility. γ-heating results in a temperature gradient of up to 11°C between the center and surface of the compact toughness specimens. The toughness is expected to be most sensitive to variations in irradiation temperature at lower fluences, when the major reduction in toughness occurs.

Hydrogen Analysis of Irradiated Material

After mechanical testing, duplicate hydrogen samples (500-mg buttons) were punched from a selection of irradiated test specimens and analyzed by hot vacuum extraction mass spectrome-

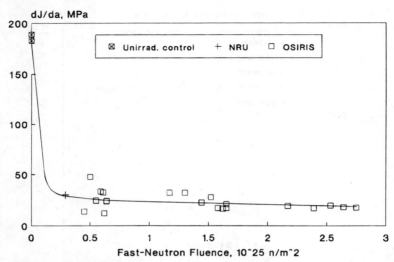

FIG. 7—*Effect of neutron fluence on crack growth toughness of Tube H737 at 240°C (preselected material irradiated in OSIRIS and NRU at 250/255°C).*

try. The maximum hydrogen concentration for all specimens irradiated in NRU was 10 ppm (0.09 atom percent), similar to that of the unirradiated material, indicating that no further hydrogen had been picked up in-reactor. However, for specimens (Tube H737) irradiated in OSIRIS, the maximum hydrogen concentration was up to 35 ppm (0.32 atom percent). The origin of this additional hydrogen is not known. The maximum total equivalent hydrogen concentration (hydrogen + 0.5 deuterium concentration) of the 40 irradiated rings removed from the PNGS A Unit 3 tubes was 30 ppm (0.27 atom percent).

The terminal solid solubility of hydrogen isotopes at 240 and 250°C is about 27 ppm (0.24 atom percent) and 31 ppm (0.28 atom percent), respectively, so that most of the hydrogen isotopes would have been in solution during mechanical testing. Therefore, the presence of hydrides (or deuterides) should not have been a factor influencing the mechanical properties of the irradiated material. This was later supported by fractographic examination that showed no evidence of hydride fracture.

Fractographic Examination

Optical and scanning electron fractographic studies were carried out on a selection of fracture toughness specimens from each test program. The fracture surfaces consisted of a central tunneled region of flat fracture, with shear lips starting to develop at the surfaces at 45° to the radial-axial plane before specimen unloading. The extent of this flat fracture zone decreased and through-thickness yielding increased with increasing toughness. Only very narrow shear lips were evident on the unirradiated fractures of some specimens, indicating that shear-lip formation was predominantly a feature of irradiated material. Failure was by ductile fibrous fracture in all cases.

For the lowest toughness tubes, the central flat zone had a very "woody" appearance, Fig. 8a, as a result of "fissures" parallel to the crack growth (axial) direction, Fig. 8b. Fissures have been observed before in both irradiated and unirradiated Zr-2.5Nb tube material and are characterized by dimple alignment in the axial or extrusion direction of the tube, with equiaxed

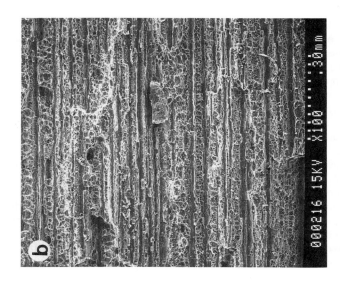

FIG. 8—*Scanning electron fractographs of fracture toughness specimen from Pickering 3 Tube 394 (Fuel Channel D18): (a) angle view of fracture surface close to fatigue crack tip showing central flat fracture and development of shear at surfaces. Note steps due to fissures on shear lips; and (b) enlarged view of fissures at midsection. (Crack growth direction is from left to right.)*

dimples of varying size in the regions between fissures [2,5,11]. The fissure density was highest in the specimen midsection, where the through-thickness stress was high and decreased towards the specimen surface where this stress was relaxed. Fissures were also observed on the faces of the shear lips, producing a series of steps, Fig. 8a, the density of such steps decreasing with increasing toughness. The shear walls between fissures and dimples on different planes, as well as the fine equiaxed dimples along shear lips, are each indicative of strain localization [11].

Fissures were observed for the unirradiated specimens, although discrimination was more difficult for higher toughness material. This was because the fissures were shorter and their width comparable to the dimple size in the remaining ligaments. The effect of irradiation was to enhance the appearance of the fissures producing the "woody" effect. The fissures appeared to be longer and more continuous than those observed for the unirradiated material, and some very long deep cracks were occasionally observed along the bottom of the fissures in the transverse-axial plane. The dimple size in the ligaments between fissures was smaller for the irradiated material than for the unirradiated material, further enhancing the effect of aligned features.

Lower toughness material had longer fissures (many hundreds of microns in length) that were more closely spaced than higher toughness material. This is illustrated in Fig. 9a and b that show the cumulative distributions of fissure length and spacing, respectively, for a lower (622-4), intermediate (C70-1), and higher (H190M-10) toughness specimen irradiated in NRU. These distributions were determined from ×100 composites at the specimen midsection, corresponding to an area of 2 mm (axial direction) and 1.5 mm (radial direction) just ahead of the fatigue crack tip; that is, the region of highest constraint characterized by the initial portion of the J-R curve. Similar distributions were obtained previously for unirradiated material at room temperature, where further details of the image analysis techniques are given [5].

A typical fissure intersecting the fatigue crack tip is shown in Fig. 10. The fissure initiates directly from the fatigue crack with no prior crack-tip blunting, indicative of a local zone of low void nucleation strain (toughness). Initiation occurs from an elongated granular zone within the fatigue crack region, showing evidence of fragmented grains. Such granular zones were frequently observed in all specimens and occasionally developed into fissures in the fracture zone. In a few cases, deep transverse-axial cracks would be associated with these regions, the room temperature cracks being extended into the high-temperature stable crack growth region and "healing up" along a fissure.

In contrast, material from the highest toughness tubes, H850 and 230, exhibited large crack-tip blunting zones with crack branching, significant through-thickness yield, and a narrow tapered zone of flat fracture at the mid-thickness, Fig. 11a. There was no evidence of fissure formation, but of equiaxed and tearing dimples of varying size, Fig. 11b. Compared to the unirradiated specimens, the dimples were generally smaller for the irradiated material, with extremely fine dimples along surface shear lips. Long deep cracks were also evident in the highly constrained region at the midsection of the irradiated material, Fig. 11a, presumably as a result of the high crack-tip strain in this region. The difference in toughness of Tubes H850 and 230, Fig. 6, appeared to be related to the extent of multiple crack nucleation (crack branching), as indicated by the raised area of material in the crack-tip blunting zone, Fig. 11a, higher toughness specimens exhibiting more crack deviation. Such material removal from one fracture surface appears to result from deformation and void nucleation along intense shear bands approximately 45° to the axial and transverse directions, as noted previously for the unirradiated control specimens from Tube H850 [5].

Recent work on the unirradiated preselected material demonstrated that fissures result from decohesion along strips of low-energy fracture on the transverse-axial plane due to microsegregation of chlorine and carbide [5]. Such species cannot be resolved in the scanning electron microscope (SEM). A correlation between the toughness and chlorine concentration of the

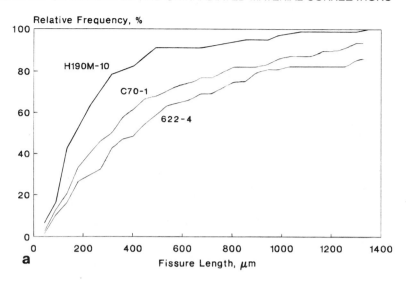

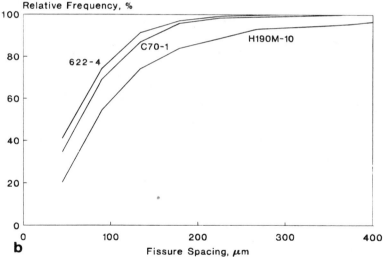

FIG. 9—*Cumulative distribution of fissures for irradiated fracture toughness test specimens tested at 240°C (preselected material irradiated in NRU at 255°C to a fast-neutron fluence of 2.1×10^{24} $n \cdot m^{-2}$): (a) fissure length and (b) fissure spacing.*

tubes, as measured by glow discharge mass spectroscopy (GDMS), supported this observation [13]. In particular, the lower toughness material had a chlorine concentration of 4 to 5 ppm (622, H737), with the higher toughness material having chlorine concentrations of 1.5 ppm (H190M) and 0.15 ppm (H850) [5].

In contrast, the chlorine concentration of the PNGS A Unit 3 material varied from 15 ppm for the low-toughness Tube 394 to 0.15 ppm for the highest toughness Tube 230 [13], consistent with the previous results. Specimens from Tube 321 (Fuel Channel G16) were unusual, since

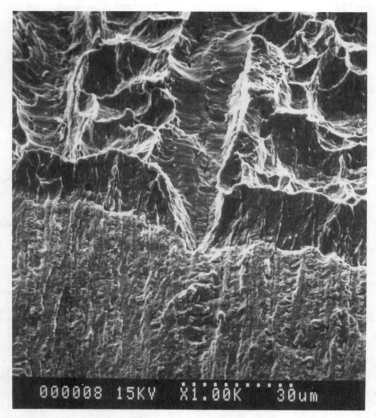

FIG. 10—*Scanning electron fractograph of fissure intersecting the blunting zone at midsection of fracture toughness specimen from Tube 622 irradiated in NRU. (Crack growth direction is towards top of the page, and the specimen is tilted 45°.)*

no fissures were evident on the fracture surfaces of the offcut specimens but a number of short fissures (about 200 μm long) were apparent on the fractures of the irradiated specimen. The chlorine concentration of this tube was about 1.5 ppm, suggesting that this represents a threshold level for fissure formation on irradiated bend-type specimens; this value is consistent with the previous result for the high-toughness Tube H190M, chlorine concentration of 1.5 ppm, that showed fissures on both irradiated and unirradiated specimens.

Particles were rarely seen on the fracture surfaces of the preselected material, the notable exception being those from the lower toughness Tube 622, for which both individual and small clusters of micron-size angular particles were observed. These particles were generally not associated with fissures. Axial alignment of particles resulted in the formation of elongated dimples that showed some similarity to a fissure, Fig. 12. Isolated particles were also observed, resulting in void nucleation and dimple formation in the ligaments between fissures, Fig. 12. Similar angular particles were observed previously on the fracture surfaces of unirradiated material from Tube 622 and were identified as zirconium carbides using scanning auger

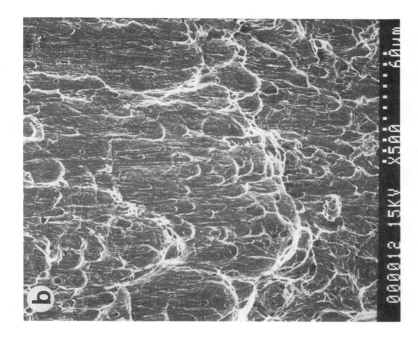
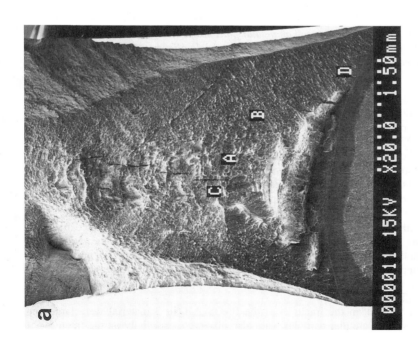

FIG. 11—*Scanning electron fractographs of fracture toughness specimen from Pickering 3 Tube 230 (Fuel Channel M21): (a) Overall view showing narrow flat fracture zone at midsection (A), crack branching and material removal at midsection of blunting zone near fatigue crack tip (B), transverse-axial cracks at midsection of fracture zone (C), and large crack-tip blunting zone near surface (D); and (b) enlargement of flat fracture zone (A) indicating equiaxed and tearing voids of varying size. (Crack growth direction is towards top of the page.)*

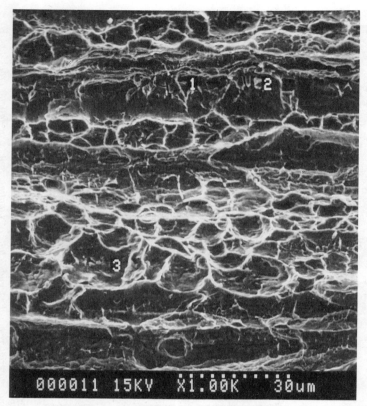

FIG. 12—*Scanning electron fractograph of angular particles at midsection of fracture toughness specimen from Tube 622 irradiated in NRU, indicating particle alignment (1 and 2) and isolated particles in ligament (3). (Crack growth direction is from left to right.)*

microscopy (SAM) [5]. In contrast to the other preselected materials, Tube 622 had a carbon concentration of 181 ppm, above the "current" specification limit of 125 ppm.

In comparison, particles were observed on all fractures of the PNGS A Unit 3 material (except for Tube 230), such particles being less evident on the irradiated specimens than the offcuts. Elongated clusters of large cracked cubic-shaped particles many microns across were seen on the fractures of the low and intermediate toughness specimens; for example, Tube 281, Fuel Channel C06 (Fig. 13). Void nucleation and growth at such particles resulted in the presence of large elongated dimples that appeared as depressions and interruptions in the fissures, effectively reducing the continuity of a fissure. For this reason, the trend of increasing fissure density with decreasing toughness was not as clear as for the preselected material. SAM analysis has shown these particles to be rich in phosphorus, with some evidence of oxygen, indicating that they are zirconium phosphides [14]. The largest clusters of cubic-shaped particles were found on specimens from tubes having the highest phosphorus concentration, as measured recently by GDMS [13]; for example, 42 to 80 ppm (Tube 321) to 97 ppm (Tube 281) to 200 ppm (Tube 296).

Most particles observed on the fractures of the PNGS A Unit 3 material were large and chunky; however, smaller angular particles were also observed, especially on the fractures of

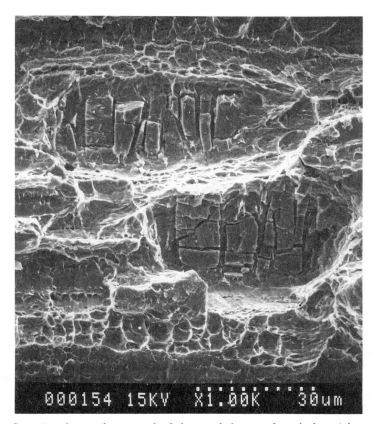

FIG. 13—*Scanning electron fractograph of elongated clusters of cracked particles at midsection of fracture toughness specimen from Pickering 3 Tube 281 (Fuel Channel C06). (Crack growth direction is from left to right.)*

some higher toughness tubes that did not exhibit fissures; for example, Tube 331 (Fuel Channel C16, see *J-R* curve in Fig. 5) and the slightly lower toughness Tube 587 (Fuel Channel H05). Small isolated particles were only rarely observed at the centers of dimples in Tube 331, but aligned zones of particles were frequently evident on the fractures of Tube 587 and were similar to the particles (carbides) observed previously in Tube 622 (Fig. 12). Other flakelike particles have been observed on lower toughness tubes (Tube 296, Fuel Channel H16). Work is continuing on the identification of these particles.

Table 2 summarizes the toughness of the unirradiated back-end offcuts of a selection of the PNGS A Unit 3 tubes and their corresponding chlorine, phosphorus, and carbon concentrations. The phosphorus concentration of these early Pickering tubes ranged from 9 to 200 ppm, considerably higher than the phosphorus levels of current production tubes of <5 ppm [*13*]. In fact, no phosphorus-rich particles were identified on the fractures of the preselected material for which the phosphorus concentration was generally ≤5 ppm (except for the early Pickering-type Tube 622 that had a phosphorus level of 12 ppm).

In summary, the fractographic examination revealed similar features on both the irradiated and unirradiated fracture surfaces. The stepped nature of the *J-R* curve was consistent with

TABLE 2—*Crack growth toughness at 250°C and chemical analysis of a selection of Pickering NGS A Unit 3 back-end offcuts.*

Pressure Tube	dJ/da, MPa	$J_{0.2}$, kJ · m^{-2}	Chlorine,[a] ppm	Phosphorus,[a] ppm	Carbon,[b] ppm
395	53	42	15	21	164
295	60	20	11	190	...
381	81	24	10	28	154
394	81	27	15	21	...
296	85	27	7	200	158
413	113	34	6	18	...
281	114	29	8	97	175
479	127	45	5	11	175
321	168	59	1.5	61	157
587	221	86	0.3	23	173
230	297	136	0.15	9	115

[a] Glow discharge mass spectroscopy, surface analytical technique.
[b] Bulk analytical technique.

material undergoing a transition from flat to shear fracture. The exceptions were material from the highest toughness Tubes H850 and 230, for which there was no extensive flat fracture zone and no detectable Stage 1 behavior on the *J-R* curve.

For the preselected material, the predominant failure mechanism was initiated by void nucleation at microsegregated chlorine and carbide, resulting in fissure formation. Lower toughness material had fissures that were longer and more closely spaced. Irradiation enhanced such fissures and increased their density. For Tube H737, there was no qualitative difference in the fissure density of specimens irradiated to the different fluences in the high (OSIRIS) and lower (NRU) flux facilities, consistent with the major reduction in toughness occurring at low fluences.

For the PNGS A Unit 3 material and the low-toughness preselected Tube 622, failure was initiated not only at microsegregated species (as revealed by the presence of fissures), but also at different distributions of particles (predominantly phosphides and carbides).

In comparison, the highest toughness tubes, H850 and 230, showed no evidence of void nucleation at microsegregated species (fissures) nor particles, but large crack-tip blunting and crack branching with equiaxed and tearing voids were evident, indicating large strain deformation.

Irradiation Damage

The microstructures of a selection of the irradiated and unirradiated specimens were studied using thin-film transmission electron microscopy (TEM) and X-ray diffraction (XRD) techniques. For XRD, samples about 1 cm^2 by 0.5-mm thick were prepared by cutting thin slices out of the tube perpendicular to each of the three principal tube axes, using a slow-speed diamond saw. Each sample was then chemically polished to remove the damaged layer introduced by the preparation (about 0.025 mm). The TEM thin-foil samples were prepared by punching 3-mm-diameter disks out of slices thinned to about 0.1-mm thickness, followed by electropolishing. TEM analysis was performed on samples whose surface was perpendicular to the axial direction of the tubes. Further details of the experimental techniques are given elsewhere [*15*].

X-Ray Diffraction

The texture of a typical tube is such that most grains have basal planes perpendicular to the transverse direction and $\{10\bar{1}0\}$-type prism planes perpendicular to the axial direction.

Analysis of the XRD peaks provides a measure of the a-component dislocation density, from prism plane line-broadening, or c-component dislocation density, from basal plane line-broadening, assuming no significant contribution from intergranular strains [16].

Calculated a-type dislocation densities have been plotted as a function of fluence for specimens from Tube H737 irradiated in the NRU and OSIRIS reactors and for samples taken from the inlets of different tubes removed from CANDU reactors, Fig. 14. Note that the a-type dislocation density is higher at the cooler inlet end (about 8×10^{14} m^{-2} at 250°C) compared to the outlet end (about 6×10^{14} m^{-2} at 290°C) [15]. There is a rapid increase in prism plane line-broadening in the early stages of irradiation ($<5 \times 10^{24}$ n · m^{-2}) that can be attributed to a-type dislocation loop formation; the total dislocation density increasing at a rate of about 1×10^{-10} n^{-1}. Any further increase is small relative to the initial transient. The basal plane line-broadening did not show any significant increase with increasing fluence and, in fact, there were indications that the lines sharpened to some extent. The latter can be attributed to the splitting of the c-component network dislocations into partials as a result of climb during irradiation [2,17].

Measurements of the α-phase and β-phase lattice parameters in the H737 material were comparable with similar measurements from the inlets of power reactors; that is, for the same operating temperature of about 250°C. For the β-phase measurements, the lattice parameters represent a composition in the range of 40 to 50% by weight niobium. This is in contrast to measurements on β-phase in material from the outlets of power reactors, for an operating temperature of about 300°C, where the β-phase composition is significantly higher (up to 70% by weight niobium) after service [15].

For the preselected materials irradiated in NRU to 2.1×10^{24} n · m^{-2}, there was little difference in the irradiation damage of the samples from the different pressure tubes. The main effect of irradiation was to increase the a-type dislocation density, measured from (20$\bar{2}$0) planes, from a range of 2.5 to 3.5×10^{14} m^{-2} to a range of 4.5 to 5.5×10^{14} m^{-2}; that is, by about 2×10^{14} m^{-2} in each case.

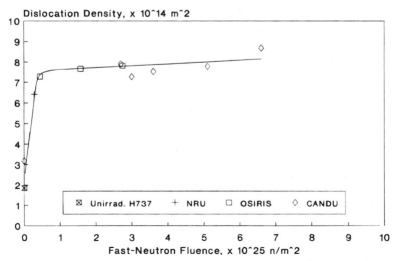

FIG. 14—*Variation of* a-*type dislocation density, measured from* (20$\bar{2}$0) *planes, as a function of fluence for Tube H737 irradiated in the NRU and OSIRIS reactors, compared with results from the inlets of CANDU reactor pressure tubes.*

TEM Examination

Micrographs showing a typical a-type and c-component dislocation structure for unirradiated and irradiated specimens of Tube H737 are shown in Fig. 15 and 16, respectively. It is clear that, whereas a-type loops are forming during irradiation, there is no apparent c-component loop formation. The c-component network dislocations, however, do show signs of splitting, Fig. 17a, and this is consistent with the line-broadening data. These micrographs for material irradiated in the OSIRIS reactor are comparable with those for samples taken from the different preselected tubes irradiated in NRU. In each case, there was little evidence for significant β-Nb precipitation in the α-phase, Fig. 17a. In contrast, the formation of β-Nb precipitates was clearly observed in reactor tube samples irradiated to higher fluences, Fig. 17b. It may be that such β-Nb precipitates are present in the preselected material irradiated to lower fluences, but are on too fine a scale to identify using TEM. It should be noted that the lattice parameter measurements indicated an expansion of the α-phase, even at low fluences (5×10^{24} n · m^{-2}) consistent with depletion of Nb (Nb being an undersized solute). These measurements are similar to those for power reactor samples containing visible β-Nb precipitates [15].

No obvious changes were observed in the morphology of the β-phase of the preselected material with irradiation, although discrimination of this complex phase is difficult with TEM. However, chemical microanalysis by energy dispersive X-ray analysis (EDXA) revealed that iron (an impurity that concentrates in the β-phase) is dispersed with irradiation, the dispersion increasing with fluence and being complete by 2.75×10^{25} n · m^{-2}; that is, the highest fluence specimens irradiated in OSIRIS. This is consistent with previous results from power reactor tubes [15].

Therefore, for the same operating temperature, there was little evidence to indicate that the different flux facilities produced significantly different microstructure evolution during irradiation.

Discussion

Transverse Tensile Strength

The present results have shown that irradiation increases the transverse UTS by about 200 MPa after a fast-neutron fluence of 2.1×10^{24} n · m^{-2}. This is similar to the value of about 190 MPa obtained by Ibrahim from ring tension tests on an early Pickering pressure tube (620) irradiated to 1.3×10^{25} n · m^{-2} [18], as well as to the scatterband of results for reduced-thickness specimens machined from different surveillance tubes [2]. These results, together with those from the high-fluence irradiation, indicate that the major increase in strength as a result of irradiation hardening occurs at low fluences, $<2 \times 10^{24}$ n · m^{-2}, with little further change with increasing fluence.

The process of irradiation hardening and the development of plastic instability in irradiated Zr-Nb alloys has been described in detail by Cheadle et al. [19]. The generation of dislocation loops is known to be the main source of hardening in irradiated zirconium alloys. For transverse tension specimens, deformation occurs initially by twinning. In irradiated material with low work hardening, this occurs in localized bands, with subsequent removal of damage by slip in the bands of reoriented material. When a sufficient number of bands are produced to form a zone, further strain is concentrated in this zone of lower flow stress. Further deformation of the defect-free zone produces localized void nucleation and fracture.

The foregoing indicates that it is the twinning process that controls the regions of irradiation-damage-free material and the transverse strength. However, the large increase in a-type dislocation density obtained for the pressure tube material for fluences $<5 \times 10^{24}$ n · m^{-2}, Fig. 14, suggests that the a-type dislocations play a significant role. It may be that the increase in such

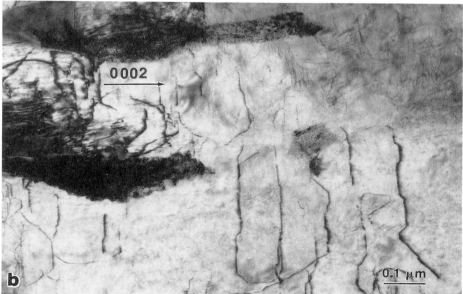

FIG. 15—*Thin-foil micrograph showing typical microstructure of unirradiated Tube H737:* (a) *high density of* a-*type network dislocations and* (b) *lower density of* c-*component network dislocations.*

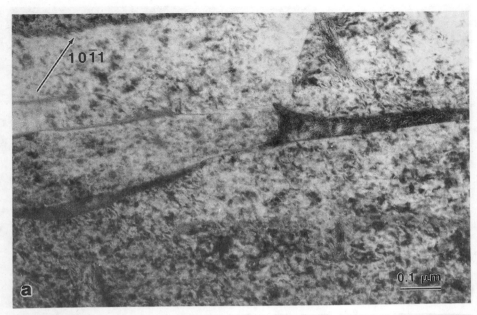

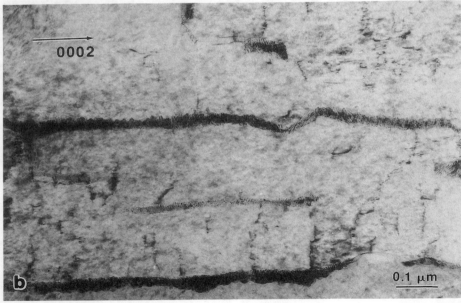

FIG. 16—*Thin-foil micrograph showing typical microstructure of specimen from Tube H737 irradiated to a fluence of 1.6×10^{25} n · m^{-2} in OSIRIS at 250°C:* (a) *high density of a-type dislocation loops in addition to the network dislocations introduced during fabrication, and* (b) *no evidence for c-component loop formation.*

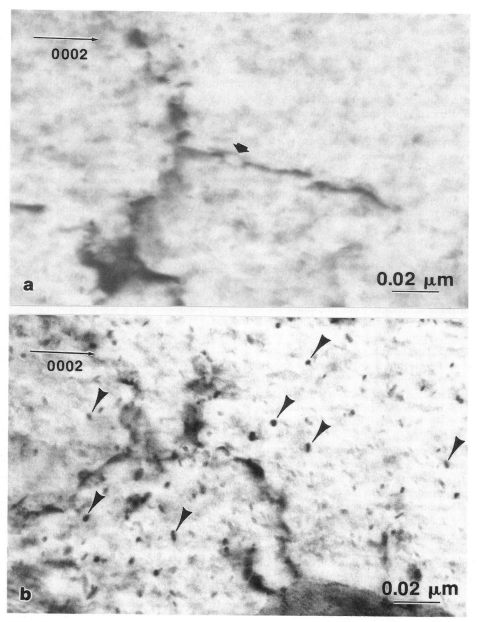

FIG. 17—*Higher magnification thin-foil micrographs showing c-component defect structures in:* (a) *Tube H737 irradiated to* 1.6×10^{25} $n \cdot m^{-2}$ *in OSIRIS at 250°C and* (b) *Pickering 3 Tube 230 (Fuel Channel M21) irradiated to* 7.3×10^{25} $n \cdot m^{-2}$ *at 257°C. The c-component network dislocations show signs of splitting into partials (arrowed in* a*).* β-*Nb (arrowed) precipitates are visible in* b, *but are not readily apparent in* a.

dislocation loops raises the stress required for prismatic slip on $\{10\bar{1}0\}$ planes to such high levels that twins are nucleated with little further increase in applied stress. For example, $\{1\bar{1}02\}$-type twins have been clearly observed in the deformation bands of transverse specimens of irradiated Zr-2.5Nb pressure tube material [2]. Any further increase in a-type dislocations to fluences of 5×10^{24} n · m^{-2}, Fig. 14, may then have little effect in further increasing the transverse yield stress or reducing the work-hardening capacity. The testing of the preselected material being irradiated in NRU to 5 to 10×10^{24} n · m^{-2} will allow this hypothesis to be investigated as the dislocation density approaches saturation.

The strength of Zr-2.5Nb pressure tubes is governed mainly by the texture [20], dislocation density (cold work) [21], grain size [4,10], and chemical composition, especially niobium and oxygen [21]. However, only the dislocation density is influenced by irradiation. For the preselected tubes irradiated in NRU, it has been shown that the increase in a-type dislocation density was similar in each case. Therefore, the relative propensity for deformation (by slip or twinning) in any grain should be similar in the unirradiated and irradiated material, accounting for the similar ranking of the strength of the tube centers before and after irradiation.

Fracture Toughness

The results from the low- and high-fluence irradiation programs have demonstrated that the major reduction in toughness of Zr-2.5Nb pressure tube material occurs early in the lifetime of a reactor at fluences $<2 \times 10^{24}$ n · m^{-2}, with little further reduction at higher fluences. Furthermore, results from the high-flux test facility suggest that saturation is complete by about 1.6×10^{25} n · m^{-2}, significantly lower than the previous best estimate of 3×10^{25} n · m^{-2} based on periodic testing of surveillance tubes [2].

Irradiation has the potential to influence the fracture toughness by increasing the number of void nucleating sites. This can occur either directly by irradiation-induced precipitation or indirectly by irradiation hardening and strain localization.

An increase in the transverse tensile strength and reduction in the work-hardening capacity of the irradiated Zr-2.5Nb matrix raises the local stress state at the crack tip and concentrates the strain in localized bands, as indicated by the localized zig-zag path of the crack [11]. This increases the probability of void nucleation (crack initiation) at any potential sites close to the crack tip. In the absence of void nucleation, further deformation of the matrix is required to raise the crack-tip stress state to higher levels over larger distances until suitable sites are activated. Microstructural features having the lowest (stress/strain) condition for void nucleation and located closest to the crack tip will be activated first.

The correlation between the toughness of the Zr-2.5Nb pressure tube material before and after irradiation is consistent with the predominant failure mechanism being governed by void nucleation at preexisting species in the as-installed tubes, rather than from additional species from reactor operation. The similarity of features on both the irradiated and unirradiated fracture surfaces, as well as an absence of fractured hydrides (deuterides) on the fractures of irradiated material, supports this hypothesis.

The presence of fissures on the majority of fractures indicates preferential void nucleation at microsegregated chlorine and carbide [5]. Such fissures reduce the through-thickness stress, allowing significant increases in the through-thickness strain in the remaining ligaments. Additional void nucleation and coalescence may then occur in the ligaments, with narrow ligaments failing by shear. Fissures are damaging in that they effectively divide the specimen into a series of thin ligaments, each of which may fail at a much lower fracture strain, resulting in a large reduction in the total fracture strain (toughness) [5]. The increase in fissure density (increase in length, reduction in spacing) with decreasing toughness for both the irradiated and unirradiated material is consistent with this. Furthermore, the increase in fissure density with irradiation is consistent with a reduction in the threshold condition for void nucleation at the microsegregated species after irradiation.

Large micron-size particles, such as phosphides and carbides, act as additional sites for void nucleation. SEM examination of metallographically polished and etched sections suggests that the larger particles or clusters were fractured before testing [22], that is during fabrication, possibly during the cold-drawing process. Such large preexisting fractured particles or voids (tens of microns across) would act as stress concentrators, producing significant increases in the local stress state across the ligaments between neighboring voids. This may lead to additional void nucleation and coalescence in the ligaments and a significant reduction in toughness. The presence of a high density of phosphide clusters in the lower toughness Tube 321 (unirradiated dJ/da of 168 MPa), compared with their absence in preselected Tube H190M (unirradiated dJ/da of 243 MPa) at a similar chlorine concentration of 1.5 ppm, supports this hypothesis. The fact that such particles were less apparent on the irradiated fractures than the offcuts again reflects a decrease in the critical condition for fissure formation with irradiation, as fewer particles closer to the fracture plane contribute to the final fracture process.

Additional evidence of irradiation hardening and strain localization reducing the condition for void nucleation and increasing the number of potential nucleation sites comes from:

(1) the reduction in dimple size in the ligaments between fissures and large particles. Such sites could be smaller preexisting particles or precipitates not resolvable in the SEM, or both. For example, the presence of silicides (<1 μm diameter) has been noted in many Zr-2.5Nb pressure tubes [2]. Recently, a large number of such inclusions were noted in bands (tens of microns long) during examination of thin foils from Tube H737 in TEM; and

(2) the presence of fine equiaxed dimples in the lower constraint region at the specimen surface, indicative of highly localized deformation. These voids were formed at 45° to the fracture plane, following the planes of maximum shear stress and resulted in shear lip formation.

Irradiation is known to enhance the precipitation of β-Nb in the α-phase of zirconium alloys [23,24]. In Zr-2.5Nb pressure-tube material, such precipitates (about 5 to 10 nm) become apparent in surveillance pressure tubes at fluences greater than about 3×10^{25} n · m^{-2} [15]. However, the good agreement between the toughness of the preselected tubes irradiated in NRU to a relatively low fluence compared with the toughness of the PNGS A Unit 3 material irradiated to much higher fluences, Fig. 6, suggests that the presence of such visible precipitates in the higher-fluence material has little effect. In fact, the influence of such small coherent precipitates, if any, should be most apparent with high-toughness material that has few other potential void nucleating sites. However, the high-toughness Tube 230 showed little evidence of any significant reduction in toughness with irradiation, indicative of an insensitivity to both irradiation hardening and the presence of any irradiation-induced precipitates. More testing of higher toughness material irradiated to higher fluences is required to investigate this further.

The present evidence indicates that the toughness of Zr-2.5Nb pressure tube material saturates early in reactor life, by a fluence of 1.6×10^{25} n · m^{-2}. However, further changes in toughness at higher exposures (fluences >1×10^{26} n · m^{-2}) cannot be ruled out. For example, early experimental results for stainless steel suggested saturation of irradiation damage occurred by a fluence of 2 to 3×10^{26} n · m^{-2} [25]. Later, further reductions in toughness were observed at fluences greater than 1×10^{27} n · m^{-2} [26], due to a change in fracture mechanism resulting from subtle changes in microchemistry [27].

A number of mechanisms might affect the toughness of Zr-2.5Nb at high fluences. These include the coarsening of β-Nb precipitates as well as a sharp increase in c-component loop formation or dislocation density under "accelerated" or "breakaway" growth conditions [28]. At present, no change in toughness has been observed that can be attributed to the presence of β-niobium precipitates. Furthermore, there is no evidence of any significant change in the c-component loop formation in Zr-2.5Nb for fluences between 1 and 10×10^{25} n · m^{-2} [15]. Recent work indicates Zr-2.5Nb is resistant to c-component loop formation compared to zirconium or Zircaloy-2 during electron radiation at 300°C [29]. This suggests that Zr-2.5Nb is likely to be more resistant to breakaway growth conditions than other zirconium alloys. However, the effect of long-term exposure to neutron radiation on the mechanical properties of Zr-2.5Nb will be investigated further by continuing the high-flux test irradiation in OSIRIS to fluences of about 4×10^{26} n · m^{-2}.

Summary

A correlation has been found between the toughness of Zr-2.5Nb pressure tube material before and after irradiation to 1×10^{26} n · m^{-2} ($E > 1$ MeV). Comparison with previous results from preselected material irradiated in the NRU reactor demonstrates that the major degradation in toughness occurs early in the life of a CANDU reactor; that is, at fluences $<2 \times 10^{24}$ n · m^{-2}. Limited results on relatively low-toughness material irradiated in the high-flux OSIRIS reactor in France indicate that the flux does not have a major effect on the irradiated toughness. These results also suggest that saturation is complete by about 1.6×10^{25} n · m^{-2}.

Fractographic and microstructural evidence has been presented that confirms that the microstructural factors controlling variations in the toughness of pressure tubes on installation also predominate after irradiation. The presence of preexisting microsegregated species (chlorine and carbide) and particles (phosphides and carbides) results in significant reductions in toughness after irradiation. This is because irradiation hardening and strain localization increase the crack-tip stress state and reduce the condition for void nucleation at preexisting species. In contrast, the toughness of material showing no evidence of microsegregated species nor significant particles is shown to be relatively insensitive to irradiation. In such cases, high crack-tip strains (toughness) can be maintained after irradiation.

Acknowledgments

These programs could not have been completed without the excellent technical assistance of many individuals, including G. E. Baas, G. R. Brady, D. D. Himbeault, J. M. Smeltzer, R. Behnke, F. J. Szostak, R. R. Bawden, A. Lockley, J. F. Watters, J. R. Theaker, J. F. Mecke, and J. E. Winegar. Particular thanks are due to C. J. Schulte (Ontario Hydro) for assistance with selection and retrieval of reactor archive material, F. J. Butcher for arranging the irradiation of material in the NRU reactor, and to C. E. Coleman and M. A. Miller for initiating and coordinating the irradiation of material in the OSIRIS reactor. Finally, the excellent work of the staff of the hot-cell facilities at Chalk River and Whiteshell Laboratories is much appreciated.

The funding of this work by the CANDU Owner's Group (COG) under Work Packages 6536, 3305, and 6506 is gratefully acknowledged.

References

[1] Cheadle, B. A., Coleman, C. E., Rodgers, D. K., Davies, P. H., Chow, C. K., and Griffiths, M., "Examination of Core Components Removed from CANDU Reactors," Report AECL-9710, AECL Research, Chalk River Labs., Chalk River, Ontario, Nov. 1988.

[2] Chow, C. K., Coleman, C. E., Hosbons, R. R., Davies, P. H., Griffiths, M., and Choubey, R., "Fracture Toughness of Irradiated Zr-2.5Nb Pressure Tubes from CANDU Reactors," *Zirconium*

in *the Nuclear Industry: Ninth International Symposium, ASTM STP 1132,* C. M. Eucken and A. M. Garde, Eds., American Society for Testing and Materials, Philadelphia, 1992, pp. 246–275.

[3] Cheadle, B. A., Coleman, C. E., and Licht, H., "CANDU-PHW Pressure Tubes: Their Manufacture, Inspection and Properties," *Nuclear Technology,* Vol. 57, 1982, pp. 413–425.

[4] Cheadle, B. A., "Fabrication of Zirconium Alloys into Components for Nuclear Reactors," *Zirconium in the Nuclear Industry (Third Conference), ASTM STP 633,* A. L. Lowe, Jr. and G. W. Parry, Eds., American Society for Testing and Materials, Philadelphia, 1977, pp. 457–485.

[5] Aitchison, I. and Davies, P. H., "Role of Microsegregation in Fracture of Cold-Worked Zr-2.5Nb Pressure Tubes," *Journal of Nuclear Materials,* Vol. 203, 1993, pp. 206–220.

[6] Simpson, L. A., Chow, C. K., and Davies, P. H., "Standard Test Method for Fracture Toughness of CANDU Pressure Tubes," CANDU Owner's Group Report COG-89-110-1, AECL Research, Whiteshell Labs., Pinawa, Manitoba, Sept. 1989.

[7] Chow, C. K. and Simpson, L. A., "Determination of the Fracture Toughness of Irradiated Reactor Pressure Tubes Using Curved Compact Specimens," *Fracture Mechanics: Eighteenth Symposium, ASTM STP 945,* D. T. Read and R. P. Reed, Eds., American Society for Testing and Materials, Philadelphia, 1988, pp. 419–439.

[8] Griffiths, M., "Displacement Energies for Zr Measured in a HVEM," "Letter to the Editors," *Journal of Nuclear Materials,* Vol. 165, 1989, pp. 315–317.

[9] Holt, R. A. and Fleck, R., "The Contribution of Irradiation Growth to Pressure Tube Deformation," *Zirconium in the Nuclear Industry: Ninth International Symposium, ASTM STP 1132,* C. M. Eucken and A. M. Garde, Eds., American Society for Testing and Materials, Philadelphia, 1992, pp. 218–229.

[10] Northwood, D. O. and Gilbert, R. W., private communication.

[11] Chow, C. K. and Simpson, L. A., "Effect of Fracture Micromechanisms on Crack Growth Resistance Curves of Irradiated Zirconium/2.5 Weight Percent Niobium Alloy," *Nonlinear Fracture Mechanics: Vol. II—Elastic-Plastic Fracture, ASTM STP 995,* J. D. Landes, A. Safena, and J. G. Merkle, Eds., American Society for Testing and Materials, Philadelphia, 1989, pp. 537–562.

[12] Davies, P. H. and Smeltzer, J. M., "On J-Controlled Crack Growth (Size Effects) in Fracture Toughness Testing of Zr-2.5Nb Pressure Tube Material," *Fracture Mechanics: Twenty-Second Symposium (Volume I), ASTM STP 1131,* H. A. Ernst, A. Safena, and D. L. McDowell, Eds., American Society for Testing and Materials, Philadelphia, 1992, pp. 93–104.

[13] Theaker, J. R., Choubey, R., Moan, G. D., Aldridge, S. A., Davies, L., Graham R., and Coleman, C. E., "Fabrication of Zr-2.5Nb Pressure Tubes to Minimize the Harmful Effects of Trace Elements," in this volume.

[14] Watters, J. F., private communication.

[15] Griffiths, M., Chow, C. K., Coleman, C. E., Holt, R. A., Sagat S., and Urbanic, V. F., "Evolution of Microstructure in Zirconium Alloy Core Components of Nuclear Reactors During Service," *Effects of Irradiation on Materials: Sixteenth Conference, ASTM STP 1175,* A. S. Kumar, D. S. Gelles, R. K. Nanstad, and E. A. Little, Eds., American Society for Testing and Materials, Philadelphia, 1993, pp. 1077–1110.

[16] Griffiths, M., Winegar, J. E., Mecke, J. F., and Holt, R. A., "Determination of Dislocation Densities in HCP Metals using X-Ray Diffraction and Transmission Electron Microscopy," *Advances in X-Ray Analysis,* Vol. 35, C. S. Barrett et al., Eds., Plenum Press, New York, 1992, pp. 593–599.

[17] Griffiths, M., Styles, R. C., Woo, C. H., Phillip, F., and Frank, W., "Study of Point Defect Mobilities in Zr during Electron Irradiation in a HVEM," *Journal of Nuclear Materials,* submitted for publication.

[18] Ibrahim, E. F., "Mechanical Properties of Cold Drawn Zr-2.5Nb Pressure Tubes After Up to 12 Years in CANDU Reactors," Paper 12, *Materials for Nuclear Reactor Core Applications,* BNES, London, 1987.

[19] Cheadle, B. A., Ells, C. E., and van der Kuur, J., "Plastic Instability in Zr-Sn and Zr-Nb Alloys," *Zirconium in Nuclear Applications, ASTM STP 551,* American Society for Testing and Materials, Philadelphia, 1974, pp. 370–384.

[20] Cheadle, B. A. and Ells, C. E., "Effect of Crystallographic Orientation on the Fracture Ductility of Zr-2.5 Wt Pct Nb (Cb) and Zircaloy-2 Tubular Products," *Transactions,* The Metallurgical Society of AIME, Vol. 233, June 1965, pp. 1044–1052.

[21] Winton, J., Murgatroyd, R. A., Watkins, B., and Nichols, R. W., "The Strength of Zr-2.5% Nb Alloy in the Annealed and Cold Worked Condition," *Transactions,* The Japan Institute of Metals, Vol. 9, Supplement, 1968, pp. 630–636.

[22] Moan, G. D., Davidson R. D., Watters, J. F., and Woo, O. T., private communications.

[23] Gilbert, R. W., Farrell, K., and Coleman, C. E., "Damage Structure in Zirconium Alloys Neutron Irradiated at 573–923 K," *Journal of Nuclear Materials,* Vol. 84, 1979, pp. 137–148.

[24] Coleman, C. E., Gilbert, R. W., Carpenter, G. J. C., and Weatherly, G. C., "Precipitation in Zr-2.5Nb During Neutron Irradiation," *Proceedings,* Phase Stability During Irradiation, Fall Meeting of AIME, Pittsburgh, Oct. 1980, pp. 587–599.

[25] Johnson, G. D., Garner, F. A., Brager, H. R., and Fish, R. L., "A Microstructural Interpretation of the Fluence and Temperature Dependence of the Mechanical Properties of Irradiated AISI 316," *Effects of Radiation on Materials—Tenth International Symposium, ASTM STP 725,* Kramer, Brager, and Perrin, Eds., American Society for Testing and Materials, Philadelphia, 1981, pp. 393–412.

[26] Hamilton, M. L., Huang, F.-H., Yang, W. J. S., and Garner, F. A., "Mechnical Properties and Fracture Behaviour of 20% Cold-Worked 316 Stainless Steel Irradiated to Very High Neutron Exposures," *Influence of Radiation on Material Properties: Thirteenth International Symposium (Part II), ASTM STP 956,* F. A. Garner, C. H. Hennger, Jr., and N. Igata, Eds., American Society for Testing and Materials, Philadelphia, 1987, pp. 245–270.

[27] Brager, H. R. and Garner, F. A., "Microstructural and Microchemical Comparisons of AISI 316 Irradiated in HFIR and EBR-II," *Journal of Nuclear Materials,* Vol. 117, 1983, pp. 159–176.

[28] Griffiths, M., Gilbert, R. W., and Fidleris, V., "Accelerated Irradiation Growth of Zirconium Alloys," *Zirconium in the Nuclear Industry: Eighth International Symposium, ASTM STP 1023,* L. F. P. Van Swam and C. M. Eucken, Eds., American Society for Testing and Materials, Philadelphia, 1989, pp. 658–677.

[29] Griffiths, M., Gilbon, D., Regnard, C., and Lemaignan, C., "HVEM Study of the Effect of Alloying Additions and Impurities on Radiation Damage in Zirconium Alloys," The Evolution of Microstructure in Metals during Irradiation, Proceedings at the International Conference on Evolution of Microstructure in Metals During Irradiation, C. E. Coleman, R. A. Holt, and R. G. Fleck, Eds., North-Holland, 1993, pp. 273–283. Muskoka, Canada, Sept. 1992.

DISCUSSION

A. M. Garde[1] (written discussion)—What is the origin of the chlorine associated with the fissures, that is, chlorine content of the bulk starting material or a surface contamination due to faulty processing?

P. H. Davies et al. (authors' closure)—The chlorine is present in the bulk starting material. It originates from the Kroll process for producing zirconium from crude zirconium sponge. Further details are given in Refs *5* and *13* cited in the paper. In billet and extruded cold-worked tubes, chlorine is evident as stringers. However, unpublished scanning Auger microscopy (SAM) and secondary ion mass spectroscopy (SIMS) work on billet and pressure tube material by I. Aitchison at AECL Research has not revealed features on the fracture faces that would be readily associated with fracture of stringers. The damaging features appear to have both different geometry and different composition: they are relatively wide (about 15 µm) and shallow (about 0.015 µm), and also contain carbon as a carbide, in a zirconium-chlorine-carbon (Zr-Cl-C) complex. These deposits are responsible for the strips of low-energy fracture on transverse-axial planes that result in the fissures observed on radial-axial planes.

C. J. White[2] (written discussion)—Could you explain the basis for using dJ/da, rather than J_{Ic}, as the parameter to describe the fracture resistance of the tested materials? At what value of Δa did you evaluate dJ/da?

P. H. Davies et al. (authors' closure)—For characterizing thin-walled pressure tube material, the use of the full crack growth resistance or *J-R* curve is preferred. This is because the crack-tip constraint undergoes a transition from high constraint (at the midsection) to lower constraint (at the surface) as the crack front develops. It is this transition in behavior that provides the high toughness of the thin-walled material. An initiation toughness parameter, such as J_{Ic}, is only characteristic of the toughness in the higher constraint midsection. A second reason for not concentrating on an initiation parameter is the difficulty of accurate determination for higher toughness materials when using d-c potential drop to measure crack growth.

The initial crack growth toughness, dJ/da, was calculated from the linear regression analysis slope to the *J-R* curve between the 0.15- and 1.5-mm offset lines. It has been found to be a very useful parameter to characterize "toughnening rate." Other crack growth toughness parameters studied were the *J*-integral at maximum load, J_{ml}, and the *J*-integral at the 1.5 mm offset, $J_{1.5}$. However, the correlation between the toughness of the irradiated and unirradiated Zr-2.5Nb pressure tube material was obtained irrespective of the toughness parameter used.

---

[1] ABB CE Nuclear Fuel, Windsor, CT.
[2] Knolls Atomic Power Laboratory, Martin Marietta Corporation, Niskayuna, NY.

Ronald G. Fleck,[1] *Joanne E. Elder,*[1] *Allan R. Causey,*[2] *and Richard A. Holt*[2]

Variability of Irradiation Growth in Zr-2.5Nb Pressure Tubes

REFERENCE: Fleck, R. G., Elder, J. E., Causey, A. R., and Holt, R. A., **"Variability of Irradiation Growth in Zr-2.5Nb Pressure Tubes,"** *Zirconium in the Nuclear Industry: Tenth International Symposium, ASTM STP 1245,* A. M. Garde and E. R. Bradley, Eds., American Society for Testing and Materials, Philadelphia, 1994, pp. 168–182.

ABSTRACT: Pressure tubes in CANDU reactors are manufactured by extrusion and cold work from the Zr-2.5Nb alloy. Historically, the details of the pre-extrusion production of the tubes have changed since the first tubes were produced (1969) for Ontario Hydro's Pickering A station. The pressure tubes exhibit variations in elongation rate during service that may partially result from these changes. Irradiation growth is a component of the elongation. This paper describes the longitudinal irradiation growth of specimens prepared from pressure tubes with a range of iron contents and in which billets were extruded either in the as-forged condition or after β-quenching.

The irradiations are performed in the Osiris reactor at Saclay, France, in two inserts with nominal operating temperatures of 280°C (Trillium 2) and 310°C (Trillium 3). Results are reported for fast fluences up to 4.3×10^{25} n/m², $E > 1$ MeV. Throughout the irradiation, the temperature of each tier of six specimens is recorded and this shows that the actual range of temperatures is 262 to 309°C. In agreement with previous studies, the temperature dependence of growth is negative, that is, an increase in temperature decreases the growth rate.

When this experimental variable is taken into account, there is still considerable variation among specimens from different pressure tubes. There is no effect of β-quenching. However, an increase in iron in the range 367 to 1060 ppm (by weight used exclusively in this paper) reduces the growth rate by about 2×10^{-29} m²/n, $E > 1$ MeV at both 280 and 310°C. If this difference extends to high fluences, it accounts for about half of the variability in elongation rates observed in pressure tubes in-service.

KEY WORDS: irradiation growth, temperature, neutron fluence, pressure tubes, elongation, iron, zirconium, zirconium alloys, nuclear materials, nuclear applications, radiation effects

The pressure tubes in CANDU reactors are manufactured by extrusion and cold work from Zr-2.5Nb. They are 6 m long and elongate several millimetres per year during service. Part of this elongation is attributed to irradiation growth. Previous studies of irradiation growth of pressure tube material at around 300°C [1–3] have examined the effects of modifications to the extrusion and post-extrusion processes, Fig. 1. These modifications were introduced with some success to change the metallurgical structure and influence the in-service deformation characteristics of the tubes.

---

[1] Unit manager and engineer, respectively, Materials Technology Unit, Ontario Hydro Research Division, Toronto, Ontario, Canada.

[2] Senior scientist and manager, respectively, Reactor Materials Research Branch, Atomic Energy of Canada Ltd. (AECL-RC), Chalk River Laboratories, Chalk River, Ontario, Canada.

MELTING

⇓

FORGING

⇓

MACHINE HOLLOW BILLETS

⇓

BETA QUENCH FROM 1010°C

⇓

EXTRUSION 11:1 AT 820°C

⇓

COLD DRAW 25-30%

⇓

AUTOCLAVE 400°C-24h

FIG. 1—*The pressure tube manufacturing route showing the location of the beta quench process.*

In each CANDU reactor, there are up to 480 pressure tubes and the tube-to-tube variability in elongation rate ($\pm$ 20% or $\pm$ 2 $\times$ 10^{-29} m^2/n, $E > 1$ MeV) has to be considered in assessing the condition of the core at a given time. The variability is due to normal production variations and also to intentional changes in details of the pre-extrusion production processes, Fig. 1, since the first tubes were made in 1969 for Ontario Hydro's Pickering A reactors.

A water quench from the β-phase prior to extrusion, was introduced to develop a finer and more uniform structure in the billets and, hence, improve the dimensional control of the product. Changes have also occurred in the chemistry of the ingot, and in particular the iron content. Early production allowed the iron to vary from about 300 up to 1500 ppm. In a few cases, special tubes had iron contents as low as 150 ppm and as high as 2000 ppm. Studies related to corrosion and hydrogen ingress in the early 1980s suggested that the iron level should be kept below 500 ppm and the specification was changed. The effects of these two variables on the irradiation growth characteristics are reported in this paper.

The results are obtained from a study focusing mainly on reaching end-of-life fluences in small specimens before the actual pressure tubes in operating CANDU reactors. The scope of the study includes both longitudinal and transverse specimens of four types of pressure tubes (one with changes in the post-extrusion manufacturing procedure) as well as specimens from another six tubes with different iron contents and pre-extrusion treatments.

However, because we are relating growth to pressure tube elongation, we will emphasize longitudinal data. Transverse growth of one material will be presented to illustrate the relationship of the current results to previous work.

Experimental

The irradiations are performed under contract in the Osiris reactor at the Commissariat a l'Energie Atomique (CEA) Centre d'Etudes Nucleaires (CEN) at Saclay, France. Two inserts, each containing a total of 42 growth specimens and seven creep capsules, are used. One insert, Trillium 2, operates at about 280°C and the other, Trillium 3, operates at about 310°C. NaK is used as the heat transfer medium, and external heaters are used to provide a uniform temperature distribution along the length of the insert and to compensate for changes in reactor power. The specimens are irradiated at a peak fast neutron flux of about 1.8×10^{18} n/m²·s, $E > 1$ MeV. The inserts comprise seven tiers in-core, each of which contains one creep capsule and six growth specimens. Figure 2 shows a cross section of one tier. The results on the creep capsules are the subject of another paper in this conference [4].

The temperature at each tier was monitored continually throughout the irradiation by one thermocouple located close to the creep capsule and one located close to one of the growth specimens. Based on the thermocouples located close to the growth specimens, the temperature

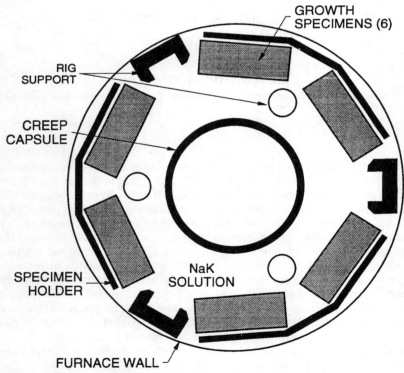

FIG. 2—*A cross section of the irradiation rigs used in the Osiris reactor that illustrates the location of the growth specimens and creep capsules in the rig. There are seven tiers in each of the Trillium 2 and 3 inserts.*

varied with time by up to ± 15°C during reactor operation. The time averaged temperature recorded for each tier during the whole irradiation varied from 262 to 295°C in Trillium 2 and from 302 to 309°C in Trillium 3.

The Zr-2.5Nb pressure tubes used in this program represent a range in production history from 1969 to 1978. The particular tubes were selected to include variations in the ingot chemistry, particularly the iron content, and pre-extrusion heat treatment, Table 1. Cold-worked Zircaloy-2 was included because its growth had been well characterized in irradiation growth studies in the DIDO reactor at Harwell, UK [5].

After machining to 38.1 by 6.4 by 1.5 or 38.1 by 6.4 by 2.5 mm, the specimens were stress-relieved for 6 or 30 h at 400°C depending upon whether the tube had already received the standard 24 h at 400°C used in production. The different thicknesses were used in an attempt to give a more uniform temperature distribution in Trillium 2.

Intermittent measurements of specimen length were made by removing the insert from the reactor, disassembling it, and transferring the specimens to a small hot cell. Before and after irradiation, the lengths of the specimens were measured using a computer-controlled system with two linear variable differential transformers. Reproducibility was verified by frequent measurement of unirradiated reference specimens during each measurement campaign. This resulted in an accuracy of 0.5 μm.

Hydrogen analysis was carried out on a few specimens removed from Trillium 2 after the first and second phases of irradiation. The results suggest that 17 to 43 ppm was picked up from the NaK in the early stages of irradiation (the highest values corresponding to specimens analyzed after the first phase).

Results

The growth strain of the Zircaloy-2 (Material G) is shown as a function of fast ($E > 1\text{MeV}$) neutron fluence in Fig. 3, which also shows the growth behavior of the same material irradiated in DIDO. Figure 4 shows the growth strain as a function of fluence for Zr-2.5Nb pressure tube material (Material A). This figure also includes data from a number of different specimens of Material A obtained from a previous irradiation growth experiment at 280°C, in the DIDO reactor [2,3]. Figure 5 shows the transverse growth data at 280°C, from Osiris, for the same material with the corresponding results from DIDO. The longitudinal growth behavior of Material A in Trilium 3 is presented in Fig. 6. For Materials B and C, the longitudinal growth

TABLE 1—*Description of the chemical content and beta treatment of the Zr-2.5Nb pressure tube materials used in this study.*

| Material | Tube Type | Alloy Contents | | | | Beta Quench |
		Nb	O	N	Fe	
A	Darlington	2.5	1150	33	676	YES
B	Pt. Lepreau	2.6	1290	49	392	NO
C	Pickering A	2.8	1240	47	990	NO
D1	Pickering B	2.5	960	26	400	YES
D2		2.6	975	38	387	
E1	Bruce B	2.5	1120	34	1060	YES
E2	Pickering B	2.5	1105	42	1035	
F1	Pickering B	2.5	1120	37	437	NO
F2		2.6	1160	44	609	
G	Zircaloy	...	...	...	...	...

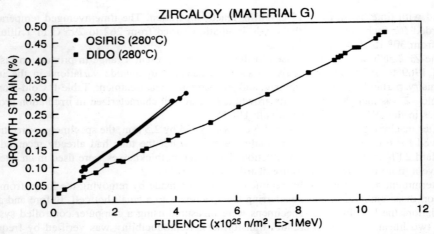

FIG. 3—*The growth characteristics of Zircaloy-2 (Material G) irradiated in Osiris and DIDO.*

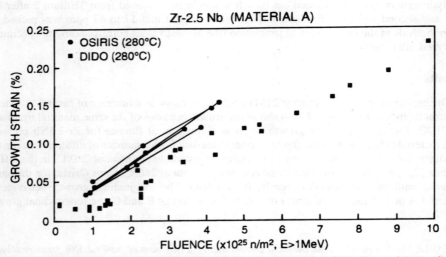

FIG. 4—*The irradiation growth, in the longitudinal direction, of Zr-2.5Nb (Material A) irradiated in Osiris (Trillium 2) and DIDO at 280°C. The DIDO data is from a number of different specimens.*

strain is shown as a function of fluence in Figs. 7 and 8, respectively. Figure 9 presents the growth data for Materials D and E representing the lowest (387, 400 ppm) and highest (1035, 1060 ppm) iron contents for tubes produced with a β-quench. For pressure tubes with low and intermediate iron contents (437, 609 ppm) produced without a beta quench, Material F, the growth results are shown in Fig. 10.

Discussion

Zircaloy-2 was included in this study to provide a comparison with previous results from DIDO where the same material showed the well-defined growth behavior of Fig. 3, with an

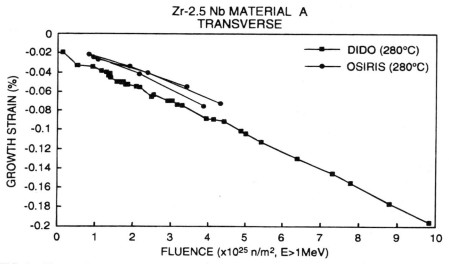

FIG. 5—*The irradiation growth, in the transverse direction, of Zr-2.5Nb (Material A) irradiated in Osiris (Trillium 2) and DIDO at 280°C. The DIDO data is from a number of different specimens.*

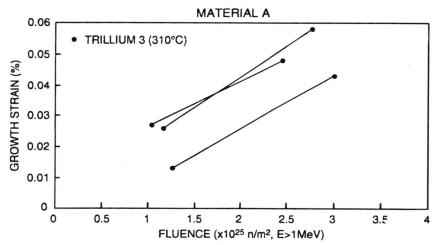

FIG. 6—*The longitudinal irradiation growth behavior of Material A irradiated in Trillium 3 in Osiris.*

initial transient and subsequent steady rate. Over the range of fluence achieved in Osiris, the material shows similar well-defined behavior, that is, linear growth extrapolating to a positive strain at zero fluence. However, the steady growth rate (per unit fluence) in Osiris is about 30% higher, suggesting that the combination of flux and spectrum in Osiris is more effective than in DIDO. Calculations based on the flux spectra suggest that the ratio of displacements to fluence in Osiris is about 10% higher than in DIDO that could account for about a third of the difference.

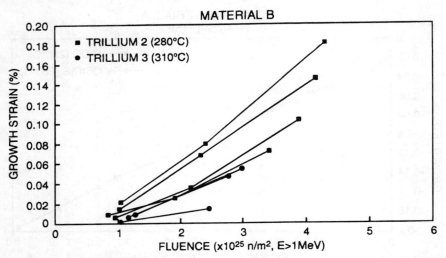

FIG. 7—*The longitudinal irradiation growth behavior of Material B irradiated in Osiris in the Trillium 2 and Trillium 3 inserts.*

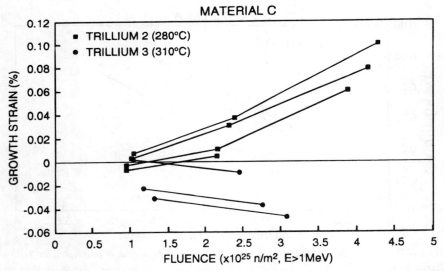

FIG. 8—*The longitudinal irradiation growth behavior of Material C irradiated in Osiris in the Trillium 2 and Trillium 3 inserts.*

Another possible contributing factor to the difference between the results from Osiris and from DIDO is the hydrogen absorbed in Osiris. This would increase the apparent growth via the volume expansion associated with hydrogen absorbtion. For the Osiris specimens, the pickup is in the 17 to 43 ppm range that corresponds to a relative volume expansion of 1.5 to 3.9×10^{-4} [6] and an isotropic linear strain of 0.5 to 1.3×10^{-4}. Since the hydrides in

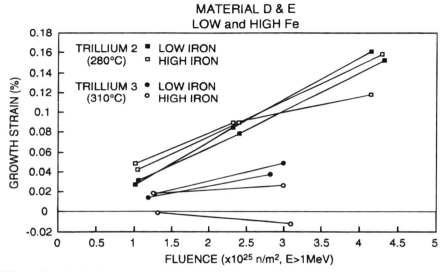

FIG. 9—*The longitudinal irradiation growth behavior of Materials D and E irradiated in the Trillium 2 and Trillium 3 inserts in Osiris.*

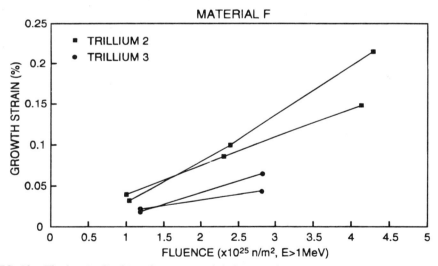

FIG. 10—*The longitudinal irradiation growth behavior of Material F irradiated in the Trillium 2 and Trillium 3 inserts in Osiris.*

these tubular materials form preferentially in the tangential plane, most of the strain would be expected to occur in the radial direction. Hence, the range just given represents an upper bound on the contribution of hydrogen to the longitudinal strain after the second phase of Trillium 2.

The growth of the Zr-2.5Nb pressure tube specimens exhibit similar effects, Figs. 4 and 5. For both longitudinal and transverse directions, the results from Osiris are higher (more positive) than those from DIDO. This reinforces the possibility of some volume expansion from the hydrogen pickup. However, the negative strains in the transverse direction are consistent with other studies of irradiation growth in Zr-2.5Nb, and indicate that the observed deformation is predominantly irradiation growth.

Allowing for a possible small contribution of hydrogen to the measured strains, the general trends in the longitudinal growth of the Zr-2.5Nb pressure tube materials are consistent with previous work, but demonstrate a wider variation than has previously been suspected.

The growth strains were not measured at low fluence but the data at 280°C are consistent with behavior reported earlier; an initial small positive strain, followed by a prolonged transient, up to 1×10^{25} n/m², $E > 1$ MeV, of low or slightly negative growth. The growth rate then increases towards a steady long-term rate. The minimum strain is negative in some cases (for example, Material C, Fig. 8), a result not observed in the narrower range of materials previously tested [2,3]. At 305°C, the growth strains and the growth rate are substantially lower over the range of 1 to 3×10^{25} n/m², $E > 1$ MeV, as observed previously [2,3]. Again, there is a wide range of behavior, with some materials exhibiting negative growth behavior only observed previously at higher temperature, or in material with a different fabrication route.

Since the long-term growth rate is of most interest technologically, a growth rate representative of a fluence of 2×10^{25} n/m², $E > 1$ MeV, was determined by averaging the rate onwards from 1×10^{25} n/m², $E > 1$ MeV. This may overestimate the rate slightly at 280°C, relative to 305°C, for Materials B and C because the growth curves are slightly concave up. These growth rates may also be lower than the long-term ones, particularly at 305°C where the growth rate is expected to increase at higher fluences. The time-averaged temperature and growth rate for each specimen are given in Table 2.

To isolate the possible effects of iron content and β-quenching on the growth of the pressure tube materials, it is necessary to eliminate the effect of temperature. The iron concentrations of the specimens fall into three clear groups, 387 to 437 ppm, 609 to 676 ppm, and 990 to 1060 ppm, each containing tubes made with and without the β-quench and each irradiated over a range of temperatures. Figure 11 shows the longitudinal growth rate for each group, plotted as a function of temperature, and the best fit line representing the decrease in growth rate with increasing temperature for each group. All three lines have similar temperature coefficients, -8×10^{-31} m²/n·°C, $E > 1$ MeV.

After accounting for the effect of temperature, there is a variation in growth rate of approximately $\pm 2 \times 10^{-29}$ m²/n, $E > 1$ MeV, about equivalent to the long-term variation in elongation rate of power reactor pressure tubes.

To determine the contributions of iron and β-quenching to this variation, the rate for each specimen was corrected to the average irradiation temperature, that is, 290°C, and the corrected rate plotted against iron content, or averaged within each group according to whether or not β-quenching had been used in manufacturing, Fig. 12 and Table 3. No effect of β-quenching is evident. However, the effect of iron is to reduce the growth rate, at least between the intermediate and high levels. The difference between the high iron and the two lower iron groups is significant at the 95% confidence level.

A study of the elongation rate of pressure tubes in Ontario Hydro's Pickering Unit 4 has revealed a correlation of iron concentration with elongation rate in tubes made from ingots from different manufacturers, Fig. 13. The magnitude of the effect (2.4×10^{-32} m²/n·ppm, $E > 1$ MeV) is very similar to that observed in our growth tests in Fig. 11 (3.4×10^{-32} m²/n·ppm, $E > 1$ MeV) suggesting that it is predominantly the contribution of growth to elongation that is affected by the iron concentration. This will have to be confirmed by higher fluence tests.

There are a number of possible mechanisms for the effect of iron on growth and elongation that are now under investigation.

TABLE 2—*The average temperature and growth rate of each specimen irradiated in the Trillium 2 and Trillium 3 inserts in the Osiris reactor.*

Material	Specimen	Insert	Temperature, °C	Strain Rate $\times 10^{29}$ m^2/n, $E > 1$ MeV
A	4A1	Trillium 2	283	3.4
	4A2	Trillium 2	288	3.1
	4A6	Trillium 2	293	3.3
	4A7	Trillium 2	286	2.9
	4AJ	Trillium 3	307	1.5
	4AK	Trillium 3	309	2
	4AL	Trillium 3	302	1.7
B	3P3	Trillium 2	280	4.2
	3P4	Trillium 2	262	4.9
	3P8	Trillium 2	277	3.3
	3P9	Trillium 2	293	2.4
	3PJ	Trillium 3	307	0.8
	3PK	Trillium 3	309	2.5
	3PL	Trillium 3	302	2.6
C	P1A	Trillium 2	280	2.4
	P1B	Trillium 2	262	2.9
	P1H	Trillium 2	277	2.1
	P1D	Trillium 3	303	−0.9
	P1J	Trillium 3	307	−0.8
	P1K	Trillium 3	309	−0.9
D	3S3	Trillium 2	283	3.7
	3T1	Trillium 2	280	4.3
	3S3	Trillium 3	302	1.8
	3T3	Trillium 3	304	1.4
E	3Q1	Trillium 2	262	3.6
	3R1	Trillium 2	280	2.2
	3Q3	Trillium 3	303	−0.6
	3R3	Trillium 3	302	0.5
F	3U1	Trillium 2	262	5.6
	3W5	Trillium 2	288	3.5
	3U3	Trillium 3	304	2.8
	3W7	Trillium 3	304	1.3
G	GVN	Trillium 2	283	6.4
	GVR	Trillium 2	288	6.1
	GVS	Trillium 2	295	
	GVU	Trillium 2	277	6.5

Initially, our interest in iron comes from results indicating that iron increases the vacancy diffusion rate in α-Zr [7]. Superficially, this suggests that the irradiation growth rate of α-Zr would increase with iron concentration. However, at this temperature, the intrinsic vacancies in zirconium are quite mobile, and the growth rate is expected to depend more on the anisotropy, than on the rate, of vacancy diffusion. The effect of iron on the anisotropy of diffusion has not been reported; depending on the sense of such an effect, it could alter the growth behavior in either direction.

One criticism of this concept is the low solubility of iron in α-Zr at these temperatures. Indeed, the iron is mostly contained in the β-phase and at α-α boundaries before irradiation [8,9]. However, about 40 ppm appears to be retained in the α-phase [10] before irradiation, and after about 1×10^{25} n/m^2, $E > 1$ MeV, most of the iron is dispersed from the β-phase

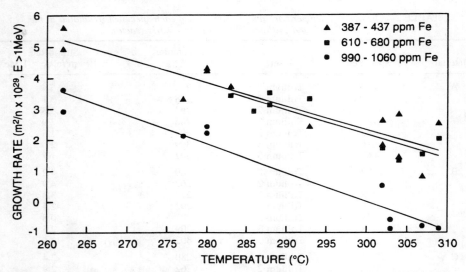

FIG. 11—*The temperature dependence of the longitudinal irradiation growth of Zr-2.5Nb with low (387 to 437 ppm), medium (610 to 670 ppm), and high (980 to 1080 ppm) iron contents.*

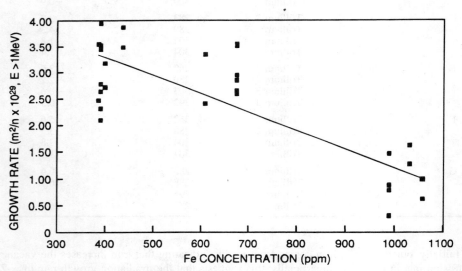

FIG. 12—*The irradiation growth of Zr-2.5Nb as a function of iron content with the growth rate normalized to the average irradiation temperature of 290°C.*

and appears to be uniformly distributed in the α-phase, with some concentration at the interfaces [*11,12*]. Thus, effects associated with iron in solution in the α-phase cannot be ruled out.

There are other possible direct effects of iron that do not require it to remain in solution. First, iron appears to stabilize vacancy defects (*a* and *c* loops) in zirconium alloys during irradiation [*13,14*]. In one theory for irradiation growth [*15*], there is a negative component

TABLE 3—*The average growth rate for pressure tube materials with different iron contents and the effect of the beta quench on the growth characteristics of these materials.*

Group	Iron Content, ppm	Average Growth Rate, 10^{-29} m²/n, $E > 1$ MeV		
		β-Quenched	Non β-Quenched	All
Low	387 to 437	2.97	3.12	3.07 ± 0.58^a
Medium	609 to 676	2.99	2.87	2.96 ± 0.52
High	990 to 1060	1.12	0.77	0.91 ± 0.42
Average		2.36	2.25	...

$^a \pm$ one standard deviation.

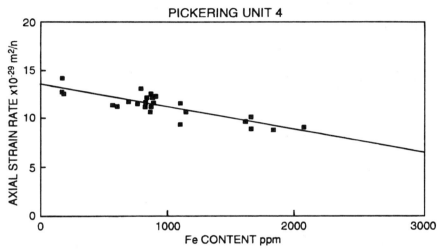

FIG. 13—*The elongation rate of pressure tubes in the Ontario Hydro CANDU reactor at Pickering Generating Station, Unit 4, as a function of iron content.*

in the axial direction, associated with the fine grain structure of Zr-2.5Nb pressure tubes and a consequent net flux of vacancies to *a* loops. Possibly, the amount of iron dispersed in the α-phase enhances this component and reduces the rate. Second, the iron may be precipitated on a very fine scale in the α-phase after irradiation, providing sites for vacancy/interstitial recombination, and, hence, suppressing growth.

There are also possible indirect effects of iron. Iron is a β-stabilizer and, hence, is expected to affect the phase distribution (α-Zr/β-Zr) during extrusion. This, in turn, could affect the texture development [*16*], grain size, and possibly the residual dislocation content after extrusion. All of these microstructural factors affect growth.

Conclusions

1. Irradiation growth of ten Zr-2.5Nb and one Zircaloy-2 pressure tube materials has been measured at temperatures between 262 and 309°C in a fast flux of 1.8×10^{18} n/m²·s, $E >$ 1MeV, to fluences of 4.3×10^{25} n/m², $E > 1$ MeV, in the Osiris reactor.

2. The growth behavior of two materials irradiated in both Osiris and DIDO are qualitatively similar, with somewhat higher strains (more positive) in Osiris. This is partly due to differences

in flux and flux spectrum, but there is probably also a small positive contribution to strain from the volume expansion caused by hydrogen picked up from the NaK in Osiris.

3. The growth rates at 2×10^{25} n/m^2, $E > 1$ MeV decrease with increasing temperature in the 262 to 309°C range with a coefficient of about -8×10^{-31} m^2/n°C, $E > 1$ MeV.

4. After correction for temperature differences, a variation of growth rate is found of similar magnitude to the tube-to-tube variation in elongation rate of pressure tubes in power reactors.

5. No effect on growth rate of β-quenching prior to extrusion is found for materials with three ranges of iron concentration.

6. The growth rates of materials with high iron contents (990 to 1060 ppm) are significantly lower than the growth rates of those with low and intermediate iron contents (387 to 437 and 609 to 676 ppm). A number of mechanisms by which the iron concentration could affect the growth rate are suggested.

7. The best-fit, straight line relating growth rate to iron concentration over the whole range of iron concentrations has a coefficient (3.4×10^{-32} m^2/n·ppm, $E > 1$ MeV) very similar to that of a plot of pressure tube elongation rate versus iron concentration for pressure tubes in Pickering Unit 4 (2.4×10^{-32} m^2/n·ppm, $E > 1$ MeV) suggesting that the effect of iron in the latter case is predominantly on the growth rate.

8. Higher fluence data are necessary to confirm these conclusions.

Acknowledgments

We would like to thank the staff of CEN Saclay, J. M. Cerles, G. Farny, M. Roche, F. Lefevre, M. Bauge, M. Thevenot, and Y. Bouilloux, for operation of the experiment and staff at OHRD and CRL for specimen preparation and measurement, M. Miller, I. Emmerton, and A. McMillan. V. Fidleris made a major contribution to the conception, design, and initiation of the experiments. This work was funded by the CANDU Owners Group.

References

[1] Fleck, R. G., Holt, R. A., Perovic, V., and Tadros, J., "Effects of Temperature and Neutron Fluence on Irradiation Growth of Zr-2.5 wt% Nb," *Journal of Nuclear Materials,* Vol. 159, 1988, pp. 75–85.

[2] Holt, R. A. and Fleck, R. G., "The Effect of Temperature on the Irradiation Growth of Cold-Worked Zr-2.5Nb," *Zirconium in the Nuclear Industry: Eighth International Symposium, ASTM STP 1023,* L. F. P. Van Swam and C. M. Eucken, Eds., American Society for Testing and Materials, Philadelphia, 1989, pp. 705–721.

[3] Holt, R. A. and Fleck, R. G., "The Contribution of Irradiation Growth to Pressure Tube Deformation," *Zirconium in the Nuclear Industry: Ninth International Symposium, ASTM STP 1132,* C. M. Eucken and A. M. Garde, Eds., American Society for Testing and Materials, Philadelphia, 1991, pp. 218–229.

[4] Causey, A. R., Elder, J. E., Holt, R. A., and Fleck, R. G., "On the Anisotropy of In-reactor Creep of Zr-2.5Nb Tubes," in this volume.

[5] Rogerson, A., "Irradiation Growth in Zirconium and Its Alloys," *Journal of Nuclear Materials,* Vol. 159, 1988, pp. 43–61.

[6] Carpenter, G. J. C., "Zirconium Hydride Precipitates in Zirconium," *Journal of Nuclear Materials,* Vol. 48, 1973, pp. 264–276.

[7] King, A. D., Hood, G. M., and Holt, R. A., "Fe-enhancement of Self-diffusion in α-Zr," *Journal of Nuclear Materials,* Vol. 185, 1991, pp. 174–181.

[8] Perovic, A., Perovic, V., Weatherly, G. C., Purdy, G. R., and Fleck, R. G., *Journal of Nuclear Materials,* Vol. 199, 1993, pp. 102–111.

[9] Griffiths, M., Phythian, W., and Dumbill, S., "Comment on Iron Distribution in Zr-2.5Nb Pressure Tube Alloy," *Journal of Nuclear Materials,* Vol. 207, 1993, pp. 353–356.

[10] Zou, H., Hood, G. M., Roy, J. A., Packwood, R., and Weatherall, V., "Solute Distribution in Annealed Zircaloy-2 and Zr-2.5Nb," *Journal of Nuclear Materials,* Vol. 208, 1994, pp. 159–165.

[11] Griffiths, M. and Holt, R. A., "Discussion on Zr-2.5Nb Alloy Pressure Tubing," *Zirconium in the Nuclear Industry: Ninth International Symposium, ASTM STP 1132,* C. M. Eucken and A. M. Garde, Eds., American Society for Testing and Materials, Philadelphia, 1991, pp. 172–175.

[12] Perovic, V., Perovic, A., Weatherly, G., Brown, L. M., Purdy, G. R., Fleck, R. G., and Holt, R. A., "Microstructure and Microchemical Studies of Zr-2.5Nb Pressure Tube Alloy," *Journal of Nuclear Materials,* Vol. 205, 1993, pp. 251–257.

[13] Griffiths, M., Gilbert, R. W., and Fidleris, V., "Accelerated Irradiation Growth of Zirconium Alloys," *Zirconium in the Nuclear Industry: Eighth International Symposium, ASTM STP 1023,* L. F. P. Van Swam and C. M. Eucken, Eds., American Society for Testing and Materials, Philadelphia, 1989, pp. 658–677.

[14] Griffiths, M., "Evolution of Microstructure in hcp Metals During Irradiation," *Journal of Nuclear Materials,* Vol. 205, 1993, pp. 225–241.

[15] Holt, R. A., "Mechanisms of Irradiation Growth of Alpha-Zirconium Alloys," *Journal of Nuclear Materials,* Vol. 159, 1988, pp. 310–338.

[16] Holt, R. A. and Aldridge, S. A., "Effect of Extrusion Variables on Crystallographic Texture of Zr-2.5 wt% Nb," *Journal of Nuclear Materials,* Vol. 135, 1985, pp. 246–259.

DISCUSSION

C. E. Coleman[1] *(written discussion)*—Have you identified the microstructural features that are responsible for the negative axial growth in some of the standard pressure tube materials?

R. G. Fleck et al. (authors' closure)—In earlier publications,[2,3] we have attributed the negative growth of a modified pressure tube material to a combination of the very fine elongated grain structure and differential anisotropic point defect diffusion (DAD). We believe that the same explanation applies (to a lesser degree) to some of the standard pressure tube materials irradiated in this study. The same model may explain the negative temperature dependence.

R. B. Adamson[4] *(written discussion)*—What is the form and location of iron in your Zr-2.5Nb alloy?

R. G. Fleck et al. (authors' closure)—Before irradiation, the iron is mostly in solution in the β-phase, but there are concentrations of iron at α-α boundaries, α-β interfaces, and on the surfaces of intermetallics.

After irradiation most of the iron from the β-phase is depleted and strongly segregated at the α/β interfaces, α-α boundaries, and intermetallics/matrix interfaces. Iron may also be dispersed in the α-matrix, but the level is below the detection limit of the high resolution X-ray spectroscopy.

We do not know the form of the iron at the interfaces or in the dispersed state.

---

[1] AECL Research, Chalk River Laboratories, Chalk River, Ontario, Canada.
[2] Holt, R. A., "Mechanisms of Irradiation Growth of Alpha-Zirconium Alloys," *Journal of Nuclear Materials*, Vol. 159, 1988, pp. 310–338.
[3] Holt, R. A. and Fleck, R. G., "The Effect of Temperature on the Irradiation Growth of Cold-Worked Zr-2.5Nb," *Zirconium in the Nuclear Industry: Eighth International Symposium, ASTM STP 1023*, L. F. P. Van Swam and C. M. Eucken, Eds., American Society for Testing and Materials, Philadelphia, 1989, pp. 705–721.
[4] GE Nuclear Energy, Pleasanton, CA.

Mitsutaka H. Koike,[1] Takashi Akiyama,[1] Kenji Nagamatsu, Itaru Shibahara[1]

Change of Mechanical Properties by Irradiation and Evaluation of the Heat-Treated Zr-2.5Nb Pressure Tube

REFERENCE: Koike, M. H., Akiyama, T., Nagamatsu, K., and Shibahara, I., "**Change of Mechanical Properties by Irradiation and Evaluation of the Heat-Treated Zr-2.5Nb Pressure Tube,**" *Zirconium in the Nuclear Industry: Tenth International Symposium, ASTM STP 1245,* A. M. Garde and E. R. Bradley, Eds., American Society for Testing and Materials, Philadelphia, 1994, pp. 183–201.

ABSTRACT: The prototype advanced thermal reactor, Fugen (165MWe), is a heavy-water-moderated, boiling-light-water-cooled, pressure tube type reactor. A Zr-2.5 Nb alloy, in the heat-treated condition to improve its high-temperature strength, is used for the pressure tubes in the Fugen. Pressure tube surveillance specimens were assembled in the inside of the special fuel assemblies from the initial stage of the operation and were exposed to irradiation. The post-irradiation examinations of the first and second surveillance specimens were performed.
The tensile strength increased by 30% when fast neutron fluence was $>2 \times 10^{25}$ n/m^2. The decreasing rate of the fracture toughness was large over the initial period of the irradiation and became smaller when fast neutron fluence was $>2 \times 10^{25}$ n/m^2. The corrosion rate was measured as -1 μm/year and the hydrogen pickup rate was measured as 1 ppm/year. Based on the results, a predictive capability for each respective property for 30 years was established, which demonstrated the integrity of the pressure tube for 30 years.

KEY WORDS: zirconium, zirconium alloys, nuclear materials, nuclear applications, radiation effects, irradiation, pressure tubes, surveillance specimens, post-irradiation examination, tensile strength, fracture toughness, corrosion, hydrogen pickup

The prototype advanced thermal reactor, Fugen (power output 165 MWe), is a heavy-water-moderated, boiling-light-water-cooled, pressure tube type reactor that entered commercial operation in March 1979 and has since performed satisfactorily [1].

Zirconium alloys are widely used as material for the pressure tubes because they possess high strength and have a low absorption rate of neutrons. A Zr-2.5Nb alloy that has been heat-treated (HT) in order to enhance its high temperature strength was chosen for the Fugen.

The fuel assemblies are placed inside the pressure tubes, and the pressure tubes are consequently exposed to a high rate of neutron irradiation. Therefore, evaluating the characteristic change of the material under irradiation is important. For this purpose, pressure tube surveillance specimens were assembled in the inside of the special fuel assemblies from the initial stage of operation and were exposed to irradiation. The first surveillance specimens were taken out

---

[1] Power Reactor and Nuclear Fuel Development Corporation, Oarai Engineering Center, 4002, Oarai, Higashi-Ibaraki, Ibaraki, 311-13, Japan.
[2] Power Reactor and Nuclear Fuel Development Corporation, Fugen Nuclear Power Station, 3, Myojin, Tsuruga, Fukui, 914, Japan.

in February 1984 for post-irradiation examinations. These specimens were transported from the Fugen to the Material Monitoring Facility in the Oarai Engineering Center in September 1984 and post-irradiation examinations that included tension tests, bend tests, burst tests, corrosion tests, hydrogen analysis, and metallographical examination were carried out [2]. In 1989, the second surveillance test was performed.

In this paper, the irradiation examination project of the pressure tubes of the Fugen, irradiation in the Fugen, and the results and evaluation of the post-irradiation examination of the first and the second surveillance specimens are described.

A concept drawing of the Fugen is shown in Fig. 1. One fuel assembly consisting of a bundle of 28 fuel rods is placed in each pressure tube. The pressure tube is approximately 5 m in length with an inner diameter of 117.8 mm and a wall thickness of 4.3 mm, which is joined to stainless steel pressure tube extensions at the top and bottom by rolled joints.

The pressure tubes are made from molten zirconium sponge to which 2.5% of niobium is added that is then cast into ingots and forged into round bar billets. The billets are formed into tubes and heat treated. Figure 2 illustrates the manufacturing processes from the billets to the final forming. The billets are hot-extruded, cold-drawn, solution-heat-treated/water-quenched, cold-drawn, and aged before the final forming stage. The distinctive feature of the

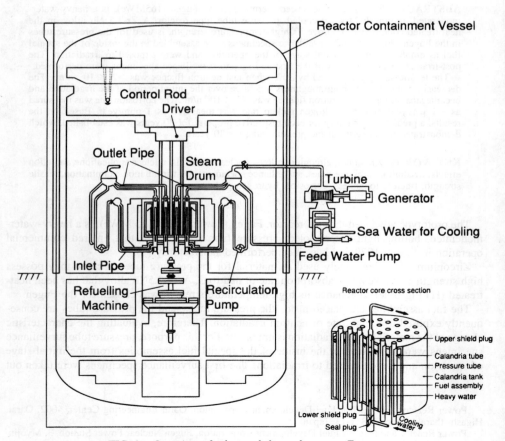

FIG. 1—*Overview of advanced thermal reactor, Fugen.*

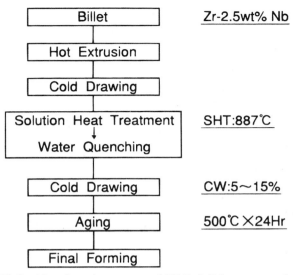

FIG. 2—*Manufacturing process of HT Zr-2.5Nb pressure tubes.*

Zr-2.5Nb pressure tubes used in the Fugen is that the solution heat treatment/water quenching process has resulted in improved strength.

Experimental Procedure

Before commencing the commercial operation of the Fugen, pressure tube specimens were irradiated at JMTR (Japan Material Testing Reactor) of Japan and SGHWR (Steam Generating Heavy Water Reactor) of England. Irradiation specimens for precursive irradiation consisted of tension specimens, fracture toughness specimens (small bend test specimen), burst test specimens, and corrosion specimens that are important in confirming the changes in the material characteristics with irradiation. These specimens are also used in the Fugen surveillance examination and by the precursive irradiation examination, the examination methods were established.

The irradiation examinations were started in 1970 with the loading and irradiation of the specimens at JMTR of the Japan Atomic Energy Research Institute and SGHWR, and the examinations were completed in 1976. The irradiation temperature and the fast neutron flux of JMTR and SGHWR were approximately equal to those of the Fugen.

The surveillance examination of the Fugen is performed in accordance with the stipulations as established by the Ministry of International Trade and Industry of Japan in the Ministerial Ordinance 62 and Notification 501 as well as in accordance with the Surveillance Examination Procedure of Nuclear Reactor Structural Material of JEAC 4201 of the Electrical Technical Regulation of the Japan Electrical Society. The types of surveillance specimens and the time when such surveillance specimens are to be taken out are established with consideration of the characteristics of the Fugen advanced thermal reactor.

Regarding the surveillance specimens, the bend test specimen is used in place of an impact test specimen as specified in Notification 501 as a procedure for evaluating the fracture toughness characteristics, with consideration for the characteristics of the pressure tube material. The as-received pressure tube material possesses adequate ductility at room temperature.

However, if sufficient hydrogen is absorbed by its material, there is the potential that it will become brittle at room temperature. This characteristic was taken into consideration, and the bend specimen was selected to evaluate the toughness of the pressure tube.

The reasons for selecting the bend test specimen to evaluate the toughness are as follows:

1. The surveillance specimens are placed in a capsule that is put into a special fuel assembly and loaded into the nuclear reactor to be irradiated. It is therefore necessary that the surveillance specimens are made as compact as possible.
2. The fracture toughness value obtained from the bend test agrees very well with fracture toughness values obtained by the standard burst test of full cross sections from actual pressure tubes as shown in Fig. 3 at a hydrogen concentration of >50 ppm. At lower hydrogen levels, the bend test gives a conservative assessment of the fracture toughness.

The bend tests consist of a bending fracture test and a bending fatigue crack propagation test that are performed by the fracture mechanics test equipment. The bend tests as well as the tension test are performed by the fracture mechanics test equipment that contains a furnace, clip gages, and a television camera. The bending fracture test determines the fracture toughness value, and the bending fatigue crack propagation test determines the crack propagation rate. The specimen is set with both ends on the jig inclined at 14.5° from the load axis to avoid out-of-plane bending at the notch area.

The bending fracture test is performed to obtain the fracture toughness value. Prior to performing the bending fracture test, the specimen is pre-cracked by fatigue. The conditions for pre-cracking and the loading rate are according to ASTM Test Method for Plane-Strain Fracture Toughness of Metallic Materials (E 399–90).

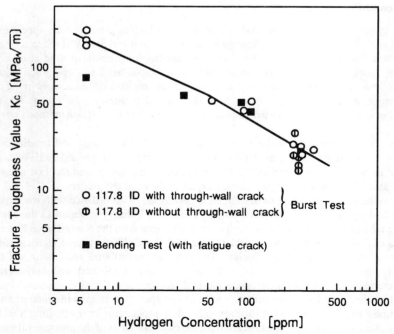

FIG. 3—*Fracture toughness values as a function of hydrogen concentration.*

The equation by Brown [3] for the three-point bending of flat plate having a crack on one side is used to calculate the stress intensity factor, K, for the bending specimen

$$K = \sigma\sqrt{\pi a} = \sigma_n\sqrt{\pi a}[1.93 - 3.07(a/H) \qquad (1)$$
$$+ 14.53(a/H)^2 - 25.11(a/H)^3 + 25.80(a/H)^4]$$

$$\sigma_n = 3PL/\{2T(H - a)^2 \cdot \cos\theta\} \qquad (2)$$

where

P = load,
L = distance between support points (200 mm),
H = height of specimen (32 mm),
a = length of crack (mm),
T = plate thickness (4.3 mm),
θ = inclination angle (14.5°), and
σ = stress.

Pressure tube surveillance examinations are performed on the surveillance specimens, shown in Fig. 4.

The surveillance specimens are assembled in a hollow cylindrical capsule and then irradiated in a special fuel assembly. The plan for removing the surveillance specimens calls for removal at two to four year intervals for a total of eight times during the reactor life (30 years) as shown in Table 1. This removing plan will be reconsidered in accordance with the Fugen operation records and the surveillance examination results.

The capsules containing the surveillance specimens are assembled into the special fuel assembly, and irradiated inside the nuclear reactor. The capsules made from HT Zr-2.5Nb alloy are assembled with surveillance specimens as illustrated in Fig. 5. The capsules also contain copper and iron flux monitors for measuring fast neutron fluence in addition to the surveillance specimens. The special fuel assembly is provided with a guide tube so that the capsule can be inserted into the fuel bundle center, and fuel elements are arranged in two circumferential layers. There are nine capsules assembled into one special fuel assembly, and as shown in Fig. 6, there are four special fuel assemblies loaded in rotational symmetry at the intermediate portion of the reactor core.

Results and Discussion

Tension Test

Shown in Fig. 7 are the results of the tension test (according to the JIS Z2241 testing method) performed at 300°C that is the approximate operating temperature of the pressure tubes of the Fugen (285°C). The changes in the ultimate tensile stress, 0.2% proof stress, reduction of area, and elongation resulting from irradiation can be seen.

The ultimate tensile stress and the 0.2% proof stress of irradiated material show values about 30% higher than those of unirradiated material. Elongation shows a small decrease with irradiation. In Fig. 7, "circumferential" and "axial" refer to specimens taken in the circumferential direction and axial direction, respectively, on the pressure tube. The circumferential specimen was taken after the pressure tube had been made into a flat plate. At room temperature, the tensile properties show the same tendency as at 300°C.

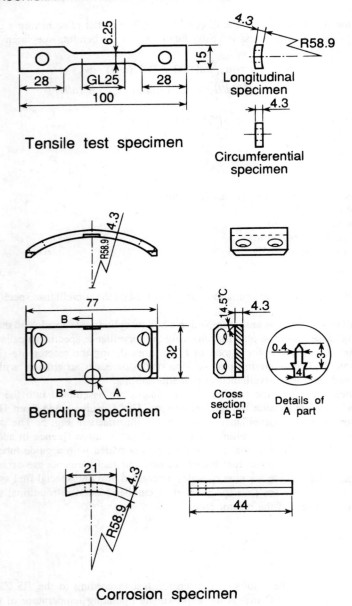

FIG. 4—*Fugen pressure tube surveillance specimens (unit:mm).*

The 0.2% proof stress and the ultimate tensile stress increased with irradiation, and therefore after the irradiation these values satisfied the allowable stress intensity of $Sm = 18.3$ kg/mm^2. The allowable stress intensity was determined as the lower value of one third of the ultimate tensile stress and two thirds of yield stress of the unirradiated material at room temperature and 300°C.

TABLE 1—*Pressure tube surveillance test program.*

	Surveillance Test Number							
	1	2	3	4	5	6	7	8
Capsule schedule (number of years after commencing commercial operation)	4	8	10	12	14	17	21	25
Number of capsules to be removed	2	2	2	2	2	2	2	2

The ultimate tensile stress increases during the initial period of irradiation and then becomes almost constant. The tensile properties approach asymptotic values at fluences about 2×10^{25} n/m², which is in agreement with CANDU reactors data [4,5]. The generation of dislocation loops is known to be the main source of hardening in irradiated zirconium alloys [5]. The ultimate tensile stress at 300°C, σ_{uts}, is predicted by extrapolating the results up to 30 years operation by the relationship

$$\sigma_{uts} = \frac{-298}{t_Y + 1.17} + 843 \tag{3}$$

where σ_{uts} is expressed in MPa and t_Y is the operating year with a reactor loading factor of 75%.

Bend Test

Fracture toughness values obtained from the bending fracture test are shown in Fig. 8. The value of fracture toughness shows the opposite tendency to tensile strength and decreases with irradiation, which is in agreement with CANDU reactors data for cold-worked Zr-2.5Nb pressure tubes [4]. During the initial period of irradiation, the decrease in room temperature fracture toughness is high, but the rate of decrease became smaller at fast neutron fluences $>1 \times 10^{25}$ n/m². The decrease in fracture toughness at 300°C is small. From the post-irradiation examination results to date, it is considered that the decrease of the fracture toughness value at operating temperature is small.

Because the hydrogen pickup is very small during the initial period of irradiation, the initial decrease of fracture toughness is considered to be caused by irradiation. The hydrogen concentration of the pressure tube after 30 years irradiation is estimated as 44 ppm [6], which is below the solubility limit of 47 to 60 ppm [7,8] at the operating temperature of 285°C. Fracture toughness at the operating temperature is considered to approach asymptotic values at fluences above 2×10^{25} n/m². At room temperature, hydrides precipitate in the pressure tube and the fracture toughness decreases according to the results as shown in Fig. 3. Considering the preceding effects, the fracture toughness is predicted by extrapolating the results up to 30 years operation by the relationship

$$K_{c(R.T.)} = \frac{52.7}{t_Y + 1.39} + 31 \tag{4}$$

$$K_{c(300°C)} = \frac{20.3}{t_Y + 1.27} + 58.9 \tag{5}$$

where K_c is expressed in MPa$\sqrt{m}$.

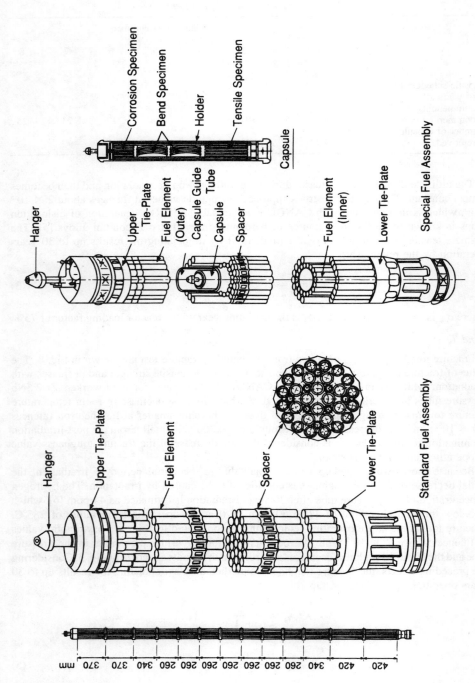

FIG. 5—*Structure of fuel assemblies.*

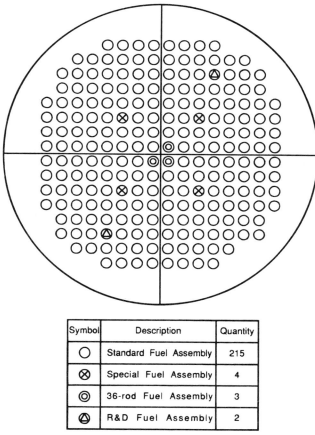

FIG. 6—*Special fuel loading position in the Fugen core (1989).*

In the bending fatigue test, the load and the length of the crack were measured. The load was continuously measured, and the length of the crack was intermittently measured by using a television camera. The intermittent measurement was performed when the propagation of the crack reached about 0.5 mm.

One example of bending fatigue crack propagation at 300°C is shown in Fig. 9. The ordinate is the crack propagation rate, da/dN (da = increment of crack length and dN = increment of fatigue cycle), and the abscissa is the stress intensity factor increment ΔK. The best-fit curve shown in Fig. 9 for unirradiated material used for the Fugen design is as follows

$$da/dN = 3.2 \times 10^{-10} \Delta K^{2.5} \tag{6}$$

At 300°C, the crack propagation rate for small flaws (when ΔK is small) is similar to that of the best-fit curve of unirradiated data, but at larger flaws (ΔK becomes larger), the propagation rate of the irradiated specimens is less than that of the unirradiated specimens. Therefore, a favorable result of the crack propagation rate decreasing with irradiation has been obtained. This tendency is the same with room temperature data.

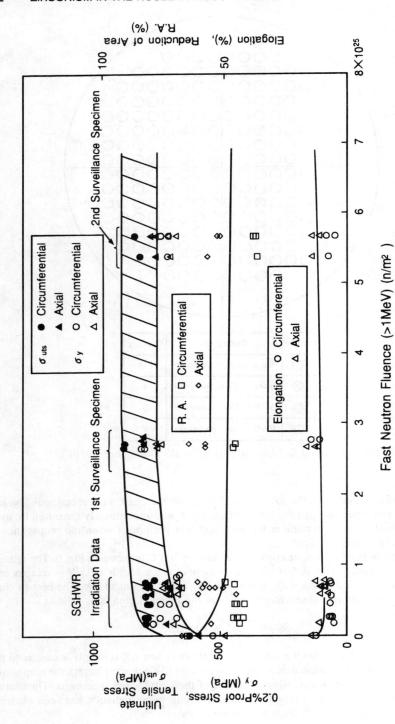

FIG. 7—Tension test results of pressure tube specimen at 300°C.

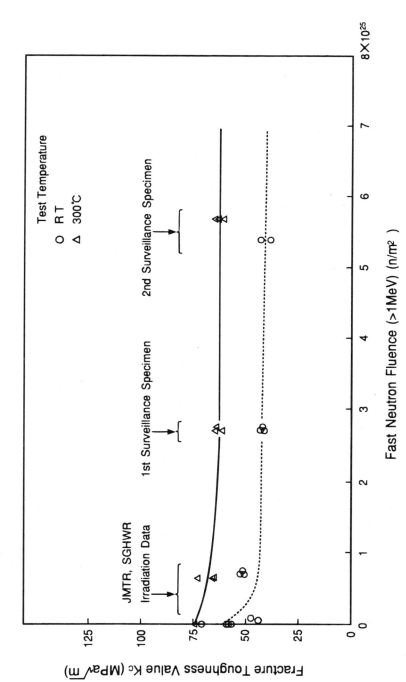

FIG. 8—*Fracture toughness test results of pressure tube specimen.*

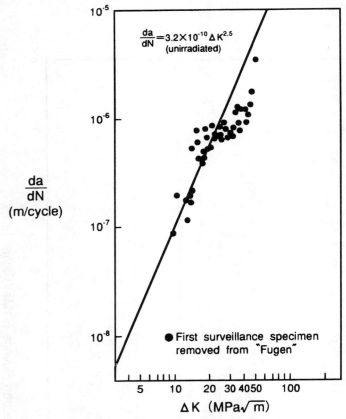

FIG. 9—*Results of bending fatigue test (300°C)*.

The evaluation of the fracture toughness of pressure tubes is made by confirming that the toughness value (critical stress intensity factor) is adequate greater than the stress intensity factor calculated on the basis of operating stress for a hypothetical crack grown from a initial flaw that can be detected by the pressure tube flaw inspection (in-service inspection).

Because the crack propagation rate of the material after irradiation is less than that of the unirradiated material, the assumed flaw dimension calculated on the basis of unirradiated data is used. The stress intensity factor during operation for an assumed flaw dimension (length = 6.3 mm, depth = 1.05 mm after 30 years, with some conservative estimate) gained by utilizing unirradiated data is 6.3 MPa$\sqrt{m}$, and the stress intensity factor during hydrostatic pressure test is 7.2 MPa$\sqrt{m}$. At this time, it is necessary that the following conditions are satisfied:

1. the stress intensity factor during operation $\times$ 2 < K_c, and
2. the stress intensity factor during hydrostatic pressure test $\times$ 1.5 < K_c.

From the present results, these conditions are satisfied with sufficient margins.

In addition, the critical crack length (CCL) for unstable fracture for a through-wall flaw and the critical crack depth for a surface crack of infinite length were estimated, and it was confirmed that there was sufficient allowance against fracture.

Corrosion Test

The corrosion test involves measuring the weight of the specimen to obtain the reduction in the wall thickness (depth of corrosion), and a comparison between the reduction in the wall thickness of the actual operational pressure tube and the design value is performed. The corrosion rate is estimated by the weight gain of the specimen before and after irradiation.

Zirconium will react with oxygen and become ZrO_2. Under the assumption that the oxide does not spall, the thickness of the corrosion film, T_f, and depth of corrosion, T_c, can be calculated by the following equations

$$T_f = 0.692 \times \Delta W/A \text{ (cm)} \qquad (7)$$

$$T_c = 0.434 \times \Delta W/A \text{ (cm)} \qquad (8)$$

where

ΔW = weight gain by corrosion (g), and
A = surface area (cm^2).

From this equation, T_c is obtained and assumed to be the thickness loss by corrosion (reduction in wall thickness).

Figure 10 shows the corrosion test results. The ordinate is the thickness loss by corrosion, and the abscissa is the effective full power days (EFPD) that is almost proportional to the fast neutron fluence. The thickness loss by corrosion is extremely small in comparison with the design value of 310 μm/30 years. The corrosion rate is very small, which is in agreement with CANDU reactors data for cold-worked Zr-2.5Nb pressure tubes [9,10].

The weight gain by corrosion as well as the thickness loss by corrosion is simply expressed with $\sqrt{t}$, where t is time. Considering the small amount of corrosion [6] and a decrease in corrosion rate during irradiation observed for Zr-2.5Nb [11], the thickness loss by corrosion, $C(\mu m)$, is predicted by extrapolating the results up to 30 years operation by the relationship

$$C = 5\sqrt{t_Y} \qquad (9)$$

Hydrogen Concentration Analysis

Hydrogen analysis was performed by cutting off hydrogen analysis specimens from the area around the fractured surfaces after testing the tension specimens, bend specimens, and burst specimens, and the absorbed hydrogen analyzed. A calibration curve had been prepared in advance using a standard sample containing a fixed quantity of hydrogen. The hydrogen analysis of the specimens was performed using a hydrogen analyzing equipment that contains a glove box, a heating tube, a quadrupole mass analyzer, and a vacuum exhauster. The accuracy of hydrogen analysis is about ±2 ppm.

If the pressure tube picks up hydrogen from the reactor water during operation, it becomes brittle as shown in Fig. 3. It is therefore important to perform hydrogen analysis of the specimens.

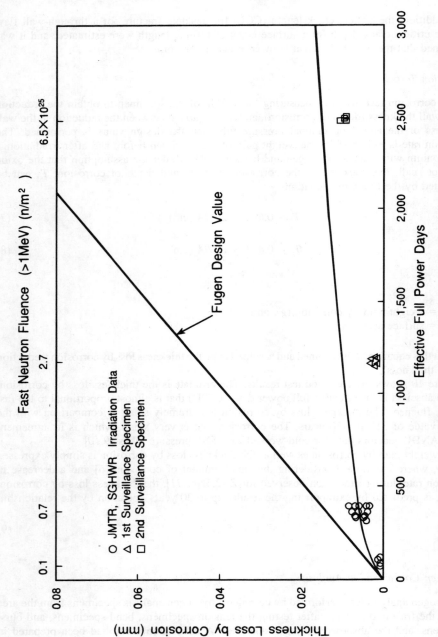

FIG. 10—*Corrosion test results of pressure tube specimen.*

Figure 11 shows the results of hydrogen concentration as a function of EFPD. There is some increase in the hydrogen concentration although the amount of increase is very small, which is in agreement with CANDU reactors data [6,9,10]. The hydrogen pickup rate is closely related with the corrosion rate [12], so that hydrogen concentration, H (ppm) is predicted with the $\sqrt{t}$ relationship by extrapolating the results up to 30 years operation by the relationship

$$H = 5.2\sqrt{t_Y} + 15 \tag{10}$$

According to this formula, hydrogen concentration after 30 years operation is estimated as 44 ppm, which is below the solubility limit of 47 to 60 ppm at the reactor operating temperature of 285°C. Therefore, the degradation of the pressure tube is very small at the operating temperature.

Metallographic Examination

The metallographic examination involves cutting off samples from the specimens that have undergone various tests and etching the surfaces of the samples. The microstructure, primary α-phase, and hydrides, were observed. As an example, a micrograph of a specimen is shown in Fig. 12.

In Fig. 12a, the acicular martensite structure can be observed. In Fig. 12b, the white parts in the photograph are the primary α-phases that are spheroidal. There is no prominent difference between the surface and the bulk of the specimen, and there is an even distribution over the entire specimen. There was also no noticeable difference in hydrides between the surface and the bulk of the specimen (Fig. 12c). In the three cases, there was no prominent difference when compared with the unirradiated material.

Conclusions

On the basis of the results for HT Zr-2.5Nb pressure tube surveillance specimens, the following conclusions are drawn:

1. The ultimate tensile stress and the 0.2% proof stress of irradiated material showed values of about 30% higher during the initial period of irradiation in comparison with those of the unirradiated material, and became almost constant at fluences $>2 \times 10^{25}$ n/m².

2. The fracture toughness at 300°C of small bend specimens showed values of about 10 to 20% lower at the initial period of irradiation in comparison with that of the unirradiated material, and became almost constant at fluences $>2 \times 10^{25}$ n/m².

3. The thickness loss by corrosion as well as the weight gain by corrosion is very small, 1 μm/year, and is simply expressed as $\sqrt{t}$.

4. The hydrogen pickup rate is very small, 1 ppm/year, and is simply expressed as $\sqrt{t}$.

5. Based on present results, the equations were developed to extrapolate the data up to 30 years. The extrapolated values indicated that the pressure tube integrity for 30 years will be maintained.

Acknowledgments

The authors are grateful to members of JMTR (Japan) and SGHWR (Great Britian) for performing a part of the tests.

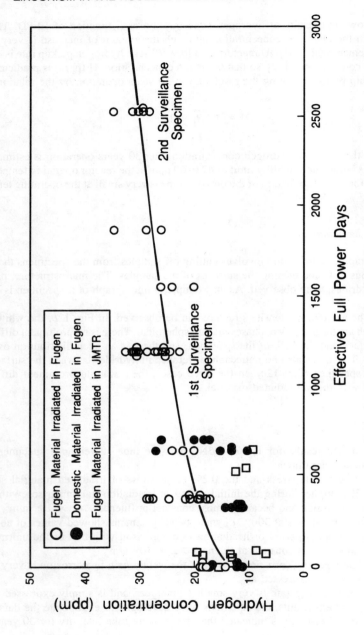

FIG. 11—*Hydrogen analysis results of pressure tube specimen.*

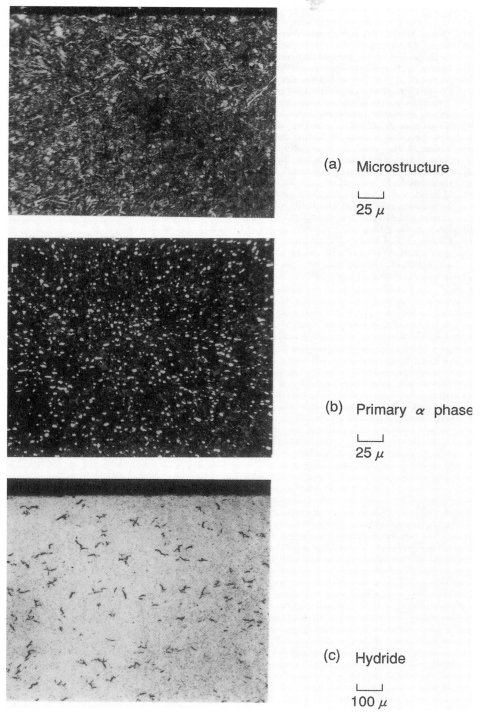

FIG. 12—*Metallographic photographs of an irradiated specimen (fluence 2.7×10^{25} n/m^2).*

References

[1] "Fugen HWR," *Nuclear Engineering International*, Vol. 24, No. 289, 1979, p. 33.
[2] Koike, M., Asada, T., Yuhara, S., and Kaneko, J. in *Collection of Summaries of Lectures*, 949th Japan Society of Mechanical Engineers Material Mechanics (in Japanese), Tokyo, 1987, p. 32.
[3] Brown, W. F. and Srawley, J. E. in *Plane Strain Crack Toughness Testing of High Strength Metallic Materials, ASTM STP 410*, American Society for Testing and Materials, Philadelphia, 1967, p. 77.
[4] Chow, C. K., Coleman, C. E., Hosbons, R. R., Davies, P. H., Griffiths, M., and Choubey, R. in *Zirconium in the Nuclear Industry: Ninth International Symposium, ASTM STP 1132*, C. M. Eucken and A. M. Garde, Eds., American Society for Testing and Materials, Philadelphia, 1991, p. 246.
[5] Cheadle, B. A., Ells, C. E., and van der Kuur, J., *Zirconium in Nuclear Applications, ASTM STP 551*, American Society for Testing and Materials, Philadelphia, 1974, p. 370.
[6] Koike, M. H., Onose, S., Nagamatsu, K., and Kawajiri, M. in *Proceedings*, The 1st JSME/ASME Joint International Conference on Nuclear Engineering, 1991, p. 533.
[7] Slattery, G. F., *Journal of Institute Metals*, Vol. 95, 1967, p. 43.
[8] Coleman, C. E. and Ambler, J. F. R., AECL-6467, Atomic Energy of Canada, Ltd., 1978.
[9] Urbanic, V. F., Cox, B., and Field, G. J., *Zirconium in the Nuclear Industry: Seventh International Symposium, ASTM STP 939*, R. B. Adamson and L. F. P. Van Swan, Eds., American Society for Testing and Materials, Philadelphia, 1987, p. 189.
[10] Price, E. G., AECL-8755, Atomic Energy of Canada, Ltd., March 1985.
[11] Urbanic, V. F., Choubey, R., and Chow, C. K., *Zirconium in the Nuclear Industry: Ninth International Symposium, ASTM STP 1132*, C. M. Eucken and A. M. Garde, Eds., American Society for Testing and Materials, Philadelphia, 1991, p. 665.
[12] Stehle, H., Kaden, W., and Manzel, R., *Nuclear Engineering and Design*, Vol. 33, 1975, p. 155.

DISCUSSION

B. D. Warr[1] (written discussion)—(1) In your plot of hydrogen content versus time, the rate of hydrogen pickup appears greater than 1 ppm hydrogen per year especially at <1500 EFPD. What is the initial hydrogen content of the samples and can you subtract them for the figure?

(2) If hydrogen pickup is related to corrosion, how can you be sure that the corrosion rate will not increase, thus increasing the hydrogen pickup rate before 30 years operation?

(3) Are you maintaining sufficiently oxidizing conditions in the annulus gas to minimize hydrogen pickup from that source?

M. H. Koike et al. (authors' closure)—(1) The average value of the initial hydrogen content is 15 ppm as shown in Fig. 11 and Eq 10. The initial hydrogen content is different from sample to sample. The specification of the hydrogen content is <25 ppm at the manufacturing stage of the pressure tube.

(2) We have data for the NPD reactor HT Zr-2.5Nb pressure tube (PNC-AECL collaboration work, unpublished). The operating time was 124 300 h that corresponded to about 20 years with a reactor loading factor of 75%. From the data, the hydrogen pickup rate (using the hydrogen equivalent concentration) was very small, 1 ppm/year. Considering these results as well as the decrease in corrosion rate during irradiation by Urbanic [*11*], our results were extrapolated up to 30 years.

(3) Yes, we are. The annulus gas of the Fugen reactor is dry CO_2 gas with 0.1% air and 100 ppm H_2O. In our corrosion experiment (10 000-h test) under the preceding condition, no hydrogen increase was observed for HT Zr-2.5Nb pressure tube materials.

D. Khatamian[2] (written discussion)—In your final stages of the production of your specimens, you state that the alloy was aged. What was the temperature and length of time for the aging process?

M. H. Koike et al. (authors' closure)—For the aging process, the temperature was 500°C and the length of time was 24 h as shown in Fig. 2.

Brian Cox[3] (written discussion)—The S-shaped nature of your fatigue crack growth curve suggests that you may have simultaneous mechanisms operating. Do you have a comparable curve obtained in vacuum? If there is an additive corrosion fatigue component in air, perhaps growth and cracking of the oxide film is providing the other component of the curve.

M. H. Koike et al. (authors' closure)—No, we do not have a comparable curve in vacuum for the irradiated material. For the unirradiated materials, we have fatigue crack growth curves in air at room temperature and 300°C, with hydrogen concentrations of 5 to 250 ppm. However, these curves do not have the S-shaped nature. The S-shaped nature of the irradiated material would be caused by the irradiation hardening of the material.

---

[1] Ontario Hydro, Toronto, Ontario, Canada M8Z 5S4.
[2] AECL Research, Chalk River Laboratories, Chalk River, Ontario, Canada.
[3] University of Toronto, Toronto, Canada.

Allan R. Causey,[1] *Joanne E. Elder,*[2] *Richard A. Holt,*[1] *and Ronald G. Fleck*[2]

On the Anisotropy of In-Reactor Creep of Zr-2.5Nb Tubes

REFERENCE: Causey, A. R., Elder, J. E., Holt, R. A., and Fleck, R. G., **"On the Anisotropy of In-Reactor Creep of Zr-2.5Nb Tubes,"** *Zirconium in the Nuclear Industry: Tenth International Symposium, ASTM STP 1245,* A. M. Garde and E. R. Bradley, Eds., American Society for Testing and Materials, Philadelphia, 1994, pp. 202–220.

ABSTRACT: Creep specimens made from cold-worked Zr-2.5Nb tubes, fabricated with two different microstructures and crystallographic textures, were irradiated in the Osiris reactor in France in a fast neutron flux of about 1.8×10^{18} n · m^{-2} · s^{-1}, $E > 1$ MeV, at 553 and 585 K. The hoop stresses from internal pressurization range from 0 to 160 MPa. The basal poles were oriented in the radial-transverse plane of the tubes, either about 30° from the radial direction or primarily in the transverse direction; these are similar to the crystallographic textures in fuel sheathing and in CANDU reactor pressure tubes. Dislocation densities were similar while grain size and shape differ significantly.

The creep specimens were irradiated to fast neutron fluences up to 7×10^{25} n · m^{-2}. Creep rates were obtained from strain versus fluence plots and creep compliances were obtained from plots of the strain rates against hoop stress for each material at each temperature. The ratio of creep rates at the nominal temperature of 583 K to those at 553 K was similar to that derived from stress relaxation results at temperatures between 523 and 568 K.

The ratio of the biaxial creep compliance in the axial direction to that in the transverse directions is different for the two test materials; 0.0 to -0.1 for the fuel sheathing texture and 0.6 to 0.7 for the pressure tube texture. The results were analyzed using a self-consistent model developed previously to account for the contributions to the creep anisotropy of the three microstructure parameters involved here and for the grain interaction effects. The model, which was normalized to other test reactor and power reactor creep data for cold-worked Zr-2.5Nb tubes, predicted the ratio of the creep compliances to be -0.26 and 0.63, respectively. Thus, in agreement with previous conclusions, the creep anisotropy of Zr-2.5Nb tubes with pressure-tube-like crystallographic texture can be adequately predicted.

KEY WORDS: zirconium, zirconium alloys, nuclear industry, creep (materials), nuclear reactors, texture (materials), tubes, nuclear materials, nuclear applications, radiation effects

The anisotropy of creep deformation of Zr-2.5Nb pressure tubes during service in CANDU reactors is related to the anisotropic physical properties of the hexagonal crystal structure of zirconium. These physical properties contribute to the development during fabrication of an anisotropic microstructure, including crystallographic textures, grain morphologies, and dislocation structures. It has been established that in-reactor deformation of zirconium alloys is the result of the combined contributions from irradiation growth, (that is, strain in the

---

[1] Senior scientist and manager, respectively; Reactor Materials Research Branch; Atomic Energy of Canada Ltd., Chalk River Laboratories, Chalk River, Ontario, Canada K0J 1J0.

[2] Engineer and unit manager, respectively, Materials Technology Unit, Ontario Hydro Technologies, Toronto, Ontario, Canada M8Z 5S4.

absence of applied stress) and creep, both thermal and irradiation induced. These strain components have strong directional anisotropy [1–5].

Several studies have attempted to relate the anisotropic deformation of the polycrystalline zirconium alloys to those of their individual grains by accounting for the microstructural features, particularly the crystallographic texture [6–8]. Application of these models to Zr-2.5Nb pressure tubes has suffered from a lack of experimental data from biaxial creep tests on materials with crystallographic texture similar to that of the pressure tubes. The experiment reported here contributes to the development of a reliable model for Zr-2.5Nb tubes by using two batches of small tubes, one of which has a crystallographic texture similar to that of the power reactor pressure tubing, while the other has a texture that is completely different. The results are analyzed in terms of the texture using a self-consistent model to account for the effects of grain interactions [8–10].

The irradiations are being conducted in the high flux Osiris reactor at Saclay, France, primarily to assess the effects of temperature and high neutron fluence on the deformation of Zr-2.5Nb. The length and diameter measurements of biaxial creep specimens reported in this paper are extended to fluences of $\approx 7 \times 10^{25}$ n · m^{-2}, $E > 1$ MeV. The previous highest fluences in creep tests that could be used to assess the creep anisotropy of Zr-2.5Nb tubes were less than 1.8×10^{25} n · m^{-2} [11]. The lead CANDU power reactors have reached fluences in excess of 11×10^{25} n · m^{-2}, while reactor end-of-life fluences are projected to be about 30×10^{25} n · m^{-2}. The small range of stress in power reactor pressure tubes precludes using that data alone for these studies.

Experimental

Test Materials

The pressurized creep capsules were fabricated from two different 10-mm-diameter Zr-2.5Nb tubes referred to as fuel sheathing (FS) and micro pressure tube (MPT). The manufacturing procedures were as shown in Fig. 1. The fabrication procedure for the micro pressure tube was similar to that used for standard CANDU pressure tubes. The crystallographic texture and microstructure, including grain size and shape, α- and β-phase structure, and dislocation densities, were determined by X-ray diffraction and electron microscopy replica techniques.

Typical microstructures for the FS, MPT, and a standard CANDU pressure tube viewed along the longitudinal and radial directions are shown in Figs. 2 through 4. The fuel sheathing in Fig. 2a shows bands of α-phase and β-phase elongated in the direction of extrusion, with the β-phase showing wider bands than the typical CANDU pressure tube, Fig. 3a. The β-phase, when viewed along the axial direction, Fig. 2b, is discontinuous and lacks the characteristic swirl pattern seen in pressure tubing, Fig. 3b. In contrast, the MPT has a grain structure similar to that of standard pressure tubes, the main difference being one of grain size, Fig. 4.

Grain sizes were measured, based on distances between intercept points of β-phase and a reference line. The average grain width in the radial direction was measured on sections with circumferential normals, and in the circumferential direction on sections with longitudinal normals. The measured widths for the FS were 0.8 and 1.1 μm, respectively. In the MPT, they were measured to be about 0.4 and 5.0 μm. In standard pressure tubing, these widths would be about 0.3 and 4.0 μm.

The $\langle a \rangle$ and $\langle c \rangle$ type dislocation densities estimated from the broadening of $(10\bar{1}0)$ and (0002) X-ray diffraction peaks [12] of the test materials are compared with those of typical pressure tubing in Table 1. The measured dislocation densities are within the range typically measured in standard cold-worked pressure tubes. The crystallographic textures of the tubes are described by the basal plane pole figures shown in Fig. 5. The resolved fraction of basal plane normals in the three principal directions of the tubing are also given in the table.

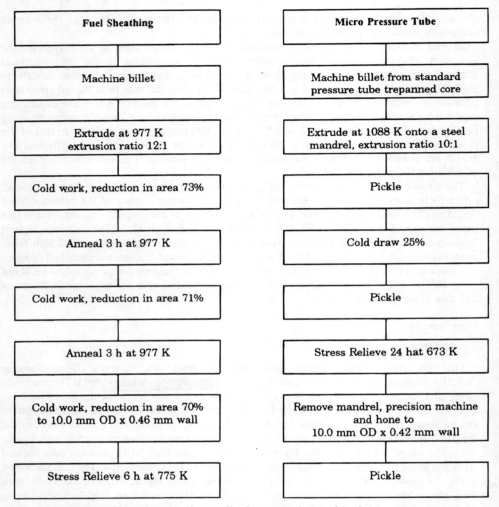

FIG. 1—*Fabrication route for small tube materials tested in Osiris reactor.*

Specimens

Figure 6 shows a photograph of a biaxial creep specimen with a gage length of 46.4 mm, outside diameter of 10.0 mm before pressurization, and a wall thickness of 0.45 mm. End-caps, machined from Zr-2.5Nb rods, were electron-beam welded to the tubes and the specimens were stress relieved 24 h at 673 K to relieve residual stresses near the welds. The total stress relief at 673 K, including 24 h during tube fabrication, was 48 h. The capsules were pressurized using high purity helium gas and sealed by tungsten inert gas welding.

Before pressurization and sealing of the capsules, their length and outside diameter were measured using a laser telemetric system. The wall-thickness of the tubes was measured using a dial gage. The internal pressure for each specimen was read from a gas gage and also estimated from the elastic deformation measured from the change in outside diameter after

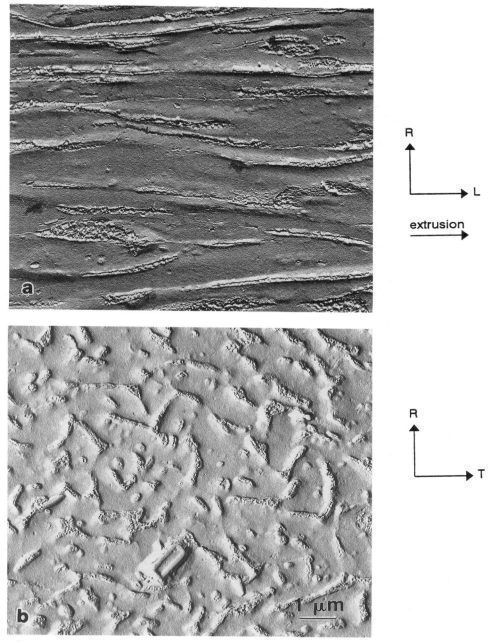

FIG. 2—*Electron replica micrograph of fuel sheathing material;* (a) *surface with circumferential normals and* (b) *surface with longitudinal (axial) normals.*

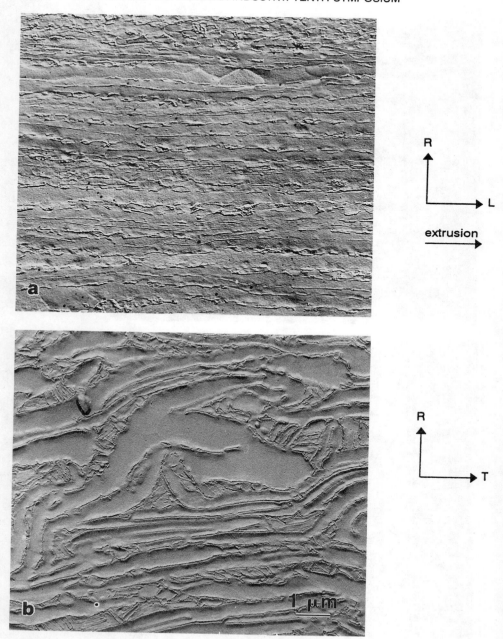

FIG. 3—*Electron replica micrograph of standard CANDU pressure tube material;* (a) *surface with circumferential normals and* (b) *surface with longitudinal (axial) normals.*

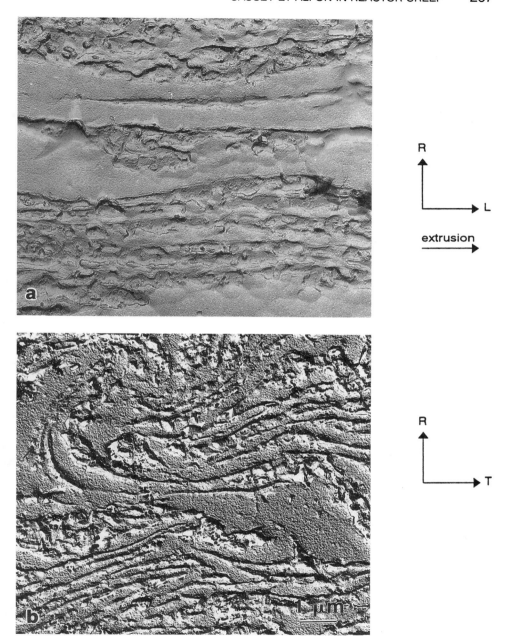

FIG. 4—*Electron replica micrograph of micro pressure tube material;* (a) *surface with circumferential normals and* (b) *surface with longitudinal (axial) normals.*

TABLE 1—*Dislocation densities and resolved basal pole fractions for Zr-2.5Nb materials.*

Material	Dislocation Density, $m^{-2} \times 10^{14}$		Resolved Basal Pole Fractions		
	$\langle a \rangle$	$\langle c \rangle$	f_R	f_T	f_L
Fuel sheathing, 10 mm OD	2.3	0.6	0.57	0.37	0.07
Micro pressure tube	3.4	0.7	0.34	0.57	0.09
Small pressure tube [16]	3.2	0.5	0.37	0.52	0.11
Fuel sheathing, 16 mm OD [11]	2.6	0.3	0.65	0.28	0.07
CANDU pressure tubes	3.0 ± 0.5	0.6 ± 0.2	0.32	0.62	0.06

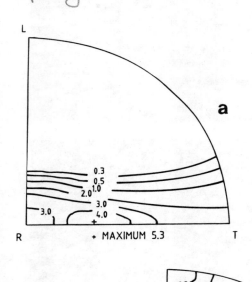

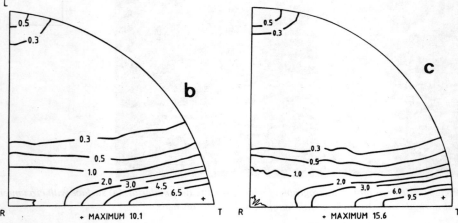

FIG. 5—*Basal pole figures for* (a) *fuel sheathing,* (b) *micro pressure tube, and* (c) *standard CANDU pressure tubes.*

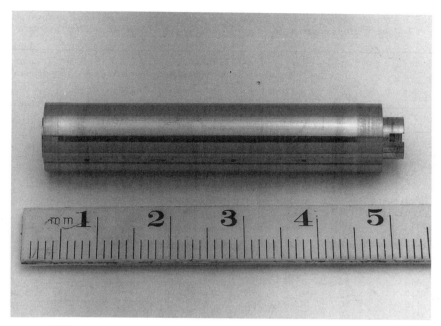

FIG. 6—*Photograph of an internally-pressurized creep test specimen.*

the capsules were pressurized. The specimen hoop stress at the nominal temperatures of the creep experiments (553 and 583 K) were calculated from the internal pressure at room temperature assuming that the helium gas behaves as an ideal gas. The capsules were pressurized at room temperature to give hoop stresses at the test temperature in the range 0 to 160 MPa (Table 2).

Before and after irradiation the specimens were measured under water in the reactor pool at a temperature of 313 K. The gaging apparatus used a linear variable differential transformer and was controlled by a computer. The diameters were measured at six positions along the capsules and the reported diameter was the average of the measurements at the three middle axial positions, each measurement being the average of 200 readings around the perimeter of the capsule. The length was obtained by averaging 200 measurements made around the specimen perimeter. A standard specimen was measured before and after each irradiated specimen, and the readings normalized to 313 K. This measurizing procedure was repeated two times on each specimen. The error in the measurement of the diameter and length of the capsules was estimated (95% confidence level) as ±2 and ±5 μm, respectively. In terms of strains, the errors are $\pm 2 \times 10^{-4}$ in diameter and $\pm 1 \times 10^{-4}$ in length.

Test Procedure

The pressurized capsules were irradiated in two NaK-filled inserts in the Osiris test reactor at Saclay, France. The reactor inserts are described in more detail in Ref *13*. The inserts were irradiated at nominal temperatures of 553 and 583 K, in experiments referred to as Trillium-2 (553 K) and Trillium-3 (583 K). The peak fast neutron flux ($E > 1$ MeV) was determined from iron wire flux monitors, to be about $1.8 \pm 0.1 \times 10^{18}$ n · m^{-2} · s^{-1} ($E > 1$ MeV). Fast

TABLE 2—*Reactor specimen loading for the four irradiation phases of Trillium-2 and two phases of Trillium-3; listing materials, specimen identity, hoop stress (MPa), and time averaged temperature (K) during each phase.*

Capsule	Stress, MPa	Trillium 2 Temperatures, K				Trillium 3 Temperatures, K	
		Phase 1	Phase 2	Phase 3	Phase 4	Phase 1	Phase 2
FS - A	0	563	554	554	554	...	...
FS - B	0	537	537	...	...	...	...
FS - C	41	558	565	...	...	...	...
FS - F	119	552	552	...	...	...	...
FS - H	78	547	541	...	...	...	...
FS - J	146	551	560	...	...	...	...
FS - K	147	551	552	551	554	...	...
FS - L16	0	...	...	...	...	583	572
FS - L17	155	...	...	...	...	580	580
MPT - M1	40	...	...	...	...	585	578
MPT - M2	160	...	...	...	...	577	576
MPT - M3	0	...	...	540	541	...	...
MPT - M4	0	...	...	...	...	582	584
MPT - M13	43	...	...	563	566	...	...
MPT - M14	82	...	...	553	565	...	...
MPT - M15	120	...	...	...	...	583	582
MPT - M16	80	...	...	...	...	589	585
MPT - M17	160	...	...	558	560	...	...
MPT - M18	122	...	...	561	567	...	...

neutron fluences of up to 6.7×10^{25} and 3.2×10^{25} n · m^{-2} ($E > 1$ MeV) were reached in Trillium-2 and Trillium-3, respectively.

The test specimen stresses and loadings for each of the four irradiation phases in Trillium-2 and the two irradiation phases in Trillium-3 are given in Table 2. The initial loading of Trillium-2 was all FS specimens; after the second phase, five of these specimens were replaced with MPT specimens, with Specimens B and K (0 and 147 MPa) continuing. The initial loading of Trillium-3 contained two FS and five MPT specimens. The temperatures were measured and controlled with thermocouples in contact with the creep specimens. The time averaged temperature of each specimen during each phase of irradiation are listed in Table 2.

Results

Strain Measurements on Fuel Sheathing

The measured transverse and axial creep strains are plotted as functions of neutron fluence in Figs. 7a and b for Trillium-2. Extrapolation of the transverse curves back to zero fluence gives intercepts within ±0.017% of the mean of 0.012%, while extrapolation of the axial curves gives positive intercepts within ±0.008% of the mean of 0.031%. The transverse strains in the specimens with no stress are negative while the axial strains are larger than those in the stressed specimens. The creep strains measured for the FS specimens in Trillium-3 are shown in Fig. 8, where the pattern appears to be similar to that of the Trillium-2 results. Strain rates have been derived for the second irradiation phases of both inserts.

The effect of stress on the transverse and axial strain rates are shown in Fig. 9. The transverse strain rates increase, while the axial strain rates decrease, with stress. The slopes of the best fit lines through the data points give creep compliances for biaxial stressing of 1.9×10^{-30}

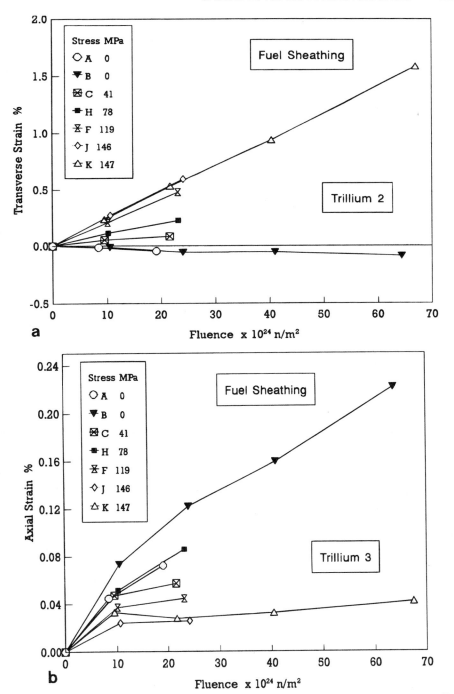

FIG. 7—*Creep strain as a function of neutron fluence for fuel sheathing specimens irradiated in Trillium-2 (nominal temperature ≈ 553 K); (a) transverse strain and (b) axial strain.*

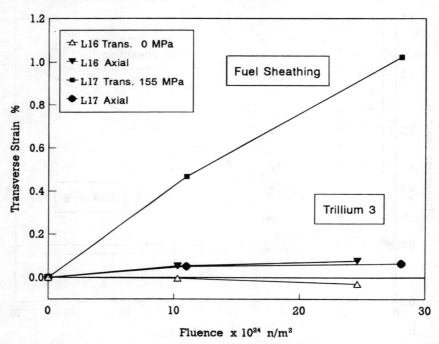

FIG. 8—*Creep strain as a function of neutron fluence for fuel sheathing specimens irradiated in Trillium-3 (nominal temperature $\approx$ 583 K).*

and -0.2×10^{-30} m$^2 \cdot$ n$^{-1} \cdot$ MPa^{-1}, in the transverse and axial directions, respectively, with a ratio of axial to transverse compliances, C_{BL}/C_{BT}, of -0.10. The zero fluence intercepts are the growth rates in transverse and axial directions, -3.9×10^{-29} and 2.9×10^{-29} m$^2 \cdot$ n^{-1}, respectively. The effect of stress on the strain rates in Trillium-3, also shown in Fig. 9, give the biaxial creep compliances of 2.2×10^{-30} and -0.3×10^{-31} m$^2 \cdot$ n$^{-1} \cdot$ MPa^{-1}, in the transverse and axial directions, respectively. The ratio, C_{BL}/C_{BT}, is approximately -0.0.

Strain Measurements on Micro Pressure Tube

The creep strains measured for the MPT specimens in Trillium-2 are shown in Figs. 10a and b, and for Trillium-3 are shown in Figs. 11a and b. The effects of stress on the strain rates derived for the second irradiation phase in each insert are shown in Fig. 12. The transverse and axial biaxial creep compliances are 0.92×10^{-30} and 0.55×10^{-30} m$^2 \cdot$ n$^{-1} \cdot$ MPa^{-1}, in Trillium-2 and 1.07×10^{-30} and 0.76×10^{-30} m$^2 \cdot$ n$^{-1} \cdot$ MPa^{-1}, in Trillium-3. The ratios, C_{BL}/C_{BT}, are approximately 0.6 and 0.7, respectively. The significant differences in the creep strains measured for the MPT specimens compared with the FS specimens are large positive axial strains and lower transverse strains in both the Trillium-2 and Trillium-3 inserts.

Discussion

The primary object of doing these in-reactor creep tests is to provide creep data at high neutron fluences at different temperatures on test materials with similar texture and microstructure to that of the cold-worked Zr-2.5Nb pressure tubes in CANDU power reactors. The MPT speci-

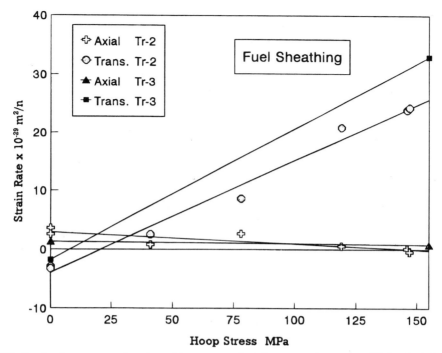

FIG. 9—*Strain rate as a function of hoop stress (axial stress = hoop stress/2) for fuel sheathing specimens.*

mens used in these experiments meet these requirements. The tests on the FS material provide a significantly different crystallographic texture on which to base modeling of the texture effects on the deformation.

The effect of temperature on the creep compliances is shown in Table 3 where the average temperature of the specimens in the two inserts are given. The ratio of the rates for the averaged temperatures of the transverse specimens in Trillium 2 and Trillium 3 averages 1.17, while the axial MPT specimens give 1.38. The value for the axial measurements of the fuel sheathing is ignored because of the uncertainty in the small values. The temperature dependence of creep has been determined previously from stress relaxation tests carried out over the range 523 to 568 K, to vary as $\dot{\epsilon} \propto e^{-Q_c/T}$, with an activation temperature, Q_c, of 2200 K [*14*]. The ratio for the transverse specimens found here is similar to that predicted from the stress relaxation data and somewhat different for the axial MPT specimens.

The growth rates in the transverse and axial directions obtained from the strain rate versus stress plots, and from the 0 stress specimens, are given in Table 4. There is good agreement between the extrapolated rates and those measured on 0 stress specimens. Comparison of the rates in Trillium 2 and Trillium 3 for MPT specimens indicates that the magnitudes of both the axial and transverse growth rates tend to decrease with increasing temperature. This is consistent with the effect of temperature on growth reported in Ref *13*.

The higher magnitudes of the transverse creep compliances in the FS materials compared with the MPT material, Table 3, clearly indicate that, while pressure tubes with radial texture may be desirable to reduce delayed hydride cracking [*15*] and axial elongation, these tubes

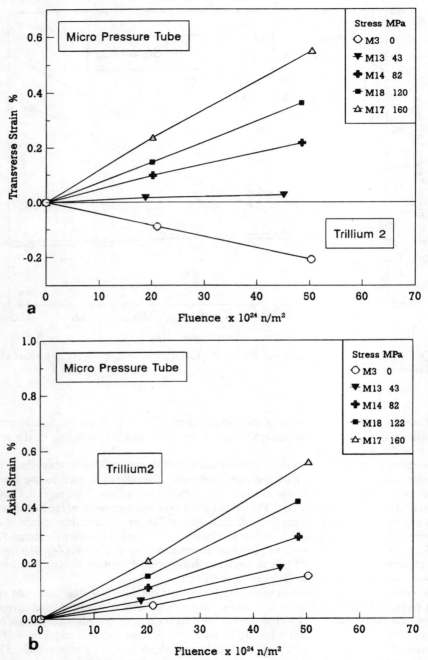

FIG. 10—*Creep strain as a function of neutron fluence for micro pressure tube specimens irradiated in Trillium-2 (nominal temperature ≈ 553 K); (a) transverse strain and (b) axial strain.*

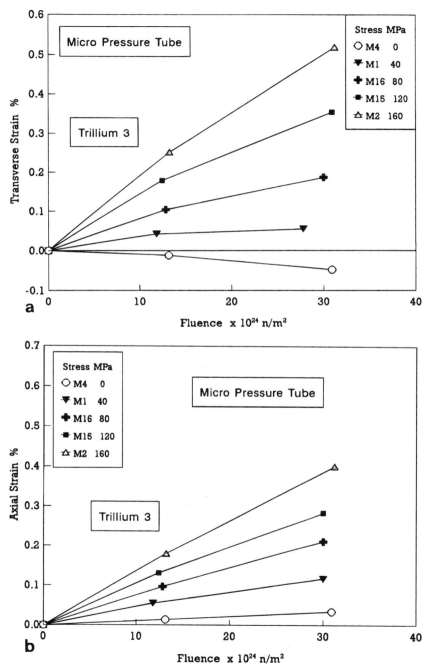

FIG. 11—*Creep strain as a function of neutron fluence for micro pressure tube specimens irradiated in Trillium-3 (nominal temperature ≈ 583 K);* (a) *transverse strain and* (b) *axial strain.*

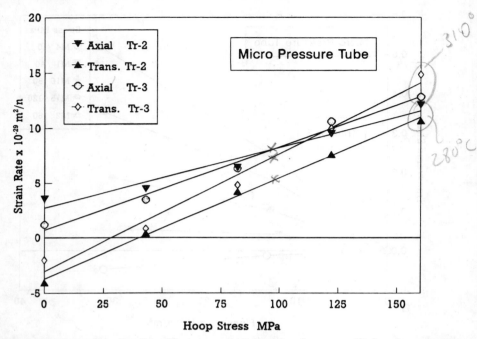

FIG. 12—*Strain rate as a function of hoop stress (axial stress = hoop stress/2) for micro pressure tube specimens.*

TABLE 3—*Effect of temperature on the biaxial creep compliances compared with the effects predicted from stress relaxation data where $\dot{\epsilon} \propto e^{-2200/T}$ [14].*

Material Orientation	Trillium 2		Trillium 3		$\dot{\epsilon}_{Trillium\ 3}/\dot{\epsilon}_{Trillium\ 2}$	
	$\dot{\epsilon}$, m²/n/MPa	Average Temperature, K	$\dot{\epsilon}$, m²/n/MPa	Average Temperature, K	Measured Ratio	Predicted Ratio
MPT axial	0.55×10^{-30}	560	0.76×10^{-30}	581	1.38	1.51
MPT transverse	0.92×10^{-30}	560	1.07×10^{-30}	581	1.16	1.15
FS axial	-0.20×10^{-30}	551	-0.03×10^{-30}	576	not valid	1.19
FS transverse	1.91×10^{-30}	551	2.24×10^{-30}	576	1.17	1.19

will have significantly higher transverse strain rates if they have the same microstructure as current tubes.

A self-consistent model has been developed by Christodoulou et al. [*10*] to calculate polycrystalline creep anisotropy from crystallographic texture and single crystal creep compliances derived from tests on polycrystals. The analysis used biaxial and shear creep compliances obtained from small pressure tube material reported in Ref *16* and normalized to measurements from the power reactors [*17*]. The single crystal compliances derived from the polycrystalline data in Ref *10* corresponded to the relative strain contributions of prism, basal, and pyramidal

TABLE 4—*Effect of temperature on the growth rates of creep specimens (in units of 10^{-29} m^2/n).*

Material Orientation	$\dot{\epsilon}_{Trillium\ 2}$		$\dot{\epsilon}_{Trillium\ 3}$	
	0 MPa Specimens	Extrapolation of $\dot{\epsilon}$ versus σ	0 MPa Specimens	Extrapolation of $\dot{\epsilon}$ versus σ
MPT axial	3.6	2.7	1.2	0.7
MPT transverse	−4.0	−3.7	−2.0	−3.0
FS axial	3.1	2.9	5.2	NA[a]
FS transverse	−3.1	−3.9	−0.4	NA

[a] NA = not available.

slip of 78, 20, and 2%, respectively. The model was then used to calculate the dependence of C_{BL}/C_{BT} on the texture parameter $(f_T - f_R)$ as shown in Fig. 13. The model is in good agreement with the measured values of C_{BL}/C_{BT} for the materials with $f_T - f_R$ values greater than 0 (that is, materials with pressure-tube-like microstructures) and poorer for values less than 0 (that is, fuel sheathing materials with microstructures different from that of a pressure tube [11]).

The C_{BL}/C_{BT} values determined for the materials tested in Osiris are shown in Fig. 13 and show the same relationship to the model as the previous results from the lower fluence tests, the MPT results falling close to the line and the FS results falling well above the prediction.

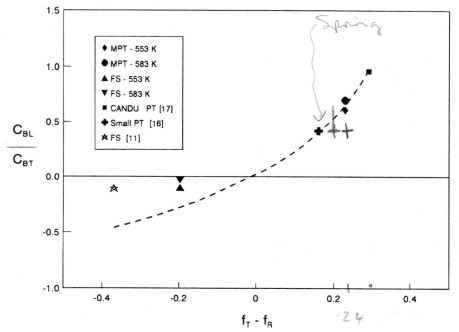

FIG. 13—*Comparison of measured biaxial creep anisotropy ratios, C_{BL}/C_{BT}, with values calculated using a self-consistent creep model (dotted line), as function of texture parameter, $f_R - f_T$, (the resolved basal pole fractions, f_d, are given in Table 1) showing good agreement for textures with $f_R > 0$ and poor agreement < 0.*

Thus, the conclusions made by Christodoulou et al. [10], that the texture alone does not accurately predict the creep behavior of all cold-worked Zr-2.5Nb materials and that a better quantitative evaluation of the microstructure is necessary, appears to be quite general and not a peculiarity of the materials, low fluence, or temperature. However, the model appears to be suitable for predicting creep anisotropy for materials with pressure-tube-like microstructures (the MPT results of the current tests).

Conclusions

Conclusions drawn from the results of the in-reactor creep experiments on cold-worked Zr-2.5Nb tubing presented in this report are as follows:

1. The axial and transverse strains of internally pressurized capsules showed a linear dependence on hoop stresses between 0 and 160 MPa for both fuel sheathing and micro pressure tube materials. The biaxial creep anisotropy ratios, C_{BL}/C_{BT}, were positive for the micro pressure tube (0.6 and 0.7) and negative for the fuel sheathing (-0.0 and -0.1), at the nominal temperatures of 553 and 583 K, respectively. These ratios confirm other results attained at much lower fluences.

2. The temperature dependence of the creep compliances are consistent with previous results and indicate an activation temperature in the Arrhenius relationship of about 2200 K in the temperature range of this experiment.

3. The contribution of creep strain to the total axial strain during irradiation is greater in specimens with CANDU pressure tube type of crystallographic texture than in specimens with fuel sheathing textures.

4. The magnitude of the transverse creep compliances in materials with fuel sheathing textures is higher than that in materials with pressure tube texture.

5. Comparison of the measured anisotropy ratios with a self-consistent model confirm previous conclusions that the creep rate cannot be entirely predicted from texture variation and that microstructural differences among the materials must be considered.

Acknowledgments

We acknowledge F. J. Butcher and S. A. Donohue for preparing the test specimens; M. Griffiths and L. G. Laurin for providing microstructural data; M. LeFevre, M. Roche, M. Mason, G. Thevenot, S. Jeggat, and Y. Bouilloux of CEN, Saclay for irradiating the specimens in the Osiris reactor and measuring the creep strains; and the CANDU Owners Group (COG) for funding the work.

References

[1] Fidleris, V., *Journal of Nuclear Materials,* Vol. 159, 1988, p. 22.
[2] Holt, R. A. and Fleck, R. G., "The Contribution of Irradiation Growth to Pressure Tube Deformation," *Zirconium in the Nuclear Industry: Ninth International Symposium, ASTM STP 1132,* C. M. Eucken and A. M. Garde, Eds., American Society for Testing and Materials, Philadelphia, 1990, p. 218.
[3] Holt, R. A., Causey, A. R., and Fidleris, V., "Dimensional Stability and Mechanical Behaviour of Irradiated Metals and Alloys," *Proceedings,* British Nuclear Society, London, 1983, p. 175.
[4] Holt, R. A., *Journal of Nuclear Materials,* Vol. 159, 1988, p. 310.
[5] Causey, A. R., Holt, R. A., and MacEwen, S. R. in *Zirconium in the Nuclear Industry: Sixth International Symposium, ASTM STP 824,* D. G. Franklin and R. B. Adamson, Eds., American Society for Testing and Materials, Philadelphia, 1984, pp. 269–288.
[6] Holt, R. A. and Ibrahim, E. F., *Acta Metallurgica,* Vol. 27, 1979, p. 1319.
[7] Holt, R. A. and Causey, A. R., *Journal of Nuclear Materials,* Vol. 150, 1987, p. 306.
[8] Woo, C. H., *Journal of Nuclear Materials,* Vol. 131, 1985, p. 105.
[9] Tomé, C. N., So, C. B., and Woo, C. H., submitted to *Philosophical Magazine,* Vol. 67, 1993, p. 917.

- [10] Christodoulou, N., Causey, A. R., Woo, C. H., Tomé, C. N., Klassen, R. J. and Holt, R. A., "Modeling the Effect of Texture and Dislocation Structure on Irradiation Creep of Zirconium Alloys," *Effects of Radiation on Materials, ASTM STP 1175*, 1994, pp. 1111–1128.
- [11] Ibrahim, E. F., *Journal of Nuclear Materials,* Vol. 102, 1981, p. 214.
- [12] Griffiths, M., Winegar, J. E., Mecke, J. E., and Holt, R. A., *Advances in X-ray Analysis,* Vol. 35, 1992, pp. 593–599.
- [13] Elder, J. E., Causey, A. R., Holt, R. A. and Fleck, R. G., in this volume.
- [14] Causey, A. R., Butcher, F. J., and Donohue, S. A., *Journal of Nuclear Materials,* Vol. 159, 1988, pp. 101–113.
- [15] Coleman, C. E., Sagat, S., and Amouzouvi, K. F., "Control Of Microstructure To Increase The Tolerance Of Zirconium Alloys To Hydride Cracking," AECL Report, AECL-9524, Chalk River, Ontario, Canada, Dec. 1987.
- [16] Causey, A. R. and Klassen, R. J., "On The Irradiation-Enhanced Creep Of Cold-Worked Zr-2.5Nb Tubes," AECL Report, AECL-10756, Chalk River, Ontario, Canada, Dec. 1992.
- [17] Causey, A. R., Fidleris, V., MacEwen, S. R., and Schulte, C. W., in *Influence of Radiation on Material Properties: 13th International Symposium (Part II), ASTM STP 956*, F. A. Garner, C. H. Henager, Jr., and N. Igata, Eds., American Society for Testing and Materials, Philadelphia, 1987, pp. 54–68.

DISCUSSION

A. T. Motta[1] *(written discussion)*—You show a negative temperature coefficient of growth strain in your experiments. Since most growth models depend on preferential migration to sinks, it could be answered that higher temperatures mean higher mobility and hence higher growth strains. Do you have any thoughts on the reason for the opposite effect being in fact observed?

A. R. Causey et al. (authors' closure)—For the type of model cited in the question, the effect of defect mobility on the growth rate depends on whether one is operating in a "recombination dominated regime" or a "sink dominated regime" (Ref *4*). In the former case, increased defect mobility decreases the proportion of recombination and increases the defect flux to sinks and, hence, the growth rate. In this case, the growth rate is proportional to (fast flux)$^{1/2}$. At 550 K, we appear to be in a sink dominated regime. In this case, most of the defects arrive at sinks and increasing the defect mobility has little effect.

Our current view is that growth in zirconium alloys at this temperature is driven by the difference in anisotropic diffusion between vacancies and interstitials (DAD). In this case, the growth rate and anisotropy are affected by changes in the anistropy of diffusion of either defect species. We have previously interpreted the decrease in the DAD effect, combined with the very fine, highly anistropic grain structure of this material[2] *[4]*. In a coarse-grained material with an isotropic grain structure, the longitudinal growth rate would increase with temperature.

Ted Darby[3] *(written discussion)*—Could you comment on the effect of as-manufactured variability (in anisotropy ratio) on irradiation creep rate variability?

A. R. Causey et al. (authors' closure)—The irradiation creep rate appears to be directly proportional to the as-manufactured variability in anisotropy.

A. T. Motta[1] *(written discussion)*—The calculations of the self-consistent model have been fitted to the pressure tube temperatures (580 K), do you intend to fit the data at calandria tube temperatures (350 K) as well? That would seem to give additional data to verify the modeling.

A. R. Causey et al. (authors' closure)—The self-consistent model has been used to fit the in-reactor creep data for Zircaloy-2 at calandria tube temperatures of about 330 K with the results described in Ref *10*. An excellent correlation was obtained between the stress relaxation data that was used to derive the single crystal creep compliances and the measured axial and transverse strain rates of small internally pressurized capsules.

---

[1] Pennsylvania State University, Department of Nuclear Engineering, University Park, PA.
[2] Holt, R. A. and Fleck, R. G., "The Effect of Temperature on the Irradiation Growth of Cold-Worked Zr-2.5Nb," *Zirconium in the Nuclear Industry: Eighth International Symposium, ASTM STP 1023*, L. P. F. Van Swam and C. M. Eucken, Eds., American Society for Testing and Materials, Philadelphia, 1994.
[3] Rolls-Royce and Associates, Ltd., Derby, UK.

James R. Theaker,[1] *Ram Choubey,*[2] *Gerry D. Moan,*[3] *Syd A. Aldridge,*[4] *Lynn Davis,*[5] *Ronald A. Graham,*[5] *and Christopher E. Coleman*[1]

Fabrication of Zr-2.5Nb Pressure Tubes to Minimize the Harmful Effects of Trace Elements

REFERENCE: Theaker, J. R., Choubey, R., Moan, G. D., Aldridge, S. A., Davis, L., Graham, R. A., and Coleman, C. E., **"Fabrication of Zr-2.5Nb Pressure Tubes to Minimize the Harmful Effects of Trace Elements,"** *Zirconium in the Nuclear Industry: Tenth International Symposium, ASTM STP 1245,* A. M. Garde and E. R. Bradley, Eds., American Society for Testing and Materials, Philadelphia, 1994, pp. 221–242.

ABSTRACT: Trace elements can reduce the fracture resistance of Zr-2.5Nb pressure tubes. The effects of hydrogen as hydrides and oxygen as an alloy-strengthening agent are well known, but the contributions of carbon, phosphorus, chlorine, and segregated oxygen have only recently been recognized. Carbides and phosphides are brittle particles, while chlorine segregates to form planes of weakness that produce fissures on the fracture face of test specimens. A high density of fissures is associated with low toughness. With long hold times in the ($\alpha + \beta$) region, oxygen partitions into the α-grains; such grains are hard and, if they survive fabrication, may reduce the toughness of the finished tube. Through a cooperative program involving AECL and the manufacturers, a series of manufacturing innovations and controls has been introduced that minimizes these harmful effects.

Hydrogen is present in the zirconium sponge as water, can be absorbed at each stage of tube fabrication, and needs to be carefully controlled, particularly during ingot breakdown and subsequent forging. Hydrogen concentrations in finished tubes have been reduced by a factor of three through the optimization of manufacturing processes and the implementation of new technology. Multiple vacuum arc melting, use of selected raw materials, and intermediate ingot surface conditioning have resulted in much improved fracture toughness through the reduction of chlorine and phosphorus concentrations. Optimum distribution of oxygen may be achieved through changes to the extrusion process cycle. An understanding of the Zr-2.5Nb-C phase diagram, particularly the solubility of carbon at low concentrations, has resulted in the specification of a lower carbon concentration.

KEY WORDS: trace elements, zirconium alloys, pressure tubes, carbon, chlorine, phosphorus, oxygen, hydrogen, fracture (materials), manufacturing process, zirconium, nuclear materials, nuclear applications, radiation effects

---

[1] Research metallurgist and branch manager, Fuel Channel Components, respectively, AECL Research, Chalk River Laboratories, Chalk River, Ontario, Canada K0J 1J0.
[2] Research metallurgist, AECL Research, Whiteshell Laboratories, Pinawa, Manitoba, Canada R0E 1L0.
[3] Senior engineer, AECL CANDU, Sheridan Park Research Community, Mississauga, Ontario, Canada L5K 1B2.
[4] President and CEO, Nu-Tech Precision Metals Inc., Arnprior, Ontario, Canada K7S 3H2.
[5] Director, Zirconium Hafnium Business Unit and technical manager of nuclear products, respectively, Teledyne Wah Chang Albany, Albany, OR 97321-0136.

The fracture toughness of cold-worked Zr-2.5Nb is very variable and remains so after irradiation [1,2]. For example, dJ/da at 240°C varied between 20 and 150 MPa after a neutron fluence of about 6.5×10^{25} n/m². Understanding the causes of the variability will allow us to fabricate components with high toughness while avoiding those with low toughness. Also, analysis of the fitness of components for service, in which we currently either use lower bound values or rely on probabilistic methods, would be simplified with tubes of consistently high toughness.

The possible causes of this variability are variations in grain size and shape, variations in distribution of the β-phase, variations in crystallographic texture, and variations in concentration and distribution of trace elements. This paper focuses on the last item.

The two principal features on a fracture surface of Zr-2.5Nb pressure tube material are ductile dimples and fissures [1] (Fig. 1). Dimples are nucleated at local inhomogeneities such as particles. Occasionally, we have observed carbide or phosphide particles at the bottom of dimples, but these are rare, except when the concentration of these elements is high. These observations provide an incentive for maintaining a low concentration of carbon and phosphorus, or distributing them in an innocuous manner. A low fracture toughness corresponds with the small spacing and large size of the fissures; examination of fracture surfaces produced inside a scanning auger microscope showed that the fissures were associated with localized concentrations of chlorine associated with carbon [3]. The fracture surface of material containing little chlorine had no fissures, and the fracture toughness was very high. This result strongly suggests that if the chlorine concentration was minimized, toughness would be increased. It also provides another reason to control the carbon concentration.

In Zr-2.5Nb pressure tubes, the oxygen concentration is maintained within the range 5200 to 7400 ppm (at) [1000 to 1300 ppm (wt)] to contribute to strength. During tube fabrication,

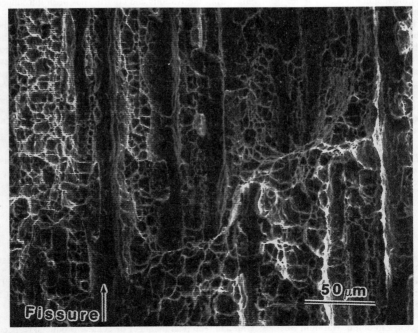

FIG. 1—*Scanning electron micrograph of the fracture surface of a Zr-2.5Nb compact specimen.*

the material is heated into the (α + β) region and the oxygen partitions between the two phases concentrating in the α-phase. This may result in hardening of the α-grains that may reduce the toughness in the finished tube. The consequence of this segregation on fracture properties is evaluated.

Hydrogen has been responsible for cracking in CANDU[6] pressure tubes [4]. Thus, we have a strong motivation to minimize the initial hydrogen concentration. A conservative guideline for determining tube lifetime is that hydrides should not be present at operating temperatures. Since the maximum pickup rate of hydrogen during operation is about 100 ppm (at) [1 ppm (wt)] per year [5], then each reduction of 100 ppm (at) [1 ppm (wt)] increases the nominal lifetime of a tube by one year. Hydrogen is picked up at several stages during manufacturing and the steps taken to minimize its concentration are described.

After briefly describing of the production of pressure tubes and their use, this paper considers each element in turn (carbon, chlorine, phosphorus, oxygen and hydrogen), identifies its source, describes the measures taken for its reduction or innocuous distribution, and then reports on the efficacy of these measures in improving tube properties. The results are summarized as a comparison with the specification of 1982 [6]. Standard test methods were used and thus experimental details are minimized. The concentrations of chlorine and phosphorus were derived from glow discharge mass spectrographic (GDMS) analysis using one machine. This technique (GDMS) has been found to give precise results for both chlorine and phosphorus, thus allowing trends to be plotted. However, due to the current absence of standards and measurement techniques at the very low concentrations involved in the investigation, the accuracy of the reported values is not known. One standard deviation for the range of concentrations measured was 8 and 4.9% for chlorine and phosphorus, respectively. The concentration of hydrogen was measured by vacuum fusion employing chromatographic detection or thermal conductivity measurement using a LECO, RH-1 Hydrogen Determinator. Fracture toughness is taken as dJ/da between 0.15 and 1.5 mm of crack growth [1].

Production of Zr-2.5Nb Pressure Tubes

Figure 2 is an abridged flow diagram showing the main steps in the fabrication of Zr-2.5Nb pressure tubes. Zircon sand is ball-milled with coke to form chemical reactor feed. The resulting compound is reacted with chlorine to form crude (zirconium + hafnium) tetrachloride that is dissolved in water and run through a counter-current MIBK (methyl-iso-butyl-ketone) solvent extraction process to remove hafnium. The zirconium aqueous solution is precipitated with NH_4OH, filtered, calcined to ZrO_2, and subjected to a second chlorination for the Kroll Process, during which the pure zirconium tetrachloride is reduced with magnesium to form a sponge donut. The $MgCl_2$ formed during reduction is intermingled with the zirconium sponge. It is removed by hot distillation under vacuum.

The resulting nuclear-grade sponge is crushed, inspected, and compacted in a press with appropriate additions of alloying elements. The compacts and recycled material are welded together, to form an electrode. Several vacuum arc melts (a process known as vacuum arc refining, VAR) are used to consolidate, refine, and homogenize the sponge, recycle material and alloying elements into a final ingot that is 585 mm (23 in.) diameter, and weighs about 6600 kg.

The ingot is initially press-forged after heating in the β-phase, conditioned by grinding, and then rotary forged in the (α + β) phase into logs about 210 mm (8¼ in.) diameter. The logs are parted and holes are trepanned, to form hollow billets that are quenched into water from

---

[6]CANDU is a registered trademark of AECL.

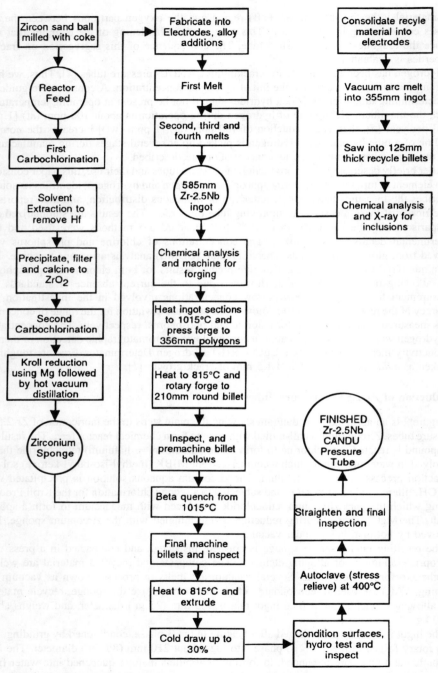

FIG. 2—*Abridged flow chart for the fabrication of CANDU Zr-2.5Nb pressure tubes.*

the β-phase and machined into extrusion billets. The billet is extruded in the (α + β) phase. The extrusion is cold-worked up to 30% and stress-relieved in the α-phase in an autoclave.

The microstructure of the cold-worked Zr-2.5Nb pressure tubes consists of flat elongated grains surrounded by a grain boundary layer of β-phase. The α-grains are usually about 0.3 to 0.5 μm thick, with dimensions in the approximate 1:5:50 ratio in the radial:transverse:longitudinal directions of the tube. The texture has basal plane normals mostly in the transverse direction of the tube. The nominal dimensions of the finished tube are length-6.1 m, inside diameter-103 mm, and wall thickness-4 mm. The pressure tubes are used in a CANDU reactor with D_2O as the heat transfer fluid at a pressure of about 10 MPa and an operating temperature range of 250 to 310°C.

Carbon

Large carbides are known to exist in some CANDU Zr-2.5Nb pressure tubes and have been associated with poor fracture toughness characteristics; they have been observed occasionally on fracture faces. To eliminate these large carbides, the role of carbon on Zr-2.5Nb alloy and its manufacturing processes was studied. The focus of the study was the carbon solubility in the Zr-2.5Nb system at the press forging soaking temperatures (1020 ± 36°C).

Zr-2.5Nb buttons meeting the chemical composition requirements for CANDU pressure tubes with a range of carbon concentrations, 350 to 2350 ppm (at) [45 to 310 ppm (wt)], were prepared by inert gas melting, chemically analyzed, rolled to thin strips, encapsulated in evacuated quartz tubing, heat-treated at temperatures ranging from 900 to 1190°C, and water quenched with the quartz tube intact, although it broke as soon as it entered the water. The presence of carbide particles was determined by metallography [7], and the maximum dimensions were recorded for these and other particles based on a visual determination of individual particles. Subsequently, these particles were characterized into three families according to the maximum dimension, optical characteristics, and the degree of confidence in their identification:

1. particles greater than 1 μm positively identified as carbides by metallography and by microprobe (wavelength dispersive);
2. particles in the range 0.25 to 0.75 μm, believed to be carbides but still to be confirmed; and
3. very small, unidentified particles.

When these particle families are inserted into a plot of carbon concentration against soaking temperature (Fig. 3), it can be seen that large carbides are not formed unless the carbon concentration is greater than about 9.5 ppm (at) [125 ppm (wt)] and the soaking temperature is below 1000°C.

Figure 3 indicates that two methods could be used to prevent the formation of large carbides in Zr-2.5Nb pressure tubes:

1. maintain the carbon concentration below 9.5 ppm (at) [125 ppm (wt)], or
2. use a solution anneal at some temperature above 1025°C (depending on the maximum carbon concentration) before either the press forging or the β-quench.

An increase in the solution annealing temperature was dismissed for several reasons, including excessive β-grain growth, a need to qualify a change in the manufacturing process, and the possibility of precipitating dissolved carbon during irradiation if a supersaturated solid

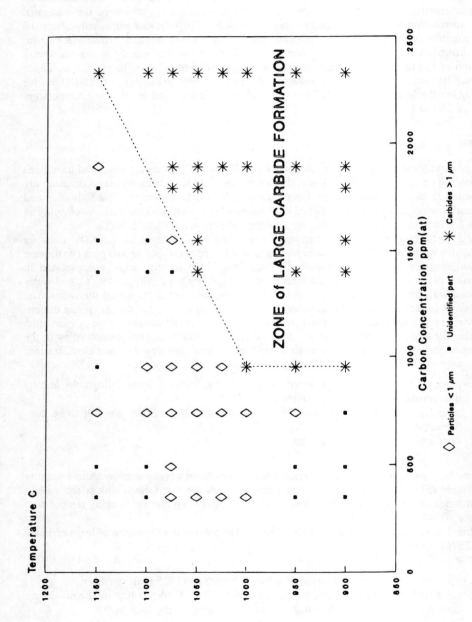

FIG. 3—*Zr-2.5Nb system showing region of carbide formation after an 8-h soaking time.*

solution was produced. Zr-2.5Nb pressure tubes manufactured before 1992 had a specified maximum concentration of 11.4 ppm (at) [150 ppm (wt)] carbon, with typical concentrations ranging from 6.1 to 11.4 ppm (at) [80 to 150 ppm (wt)]. Thus, lowering the maximum concentration was the most effective way to eliminate the large carbides, and it required minimal manufacturing changes.

Source and Removal

The primary processes that introduce carbon into zirconium metal occur during the second "pure" (that is, hafnium free) carbochlorination and subsequent magnesium reduction and the melting of Zr-2.5Nb recycle.

The hafnium-free zirconium oxide is blended with carbon (coke) to help reduce the zirconium oxide during the second carbochlorination. The control of this carbon is accomplished by filtration of fine residual carbon particles from the carbochlorinated feedstock, but some residual carbon is carried through the process and remains with the sponge. A secondary source of carbon comes from entrapped CO_2, CO, and $COCl_2$ gases in the $ZrCl_4$, formed during carbochlorination of the pure zirconium oxide. A tertiary source of carbon that must be monitored and controlled is carbon contained in the magnesium used for the reduction process.

Carbon can also enter the process during the processing of recycled material. Machining chips can contain residual oily cutting fluids that have high carbon concentrations, and it is important to use only machining fluids that can be readily washed from the chips. Residual carbon in recycled material made from chips and other recycled forms must be analyzed. Material with various carbon concentrations can be blended through careful assembly of the melt electrode, to produce a finished ingot with the desired concentration.

Implementation

To avoid the large detrimental carbides, a process using only selected zirconium sponge, based on chemical composition and material inspection, and recycled material has been implemented in the ingot manufacturing cycle, and a new specification limit of 9.5 ppm (at) [125 ppm (wt)] carbon has been introduced. Typical carbon concentrations of recently manufactured pressure tubes have been in the range of 6.8 to 8.3 ppm (at) [90 to 110 ppm (wt)], and no large carbides have been observed, either on the fracture faces of test specimens or in the metallographic examination of samples from finished tubes.

Chlorine

As the chlorine concentration increases, the fracture toughness of Zr-2.5Nb pressure tubes decreases (Fig. 4).

Examination of the pressure tube manufacturing history showed that all the tough tubes had been produced from ingots made from 100% recycled trepanned cores (the material that is removed from the center of a solid billet so that it can be extruded into a tube). Therefore, the material used to manufacture these pressure tubes had been melted at least four times; that is, twice when the original material was produced and twice more when the cores from the original material were remelted. Effectively, these pressure tubes were manufactured from ingots that had been subjected to quadruple VAR.

Cross-sectional chemical analyses at several positions from top to bottom of double-melted ingots showed a chlorine concentration at the ingot tops of 0.3 to 0.6 ppm (at) [0.1 to 0.2 ppm (wt)], the same as the concentration of pressure tubes made from 100% recycled trepanned cores. The other positions in the ingot had chlorine concentrations of 14 to 20.5 ppm (at) [5.5 to 8 ppm (wt)]. A hot topping practice, utilizing a reduced melting rate, is applied during the

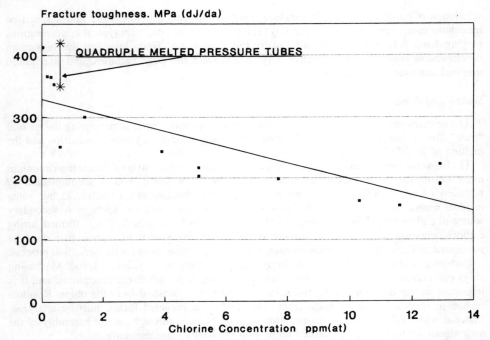

FIG. 4—*Fracture toughness [8] as a function of chlorine concentration for pressure tubes with phosphorus concentrations of <30 ppm (at) [10 ppm (wt)].*

final stages of ingot melting to minimize piping and segregation. This practice results in a slow transfer of liquid particles in the arc, and an increased retention time of the liquid metal at the top of the ingot, thus increasing the time of exposure to low pressure, where the vacuum system operates most efficiently. This is further evidence that the chlorine concentration is lowered by VAR.

Source and Removal

The source of the chlorine is the magnesium chloride formed during the Kroll process in manufacturing zirconium sponge.

$$ZrCl_{4(gas)} + 2Mg_{(liq)} \longleftrightarrow Zr_{(sol)} + 2MgCl_{2(liq)}$$

To date, the standard practice has been to manufacture Zr-2.5Nb pressure tubes from ingots that were melted two times, and quadruple melting can be used to enhance the chemical refinement of Zr-2.5Nb. It is especially effective in removing volatile species such as hydrogen, chlorine, and magnesium. The quadruple melting process starts by electron beam welding sponge compacts to form a 280-mm-diameter first-melt electrode. Four 355-mm-diameter first-melt ingots are vacuum arc refined. The 355-mm-diameter first-melt ingots are cut, and recycle billets that have been previously melted into a 355-mm-diameter ingot and cut into short lengths are inserted into the second-melt electrodes. The 355-mm-diameter ingots are inverted during assembly and form three second-melt electrodes that are cast into three 430-mm-diameter second-melt ingots. The second-melt ingots are cut, inverted, and welded together to form two electrodes that are cast into two 510-mm-diameter triple-melt ingots. The triple-melt ingots are inverted, welded, and VAR into a 585-mm (23-in.)-diameter quadruple-melted ingot with a final weight of about 6600 kg.

During melting, the melt rate is carefully controlled by adjusting the furnace voltages and current. Vacuum levels are established and maintained to provide optimum removal of volatile species while maintaining manufacturing control.

Figure 5 shows the effect of the multiple VAR on ingot chlorine concentrations. Geiger and Poirier [9] showed that the fraction retained, f_r, of a volatile compound in a melt exposed to vacuum for time, t, in seconds, is given by

$$Ln \cdot f_r = -(A_s/V)Kt \qquad (1)$$

where

A_s = the surface area of melt (m²),
V = the volume of the melt (m³), and
K = the overall mass transport rate constant (m/s).

K in turn is related to constants for intra-liquid transport, K_{liq}, and evaporation transport, K_{evap}, as

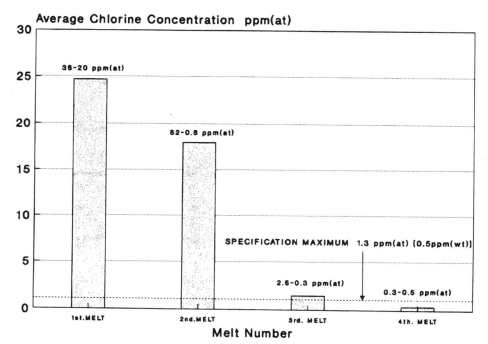

FIG. 5—*Chlorine concentration change by melt number in a quadruple-melted ingot. The initial chlorine concentration of the sponge was approximately 193 ppm (at) [75 ppm (wt)]. The numbers at the top of each bar represent the maximum and minimum concentrations measured.*

$$1/K = 1/K_{liq} + 1/K_{evap} \tag{2}$$

Now

$$K_{liq} = 2\sqrt{(D/\pi\theta)} \text{ m/s} \tag{3}$$

where

D = the liquid diffusion coefficient ($\sim 10^{-8}$ to 10^{-9} m²/s) and
θ = a time parameter equal to the average distance from the center of the melt to the edge of the crucible divided by the average velocity of the melt at the surface, usually estimated as ≈ 1 s.

K_{evap} comes from the Knudsen evaporation equation and is given by

$$K_{evap} = \gamma p^\circ / \rho \sqrt{(2\pi MRT)} \tag{4}$$

where

p° = the vapor pressure of pure vaporizing species ($\sim 10^8$ kg/m-s²),
γ = the activity coefficient taken as unity,
ρ = the molar density at melting ($\sim 7 \times 10^{-8}$ mol/m³ for zirconium),
M = the molecular weight of the vaporizing species ($\sim 10^2$ for ZrCl or ZrCl$_4$),
R = the gas constant (8.31 J · mol^{-1} · K^{-1}), and
T = about 2200 K.

Substituting appropriate values in both Eqs 3 and 4, it can be shown that

$$\frac{1}{K_{evap}} \ll \frac{1}{K_{liq}}$$

therefore

$$\frac{1}{K} \approx \frac{1}{K_{liq}}$$

The time-at-liquid, t, is calculated from

$$\frac{P}{\phi}$$

where

P = the pool depth at the center and
ϕ = the fill rate.

and

$$\phi \propto \mu/A_s$$

where

μ = the melt rate, and
A_s = the surface area of the pool.

Time-at-liquid ranges between 2000 and 3000 for the melt sequences previously described.

Figure 6, derived from data obtained during this work, shows the plot of f_r versus $e^{-(A/V)Kt}$ showing that Eq 1 describes the elimination of chlorine during multiple vacuum arc melting operations. The value of K that provides the best fit is approximately 1×10^{-4} m/s. This compares well with theoretically calculated values of K from Eq 3, giving a range of 10^{-4} to 10^{-5} m/s.

Examination of ingots produced from the first melt showed a partial layer of residual magnesium chloride salts on the outside surface. This layer was removed by machining.

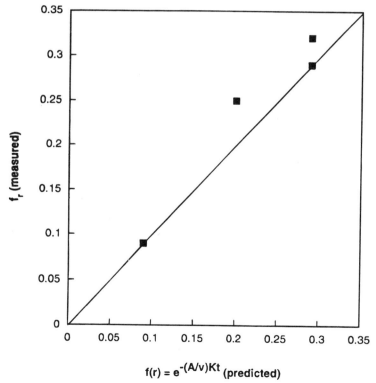

FIG. 6—*Fraction of chlorine retained: measured versus predicted during four VAR melts.*

Implementation

A melting practice, including ingot conditioning (removal of outside surface) and four melting cycles, has been established and a maximum chlorine concentration of 1.3 ppm (at) [0.5 ppm (wt)] has been introduced into the specification. This value was chosen based on the concentrations measured by an uncalibrated GDMS system and estimates of chlorine concentrations that would be achieved by four melting cycles.

CANDU Zr-2.5Nb pressure tubes, conforming to all the specification requirements, have been produced with chlorine concentrations of 0.3 to 0.6 ppm (at) [0.1 to 0.2 ppm (wt)] and resultant fracture toughness values are in the range 350 to 419 MPa (Fig. 4) for two ingots. Other mechanical properties (Table 1) are similar to those of pressure tubes made from double-melted material, although the tensile elongation and the ultimate strength appear higher for the material melted four times. Examination of the chlorine concentrations of successive melts has shown that at least four melting cycles are required to reduce the chlorine concentration in Zr-2.5Nb to <1.3 ppm (at) [<0.5 ppm (wt)] throughout the ingot. When this is done, pressure tubes with consistently high fracture toughness characteristics can be obtained. Quadruple melting practice has now been adopted for the production of Zr-2.5Nb pressure tubes for CANDU.

Phosphorus

A number of pressure tubes manufactured from 100% recycled trepanned cores had low chlorine concentrations, but their fracture toughness values were lower than expected (based on the chlorine concentration). The phosphorus concentrations of these tubes correlated with the fracture toughness; as the phosphorus concentration increased, the fracture toughness decreased (Fig. 7). Pressure tubes made from double-melted material exhibited a similar but weaker correlation, the effect of phosphorus being masked by the effect of chlorine. In addition, large phosphorus-containing particles were found on the fracture faces of samples with phosphorus concentrations of 124 ppm (at) [42 ppm (wt)] (Fig. 8).

Source and Removal

The source of phosphorus is primarily the zircon sand. An upper specification limit for phosphorus is applied to the sand, but this limit is generally well above the concentration of

TABLE 1—*Mechanical properties of Zr-2.5Nb pressure tubes made from quadruple-melted material (tested in the longitudinal direction at 300°C).*

	Quadruple-Melted Ingots		
	Ultimate Tensile Strength, MPa	Yield Strength, MPa	Elongation, %
Mean (6 results)	550	412	19
	COMPARISON DATA		
	STATION A		
Mean (2058 results)	513	393	14.7
Standard deviation	14.8	23.4	1.7
	STATION B		
Mean (2007 results)	512	379	14.8
Standard deviation	17.2	20.7	1.9

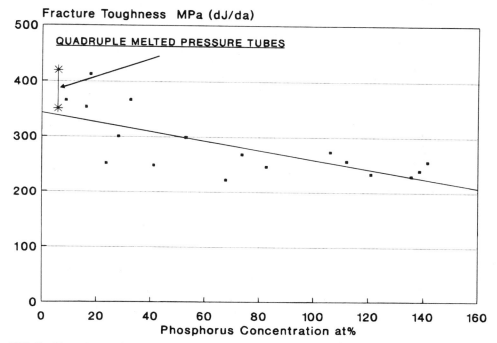

FIG. 7—*Fracture toughness* [8] *as a function of phosphorus concentration for pressure tubes with chlorine concentrations <1 ppm (at) [0.4 ppm (wt)].*

phosphorus desired in the final product. The phosphorus is primarily removed during the chlorination process, but can be further lowered through the additional step of subliming the pure zirconium chloride. Since there is some variability in the phosphorus concentration in the zirconium tetrachloride, it is necessary to sample the chloride on a lot-by-lot basis and select lots that have phosphorus concentrations below the specification maximum, or to blend lots with variable concentrations to meet the required phosphorus concentration.

Results

As part of the quadruple melting practice, a phosphorus concentration limit of 30 ppm (at) [10 ppm (wt)] has been introduced and, using selected zirconium sponge, pressure tubes with a concentration of 2.9 to 5.9 ppm (at) [1 to 2 ppm (wt)] have been produced from two quadruple-melted ingots (the same ones used in the chlorine reduction trial referenced earlier). These pressure tubes have a high fracture toughness (Fig. 7).

Implementation

To consistently manufacture CANDU Zr-2.5Nb pressure tubes with high fracture toughness, it is not only necessary to reduce the chlorine concentration, but maintain the phosphorus concentration below 30 ppm (at) [10 ppm (wt)]. This is achievable through the use of selected zirconium sponge.

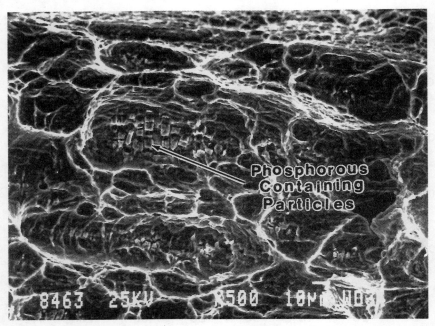

FIG. 8—*Secondary electron micrograph of cracked phosphorus-containing particles in the fracture face of a Zr-2.5Nb compact specimen containing 124 ppm (at) [42 ppm (wt)] phosporus.*

Oxygen

In CANDU Zr-2.5Nb pressure tubes, oxygen is used as an alloying element to increase the short-term mechanical strength. However, because it is used in small concentrations, 5000 to 7400 ppm (at) [1000 to 1300 ppm (wt)], it is defined here as a trace element. For the extrusion process, the Zr-2.5Nb hollow billets are heated and held in the ($\alpha + \beta$) phase. However, oxygen is a strong alpha stabilizer and secondary ion mass spectrometry (SIMS) has shown that oxygen concentration in some α-grains has increased after soaking at these temperatures. Assuming that the oxygen is distributed evenly after the β-soak and water quench, even short soaking times result in a partitioning of the oxygen to some α-grains (Table 2). As the soaking time is extended, the amount of partitioning increases. Microhardness measurements on the α-grains show that these oxygen enriched grains are harder (HV 380) than the matrix (HV 280 to 300).

On cooling, after extrusion, the microstructure consists of the oxygen-enriched α-grains in a network of oxygen-lean α-grains transformed from the β-phase. If these prior α-grains contain different oxygen concentrations due to different soak times, then it is possible that the resultant Zr-2.5Nb pressure tubes could have variable mechanical properties, including fracture toughness.

Investigation

Trials were run on hollow billets from the same double-melted ingot to examine the effect of soaking time at extrusion temperature. The extrusion process variables, mechanical property variations, including fracture toughness, and microstructural differences of pressure tubes

TABLE 2—Oxygen-partitioning in Zr-2.5Nb material heat-treated in the (α + β) phase field.

Heat-Treatment Temperature, °C	Heat-Treatment Time, h	Average Oxygen Concentration, ppm (wt)		
		Large α-Grains (a)	β-Matrix (b)	Ratio a/b
8¼ in. Forged Billet				
850	¼	2076	697	3.0
850	2½	2354	668	3.4
850	12	2422	422	5.7
Pressure Tube				
815	¼	1014	647	1.6
815	2¾	1685	728	2.3

TABLE 3—Mechanical properties in the longitudinal direction as a function of pre-extrusion soaking time.

Tube No.	Soak Time, h	Yield Strength, MPa	Ultimate Tensile Strength, MPa	Elongation, %	dJ/da, MPa
Tube 1	2¾	383	521	15	174
Tube 2	2¾	381	507	18	160
Tube 3	¼	384	517	18	182
Tube 4	¼	384	518	17	178

manufactured with pre-extrusion soaking times of 2¾ h, were compared with pressure tubes made with a normal soaking time of ¼ h. No unusual extrusion variables were observed, and there were no significant mechanical property variations. SIMS analyses did indicate partitioning of oxygen to the larger prior α-grains (Table 2), and there is some indication that the fracture toughness of the material soaked for 2¾ h may be lower than that of the material soaked at the shorter time (Table 3). The microstructure of the pressure tubes soaked at the shorter time appears more uniform and contains fewer large prior α-grains (Fig. 9) than the material made after the longer pre-extrusion soak time.

Results

Based on the microstructure and the oxygen concentration gradient between the prior α-grains and the matrix, the absence of clear changes in pressure tube attributes for increased soaking time is perhaps surprising. However, the chlorine concentration of this material is 18 ppm (at) [7 ppm (wt)] and it is suspected that this high value may mask the effect of any oxygen segregation. The effect of soaking temperature and time is being investigated on low chlorine, quadruple-melted material.

Implication

For double-melted Zr-2.5Nb pressure tubes, long soaking times result in oxygen segregation to the prior α-grains, and a less uniform microstructure, but there is no significant effect on the mechanical properties and only a possible reduction in fracture toughness. The mechanical properties, including fracture toughness, appear independent of extrusion soaking times,

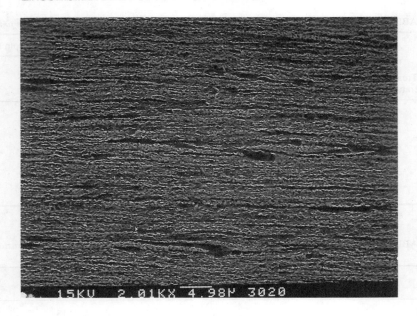

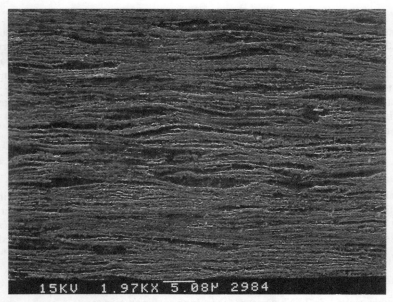

FIG. 9—*Typical microstructures of pressure tubes made from a double-melted ingot soaked at 815°C for different times before extrusion:* (a) $1/4$-h soak time and (b) $2^3/_4$-h soak time.

although this remains to be confirmed on quadruple-melted material with low chlorine concentrations.

Hydrogen

Hydrogen is responsible for a cracking mechanism in zirconium alloys called delayed hydride cracking (DHC) [4,10–12]. For DHC to occur in Zr-2.5Nb pressure tubes, the following must be present at the same time:

1. a large tensile stress,
2. a crack initiator, and
3. hydrides, the presence of which indicates that the hydrogen concentration is greater than the solubility limit of hydrogen in zirconium at the temperature of interest.

During reactor operation, the hydrogen concentration in pressure tubes gradually increases, primarily from the pickup of some of the deuterium that is released in the corrosion reaction with the D_2O coolant at the inside surface. It is therefore desirable that new pressure tubes have low initial hydrogen concentrations to ensure that the equivalent hydrogen concentrations that would cause hydrides to be present at operating temperatures are not reached throughout the design life.

Source and Removal of Hydrogen

Hydrogen found in zirconium alloys may have entered the material at many stages during the manufacturing process. The principal source is the manufacture of the sponge. The pure $ZrCl_4$ is reduced by magnesium in the Kroll Process, and it is essential to remove the $MgCl_2$ by vacuum distillation. Some residual $MgCl_2$ remains in the sponge after the distillation, and it can absorb moisture as $MgCl_2 \cdot 5H_2O$. The sponge must be handled so that moisture is not absorbed by the $MgCl_2$ or by the reactive sponge surfaces. The residual $MgCl_2$ is distilled out during the VAR, but furnace variables such as melt rate and vacuum levels must be controlled so that the associated hydrogen and residual moisture are also removed. Continued VAR will reduce the hydrogen concentration, but with diminishing returns.

Hydrogen pickup can also occur during high-temperature heat treatments such as those used for forging, β-quenching, and extrusion. Gas-fired furnaces with reducing atmospheres have been eliminated and replaced by electric furnaces for these operations. Holding times and temperatures have been modified to keep them as low as possible while maintaining manufacturing efficiency and product integrity.

A study was carried out to examine the relative importance of the different processing stages in the pickup of hydrogen in new CANDU Zr-2.5Nb pressure tubes. Material was collected during the processing of tubes from two double-melted ingots. The material sampled included sponge and slices from the ingots, the forgings, the billets before trepanning, the extrusions, the cold-drawn tubes, and the finished tubes. Several thousand samples were prepared, so that the hydrogen concentrations and their distributions in the material after the different processing stages could be determined.

Results

The results of the hydrogen analyses are summarized in Table 4. They show that the sponge had the highest hydrogen concentrations and a large variability. The concentrations in the ingot were generally 400 or 500 ppm (at) [4 or 5 ppm (wt)] after double vacuum arc melting, but values of up to 1300 ppm (at) [14 ppm (wt)] were measured at locations close to the ingot

TABLE 4—*Summary of hydrogen concentration at each stage of production.*[a]

Sponge	9 samples tested; range = 0.12 to 0.59 atom % (13 to 66 ppm); mean = 0.32 atom % (35 ppm)
Ingots	Most data 0.04 to 0.05 atom % (4 to 5 ppm); some surface values up to 0.09 to 0.13 atom % (10 to 14 ppm)
Billets	Most data 0.04 to 0.05 atom % (4 to 5 ppm), but at forged surfaces, most surface values 0.1 atom % (11 ppm), maximum 0.36 atom % (40 ppm)
Extrusions	Mean = 0.05 atom % (5 ppm); standard deviation = 0.007 atom % (0.8 ppm)
Drawn tubes	Mean 0.04 atom % (4.3 ppm); standard deviation 0.005 atom % (0.6 ppm)
Autoclaved tubes	Mean 0.05 atom % (4.8 ppm); standard deviation 0.006 atom % (0.7 ppm)

[a] Summary of the hydrogen concentrations measured in several thousand samples prepared during the processing of the pressure tubes from two ingots. The data show that the highest hydrogen concentrations were found in the sponge, that after the double vacuum arc melting of the ingot, most of the data were close to 0.05 atom % (5 ppm), and that there was a higher concentration close to the forged surfaces. The original data were reported as ppm (wt).

surface. The concentration in the billets was normally 400 or 500 ppm (at) [4 or 5 ppm (wt)]. However, all the billet samples showed an increasing hydrogen concentration gradient from the center to the outside surface, with some values as high as 3600 ppm (at) [40 ppm (wt)] just below the billet surface. Cross-sectional samples from the extrusions, drawn tubes, and autoclaved tubes normally had mean concentrations of 400 or 500 ppm (at) [4 or 5 ppm (wt)], and standard deviations of less than 100 ppm (at) [1 ppm (wt)].

The hydrogen concentration changes during the quadruple melting process were monitored in recent ingots. After the first melt, the hydrogen concentration varied between a high of 1600 ppm (at) [18 ppm (wt)] and a low of less than 300 ppm (at) [3 ppm (wt)]. As the material progressed through the other three melting cycles, the hydrogen concentration decreased to be consistently less than 300 ppm (at) [3 ppm (wt)] in the final ingot with the same concentrations in the pressure tubes.

The melting process clearly makes the largest contribution to the reduction of the hydrogen concentration. Hydrogen absorbed during the heating for forging, β-quenching, and extrusion operations remained close to the surface. By modifying the machining practice to remove the appropriate amount of stock and ensuring that the billets are correctly centered, most of this absorbed hydrogen can be removed.

Implementation

The average hydrogen concentration in a batch of tubes made in the 1980s was 950 ppm (at) [10.3 ppm (wt)], the standard deviation was 300 ppm (at) [3.1 ppm (wt)], and some values were up to 1600 ppm (at) [18 ppm (wt)] (Fig. 10). Process changes and system improvements, such as the use of new vacuum systems for the vacuum arc furnaces and electric furnaces for heating material for forging and extrusion, and a modified machining practice, have been implemented, and hydrogen concentrations in a recent batch of tubes have a mean concentration of 260 ppm (at) [2.6 ppm (wt)], with a standard deviation of 40 ppm (at) [0.4 ppm (wt)] and a maximum value of 500 ppm (at) [5 ppm (wt)]. This is a real reduction as a result of the modified practice, and it has allowed the specified maximum hydrogen concentration of CANDU Zr-2.5Nb pressure tubes to be lowered from 2300 to 500 ppm (at) [25 to 5 ppm (wt)].

Summary

This paper is summarized in Table 5, which lists the elements, the mechanism by which they reduce toughness, their source during tube fabrication, and the steps taken to minimize

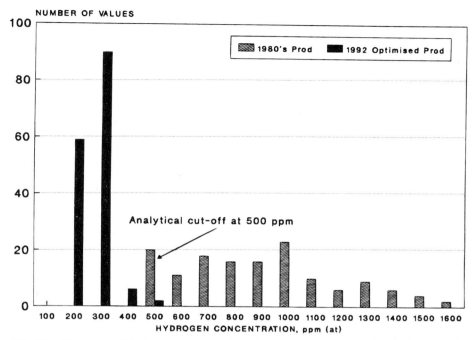

FIG. 10—*Histogram of hydrogen concentrations for pressure tubes manufactured in the 1980s and 1992. The analytical cutoff at 550 ppm (at) for the pressure tubes manufactured in the 1980s represents the analytical capability of the equipment used at that time to measure the hydrogen concentration.*

TABLE 5—*Minimizing the effects of five trace elements on fracture properties of Zr-2.5Nb.*

Element	Mechanism	Effect	Principal Source	Mitigation of Deleterious Effect
C	Forms brittle ppts and associates with embrittling Cl	reduces fracture toughness	Zr sponge and scrap	use selected sponge and recycled material
Cl	Forms fissures	reduces fracture toughness	Zr sponge	quadruple melting
P	Forms ppts that initiate microvoids	reduces fracture toughness	zircon sand	use selected sponge
O	Partitions to the α-phase during heating in (α + β) regions	may reduce fracture toughness	Zr sponge	control extrusion variables
H	Forms brittle hydrides	leads to DHC and brittle fracture	Zr sponge and forging	control melting and forging

TABLE 6—*Element changes, ppm (wt), (as defined by the pressure tube specification).*

Element	1982	1992
C	270 max	125 max
Cl	not covered	0.5 max
P	not covered	10 max
O	900 to 1300	1000 to 1300
H	25 max[a]	5 max[a]

[a] Finished tube.

their effect. The consequence of these studies has been a change in the specification of the chemical composition of CANDU pressure tubes. Table 6 compares the relevant parts of the new specification with that used in 1982 [6]. Noteworthy is the reduction in allowable hydrogen concentration by a factor of five. If hydrogen is picked up during operation at 100 ppm (at) [1 ppm (wt)] per year, then the period without hydrides present (at the operating temperature) has been increased by 20 years in a tube containing the maximum concentration allowed by the new specification. Also of significance is the introduction of a specified maximum value for chlorine and phosphorus, as well as a reduction in the maximum carbon concentration permitted. All the low concentrations are challenging the resolution of the chemical analytical techniques, and additional independent confirmation may be required to assure compliance with the specification.

Since tubes made from 100% recycled material (that is, melted four times) have already been in service in power reactors, we have examined their other properties to ensure that they have not been sacrificed to improve fracture toughness. No changes were observed that could be attributed to the differences in purity:

1. the characteristic properties of delayed hydride cracking, crack velocity, and K_{IH} were measured on material at the extremes of measured fracture toughness, and no difference was noted;
2. measurements of oxidation and deuterium pickup in tubes removed from Pickering Unit 3 revealed no difference at any position along the tubes between tubes made from double-melted (18 tubes) and 100% recycled material (seven tubes); and
3. within 10% of the mean values, the deformation of the pressure tubes in Pickering Unit 3 were indistinguishable.

Acknowledgments

This work could not have been done without the assistance and technical support of many individuals, including I. Aitchison, A. M. Babayan, A. R. Causey, C. K. Chow, J. Watters, M. Griffith, P. H. Davies, E. F. Ibrahim, D. P. Nguyen, R. A. Ploc, S. Sagat, and V. F. Urbanic of AECL; P. T. Finden, J. P. Tosdale, J. A. Sommers, C. M. Eucken, and B. A. McClanahan of Teledyne Wah Chang Albany; and C. J. Schulte of Ontario Hydro. Special thanks are due to R. Hall, of Teledyne Wah Chang Albany for the logistics support.

The funding of portions of this work by the CANDU Owners Group (COG) under Work Packages 6583 and 6465 is gratefully acknowledged.

References

[1] Chow, C. K., Coleman, C. E., Hosbons, R. R., Davies, P. H., Griffiths, M., and Choubey, R., "Fracture Toughness of Irradiated Zr-2.5Nb Pressure Tubes from CANDU Reactors," *Zirconium*

in the Nuclear Industry: Ninth International Symposium, ASTM STP 1132, C. M. Eucken and A. M. Garde, Eds., American Society for Testing and Materials, Philadelphia, 1991, pp. 246–275.

[2] Davies, P. H., Hosbons, R. R., Griffiths, M., and Chow, C. K., "Correlation between Irradiated and Unirradiated Fracture Toughness of Zr-2.5Nb Pressure Tubes," in this volume.

[3] Aitchison, I. and Davies, P. H., "Role of Microsegregation in Fracture of Cold-Worked Zr-2.5Nb Pressure Tubes," *Journal of Nuclear Materials*, Vol. 203, 1993, pp. 206–220.

[4] Cheadle, B. A., Coleman, C. E., and Ambler, J. F. R., "Prevention of Delayed Hydride Cracking in Zirconium Alloys," *Zirconium in the Nuclear Industry: Seventh International Symposium, ASTM STP 939*, R. B. Adamson and L. F. P. Van Swam, Eds., American Society for Testing and Materials, Philadelphia, 1987, pp. 224–240.

[5] Urbanic, V. F., Warr, B. D., Manolescu, A., Chow, C. K., and Shanahan, M. W., "Oxidation and Deuterium Uptake of Zr-2.5Nb Pressure Tubes in CANDU-PHW Reactors," *Zirconium in the Nuclear Industry: Eighth International Symposium, ASTM STP 1023*, L. F. P. Van Swam and C. M. Eucken, Eds., American Society for Testing and Materials, Philadelphia, 1989, pp. 20–34.

[6] Cheadle, B. A., Coleman, C. E., and Licht, H., "CANDU-PHW Pressure Tubes: Their Manufacture, Inspection and Properties," *Nuclear Technology*, Vol. 57, June 1982, pp. 413–425.

[7] Wang, C. T. and Danielson, P. E., "Use of Anodization Techniques for Optical Microscopy Determination of Solubility Limits for Carbon in Zircaloy," *Materials Characterization*, Vol. 24, No. 87, 1990.

[8] Davies, P. H., private communication.

[9] Geiger, G. H. and Poirier, D. R., *Transport Phenomena in Metallurgy*, Addison-Wesley, 1973, pp. 560–567.

[10] Coleman, C. E. and Ambler, J. F. R., "Delayed Hydride Cracking in Zr-2.5Nb Alloy," *Reviews of Coatings and Corrosion*, Vol. 3, 1979, pp. 224–240.

[11] Sagat, S., Coleman, C. E., Griffiths, M., and Wilkins, B. J. S., "The Effect of Fluence and Irradiation Temperature on Delayed Hydride Cracking in Zr-2.5Nb," in this volume.

[12] Dutton, R., Nuttall, K., Puls, M. P., and Simpson, L. A., "Mechanisms of Hydride-Induced Delayed Cracking in Hydride Forming Materials," *Metallurgical Transactions A*, 8A, 1977, pp. 1553–1562.

DISCUSSION

D. G. Franklin[1] *(written discussion)*—In changing to quadruple-melted ingots, did you determine the effects on other critical properties, for example, corrosion, hydrogen, pickup, and margin to nodular corrosion?

J. R. Theaker et al. (authors' closure)—Since tubes made from 100% recycled material (that is, melted four times) are already in service in power reactors, the data from these tubes can be examined. For example, see Fig. 11 examining in-reactor pressure tube elongation. No changes from double-melted material have been observed. This subject is discussed in the summary section of the paper, although it was not covered during the conference presentation.

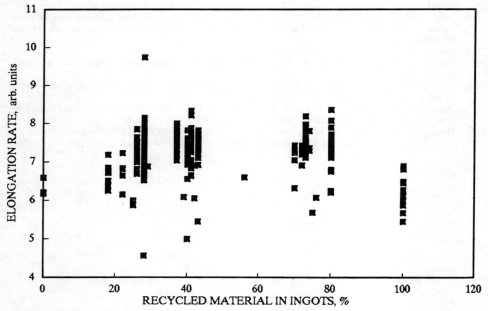

FIG. 11—*Graph showing the effect of the amount of recycled material in the Zr-2.5Nb ingots in the in-reactor axial elongation rate of pressure tubes after approximately eight years of operation.*

---

[1] Westinghouse Bettis Atomic Power Laboratory, West Mifflin, PA.

Fabrication

Fabrication

Jean Luc Aubin,[1] *Emmanuel Girard,*[2] *and Pierre Montmitonnet*[3]

Modeling of Damage in Cold Pilgering

REFERENCE: Aubin, J. L., Girard, E., and Montmitonnet, P., "**Modeling of Damage in Cold Pilgering,**" *Zirconium in the Nuclear Industry: Tenth International Symposium, ASTM STP 1245*, A. M. Garde and E. R. Bradley, Eds., American Society for Testing and Materials, Philadelphia, 1994, pp. 245–263.

ABSTRACT: Surface defects may occur in cold pilgering of Zircaloy-4 tubes caused by the stress and strain history. Because the low frequency of these defects makes their direct study difficult, a model has been built to describe the mechanical conditions that prevail during the pass, as a function of tool design, rolling parameters, and friction. It is based upon the geometrical definition of strains that is transformed into stresses via the Von Mises plastic flow rule and a slab method. Rolling forces are deduced and compared with experiments, and the damage function is computed. A number of experimental observations has been performed (axial movements of tube and mandrel, bulge wave propagation, contact length, shearing strains, etc.) that provides the correct hypotheses for the model. A correlation is established between the computed damage function and the frequency and severity of the studied defects.

KEY WORDS: zirconium, zirconium alloys, nuclear materials, nuclear applications, radiation effects, cladding tubes, cold pilgering, mechanical modeling, surface defects, damage (materials)

Nomenclature

$D(X)$ Tube average diameter at Position X, m
$e(X)$ Tube thickness at Position X, m
Δe Tube thickness reduction during one stroke, m
E, ν Young's modulus (GPa) and Poisson's ratio of roll
F Vertical roll separation force, N
F_{ax} Horizontal, axial force on mandrel and preform, N
F_x Distance between mandrel center and groove center in a cross section, m
G Roll gap due to the roll separation force, m
Ld Roll/tube contact length, m
m Feed length (by which the preform is pushed forward after each stroke), m
R_i, R_e Internal and external tube radii (at a given Position X), m
R_m Mandrel radius (at a given Position X), m
R_{GB} Longitudinal roll radius at groove bottom, m
R_{pg} Gear pinion radius, m
R_x Groove cross section radius, m
$S(X)$ Tube cross section at Position X, m^2

---

[1] Docteur ingénieur, adjoint au service production, Zircotube, BP 21-44560 Paimboeuf, France.
[2] Agrégé génie mécanique, I. U. T. de Saint-Nazaire, L. A. M. M., BP 420-44606 Saint-Nazaire, France.
[3] Chargé de Recherche, Ecole des Mines de Paris–CEMEF–UA CNRS 1374, BP 207-06904 Sophia-Antipolis, Cedex, France.

t	Time, s
U	Axial displacement of an element of the system (indicated by subscript), m
Ur, Uf	Mandrel displacement during the return (respectively forward) stroke, m
X	Axial position along the mandrel; $X = 0$ is the mandrel end, m
$\delta(X)$	Mandrel half conicity angle at Position X, rad
$\psi(X)$	Roll groove half conicity angle at Position X, rad
ϵ	Global strains
$\Delta\epsilon$	Strain increment tensor
$\bar{\epsilon}$	Equivalent strain (see definition Eq 23)
$\dot{\epsilon}$	Strain rate tensor
$\dot{\bar{\epsilon}}$	Equivalent strain rate, s^{-1}
θ	Angular position around mandrel, tube, or roll groove cross section, rad
σ	Stress tensor
σ_0	Yield stress, MPa
Subscript k	"At station k." Mandrel station number, k, is defined as the ratio, X/Lm, where Lm is the total mandrel work zone length: $k = 0$ is the mandrel downstream extremity, $k = 1$ is the upstream extremity. A similar definition exists for the roll, so that Station k of the roll should face Station k of the mandrel, if there was no elastic tool deformation.

This study discusses cold pilgering of cladding tubes in zirconium alloy. Such tubes require a high level of quality, so it is important to prevent surface defects (an example is shown in Fig. 1) that may have critical consequences. Such defects occur on a very small proportion of tubes (say <1%), so that an experimental study of the origin and influential factors is extremely difficult. This is why a mechanical model of cold pilgering was built to determine the stress and strain history of metal particles, and to try and relate it to the observed failures and help design defect-free tools.

A metal particle undergoes a very complex stress and strain history in a pilger mill. In the third pass of zirconium tube pilgering that we studied here, a material element will go through

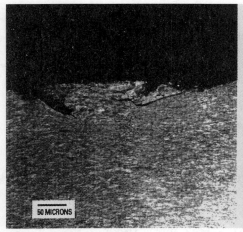

FIG. 1—*A typical external surface defect studied:* (left) *longitudinal cross section of the tube and* (right) *view of the defective surface.*

a series of about 100 strokes as it travels from entry to exit of the mandrel. Each stroke imposes only a very small strain, $\Delta\bar{\epsilon} \approx 0.02$ or 0.03. Different stress states prevail at different orthoradial positions, θ, around the mandrel cross-section and the metal particle experiences all of those states of stress as it rotates after each back-and-forth movement of the rolls.

In the first phase of the research, we investigated the influence of pilgering conditions on internal surfaces [1]. In particular, we built a "two dimensional model" disregarding rotation and changing the tube to a sheet. This model was derived from the classic analysis by Neumann and Siebel [2,3] that emphasized the normal stress in the groove bottom and the roll separation force. In the present paper, we investigated a more involved "three dimensional model" that includes rotation and determines the complete stress tensor. The complexity of the analysis of pilgering arises from at least three conditions:

1. First, the basic tool geometry is complex, that is, mandrel radius evolution and groove profile follow a parabolic law; and the transverse profiles with side relief are difficult to account for and represent a large number of parameters.
2. The process dramatically depends on tool elastic deformation, that is, the axial movement of the mandrel separates the roll and mandrel stations, the roll separation force creates the roll gap so that the evolution of the diametral and thickness reduction is disturbed and roll flattening in the groove bottom strongly influences the state of stress.
3. Finally, dynamic effects may arise from the accelerations caused by the back-and-forth movement of the rolls (up to 260 cpm).

The finite element method (FEM) is now a well established tool for analyzing the plastic flow and forming processes [4,5]. However, because of the complex three-dimensional geometry, the need to consider tool deformation, along with the non-steady-state kinematics of the process, pilgering precludes the use of FEM except for rough estimates [6] because it would require enormous meshes and central processing unit (CPU) time. Here, it will only help us study specific local phenomena with simplified geometries. The global model is based on the slab method and similar analyses, where the hypotheses will be derived from the experiments and observations on the pilger mill. The basic ideas of Hayashi [7] have been updated, and we have elaborated on the previous work of Farrugia [8] that resulted from the thorough measurement of forces and displacements for an instrumented industrial rolling mill.

Experimental Procedures

All of the experiments were conducted in rolling tube preforms on an instrumented pilger mill (Mannesmann 25 VMR) down to inner/outer diameters of 8.3/9.5 mm. The transducers presented in Fig. 2 were specially designed to resist the aggressive chemical (lubricant) and electrical environment. The roll separation force (vertical) and the axial (horizontal) forces applied to both the mandrel and preform were measured using four strain-gage transducers. The preform displacement, the tube end, and the mandrel end displacement were measured using linear voltage differential transformer (LVDT) transducers.

Analysis of Displacements

The axial displacements are pictured in Fig. 3. Positive displacements represent a movement forward (or downstream). The mandrel, tube, and preform displacements are compared in Fig. 3a, and the elongation (tube-preform) and sliding (mandrel-preform) are thrown in Fig. 3b. Zero mandrel displacement is assumed when the force is zero. The origins of tube and preform

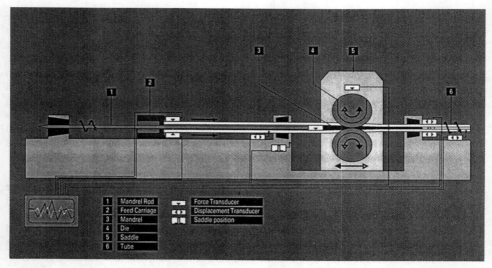

FIG. 2—*Schematic view of the location of the transducers on the pilger mill for the experimental study.*

displacements (being more or less arbitrary) were chosen at the end of the forward pass. Note that the displacement curves are quite sensitive to the dynamic effects (that is, the number of strokes per minute).

For the tooling, the mandrel undergoes a tensile force during the forward stroke and a compressive force during the return stroke. From the similarity between the mandrel displacement and force evolution, the stiffness of the mandrel-mandrel rod system can be determined to be 3000 N/mm during the forward stroke. It is smaller during the return stroke (1500 N/mm) due to elastic buckling and bending of the mandrel rod. The mandrel displacement is important because it changes the tool design; therefore, the measured curve is introduced into the model.

The preform displacement curve reflects the competition between tube elongation and feeding on one hand (forward movements, partly hindered by the tube sticking on the mandrel) and the axial forces (forward or backward oriented) applied by the rolls on the other hand. Figure 3b shows that, for the tool considered, the preform first slips forward ($U_{mandrel}-U_{preform}$ decreases) during the forward stroke, then slips backward during the return stroke. Comparing the preform displacement and force, the preform/carriage stiffness is found to be 2500 N/mm during the forward stroke and 3000 N/mm for the return stroke. The mandrel/tube and mandrel/preform sliding curves are important because their signs determine the direction of the friction forces.

The relative movement of the tube along the mandrel is complex. The wave propagation of the tube deformation (Fig. 4) is mentioned by practically all researchers [2,3,7,9], but no measurement has ever been reported in the literature. We have conducted some interrupted rolling, stopping the rolls at chosen stations during a forward stroke. Figure 4a shows the wave at Station 0.55. Clearly, a bulge is seen downstream of the roll; a sketch of the phenomenon is shown in Fig. 4b. This bulge is about 0.1 mm in diameter and is observed in both the internal and external surfaces: it is not a thickened, but a bent area. This phenomenon has

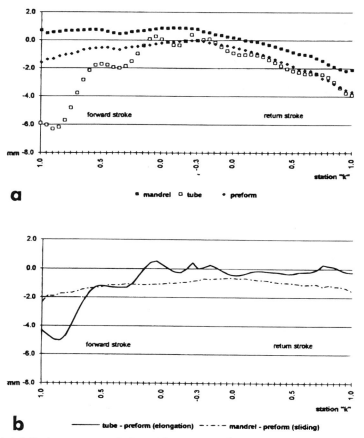

FIG. 3—*Axial displacements, evolution with station number:* (a) *mandrel, preform, and tube end displacements and* (b) *mandrel-preform (sliding) and tube-preform (elongation).*

been reproduced by the FEM (simplified axisymmetric geometry, Code FORGE2®).[4] This approach showed that the existence of bulges and wave propagation requires a restriction to the sliding capabilities of the tube, for instance, sticking of the tube on the mandrel, or blocking of the preform by the feed carriage. Such a phenomenon has been understood and studied qualitatively, along with its consequences on the state of stress, but this local feature cannot be included at present in the global model that we have built.

The Vertical Roll Separation Force and Its Consequences

A typical experimental curve is shown in Fig. 5 for both forward and return strokes Standard deviation is about 2.10^3 N, indicating good reproducibility. The force never falls down below 3.10^4 N, that is, the pre-constraint force imposed before the beginning of rolling. Measurement of the force is only possible if the tube/contact force is larger than this pre-constraint force;

[4] Transvalor, Sophia-Antipolis, France.

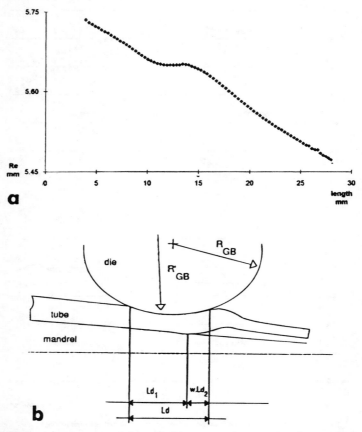

FIG. 4—*The wave downstream of the roll:* (a) *measured external radius and* (b) *a schematic view of the contact area.*

under this condition, a roll gap, G, exists (refer to Fig. 7b later) [1]. It has been measured by interposing an indium wire between the lateral treads of the two rolls during a single forward or return stroke. The measurement of its thickness gave the roll gap evolution under quasi-production conditions (Fig. 5a). Comparing gap and roll force curves, measured for several rolling conditions and tools, a stand stiffness of $7.4 \cdot 10^5$ N/mm has been deduced.

Another consequence of the roll force and contact stresses is the roll elastic deformation, specifically, the roll flattening that significantly influences the contact length, Ld, and stresses. Figure 5b shows a comparison of measured contact length, a classic formula by Neumann and Siebel [2], and a corrected formula accounting for roll flattening. Measurement consisted of spreading paint at a given station on a partly rolled preform, then closing the stand with the force corresponding to this station; the area from which the paint had been squeezed out was taken as the contact area. It was shown that, as measured by Yoshida et al. [3], the contact length is only slightly dependent on the orthoradial position around the mandrel. The measured contact length is significantly larger than the one calculated from the formula of Siebel and Neumann [2]

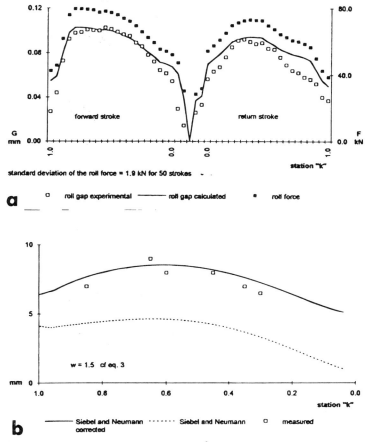

FIG. 5—*Roll separation force and its consequences:* (a) *evolution of the roll force and roll gap with station number and* (b) *comparison of experimental contact length* Ld, *with theory (Eqs 1 and 3).*

$$Ld = Ld_1 + Ld_2 = \sqrt{2mR_{GB}\frac{S_1}{S_k}\{\psi_k - \delta_k\}} + \sqrt{2mR_{GB}\frac{S_1}{S_k}\delta_k} \qquad (1)$$

where Ld_1 corresponds to a double contact mandrel/tube/roll, whereas Ld_2 is an area of single tube/roll contact, that is a kind of "tube-sinking" area undergoing much smaller normal stress.

The difference between Eq 1 and our measurement is mostly connected with roll flattening, a well-known phenomenon in strip rolling. Therefore, we corrected Eq 1 using Hitchcock's formula [*10*] that gives the deformed, equivalent roll radius

$$R'_{GB} = R_{GB}\left(1 + \frac{16(1-\nu^2)}{\pi E}\frac{F}{\Delta e}\right) \qquad (2)$$

Also, a factor ($w\ Ld_2$) for the increase of the contact caused by the wave needs to be inserted into the equation, where w is the wave factor. This fitting factor is greater than 1 and has been occasionally found equal to 4; the normal value is about 1.5 to 2. Therefore, the contact length can be expressed by the following formula

$$Ld = \sqrt{2mR'_{GB} \frac{S_1}{S_k} \{\psi_k - \delta_k\}} + w\sqrt{2mR_{GB} \frac{S_1}{S_k} \delta_k} \qquad (3)$$

The analysis of the results showed that Hitchcock's formula and the wave factor could be considered sufficient as a first approach (Fig. 5b), although numerical tests with a more sophisticated strip rolling model [11] showed it to be only approximate. Under real pilgering conditions, roll flattening can multiply the contact length by a factor greater than 2. This is therefore a very important effect that should never be overlooked when analyzing the plastic deformation of the tube.

Measurement of Shearing Strains

Torsion of the tube (pilgering helix) is known to occur during pilgering. It is also known that a radial marker in the preform will be oblique at the exit. These two observations show that shear strains are developed. Their prediction is extremely difficult under real pilgering conditions, therefore their measurement is a prerequisite for their inclusion in the model. Special preforms have been prepared by drilling holes (0.4 mm diameter) every 4 mm, with four holes every 90° on a section, filled with a gold-silver-copper alloy. These preforms have been partially rolled so as to observe the evolution of the location and shape of the inserts that can be revealed by etching and microscopic observation. First, the evolution of the orthoradial position of the holes was measured. The initial straight line changes to an helix with a pitch of 0.6 to 0.9 m for the reference tool. This torsion corresponds to a very small strain

$$\epsilon_{\theta z} \approx \frac{2}{\sqrt{3}} \pi \frac{N}{L} R \approx 0.02 \text{ to } 0.03 \qquad (4)$$

where N (revolutions)/L (length) is the inverse pitch and R is the average finished tube radius. Although the torsion phenomenon is conspicuous, its effect in the model is negligible. On the contrary, the ϵ_{rz} shear shown in Fig. 6 is quite large. The angle, γ, after pilgering is 40° to 55° for the reference tool, giving

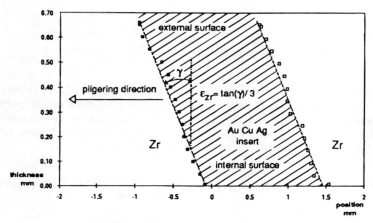

FIG. 6—*Measurement of ϵ_{rz} shear.*

$$\epsilon_{rz} \approx \frac{1}{\sqrt{3}} \tan(\gamma) \approx 0.5 \text{ to } 0.8 \tag{5}$$

a non-negligible value with respect to the other strain components. A numerical approach by the FEM (code FORGE2), in a simplified, axisymmetric geometry, has shown that the incremental strain, $\Delta\epsilon_{rz}$, is proportional to the elongational strain increment, $\Delta\epsilon_{zz}$, and also depends on the friction coefficients (internal and external) and the sliding factor that are determined by tool design. It is also quite sensitive to anything that hinders a tendency to a movement forward (tube sticking on mandrel) or backward (restriction by the feed carriage)

Development of the Model

The first step in the development of the model exploits the geometric evolution of the tube imposed by the tool design (corrected by the experimental sliding and roll gap) and derives the increment of strain due to a single stroke (we will hereafter focus on the forward stroke). Then, the Von Mises plastic flow rule determines the deviant stress components. Finally, the slab method is used to compute the axial stress, σ_{zz}, that completes the determination of stress components. Note that the shear stresses can be determined by this approach only if experimental shear strain components are introduced at the beginning of the computation.

The stresses are then integrated to give the roll separation force and the axial force. Also, a damage function is derived from the stresses.

Geometric Analysis of Strain

Axial Movements—Figure 7a shows schematically the displacement of the point of the tube during the *n*th forward and return stroke:

1. After feeding (feed length, *m*), the particle coming from H_{n-1} is at point K_n; the tube section there is SK_n. S_1 is the initial tube section

$$XK_n = XH_{n-1} - m \tag{6}$$

2. After the forward pass, the length, *m*, has been elongated by a factor, S_1/SK_n, so that the particle reaches Point J_n

$$XJ_n = XK_n - m(S_1/SK_n - 1) + Uf(XK_n) = XH_n + Uf(XK_n) \tag{7}$$

accounting for the mandrel movement of the forward stroke. *Uf* (the measured value, see Fig. 3, is introduced into the model that could not be predicted until now).

3. After the return stroke, where almost no elongation occurs, only the mandrel position has been altered (mandrel movement, *Ur*, is referred to the equilibrium position)

$$XI_n = XH_n + Ur(XK_n) \tag{8}$$

4. Just before feeding, the tube and mandrel are unloaded and the point comes to XH_n.

Strain Increments—The strain increments (for the forward stroke) will be considered homogeneous around a cross section:

1. The axial strain increment

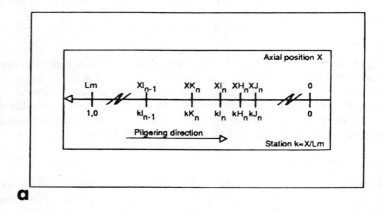

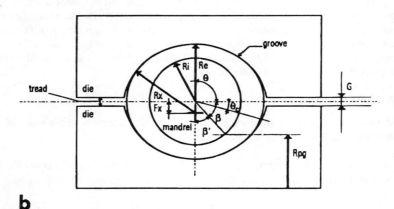

FIG. 7—*Geometrical variables involved in the model: (a) definition of the axial position involved in the computation of strains and (b) cross section of the tube/mandrel/roll groove system.*

$$\Delta\epsilon_{zz} = \ln\left(\frac{SI_{n-1}}{SJ_n}\right) \quad \text{(elongation)} \tag{9}$$

2. The orthoradial strain increment

$$\Delta\epsilon_{\theta\theta} = -\ln\left(\frac{DI_{n-1}}{DJ_n}\right) \quad \text{(diametral evolution)} \tag{10}$$

where D, the average diameter, comes from an analysis described later. In this equation, proper account is taken of the tube rotation between stroke $n - 1$ and n, and of the variations of the average diameter around the section.

3. The radial strain increment

$$\Delta\epsilon_{rr} = -(\Delta\epsilon_{zz} + \Delta\epsilon_{\theta\theta}) \quad \text{(plastic incompressibility)} \tag{11}$$

4. The strain tensor is completed by the measured shear strain increment, the global value, ϵ_{rz}, given previously is introduced in the following formula so that

$$\Delta\epsilon_{rz} = \epsilon_{rz} \frac{\Delta\epsilon_{zz}}{\epsilon_{zz}} (\beta' - (\beta - \theta_c)) \tag{12}$$

where ϵ_{zz} is the global axial strain. The β, β', and θ_c angles are defined in Fig. 7b and represent the sliding contribution to the shear strain increment, ϵ_{rz}.

Computation of Cross Sections and Average Diameters—Figure 7b presents the cross section of the system. The mandrel is circular and the groove is essentially a circle with a center shifted with respect to the mandrel center. The roll gap due to the separation force modifies the shift. Moreover, the tube has been shown experimentally to be in contact only partially (Angle θ_c) [1]; in the non-contacting zone, the tube thickness is taken constant, equal to the thickness at θ_c:

1. External surface equation

$$R_e(X, \theta) \approx R_x(X) - (F_x)(X) - G(X)/2)\cos\theta \tag{13}$$

2. Internal surface equation

$$R_i(X, \theta) = R_m(X) \quad \text{if } 0 \leq \theta \leq \theta_c \tag{14a}$$

$$R_i(X, \theta) \approx R_m(X) + (F_x(X) - G(X)/2)(\cos\theta_c - \cos\theta) \quad \text{if } \theta_c \leq \theta \leq \pi/2 \tag{14b}$$

The cross section and the average diameter are given by

$$S(X) = 4 \int_0^{\pi/2} (R_e^2 - R_i^2) d\theta \tag{15}$$

$$D(\theta, X) = R_e + R_i = R_m + \sqrt{R_x^2 - \sin^2\theta(F_x - G/2)^2} - (F_x - G/2)\cos\theta \tag{16}$$

Application of the Von Mises Plastic Flow Rule

This equation relates the plastic strain rate, $\dot{\epsilon}$, to the stress deviator, s

$$\dot{\epsilon} = \frac{3}{2} \frac{\bar{\dot{\epsilon}}}{\sigma_0} s \tag{17}$$

where σ_0 is the yield stress and $\bar{\dot{\epsilon}}$ is the equivalent strain rate. From Eq 17, stress components are deduced

$$\sigma_{\theta\theta} - \sigma_{zz} = \frac{2}{3} \frac{\dot{\epsilon}_{\theta\theta} - \dot{\epsilon}_{zz}}{\dot{\epsilon}} \sigma_0 = \sqrt{\frac{2}{3}} \frac{\Delta\epsilon_{\theta\theta} - \Delta\epsilon_{zz}}{\sqrt{\Delta\epsilon_{rr}^2 + \Delta\epsilon_{\theta\theta}^2 + \Delta\epsilon_{zz}^2 + 2\Delta\epsilon_{rz}^2}} \sigma_0 \quad (18a)$$

$$\sigma_{rr} - \sigma_{zz} = \frac{2}{3} \frac{\dot{\epsilon}_{rr} - \dot{\epsilon}_{zz}}{\dot{\epsilon}} \sigma_0 = \sqrt{\frac{2}{3}} \frac{\Delta\epsilon_{rr} - \Delta\epsilon_{zz}}{\sqrt{\Delta\epsilon_{rr}^2 + \Delta\epsilon_{\theta\theta}^2 + \Delta\epsilon_{zz}^2 + 2\Delta\epsilon_{rz}^2}} \sigma_0 \quad (18b)$$

$$\sigma_{rz} = \frac{2}{3} \frac{\dot{\epsilon}_{rz}}{\dot{\epsilon}} \sigma_0 = \sqrt{\frac{2}{3}} \frac{\Delta\epsilon_{rz}}{\sqrt{\Delta\epsilon_{rr}^2 + \Delta\epsilon_{\theta\theta}^2 + \Delta\epsilon_{zz}^2 + 2\Delta\epsilon_{rz}^2}} \sigma_0 \quad (18c)$$

so that the state of stress is now determined if one of the diagonal components can be explicitly computed; this limitation is a consequence of the incompressible character of plastic deformation.

Computation of σ_{zz} by the Slab Method

The equilibrium of forces applied to an elementary slab of the tube gives a first-order ordinary differential equation that is easily integrated to give

$$\sigma_{zz}(x) = \frac{K_0 + \int A \cdot D \left[B - \frac{de}{dx} \right] dx}{De} \quad (19)$$

where D is the average diameter; e is the thickness; A, B, and K_0 are constants computed from friction coefficients, contact angles ($\approx 80°$) on internal and external surfaces, respectively, and the mandrel and roll local cone angles. The integral is to be taken on an interval of length, Ld, given by Eq 3. In fact, the main difficulty is in the choice of the boundary conditions through K_0. At the present time, a boundary condition at the downstream end of the contact is deduced from the restricting force imposed by the exit chuck (which has been measured to be about 200 N). However, the finite element study has shown how the carriage and stuck areas, blocking the tube and interacting with its forward or backward movement, could generate different patterns of axial stress.

Finally, an average value of σ_{zz} along the contact length, Ld, is taken to complete the stress tensor.

Computation of the Global Forces

Integration of the vertical projection of σ_{rr} over the contact surface gives the separation force. Due to the very small angles, the contributions of the axial component and the friction stress are ignored

$$F = \sigma_{rr} \cdot Ld \cdot D \quad (20)$$

The global axial force is given by the axial stress at the entry of the contact (Position $X + Ld$)

$$F_{ax} = S(X + Ld) \cdot \sigma_{zz}(X + Ld) \quad (21)$$

and is to be compared with the sum of the experimental mandrel and preform forces. The partition of the computed axial force between mandrel and preform is not possible at the present time, because it depends on dynamic and elastic characteristics of the system.

Damage Function [12]

Once the stress state and the strain increments are computed, the damage function, LC, is readily derived by

$$LC = \sum_{stroke} \text{Max}(0, \sigma_I, \sigma_{II}, \sigma_{III}) \Delta \bar{\epsilon} \quad (22)$$

where σ_I, σ_{II}, and σ_{III} are the principal stresses (supposed to be active only if tensile) and $\Delta \bar{\epsilon}$ is the incremental equivalent strain

$$\Delta \bar{\epsilon} = \sqrt{\frac{2}{3}(\Delta \epsilon_{zz}^2 + \Delta \epsilon_{\theta\theta}^2 + \Delta \epsilon_{rr}^2 + 2\Delta \epsilon_{rz}^2)} \quad (23)$$

Such a damage function is not supposed to predict accurately the occurrence of a given type of defect, but it should make possible the comparison between different rolling conditions.

Confrontation Model—Experiment

Stress Computation

Figure 8 shows the results of the calculation of stresses applied to one particle at each stroke during its motion from entry to exit of the mandrel. We can observe that σ_{rr} and $\sigma_{\theta\theta}$ are very compressive with mean values of about -870 MPa and -720 MPa respectively. σ_{zz} is lightly compressive with a mean value of about -20 MPa. The shear stress, σ_{rz}, is positive and its mean value is about 100 MPa. It should be noted that the effect of tube rotation generates a variation of stress from one stroke to another of about 500 MPa for $\sigma_{\theta\theta}$. This is due to the nonuniformity of the wall thickness around a cross section.

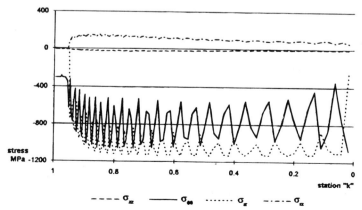

FIG. 8—*Evolution of the stress components along the stroke.*

Roll Separation Force

In order to verify the preceding calculation, some comparisons between experiment and computation are made. Figure 9a shows the results of Case (a) that we have named the reference case. Figure 9b shows the results of Case (b) where a higher Qp factor (that is, $Qp = 1.43\ Qp_{\text{ref}}$) is used. Qp is defined by the following expression

$$Qp = \frac{\dfrac{e_1 - e_0}{e_1}}{\dfrac{D_1 - D_0}{D_1}} \tag{24}$$

where e_1 and e_0 are respectively the wall thickness of the preform and of the finished tube and D_1 and D_0 are respectively the outside diameter of the preform and the finished tube. Figure 9c shows the results of Case (c) where the feed length (that is, $m = 1.36\ m_{\text{ref}}$) has been changed. Figure 9d shows the results of Case (d) for another gear pinion radius (that is, $R_{\text{pg}} = 0.98\ R_{\text{pgref}}$).

In all of these cases, we have a reasonably good agreement between experiment and calculation. We can observe that for the lower station (that is, 0.4 to 0) the computed forces are lower than the experimental forces. Our interpretation is that this deviation is due to elastic components that are ignored in this model.

To perform these calculations, we had to introduce three experimental coefficients:

1. mandrel motion, U;
2. wave factor, w; and
3. shear strain, ϵ_{rz}.

In the reference case, measurements of these parameters were made as described earlier and introduced into the computation (Fig. 9a, $\epsilon_{rz} = 0.6$). In other cases, w and ϵ_{rz} were determined by adjusting the computed force curve to the experimental one. In Fig. 9a, we have also drawn the curve for the reference case with its fitting value, $\epsilon_{rz} = 0.44$, included in order to show the sensitivity of the model to this parameter.

Among these different cases, only the change of gear pinion radius notably modifies the roll separating force. Under this particular condition, the elongation is distributed between the forward and the return stroke. The consequence is to decrease the maximum separating force from 110 to 85 kN.

Damage Factor

Figure 10 shows the evolution of the cumulative damage function along the stroke in the different cases described earlier. We can observe that contrary to roll separation forces, large differences are found between the tools. The values range from 68 MPa for the reference tool, that is Case(a) up to 163 MPa for the lower R_{pg} that is Case (d). We can also observe that the contribution for the return stroke is very large and that the increase of the feed length and of Qp increases the damage function up to 87 MPa and 142 MPa, respectively. These differences are due to different shear strain, ϵ_{rz}, and stress, σ_{rz}. It is noted that ϵ_{rz} is deduced by fitting experiment and computed vertical roll forces. Physically, the differences in ϵ_{rz} can be explained by variations of internal friction and pilgering conditions: higher pressure, for example, Case (b), higher local temperature, for example Case (c), or poorer circulation of internal lubricant, for example, Case (d).

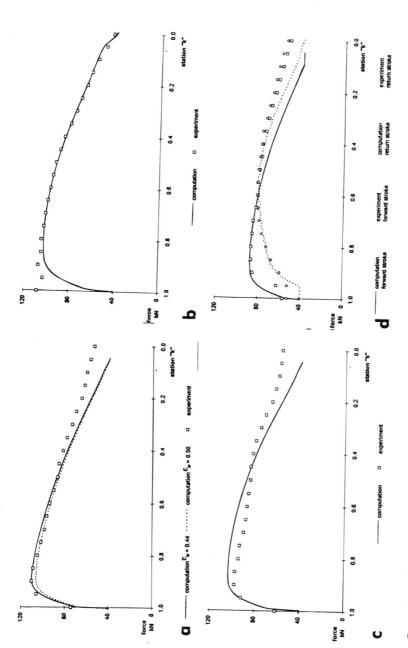

FIG. 9—*Comparison of experimental and theoretical roll separation force curves: (a) comparison of computed and measured force curves for reference condition, (b) comparison of computed and measured force curves for a higher Qp, (c) comparison of computed and measured force curves for a higher feed length, and (d) comparison of computed and measured force curves for a lower gear pinion radius.*

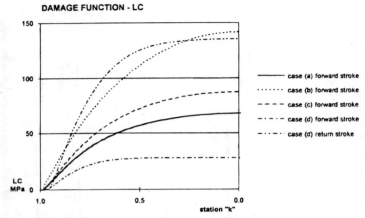

FIG. 10—*Evolution of damage function, LC, along the stroke.*

Relationships Between Damage Factor and Defects

A natural validation of this model can be shown by verifying that the damage function value is correlated to the frequency and severity of defects. For this purpose, two lots of about 400 tubes were pilgered. For one lot, the standard production tools and conditions were used (different from the reference tools described earlier), giving an LC = 79 MPa. The other lot was pilgered with an unacceptable internal lubricant, a higher feed length, and with other tools giving a higher Qp factor in order to increase LC; this model gave an LC = 315 MPa. Both lots were then checked by ultrasonic inspection, and rejected tubes were visually inspected to sort out defects of the type pictured in Fig. 1.

Frequency $f1$ is the ratio between the number of tubes with at least one defect to the total number of tubes of the lot (Table 1). Obviously, $f1$ increases with LC. However, the relationship is more clearly shown using modified Frequency $f2$ (number of defects per kilometre of pilgered tubes). Finally, a third indicator of the severity of the defect is its maximum depth (Table 1) and the depth distribution (Fig. 11) of defects.

This correlation shows that a mechanical damage function such as Eq 22 is a correct and sensitive indicator of the severity of this type of defect. The model shows that LC dramatically increases with shear strain and stresses. Preliminary tests have succeeded in decreasing ϵ_{rz} and therefore LC, and we plan to check that the frequency and severity of defects has been decreased.

Conclusion

This model has made it possible to calculate the strains and stresses applied to a particle of tube at each stroke during its motion, from the entry to the exit of the mandrel. Then the

TABLE 1—*Analysis of the frequency and the harmfulness of defects using the damage function.*

Damage Function (LC), MPa	Number of Tubes	Frequency, $f1$ (%)	Frequency, $f2$ (defects/km)	Maximum Depth, μm
79	429	1.40	4.4	35
315	438	3.42	154.5	80

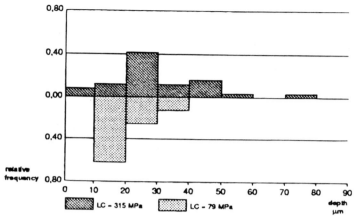

FIG. 11—*Comparison of the distribution of the defects' depth for a low and a high damage function.*

roll separation force for the forward and return stroke has been calculated. A damage function has been computed from strains and stresses. We have chosen the Latham and Cockroft (LC) criterion that correlates well with the occurrence of surface defects and their severity. The predictive aspect of this model is limited by the necessity of separately evaluating the mandrel motion, the wave factor, and the shearing strain. Nevertheless, these different factors can be evaluated by the described experiments on the pilger mill and on the tube. The next step may consist of processing experimental data into empirical law.

This study shows that it is not only possible to predict the severity of defects, but also to control the pilgering conditions in order to avoid surface defects, such as those shown in Fig. 1, through factors that influence the shearing strain. The improvement can be obtained by optimizing the lubricant and pilgering conditions and trying to design a "defect free tool."

From a more general point of view, the set of measurements that has been designed and implemented on the pilger mill is able to provide an in-depth knowledge of the pilgering process, which itself is a source of future improvements.

References

[1] Montmitonnet, P., Farrugia, D., Aubin, J. L., and Delamare, F., "Internal Surface Roughness of Cold Pilgered Zircaloy Tubes," *Wear,* Vol. 152, 1992, pp. 327–342.
[2] Siebel, E. and Neumann, F. W., "Das Kaltpilgern von Rohren—Versuch Ergebnisse aus Untersuchungen über dem Walzvorgang," *Stahl und Eisen,* Vol. 74, No. 3, 1954, pp. 139–145.
[3] Yoshida, H., Matsui, T., Otani, T., and Mandai, K., "Experimental Investigation of the Cold Pilgering of Copper Tubes," *Annals CIRP,* Vol. 24, 1975, pp. 191–197.
[4] Kobayashi, S., Oh, S. I., and Altan, T., *Metal Forming and the Finite Element Method,* Oxford University Press, Oxford, 1989.
[5] *Proceedings,* NUMIFORM'92, J. L. Chenot, O. C. Zienkiewicz, and R. D. Wood, Eds., Balkema, Rotterdam, 1992.
[6] Osika, J. and Libura, W., "Mathematical Model of Tube Cold Rolling in Pilger Mill," *Journal of Materials Processing Technology,* Vol. 34, 1992, pp. 325–332.
[7] Furugen, M. and Hayashi, C., "Application of the Theory of Plasticity to the Cold Pilgering of Tubes," *Journal of Mechanical Working Technology,* Vol. 10, 1984, pp. 273–286.
[8] Farrugia, D., "Etude mécanique et tribologique du laminage à pas de pèlerin," thesis, Ecole des Mines de Paris, 1990 (in French).
[9] Haleem, A. S., Cole, I. M., and Sansome, D. H., "Measurement of Contact Pressures in Tube Rolling," *Journal of Mechanical Working Technology,* Vol. 1, 1977, pp. 153–168.

[10] Hitchcock, J. H., "Roll Neck Bearings," Report of ASME Research Committee, 1935.
[11] Montmitonnet, P., Wey, E., Delamare, F., Chenot, J. L., Fromholz, C., and de Vathaire, M., "A Mechanical Model of Cold Rolling. Influence of the Friction Law on Roll Flattening Calculated by a Finite Element Method," *Proceedings*, 4th Steel Rolling Conference, Deauville, France, June 1987.
[12] Latham, D. J. and Cockroft, M. G., "Ductility and Workability of Metals," *Journal*, Institute of Metals, Vol. 96, 1968, pp. 33–39.

DISCUSSION

B. Narayan[1] (*written discussion*)—Will you define "defect-free tool?"

J. L. Aubin et al. (authors' closure)—A defect-free tool is a set of tools (dies and mandrel) specially designed in order to improve the internal friction conditions. This is done mainly by optimizing the side relief.

[1] Westinghouse Electric Corporation, Blairsville, PA.

Christopher E. Coleman,[1] George L. Doubt,[1] Randy W. L. Fong,[1] John H. Root,[1] John W. Bowden,[1] Stefan Sagat,[1] and R. Terrence Webster[2]

Mitigation of Harmful Effects of Welds in Zirconium Alloy Components

REFERENCE: Coleman, C. E., Doubt, G. L., Fong, R. W. L., Root, J. H., Bowden, J. W., Sagat, S., and Webster, R. T., "**Mitigation of Harmful Effects of Welds in Zirconium Alloy Components,**" *Zirconium in the Nuclear Industry: Tenth International Symposium, ASTM STP 1245,* A. M. Garde and E. R. Bradley, Eds., American Society for Testing and Materials, Philadelphia, 1994, pp. 264–284.

ABSTRACT: Welding produces local residual tensile stresses and changes in texture in components made from zirconium alloys. In the heat-affected zone in tubes or plates, the basal plane normals are rotated into the plane of the component and perpendicular to the direction of the weld. Thin-walled Zircaloy-2 tubes containing an axial weld do not reach their full strength because they always fail prematurely in the weld when pressurized to failure in a fixed-end burst test. Reinforcing the weld by increasing its thickness by 25% moves the failure to the parent metal and improves the biaxial strength of the tube by 20 to 25% and increases the total elongation by 200 to 450%. In components made from Zr-2.5Nb, the texture in the heat-affected zone promotes delayed hydride cracking (DHC) driven by tensile residual stress. Although the texture is not much affected by heat-treatments below 630°C and large grain interaction stresses remain as a result of mixed textures, macro-residual tensile stresses can be relieved by heat treatment to the point where the probability of cracking is very low.

KEY WORDS: welds, heat-affected zone, burst strength, delayed hydride cracking, residual stress, post-weld heat treatment, zirconium, zirconium alloys, nuclear materials, nuclear applications, radiation effects

Welding zirconium produces residual stresses and local changes in microstructure and crystallographic texture. Any use of a welded component has to take these features into account because they affect properties and thus behavior. Many nuclear components made from zirconium alloys, such as fuel and support structures, are welded and perform with few problems. In this paper, we describe two applications:

1. the calandria tubes in the CANDU reactor have given exemplary service, but to increase margins we would like to improve their strength; and
2. some pressure tubes are welded and must be protected from delayed hydride cracking (DHC).

---

[1] Research scientists, Atomic Energy of Canada Ltd., AECL Research, Chalk River Laboratories, Chalk River, Ontario, Canada K0J 1J0.
[2] Metallurgical consultant, Scio, Oregon 97374.

Zircaloy-2 Calandria Tube

In the fuel channel of a CANDU reactor, the calandria tube surrounds the hot pressure tube and isolates it from the cool heavy-water moderator. The gap between the two tubes is filled with circulating dry CO_2 that acts as an insulator and, in the event of a through-wall crack in the pressure tube, provides a leak detection system. The calandria tube controls the sag of the fuel channel; as demonstrated in Unit 2 of Pickering NGS, the calandria tube may also contain the pressure of the heat-transport fluid when a pressure tube ruptures [1]. We are exploring ways to strengthen the calandria tubes to increase the margin on the consequences of a failed pressure tube. Thus, the strength being addressed is the burst strength of the tube when it is internally pressurized rapidly rather than the creep strength.

To manufacture calandria tubes, annealed Zircaloy-2 strip is brake-formed and seam-welded [1]. The tubes are 6 m long, have an inside diameter of 130 mm, and a wall thickness of 1.4 mm. They normally operate at a temperature of about 70°C but, when exposed to the heat-transport fluid, the average temperature is raised to about 170°C. The tubes are firmly attached to the ends of the reactor structure and when pressurized are stressed in fixed-end mode. The importance of fixed-end biaxial stressing is that it provides a large amount of texture strengthening. The basal plane normals are highly concentrated in the plane perpendicular to the axial direction and, depending on the fabrication route, are localized between 0° and 40° from the thickness direction in the sheet from which the tubes are made. As a result, the material is highly anisotropic. By applying Hill's analysis [2] to fixed-end stressing, we find [3] that the biaxial strength, σ_H, is predicted to be related to the uniaxial strength, U, through

$$\sigma_H = U_A \sqrt{\frac{(1 + R)^2 P}{R(1 + R + P)}} \quad (1)$$

$$= U_T \sqrt{\frac{(1 + R)(1 + P)}{(1 + R + P)}} \quad (2)$$

Subscripts A and T denote the axial and transverse directions, respectively. The anisotropy factor, R, is determined from the dimensions of longitudinal tension specimens

$$R = \frac{\ln(w/w_0)}{\ln(t/t_0)}$$

The width and thickness of the tensile specimen are written w and t; subscript 0 indicates an initial dimension. The anisotropy factor, P, is similarly determined from transverse tension specimens. With R and P in the range 2 to 7, we expect a biaxial strength between 35 and 80% greater than the uniaxial strength.

Tubes always fail along the weld in fixed-end burst tests (Appendix I) on both irradiated and unirradiated tubes. Comparison of uniaxial and biaxial strengths (Table 1) shows that the full texture strengthening is not realized, even though the weld has a higher strength than the main body of the tube. The texture of the welds has been evaluated by neutron diffraction (Appendix II). The high concentration of basal plane normals in the radial direction, the source of the texture strengthening, is much reduced in the weld, Fig. 1, and in biaxial stressing provides the weak link by allowing wall thinning that leads to rupture. We have examined methods of strengthening calandria tubes and concluded that to maintain the minimum quantity of parasitic material without destroying the current excellent properties we have to concentrate on eliminating the effect of the weld. The potential improvement in strength can be estimated

TABLE 1—*Development of biaxial strength in Zircaloy-2 calandria tubes: mean value (range).*

Material	Ultimate Tensile Strength, MPa			R	P	Burst Strength, MPa			
	Transverse		Longitudinal				Predicted from		Weld Reinforcement: break in parent metal
	Parent Metal	Weld	Parent Metal			As-Received	Eq 1	Eq 2	
Sample 1: basal planes 40° from radial direction, oxygen concentration 0.74 at% (1300 ppm)	320	350	350	2.5 (2 to 3)	4.4 (3.5 to 5.3)	535 (510 to 560)	577 (545 to 610)	493 (461 to 525)	675 (640 to 710)
Sample 2: basal planes in radial direction, oxygen concentration 0.69 at% (1200 ppm)	305	360	320	5 (4 to 6)	5.5 (4.4 to 6.6)	550 (520 to 580)	595 (550 to 640)	560 (520 to 600)	665 (630 to 700)

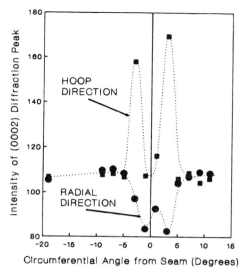

FIG. 1—*Variation in the density of (0002) poles in the radial and transverse directions, near the seam weld in a Zircaloy-2 calandria tube.*

as the ratio of the full texture strengthening to the biaxial strength of the weld, assuming it to be isotropic

$$\text{potential improvement} = \frac{\sigma_H}{1.15\sigma_u^w} \quad (3)$$

where σ_u^w is the uniaxial strength of the weld. Applying Eq 3 to the data in Table 1 suggests that if the tube fails outside the weld an improvement in strengthening of at least 10% should be achievable, with values as high as 40% being possible.

Two potential methods for achieving this extra strength are to thicken the weld or make seamless tubes. To produce a thick weld, we have chosen to use uniform strip that is thicker than that currently used, to fabricate the tube with present methods, then to remove the excess material to the desired thickness except for the weld. The aim of the test program was to measure how thick the weld had to be to ensure failure in the main body of the tube then evaluate the benefit provided by the change in fabrication route. Prototype seamless tubes have been made by roll-extrusion [4,5].

Welded Tubes

The material and the tubes used in these tests were made at four different times by two strip suppliers but a single tube fabricator. Three orders of tubes had a concentration of basal plane normals about 40° from the radial direction and an oxygen concentration of 0.74 atomic percent (1300 ppm by weight); this material is called Sample 1. In the fourth order of tubes, the basal plane normals were concentrated close to the radial direction but the oxygen concentration was 0.69 atomic percent (1200 ppm); this material was called Sample 2.

Test specimens were made from standard calandria tubes by dissolving material, in a pickling solution of 3% HF, 15% HNO_3, and 82% H_2O, from the outside of the wall everywhere except along a 50-mm band centered on the longitudinal seam weld. Thus, a weld of thickness, w_w, was reinforced with respect to the rest of the tube of reduced thickness, w_r. The reinforcement, $(w_w/w_r - 1) \times 100$, was up to 40%. The specimens were pressurized to rupture in a fixed-end biaxial test (Appendix I) at 170°C. The hoop stress and strain at burst are plotted as a function of weld reinforcement in Fig. 2 for Sample 1. The lowest reinforcement that caused some specimens to rupture in the parent material was 17%, while specimens with 22% reinforcement could still fail in the weld. On average, specimens that ruptured in the parent material reached a hoop stress about 25% higher than specimens with no weld reinforcement (Table 1) confirming the promise of Eq 3. They also reached a hoop strain almost three times as high as that obtained with uniform wall tube. The strain results were variable and this is attributable to increasing sensitivity to small flaws at high strains and variation in the wall thickness from pickling.

A smaller number of tests was done on the tubing made from Sample 2. This material had a sharper texture than Sample 1, but its maximum strengths (Table 1), with or without weld reinforcement, were similar to those in Fig. 2. The average improvement to the strength by thickening the weld of these calandria tubes was about 21%. The hoop strains to rupture were much different, as shown in Table 2. The low strain values for the unreinforced tubes in Sample 2 are a result of the large difference in texture between the weld and parent material. With weld reinforcement, the hoop strain increased 450% in Sample 2, but it was still 55% lower than that of Sample 1.

The maximum hoop stresses generated in fixed-end burst tests were greater than the transverse uniaxial tensile strength by about 110% in Sample 1 and 120% in Sample 2. A similar comparison for the axial uniaxial tensile strength gives values of 93% and 107% for Sample 1 and Sample 2, respectively. These factors are greater than predicted by Hill's analysis, but we have not determined whether the cause of the discrepancy lies in the limitations of theory

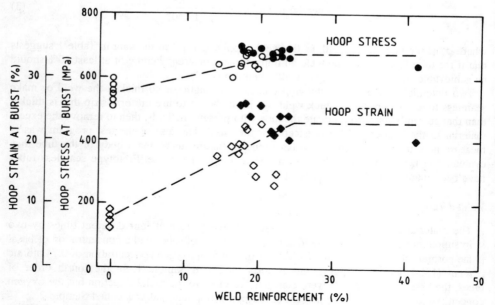

FIG. 2—*Dependence of hoop stress and strain at burst on the thickness of the seam weld in Zircaloy-2 calandria tubes. Filled symbols indicate rupture in the parent material.*

TABLE 2—*Average hoop strain at burst (number of specimens in parentheses).*

	Sample 1 (Fig. 2)	Sample 2 (sharp radial texture)
Unreinforced, rupture in weld	7.9 (14)	1.9 (12)
Reinforced, rupture in parent material	23.2 (11)	10.4 (3)

when applied at such large strains, whether the values of R and P are large enough, or whether the assumed stress state is fully applicable for the experimental method.

The weld reinforcement, by postponing the tube rupture until the parent material fails, has allowed realization of the full texture strengthening. From the results of the experiments, a difference in thickness between the weld and parent metal of 25% is recommended. The increase in hoop strain obtained by weld reinforcement is a bonus since it can be of substantial practical benefit. It improves tolerance to the waterhammer-type loading that may occur in certain postulated accidents and creates a large margin for loss of ductility from neutron irradiation.

Seamless Tubes

Prototype seamless tubes have been made by roll-extrusion. The fabrication sequence begins with extrusion of a hollow billet at 650°C followed by externally roll-extruding the annealed hollow, vacuum annealing at 700°C for 1.5 h, then internally roll-extruding the tube to final dimensions. The tubes were given a final vacuum anneal of 750°C for 0.5 h. The finished tube contained equiaxed grains (6 to 9 µm) with their basal plane normals predominately aligned along the radial direction of the tube.

The effect of a weld on burst strength was determined by testing two samples taken from the seamless tube, where one sample contained a simulated longitudinal weld and the other was seamless. Results from fixed-end burst tests at 170°C showed that the tube with the simulated weld had a burst strength of about 510 MPa and a hoop strain of about 1% at burst, with the rupture being confined to the weld. The seamless tube withstood a hoop stress greater than 570 MPa and a hoop strain greater than 2% without rupture. The tube did not burst but failed at the weld between the tube and one flange in the fixed-end burst test fixture. These results confirm that the weld is the weakest part of the tube in biaxial stressing, and support the idea that the burst strength will be increased with a seamless tube.

Welds In Zr-2.5Nb

Delayed Hydride Cracking (DHC)

To prevent DHC in Zr-2.5Nb, one of the following risk factors must be absent [6,7]:

1. hydrides—the hydrogen concentration must be less than the solubility limit so that brittle hydrides do not form,
2. tensile stress—the total stress and its amplification by flaws must be less than the critical value to fracture hydrides, and
3. time—the duration of exposure to the first two risk factors must be minimized.

The critical stress and the susceptibility to DHC are characterized by K_{IH}, the threshold stress intensity factor. The microstructure of the material affects DHC properties but K_{IH} appears to be dominated by crystallographic texture [8]. A common habit plane for hydride precipitates is close to the basal plane, and this tends to be the cracking plane. Thus, high tensile stresses parallel to basal plane normals should be avoided.

Experience with Welds

Experience with welded Zr-2.5Nb (ASTM Grades R60705 and R60901) has been mixed:

1. The welds between the end caps and fuel sheathing in experimental fuel elements cracked in the heat-affected zone (HAZ) during storage at room temperature [9]. A heat treatment of 530°C for 15 min and a modification to the design to reduce stress risers appeared to solve the problem.
2. Similar cracking at room temperature was experienced in gas tungsten arc (GTA) welds in structures used in chemical processing [7]. In one example, the cracking was initiated by an oxygen-rich layer within five months of fabrication while in another, cracking was discovered about 18 months after welding. Cracks followed the boundary between the HAZ and the fusion zone (FZ) of the weld.
3. In 1986, the Tennessee Valley Authority installed two welded pressure vessels made from Zr-2.5Nb for the hydrolysis of wood fibers to produce ethanol. Both vessels were heated to 565 ± 10°C for 1 h within 14 days after welding. The closure door on one vessel could not be heat treated because of possible warpage. Consequently, the inside face was loosely clad with commercially pure zirconium (ASTM Grade R60702) and held in place with zirconium stud bolts that were sealed by welding. The units were batch operated at 215 ± 10°C and 2.8 MPa for up to 20 h. After 40 cycles of operation and about 1.5 years after welding, the closure door began to leak. The stud bolt seal welds had cracked by DHC. The zirconium bolts were Zr-2.5Nb rather than pure zirconium. Because the welds had not been stress relieved, they were subject to DHC. The bolts were replaced with pure zirconium and the vessel went back into service. All of the equipment has been in use for five years without further failure in any of the welds.
4. A loop tube for WR-1, a research reactor at Whiteshell Laboratories, made from Zr-2.5Nb contained five GTA welds. Within a few days of welding being completed, but over a month after the first weld was done, the tube was heat-treated at 400°C for 24 h and within five months of welding the loop was operating at 230°C. Subsequent operation was at temperatures between 250 and 300°C. The tube functioned without problems for the next eleven years during which time the tube was at room temperature for 33 000 h. The welds received a low dose of neutrons, that is, about 10^{23} n/m². No cracking was observed during post-irradiation examination of the welds.
5. Electron-beam welding appears less harmful than GTA welding. Seven loop tubes for NRU, a research reactor at Chalk River Laboratories, contained 20 welds between Zr-2.5Nb and itself, Zircaloy-2 or Excel. These tubes were stored at room temperature for times up to 22 months before operating successfully at temperatures in the range 260 to 300°C. Again, no cracking was detected during post-irradiation examination.
6. A similar tube for an out-reactor loop contained two circumferential electron-beam welds one of which cracked in the axial direction 16 months after welding. The weld was redone, the tube was operating at 250°C within two months of welding and performed without incident for eight months.

7. The end caps of biaxial creep capsules were attached by electron beam welding then heat treated at 400°C for 24 h within a few hours of welding [10]. Despite internal pressurization, none of the capsules has failed by DHC.
8. None of the 50 000 electron beam welds between the diffusion joints and the main body of the pressure tube in reaktor bolshoy moshchnosty kipyashchiy (R,B,M, and K) reactors has suffered DHC after up to 16 years of reactor operation [11]. These welds were heat treated at 580°C for 10 h within a few days of welding. To date, these welds have been irradiated to about 5×10^{22} n/m².

These examples illustrate the application of the risk factors. The failures occurred after extended periods at room temperature after welding. Hydrides are always present during storage at room temperature because the solubility limit is less than 0.01 atomic percent (1 ppm). Reducing the hydrogen concentration below the solubility limit at room temperature is impractical, and thus this risk factor is always present. Once a component is operating at high temperatures, this risk factor disappears because the hydrogen is all in solution. For example, a component with an initial hydrogen concentration of 0.10 atomic percent (10 ppm) will contain no hydrides at temperatures above 170°C. Also, the risk factor associated with stress will decline because residual stresses will start to relax at elevated temperatures. Heat-treatment is effective in reducing residual stresses. No weld that was heated above 400°C soon after welding has failed. Thus, an effective antidote to the stress risk factor is to stress relieve by heat treatment before cracking can start.

Stress Relief

To estimate the time at temperature required to minimize cracking, we need to know:

1. the maximum depth of surface flaw, *a*, that can be tolerated, which is estimated from Ref *12*

$$a = C\left(\frac{K_{IH}}{\sigma}\right)^2 \tag{4}$$

where the shape factor, *C*, is about 0.4, and
2. the potential reduction in stress by stress relaxation.

As an illustration, we assume the as-welded residual tensile stress is the yield stress of Zr-2.5Nb annealed at 730°C for 1 h and calculate the change in stress and critical flaw size from the unrelaxed stress ratio[3] and Eq 4 assuming that K_{IH} is 4.5 MPa$\sqrt{m}$. The results, summarized in Table 3, show that a heat-treatment at a moderate temperature, 400 to 450°C, can increase the critical flaw size by factors of 3 to 10 while an extra 50°C reduces the residual stress to 15% of its initial value and increases the critical flaw size by a factor of over 40. These results agree with experience. To confirm the analysis and understand why the cracks initiate in the HAZ, we have evaluated by neutron diffraction the texture and residual strains in welded Zr-2.5Nb plates.

---

[3] A. R. Causey, private communication, Atomic Energy of Canada Ltd., Chalk River Laboratories, Chalk River, Ontario, Canada, 1993.

TABLE 3—*Stress relaxation and critical flaw size for cracking.*[a]

Temperature °C	Time, h	Unrelaxed Stress Ratio	Stress, MPa	Critical Flaw Size, mm
...	...	1.0	380	0.06
400	6.0	0.55	209	0.19
450	3.0	0.3	114	0.62
500	1.0	0.15	57	2.49
550	0.5	0.10	38	5.6

[a]Zr-2.5Nb annealed at 730°C for 1 h. At room temperature, the yield stress is 380 MPa.

Measurement of Texture and Residual Stress

Plates of Zr-2.5Nb, 6 mm thick, 190 mm wide, and 300 mm long were made by hot-rolling at 750°C followed by annealing at 730°C for 1 h. The plates were welded together in pairs perpendicular to the rolling direction using GTA welding. One set was left as-welded while three others were heated to either 530, 590, or 650°C for 1 h.

Texture—To measure texture, specimens were cut from the fusion zone (FZ) and from the region of the HAZ with large prior-β grains, Fig. 3. The specimens had dimensions of 30 mm parallel to the long axis of the weld, 6 mm normal to the plate, and 4 mm transverse to the weld. A third sample of base metal (BM) was obtained far from the weld.

The data are presented as pole figures in Fig. 4. There are remarkable differences in the textures observed in the BM, HAZ, and FZ, Figs. 4a, b, and c, respectively. In the HAZ, there is a pronounced maximum in the density of basal plane normals in the rolling direction. Hydride platelets will tend to precipitate on the normal-transverse plane, which accounts for the susceptibility of the HAZ to cracking parallel to the weld in the HAZ.

Heating for 1 h at 650°C effected a qualitative change in the texture of the FZ; it had no effect on the textures of the HAZ and BM. In Fig. 4c, eight markers ($\otimes$) indicate the ideal positions of the {110} family of reflections in a cube texture. The cube texture is expected to develop during solidification from the melt into the high-temperature (bcc) phase of zirconium [14]. The [110] direction transforms into the [0002] direction on subsequent cooling into the low-temperature (hcp) phase [15]. In the as-welded FZ, all of the transformation variants are present, but in the FZ that was heat-treated at 650°C, the variants at the circumference of the pole figure ($\chi = 90°$) have vanished, Fig. 4d. Unlike the heat-treatment at 650°C, those at the two lower temperatures had no effect on the weld textures.

In Figs. 4e through g, the ($10\bar{1}0$) pole figures are presented from the BM, HAZ, and FZ of the as-welded plate, respectively. In the HAZ, it can be seen that wherever there is a high intensity in the (0002) pole figure, Fig. 4b, there is a corresponding high intensity in the ($10\bar{1}0$) pole figure, Fig. 4f. Grain interaction stresses are expected in the HAZ [16].

Spatial Mapping of Texture Variations and Lattice Strains—Because of the variation in texture associated with the weld, no single diffraction peak could be chosen to determine the lattice strains over the area of interest. As the base metal has a strong transverse texture, Fig. 4a, the strains in the rolling (RD), transverse (TD), and normal (ND) directions were calculated from the respective shifts in ($10\bar{1}0$), (0002), and ($11\bar{2}0$) peaks. In the HAZ and FZ, different diffraction peaks were required to obtain measurable signals. For the HAZ of the as-welded plate, the intensities of reflections from ($10\bar{1}0$) and (0002) planes normal to the RD are presented in Fig. 5. The ($10\bar{1}0$) intensity drops precipitously on entering the HAZ, which

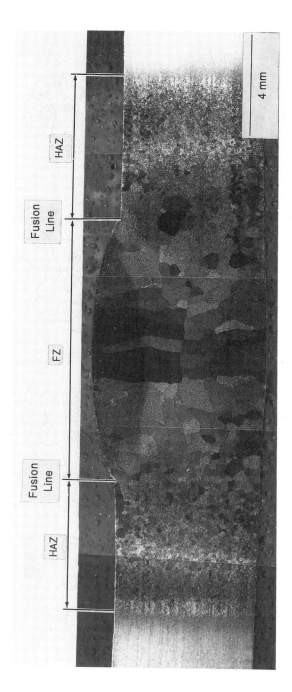

FIG. 3—*Macrostructure of a GTA weld between Zr-2.5Nb plates.*

274 ZIRCONIUM IN THE NUCLEAR INDUSTRY: TENTH SYMPOSIUM

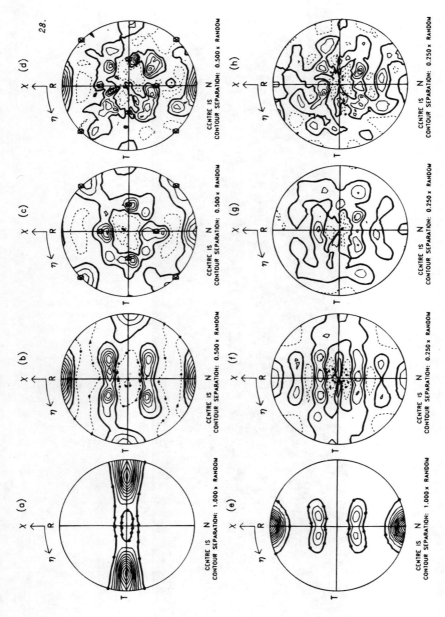

FIG. 4—*Pole figures of GTA welds between Zr-2.rNb plates: (002) as-welded: (a) base metal [13]; (b) Heat-affected zone [13], and (c) fusion zone [13]; welded and 650°C for 1 h: (d) fusion zone; (1010) as-welded: (e) base-metal, (f) heat-affected zone, (g) fusion zone; and welded and 650°C for 1 h: (f) fusion zone.*

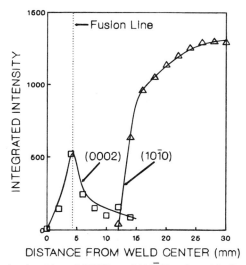

FIG. 5—*Variation in the densities of (0002) and (10$\bar{1}$0) poles with distance from the center line of a GTA weld between Zr-2.5Nb plates.*

extends to a distance of about 13 mm from the center of the weld. The (10$\bar{1}$0) intensity is too low to determine a strain value for distances less than 12 mm from the center of the weld. However, the intensity of the (0002) peak increases on entering the HAZ and reaches a maximum at the fusion line. Thus, RD strains within the HAZ could be determined from shifts in the (0002) diffraction peak. In the FZ, intensities of Bragg reflections exhibited an erratic dependence on position and the calculated strains displayed large fluctuations. Therefore, only the strains obtained in the base metal and HAZ are presented in this paper.

In Fig. 6, the intensity of (0002) reflections is plotted against position. The intensities for the heat-treated plates have been normalized to coincide with the average of the intensities for the as-welded plate at 8, 10, and 12 mm from the weld center. The normalized profiles are represented reasonably well by a single curve and thus, heat treatment does not appear to influence the density of grains with (0002) normal to the RD in the HAZ. The volume-averaged (0002) texture of the HAZ, Fig. 4b, revealed the undesirable tendency for basal plane normals to be oriented perpendicular to the weld direction; but now it can be concluded that this crystallite orientation is pronounced at the boundary between the HAZ and FZ, where the cracks develop in a weld.

Heat treatment is very effective in relieving the macroscopic residual stresses. This is evident in Fig. 7, where the TD, RD, and ND strain components in the base metal are compared for the plates (a) as-welded and (b) heat-treated at 530°C for 1 h. In the as-welded plate, the dominant strain is parallel to the long dimension of the weld. The TD strain profile, which exhibits a high tensile value near the weld and an oscillation through a complementary compressive minimum about 30 mm from the weld center, is expected to arise from the constrained shrinkage of the hot zones of the weld against the remainder of the base metal [17] and has been observed previously by neutron diffraction in a steel weld [18]. The ND and RD strains are approximately equal to each other and are mainly the Poisson's-ratio responses to the underlying TD stress field. After heat-treatment, the amplitude of the TD strain profile is reduced by about a factor of three, showing that the residual stress due to welding is relieved to a large extent.

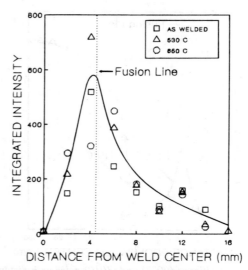

FIG. 6—*Variation in the density of (0002) poles with distance from the center line of GTA welds between Zr-2.5Nb plates, before and after heat treatment at 590 and 650°C.*

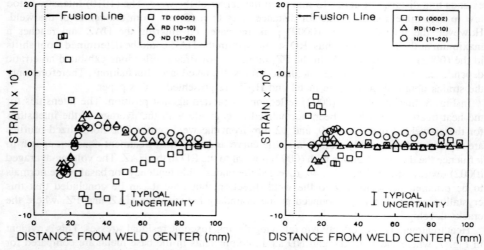

FIG. 7—*Distribution of residual strains in the transverse, rolling, and normal directions in GTA welds between Zr-2.5Nb plates: (a) as-welded, and (b) after heat treatment of 530°C for 1 h.*

Heat treatment at 530°C did not reduce the component of strain in the RD in the HAZ, Fig. 8. The pole figures of the HAZ show that in neighboring grains [0002] and $\langle 10\bar{1}0 \rangle$ may be parallel, Figs. 4b and f. The coefficient of thermal expansion (CTE) along [0002] is about twice as great as that along $\langle 10\bar{1}0 \rangle$. When the HAZ cools to room temperature, the difference in CTEs leaves [0002] in tension and $\langle 10\bar{1}0 \rangle$ in compression. The resulting stresses, called grain-interaction stresses, are not relaxed during heat treatment; they will remain at a level

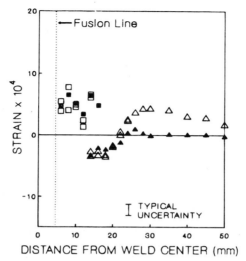

FIG. 8—*Distribution of residual strains in the rolling direction associated with GTA welds between Zr-2.5Nb plates: (0002) □ as-welded and ■ heat treated for 1 h at 530°C and (10$\bar{1}$0) △ as-welded and ▲ heat-treated for 1 h at 530°C.*

that depends on the relative volume fractions of the grains of each orientation. It follows that the component of strain in the RD results from grain-interaction stresses.

Application

To achieve approval for the use of ASTM Grade R60705 (Zr-2.5Nb) in Section VIII of the American Society of Mechanical Engineers (ASME) Boiler and Pressure Vessel Code, a test program was initiated for the mitigation of DHC in welds. Four tests were conducted on samples cut from 9-mm-thick plate of ASTM Grade R60705 annealed at 700°C for 1 h and filler welded into T-test plates. Two samples were heat treated at 566 ± 10°C for 1 h at temperature 14 days after welding. The T-section samples, Fig. 9, were subjected to static

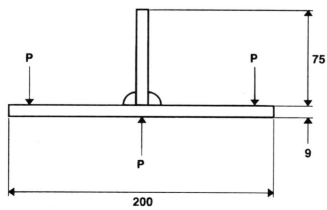

FIG. 9—*Welded T-sections used to test efficacy of heat treatment to prevent DHC in Zr-2.5Nb (dimensions are in mm).*

bending at room temperature in a three-point-bend fixture. The outer fiber stress at the welds was estimated to be 80% of the yield strength of the plate material. The first sample to break was non-heat-treated and ruptured after 35 days and the second non-heat-treated plate cracked after 14 months. The two heat-treated samples have not cracked in seven years of testing.

On the basis of the cracking experience, the various analyses, and experiments, the Boiler and Pressure Vessel Code of the ASME now provides guidelines for time at temperature within a time after welding, as well as cooling rates to avoid large temperature gradients in thick-walled structures [19].

> Within 14 days after welding, all products of zirconium Grade R60705 shall be heat-treated at 1000°F–1100°F (538°C–593°C) for a minimum of 1 h for thicknesses up to 1 in. (25.4 mm) plus $1/2$ h for each additional inch of thickness. Above 800°F (427°C), cooling shall be done in a closed furnace or cooling chamber at a rate not greater than 500°F/h (260°C/h) divided by the maximum metal thickness of the shell or head plate in inches but in no case more than 500°F/h. From 800°F, the vessel may be cooled in still air.

For some applications, this practice may be too restrictive. On a large vessel, insufficient time may be available to meet the requirement of 14 days; a lower temperature of heat-treatment may be more suitable to protect other metallurgical properties; or an electron beam weld may be less susceptible to DHC than a GTA weld. Future work will be aimed at reducing the conservatism built into the Code.

Conclusions

1. In the heat-affected zone of welds in zirconium alloy tubes or plates, the basal plane normals are rotated into the plane of the component and perpendicular to the direction of the weld.

2. The consequence for thin-walled Zircaloy-2 tubing containing an axial weld is that the tube always fails in the weld when pressurized to failure in a fixed-end burst test. Reinforcing the weld by increasing its thickness by 25% moves the failure to the parent metal and improves the biaxial strength of the tube by 20 to 25% and increases the total elongation by 200 to 450%.

3. In components made from Zr-2.5Nb, the texture in the heat-affected zone promotes DHC. Macro-residual tensile stresses can be relieved by heat treatment, for example, 530°C for 1 h, and thus much reduce the probability of cracking. Large grain interaction stresses remain as a result of mixed textures.

Acknowledgments

We would like to thank C. D. Cann, A. J. White, and R. M. Lesco for information of the performance of welds in tubes and C. E. Ells for useful discussion. We would also like to thank K. V. Kidd, K. McCarthy, R. M. Condie, and M. Cross for assistance with the experiments.

Funding for portions of this work by the CANDU Owners Group through Working Parties 22 and 33 is gratefully acknowledged.

APPENDIX I

Calandria Tube Fixed End Burst Test

Fixed-end burst tests are conducted at 170°C to simulate reactor conditions in the event of pressure tube rupture. In Fig. 10, a burst test assembly is shown with both ends of the tube

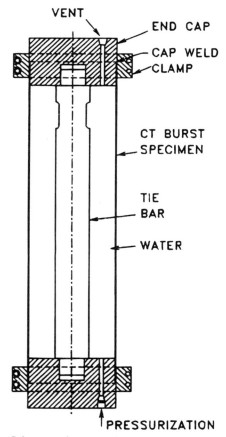

FIG. 10—*Schematic diagram of fixed-end burst test apparatus.*

secured with a tie bar to restrict axial movement, usually contraction, of the internally pressurized tube. End-caps are welded to the tube to seal the ends, and clamps prevent the welded ends from breaking when the tube bulges. To measure axial and hoop strains, foil strain gages are bonded to the tube at mid-length. Temperature is measured with type-K thermocouples attached to the tube. Burst tests were done in a steel bell jar. The sample is placed vertically inside the bell jar, heated to 170°C, and rapidly pressurized to failure. The pressure and circumference of the tube is measured at burst. Burst stress, σ_H, is given by

$$\sigma_H = \frac{PD}{2t} \qquad (5)$$

where P is the maximum pressure in MPa, and D and t are respectively the instantaneous inside diameter and initial wall thickness of the tube in mm. The maximum hoop strain, e_H, in percent at burst is written as

$$e_H = \frac{c_f - c_i}{c_i} \times 100 \tag{6}$$

where c_i and c_f are the initial and final circumferences of the tube, respectively.

APPENDIX II

Neutron Diffraction as a Probe of Texture and Strain Distributions in Zirconium-Alloy Welds

Neutrons penetrate easily through zirconium. A neutron beam is attenuated by 30% on traversing a thickness of 12 mm. Therefore, neutron diffraction can be used for nondestructive examination of engineering components. Two techniques are presented in this paper.

The first technique is the quantitative analysis of the crystallographic texture. To characterize the texture of a rolled plate, a specimen is made with one axis parallel to the normal direction. The angle, χ, is measured from the normal direction (N), while the angle η is measured counterclockwise from the rolling direction (R). An Eulerian cradle tilts the specimen through 90° in χ and rotates the specimen about its axis by 360° in η. For each direction in the specimen, (χ, η), measurements are made of the intensities of at least five neutron diffraction peaks of $(10\bar{1}0)$, (0002), $(10\bar{1}1)$, $(10\bar{1}2)$, and $(11\bar{2}0)$. The intensity is proportional to the volume fraction of crystallites with plane normals parallel to the direction (χ, η). The intensity variation of a particular diffraction peak is plotted on a stereogram. Contour lines connect points of equal intensity. Dashed lines indicate intensities less than those of a random distribution of crystallite orientations; continuous lines indicate intensities greater than those of a random distribution. A thick line indicates the intensity that would be observed if there were a completely random distribution of crystallite orientations in the material. The contour interval is quoted in multiples of the intensity of a random distribution. Because neutrons penetrate easily through the material, complete pole figures can be obtained with a single setting of a specimen on the Eulerian cradle. In all orientations, the full volume of the specimen is sampled by the neutron beam. Since neutrons probe the full volume of the specimen, many more grains are sampled than would occur in X-ray diffraction, which probes only a thin surface layer.

The second technique is the spatial mapping of texture and lattice strain. Both the incident and diffracted neutron beams can be shaped by rectangular slits in neutron-absorbing cadmium masks. The intersection of the incident and diffracted beams defines the sampling volume, typically of the order of $2 \times 2 \times 2$ mm^3, which can be set at various locations within a welded zirconium-alloy plate. The plate is moved by computer-controlled XYZ translators. At each location, a diffraction peak is measured. The raw data, neutron counts versus scattering angle (2θ), are fitted with a model function, a Gaussian on a linear background, to obtain accurate measures of the integrated peak intensity and the mean value of 2θ. The peak intensity varies with location in a welded plate because the texture is strongly affected by the thermal history of each weld zone. The mean value of 2θ also varies with location because welding produces a residual stress field that creates strains in the crystal lattices of the grains. By comparing the value of 2θ obtained at a particular location within the weld with the $2\theta_0$ value obtained from a specimen that is stress-free, the lattice strain, ϵ, can be deduced from the relationship

$$\epsilon = \frac{\sin \theta_0}{\sin \theta} - 1 \tag{7}$$

Outside the FZ, the principal axes of the residual stress field are expected to be the symmetry directions of the rolled plate, the rolling (RD), transverse (TD), and normal (ND) directions. To determine the corresponding components of strain, these directions must each be placed parallel to the bisector, as illustrated in Fig. 11. Because neutrons easily penetrate through zirconium alloys, direct measurements of lattice strain can be made without destroying the weld.

The wavelength of the neutron beam was 0.190 05 nm, calibrated against a silicon powder specimen obtained from the National Institute of Standards and Technology. Strains were calculated from scattering angles at locations within the weld and those obtained from a reference specimen, a small rectangular coupon cut from a corner of the plate heat treated at 650°C for 1 h.

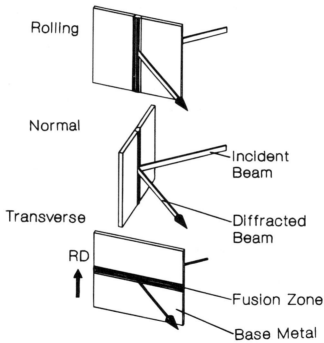

FIG. 11—*Orientation of welded plate for the measurement of strain in rolling, normal, and transverse directions.*

References

[1] Ells, C. E., Coleman, C. E., and Chow, C. K., "Properties of a CANDU Calandria Tube," *Canadian Metallurgical Quarterly*, Vol. 24, 1985, pp. 215–223.
[2] Hill, R., "A Theory of the Yielding and Plastic Flow of Anisotropic Metals," *Proceedings*, Royal Society, Vol. A 193, 1948, p. 281.
[3] Ells, C. E., Coleman, C. E., Hosbons, R. R., Ibrahim, E. F., and Doubt, G. L., "Prospects for Stronger Calandria Tubes," AECL-10339, Atomic Energy of Canada Ltd., Chalk River Laboratories, Chalk River, Ontario, Canada, Dec. 1990.
[4] Ernestus, A. W., U. S. Patent No. 3, 222,905, 14 Dec. 1965.
[5] Ernestus, A. W., U. S. Patent No. 3, 411,334, 19 Nov. 1968.

[6] Coleman, C. E., Cheadle, B. A., Ambler, J. F. R., Lichtenberger, P. C., and Eadie, R. L., "Minimising Hydride Cracking in Zirconium Alloys," *Canadian Metallurgical Quarterly,* Vol. 24, 1985, pp. 245–250.

[7] Cheadle, B. A., Coleman, C. E., and Ambler, J. F. R., "Prevention of Delayed Hydride Cracking in Zirconium Alloys," *Zirconium in the Nuclear Industry: Seventh International Symposium, ASTM STP 939,* R. B. Adamson and L. F. P. Van Swam, Eds., American Society for Testing and Materials, Philadelphia, 1987, pp. 224–240.

[8] Coleman, C. E., "Effect of Texture on Hydride Reorientation and Delayed Hydrogen Cracking in Cold-Worked Zr-2.5Nb," *Zirconium in the Nuclear Industry: Fifth International Symposium, ASTM STP 754,* D. G. Franklin, Ed., American Society for Testing and Materials, Philadelphia, 1982, pp. 393–411.

[9] Simpson, C. J. and Ells, C. E., "Delayed Hydrogen Embrittlement in Zr-2.5wt%Nb," *Journal of Nuclear Materials,* Vol. 52, 1974, pp. 289–295.

[10] Causey, A. R., Holt, R. A., Christodoulou, N. C., Elder, J. E., and Fleck, R. G., "On the Anisotropy of In-Reactor Creep of Zr-2.5Nb Tubes," in this volume.

[11] Platonov, P. A., Ryazantseva, A. V., Frolov, I. A., Rodchenkov, B. S., Sinelnikov, L. P., and Eperin, A. P., "A Review of Investigations of RBMK Fuel: Control and Protection of Rod Channels for the 15 Years of Operation," abstract submitted to this conference.

[12] Tiffany, C. F. and Masters, J. N., "Applied Fracture Mechanics," *Fracture Toughness Testing and its Applications, ASTM STP 381,* W. F. Brown, Jr., Ed., American Society for Testing and Materials, Philadelphia, 1965, pp. 249–277.

[13] Root, J. H. and Salinas-Rodriguez, A., "Neutron Diffraction Measurement of Texture Variations near a Weld in a Zr-2.5Nb Plate," *Textures and Microstructures,* Vol. 14–18, 1991, pp. 989–994.

[14] Davies, G. J. and Garland, J. G., "Solidification Structures and Properties of Fusion Welds," *International Metallurgical Reviews,* Vol. 20, 1975, p. 83.

[15] Burgers, W. G., "On the Process of Transition of the Cubic-body-centered Modification into the Hexagonal-close-packed Modification of Zirconium," *Physica,* Vol. 1, 1934, pp. 561–586.

[16] MacEwen, S. R., Tome, C., and Faber, J., "Residual Stresses in Annealed Zircaloy," *Acta Metallurgica,* Vol. 37, 1989, p. 979.

[17] Wohlfahrt, H., "Residual Stresses due to Welding: their Origin, Calculation, and Evaluation," *Residual Stresses,* E. Macherauch and V. Hauk, Eds., Informationsgesellschaft Verlag, Oberursel, Germany, 1986, pp. 81–112.

[18] Root, J. H., Holden, T. M., Schroeder, J., Hubbard, C. R., Spooner, S., Dodson, T. A., and David, S. A., "Residual Stresses in a Multipass Ferritic Weldment," *Materials Science and Technology,* in press.

[19] *The Boiler and Pressure Vessel Code,* American Society of Mechanical Engineers, Section VIII, Division I, Subsection C, UNF-56, (d), 1992, p. 183.

DISCUSSION

Keith N. Woods[1] (written discussion)—What process is used to reduce the wall thickness of the calandria tube sheet material? Please comment on the apparent inefficiency of this fabrication process that loses up to 25% of the material.

C. E. Coleman et al. (authors' closure)—The wall is reduced by pickling. We found that this method required the least development, although we agree that it is wasteful and inefficient. This inefficiency is one of the reasons why we would like to develop a seamless calandria tube.

Brian Cox[2] (written discussion)—In order to roll a ridged calandria tube into the end shields, you presumably have to grind off the ridge. Has this caused any problems with leak tightness of the rolled joint?

C. E. Coleman et al. (authors' closure)—At the rolled joint, the tube has a uniform wall thickness because the thinning process is not extended the full length of the tube. Rolled joints have been made with the thicker tube material, and they behaved satisfactorily and did not leak.

C. J. White[3] (written discussion)—You have indicated that 30-min treatment at 530 to 560°C will substantially lower the residual stresses in the Zr-2.5Nb weld and heat-affected zones and significantly improve the DHC behavior of this material. Is the improved DHC behavior due solely to a reduction in weld residual stress or is there a microstructural change brought on by the heat treatment that also acts to improve DHC resistance. Specifically, does the heat treatment promote beta zirconium decomposition to discrete beta niobium precipitates therefore inhibiting diffusion of hydrogen to the crack front? Secondly, have you evidence that the 530 to 560°C/30-min heat treatment has in fact reduced DHC susceptibility or have you assumed this from your measurements which show reduced residual stress?

C. E. Coleman et al. (authors' closure)—The reduction in residual stress is the sole source of improvement. Any beta-phase is discretely distributed in the weld region and no decomposition of continuous beta-phase is involved (see S. Sagat, C. E. Coleman, M. Griffiths, and B. J. S. Wilkins, in this volume).

DHC susceptibility is characterized by K_{IH}. K_{IH} is controlled mostly by crystallographic texture but texture is not affected by the heat treatment. Thus, the improved performance and lack of cracking is because the tensile stresses have been reduced not because K_{IH} has been increased.

Mitsutaka H. Koike[4] (written discussion)—The textures of the parent metal and the heat-affected zone of the weld are different. Irradiation creep and growth phenomena are inhomogeneous around the weld portion. Is there any problem with the irradiation creep and growth for reactor use under irradiation?

C. E. Coleman et al. (authors' closure)—The calandria tubes in CANDU reactors are subjected to a high flux of fast neutrons at about 80°C. To date, we have detected no problems that can be attributed to irradiation-induced deformation. The reasons are as follows. The growth strains are very small. Possible differential strain between the body of the tube (α-annealed) and the weld (β-heat treated) is about 10^{-3} at a fluence of 10^{26} n/m².[5] With no stress relaxation, such a strain difference would generate a stress of about 100 MPa. If this

---

[1] Siemens Power Corporation, Richland, WA.
[2] University of Toronto, Toronto, Ontario, Canada.
[3] Knolls Atomic Power Laboratory, Niskayuna, NY.
[4] Power Reactor and Nuclear Fuel Development Centre, Oarai Engineering Center, 4002, Oarai, Higashi-Ibaraki, Ibarakiken, 311-31 Japan.
[5] Fidleris, V., Causey, A. R., and Holt, R. A., "Factors Affecting In-core Dimensional Stability of Zircaloy-2 Calandria Tubes," *Optimising Materials for Nuclear Applications*, F. A. Garner, D. S. Gelles, and F. W. Wiffen, Eds., Metallurgical Society of AIME, Los Angeles, CA, 1985, pp. 35-50.

strain were to be generated linearly with time, the average stress would be 50 MPa. Using a creep compliance of 5×10^{-31} m^2/n·MPa produces a creep strain of about 2×10^{-3}, similar to that from growth, thus any residual stresses would be relaxed. A second mitigating factor is that the weld is only a small fraction of the whole circumference.

Similar arguments may be made for the pressure tubes. The welds in the Zr-2.5Nb pressure tubes in RBMK reactors and our research reactors receive only a low flux of fast neutrons and thus little effect is expected. No deformation problems have been reported from the welded Zircaloy-2 tubes used in the Hanford Site N Reactor.

A. Strasses[6] (written discussion)—Is the calandria tube made with one or two longitudinal welds? If one weld, could the differential growth between the asymmetric weld and the tube result in stresses (or dimensional changes)? The thickness weld section and the accelerated growth rate of annealed material at extended exposures could magnify such an effect.

C. E. Coleman et al. (authors' closure)—Calandria tubes are made with one longitudinal weld. The question of differential growth is answered in the previous question. If the growth rate accelerates, it will only do so late in life. Since the growth strain is very small, we are not anticipating any problems.

Ted Darby[7] (written discussion)—An adverse weld HAZ texture in the zirconium-niobium originated from a very strong parent material texture. Could you comment on the type of HAZ texture that would develop from a weakly textured parent material such as the (heat-treated) Fugen tube?

C. E. Coleman et al. (authors' closure)—We have not yet measured the texture of the HAZ of a heat-treated Zr-2.5Nb pressure tube; however, we can try to predict it from the texture of the HAZ of the Zr-2.5Nb sheet. The axial, transverse, and radial directions of a pressure tube correspond to the rolling, transverse, and normal directions of sheet. Like the sheet, a heat-treated pressure tube has a high density of grains with (0002) poles in the transverse direction and $\{11\bar{2}0\}$ poles between the radial and axial directions. Such grains were the source of the undesirable concentration of (0002) poles in the axial direction in the HAZ of the sheet. We anticipate similar behavior in the HAZ of a heat-treated pressure tube. It must also be noted that even before being welded, a heat-treated pressure tube has a high density of (0002) poles in the axial direction. Although a heat-treated Zr-2.5Nb pressure tube has a weaker texture than the Zr-2.5Nb sheet, it will still develop an unfavorable texture in the HAZ during welding.

[6] S. M. Stoller Corporation, Pleasantville, NY.
[7] Rolls-Royce and Associates, Ltd., Derby, UK.

Hideaki Abe,[1] *Katsuhiko Matsuda*,[2] *Tadao Hama*,[1] *Takao Konishi*,[2] *and Munekatsu Furugen*[2]

Fabrication Process of High Nodular Corrosion-Resistant Zircaloy-2 Tubing

REFERENCE: Abe, H., Matsuda, K., Hama, T., Konishi, T., and Furugen, M., "**Fabrication Process of High Nodular Corrosion-Resistant Zircaloy-2 Tubing,**" *Zirconium in the Nuclear Industry: Tenth International Symposium, ASTM STP 1245*, A. M. Garde and E. R. Bradley, Eds., American Society for Testing and Materials, Philadelphia, 1994, pp. 285–306.

ABSTRACT: A new fabrication process of high nodular corrosion-resistant Zircaloy-2 tubing for boiling water reactors (BWRs) water rods has been developed. The main features and findings are:
  1. Development of a new quenching machine and method, by which tube shell can be $\alpha + \beta$ through-wall quenched at a rapid cooling rate of >50 K/s from 1173 to 1073 K.
  2. High nodular corrosion-resistant tube shells, quenched at both hollow billet and tube shell stages, have less deformability, especially in the circumferential compression direction.
  3. Reduction in area in combination with Q-value affect crack sensitivity during pilgering. Area reductions greater than 80% at high Q-values were proved possible with minimal or no cracking.
  4. Consequently, a new fabrication process, consisting of two cold reduction passes from tube shell to final product, has been successfully developed. The final tubing showed high nodular corrosion resistance, having no nodules in the high temperature (773 K) corrosion test.

KEY WORDS: zirconium, zirconium alloys, water rods, nodular corrosion, quenching, cold pilgering, reduction in area, Q-value, cracks (materials), tube shells, nuclear materials, nuclear applications, radiation effects

Recently, high nodular corrosion resistance has been required for water rods as well as cladding tubes in boiling water reactors (BWRs). Many investigaters have already reported the effect of chemical composition and thermomechanical processes on corrosion behavior [1–6]. Their findings showed that controlled-chemistry Zircaloy-2, which had low tin, high iron, and high nickel content, exhibited better nodular corrosion resistance than conventional Zircaloy-2, and low accumulated annealing parameters (ΣA_i) tubing exhibited better nodular corrosion resistance than high ΣA_i tubing. In addition, some practical ideas regarding heat treatment for improving nodular corrosion resistance have been proposed [7,8].

These findings already have been applied for cladding tubing that has subsequently shown good resistance to nodular corrosion in the reactor. On both the inner and outer surfaces, corrosion resistance is required for water rods, while for cladding tubes only the outer surface requires corrosion resistance.

---

[1] Assistant manager and engineer, respectively, Precision Tube Making Plant; Steel Tube Works, Sumitomo Metal Industries, Ltd., Amagasaki, Hyogo, 660, Japan.
[2] Managers, respectively, Technical Department, Steel Tube Works, Sumitomo Metal Industries, Ltd., Amagasaki, Hyogo, 660, Japan.

The paper describes a new method of heat treatment that can be applied to improve nodular corrosion resistance for both surfaces of the tubes, and a new cold work process for Zircaloy-2 tubing for water rods.

Material and Process Flow

The materials were typical controlled-chemistry Zircaloy-2, used for nodular corrosion resistance. The chemical composition of the material is shown in Table 1. A conventional tube shell is for reference and was made from conventional-chemistry Zircaloy-2.

The tube-making process is as follows: the tube shell is extruded from the hollow billet, which is β-quenched, and the tube is reduced to final tubing, as shown in Fig. 1.

Heat Treatment to Improve Corrosion Resistance

Hollow Billet β-Quenching

The hollow billet is β-quenched and, after billet conditioning, extruded into a tube shell for nodular corrosion resistance. Usually, quenching of the hollow billet is performed by dropping into a water tank immediately after induction heating. The temperature during extrusion is controlled below 923 K. Lower temperature (lower than 950 K) annealing is performed after extrusion.

This method has been applied for almost all Zircaloy-2 for BWRs, including the cladding tube.

Tube Shell Heat Treatment

A schematic diagram of the heat treatment equipment is shown in Fig. 2. The equipment is the vertical type. The tube shell, 63.5 mm outside diameter (OD) by 10.9 mm wall thickness (WT), is induction heated (through-wall) to 1213 K ($\alpha + \beta$) and instantly water quenched. Subsequently, the tube shell is stress relieved at 853 K.

The operation parameters are shown in Table 2. This method applies high-frequency induction heating to the surface of the tube shell as the tube shell is rotated and transported through the induction coil. During the heat treatment, the temperature just below the coil is monitored with a two-color pyrometer. To ensure the circumferential uniformity of the heating, the tube

TABLE 1—*Chemical composition of the material (%).*

	Sn	Fe	Cr	Ni	C	O	Si	Zr
Controlled-chemistry Zircaloy-2	1.30	0.19	0.10	0.07	0.012	0.110	0.01	balance
Conventional Zircaloy-2	1.52	0.15	0.10	0.05	0.018	0.120	0.01	balance

Melting → Forging → Machining → β-Quenching → Extruding → Annealing
63.5φ × 10.9t

α+β-Quenching → Stress Relieving → Cold-Pilgering → Annealing → Condition
34φ × 1t

(Repeat)

FIG. 1—*Process flow of high-corrosion-resistant tubing.*

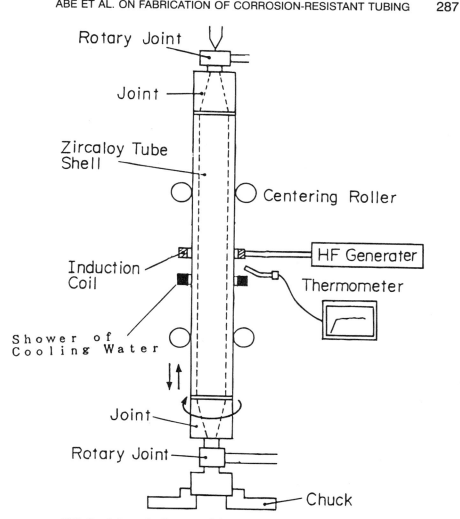

FIG. 2—*Schematic diagram of the new heat-treatment method.*

TABLE 2—*α + β-quenching parameters of the tube shell.*

Heating temperature,	1213 K
Frequency of induction coil,	2 kHz
Feeding rate,	4 mm/s
Revolution speed,	10 rpm
Cooling water flow rate,	46 L/min

is rotated at 10 rpm. Just after induction heating, the tube is rapidly through-wall quenched using a water spray on the outer surface.

The results of temperature measurement during heat treatment are shown in Fig. 3. The measured points on the tube shell were at the outer surface, the middle wall, and near the inner surface. The cooling rate at all three measured points was more than 50 K/s from 1173 to 1073 K. During this measurement, the tube shell was not rotated.

This method is easier to maintain for both circumferential and axial uniformity during through-wall quenching than other methods. Further, it is more practical and economical, since heat treatment is performed on large tube shells (63.5 mm OD by 10.9 mm WT), compared to past practice of heat treatment of smaller diameter tubes.

Evaluation of the New Heat Treatment

Corrosion tests were performed on the heat-treated tube shell at high temperature (773 K, 24 h, 10.3 MPa). A conventional tube shell was also tested for reference. A conventional tube shell was extruded from α-forged billet after β-quenching and was annealed. The corrosion test results are shown in Table 3.

The heat-treated tube shell had no nodules on either surface and had very low weight gains in the corrosion test, while the conventional tube shell had nodules and higher weight gains.

By means of scanning electron microscopy (SEM), the size distribution of the precipitates was observed. The photographs are shown in Fig. 4. The heat-treated tube shell exhibited very fine precipitates.

This heat treatment, including β-quenching at hollow billet and α + β-quenching at the tube shell, improved nodular corrosion resistance.

Properties of High Nodular Corrosion-Resistant Tube Shells

In order to choose the appropriate cold pilgering schedule for tube shells that were heat treated with this new method, the fundamental properties of the heat-treated tube shell were investigated. Properties tested included microstructure, hardness, texture, tensile property, flattening property, charpy impact, and compressive property. The descriptions of the tube shell investigated are shown in Fig. 5 and Table 4. The dimensions of tube shells are all 63.5 mm in diameter and 10.9 mm in wall thickness. The sampling method is illustrated in Fig. 6.

It is desirable to reduce the cold pilgering pass for lowering the accumulated annealing parameter, ΣA_i. And the target is implemented by high reduction (>80%) pilgering for reducing the pass.

Microstructure

The microstructure of the tube shells is shown in Fig. 7.

Sample A had duplex microstructure due to lower temperature extrusion and low temperature annealing after extrusion. This result corresponds with the previous study showing the influence of extrusion conditions [9]. Sample B from α + β-quenched tube shell showed much finer grains in the α grains. The finer grains are caused by α + β-quenching of the tube shell. However, the remarkable change in microstructure could not be seen after quenching. Sample C had a little more uniform microstructure.

Hardness

The hardness results are shown in Fig. 8.

There are little differences between the four types of tube shells (A, B, C, and S). Both β-quenching at billet and α + β-quenching at the tube shell had little influence on hardness.

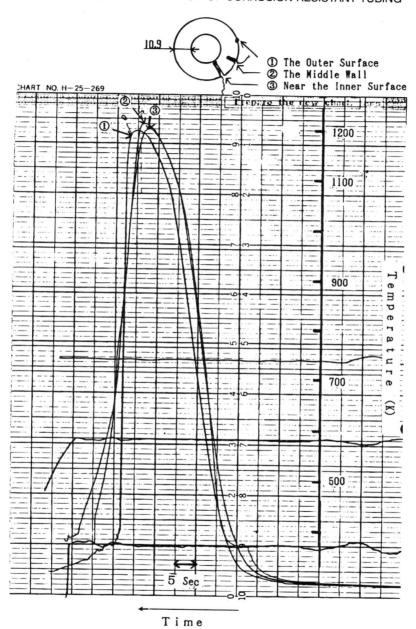

FIG. 3—*The result of temperature measurement in $\alpha + \beta$-quenching of the tube shell.*

TABLE 3—*The results of 773 K corrosion test for the tube shell.*

Tube Type	Number of Specimens	Appearance	Weight Gain, mg/dm^2
The heat-treated tube shell	5	no nodules	42, 43, 41, 43, 42
Conventional (reference) tube shell	2	many nodules	850, 550

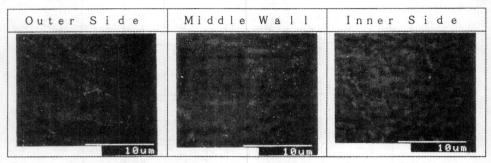

FIG. 4—*Distribution of precipitates in the heat-treated tube shell.*

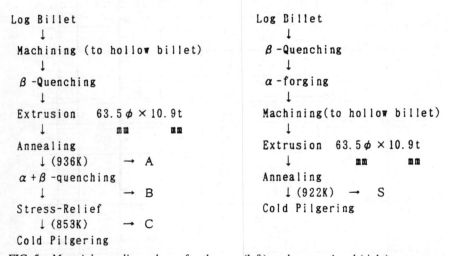

FIG. 5—*Material sampling scheme for the new* (left) *and conventional* (right) *processes.*

Texture

The tube shell texture at different metallurgical conditions is shown in Fig. 9.

There is no appreciable difference between the three types of tube shells (A, C, and S). After extruding, the influence of prior β-quenching at billet could not be seen, and the α + β-quenching at the tube shell had little influence on the texture.

TABLE 4—*Tube shell investigated.*

Symbol	Description
A	Extruded and annealed tube shell from β-quenched hollow billet
B	As α + β-quenched tube shell by induction heating
C	Stress relieved (at 853 K) tube shell after the α + β-quenching
S	Conventional tube shell (extruded from α-forged billet and annealed)

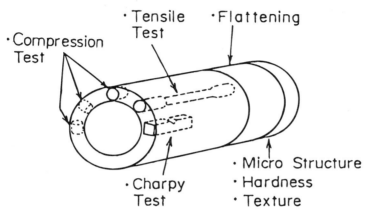

FIG. 6—*Sampling method of specimens.*

Tensile Property

The tension test results in the axial direction of the tube shell are shown in Table 5.

There are few differences in the ultimate strength between the four types of tube shells. The result shows that neither β-quenching at hollow billet nor α + β-quenching at the tube shell significantly affects tensile property in the axial direction of tube shell.

Flattening Property

The flattening test results of the tube shells are shown in Table 6. Materials B and C had less flattening rate than the others (A and S). This result shows that α + β-quenching affects the flattening property, even after stress relief.

Charpy Impact

The charpy impact test results of the tube shells are shown in Table 7. The specimens were 7.5 mm in width, 10 mm in height, 55 mm in length, and had a V-notch. The data shows that the tube shell after α + β-quenching and stress relief annealing has the lowest value.

Compressive Property

For compression tests, cylindrical specimens in three directions were machined from the tube shells, as shown in Fig. 10. The specimens were 4 mm in diameter and 8 mm in height. Compression testing was performed on an Instron testing machine at room temperature. The specimens were grooved on both ends and lubricated in oil. The compression speed was 5

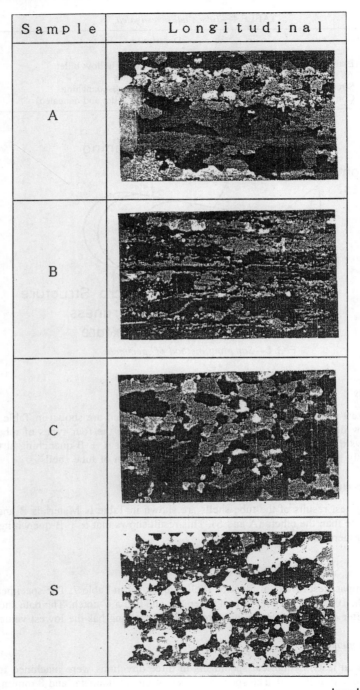

FIG. 7—*The microstructure of the tube shell.*

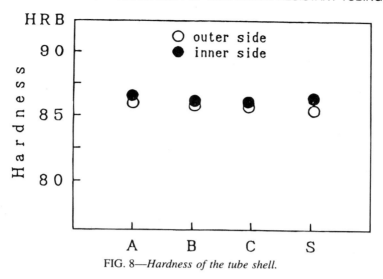

FIG. 8—*Hardness of the tube shell.*

	A	C	S
(0002) Pole Figure	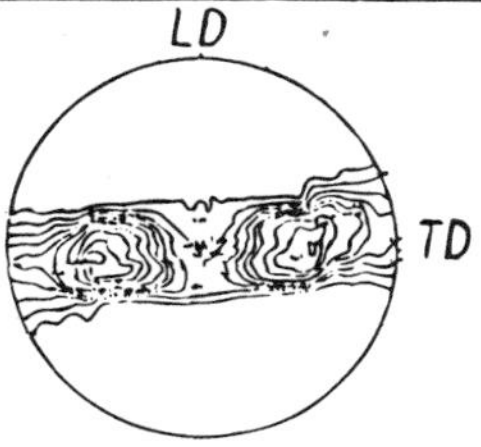 LD / TD	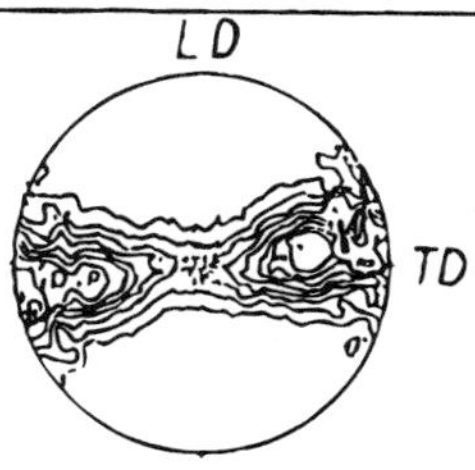 LD / TD	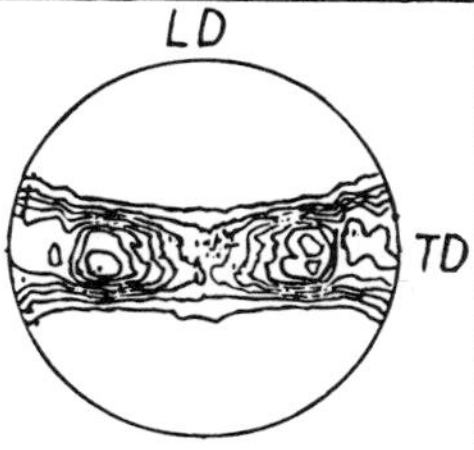 LD / TD
fr	0.50	0.47	0.43
ft	0.46	0.47	0.46
fl	0.05	0.06	0.05

FIG. 9—*The texture of the tube shell.*

mm/min. The load and the height of specimens during testing were continuously recorded in a chart. When cracks were initiated on the surface of the specimen, the recording load started to decrease. Reduction in height was calculated by the following equation.

$$\text{reduction in height } (\%) = \frac{H - h}{H} \times 100$$

where

TABLE 5—*Tensile property (at room temperature).*

	A, extruded + annealed	B, α + β-quenching	C, α + β-quenching and stress relief	S, conventional
Ultimate tensile strength, MPa	544	530	533	549
0.2% Offset yield strength, MPa	377	315	328	386
Elongation in 50 mm, %	23.8	22.0	22.5	23.8

TABLE 6—*Flattening property.*

	A, extruded + annealed	B, α + β-quenching	C, α + β-quenching and stress relief	S, conventional
Flattening rate, %	41	13	15	47

TABLE 7—*Charpy impact test.*

	A, extruded + annealed	C, α + β-quenching and stress relief	S, conventional
Charpy impact, J/cm²	27	17	28

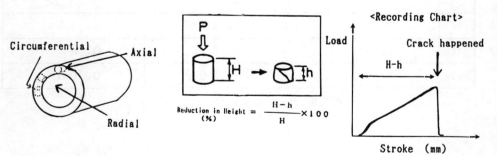

FIG. 10—*Compression test method.*

H = the height of specimen before testing, and
h = the height when cracks happened.

The compression test results are shown in Fig. 11. Materials A and C showed less compressive property than Material S. In addition, Material C showed the least reduction in the tangential

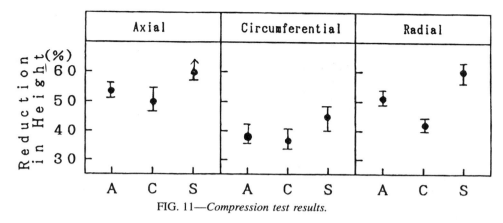

FIG. 11—*Compression test results.*

direction. The tangential compressive property is important because circumferential compressive deformation during cold pilgering is large.

This result showed that both α + β-quenching at the tube shell and β-quenching at the hollow billet significantly affected the compressive property of the tube shell. The reason might be the prior β-grains, residual strain, and precipitates in the grain boundary. Therefore, Materials A and C (and B) might have less deformability in cold pilgering.

The reason for the large difference in compressive property between three directions of the tube shell is the texture. As shown in Fig. 9, the tube shells are highly textured in the circumferential direction. Therefore, compressive deformability in the circumferential direction is less than the axial direction, because of the inherent nature of the basal pole [10].

The Influence of Annealing on Properties of the Tube Shell

As shown earlier, β-quenching at hollow billet and α + β-quenching at the tube shell significantly affects the properties of the tube shell, including microstructure, flattening, charpy impact, and compressive property. These results show that both β-quenching at the hollow billet and α + β-quenching at the tube shell lowered the deformability of the tube shell.

In order to find out the reason for the lower deformability of the tube shell, the influence of annealing temperature on the tube shell properties including microstructure, the compressive property, and residual stress were investigated.

Annealing was performed for both the extruded tube shell (A) and the α + β-quenched tube shell (B) at five temperatures, that varied from 853 to 1123 K. The microstructures are shown in Fig. 12. As the annealing temperature was raised, the microstructure was more recrystallized. At the highest annealing temperature (1123 K) grains were coarser.

The compression test results given in Fig. 13 show that as the annealing temperature is increased, the reduction in height is increased in the circumferential and axial directions.

In order to improve the deformability of the tube shell, a low-temperature heat treatment after quenching was decided upon. However, as shown in Fig. 14, the influence of quenching at the tube shell on residual stress could be measured. Even after stress relief anneal at 853 K, the residual stress was higher in the circumferential and axial directions for the α + β quenched material than the as-extruded tube shell.

Although high temperature (from 973 to 1073 K) annealing improved deformability of the tube shell, the heat treatment naturally could not be adopted because of degradation in the

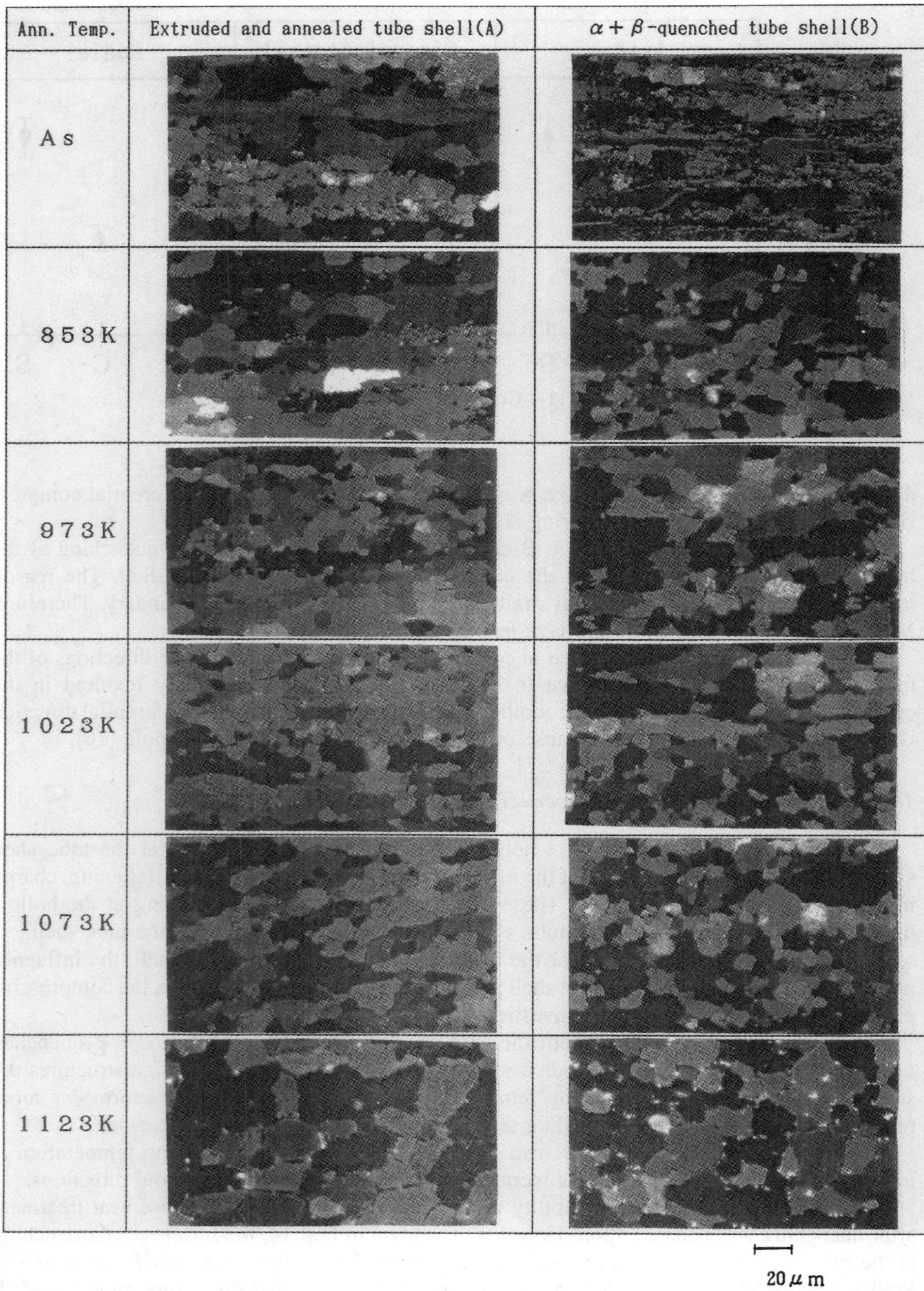

FIG. 12—*Microstructure of the tube shell annealing tested (longitudinal).*

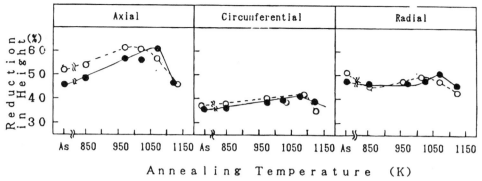

FIG. 13—*Tube shell compression test results after annealing at various temperatures:* ○ *extruded tube shell (A) and* ● α + β − Q *tube shell (B).*

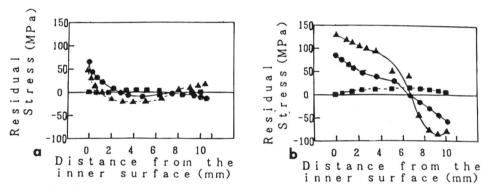

FIG. 14—*The results of residual stress measurement of the tube shell: (a) extruded tube shell (A) and (b)* α + β − Q *and stress relieved tube shell (B), where* ● = *axial,* ▲ = *circumferential, and* ■ = *radial.*

nodular corrosion resistance. Hence, it was necessary to develop an alternate cold work process for a high-corrosion-resistant tube shell.

Deformability of the Tube Shell in Cold Pilgering

Cold Pilgering Test

In some cold pilgering schedules for the high-corrosion-resistant tube shell, some cracks could be seen on the surface of the reduced tubes, since, as previously discussed, the high-corrosion-resistant tube shell has less deformability than conventional tube shells.

In order to determine the appropriate cold pilgering schedule for the high-corrosion-resistant water rods made from α + β-quenched and stress relieved tube shells, the cold pilgering parameters were widely varied and the surface of the reduced tubes investigated.

The results of cold pilgering tests are given in Table 8. The tests were performed on a typical 75 VMR cold pilger mill. After cold pilgering, the reduced tubes were inspected by ultrasonic flaw testing. The pilgering parameters are defined by the following equations

$$\text{Reduction in area (Rd)} = \frac{(A - a)}{A} \times 100 = 1 - \left[\frac{t}{T} \times \frac{d}{D}\right] \times 100$$

and

$$Q\text{-value (QE)} = \frac{Ln(t/T)}{Ln(d/D)}$$

where

A = cross-sectional area of a mother tube,
a = cross-sectional area of a reduced tube,
T = wall thickness of a mother tube,
t = wall thickness of a reduced tube,
D = mean diameter of a mother tube, and
d = mean diameter of a reduced tube.

The test results are shown in Fig. 15. The test proved that the combination of the reduction in area and Q-value significantly affect the surface of the reduced tubes. At low reduction in area, pilgering with no cracks (c, d) are possible at low Q. However, at higher reduction, it is necessary to increase the Q-value to prevent cracks on the reduced tube, and higher reductions (>80%) with higher Q-value (>2) were possible. The appropriate cold pilgering range to produce crack-free surfaces is shown in Fig. 15.

Analysis of Cold Pilgering

In order to understand the influence of cold pilgering conditions on crack sensitivity, a theoretical analysis of cold pilgering tubes at different parameters was performed according to the method of the plastic deformation model [11,12]. The analytical model is shown in Fig. 16.

By means of this analytical model, the plastic strains of the tube during cold pilgering in the six schedules (shown in Table 8) were calculated. The calculation results for Schedules a and f, with almost the same area reduction, are shown in Fig. 17. The plastic strain (ϵ_r, ϵ_θ, ϵ_z) shows radial, circumferential, and axial strain, respectively. In Schedule a, which had cracks on the surface of the reduced tube, the circumferential compressive strain is larger; and the

TABLE 8—*Cold-pilgering test parameters and results.*

Symbol	Pilgering Schedule	Rd	QE	Test Results
a	63.5 φ × 10.9t → 25.5 φ × 3.5t	86.6	1.3	cracks
b	63.5 φ × 10.9t → 34.0 φ × 4.6t	76.5	1.5	cracks
c	63.5 φ × 10.9t → 44.4 φ × 7.6t	51.2	1.0	no cracks
d	63.5 φ × 10.9t → 40.0 φ × 6.0t	64.4	1.4	no cracks
e	63.5 φ × 10.9t → 34.0 φ × 2.8t	84.8	2.6	no cracks
f	63.5 φ × 10.9t → 37.0 φ × 2.2t	86.3	3.8	no cracks

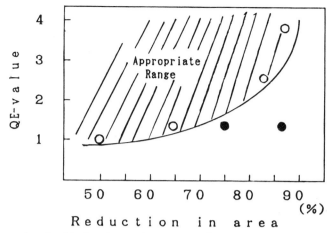

FIG. 15—*The results of cold-pilgering testing:* ○ = *no flaws occurred and* ● = *flaws occurred.*

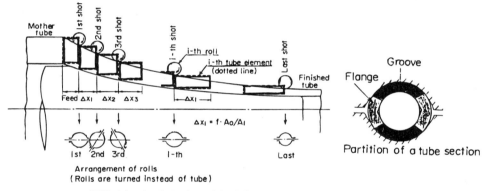

FIG. 16—*Analytical model of the cold-pilgering process.*

radial tensile strain is also larger, especially on the flange part. While in Schedule f, which had no cracks, these strains are lower.

Furthermore, the difference of plastic strain on the flange part between the six pilgering schedules is shown in Fig. 18. In Schedules a an b, both circumferential (compression) and radial (tensile) strain are larger, and the ratio of radial strain to circumferential strain ($-\epsilon_r/\epsilon_\theta$) is larger. While, in the other schedules, the ratio ($-\epsilon_r/\epsilon_\theta$) is smaller. In other words, in the case of good schedules, the increase of wall thickness on the flange part, caused by large circumferential compression, is very small even at high area reduction pilgering and, consequently, the frequency of crack occurrence is lower. This crack mechanism is illustrated in Fig. 19.

Thus, in the tube shell with less circumferential deformability, larger circumferential compression strain during cold pilgering tends to produce cracks in the reduced tube surface. It is, therefore, necessary to choose for less deformable tube, a higher Q schedule to lower the radial tensile strain on the flange part during cold pilgering.

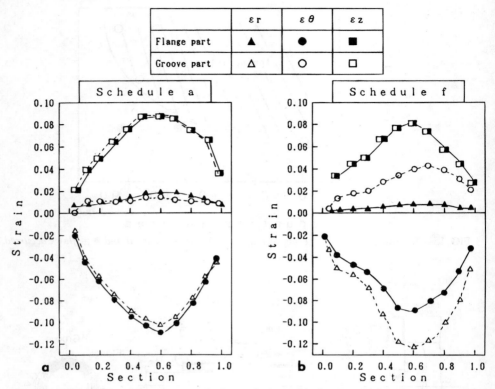

FIG. 17—*The results of plastic strain calculation during cold pilgering:* (a) *low*-Q *pilgering and* (b) *high*-Q *pilgering.*

Fabrication of High Nodular Corrosion-Resistant Zircaloy-2 Tubing

Fabrication Process

The cold pilgering schedule was established in accordance with the preceding investigation, and is given in Fig. 20. A fabrication process, that has only two cold reduction passes from the tube shell to final product was achieved. Intermediate and annealing parameters were 853 K (2 h) and 843 K (3 h). The accumulated annealing parameter (ΣA_i) for this new process was 2.8×10^{-20} h, with a $Q = 80\ 000$ calories, a very low value.

The dimension of the final product is 34 mm OD and 1 mm WT. The ratio of wall thickness to diameter is smaller (3%) than cladding tube, and the final cold work is performed by means of the three-roll-type cold-pilger mill (HPTR). Though the three-roll mill can not have very high reduction in area, the mill is suited for cold working thin wall tubing with better dimensional precision.

Evaluation of the Final Product

The evaluation of the final products were performed for nodular corrosion resistance, dimensional precision, and mechanical properties.

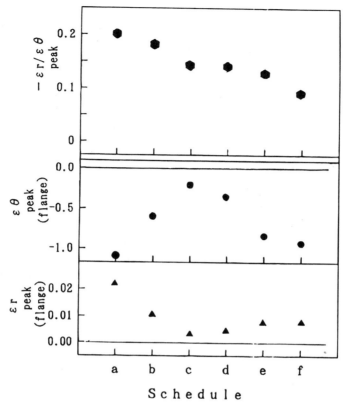

FIG. 18—*Comparison of plastic strain between the six pilgering schedules.*

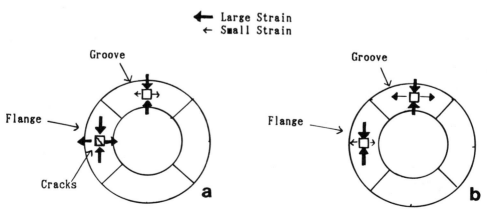

FIG. 19—*Schematic diagram of crack mechanism:* (a) *bad pilgering and* (b) *good pilgering.*

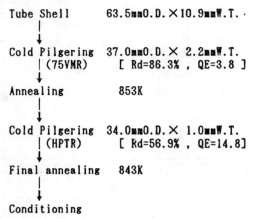

FIG. 20—*The newly developed fabrication process flow.*

The 773 K corrosion test results are shown in Table 9. No nodules could be seen on both outer and inner surfaces, and the weight gains were low, indicating that the developed tubing had high nodular corrosion resistance.

The histograms of the dimensions (outer diameter, inner diameter, and wall thickness) are shown in Figs. 21*a*, *b*, and *c*, respectively. All dimensional data were obtained by ultrasonic method in full length (3.3 m) of the final products. The range of outer diameter was within 0.10 mm. The range of inner diameter was within 0.15 mm. The range of wall thickness was within 0.1 mm. Eccentricity was less than 8% of nominal value.

TABLE 9—*The results of a two-step corrosion test of the final products.*

Number of Tested Pieces	Appearance	Weight Gain, mg/dm^2
5	no nodules	45, 46, 45, 48, 47

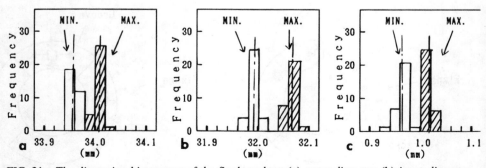

FIG. 21—*The dimension histograms of the final product:* (a) *outer diameter,* (b) *inner diameter, and* (c) *wall thickness.*

TABLE 10—*The results of properties of the final products.*

Tensile property at room temperature	
Ultimate tensile strength, MPa	545
0.2% offset yield strength, MPa	434
Elongation in 50 mm, %	53
Tensile property at 616 K	
Ultimate tensile strength, MPa	267
0.2% offset yield strength, MPa	173
Elongation in 50 mm, %	38
Grain Size (ASTM No.)	
Transverse	12.9
Longitudinal	12.8
Texture (Kearn's Number)	
f_r	0.68
f_t	0.27
f_l	0.05

The other properties of the products are shown in Table 10. The data are consistent with standard Zircaloy-2 cladding tube for BWRs.

Conclusion

A new fabrication process of high nodular corrosion-resistant Zircaloy-2 tubing for BWR water rods has been developed. The main features and findings were as follows:

1. A new quenching machine and method where the tube shell can be $\alpha + \beta$ through-wall quenched at a rapid cooling rate (>50 K/s from 1173 to 1073 K) has been developed.
2. High nodular corrosion-resistant tube shells quenched at both hollow billet and tube shell stages, have less deformability, especially in the circumferential compression. This compression test proved to be a useful measure of deformability of the tube shell in cold pilgering.
3. Crack sensitivity in cold pilgering can be correlated to a combination of the reduction in area and Q-value. Higher reduction (>80%) pilgerings with higher Q-values (>2) are proved possible.
4. A new fabrication process, which has only two cold reduction passes from the tube shell to the final product, has been successful. The new two-step reduction process allows fabrication of tubing with a low annealing parameter ($\Sigma A_i = 2.8 \times 10^{-20}$).
5. The tube shells and final tubing using the new fabrication process showed high nodular corrosion resistance and had no nodules in the 773 K corrosion tests.

References

[1] Weidinger, H. G., et al. in *Zirconium in the Nuclear Industry: Seventh International Symposium, ASTM STP 939*, R. B. Adamson and L. F. P Van Swam, Eds., American Society for Testing and Materials, Philadelphia, 1987, pp. 364–386.
[2] Eucken, C. M., et al. in *Zirconium in the Nuclear Industry, Eighth International Symposium, ASTM STP 1023*, L. F. P. Van Swam and C. M. Eucken, Eds., American Society for Testing and Materials, Philadelphia, 1989, pp. 113–127.
[3] Charquet, D., et al. in *Zirconium in the Nuclear Industry, Eighth International Symposium, ASTM STP 1023*, L. F. P. Van Swam and C. M. Eucken, Eds., American Society for Testing and Materials, Philadelphia, 1989, pp. 374–391.

[4] Moulin, I., et al. in *Zirconium in the Nuclear Industry, Eighth International Symposium, ASTM STP 824*, D. G. Franklin and R. B. Adamson, Eds., American Society for Testing and Materials, Philadelphia, 1984, pp. 225–243.

[5] Wang, C. T., et al. in *Zirconium in the Nuclear Industry, Ninth International Symposium, ASTM STP 1132*, C. M. Eucken and A. M. Garde, Eds., American Society for Testing and Materials, Philadelphia, 1991, pp. 319–345.

[6] Bradley, E. R., et al. in *Zirconium in the Nuclear Industry, Ninth International Symposium, ASTM STP 1132*, C. M. Eucken and A. M. Garde, Eds., American Society for Testing and Materials, Philadelphia, 1991, pp. 304–316.

[7] Andersson, T., et al. in *Zirconium in the Nuclear Industry, Fifth International Symposium, ASTM STP 754*, D. Franklin, Ed., American Society for Testing and Materials, Philadelphia, 1982, pp. 75–95.

[8] Sabol, G. P., et al. in *Zirconium in the Nuclear Industry, Seventh International Symposium, ASTM STP 939*, R. B. Adamson and L. F. P. Van Swam, Eds., American Society for Testing and Materials, Philadelphia, 1987, pp. 168–186.

[9] Chkravartty, J. K., et al. in *Zirconium in the Nuclear Industry, Ninth International Symposium, ASTM STP 1132*, C. M. Eucken and A. M. Garde, Eds., American Society for Testing and Materials, Philadelphia, 1991, pp. 48–61.

[10] Tenckhoff, E. in *Zirconium in the Nuclear Industry, Fifth International Symposium, ASTM STP 754*, D. Franklin, Ed., American Society for Testing and Materials, Philadelphia, 1991, pp. 5–25.

[11] Furugen, M., et al. *Journal of Mechanical Working Technology*, Vol. 10, 1984, pp. 273–286.

[12] Abe, H., et al. in *Zirconium in the Nuclear Industry, Ninth International Symposium, ASTM STP 1132*, C. M. Eucken and A. M. Garde, Eds., American Society for Testing and Materials, Philadelphia, 1991, pp. 35–47.

DISCUSSION

Helmut Sprenger[1] (written discussion)—Your results of term autoclave tests showed a significant decrease in weight gain. It was mentioned yesterday that a decrease in weight gain may increase hydrogen uptake. Can you comment on hydrogen uptake in your type of cladding?

H. Abe et al. (authors' closure)—In the hydrogen uptake test for the final products, we could not see the difference between the developed tubing and the conventional tubing.

Satya R. Pati[2] (written discussion)—What was the heat-treatment temperature before the introduction of heat treatment in the $\alpha + \beta$ region?

H. Abe et al. (authors' closure)—The temperature was 893 to 923 K.

H. G. Weidinger[3] (written discussion)—It is very interesting that you show that $\alpha + \beta$-quenching leads to pilgering problems to manufacture tubes. You overcame that problem by high amounts of cold work and Q-faction. This should lead to a rather sharp radial texture in the tube wall, and I wonder if you determined the texture. Since it is known that radial textures improve the nodular corrosion, my first question is: did you test the nodular corrosion behavior in autoclave tests on materials with ΣA-values higher than the one you mentioned as 2.8×10^{-20}? Did you test the uniform corrosion of your tubes as fabricated by the new process using $\alpha + \beta$-quenching with $\Sigma A = 2.8 \times 10^{-20}$, for example, by a 400°C test on the loupe exposure time on the standard ASTM test.

H. Abe et al. (authors' closure)—As for the nodular corrosion, in the controlled-chemistry Zircaloy-2, we had no experimental data of higher ΣA_i tubing. However, in the conventional (1.5Sn) Zircaloy-2, we found that the higher ΣA_i affected nodular corrosion resistance, even if $\alpha + \beta$-quenched tube shell.

And for the uniform corrosion, the final products of the new process were evaluated in the 673 K, 72 h autoclave test. Consequently, there were no differences between the product of the new process and one of the conventional process.

Nobuaki Yamashita[4] (written discussion)—(1) How high is the frequency of the high-frequency generator for heat treatment?

(2) How long does the tube shell stay in the heat-treatment coil?

H. Abe et al. (authors' closure)—(1) The frequency is 2 kHz.

(2) The tube shell stayed in the coil for 25 s.

C. D. Williams[5] (written discussion)—(1) The speaker referred to flaws in the tubing after the use of through-wall $\alpha + \beta$ heat treatment. Were the flaws similar in form to those shown in the Zircotube paper presented by M. Aubin?

(2) After development of the high-Q reduction process steps, do you see flaws of the type described by M. Aubin?

H. Abe et al. (authors' closure)—(1) The flaws were different from the Zircotube paper's flaws. Our objective flaws occurred on the inner surface, while Zircotube's were on the outer surface.

(2) We had no experiences of the type of flaws described by M. Aubin.

David W. White[6] (written discussion)—In your process flow outline, a stress-relief anneal was mentioned that followed the $\alpha + \beta$ quench. What is your opinion of the value of the anneal in reducing cracking?

---

[1] TUEV Hannover/Sachsen-Anhalt, Am TUEV 1, 3000 Hannover 81, Germany.
[2] ABB CENF, Windsor, CT.
[3] The S. M. Stoller Corp., Erlangen, Germany.
[4] General Electric Company, Wilmington, NC.
[5] GE Nuclear Energy, Wilmington, NC.
[6] GE Nuclear Energy, Wilmington, NC.

H. Abe et al. (authors' closure)—We think that the stress relief at 853 K is not so effective for reducing cracking. It might be possible to delete the stress relief after α + β quenching.

Bo Cherry[7] (written discussion)—Is α + β quenching of the tube shell necessary to control nodular corrosion? Can this process be eliminated if care is given to reduce the billet extrusion temperature following billet (β) quenching?

H. Abe et al. (authors' closure)—We think that lower extrusion temperature after β-quenching at billet is effective for nodular corrosion resistance.

We also think that our new heat treatment at the tube shell can secure nodular corrosion resistance because of elimination of thermomechanical history.

R. B. Adamson[8] (written discussion)—(1) Have you seen evidence of phase formation on the surface of the tube shells heat treated at the nominal temperature of 940°C?

(2) Have you noted any texture difference by the heat treatments at 940°C or by the high-reduction, high-Q fabrication process?

H. Abe et al. (authors' closure)—(1) We could not see the drastic change of microstructure after the heat treatment. However, we think that β-phase formation is possible.

(2) We could not see the change of texture by the heat treatment at the tube shell. The texture of final tubing is affected by the final cold-working parameter. We could also obtain the radial texture at the high-Q (14.8) pilgering.

---

[7] Electric Power Research Institute (EPRI), Palo Alto, CA.
[8] General Electric, Pleasanton, CA.

Hiroyuki Anada,[1] Ken-ichi Nomoto,[1] and Yoshiaki Shida[1]

Corrosion Behavior of Zircaloy-4 Sheets Produced Under Various Hot-Rolling and Annealing Conditions

REFERENCE: Anada, H., Nomoto, K., and Shida, Y., "**Corrosion Behavior of Zircaloy-4 Sheets Produced Under Various Hot-Rolling and Annealing Conditions,**" *Zirconium in the Nuclear Industry: Tenth International Symposium, ASTM STP 1245,* A. M. Garde and E. R. Bradley, Eds., American Society for Testing and Materials, Philadelphia, 1994, pp. 307–327.

ABSTRACT: The uniform corrosion resistance of Zircaloy-4 sheets produced under various hot-rolling and annealing conditions was investigated. The uniform corrosion resistance in static steam at 673 K improved with an increasing cumulative annealing parameter (ΣA). However, to clarify the physical meaning of the annealing parameter and the different combinations of various heating stages, the effects of individual manufacturing conditions that influence the corrosion behavior and ripening behavior of the precipitates were investigated. As a result, the annealing temperature after hot-rolling influenced the uniform corrosion more significantly than the intermediate annealing and the final annealing temperatures. Size distributions of the intermetallic compound precipitates that were observed by means of a scanning electron microscope (SEM) with an image analyzer were examined for a correlation with weight gain. The percentage of small precipitates below 100 nm in size was important in determining corrosion resistance and had a good correlation with the annealing parameter; that is, if the percentage was lower than 30%, the corrosion resistance was satisfactory. Also, the frequency of precipitates under 100 nm in size decreased sharply with increasing annealing temperature after hot-rolling. The intermetallic compounds that were incorporated into the ZrO_2 oxide scale and the scale/metal interface were observed by a transmission electron microscope (TEM). The incorporated intermetallic precipitates were observed to have changed their composition; that is, the iron content decreased and the iron/chromium (Fe/Cr) ratio for the zirconium-chromium-iron intermetallic precipitate was approximately 1 compared to approximately 2 in the Zircaloy matrix. During oxidation, iron from the intermetallic compound precipitates scattered more into the oxide than chromium. These results suggest that the iron that diffused into the oxide from the incorporated precipitates may play an important role in the corrosion process. It was observed that a stress-relieved microstructure of the Zircaloy-4 just beneath the growing front of the oxide was transformed into a recrystallized microstructure. This was especially clear in a high ΣA material compared to a low ΣA material. This seems to be one of the reasons that a higher ΣA Zircaloy-4 is more corrosion resistant under uniform corrosion conditions.

KEY WORDS: zirconium alloys, zirconium, uniform corrosion, intermetallic compounds, manufacturing conditions, cumulative annealing parameter, oxide scale, transmission electron microscopy, autoclave tests, nuclear materials, nuclear applications, radiation effects

In recent years, the manufacturing process for Zircaloy-4 cladding tubes has been controlled to maintain superior in-pile corrosion resistance. It has been reported that each manufacturing step affected the uniform corrosion, for example, recrystallization and cold rolling [1]. The

---

[1] Research engineers and general manager, respectively, Advanced Technology Research Laboratories, Sumitomo Metal Industries, Ltd., 1–8, Fuso-cho, Amagasaki, Hyogo, Japan.

cumulative annealing parameter (ΣA) [2–5] was used to correlate the uniform corrosion resistance of Zircaloy-4 with its processing history, and was generally successful. However, the precise role of individual heating has yet to be understood.

It was also reported that the size and distribution of precipitated intermetallic compounds influence the uniform corrosion resistance [3,5]. The intermetallic compound ripening parameter was proposed as the secondary ordered cumulative annealing parameter (SOCAP) [6]. However, the relationship between the size distribution and the corrosion has not been clarified and explained quantitatively.

In this paper, the effect of each heating step in the Zircaloy-4 manufacturing process on its corrosion behavior was investigated in an attempt to obtain the physical meaning of the annealing parameter, ΣA_i; that is, $\Sigma A_i = \Sigma t_i \exp(40\,000/T)$, where t = time (h) and T-temperature (K). These results are quantitatively correlated with the size distribution data for the intermetallic precipitates.

Experimental Procedure

Materials

Sheet specimens of Zircaloy-4 (chemical composition is shown in Table 1) were manufactured using several rolling and annealing steps. Figure 1 depicts the conditions where particular annealing temperatures varied in the rolling process, to explain the effect of each heating process, that is, hot rolling temperature, annealing after hot rolling and cold rolling, and final stress-relief annealing. As a result, ΣA varied from 5.2×10^{-20} to 1.8×10^{-16} (h).

After a beta solution treatment followed by quenching into water, a plate billet 25 by 200 by 400 mm in size was sliced into small plates 10 by 50 by 150 mm in size. They were hot-rolled and annealed at several temperatures. They were also cold-rolled and annealed three times to obtain 1-mm-thick sheet materials. The intermediate annealing during cold rolling was conducted twice at a same temperature from 853 to 1053 K for 2 h. The final annealing temperature varied from 698 to 773 K.

Corrosion Test

Two 1 by 40 by 40-mm specimens were cut from each sheet for the corrosion test. The specimens were abraded using No. 800 grit silicon-carbide paper and degreased in acetone before measuring their weight.

Static steam autoclave tests, (673 K, 10.3 MPa, for up to 210 days) were conducted to compare their uniform corrosion resistances by measuring their weight gains.

Observation of Intermetallic Compounds and Scales

It is well-known that the corrosion resistance of Zircaloy-4 depends on the precipitation behavior of the intermetallic compounds. Therefore, the intermetallic precipitates were observed

TABLE 1—*Chemical composition of the Zircaloy-4 sheet (percent by weight).*

Sn	Fe	Cr	Si	C	O
1.53	0.21	0.10	0.0026	0.0089	0.120

```
β-solution treatment : 1323K → Water quenched
          ↓
Hot rolling : 873,923,973,and 1023K for 2hr
              Reduction 50%
          ↓
Annealing : 853,908,953,973,1003,and 1053K for 2hr
          ↓
Cold rolling : Reduction 60%
          ↓
Intermediate annealing
Group-A                              | Group-B ,-C
853 | 923 | 973 | 1013 | 1053K       | 923K for 2hr
          ↓
Cold rolling : Reduction 60%
          ↓
Intermediate annealing
Group-A and Group-B                  | Group-C
853 | 923 | 973 | 1013 | 1053K       | 923K for 2hr
          ↓
Cold rolling : Reduction 70%
          ↓
Final stress relieving : Group-A,-B  723 for 2hr
                         Group-C
                         673,693,723,748,773K for 2hr
```

FIG. 1—*Processing conditions for Zircaloy-4 sheet specimens.*

by a TEM to analyze their chemical composition and crystal structure. Particle size distributions were also measured by means of a scanning electron microscope (SEM) (JEOL S510) equipped with an image analyzer. Size distributions were obtained by measuring more than 200 intermetallic precipitates for each specimen in the SEM images magnified at ×10 000.

The surface oxide, the oxide/Zircaloy-4 matrix interface, and the intermetallic compounds incorporated into the oxide were examined by a transmission electron microscope (TEM) (HITACHI,H-8100,HS-700S).

Results

Weight Gain

1. Effect of Hot Rolling and Annealing Immediately after Hot Rolling—The heating temperature effect before hot-rolling is depicted in Fig. 2. In these specimens, the intermediate annealing and the final stress relief annealing after cold rolling were conducted at 823 and 623 K, respectively. The weight gain of the specimen heated at lower the temperature increased with time significantly. The sample heated at 1023 K for 2 h showed only a slight increase in weight. These results indicate that high hot-rolling temperatures are effective for improving corrosion resistance.

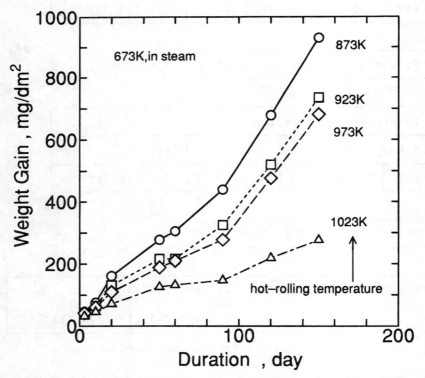

FIG. 2—*Effect of the heating temperature prior to hot rolling on the weight gain in steam at 673 K, Group A specimens.*

The annealing temperature after hot-rolling also significantly affected the weight gains, particularly when hot-rolling was performed at low temperatures as shown in Fig. 3. In these specimens, the intermediate annealing and the final stress relief annealing after cold rolling were conducted at 823 and 623 K, respectively. When the hot-rolling temperature was lower than 923 K, the weight gain-annealing condition curve was different from the specimens with higher hot-rolling temperatures. With lower hot-rolling temperatures, the weight gain showed a maximum and decreased sharply with increasing annealing temperature. When the hot-rolling temperature was higher than 1023 K, the annealing after hot-rolling affected the weight change less significantly. It is clear that the weight gain data for the samples that were hot-rolled at low temperatures show a strong dependence on ΣA. This indicates that the corrosion resistance of Zircaloy-4 under low ΣA conditions can show a large scatter due to the variations in annealing and hot-rolling temperatures.

2. Effect of Intermediate Annealing after Cold-Rolling—After the hot-rolling and annealing, Group-A specimens, as shown in Fig. 1, were cold-rolled and intermediate annealed twice at the same temperature at 853 to 1053 K and finally cold-rolled and stress relieved at 723 K. For Group-B specimens, as shown in Fig. 1, the first intermediate annealing was conducted at 923 K for 2 h, and the temperature for the second intermediate annealing was varied.

The weight gain in Group-A specimens versus the intermediate annealing temperature relationships are shown in Fig. 4a. In Fig. 4a, a steep decrease in the weight gain-temperature

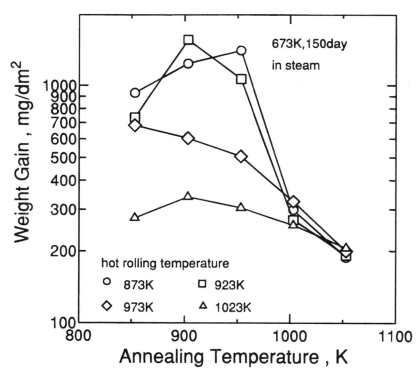

FIG. 3—*Effect of hot rolling temperature and annealing after hot rolling on the weight gain, Group A specimens.*

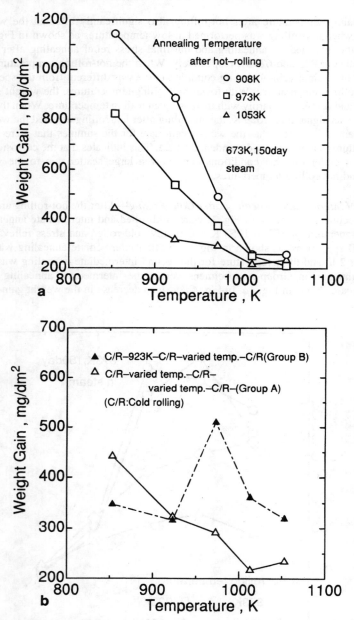

FIG. 4—(a) *Effect of intermediate annealing temperature after cold-rolling on the weight gain, Group A specimens.* (b) *Effect of annealing and cold rolling on the weight gain, Group A and Group B specimens.*

curve was observed by increasing the intermediate annealing temperature. And in Fig. 4b, the weight gain of Group-A specimens decreased when increasing the intermediate annealing temperature, however, the maximum in the Group-B curve was observed. This maximum might be same as for the specimen annealed immediately after hot rolling at a lower temperature, as shown in Fig. 3. The cold rolling and intermediate annealing gradually improved the corrosion behavior.

3. Effect of Final Stress-Relief Annealing—The final stress-relief annealing varied from 673 to 773 K. As shown in Fig. 5, the weight gain decreased when increasing the annealing temperature. In this range, ΣA hardly changed, however, the weight gain depended on the final annealing temperature. Recrystallization occurred in this temperature range, which might result in a decreased weight gain [1].

4. Accumulative Annealing Parameter and Weight Gain—Relationships between weight gain data and ΣA that were divided into the effect of the each heating and the annealing stages are depicted in Fig. 6. Generally, the weight gain decreases with increasing ΣA. It may be more convenient to study the data in detail by dividing it into three groups; Group 1 is the hot-rolling temperature and intermediate annealing temperature after cold rolling, Group 2 is the annealing temperature immediately after hot-rolling, and Group 3 is the final stress-relief annealing. These three groups that are shown by different symbols in Fig. 6, were characterized as follows.

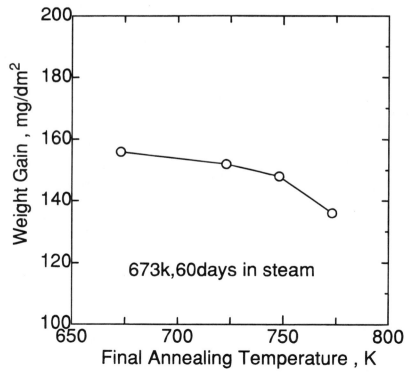

FIG. 5—*Weight gain versus final annealing temperature plots, Group C specimens.*

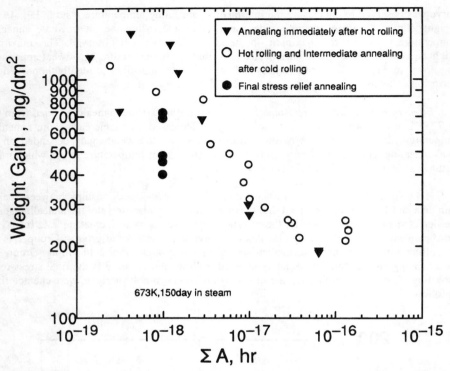

FIG. 6—*Weight gain and ΣA plots, showing the effectiveness of the different annealing stages.*

Group 1 shows that the weight gain decreased with increasing ΣA. These were the specimens that were cold-rolled and intermediate annealed several times.

Group 2 shows the maximum in the curve. At low ΣA, the weight gain was greater than that of Group 1. The weight gain of Group 2, however, decreased more steeply with increasing ΣA than that of the Group 1, which made Group 2 the lowest weight gain group. These observations suggest an annealing or heating effect on the microstructure, for example, the growth of intermetallic precipitates was different for each process.

The ΣA in Group 3 specimens were hardly changed, however, the weight gain difference among the specimens was significant.

Observation of Intermetallic Compounds

Intermetallic compounds that precipitated in the Zircaloy-4 matrix were examined by TEM. The $ZrCr_2$ type intermetallic compounds were precipitated in every specimen, and iron forms a solid solution with the intermetallic precipitates. As a form of $Zr(Cr,Fe)_2$, the iron/chromium (Fe/Cr) ratio was ~2 in the solid solution irrespective of the intermediate annealing temperature.

Size distribution of the intermetallic particles depended significantly on the annealing conditions. Representative histograms are shown in Fig. 7. Figure 8 shows the relationship between

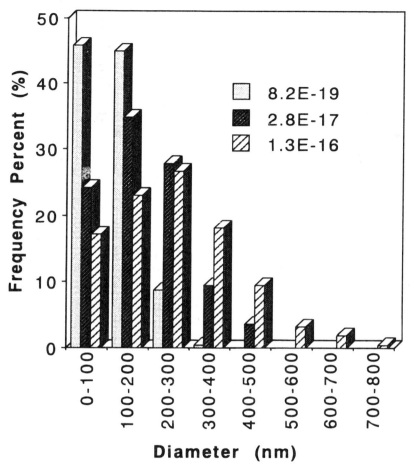

FIG. 7—*Diameter histograms of precipitated intermetallic compounds in Zircaloy-4 under representative ΣA observed by means of SEM images magnified to ×10 000, Group A specimens.*

ΣA and average diameters that were derived from the size distribution such as Fig. 7. The average diameter of an intermetallic compound grew gradually with increasing ΣA, up to 200 nm when ΣA was 1.8×10^{-16}. Because the relationship made it difficult to understand the growth of the intermetallic precipitates, and the diameter distributions of the intermetallic precipitates were divided into four ranges; that is, (*a*) <100 nm, Range A; (*b*) 100 to 200 nm, Range B; (*c*) 200 to 300 nm, Range C; and (*d*) >300 nm, Range D. Each frequency range is depicted in Fig. 9. The intermetallic precipitates in Range (*a*) grew with increasing ΣA, and its frequency percent decreased sharply. On the contrary, the frequency percent for Range D increased. Above 3×10^{-17}, the curves for Range A and Range C intersected each other. In Range B, however, the ΣA and the intermetallic diameter relationship was different. The frequency percent in Range B increased once up to 3×10^{-18} in ΣA and then decreased slightly. These results suggest that the intermetallic precipitates in Range A may have ripened faster than those in the other ranges, because Range B frequency may increase in the early stage.

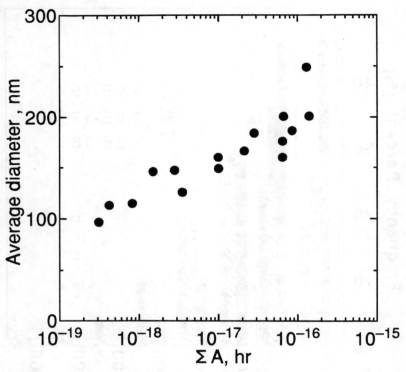

FIG. 8—*Average diameter of intermetallic precipitates ripening with increasing ΣA, Group A specimens.*

Discussion

Relationship Between Diameter Distribution of Intermetallics and Weight Gain

It has been reported that weight gains in Zircaloy may be interrelated with the size of intermetallic compounds. However, it has not been quantitatively understood which size influences the corrosion resistance significantly. Therefore, the diameter distribution was divided into four ranges as shown in Fig. 9 and was correlated with the weight gain as shown in Fig. 10. Note in Fig. 10a that the weight gain was small when the frequency percent in diameter below 100 nm remained less than ~30% or the frequency percent above 200 nm exceeded more than ~25% by ripening the intermetallic precipitates during the annealing. This suggests that the fine intermetallic precipitates below 100 nm in diameter influence the uniform corrosion resistance.

Also, when the frequency in Range C increased above 25%, the weight gain apparently decreased. An estimation for the frequency above 200 nm or below 100 nm may be utilized as a conventional measure to evaluate the degree of uniform corrosion resistance in Zircaloy-4.

Difference in Growth of Intermetallic Particles

As mentioned earlier, the annealing temperature after the hot rolling influenced the corrosion resistance significantly more than the intermediate annealing temperature during repeated cold

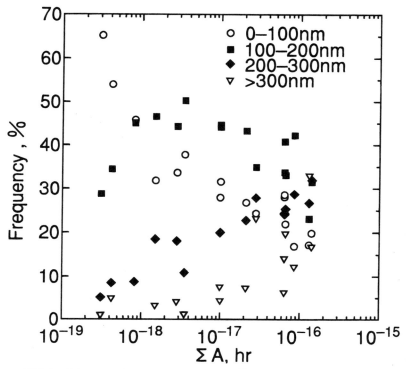

FIG. 9—*Diameter ripening with increasing ΣA, divided into four ranges.*

rolling. Also, the size effect of intermetallic compounds was studied in correlation to the annealing temperature and the weight gain. As a result, the frequency percentage of the intermetallic precipitates below 100 nm in diameter was important for evaluating the corrosion resistance. Therefore, one may expect that the finest range of intermetallic precipitates might be decreased rapidly and effectively during high temperature annealing after hot rolling. The frequencies below 100 nm in diameter are compared to the annealing after the hot-rolling and the intermediate annealing after the cold rolling, as shown in Fig. 11. The frequency percent descended more rapidly, below 30% for the hot-rolling annealing, than for the intermediate annealing. This seems to be the reason for the difference in the corrosion resistance between them.

The intermetallic particles in the hot-rolling annealing-modified specimens were observed to mainly precipitate at the grain boundaries and grew faster than those precipitated in the grains, as shown in Figs. 12 and 13. After repeated cold rolling several times, such a grain boundary may diminish and the preferential ripening of particular precipitates at the boundaries may be suppressed.

Intermetallic Particles Incorporated into Oxide Scale

The intermetallic compound precipitation and its growth are compared to corrosion behavior as mentioned earlier. The intermetallic compounds may be incorporated within the ZrO_2 scale

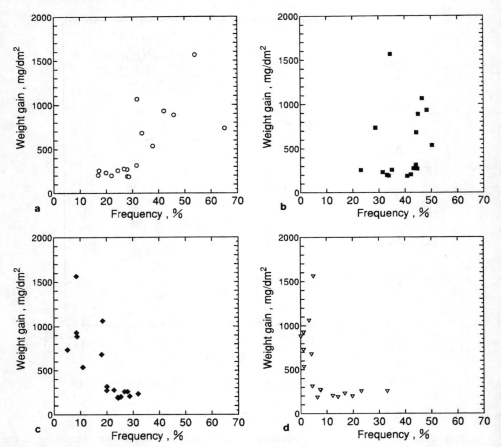

FIG. 10—(a) *Relationship between the weight gain and intermetallic precipitates diameter, under 100 nm.* (b) *Relationship between the weight gain and intermetallic precipitates diameter, 100 to 200 nm.* (c) *Relationship between the weight gain and intermetallic precipitates diameter, 200 to 300 nm.* (d) *Relationship between the weight gain and intermetallic precipitates diameter, above 300 nm.*

and oxidized in the ZrO_2 oxide when the Zircaloy-4 is oxidized in steam [7]. The oxidation behavior of the intermetallic compound was described in two publications [8,9].

The oxide scales on the Zircaloy-4, ($\Sigma A = 1.4 \times 10^{-17}$) were observed by means of TEM in cross section, as depicted in Fig. 14.

Figure 15 shows the interface region between the base metal and the scale for the two samples with representative ΣA. Particular attention was paid to the direct observation of the intermetallic compound incorporated into the oxide and the morphology of growing front.

The intermetallic precipitate incorporated into the ZrO_2 was observed as shown in Fig. 15a. The crystal structure in the middle of the ZrO_2 intermetallic precipitates was analyzed to be the $ZrCr_2$ hexagonal type, as in the base metal. Its chemical composition was analyzed at several points from center to edge, results of which are shown in Fig. 16. The Fe/Cr ratio of the incorporated intermetallic precipitate was reduced to be ~1, which is much smaller than

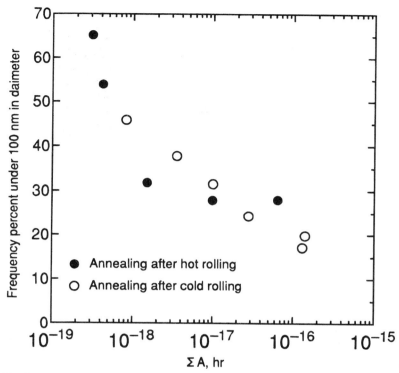

FIG. 11—*Ripening intermetallic precipitates under 100 nm in diameter compared with annealing after hot rolling and cold rolling.*

2 in the Zircaloy-4 base metal. This suggests that a small amount of iron and chromium may diffuse into the surrounding ZrO_2 due to oxidation of the intermetallic precipitates, and in particular, dissolution of iron must be much greater than chromium. A solid solution of Fe^{3+} (or Cr^{3+}) into ZrO_2 and substitution for Zr^{4+} may induce an oxygen vacancy in ZrO_2 because of the charge balance in the ZrO_2.

If the iron (or chromium) that are diffused into the oxide from the intermetallic precipitates influence the corrosion behavior, as is believed so far, the surface area should be important. Therefore, the surface areas were derived from the diameter distribution data and were correlated with the weight gain data as shown in Fig. 17. The specimen weight gains with different intermediate annealing temperatures after cold rolling, increased with the increasing intermetallic particle surface area. However, the relationship between the weight gain and the particle surface area for specimens with different annealing temperatures after hot-rolling was different.

It was remarkable, as shown in Fig. 15, that a few recrystallized grains, 100 to 500 nm in diameter, were observed in the higher ΣA Zircaloy-4 matrix just beneath the growing front of the oxide. The recrystallized grains aligned along the interface forming a 600 to 800-nm–wide layer. In the lower ΣA Zircaloy-4, similar recrystallization grains were also observed, however, there were fewer of the recrystallized grains than in the high ΣA Zircaloy-4. The microstructure beneath the recrystallized layer remained as a Zircaloy-4 stress-relived structure.

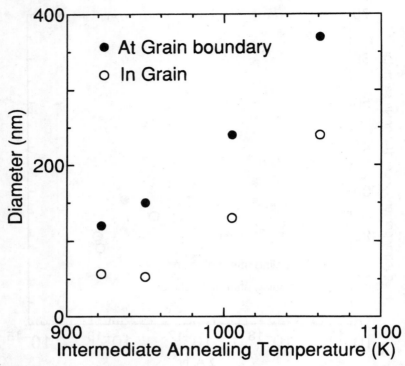

FIG. 12—*Rapid ripening of intermetallic precipitates at the grain boundary by annealing after hot rolling.*

Generally, it is known that recrystallization improves uniform corrosion resistance. The recrystallization that occurred during corrosion, primarily in the higher ΣA Zircaloy-4, might be a cause for the improved corrosion resistance of those materials. Recrystallization may also be caused by a stress that was induced by volume expansion of ZrO_2 when zirconium transformed into ZrO_2. In the weight gain curve for the higher ΣA Zircaloy-4, transition of oxidation was observed repeatedly, for example, see Fig. 2. However in the weight gain curve for the lower ΣA Zircaloy-4, the transition was not clearly repeated after the first transition.

Conclusion

In this study, the relationship between uniform corrosion resistance and intermetallic compound precipitation behavior in the Zircaloy-4 sheets produced under various hot-rolling and annealing conditions were investigated and the following conclusions were drawn.

1. Uniform corrosion resistance was improved with an increasing cumulative annealing parameter, however, the annealing temperature after hot-rolling had greater impact compared to the intermediate annealing and the final annealing temperatures.
2. When a frequency of the intermetallic compounds under 100 nm in diameter decreased below 30%, the corrosion resistance was improved. The frequency for particles under

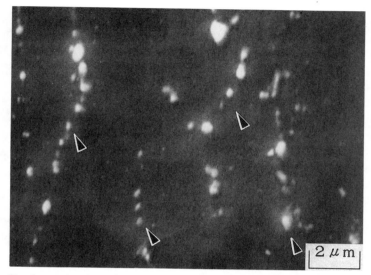

FIG. 13—*Dominant ripening of intermetallic precipitates at the grain boundary by annealing after hot rolling.*

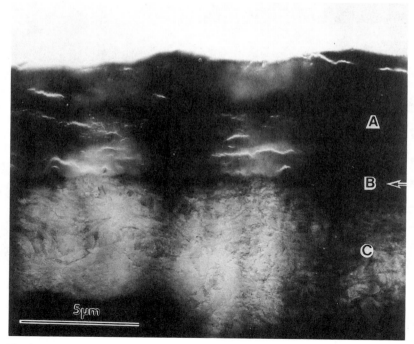

FIG. 14—*Scale on Zircaloy-4 under 1.4×10^{-17} in ΣA observed by TEM in cross section (673 K, 10.3 MPa, 150 days in steam): (A) ZrO_2, (B) interface, and (C) Zircaloy.*

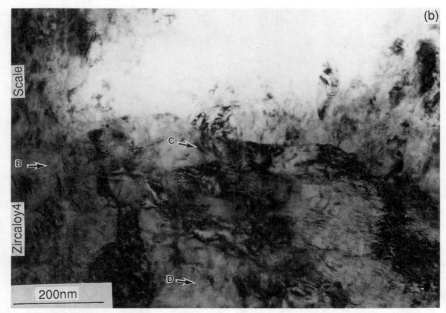

FIG. 15—*Intermetallics incorporated into the oxide and interface structure in the vicinity of the scale and Zircaloy-4 matrix (673 K, 10.3 MPa, 150 days in steam):* (a) $\Sigma A = 1.4 \times 10^{-17}$ *and* (b) $\Sigma A = 8.9 \times 10^{-19}$. *(A) intermetallic compound, (B) interface, (C) recrystallized grain, and (D) stress-relieved structure.*

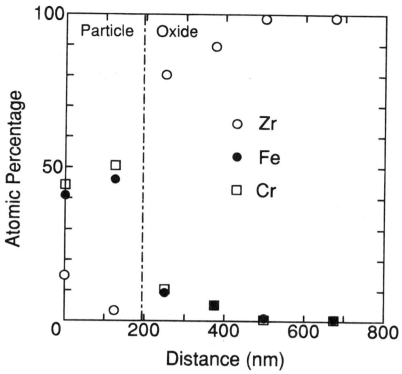

FIG. 16—*Iron and chromium distributions in the intermetallic precipitates and in the surrounding oxide analyzed by means of TEM with EDS.*

100 nm in diameter decreased sharply, especially when increasing the annealing temperature after the hot rolling.
3. The intermetallic compounds incorporated into the ZrO_2 oxide were observed by TEM, and it was confirmed that primarily iron in the intermetallic precipitates diffused out into the oxide and the Fe/Cr ratio of the precipitate changed to ~1 in the oxide compared to ~2 in the base metal.
4. Recrystallization had occurred in the Zircaloy-4 matrix just beneath the growing oxide, and the degree of recrystallization in the higher ΣA Zircaloy-4 was higher than that in the lower ΣA Zircaloy-4. This might affect the corrosion behavior significantly.

Acknowledgment

The authors thank their colleagues in Sumitomo Metal Industries, Ltd., particularly, Mr. T. Konishi for his guidance, Mr. H. Nakamura for autoclave testing, and Mr. Y. Kohzuki for particle size measurements.

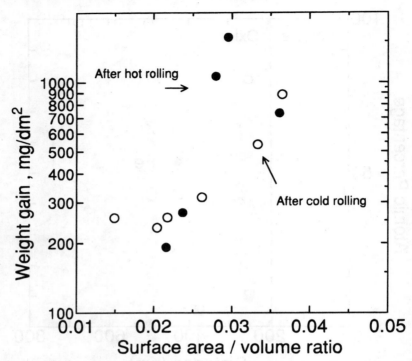

FIG. 17—*Surface area/volume ratio of intermetallic precipitates depending on annealing conditions.*

References

[1] Schemel, J. H., Charquet, D., and Wadier, J. F., "Influence of the Manufacturing Process on the Corrosion Resistance of Zircaloy-4 Cladding," *Zirconium in the Nuclear Industry: Eighth Symposium, ASTM STP 1023,* L. F. P. Van Swam and C. M. Eucken, Eds., American Society for Testing and Materials, Philadelphia, 1989, pp. 141–152.

[2] Charquet, D., "Improvement of the Uniform Corrosion Resistance of Zircaloy-4 in the Absence of Irradiation," *Journal of Nuclear Materials,* Vol. 160, 1988, pp. 186–195.

[3] Garzarolli, G., Steinberg, E., and Weidinger, H. G., "Microstructure and Corrosion Studies for Optimized PWR and BWR Zircaloy Cladding," *Zirconium in the Nuclear Industry: Eighth Symposium, ASTM STP 1023,* L. F. P. Van Swam and C. M. Eucken, Eds., American Society for Testing and Materials, Philadelphia, 1989, pp. 202–212.

[4] Anderson, T., Thorvaldsson, T., Wilson, A., and Wardle, A. M., "Influence of Thermal Processing and Microstructure on the Corrosion Behavior of Zircaloy-4 Tubing," IAEA-SM-288/59, International Atomic Energy Agency, Vienna, 1986.

[5] Foster, J. P., Dougherty, J., Burke, M. G., Bates, J. F., and Worcester, S., "Influence of Final Recrystallization Heat Treatment on Zircaloy-4 Strip Corrosion," *Journal of Nuclear Materials,* Vol. 173, 1990, pp. 164–178.

[6] Gros, J. P. and Wadier, J. F., "Precipitate Growth Kinetics in Zircaloy-4," *Journal of Nuclear Materials,* Vol. 172, 1990, pp. 85–96.

[7] Kubo, T. and Uno, M., "Precipitate Behavior in Zircaloy-2 Oxide Films and Its Relevance to Corrosion Resistance," *Zirconium in the Nuclear Industry: Ninth International Symposium, ASTM STP 1132,* C. M. Eucken and A. M. Garde, Eds., American Society for Testing and Materials, Philadelphia, 1991, pp. 476–498.

[8] Weidinger, H. G., Ruhmann, H., Cheliotis, G., Maguire, M., and Yau, T.-L. "Corrosion–Electrochemical Properties of Zirconium Intermetallics," *Zirconium in the Nuclear Industry: Ninth International Symposium, ASTM STP 1132*, C. M. Eucken and A. M. Garde, Eds., American Society for Testing and Materials, Philadelphia, 1991, pp. 499–535.

[9] Pecheur, D., Lefebvre, F., Motta, A. T., Lemaingnan, C., and Charquet, D., "Oxidation of Intermetallic Precipitates in Zircaloy-4: Impact of Irradiation," in this volume.

DISCUSSION

J. B. Bai[1] (written discussion)—Your results showed an improvement of uniform corrosion resistance by increasing the accumulative annealing after hot rolling. Which do you think contributes the most; the microstructure of the Zircaloy-4 or the intermetallic compounds to the corrosion resistance improvement, and why?

H. Anada et al. (authors' closure)—I think both of them contribute to corrosion resistance. We compared corrosion resistance with frequency percent, under 100 nm in size distribution, for the intermetallic compound. In this correlation, we can not separate the contributions of the Zircaloy-4 microstructure and of the intermetallic compounds.

B. Warr[2] (written discussion)—You state that your improved corrosion resistance is due to the recrystallized layer just below the oxide. Is the improvement due to (1) a different oxide forming in the recrystallized matrix or (2) the recrystallized layer itself. If (1) is the answer, what is the difference you see in the oxide that leads to the improved corrosion performance?

H. Anada et al. (authors' closure)—We mentioned that the recrystallized layer may play a role in improving corrosion resistance. But we think that a difference in the size distribution of the intermetallic compounds would influence corrosion resistance more. We have not observed any difference in the crystal structure of the oxide. Most of our attention was given to the intermetallic compound incorporated in the oxide. In future work, we will consider the relationship between the oxide structure and the recrystallized layer during corrosion.

S. K. Yagnik[3] (written discussion)—The recrystallized zone just beneath the growing oxide front is not always seen in other investigations. What is the origin of such a zone?

H. Anada et al. (authors' closure)—We assume that recrystallization might be caused by a stress induced by volume expansion during transformation of zirconium into ZrO_2. In the weight gain-test duration curve of a low corrosion-resistant Zircaloy-4, transition was not obvious after the first transition occurred. However, for a high corrosion-resistant Zircaloy-4, several transitions were apparently observed. These indicate the possibility of dense oxide formation and of high induced stress.

Bo Cheng[4] (written discussion)—Your data suggest that ΣA can not or may not predict the corrosion rate of Zircaloy in 400°C steam, where annealing is also important in the processing sequence. You suggest that the precipitate size distribution is important. With this data, can you provide an insight into the optimum processing sequence for uniform corrosion resistance?

H. Anada et al. (authors' closure)—Generally speaking, the total characteristics for the optimization of Zircaloy-4 cladding tubes should be considered. Therefore, we can not propose an optimum process for Zircaloy-4 tube using the data based on only this study. For corrosion behavior only, we believe that the condition in hot extrusion process or in annealing or both after hot rolling are the most important for improving corrosion resistance.

Young S. Kim[5] (written discussion)—(1) You mentioned that diffusion of iron the chromium from the precipitates into the oxide caused oxygen vacancy leading to corrosion enhancement. Does that mean that lattice diffusion of oxygen is the main controlling process in determining corrosion? If grain boundary diffusion of oxygen is the one that controls the corrosion rate, the vacancy in the oxide does not affect the corrosion at all. Please comment on this point.

(2) Can you tell what the phase change of the oxide is across the oxide layer, if there is?

---

[1] Lab. MSS/MAT, CNRS URA 850, Ecole Centrale Paris, 92295 Chatenay Malabry, France.
[2] Ontario Hydro, Toronto, Ontario, Canada.
[3] Electric Power Research Institute (EPRI), Palo Alto, CA.
[4] Electric Power Research Institute (EPRI), Palo Alto, CA.
[5] Korea Atomic Energy Research Institute, Daeduk-Danji, Daejom, Korea, 305-606.

(3) Comment on whether the precipitates enclosed in the oxide were oxidized or not? If yes, how are you sure of it?

H. Anada et al. (authors' closure)—(1) It was not apparent which diffusion process, a bulk diffusion or a short circuit diffusion, controlled the oxidation reaction. If the bulk diffusion was predicted as a controlling process, vacancies of iron or chromium by diffusion might be important. If a short circuit diffusion was predicted, the oxide structure (that is, grain size, grain shape, etc.) might be important. However, the difference in the oxide structure was not apparent in this study. Therefore, we assumed the bulk diffusion model enhanced by iron or chromium diffusion. In our future work, it will be necessary to explain the relationship of the oxide structure and corrosion behavior.

(2) A monoclinic ZrO_2 was observed in the vicinity of the oxide and the metal interface.

(3) The electron diffraction technique showed that the intermetallic compound that was incorporated into the oxide was oxidized gradually and decomposed into small ZrO_2 particles.

Jean-Paul Mardon,[1] Daniel Charquet,[2] and Jean Senevat[3]

Optimization of PWR Behavior of Stress-Relieved Zircaloy-4 Cladding Tubes by Improving the Manufacturing and Inspection Process

REFERENCE: Mardon, J.-P., Charquet, D., and Senevat, J., "**Optimization of PWR Behavior of Stress-Relieved Zircaloy-4 Cladding Tubes by Improving the Manufacturing and Inspection Process,**" *Zirconium in the Nuclear Industry: Tenth International Symposium, ASTM STP 1245,* A. M. Garde and E. R. Bradley, Eds., American Society for Testing and Materials, Philadelphia, 1994, pp. 328–348.

ABSTRACT: With the aim of optimizing the basic properties of stress-relieved Zircaloy-4 cladding tubes, particularly those that make it possible to push back the initial technological limits that may be encountered, and of reducing the scatter of those properties and enhancing tube quality, the role of the main parameters involved in manufacturing the ingot, Trex, and cladding tube has been evaluated on an industrial scale. A series of large-sized tube lots were produced under controlled manufacturing conditions, then characterized by out-of-pile test results (short- and long-term corrosion, stress corrosion cracking (SCC), creep, mechanical, and structural properties) on finished tubes. For the investigated parameters (chemical composition, number of melt, quench rate, accumulated annealing parameter, the ΣA factor, surface condition (outside and inside diameters), and finished tube quality), this role is indeed important but complex due to the highly interactive nature of the variables investigated.

Adjustment of the chemical composition within ASTM limits enables generalized corrosion resistance to be enhanced and irradiation growth to be minimized. A significant decrease of the observed scatter in corrosion and mechanical properties is obtained by optimization of the ΣA range, the quenching rate, and the final heat treatment. The optimum seems to be reached for a final treatment at the highest possible temperature compatible with the stress-relieved state, corresponding to an average precipitate size and ΣA. Moreover, by adding anneals upstream in the process, a further increase in this ΣA no longer seems to have a significant effect on generalized corrosion.

Finally, extensive efforts have been employed in the pilgering, surface preparation (outside diameter polishing, flush-pickling), and examination method (UT, EC) leading to a sizable improvement in SCC resistance and to a reduction in scatter for finished tubes.

The result of these optimizations has been implemented in the current AFA-2G, that shows that under irradiation a 30% corrosion gain is reached after three cycles, without degrading creep strength or growth. The intrinsic effect of tin on generalized corrosion resistance under irradiation was also confirmed on this occasion.

KEY WORDS: zirconium, zirconium alloys, cladding tubes, process outling, uniform corrosion, quenching, annealing parameter, precipitates, nondestructive testing, in-reactor behavior, nuclear materials, nuclear applications, radiation effects

---

[1] Consulting engineer, FRAMATOME Nuclear Fuel Division, FRAGEMA, 69006, Lyon, France.
[2] Senior scientist, CEZUS, Centre de Recherche d'Ugine, 73400 Ugine, France.
[3] Manager, Research and Development, ZIRCOTUBE, BP 21, 44560 Paimboeuf, France.

Zircaloy-4 has been used as the cladding tube material in pressurized water reactors (PWRs) for more than 30 years, and its in-core uniform corrosion behavior has been demonstrated to be acceptable for limited discharge burnups and standard operating conditions. To meet the utility need to increase burnups, the designers and manufacturers have been led to give priority to the improvement of Zircaloy-4 tube corrosion resistance. For this purpose, several studies on the corrosion behavior of Zircaloy-4, particularly uniform corrosion, have been conducted on the basis of out-of-pile and in-pile tests. The results of these tests show that Zircaloy-4 corrosion behavior is significantly affected by the ΣA parameter and by the morphology of the intermetallic particles [1–5] and the quenching rate [6–8]. It is also clear that adaptations of the chemical composition (that is, reduction of tin content, increase of iron and chromium), while remaining within the ranges proposed by ASTM Specification for Wrought Zirconium and Zirconium Alloy Seamless and Welded Tubes for Nuclear Service (B 353-83), may lead to enhanced waterside corrosion resistance [9–12].

To meet the needs of users for the 1990s 50 to 55 GWd/t maximum burnup, we have run a Zircaloy-4 optimization program that aims not only to strive towards higher fuel cycle cost savings through increased burnup, but also to cater for the large maneuverability needs of the reactors, and to continue improving fuel reliability.

To achieve these goals, we have optimized Zircaloy-4 in two ways: one aiming to optimize the basic properties of the material and particularly those making it possible to push back the first technical limits that may be encountered by the product under irradiation (creep, growth, stress corrosion cracking (SCC), generalized corrosion, hydriding) and the other involving both reduction of the scatter in these properties to restore margins and improvement in tube quality.

This study closely examines the effects of several parameters from the ingot to the tube on the properties of the finished tube, through many cladding tube industrial scale production runs. On the strength of the out-of-pile characterization of test results (corrosion, SCC, creep, mechanical properties) and of the in-pile results (corrosion, growth, creep), the most promising improvements have been implemented in the current AFA-2G product proposed for burnups just quoted.

Experimental Procedure

Program

In order to optimize the manufacturing and inspection process outline for stress-relieved Zircaloy-4 tubes, the manufacturing parameters with a potential influence on the finished tube basic properties (corrosion, creep, etc.) were identified, then a test program detailed in Table 1 was devised to highlight not only the effect of each parameter, but also their interactions, if any. The main variables studied are: chemical composition, number of melts, quench rate, the ΣA factor, the tube pilgering process outline, the final stress-relieving anneal, and the nondestructive examination techniques.

Materials

These experiments were run on an industrial scale on 37 ingots of 3 or 6 tonnes, leading to more than 450 lots of 500 to 600 tubes, manufactured according to the standard forged, β-quenched, hot-extruded process, followed by several cold-working/annealing cycles, of which the last is a stress-relieving treatment. Each of the ingots has a homogeneous chemical composition close to the nominal values (Table 1), so it is representative of everyday production.

TABLE 1—*Process outline parameters.*

Parameter	Variation Range
Chemical composition	
Sn, % wt	1.23 to 1.40
Fe, % wt	0.19 to 0.24
Cr, % wt	0.09 to 0.13
O_2, % wt	0.108 to 0.135
C, ppm	100 to 174
Si, ppm	77 to 113
Number of melts	2, 3, and 4
Quench rate V_{max}[a]	5 to 100°C s^{-1}
Annealing parameter[b]	1.10^{-18} to 2.10^{-16} h
Intermediate anneals	after quench, after drawing, and on shell and Trex.
Final anneal	T (450, 530°C) t (2 to 4 h)
Tube surface preparation	belt polishing + OD polishing wheel, and ID surface pickling
Nondestructive testing	automatic visual OD examination, and ID eddy current examination

[a]V_{max} = rate equivalent to the rate obtained at the surface of the draw-ready billet during the $\beta \to \alpha$ transformation (950°C → 800°C).
[b]Calculated with $Q/R = 40\,000$ K^{-1}.

Methods and Tests

The experimental techniques are:

1. Autoclave corrosion tests: short-term three days at 400°C as per ASTM Practice for Preparing, Cleaning, and Evaluating Corrosion Test Specimens (G 2-88) on a non-pickled specimen and long-term tests in steam at 400°C on pickled specimens.
2. Measurement of the diameter of the intermetallic precipitates by transmission electron microscopy (TEM) or scanning electron microscopy SEM [*13*].
3. Internal pressurization tests with and without iodine at 350°C and slow tension tests (1.7×10^{-5} s^{-1}) on a ring at room temperature in inert atmosphere and in iodine.
4. Cold and hot tensile mechanical tests according to ASTM Test Methods for Tension Testing of Metallic Materials (E-893) and ASTM Test Methods for Elevated Temperature Tension Tests for Metallic Materials (E 21-92,) and thermal creep tests (400°C, 130 MPa, 240 h) on pre-pressurized specimens.

Results of Optimization Studies and Comments

The manufacturing parameters, whose influence on the finished tube properties is analyzed and discussed later, are mostly interactive and their effects are not necessarily cumulative.

The effects observed in this study cannot be considered separately or taken out of the general background in which they were investigated, that is, the complete ingot to tube process, otherwise their role will be inhibited or reverse effects will even occur. We therefore consider that the effects obtained here are quantified and validated solely for the qualified ranges of

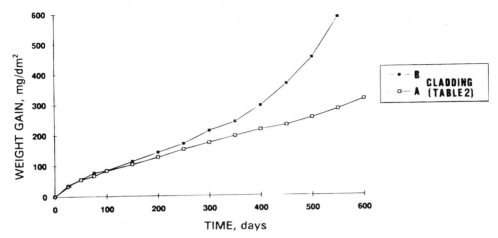

FIG. 1—*Effect of tin content on long-term corrosion resistance in autoclave at 400°C of stress-relieved Zircaloy-4.*

our Trex and tube process outlines, leading to an AFA-2G low-tin stress-relieved Zircaloy-4 cladding tube; extrapolation to another manufacturing context or another metallurgical state calls for experimental validation.

Chemical Composition

Our studies for optimization of Zircaloy 4 chemical composition, while remaining within the standard ASTM ranges, show the beneficial effect on autoclave corrosion at 400°C of lowering tin content down to 1.2% (Fig. 1, Table 2) and the existence of optimum values for

TABLE 2—*Irradiation characteristics and conditions of A and B cladding for stress-relieved Zircaloy-4.*

Assembly	Zircaloy-4 CWSR	Burnup (MWd/tM) Fast fluence (10^{-21} n·cm^{-2}) $E > 1$ MeV		
		EOC 1	EOC 2	EOC 3
1	Sn (1.20 to 1.29%)	9391	23 513	35 243
	A CLADDING			
	132 fuel rods/ assembly	1.6	4.1	6.3
2	Sn (1.38 to 1.50%)	9378	23 728	34 550
	B CLADDING			
	132 fuel rods/ assembly	1.6	4.1	6.1

silicon in the 50 to 90 ppm range (Fig. 2), a result that supplements that obtained by Fuchs et al. [14], and for an iron/chromium (Fe/Cr) ratio close to 1.8 to 1.9 (Fig. 3).

Carbon must be maintained at a relatively low level, less than 200 ppm, because of its influence on growth (Fig. 4).

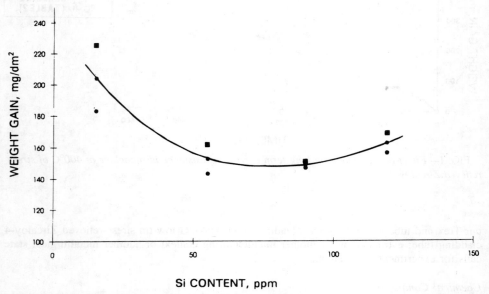

FIG. 2—*Influence of silicon content on 340 days corrosion resistance in autoclave at 400°C of recrystallized Zircaloy-4.*

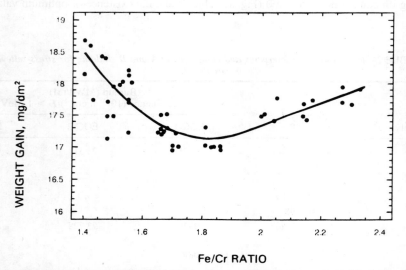

FIG. 3—*Weight gain after ASTM G 2-88 three-day, 400°C test, versus Fe/Cr ratio of stress-relieved low tin Zircaloy-4.*

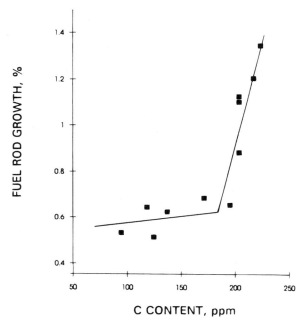

FIG. 4—*Influence of carbon content on Zircaloy-4 stress-relieved fuel rod irradiation growth.*

Production of the ingot in three melts also completely rules out stringers, a sign of the presence of volatile elements [8,15,16].

Quench Rate

On a number of ingots, the quenching conditions were controlled to cover a wide quench rate range (V_{max} 5°C s^{-1}, >100°C s^{-1}). The quenched billets then followed the standard manufacturing process outlines up to the finished tube stage. The set of corrosion results obtained for these lots is given in Fig. 5 and shows the existence of a corrosion optimum. Parametric tests in a narrower range confirm the existence of this minimum and show that scatter can be minimized for a proper rate.

Accumulated Annealing Parameter: ΣA and Precipitate Size

To cover a wide ΣA range (10^{-18} to 10^{-16}) calculated with $Q/R = 40\,000$°K^{-1} (Table 1), we acted upon the temperatures of the intermediate recrystallization anneals of the Trex and stress-relieved Zircaloy-4 process outlines; the results obtained for two tin contents are illustrated in Fig. 6. Note a strong interaction between tin content and ΣA where the reduction in tin content leads to:

1. less corrosion susceptibility for values under 10^{-17},
2. a change of slope shifted towards the high ΣA values, and
3. a plateau that is logically lower.

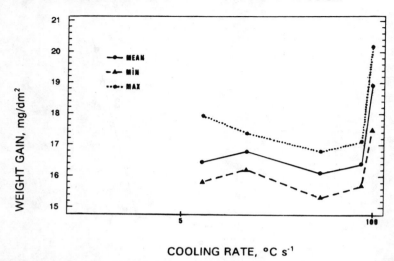

FIG. 5—*ASTM G 2-88 three-day, 400°C test weight gain versus cooling rate, V_{max} (Table 1), of stress-relieved low-tin Zircaloy-4.*

Also, no rise, even slight, of the curves was observed for the large values involved (2×10^{-16}) unlike the findings of Fuchs et al. [*14*].

The ΣA of the AFA 2G process outline, from 2.5 to 4.4×10^{-17}, is an optimum level for generalized corrosion. To assess the influence of the annealing position on the corrosion resistance of the tube, we performed anneals on a needle structure or on a cold and hot worked structure or both: the results of the 400°C long-term corrosion tests show that all the configurations are equivalent and do not improve corrosion resistance.

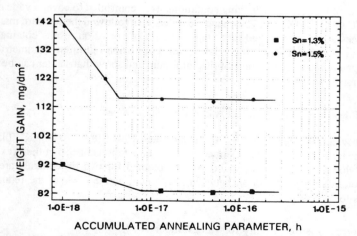

FIG. 6—*Influence of ΣA and tin content on weight gain after 112 days at 400°C of stress-relieved Zircaloy-4.*

A close interaction was also found between the size of the intermetallic precipitates and the tin content (Fig. 7); the corrosion plateau is reached for precipitates of 0.18 to 0.20 μm in diameter in the case of stress-relieved Zircaloy-4. With the chosen ΣA range, we obtain particles of about 0.25 μm for AFA-2G that locates it confortably on the plateau.

Final Stress-Relieving Heat Treatment

A significant increase in autoclave corrosion resistance can also be obtained by optimizing the final heat treatment. Figure 8 shows long-term corrosion weight gain trends for different time/temperature ratings on a product obtained at a high quench rate that behaved poorly in the three day, 400°C ASTM G 2-88 corrosion test.

It is clear that, given the fact that 500°C must not be exceeded to remain in the stress-relieved condition for this type of tube, the optimum is located as close as possible to this temperature for the final heat treatment, a temperature that was therefore chosen for the AFA-2G tubes.

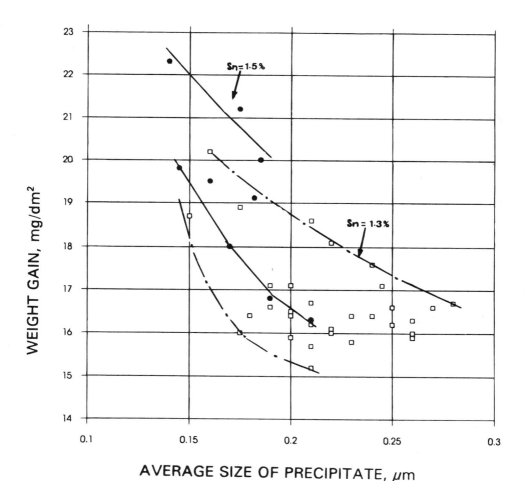

FIG. 7—*Influence of precipitate size and tin content on weight gain after three days at 400°C.*

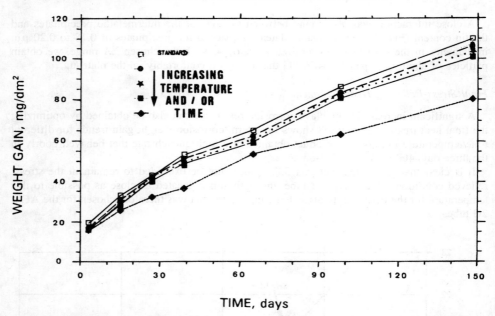

FIG. 8—*Weight gain at 400°C for different final anneal points (t, T) of stress-relieved low-tin Zircaloy-4.*

Later tests confirmed that the scatter in the ASTM G 2-88 corrosion-results was significantly reduced at this temperature.

Surface Preparation

Inside Diameter (ID) Surface—It is acknowledged that a wise choice of cladding ID surface finish process provides an improvement in its SCC behavior, one of the main phenomena contributing to pellet cladding interaction (PCI). On the strength of the progress made in understanding SCC-induced Zircaloy-4 failure mechanisms [17,18], the decision was taken to adopt ID surface treatment by forced circulation of an acid solution.

Three types of ID surface finish are commonly employed: sand-blasting, silicone carbide (SiC) blasting, and flush-pickling. The SCC behavior with these three techniques was investigated through pressurization and slow tension tests on rings at room temperature, with and without iodine.

The results obtained with these three surface finishes for the pressurization tests show (Fig. 9) that the lifetime reduction due to iodine is only by a factor of 15 for flush-pickled tubes, whereas it is by 100 for sand-blasted claddings.

Likewise, the slow tension tests (Table 3) show that the loss of relative ductility is significantly lower for the flush-pickled state, thus confirming the lesser SCC susceptibility of this surface condition.

Outside Diameter (OD) Surface—Concurrently with this effort to upgrade the ID surface, significant advances were also made in OD polishing. Two areas were investigated:

1. belt polishing on the three heads, and

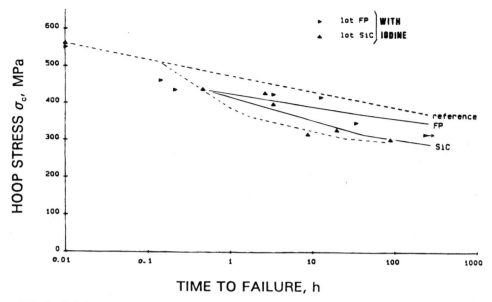

FIG. 9—*Lifetime curves at pressurization for the different types of ID surface finish.*

2. belt polishing on the first two heads and by fine abrasive wheels on the third head.

The latter brings a perceptible improvement in roughness (Table 4) that translates into a gain on the three-day, 400°C corrosion results, together with a reduction of the scatter.

Quality Improvement

Independent of the previously mentioned actions, a full-scale effort at each tube fabrication process was pursued to reduce the scatter of tube properties and to increase the quality of the finished tubes. This effort, focused on:

1. tighter control of pilgering and tooling parameters [19] to obtain the required level of tube properties and dimensions and reduce their scatter,
2. improvement of lubrication during the last pilgering pass to reduce the defect rate of the tube inner surface, and

TABLE 3—*Loss of ductility in iodine-containing alcohol during a slow tension test (1.7×10^{-5} s^{-1}).*

	Relative Loss of ductility due to Iodine, %
Standard finish	60
Flush-pickling finish	40

TABLE 4—*Weight gain at ASTM G 2-88 test and roughness for two OD polishing configurations.*

	Roughness, μm						ASTM G 2-88, three days at 400°C	
	R_a		R_z		R_m		ΔP, mg/dm²	
	$\overline{m}$	σ	$\overline{m}$	σ	$\overline{m}$	σ	$\overline{m}$	σ
Standard lots, belt-polished only	0.256	0.036	2.89	0.39	3.87	0.76	17.88	0.80
Lots polished with belts and wheels	0.190	0.023	2.20	0.215	2.84	0.35	16.21	0.34

3. the use of heat treatments with argon blanket to homogenize the temperature within the furnace at each anneal treatment, and thus reduce the scatter of the tube properties (corrosion, etc.).

The improvement made by these advances can be illustrated by a number of relevant indicators of tube quality:

1. From 1985 to 1992, we noted a spectacular decrease in the defect rates on the OD surface and above all the ID surface.
2. During the same period, CSR, which is a good indicator of the quality of pilgering and of the overall thermomechanical process, saw its scatter improve by a factor 1.4.
3. Finally, the scatter of the dimensional characteristics was also reduced by a factor 1.3 to 2, depending on the characteristic concerned.

Concurrently, the introduction of two new higher-performance, nondestructive examination processes, that is, eddy current examination of the ID surface [20] and automatic visual examination, was a contributor to this quality upgrade effort.

Optimized AFA-2G Zircaloy-4

To show the progress made by the introduction of the process optimizations, we decided to follow the weight gain trend resulting from the ASTM G 2-88 three day, 400°C test over a long manufacturing period (Fig. 10). It was noted that at the end of the optimization phase, the AFA-2G product reaches an optimum from the corrosion standpoint and that it stabilizes at this floor value.

This result is confirmed by long-term 400°C corrosion tests (Fig. 11).

PWR Performances of AFA-2G Zircaloy-4

Properties in Use as Design Inputs

The reduction in tin content, within the ASTM standards, has a negligible effect on free growth, does not modify the unirradiated material endurance limit of 350 MPa at 350°C and 1 Hz [21], does not affect the elastic and plastic behavior under uniaxial and biaxial loadings, and causes a slight decrease in creep resistance under high stress and under neutron flux.

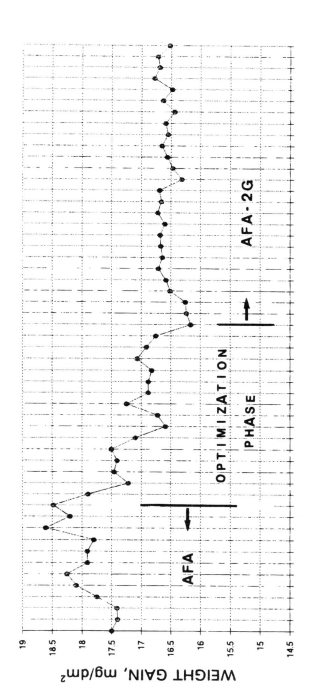

FIG. 10—*Monthly average weight gains after ASTM G 2-88, three days at 400°C of stress-relieved Zircaloy-4 cladding tube.*

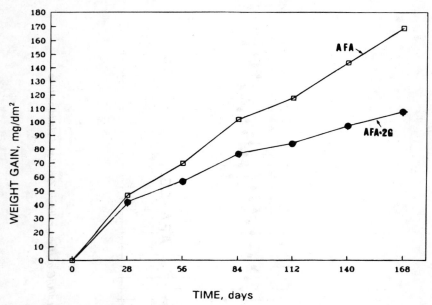

FIG. 11—*Weight gain at 400°C for a Zircaloy-4 AFA (Sn = 1.55%) and an AFA-2G (Sn = 1.32%).*

Overall, the effect of improving the manufacturing process outline outweighs the few possible negative effects due to the slight variation in chemical composition for the set of analyzed properties, as subsequently illustrated.

Irradiation Behavior in PWRs

Zircaloy-4 cladding tubes manufactured according to an optimized process and only differing in their tin content (Table 2) were irradiated in PWR conditions in 1988; the irradiation experience built up to date is three cycles (35 000 MWd/T).

The very substantial gain in corrosion resistance under irradiation (Fig. 12) is on the order of 30% average oxide thickness after three cycles, and the highest values observed on the high-tin claddings are eliminated with the optimized cladding. On the basis of the corrosion kinetics observed to date, it is expected that this gain will expand with increasing burnup.

The irradiation growth (Fig. 13) and diametral creep (Fig. 14) of the fuel rods are unaffected by the decrease in tin content.

A comparison with the whole data base shows that for growth (Fig. 15), as for corrosion, scatter is reduced by clipping the high values, and the average values are reduced.

This excellent behavior of AFA-2G confirms that we have indeed reached the optimum level with regard to the in-reactor properties of stress-relieved Zircaloy-4, by combining the different plateaus and optimum values defined in the previously described parametric study.

Conclusions

This paper presents the results of Zircaloy-4 optimization, in accordance with ASTM specifications.

The first optimization criterion is based on corrosion behavior:

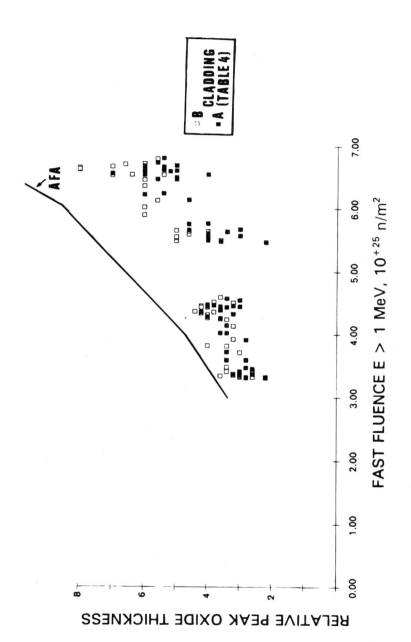

FIG. 12—*Maximum zirconia layer thickness measured at Span 6 on Zircaloy-4 stress-relieved peripheral rods, versus fluence (E > 1 MeV).*

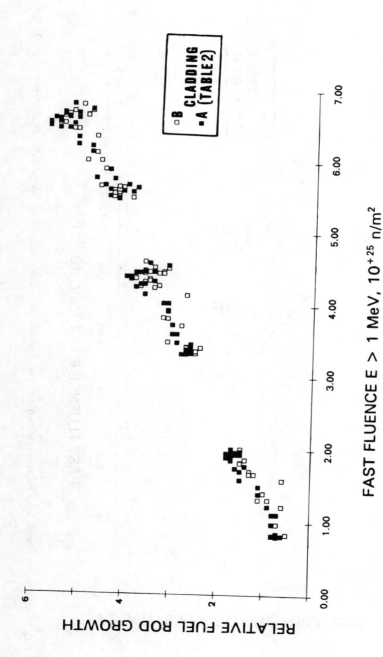

FIG. 13—*Growth of stress-relieved Zircaloy-4 peripheral rods, versus fluence (E > 1 MeV).*

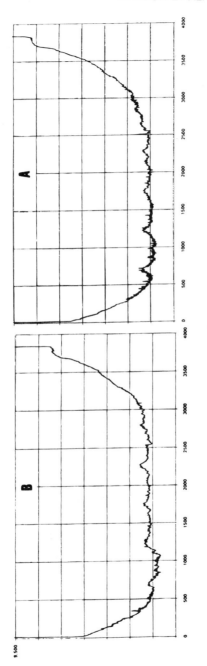

FIG. 14—*Average diameter corrected for zirconia layer after one irradiation cycle stress-relieved Zircaloy-4:* (a) *standard tin and* (b) *low tin.*

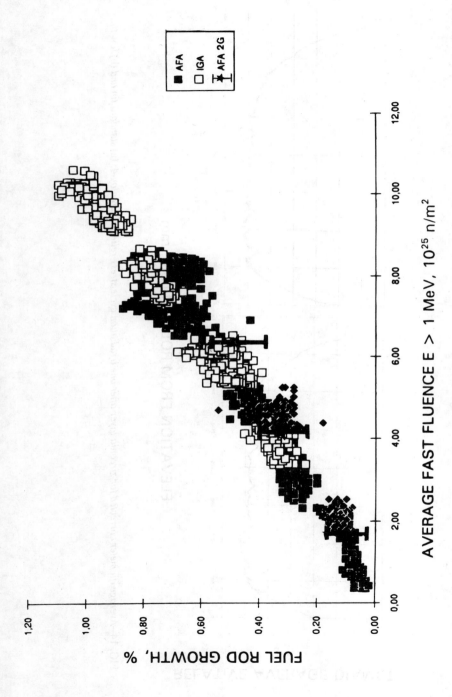

FIG. 15—*AFA fuel rod growth* [22].

1. For some factors, an optimum or plateau was found, regarding their impact on this property. This is the case for the Fe/Cr ratio close to 1.8 to 1.9, for silicon in the 50 to 90 ppm range, for the quench rate, for ΣA from 10^{-17}, and for precipitate size from 0.18 µm. At these optimum levels, together with the improvement in average corrosion values, scatter is minimized while the complex interactive phenomena observed for the extreme values are avoided. The choice of these factors can practically be made by considering only corrosion, since their influence on the other properties is negligible.
2. Other factors like tin content or final annealing temperature have a monotonic influence on corrosion. These factors also modify other major properties like creep, for example; this means that their choice must be the result of a compromise. A tin content of 1.3% and a stress-relieving treatment at the highest possible temperature compatible with the cold-worked stress relieved (CWSR) state is an acceptable tradeoff. The decrease in tin content, together with other factors, does not change behavior under loss of coolant accident (LOCA) conditions.

Another goal is improvement of other aspects of in-reactor behavior and quality enhancement. The importance of surface preparation, particularly the benefits of ID flush pickling on SCC, has been convincingly demonstrated. To limit growth, carbon has to be held at a relatively low level, under 200 ppm. More generally, the advances made in the manufacturing and inspection process outline have led to a further improvement in tube quality.

These optimizations are only of interest if they pay off in enhanced in-core cladding behavior. This has clearly been the case for the AFA-2G optimized product, since after three irradiation cycles, a corrosion gain of 30% was recorded, with no degradation of creep or growth.

The first AFA 2G reloads were delivered by FRAMATOME in 1992 and since then more than 600 assemblies have been supplied; this design will enter into general use in 1995.

However, Zircaloy-4 is reaching its optimization limits and only a few minor improvements can still be considered for the future. To reach even higher performance levels, more advanced technological solutions must be envisaged.

Acknowledgments

The authors thank Mrs. I. Schuster and Mr. T. Forgeron of the CEA (Grenoble and Saclay) for their kind cooperation in running SCC and high-temperature creep tests.

References

[1] Anderson, T., Thorvaldsson, T., Wilson, A., and Wardle, A. M., "Influence of Thermal Processing and Microstructure on the Corrosion Behavior of Zircaloy-4 Tubing," *Improvements in Water Reactor Fuel Technology and Utilization,* IAEA-SM-288/59, International Atomic Energy Agency, Vienna, 1987, pp. 435–449.

[2] Maussner, G., Ortlieb, E., and Weidinger, H. G., "Basic Properties of Zirconium Alloys with Respect to Mechanical and Corrosion Behavior," *Materials for Nuclear Reactor Core Applications,* British Nuclear Energy Society (BNES), London, 1987, pp. 49–55.

[3] Charquet, D., Steinberg, E., and Millet, Y., "Influence of Variations in Early Fabrication Steps on Corrosion, Mechanical Properties, and Structure of Zircaloy-4 Products," *Zirconium in the Nuclear Industry: Seventh International Symposium, ASTM STP 939,* R. B. Adamson and L. F. P. Van Swam, Eds., American Society for Testing and Materials, Philadelphia, 1987, pp. 431–437.

[4] Thorvaldsson, T., Anderson, T., Wilson, A., and Wardle, A., "Correlation Between 400°C Steam Corrosion Behavior, Heat Treatment, and Microstructure of Zircaloy 4 Tubing," *Zirconium in the Nuclear Industry: Eighth International Symposium, ASTM STP 1023,* L. F. P. Van Swam and C. M. Eucken, Eds., American Society for Testing and Materials, Philadelphia, 1989, pp. 128–140.

[5] Garzarolli, F., Steinberg, E., and Weidinger, H. G., "Microstructure and Corrosion Studies for Optimized PWR and BWR Zircaloy Cladding," *Zirconium in the Nuclear Industry: Eighth Interna-*

tional Symposium, ASTM STP 1023, L. F. P. Van Swam and C. M. Eucken, Eds., American Society for Testing and Materials, Philadelphia, 1989, pp. 202–212.

[6] Kass, S. in *Corrosion of Zirconium Alloys, ASTM STP 368*, American Society for Testing and Materials, Philadelphia, 1963, pp. 3–27.

[7] Garzarolli, F., "Long Time Out of Pile Corrosion of Zircaloy 4 in 350°C Water," IAEA Technical Committee Meeting on External Cladding Corrosion in Water Power Reactor, Cen Cadarache, Oct. 1985, pp. 66–69.

[8] Charquet, D., "Improvement of the Uniform Corrosion Resistance of Zircaloy 4 in the Absence of Irradiation," *Journal of Nuclear Materials*, Vol 160, 1988, pp. 186–195.

[9] Harada, M., Abe, K., Furuya, T., Kakuma, T., and Onta, S., "Effect of Allowing and Impurity Elements on Nodular Corrosion Resistance of Zirconium Alloys," presented at Seventh International Symposium on Zirconium in the Nuclear Industry, Strasbourg, France, June 1985, unpublished.

[10] Weidinger, H. G. and Lettau, H., "Advanced Material and Fabrication Technology for LWR Fuel," *Improvements in Water Reactor Fuel Technology and Utilization*, IAEA-SM-288/27, International Atomic Energy Agency, Vienna, 1987, pp. 451–467.

[11] Eucken, C. M., Finden, P. T., Trapp-Pritsching, S., and Weidinger, H. G., "Influence of Chemical Composition on Uniform Corrosion of Zirconium-Base Alloys in Autoclave Tests," *Zirconium in the Nuclear Industry: Eighth International Symposium, ASTM STP 1023*, L. F. P. Van Swam and C. M. Eucken, Eds., American Society for Testing and Materials, Philadelphia, 1989, pp. 113–127.

[12] Eucken, C. M. et al., in *Zirconium in the Nuclear Industry: Seventh Internationl Symposium, ASTM STP 939*, R. B. Adamson and L. F. P. Van Swam, Eds., American Society for Testing and Materials, Philadelphia, 1987, pp. 113–127.

[13] Gros, J. P. and Wadier, J. F., "Precipitate Growth Kinetics in Zircaloy 4," *Journal of Nuclear Materials*, Vol. 172, 1990, pp. 85–96.

[14] Fuchs, H. P. et al., "Cladding and Structural Material Development for the Advanced Siemens PWR Fuel "FOCUS"," American Nuclear Society (ANS), Avignon, 1991.

[15] Charquet, D. et al., in *Zirconium in the Nuclear Industry: Seventh International Symposium, ASTM STP 939*, R. B. Adamson and L. F. P. Van Swam, Eds., American Society for Testing and Materials, Philadelphia, 1987, p. 284.

[16] Charquet, D. et al., "Heterogeneous Scale Growth During Steam Corrosion of Zircaloy 4 at 500°C," *Fracture Mechanics: Perspectives and Directions (Twentieth Symposium), ASTM STP 1020*, R. P. Wei and R. P. Gangloff, Eds., American Society for Testing and Materials, Philadelphia, p. 374.

[17] Brunisholz, L., "Caractérisation de la CSC du Zircaloy en milieu iodé," *Thèse*, Ph. D, Grenoble, 1985.

[18] Brunisholz, L. and Lemaignan, C., "Iodine Induced Stress Corrosion of Zircaloy Fuel Cladding Initiation and Growth," Strasbourg, June 1985.

[19] Senevat, J. et al., "Establishing Statistical Models of Manufacturing Parameters in Zircaloy-4 Cladding Tube Properties," *Zirconium in the Nuclear Industry, Ninth International Symposium, ASTM STP 1132*, C. M. Eucken and A. M. Garde, Eds., American Society for Testing and Materials, Philadelphia, 1990, pp. 62–79.

[20] Senevat, J. et al., "Eddy Current Examination Technique During Manufacturing of Zircaloy-4 Fuel Cladding Tubes," *Journal of Nuclear Materials*, Vol. 178, 1991, pp. 315–320.

[21] Soniak, A. et al., "The Influence of Frequency on Fatigue Behavior of Unirradiated Zircaloy 4 Cladding Tubes," *11th International Conference on SMIRT*, Tokyo, 1991.

[22] Morel, M. et al., "Irradiation Experience Leads FRAGEMA to New Fuel Designs," *Nuclear Engineering International*, Oct. 1992, pp. 33–35.

DISCUSSION

C. D. Williams[1] (written discussion)—Eddy current inspection and automated visual inspection were mentioned as means to improve quality. Were these used on intermediate-sized tubing during tube manufacture or on final-sized tubing?

J.-P. Mardon et al. (authors' closure)—Eddy current examination of the inside-diameter surface and visual examination of the outside-diameter surface are performed on the finished tube.

Nobuaki Yamashita[2] (written discussion)—What is the main reason for the improvement of corrosion resistance by adding wheel polish, surface, or amount of removal?

J.-P. Mardon et al. (authors' closure)—Improvement of corrosion behavior by adding an abrasive wheel during final polishing of the outside-diameter surface is not due to material removal since we have not observed a weight gain gradient across the tube thickness. Rather, it corresponds to an overall quality upgrade of the outside-diameter surface, and particularly of roughness, which mainly pays off in a reduction in scatter.

R. A. Ploc[3] (written discussion)—In a number of presentations, including yours, it is implicitly assumed that the hydrogen pickup is directly related to the corrosion rate. In view of your data that the corrosion rate is related to the tin and silicon contents, can you also tell us if the hydrogen pickup data shows the same dependence as the oxide film growth?

J.-P. Mardon et al. (authors' closure)—The hydrogen pick-up fraction is of the same order of magnitude (0.13 on average) for Zircaloy-4 AFA (standard tin) and for AFA 2G (low tin) irrespective of ZrO_2 thickness, at least up to 90 μm. Consequently, a decrease in the oxide thickness formed under irradiation necessarily results in a smaller quantity of H_2 in ppm, in proportion to the decrease.

G. D. Moan[4] (written discussion)—In Fig. 3, the weight gain in the corrosion test plotted against the Fe/Cr ratio showed a minimum at a Fe/Cr ratio of about 1.8.

The scatter in the data for Fe/Cr ratio less than 1.8 was much greater than the Fe/Cr ratio greater than 2.

What are the reasons for the difference in the scatter in the two Fe/Cr areas?

J.-P. Mardon et al. (authors' closure)—The wider scatter observed for the Fe/Cr ratio values less than 1.8 to 1.9 is due to the combination of the following two factors:

1. the iron and chromium chemical composition fluctuations between $Zr(Fe,Cr)_2$ precipitates, and
2. the more pronounced slope of the weight gain curve $= f(Fe/Cr)$ in this Fe/Cr ratio range.

M. Limback[5] (written discussion)—In your presentation you mentioned that AFA 2G cladding material has optimized creep performance. In what respect do you regard the creep behavior as optimized? (When regarding pellet cladding interaction, one can assume that a lower creep role, resulting in longer time to gap closure, would be preferable. However, when considering fission gas release, a short time to gap closure results in lower fuel temperature and thereby to lower fission gas release. Meaning that, it is not an easy task to determine the optimum creep behavior.) Perhaps you are showing that low tin (1.3% by weight) AFA 2G Zircaloy-4 cladding has creep performance similar to normal tin (1.5% by weight).

---

[1] G. E. Nuclear Energy, Wilmington, NC.
[2] G. E. Nuclear Energy, Wilmington, NC.
[3] AECL Research, Chalk River, Ontario, Canada.
[4] AECL CANDU, Mississauga, Ontario, Canada.
[5] ABB Atom AB., Vasteras, Sweden.

J.-P. Mardon et al. (authors' closure)—The decrease in tin content between AFA and AFA 2G results in a slight reduction in thermal creep resistance, which was partially offset by optimizing the thermal/mechanical process outline. Moreover, the process upgrades enabled the scatter in AFA 2G creep behavior to be drastically reduced. The result of these upgrades was that a marked similarity in irradiation creep behavior was observed between the two products after one cycle in a PWR (Fig. 14).

The optimum creep value we have chosen for AFA 2G is the result of a tradeoff providing optimum fuel rod irradiation behavior in terms of high-stress primary creep and secondary creep for standard operating conditions.

R. Holt[6] *(written discussion)*—Concerning the effects of carbon: (1) would you care to speculate on why carbon affects fuel rod growth and (2) is the effect of carbon on creep based on in-reactor or out-of-reactor testing?

J.-P. Mardon et al. (authors' closure)—We have observed an influence of carbon content on fuel rod growth (axial creep + free growth) under PWR irradiation conditions. Carbon may have an impact, non-qualified to date, on the two properties involved in fuel rod growth.

---

[6] Chalk River Laboratories, Chalk River, Ontario, Canada.

Lithium Effects and Second-Phase Particles

Philippe Billot,[1] *Jean-Charles Robin,*[1] *Alphonse Giordano,*[1] *Jean Peybernès,*[1] *Joël Thomazet,*[2] *and Hélène Amanrich*[3]

Experimental and Theoretical Studies of Parameters that Influence Corrosion of Zircaloy-4

REFERENCE: Billot, P., Robin, J.-C., Giordano, A., Peybernès, J., Thomazet, J., and Amanrich, H., "**Experimental and Theoretical Studies of Parameters that Influence Corrosion of Zircaloy-4,**" *Zirconium in the Nuclear Industry: Tenth International Symposium, ASTM STP 1245,* A. M. Garde and E. R. Bradley, Eds., American Society for Testing and Materials, Philadelphia, 1994, pp. 351–377.

ABSTRACT: Waterside corrosion of Zircaloy cladding in pressurized water reactors (PWRs) is largely dependent upon the operating parameters and microstructure of the zirconium alloys. The impact of these parameters on the corrosion kinetics of Zircaloys is investigated on the basis of empirical data and experiences that can be interpreted using existing corrosion models.

The influence of thermo-hydraulic data, heat flux, local boiling conditions, and of the growing oxide films has been studied from corrosion tests performed in static autoclaves or in out-of-pile loops. These parametric investigations are described as well as the models that were developed.

The impact of microstructure is studied from the comparison of the corrosion behavior of different Zircaloy-4 specimens corroded in out-of-pile tests. In particular, a poor corrosion resistance of an experimental Zircaloy-4 material is analyzed as a function of the microstructure close to the metal/oxide interface.

The impact of the alloy composition and primary coolant chemistry on the corrosion kinetics of Zircaloy-4 is modeled empirically or uses a mechanistic approach that proposes a series of chemical equations with a mathematical representation of the kinetics.

These proposed models are then used to investigate the corrosion behavior of Zircaloy-4 cladding in 17 by 17 plants for rods irradiated at high burnups. Higher PWR operating cycles, core average coolant temperature, power, and elevated primary coolant lithium concentrations (3.5 to 4 ppm) are then simulated and discussed in terms of Zircaloy corrosion resistance considerations.

KEY WORDS: corrosion, autoclaves, loops, power reactors, boiling conditions, lithium, microstructure, zirconium, zirconium alloys, nuclear materials, nuclear applications, radiation effects

Currently developed corrosion models produce satisfactory assessments of the oxidation behavior of Zircaloy-4 cladding in pressurized water reactors [1]. The large number of results from autoclave, out-of-pile loop and pressurized water reactor (PWR) tests can be used to interpret the effect of the various parameters involved in the corrosion process but also reveal the inadequacies of the models as soon as more severe chemistry and thermo-hydraulic conditions (boiling, high thermal flux, and pH) or different design materials are involved.

---

[1] Senior research engineer and research engineers, respectively, Commissariat à l'Energie Atomique, Bat 224, C. E. Cadarache, 13108 Saint Paul lez Durance, France.
[2] Senior research engineer, FRAMATOME, Division Combustible, 69006 Lyon, France.
[3] Senior research engineer, EDF/SEPTEN, Lyon, France.

Current models are made up of two modules:

1. the first module determines the thermo-hydraulic characteristics of the fuel cladding environment (coolant temperature, wall temperature, void fraction, etc.); and
2. the second module calculates external axial and azimuthal corrosion rates of the fuel claddings using kinetics laws of oxidation.

Zircaloy-4 corrosion kinetics are characterized by the succession of two laws: the pre-transition and the post-transition regimes. As a first approximation, the base mechanisms are represented by a cubic law in pre-transition phase and a linear law in post-transition phase [2].

The out-of-pile loop corrosion data and characterizations of oxide layers formed at various boiling rates have led to a better understanding of the impact of two-phase flow regimes on Zircaloy-4 corrosion. It has also been possible to develop the bases for modeling the lithium enrichment processes in the oxide layers formed under boiling conditions.

Interpretation of out-of-pile and in-pile results about the corrosion behavior of varied zirconium base alloys leads to establish correlations between the nature of the cladding material used and its resistance to corrosion.

A more mechanistic approach, using chemical equilibrium, is undertaken in order to explain the influence of the alloying elements and primary fluid chemistry on the corrosion behavior of various cladding materials.

In addition, the semi-empirical approach to corrosion processes is currently improved in order to lead to a phenomenological model. This model makes it possible to integrate the effects due to heat flux and primary fluid chemical conditions into the oxidation model developed by the CEA (COCHISE code). Other parameters such as irradiation and microstructure impacts are under development.

Finally, the COCHISE corrosion model is used to analyze the corrosion behavior of varied materials in power reactors, then is applied to new PWR operating conditions.

Influence of Two-Phase Flow Regimes on Zircaloy Corrosion

Out-of-Pile Loop Tests

The possible two-phase flow regimes that appear locally in the hot channels of PWRs undergoing more severe thermo-hydraulic conditions led us in recent years to study the mechanisms of zirconium-based alloys oxidation under boiling [1–3].

The experimental results indicate an increase in cladding corrosion that is caused by lithium enrichment in the oxide layers formed [2,4]. Overconcentration mechanisms under boiling are linked to the thermo-hydraulic conditions in a region close to the wall as well as to the porous structure of zirconia.

The parameter generally used to correlate the oxidation phenomena observed in the presence of the vapor phase is void fraction, which is defined as the percentage of surface occupied by the vapor phase in the section of the hot channel. The void fraction determination makes it possible to identify the different boiling regimes to which the claddings are subjected: partial nucleate boiling, fully developed nucleate boiling, and bulk boiling.

The influence of boiling on Zircaloy-4 corrosion varies according to the oxidation phase reached. In fact, the various boiling regimes (nucleate or generalized boiling) only have a slight influence on cubic kinetics in the pre-transition regime. At this stage of oxidation, the corrosion profile is therefore practically independent of void fraction variations.

Analyses of pre-transition oxide layers by secondary ion mass spectrometry (SIMS) reveal lithium enrichment at the oxide/water interface. The morphology of oxide formed before the transition time prevents significant diffusion of lithium in the more protective oxide films (Fig. 1). The enrichment at the wall of low-volatility chemical species, such as lithium, occurs in a micro-layer present below the forming vapor bubbles [1]. Lithium concentration at the surface is a function of the mass evaporation rate as well as the bubble density at the wall. Boiling conditions in out-of-pile loop tests provide an almost constant assessment of the overconcentration factor of around 2 to 3 for void fractions at the wall varying from 5 to 35%.

On the other hand, beyond the transition thickness, corrosion kinetics depend greatly on the boiling regimes and void fraction. In particular from out-of-pile results, it was possible to correlate the axial and azimuthal profiles of the corrosion rates with the axial and circumferential variations of void fractions due to the geometry of the elementary cells surrounding the rods [1].

At the same time, various analyses have shown an increase in lithium concentration in oxide films subjected to increasing void fractions (Fig. 2). The lithium increase in the oxide that has become porous in the post-transition regime leads to a uniform lithium distribution from the metal/oxide interface to the oxide/water interface (Fig. 1). It was possible to define an acceleration factor of post-transition kinetics as a function of the void fractions applied to the claddings

$$Fa(t) = DS/DS_0$$

where

DS = increase in oxide thickness at time, t, under boiling regime (wall temperature = T_0) and
DS_0 = increase in oxide thickness at time, t, in one phase flow heat transfer condition (wall temperature = T_0).

The acceleration factors measured are given in Fig. 3. They show the influence of the two-phase flow regimes on post-transition corrosion kinetics under various chemical and thermohydraulic conditions for 60-day tests. The corrosion profiles as a function of void fraction are similar for each test. The acceleration factor increases gradually up to a void

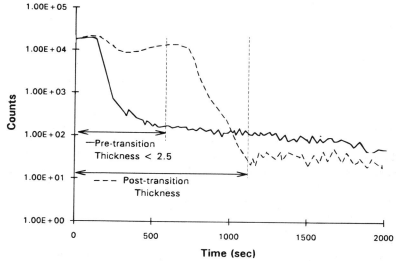

FIG. 1—*Usual lithium profiles in the oxide scales using the SIMS technique.*

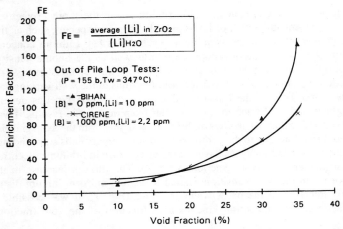

FIG. 2—*Lithium enrichment in the oxide layer as a function of void fraction (out-of-pile loop tests).*

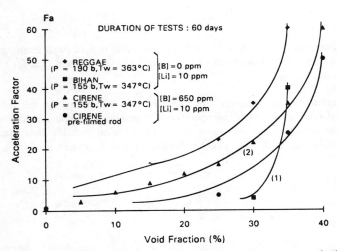

FIG. 3—*Corrosion enhancement factor versus boiling conditions (out-of-pile tests).*

fraction of 30%, then increases more significantly beyond this value. In addition, the acceleration factor varies as a function of the chemical and thermal conditions imposed during the tests.

The deviations observed between corrosion profiles, (1) and (2) of Fig. 3, are explained by different chemistry conditions. We suppose that boron at the oxide/water interface hinders the incorporation of lithium into the oxide layer and thus diminishes the effect of lithium enrichment at the wall.

The more severe thermohydraulic conditions of the tests carried out in the REGGAE loop (363°C, 191 bar) result in higher acceleration factors due to an intensification of the boiling processes or the synergetic effect of lithium and temperature or both. Finally, claddings that have been subjected to prefilming in the pre-transition phase under one-phase flow heat transfer conditions show lower acceleration factors. The pre-transition oxide layer formed seems therefore to have a protective nature with regard to boiling processes. It is shown that the enhancement of corrosion rates is not only a function of thermo-hydraulic conditions, such as void fraction, but also appears after a modification of the oxide structure in the post-transition phase (creation of pores and cracks).

Corrosion Processes under Boiling Conditions

Pores and cracks created during the post-transition phase allow the lithium to diffuse more easily in the zirconia films that result in a flat profile of lithium across the oxide scales (Fig. 1). In addition, the incorporation of lithium into the oxide leads to a growth in the porous network, the mechanism of which is not well established [5,6]. According to the hypothesis proposed by Cox, the generation of pores by lithium is caused by a process in which the tetragonal zirconia dissolves and reprecipitates as monoclinic zirconia. In the two-phase flow regime, the development of microcracks and pores at the surface of the oxide influences the distribution and density of the nucleation sites where bubbles can grow. The lithium enrichment in the oxide films originates at these sites. Given that the incorporated lithium can develop and enlarge the porosity network, the result is an intensification of the boiling process. In other words, the liquid is absorbed by capillary action into the porous network then is vaporized through the widest pores. These mechanisms, now quite familiar [7–9], have the effect of increasing the mass and heat transfer to the wall and of gradually increasing the enrichment factor as the oxide layer develops.

These processes are largely responsible for the corrosion acceleration phenomena observed from the loop tests conducted under boiling conditions (Fig. 3).

Understanding the mechanisms of vaporization and lithium enrichment in the pores of the oxide layers requires experimental studies in the following fields:

1. porosimetry measurements of the oxide films and
2. visual examination of the boiling process.

Porosity Measurements—Currently developed porosity measurements are based on the Brunauer-Emmet-Teller method (BET). This method uses krypton adsorption and desorption in the oxide pores and produces an assessment of the open pore fraction and the size and volume distribution of pores with a diameter less than 0.1 µm. Additional measurements using mercury porosimetry show the distribution of wider pores in the zirconia films (diameter > 0.1 µm).

BET porosity measurements carried out on various oxide layers in the post-transition phase reveal the presence of interconnected pores with a diameter of around 1 to 10 nm. Joint measurements using mercury porosimetry and the BET technique confirm the presence of these pores but also reveal the appearance of wider cracks whose size and volume increase with the oxide thickness. This increase of the cracks is accompanied by a rise in the open porosity. Figure 4 shows the development of the mean size of cracks that join onto the pore network obtained on oxide samples of 10 to 100 µm formed in out-of-pile loops and measured by mercury porosimetry.

Visual Examination of the Boiling Process—Direct observation of the cladding surfaces subjected to nucleate boiling conditions provides information about the development of boiling mechanisms in relation to the morphology of the oxide layers (porosity, thickness, roughness, etc.). Observations of the thick oxide films at atmospheric pressure and in static conditions revealed the tendency for vapor bubbles to appear from the porosity of oxide layers despite subsaturation of the water coolant (Fig. 5). Moreover, the intensification of the vaporization process in the thickest layers of zirconia is shown.

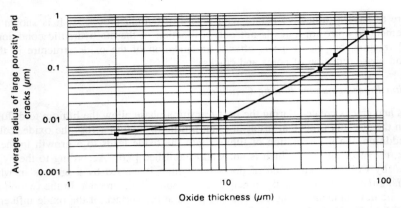

FIG. 4—*Large porosity and cracks as a function of thickness of the oxide film formed in the out-of-pile loop (mercury measurements).*

In addition, the fitting of a window on the REGGAE out-of-pile loop provides optical access to the two-phase flow regime under PWR operating conditions, allowing for the characterization of size, velocity, and the concentration distribution of bubbles, that is, three parameters at the origin of the chemical species hideout process.

Lithium Enrichment Mechanisms in Thick Layers

In the post-transition regime, the porous network allows the lithium concentrated at the oxide/water interface to diffuse rapidly to the metal/oxide interface, leading to an increase in the corrosion rate.

Moreover, for thick oxide layers, the creation of large pores and microcrack network allows water to circulate into the oxide scale. From the Clausius-Clapeyron relationship, the mean radius required for nucleation sites to create vapor bubbles on the surface of a cladding subjected to PWR thermo-hydraulic conditions must be greater than 20 nm. In other words, the cracks and pores that open on the surface with dimensions promoting the nucleation of bubbles will trigger vaporization mechanisms across the oxide films. Porosity measurements (Fig. 4) show that pores and cracks at the origin of these vaporization mechanisms can appear beyond an oxide layer of about 20 μm. Water that vaporized in the pores and evolved out of the oxide creates suction movements in the neighboring pores. Liquid phase circulation is maintained in the pores by the pressure gradient caused by the cyclic growth and departure of the bubbles in the bulk of the coolant. The mass flow rate of liquid into the pore network is governed by viscosity forces and can be suitably described using Darcy's law. This law uses the permeability parameter that characterizes the ability of the porous environment to transport the liquid. This parameter is a function of the open porosity, the interconnections between pores, and the morphology of the porous environment. Various semi-empirical relationships can be used to assess the permeability as a function of the open pore fraction and the characteristics of the porous environment (fissured media, granular media, etc.)[*10*].

The chemical species that are contained in the water at a concentration, C_0, and drawn into the porous network are concentrated in the pores and diffuse into the oxide. The enrichment factor is therefore linked to the mass evaporation rate and the diffusion coefficient of the species in zirconia.

FIG. 5—*Boiling due to the thick oxide films formed (pool conditions).*

The mass evaporation rate in the oxide can then be estimated by iterative calculation using Darcy's law, change of phase law, and thermal and mass balances.

Lithium, because of its low volatility, is only considered in the liquid phase, that is, $\Gamma = C_v/C_l \sim 0$, where C_v and C_l are concentrations of the chemical species in the vapor and liquid phases, respectively.

The chemical balance of species in the oxide also depends on mass transfer between the oxide film and the two-phase flow. The corresponding mass transfer coefficient, K_d, is therefore a function of the eddy diffusion coefficients due to the liquid and the bubbles at the wall.

The steady-state diffusion equation can be expressed in a simplified monodimensional formulation as follows

$$D \frac{d^2C}{dx^2} - \frac{(1-\Gamma)}{\rho_l} C_0 \frac{d\dot{m}_v}{dx} = 0 \qquad (1)$$

where

C = concentration in the oxide (kg/m³),
D = diffusion coefficient in the oxide of the species (m²/s),
$\frac{d\dot{m}_v}{dx}$ = mass evaporation rate per unit of volume and time (kg/m³·s), and
x = distance into porous oxide, from oxide/water interface as origin (m).

Transfers at both interfaces impose the following limit conditions:

1. oxide/water interface, $x = 0$

$$D\frac{dC}{dx} = K_d (C - C_0)$$

where

K_d = mass transfer coefficient (m/s), and
C_0 = concentration of chemical species in the coolant (g/m³).

2. metal-oxide interface, $x = \delta$

$$D\frac{dC}{dx} = 0$$

(the chemical species do not diffuse in the metal).

Resolution of Eq 1 gives the lithium concentration in the oxide layers as a function of the lithium level in the primary water and of the mass evaporation rate through the oxide, depending on the thermo-hydraulic conditions at the wall of the fuel claddings. This model predicts the lithium contents measured in the thick porous oxide layers formed under boiling during out-of-pile loop tests, assuming an open porosity fraction around 10%. Viewing two-phase flows under PWR operating conditions should lead to the assessment of the parameters involved in the mass transfers and an improvement in modeling.

In the reactor, vapor formation in the porous oxide can also take place in spite of the thermo-hydraulic conditions thought to prevent local boiling at the cladding wall. In this case, enrichment of the lithium in the oxide can thus occur.

In particular, boiling in the oxide pore network will occur if the following conditions are combined:

1. water penetration in the oxide,
2. sufficient heating of the water to reach vaporization temperature, and
3. sufficient oxide porosity to allow for vapor formation and evacuation.

Effect of Materials

Corrosion studies have been conducted for a number of years to optimize Zircaloy-4 and to propose new alloy solutions. These studies were conducted to support the industrial developments described in Ref 11. Before products are put into a power reactor, tests are conducted in static autoclaves and out-of-pile loops and characterizations are made.

Tests performed at 360°C in autoclave, with high lithium levels in the coolant as lithium hydroxide (0.01 to 0.2 M), show erratic results, together with large-scale hydriding that led us to postpone the tests under these conditions.

Impact of Microstructure

A microstructure study was carried-out for two specific materials. These alloys were developed from the same ingot (standard Zircaloy-4 chemistry), with intermediate recrystallization treatment chosen so that these alloys have very different annealing factor values ($\Sigma A = \Sigma\, t_i \exp(-Q/RT_i)$ where t_i and T_i are respectively the duration (h) and temperature (K) of heat treatment No. i, and $Q/R = 40\,000$ K):

1. Alloy A: $\Sigma A = 5.7\ 10^{-18}$ h and
2. Alloy B: $\Sigma A = 9.7\ 10^{-19}$ h.

These two materials underwent final stress-relieving treatment and were then mechanically polished.

The different material behaviors were first revealed under ASTM Practice for Aqueous Corrosion Testing of Samples of Zirconium and Zirconium Alloys (G 2-88) conditions, at 400°C. After three days in autoclave, the respective weight increases are 17.71 mg/dm² for Alloy A, and 20.84 mg/dm² for Alloy B. When the test is continued, the behavior of these alloy becomes more and more different, as shown in Table 1.

Material characterization confirmed the following observations:

1. The precipitates in Alloys A and B are mainly composed of $Zr(Fe, Cr)_2$ phases that have a hexagonal structure in Alloy A but are predominantly cubic in Alloy B. The mean size of the precipitates is greater and more dispersed in Alloy A, but their density is greater in Alloy B

2. A significant difference is observed for the hydride distribution, as seen in Fig. 6, for cladding samples corroded at the same level of oxide thickness (~6 μm after 100 days for Product A and ~7 μm in less than 50 days for Alloy B). Alloy B, which corrodes more rapidly, shows very few hydrides that are circumferencially oriented. These may be high-stress zones. On the other hand, Alloy A shows more numerous and better distributed hydrides.

Impact of the Last Thermal Treatment

Three Zircaloy-4 products from the same ingot with optimized chemistry (including a reduction in tin content around 1.3%) were produced in an identical process except for the final heat treatment that was characterized by:

TABLE 1—*Corrosion results from autoclave test (400 °C, vapor phase).*

Effective Full Power Days	3	14	50	100
A, mg/dm²	17.71	30.2	61	98
B, mg/dm²	20.84	34	160	425

360 ZIRCONIUM IN THE NUCLEAR INDUSTRY: TENTH SYMPOSIUM

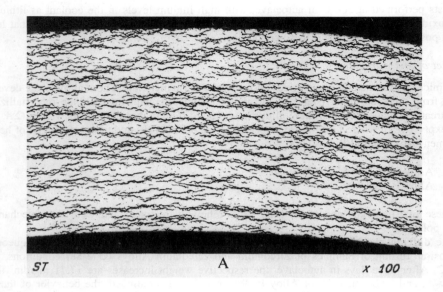

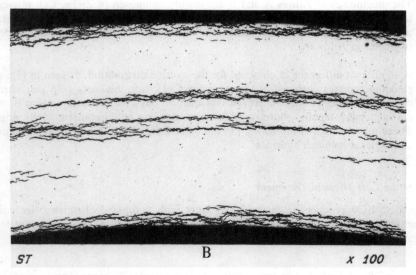

FIG. 6—*Distribution precipitates in Materials A and B after the autoclave test.*

1. stress-relief annealing (SRA),
2. recrystallization annealing (RXA), and
3. β-range quenching treatment (β-quench).

A fourth material in the stress-relieved state that differs from SRA product by having standard Zircaloy-4 (about 1.5% tin) chemistry and optimized microstructure was added to this study. This product is known as standard (STD).

These four cladding materials were studied in out-of-pile conditions (static autoclaves and loop) and in a power reactor.

Out-of-Pile Tests—Tests were performed in static autoclaves under three conditions:

1. at 360°C, 180 bar, in a solution of boric acid (650 ppm) and lithium hydroxide (1.5 ppm);
2. at 400°C, 100 bar, in vapor phase; and
3. at 415°C, 105 bar, in vapor phase.

Oxidation kinetics in isothermal conditions are shown in Figs. 7a (360°C), b (400°C), and c (415°C).

Two corrosion tests were carried out in the out-of-pile loop. The first test, under thermohydraulic and chemical conditions close to those of the pressurized reactors ([Li] = 5 ppm, [B] = 650 ppm, T_{WALL} = 347°C, void fraction <5%) resulted in zirconia thicknesses of around 5 μm. The second loop test was conducted under severe chemical conditions (average lithium content in the fluid = 210 ppm, T_{WALL} = 363°C, no void fraction) with the objective of obtaining corrosion data up to 80 μm on average for the standard material. In order to compare the corrosion behavior of the different variants independent of the corrosion conditions, the measured oxide thicknesses versus predicted oxide thicknesses using the CEA corrosion model (COCHISE) applied to standard Zircaloy-4 are studied (Fig. 8). The average corrosion data obtained from metallographic measurements are presented in Table 2.

Under all of the out-of-pile conditions tested (isothermal or under thermal gradient), a beneficial effect in corrosion in the 20 to 40% range results in reducing the tin content in the 1.5 to 1.3% range (see Figs. 7 and 8). In addition, from the vapor phase results, tin reduction appears to improve the resistance of the material to nodular corrosion.

From the autoclave tests, the RXA treatment significantly improves the corrosion behavior of the material. On the other hand, the beneficial effect of the RXA treatment disappears in the out-of-pile loop, under thermal gradient (see significantly higher corrosion kinetics of RXA material compared to SRA material in Fig. 8).

In every isothermal condition tested, the corrosion behavior of β-quenched material is better compared to the STD material (Figs. 7a to c). This corrosion behavior is also observed at the end of the first out-of-pile loop test (Fig. 8). At this stage of material oxidation (low oxide thickness), β-range quenching treatment significantly improves the corrosion behavior of Zircaloy-4. However, for the higher oxide thicknesses obtained at the end of the second out-of-pile loop test, the β-quenched material shows scattered corrosion rates, good behavior alternating with poor behavior, and spalling (Fig. 8).

Tests in Pressurized Water Reactors—Test assemblies, containing the four material variants described, were placed in a French 900 MWe power reactor. To compare the behavior of the different variants independent of the varied burnups, Fig. 9 presents oxide thickness measurements versus oxide thicknesses predicted by the CEA corrosion model (COCHISE) for standard Zircaloy-4.

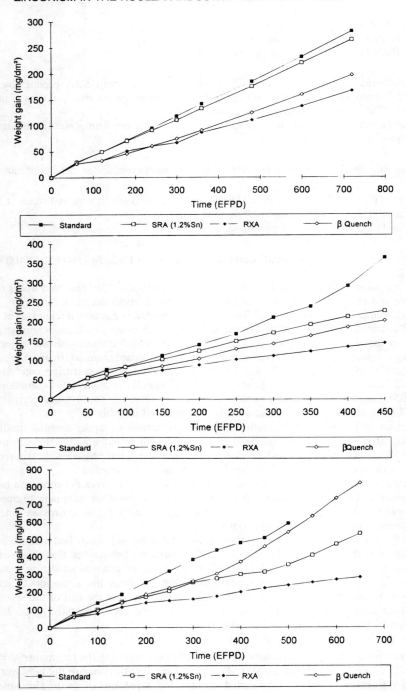

FIG. 7—*Autoclave corrosion data* (a) 360°C, 180 bar, [Li] = 1.5 ppm, and [B] = 650 ppm; (b) 400°C, 100 bar, and vapor phase; and (c) 415°C, 105 bar, and vapor phase.

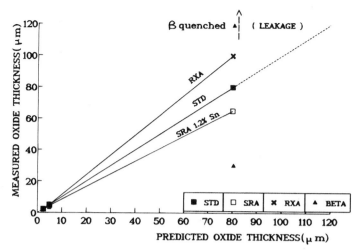

FIG. 8—*Comparison of the calculated and measured oxide thickness using CEA/COCHISE model; microstructure impact in the out-of-pile loop.*

TABLE 2—*Corrosion results from out-of-pile loop test ($T_w = 363°C$).*

Material Variant	Thickness Measurement, μm
Stress relief annealed (reference)	65
Standard	80
Recrystallized	100
β-quenched	30 to 800 (leakage)

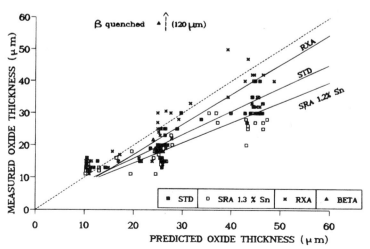

FIG. 9—*Comparison of the calculated and measured oxide thickness using CEA/COCHISE model; microstructure impact in the pressurized water reactor.*

Following the first two irradiation cycles, widespread corrosion results were observed for the β-quenched material. Corrosion measurements, carried out using the eddy current technique, lead to thicknesses as great as 160 μm (without cladding defect), coexisting with cladding zones showing low corrosion rates (thicknesses between 20 and 30 μm), as observed at the completion of the out-of-pile loop tests.

The conclusions concerning the material behaviors are:

1. Reducing the tin content leads to a gain in corrosion of about 25%. This improved corrosion behavior under thermal gradient is less than that assessed in isothermal conditions.
2. The beneficial effect of RXA treatment, observed in isothermal conditions (autoclave), disappears in pressurized water reactors, as it does in the out-of-pile loop under thermal gradient.
3. The beneficial effects observed for the β-quench material under isothermal conditions disappears under heat flux in the out-of-pile loop and in power reactors.

The Oxidation Mechanism

Alloy Chemistry

A more phenomenological approach to Zircaloy-4 with different variants of zirconium-based alloy corrosion is conducted parallel to the semi-empirical studies that are based mainly on the autoclave analyses, out-of-pile loop, and PWR corrosion data under various thermohydraulic and chemical conditions. The purpose of this approach is to determine the reaction mechanisms that describe the corrosion processes.

In order to take the crystalline network of zirconia into account, Kroger's notation will be used. Using this formulation, zirconia is mainly made up of:

1. oxygen in the oxygen sites, $(O_{O^{2-}}^{2-})^x$;
2. zirconium in the zirconium sites, $(Zr_{Zr^{4+}}^{4+})^x$; and
3. oxygen vacancies $(V_O^{2-})^{\circ\circ}$.

Table 3 summarizes the different species used hereafter.
Zirconium corrosion by water can be represented according to four main stages:

1. dissociative water adsorption at the oxide/water interface

$$H_2O + (V_{O^{2-}})^{\circ\circ} + (O_{O^{2-}}^{2-})^x === {}_2(OH_{O^{2-}}^-)^\circ$$

2. oxygen and hydrogen diffusion via the oxygen vacancies to the grain boundaries,
3. oxygen dissolution into the interstitial sites of the metal to form solid solution (Zr,O)

$$(e_i^-)' + (OH_{O^{2-}}^-)^\circ + (V_i)^x === (O_{O^{2-}}^{2-})^x + (H_i)^x$$

$$(O_{O^{2-}}^{2-})^x + (V_i)^x === (V_{O^{2-}})^{\circ\circ} + (O_i)^x + {}_2(e_i^-)'$$

TABLE 3—*Notations for mechanistic model.*

Notation	Chemical species
H_2O	water
LiOH	lithium hydroxide
ZrO_2	zirconia
$(O_{O^{2-}})^x$	oxygen ion in oxygen ion site
$(OH_{O^{2-}}^-)^\circ$	hydroxil group in oxygen ion site
$(V_O^{2-})^{\circ\circ}$	oxygen ion vacancy
$(Zr_{Zr^{4+}}^{4+})^x$	zirconium ion in zirconium ion site
$(FE_{Zr^{4+}}^{3+})'$	iron ion in zirconium site
$(V_{Zr}^{4+})''''$	zirconium ion vacancy
$(Li_{Zr}^{+4+})'''$	lithium ion in zirconium site
$(e_i^-)'$	electron
$(Zr_{Zr})^x$	zirconium metal in zirconium metal site
$(O_i)^x$	oxygen in interstitial site
$(H_i)^x$	hydrogen in interstitial site
$(V_i)^x$	vacancy in interstitial site
$(Fe_M)^x$	iron in metallic site (from precipitate)

4. zirconia production

$$(Zr_{Zr})^x + {}_2(O_i)^x === (Zr_{Zr^{4+}}^{4+})^x + {}_2(O_{O^{2-}}^{2-})^x$$

Part of the hydrogen produced by the corrosion reaction remains in solution in the metal, and precipitates in the form of ZrH_2 when the solubility limit is reached, while another part is dissolved in the water via the porous network.

It has already been observed that iron seems to diffuse towards the water/oxide interface, with the iron/chromium (Fe/Cr) ratio in the intermetallic precipitates moving from 2 to 0.5 [*12*]. Furthermore, a SIMS analysis shows that chromium enriched zones are present in the oxide film, while iron peaks disappear (Fig. 10). This observation can be described by the following processes:

1. iron oxidation/dissolution in the zirconia

$$(Fe_M)^x + {}_2(O_i)^x + (e_i^-)' === (FE_{Zr^{4+}}^{3+})' + {}_2(O_{O^{2-}}^{2-})^x$$

2. iron diffusion in the matrix, and
3. iron oxidation in the form, Fe_2O_3

$$_2(FE_{Zr^{4+}}^{3+})' + {}_3(O_{O^{2-}}^{2-})^x === Fe_2O_3 + {}_2(V_{Zr^{4+}}^{4+})'''' + {}_3(V_{O^{2-}})^{\circ\circ}$$

Chromium, on the other hand, does not seem to diffuse so easily in the zirconia. Zones with

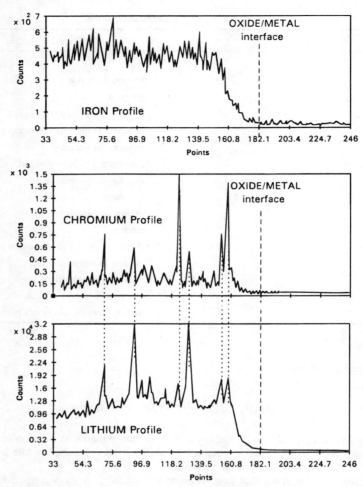

FIG. 10—*Secondary ion mass spectroscopy profiles of iron, chromium, and lithium elements in the oxide films developed in the out-of-pile loop.*

high concentrations of chromium are still observed in the oxide layers (Fig. 10), corresponding to the intermetallic precipitates.

The complete model (including water adsorption) therefore results in the formation of anionic and cationic vacancies $(V_{Zr^{4+}})''''$ and $(V_{O^{2-}})^{\circ\circ}$, that is, zirconia dissolution. This dissolution, more active close to the intermetallic precipitates due to iron diffusion, could be the cause of part of the porous network observed in the zirconia films that are formed, as well as the irregular features generally observed along the metal/oxide interface.

Primary Fluid Chemistry

The measurements of lithium across the oxide films using SIMS technique show two types of lithium distribution:

1. a diffusion profile for thin protective oxide films in the pre-transition regime and
2. a flat profile from the oxide/water interface to the metal/oxide interface for thick oxide layers with higher porosity in the post-transition phase (see Fig. 1).

Using porosity measurements from the oxide layers formed in out-of-pile loop conditions, the development of highly porous oxide was correlated with the high lithium levels in the oxide films. As suggested by Cox [6], tetragonal zirconia could be dissolved by lithium hydroxide at the bottom of the pores, before being reprecipitated on the surface of the oxide in the monoclinic form. This mechanism could be described as follows:

1. at the bottom of the pore, ZrO_2 is replaced by LiOH

$$LiOH + (Zr^{4+}_{Zr^{4+}})^x + {}_2(O^{2-}_{O^{2-}})^x$$
$$=== (Li^+_{Zr^{4+}})''' + (OH^-_{O^{2-}})^\circ + (V_{O^{2-}})^{\infty\infty} + (ZrO_2)_{dissolved}$$

2. diffusion of Zr^{4+} throughout the porosity, and
3. reprecipitation of zirconia at the water/oxide interface

$$(ZrO_2)_{dissolved} === (Zr^{4+}_{Zr^{4+}})^x + {}_2(O^{2-}_{O^{2-}})^x$$

This schematic reaction leads to the incorporation of lithium into the matrix and to the reprecipitation of zirconia towards the oxide/water interface. The lithium incorporated into the oxide film then reacts with the oxygen ions to form the lithium oxide that is soluble in water

$$_2(Li^+_{Zr^{4+}})''' + (O^{2-}_{O^{2-}})^x === (Li_2O)_{dissolved} + {}_2(V_{Zr^{4+}})'''' + (V_{O^{2-}})^{\infty\infty}$$

The result in this case is the formation of anionic, $(V^{2-}_O)^{\infty\infty}$, and cationic, $(V^{4+}_{Zr})''''$, vacancies at the bottom of the pore, that is, the development of the porous network.

In addition, recent SIMS examinations show that the chromium-enriched zones (precipitates) are also enriched in lithium (Fig. 10). Furthermore, as the iron solubility in the water increases with the lithium content (pH increase of the fluid), it is suggested that higher iron dissolution from the precipitates can be related to the incorporation of lithium into the oxide and to an increase in anionic vacancies and porosity development. This leads to the following mechanism:

1. iron substitution by lithium close to the oxide porosity or the oxide/water interface

$$_2LiOH + {}_2(Fe^{3+}_{Zr^{4+}})' + {}_3(O^{2-}_{O^{2-}})^x === {}_2Fe^{3+} + {}_3O^2 + {}_2(Li^+_{Zr^{4+}})''' + {}_2(OH^-_{O^{2-}})^\circ + (V_{O^{2-}})^{\infty\infty}$$

2. iron transfer and hematite (Fe_2O_3) reprecipitation at the oxide/water interface,
3. soluble lithium oxide (Li_2O) formation and further increase in porosity

$$_2(Li^+_{Zr^{4+}})''' + (O^{2-}_{O^{2-}})^x === (Li_2O)_{dissolved} + {}_2(V_{Zr^{4+}})'''' + (V_{O^{2-}})^{\infty\infty}$$

4. diffusion of the remaining lithium in the oxide towards the oxide/metal interface.

The overall balance of this mechanism therefore leads to:

1. substitution of a part of the iron by lithium,
2. development of the porous network in the zirconia, and
3. dissolution of precipitates, capable of promoting the transformation of the tetragonal phase to monoclinic zirconia [13].

The CEA/COCHISE Corrosion Model

Development of the COCHISE Model

Analysis of the corrosion results from the static autoclave and the out-of-pile loops has made it possible to show the effect of important parameters influencing the oxidation kinetics of zirconium-based alloy claddings:

1. primary fluid chemistry (lithium, boron) and
2. heat flux and boiling conditions at the wall of the fuel elements.

Effect of Primary Coolant Chemistry—The influence of chemistry parameters is modeled using the usual structure of the CEA/COCHISE corrosion code [14], based on the thermo-hydraulic model, that provides the temperature and the void fraction at the wall of the fuel cladding. Corrosion kinetics are then divided into two stages:

1. pre-transition corrosion regime

$$\frac{ds^3}{dt} = K_{pre} e^{-\frac{Q_{pre}}{RT_i}} \tag{2}$$

2. post-transition corrosion regime

$$\frac{ds}{dt} = K_{post} e^{-\frac{Q_{post}}{RT_i}} \tag{3}$$

where

$$T_i = T_{wall} + \frac{\Phi s}{\lambda}$$

and

s = oxide thickness (m),
t = time (s),
K_{pre}, K_{post} = frequency factors for pre- and post-transition kinetics (m³/s and m/s),
Q_{pre}/R, Q_{post}/R = activation energies for pre- and post-transition kinetics (K),
T_i = metal/oxide interface temperature (K),
T_{wall} = wall temperature (K),
Φ = heat flux (W/m²),
λ = thermal conductivity of the zirconia (1.6 W/m · K), and
R = perfect gas constant (J/mol · K).

Lithium effect—Using results obtained in isothermal conditions (static autoclave), under chemical conditions ranging from the usual PWR environment (lithium concentration [Li] = 2.2 ppm, boron concentration [B] = 650 ppm) to high lithium content ([Li] = 700 ppm), two types of correlation are established: pre- and post-transition activation energies are described by linear laws as a function of the lithium content in the primary fluid. The frequency factors are described by exponential laws, as a function of the lithium content at the oxide/water interface, taking into account the enrichment factor in boiling conditions. The following two kinetics laws are therefore obtained that are valid in isothermal conditions:

$$\frac{ds^3}{dt} = e^{(a+b\,[Li])}\, e^{-\frac{c+d[Li]}{RT}} \qquad (4)$$

and

$$\frac{ds}{dt} = e^{(\alpha+\beta[Li])}\, e^{-\frac{\gamma+\delta\times[Li]}{RT}} \qquad (5)$$

where

a, b, c, d = coefficients of K_{pre} and Q_{pre} correlations, and
$\alpha, \beta, \gamma, \delta$ = coefficients of K_{post} and Q_{post} correlations.

Boron effect—It has been confirmed experimentally that the presence of boric acid in the coolant slows down Zircaloy corrosion kinetics. The decrease in corrosion kinetics does not appear to be proportional to the boron content in the fluid, hence, no pH effect can be estimated. Furthermore, SIMS analyses have shown that boron remains mainly at the oxide/water interface, forming a protective film preventing oxidizing species from entering the oxide film. This phenomenon has therefore been expressed in the form of a multiplicative coefficient that can be applied to the frequency factors:

1. in the absence of boron (out-of-pile test), $factor_{boron} = 1$, and
2. in the presence of boron (out-of-pile and in-pile test), $factor_{boron} = 0.64$.

Heat Flux Effect—As already presented in this document and at other conferences [14], the influence of a thermal gradient through the cladding on the corrosion kinetics is not well represented by simply taking in account the temperature increase at the metal/oxide interface. Under heat transfer, the flux of anionic species is represented by the Fick's first law, taking the thermo-diffusion processes into account.

Relating to this law, it was chosen to model the effect of the heat flux by modifying the frequency factor in the form of a multiplicative factor, as a function of both heat flux parameters, and the lithium content in the coolant. The coupling energy, Q^*, that correlates the mass and heat transfer is assumed to be a function of the cladding material and the chemical additives of the coolant.

Finally, the following two kinetics laws are suggested:

1. the pre-transition regime

$$\frac{ds^3}{dt} = e^{a+b\,[Li]}\left(1 + \frac{Q^*_{pre}([Li])\, s}{RT^2}\, \frac{\Phi}{\lambda}\right) factor_{Boron}\, e^{-\frac{c+d[Li]}{RT}} \qquad (6)$$

2. the post-transition regime

$$\frac{ds}{dt} = e^{\alpha+\beta\,[\mathrm{Li}]}\left(1 + \frac{Q^*_{\mathrm{post}}([\mathrm{Li}])s}{RT^2}\frac{\Phi}{\lambda}\right)\mathrm{factor}_{\mathrm{Boron}}e^{-\frac{\gamma+\delta[\mathrm{Li}]}{RT}} \qquad (7)$$

Prediction of Out-of-Pile Corrosion Kinetics—The corrosion model described before was used to predict the corrosion kinetics of Zircaloy-4 in static autoclave and out-of-pile loop in various fluid chemical conditions. The results are reported on Fig. 11. It can be seen that, up to a thickness of 10 μm, there is no marked effect of heat flux on corrosion kinetics. Beyond this oxide thickness, the heat flux effect appears to be well predicted by the mathematical law [7] that can be extrapolated to isothermal conditions.

Prediction of In-Pile Corrosion Kinetics—The corrosion model described earlier was used to predict the corrosion kinetics of Zircaloy-4 claddings in some pressurized water reactors, using real power history.

The analysis of predicted corrosion data in both out-of-pile loop and PWR conditions led us to take into account the irradiation effect on corrosion by using the following frequency factor in the post-transition phase

$$K_{\mathrm{post}}([\mathrm{Li}], \Phi, s)_{\mathrm{in\text{-}pile}} = 2.15\, K_{\mathrm{post}}([\mathrm{Li}], \Phi, s)_{\mathrm{out\text{-}of\text{-}pile}}$$

Calculations and measurements of oxide thicknesses were performed on rods selected to meet the following irradiation conditions:

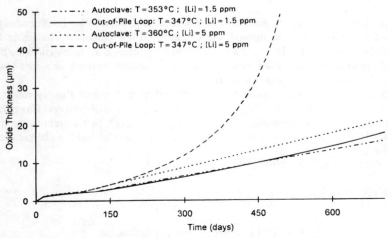

FIG. 11—*Predicted corrosion kinetics: lithium and heat flux effect.*

1. rods enriched with 3.25% of ^{235}U after three irradiation cycles in a 900 MWe PWR, corresponding to a burnup of 37 500 MWd/tU (Fig. 12a);
2. rods enriched with 4.5% of ^{235}U after five irradiation cycles in a 900 MWe PWR, corresponding to a burnup of 58 000 MWd/tU (Fig. 12b); and
3. rods after three irradiation cycles in a 1300 MWe PWR, corresponding to a burnup of 34 000 MWd/tU (Fig. 12c),

The oxide thickness measurements used for this comparison were obtained by the eddy current technique. If available, metallographic examinations were also used. In spite of some underpredicted data not completely elucidated, a good agreement is generally observed between calculated and measured corrosion rates.

The Impact of $[Li]_{coolant} = 3.5$ *ppm on Oxidation Kinetics in PWRs*—As seen earlier, lithium used in PWRs increases the corrosion rate of zirconium-based alloys. However, the new types of fuel management (longer cycle and higher burnup) will cause a rethinking of the primary coolant chemistry, for example, an increase of the boron content in the coolant. In order to maintain the pH$_{300}$ of the primary coolant throughout the cycle between the recommended values of 7.2 and 7.4, it will be necessary to increase the lithium content in the fluid. This will have consequences in two fields: dose rates and cladding corrosion.

Studies have shown that a pH increase in the primary coolant, which increases metallic ion solubility, could result in a significant reduction in circuit contamination by corrosion products, thus reducing dose rates [*15*]. These studies made it possible to determine the optimum pH of the coolant between 7.2 and 7.4. In France, where pH$_{300}$ was maintained at 7.0 up to now, circuit contamination was controlled in six pressurized water reactors operated by the EDF with the pH$_{300}$ of the primary coolant maintained at 7.2 throughout the cycle. The conclusion is that no significant impact of this raised pH is detected on the dose rates (see Table 4). The impact of the pH increase on dose reduction of pressurized water reactors is therefore probably less than expected and often less important compared to the effect of other factors.

On the other hand, an attempt is made using the CEA corrosion model (COCHISE) to determine the influence of increasing the lithium content in the primary coolant (coordinated chemistry at constant pH = 7.4, with lithium content of 3.5 ppm early in the cycle) on fuel cladding corrosion. Predictions are made for two types of fuel management:

1. 900 MWe PWR: five 12-month cycles, up to a final burnup of 60 000 MWd/tU and
2. 1300 MWe PWR: three 18-month cycles, up to a final burnup of 50 000 MWd/tU.

Both fuel managements led to a calculated increase in the corrosion data of around 30%, compared to the oxide thicknesses formed under current operating conditions for French PWRs, that is, coordinated chemistry with maximum lithium content of 2.2 ppm (Figs. 13a and b).

Conclusions

The analysis of the corrosion results obtained from static autoclaves, out-of-pile loops, and pressurized water reactors led to the discussion of the main parameters that influence the oxidation processes.

It was possible to correlate the corrosion acceleration of Zircaloy-4 claddings subjected to boiling conditions with the lithium enrichment of oxide films. The enrichment mechanisms under boiling were related to the thermo-hydraulic conditions in a zone close to the cladding wall, but also to the porous structure of the oxide. Furthermore, as porosity increased due to lithium hideout, there was a resulting intensification of the boiling process and a gradual

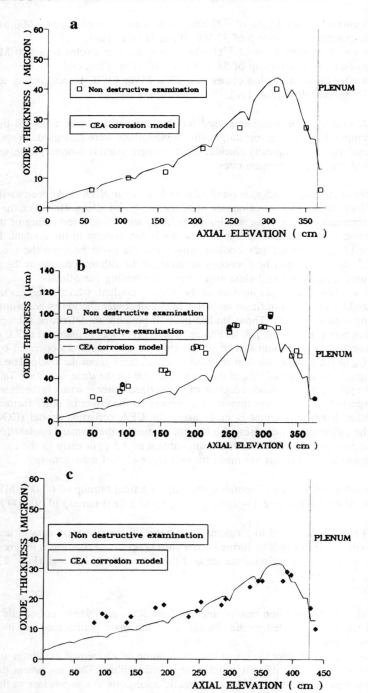

FIG. 12—*Comparison of measured and predicted corrosion data: (a) three cycles, 900 MWe (3.25% ^{235}U), (b) five cycles, 900 MWe (4.5% ^{235}U), and (c) three cycles, 1300 MWe.*

TABLE 4—*Dose rate from PWRs as a function of pH in the coolant.*

	Mean Value of the Dose Index, mRem/h	Standard Deviation, mRem/h
Operation at pH = 7.0	59	19
Operation at pH = 7.2	66	21

increase in the overconcentration factor during oxide growth. Modeling was developed in order to assess the lithium enrichment in the zirconia as a function of the mass evaporation rate through the oxide films.

Corrosion studies conducted on various zirconium-based alloys subjected to different final heat treatments give the following conclusions:

1. the relative behavior of the materials varied according to the corrosion conditions. In

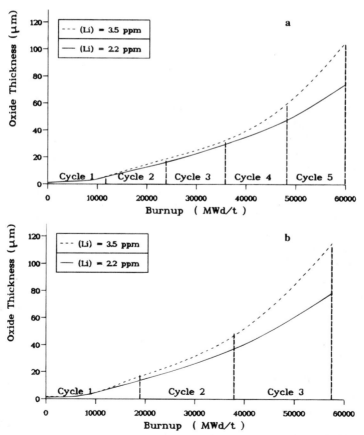

FIG. 13—*Simulation of lithium effect using CEA corrosion model; (a) PWR 900 MWe, 12-month cycle (Φ_{ave} = 57 W/cm^2) and (b) PWR 1300 MWe, 18-month cycle (Φ_{ave} = 57 W/cm^2).*

particular, the behavior of materials corroded in isothermal conditions (static autoclave) and under thermal gradient (out-of-pile loop test and in-pile conditions) included significant differences: the beneficial effect of recrystallization observed in isothermal conditions disappears in the presence of heat flux at the wall of the cladding tube; and

2. the reduction in tin content led to reduced corrosion rates of at least 20%, based on studies of Zircaloy-4 alloys.

These studies led to a mechanistic approach to Zircaloy-4 corrosion processes in PWRs, taking into account the chemical composition of the material as well as the effect of chemical additives of the primary fluid. A proposal concerning the roles of iron and lithium in corrosion mechanisms is presented.

A phenomenological approach to corrosion processes was developed that explicitly takes into account the effect of the primary fluid chemistry and the heat flux on Zircaloy-4 corrosion kinetics. Irradiation, and microstructural impacts are currently in progress. This model was used to predict corrosion kinetics for burnups close to 50 000 MWd/tU.

A predictive study of the impact of increasing lithium content in the primary coolant is carried out and leads to the observation of a non-negligible increase in corrosion kinetics. On the other hand, it is shown that there is little evidence to suggest that the pH increase of the primary coolant has a dose-reducing impact.

References

[1] Billot, P. et al., "Development of a Mechanistic Model to Assess the External Corrosion of the Zircaloy Cladding in PWRs," *Zirconium in the Nuclear Industry, Eighth International Symposium, ASTM STP 1023*, L. F. P. Van Swam and C. M. Eucken, Eds., American Society for Testing and Materials, Philadelphia, 1989, pp. 165–184.

[2] Billot, P. and Giordano, A., "Comparison of Zircaloy Corrosion Models from the Evaluation of In-Reactor and Out-of-Pile Loop Performance," *Zirconium in The Nuclear Industry, Ninth International Symposium, ASTM STP 1132*, C. M. Eucken and A. M. Garde, Eds., American Society for Testing and Materials, Philadelphia, 1991, pp. 539–565.

[3] Billot, P., Beslu, P., and Frejaville, G., "Justification of In-Plant Waterside Corrosion of Zircaloy Clad on the Basis of Thermal and Hydraulic Calculation," *Proceedings*, IAEA Technical Meeting on External Cladding Corrosion in Water Power Reactor, Cadarache, France, 14–18 Oct. 1985.

[4] Noe, M., Beslu, P., and Thomazet, J., "External Cladding Corrosion in Water Power Reactors," International Atomic Energy Agency, Vienna, 1986, pp. 70–80.

[5] Ramasubramanian, N., "Lithium Uptake and the Corrosion of Zirconium Alloys in Aqueous Lithium Hydroxide Solutions," *Zirconium in the Nuclear Industry, Ninth International Symposium, ASTM STP 1132*, C. M. Eucken and A. M. Garde, Eds., American Society for Testing and Materials, Philadelphia, 1991, pp. 613–627.

[6] Cox, B. and Wu, C., "Dissolution of Zirconium Oxide Films in 300°C LiOH," *Journal of Nuclear Material*, Vol. 199, No. 3, Feb. 1993.

[7] Afgan, N. H. et al., "Boiling Heat Transfer from Surfaces with Porous Layers," *International Journal of Mass Transfer*, Vol. 28, No. 2, 1985, pp. 415–422.

[8] Styrikovich, M. A., Leont'ev, A. I., and Molyshenko, S. P., "Transport Mechanism of Nonvolatile Impurities in Boiling on Surfaces Coated with Porous Structure," *High Temperature*, Vol. 14, 1976, pp. 886–893.

[9] Pan, C., Jones, B. G., and Machiels, A. J., "Concentration Levels of Solutes in Porous Deposits with Chimneys Under Wick Boiling Conditions," *Proceedings*, Third International Topical Meeting on Reactor Thermal Hydraulics, Vol. 2, Chong, Chivard, Gilbert, Brown, Eds., Newport, RI, Oct. 1985.

[10] Dagan, G., *Flow and Transport in Porous Formations*, Springer-Verlag, Berlin Heidelberg, 1989.

[11] Mardon, J. P., Charquet, D., and Senevat, J., "Optimization of PWR Behavior of Stress Relieved Zircaloy-4 Cladding Tubes by Improving the Manufacturing and Inspection Process," in this volume.

[12] Pecheur, D. et al., "Precipitates Evolution in the Zircaloy-4 Oxide Layer," *Journal of Nuclear Material*, Vol. 189, No. 3, Aug. 1992.

[13] Godlewski, J., "Oxydation d'Alliages de Zirconium en Vapeur d'Eau: Influence de la Zircone Tetragonale sur le Mécanisme de Croissance de l'Oxyde," thesis, Université Technologique de Compiègne, France, 1990.
[14] Thomazet, J., Robin, J. C., Billot, P., and Girard, M., "Integration of the Mechanistic Modelling with the Empirical Aspects of Zircaloy Corrosion," presented at the poster session of Zirconium in The Nuclear Industry, Ninth International Symposium, 1991.
[15] Anthoni, S. et al., "Effects of pH of Primary Coolant on PWR Contamination," *Water Chemistry of Nuclear Reactor Systems,* British Nuclear Energy Society (BNES), Bournemouth, UK, 1992.

DISCUSSION

Y. S. Kim[1] (written discussion)—Please comment on why LiOH enhances the corrosion rate. Unless it is clear, it is very difficult to predict the effect of lithium on the in-reactor corrosion based only on the out-of-pile data.

P. Billot et al. (authors' closure)—It seems that lithium (as LiOH or Li^+) incorporation into the oxide largely increases the porosity of the oxide layers. Therefore, the access of water in the oxide through the pores and cracks increases the short-circuit of diffusion from the oxide/water interface to the metal/oxide interface. In addition, the increase in Li^+ is at the origin of the increase in anionic vacancies with a subsequent enhancement of corrosion rates.

The effect of lithium on the in-reactor corrosion is based on the out-of-pile loop results knowing that the thermo-hydraulic and chemistry conditions maintained in the loop are equivalent to those in PWRs and that irradiation certainly has no impact on reducing this lithium effect.

Bo Cheny[2] (written discussion)—Your corrosion model incorporated an effect of lithium that predicts a corrosion enhancement of up to 50% by lithium. If the model is used for fuel management, it may result in the premature discharge of fuel rods being operated under 3.5 ppm lithium. It is therefore important that the model be validated using actual PWR fuel data. Do you have a plan to do this?

P. Billot et al. (authors' closure)—The maximum corrosion enhancement calculated by our model is very close to 30% and depends on the fuel management (power history and burnup).

Until now, the model has been validated in the out-of-pile loop corrosion data collected from tests conducted under higher lithium content in the coolant. Because it can be reasonably assumed that the lithium influence on corrosion kinetics is not really modified (specially reduced) by irradiation, we are confident in our validation can be considered in some respects easier than those conducted in power reactors, where often many other parameters (materials, etc.) are modified with lithium. Nevertheless, we plan to validate our model on available PWR data.

Brian Cox[3] (written discussion)—(1) You claim a heat flux effect in additional to and independent of the effect on the temperature gradient in the oxide. I am unaware of any physical phenomenon that could lead to such an independent effect. Can you explain the physical basis for this effect?

(2) You argue that iron is preferentially dissolved from the oxide in lithium solutions, yet you appear to have no clear evidence that the iron in the oxide was originally in zirconium rather than in the loop water. I would be happier about this argument if you had tagged the iron in the specimen (for example, with ^{57}Fe) so that its origin was clearly distinguishable.

P. Billot et al. (authors' closure)—(1) We claim an additional heat flux effect on corrosion kinetics independent of its effect on interface temperature increase ($T_i = T_w + \Phi S/\lambda$). It appears that the heat flux through thermal gradient ($\vec{\nabla} T$) acts as an additional driving force causing the acceleration of the flux of anionic species, along grain boundaries or through anionic vacancies. In other words, the interface temperature increases ($T_w + \Phi S/\lambda$) and $\vec{\nabla} T(\Phi/T)$ have their respective influence.

(2) As we observed using the SIMS technique, the chromium-enriched zones are also enriched with lithium, and while the iron peaks disappear, it can be considered that the iron dissolution from the $Zr(Fe,Cr)_2$ precipitates are increased due to the iron solubility increase

---

[1] Korea Atomic Energy Research Institute, Daljom, Korea.
[2] Electric Power Research Institute (EPRI), Palo Alto, CA.
[3] University of Toronto, Toronto, Ontario, Canada.

with pH. It is then suggested that the higher dissolution of iron from the precipitates can be related to the incorporation of lithium into the oxide.

B. P. Sabol[4] (written discussion)—(1) In the loop tests, accelerated corrosion occurred only after an appreciable amount of sub-cooled boiling, about 10%. Is sub-cooled boiling required for lithium-accelerated corrosion, and if so, how much boiling is required?

(2) You indicate that the porosity of post-transition films is sufficient to concentrate the soluble lithium to levels sufficient to cause accelerated corrosion. What value of porosity is used for the oxide film? What is the expression for the concentration factor for lithium?

(3) You indicate that boron at ≥ 100 ppm boron mitigates the corrosion acceleration due to lithium. Because boron is above this value for most of the fuel residence time, how do you value for most of the fuel residence time, and how do you account for higher corrosion with 3.54 ppm lithium versus 2.0 ppm lithium?

P. Billot et al. (authors' closure)—(1) In order to observe lithium enhanced corrosion in corrosion tests of a short duration (three to six months), a significant void fraction of about 10% is needed. Nevertheless, for longer corrosion tests, an acceleration of corrosion kinetics has been shown with lower void fractions (<5%) when thick oxide films (>30 μm) are subjected to these conditions. The reason can be attributed to the increase in porosity for thicker oxide layers that leads to an increase in the lithium incorporated in the oxide films with a possible enrichment due to low void fraction.

(2) The oxide film porosity in the post-transition phase was assessed to be in the 3 to 15% range according to the thermo-hydraulic and chemistry conditions of the corrosion tests. Severe chemistry conditions (lithium at 10 ppm with boiling) led to very porous oxide layers (open porosity of 15%). The lithium enrichment factor in the oxide films can be expressed as a function of the mass evaporation rate of water, the lithium diffusion coefficient, and the mass transfer at the oxide/water interface. These parameters depend on the open porosity fraction of the oxide films. Lithium concentration in the oxide films is derived from the resolution of a differential equation integrating all of these parameters.

(3) It has been experimentally shown that boric acid at >100 ppm in the coolant slows down corrosion kinetics. Since corrosion rates were observed to be independent of boron content in the 100 to 1000 ppm range, a mitigating factor of 0.64 is applied that is constant for the whole cycle to taking into account the boron effect. No difference in the boron effect was observed in the 2.2 to 5 ppm lithium range. Therefore, the same factor is applied.

[4] Westinghouse NMD, Energy Center, Pittsburgh, PA.

Natesan Ramasubramanian[1] *and Poyilath V. Balakrishnan*[2]

Aqueous Chemistry of Lithium Hydroxide and Boric Acid and Corrosion of Zircaloy-4 and Zr-2.5Nb Alloys

REFERENCE: Ramasubramanian, N. and Balakrishnan, P. V., "**Aqueous Chemistry of Lithium Hydroxide and Boric Acid and Corrosion of Zircaloy-4 and Zr-2.5Nb Alloys**," *Zirconium in the Nuclear Industry: Tenth International Symposium, ASTM STP 1245*, A. M. Garde and E. R. Bradley, Eds., American Society for Testing and Materials, Philadelphia, 1994, pp. 378–399.

ABSTRACT: The chemistry of the aqueous solution, surface chemistry of zirconium oxide (ZrO_2), and the physical structure of the corrosion film have to be considered for an understanding of the mechanism of corrosion of zirconium alloys in aqueous solutions. Based on information available in all these areas, we are proposing a model for oxide growth on Zircaloy-4 fuel cladding and Zr-2.5Nb pressure tube material, in lithium hydroxide (LiOH) solutions with and without added boric acid (H_3BO_3).

Corrosion exposures were at 360°C and were short term of four-day duration. Concentration of lithium covered the range 0.7 to 3500 ppm and boron was added at 300, 600 and 1200 ppm. Weight gain, Fourier Transform InfraRed (FTIR) spectroscopy, and Secondary Ion Mass Spectrometry (SIMS) were used to characterize the oxide films. Potentiodynamic polarization measurements were made in separate tests at 315°C. The chemistry of $LiOH$-H_3BO_3 system at 300 to 360°C was evaluated from the ionization constants of water, LiOH, and H_3BO_3.

There is no simple relationship between pH and corrosion. In the absence of boron acceleration in corrosion, in these short-term tests, is observed at concentrations >350 ppm Li for Zircaloy-4 and at >60 ppm Li for Zr-2.5Nb. This is attributed to the modification of normal oxide growth by surface OLi groups formed by reaction with undissociated LiOH in the solutions. In the case of Zircaloy-4, the inhibition of accelerated corrosion by H_3BO_3 is greater than what can be accounted for by its neutralizing action on LiOH. The mechanism suggested is the removal of surface OLi groups by reaction with non-ionized H_3BO_3. Boric acid at high concentrations has a detrimental effect on the accelerated corrosion of Zr-2.5Nb alloy.

According to the model proposed for oxide growth, the corrosion behavior can be classified into two categories: (1) growth of post-transition type films under acceleration conditions where solution had access into the oxide and (2) growth of pre-transition films under non-acceleration conditions where the solution had not gained access into the oxide. For Case 1, the pores have to be >2 nm in diameter so that interfacial double layer requirements are met for solution incursion in the oxide and the unleachable lithium and boron on the surfaces of oxide grains are from ion exchange and reactions with LiOH and H_3BO_3. In Case 2, the outer surfaces of oxide grains, which were exposed to the solution, have lithium and boron similar to that described for Case 1. The lithium and boron in the bulk of these pre-transition films are ascribed to LiOH and H_3BO_3 hydrogen-bonded to water and adsorbed on the surfaces of oxide grains.

KEY WORDS: zirconium oxide surface chemistry, lithium hydroxide-boric acid aqueous chemistry, corrosion acceleration, lithium, corrosion inhibition, boron, zirconium, zirconium alloys, nuclear materials, nuclear applications, radiation effects, lithium hydroxide, boric acid

---

[1] Scientist–metallurgical, Ontario Hydro, Research Division, Toronto, Ontario M8Z 5S4, Canada.
[2] Research scientist, Chalk River Laboratories, Chalk River, Ontario KOJ 1JO, Canada.

Zircaloy-4 fuel cladding and Zr-2.5Nb alloy pressure tubes are exposed to lithiated water in pressurized water reactors (PWRs). In light water-cooled and -moderated reactors boric acid H_3BO_3 is also added as chemical shim. Zirconium oxide (ZrO_2), under hydrated conditions, can function both as a cation and as an anion exchanger [1]. The possibility thus exists for both lithium and boron to participate in the growth of the oxide film and affect the corrosion of the zirconium alloy components.

In out-reactor tests, the detrimental effect of high concentrations of lithium hydroxide (LiOH) in the corroding aqueous environment on the corrosion of zirconium alloys is well documented [2–5]. In the case of pressure tube materials, acceleration in corrosion at a lower lithium concentration is reported for Zr-2.5Nb alloy compared to that for Zircaloy-2 [5]. Similarly, the ameliorating effect of H_3BO_3 addition on the corrosion of Zircaloy-4 in concentrated LiOH solutions has been demonstrated in autoclave tests [6]. Acceleration of corrosion is observed when lithium in the oxide is above about 100 ppm [2]. For inhibition by boron, a minimum concentration of about 50 ppm in the corroding solution seems to be necessary [6].

The PWR coolant chemistry, typically 2.5 ppm lithium as LiOH and 1000 ppm boron as H_3BO_3 added at the start of a cycle in the case of light water reactors, is not generally considered to be life limiting from the corrosion point of view. However, situations may be anticipated where a hideout of lithium in the oxide may lead to an enhancement of corrosion. For example, a move towards increased fuel burn-up, higher fuel temperatures, and local boiling can cause concentrating of lithium in the oxide. With H_3BO_3 in the coolant, a hideout of boron in the oxide can also be expected. Whereas long-term tests at conditions close to those in operating reactors are essential to achieve a predictive capability, short-term tests at accelerated test conditions can serve the purpose of investigating relationships between the chemistry of the aqueous solutions and the effects of a possible hideout of lithium and boron in the oxides. We report here short-term laboratory tests on the corrosion behavior of Zircaloy-4 fuel cladding and Zr-2.5Nb pressure tube material in LiOH solutions with and without added H_3BO_3. The total system chemistry, that is, the equilibria in water calculated from the appropriate dissociation constants of LiOH and H_3BO_3 and the surface chemistry of ZrO_2, is taken into consideration in interpreting the results.

Experimental Procedures

Zircaloy-4 fuel cladding specimens, 2.5-cm long and 22 cm² in area, were cut from standard cold-worked and stress-relieved tubing 1.5 cm in diameter and 0.04 cm in wall thickness. Zr-2.5Nb specimens were obtained from CANDU (CANada Deuterium Uranium) pressure tube material, that is, cold-worked and stress-relieved tube 10 cm in diameter and 0.4 cm in wall thickness. The specimens were 2.5 by 1 cm and 7.5 cm² in area. The surfaces were prepared following established procedures of mechanical polishing and pickling in acid baths. All the corrosion exposures, four days at 360°C, were carried out in 1-L Hastelloy autoclaves. Hydrogen in argon was bubbled through the solution at room temperature and the autoclave was charged with the gas to 0.7 MPa that corresponded to a hydrogen partial pressure of 0.18 MPa. In each experiment, a pair of cladding and a pair of pressure tube specimens were corroded to obtain weight gain data points on the kinetic curves. Concentration of LiOH covered the range 0.7 to 3500 ppm lithium and boric acid was added at 300, 600, and 1200 ppm boron.

Infrared spectra of the corrosion films were obtained in the reflection mode on a Bio-Rad Model FTS-40 Fourier Transform InfraRed (FTIR) spectrometer. The microscope accessory was used to obtain spectra using a beam focussed to 300 μm in diameter. Oxide thicknesses were estimated from the spacing of the interference peaks [7]. The presence of molecular functional groups, such as hydroxyl and carbonate, in the oxides was evidenced by their characteristic infrared absorption bands in the spectrum [8]. A Cameca IMS-3f ion microscope was used to obtain concentration profiles of lithium and boron isotopes in the oxides by Secondary Ion Mass Spectrometry (SIMS) [9]. Lithium and boron implants, in 1-μm-thick corrosion films grown in steam on pressure tube specimens, were used for quantification of secondary ions in SIMS analyses. The quantification is expressed as a volume concentration, that is, number of atoms detected in a unit volume of oxide analyzed. It can be converted to ppm by using the formula, $10^6 \, m \, CN^{-1} \, \sigma^{-1}$, where m is the atomic mass of the isotope, C is the concentration in atoms cm^{-3}, N is Avagadro's number, and σ is the density of zirconia. The oxide was mainly monoclinic zirconia and the implantation depth was 0.3 μm. In using the calibration for oxides on Zircaloy-4 and Zr-2.5Nb alloy, differences in matrix effects, if any, are not taken into consideration.

In separate two-day exposure tests at 316°C, electrochemical potential and polarization measurements were made. An Ag/AgCl (0.01 M KCl) reference electrode, a platinum gauze counter electrode, and a scan rate of 1.2 V/hour were used for potentiodynamic polarization measurements.

The concentrations of the various species in the system H_2O-LiOH-H_3BO_3 at 300° to 360°C, were calculated from the equilibrium constants reported for the ionization of water [10], the dissociation of LiOH [11], and the acid ionization of H_3BO_3 [12]. Appropriate equations of the form

$$\log k = A/T + B + CT + D \log T + E/T^2$$

where k is the equilibrium constant, were fitted to the experimental data reported in the range 50 to 290°C [11,12] and the A to E parameters were evaluated. The equilibrium constants at temperatures of interest were then calculated from these equations. A FORTRAN routine to solve chemical equilibria was used to calculate the concentrations of the species in water for various concentrations of LiOH and H_3BO_3.

Results

Kinetics of Corrosion

The average weight gain for a pair of specimens, exposed in each experiment, is plotted as a function of lithium concentration in Figs. 1 and 2. A maximum variation of 10% was observed between the weight gained by the two specimens. In Fig. 1, weight gain data for Zircaloy-4 are shown. With increase in LiOH concentration, corresponding to 0.7 and 350 ppm lithium, a small decrease and a slight increase in weight gain are seen. The decrease in weight gain with increase in lithium, observed at low lithium concentrations has also been reported by others [5]. Its significance, however, is not understood. Acceleration in corrosion is observed above a concentration of 350 ppm lithium. Inhibition by H_3BO_3 is clearly evidenced when lithium concentration is higher than about 100 ppm. At lithium concentrations less than 70 ppm the effect of added H_3BO_3 is difficult to discern. The weight gain data for Zr-2.5Nb alloy are plotted in Fig. 2. With increase in lithium concentration from 0.7 to 40 ppm, small changes in weight gain, similar to those observed for Zircaloy-4, are seen. However, acceleration in corrosion is observed at about 60 ppm lithium. The effect of added boric acid at 1200 ppm boron, was also different from that in the case of Zircaloy-4. Although an inhibiting effect was evident at lithium concentrations less than 350 ppm, spalling of the oxide in the form of a fine powder was observed at higher lithium concentrations. Some of the specimens showed a weight loss. In the absence of added H_3BO_3, the corrosion films were compact in spite of high weight gains.

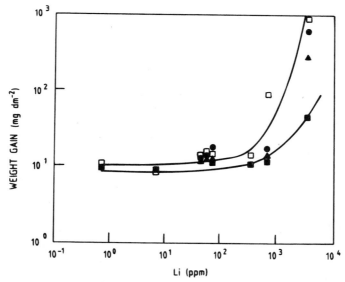

FIG. 1—*Corrosion of Zircaloy-4 fuel cladding in $LiOH$-H_3BO_3 solutions; four-day weight gains at 360°C; boron (ppm):* □ = 0, ● = 300, ▲ = 600, *and* ■ = 1200.

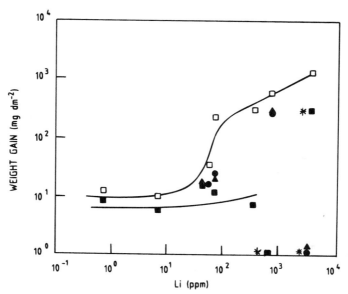

FIG. 2—*Corrosion of Zr-2.5Nb pressure tube alloy in $LiOH$-H_3BO_3 solutions; four-day weight gains at 360°C; boron (ppm):* □ = 0, ● = 300, ▲ = 600, ■ = 1200, *and* * = *oxide spalling.*

Infrared Spectra of Corrosion Films

The reflection spectra for films grown on Zircaloy-4 in concentrated LiOH solutions are shown in Fig. 3. Spectrum, Fig. 3a, is for a specimen corroded in 3500 ppm lithium solution that had a weight gain of 900 mg dm^{-2}. The spectral characteristics to be noted are: an absorption peak in the O-H (hydroxyl) stretching region at 3600 cm^{-1}, well-defined interference peaks in the mid-infrared region from 2800 to 1800 cm^{-1}, absorption peaks in the 1600 to 1400 cm^{-1} region, and a rapid increase in absorbance near 1000 cm^{-1}. From the spacing of the interference peaks, the oxide thickness is calculated as 60 μm, which is in good agreement with the weight gain data. When H_3BO_3, 1200 ppm boron, was added to the 3500 ppm lithium solution, the average weight gain was 45 mg dm^{-2}, which corresponds to a uniform oxide of 3 μm thickness. The spectrum obtained for the oxide on the outside surface of the cladding is shown in Fig. 3b. From the two interference peaks, an oxide thickness of 1.2 μm is calculated. The spectra from the inside surface showed variations in film thickness from 1 to 5 μm. Thus, the weight gain represents an average of oxide thicknessess on the two surfaces of the cladding specimen.

In Fig. 4, spectra are shown for Zr-2.5Nb specimens corroded in 700 ppm lithium solutions with 300 and 1200 ppm boron added. The specimens corroded in the solution containing 1200 ppm boron showed weight loss. Although an oxide film was visible on these specimens, their spectra were featureless. A typical spectrum is shown in Fig. 4a. Absorption peak due to OH groups in the oxide at 3600 cm^{-1}, interference peaks in the 3200 to 1000 cm^{-1} and absorption peaks in the 1600 to 1400 cm^{-1} regions, and an increase in absorbance near 1000 cm^{-1} are seen in Fig. 4b spectrum for the specimen corroded in the solution containing 300 ppm boron. Thickness calculated from the spacings of the interference peaks, 16.5 μm, compares well with the weight gain of 260 mg dm^{-2}. When compared to the spectrum for Zircaloy-4 in Fig. 3a, the absorption peak for O-H stretching vibration at 3600 cm^{-1} is more intense although the oxide is only one fourth in thickness.

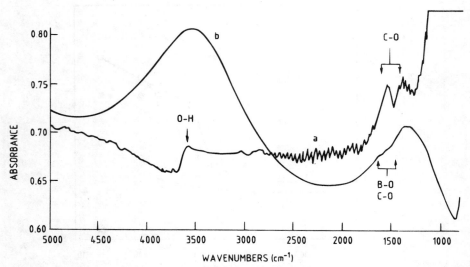

FIG. 3—*Fourier Transform InfraRed (FTIR) reflection spectra of corrosion films on Zircaloy-4; four-day exposures at 360°C in LiOH (3500 ppm Li) solutions: (a) no H_3BO_3 added, weight gain 900 mg dm^{-2} and (b) H_3BO_3 (1200 ppm B) added, weight gain 45 mg dm^{-2}.*

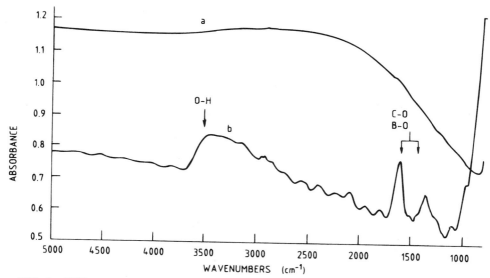

FIG. 4—*FTIR spectra of corrosion films on Zr-2.5Nb alloy exposed four-days at 360°C in LiOH (700 ppm Li) solutions containing H_3BO_3 (a) 1200 ppm B, weight loss (oxide spalling had roughened the surface and the resulting scattering losses may be the cause for the featureless spectrum) and (b) 600 ppm B, weight gain 260 mg dm^{-2}.*

When specimens of Zircaloy-4 and Zr-2.5Nb that have weight gains greater than 150 mg dm^{-2} were leached in dilute nitric acid, changes in the spectra occurred at the 1600 to 1400 cm^{-1} region that correspond to C-O and B-O vibrations. The intensity of the absorption peaks in this region decreased and absorption at 1500 cm^{-1}, characteristic of carbonate species, was eliminated. Carbonate as an impurity was likely present in the LiOH stock material or was formed on exposure of the corroded specimen to atmospheric air [8].

SIMS Profiles of Corrosion Films

A typical set of profiles for thick oxide films grown in concentrated LiOH solution with added H_3BO_3 is shown in Figs. 5a and b. The oxide was grown on Zircaloy-4 in 3500 ppm lithium solution with 600 ppm boron added. A thickness of 15 to 17 µm (estimate is based on the start of the decrease in ZrO signal at the oxide-alloy interface) indicated by the profiles compares well with a weight gain of 300 mg dm^{-2}. In the unleached oxide (Fig. 5a), the concentration of lithium shows changes in the outer layers, it is almost constant in the interior layers and it rapidly decreases near the oxide-alloy interface. In the bulk of the oxide, the concentration is 820 ppm lithium. The concentrations of the boron isotopes are almost the same in the outer layers and in the bulk oxide, and they rapidly decrease at the oxide-alloy interface. The concentration in the bulk oxide is 320 ppm boron. When the oxide was leached in warm dilute nitric acid, changes in the profiles, shown in Fig. 5b, were observed only for the lithium isotopes. The concentration decreased to an almost constant level of 100 to 120 ppm lithium in the surface and bulk oxide, and the profile shapes were comparable to those of boron. The results obtained for films >3 µm thick, grown on Zircaloy-4 cladding and Zr-2.5Nb pressure tube material, were similar to those shown in Figs. 5a and b.

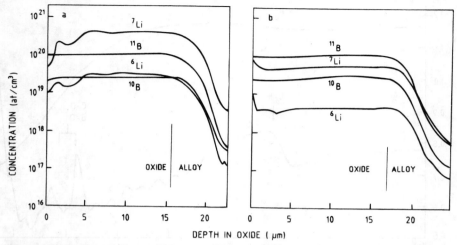

FIG. 5—*Depth concentration profiles for lithium and boron isotopes, obtained by SIMS, in oxide grown on Zircaloy-4 cladding corroded four-days at 360°C in LiOH (3500 ppm Li) and H_3BO_3 (600 ppm B) solution; oxide-alloy interfaces identified with start of the decrease in ZrO signal: (a) before and (b) after leaching in acid.*

In Fig. 6, profiles are shown for thin oxide films grown in 0.7 ppm lithium solution without added H_3BO_3 and in 3500 ppm lithium solution with H_3BO_3 added at a concentration of 1200 ppm boron. The oxide grown in the dilute solution is 0.65 μm thick and that grown in the concentrated solution is 1.2 μm thick (see Fig. 3b). The concentration of lithium is higher in the film grown in the concentrated solution by an order of magnitude. In the case of the film grown in the dilute solution, an acid leach reduced the lithium concentration only on the surface to 80 ppm from 200 ppm lithium. For both the oxide films, the lithium concentrations rapidly decrease through the oxide thickness, a characteristic similar to that observed near the oxide-alloy interface for thick oxides shown in Fig. 5. However, the decrease in the oxide up to the oxide-alloy interface is nearly two orders of magnitude compared to a factor of at the most only 2 in the case of the thick oxides. The decrease in the the case of thick oxides, shown in Fig. 5, is after the ZrO signal starts to diminish.

Electrochemical Polarization Measurements

A typical set of potentiodynamic polarization curves traced for Zr-2.5Nb alloy specimens corroding in dilute (2.5 ppm lithium) and concentrated (7000 ppm lithium) LiOH solutions is shown in Fig. 7. The reverse cathodic and forward anodic portions of the traces are plotted in the figure; a large hysteresis of greatly reduced current during the reverse anodic trace was present in the case of polarizations in the concentrated solution. Extrapolation of the linear anodic portions to the rest potentials gives corrosion currents of 2.3 μA for the 0.5-μm-thick film in the dilute and 130 μA for the 0.7-μm-thick film in the concentrated solution. The slopes of these linear portions are comparable; 8.23 V in the dilute and 8.7 V in the concentrated solution, for an order of magnitude change in current.

Aqueous Solution Chemistry

The ionization of water, the dissociation of LiOH, and the four stages of acid ionization of H_3BO_3 were considered in establishing the aqueous chemistry of the LiOH-H_3BO_3 system. The constants for the equilibria and the four ionization stages of H_3BO_3 are listed in Table 1

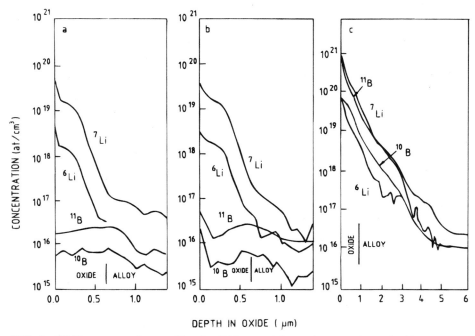

FIG. 6—*SIMS concentration profiles for lithium and boron isotopes in oxide films grown on Zircaloy-4 cladding following four-day exposures at 360°C: (a) 0.7 ppm Li as LiOH; (b) same as (a) after leaching in dilute nitric acid; and (c) 3500 ppm Li as LiOH and 1200 ppm B as H_3BO_3 (see Fig. 4(a); oxide-alloy interface corresponds to start of the decrease in ZrO signal.*

and the calculated concentrations of the species are listed in Table 2 for the three selected temperatures. The first column in Table 2 lists the total concentration of LiOH in ppm lithium. Under the four sections grouped according to the concentration of added H_3BO_3 (expressed in ppm B) are listed the pH of the solution and the amounts of undissociated LiOH in ppm Li and non-ionized H_3BO_3 in ppm B, respectively. It is readily seen that the dissociation of LiOH, with and without added H_3BO_3, decreases with increase in the concentration and temperature. This effect of temperature is also reflected in the ionization of H_3BO_3. Although the ionization of H_3BO_3 increases with an increase in LiOH concentration, with an increase in temperature, it decreases for a combination of concentrations of the alkali and the acid.

In Fig. 8, weight gain data for Zircaloy-4 are plotted as a function of the concentration of undissociated LiOH in the solution. The similarity to Fig. 1, where total LiOH concentration is used, is obvious. The plots of weight gain versus concentration of undissociated LiOH for Zr-2.5Nb were similar to the plots shown in Fig. 2. When pH of the solution is used, instead of the LiOH concentrations, the plots were again similar to those in Figs. 1 and 2. From Fig. 8, it is evident that for the same concentration of undissociated alkali, the weight gains at high alkali concentrations (>100 ppm for Zircaloy-4 and >30 ppm for Zr-2.5Nb) are greatly reduced when H_3BO_3 is also present. The concentration of dissolved hydrogen at the corrosion conditions is calculated to be 65 ppm. What effect this had on oxide growth is not discernible from the results reported here. However, the dissolved hydrogen had not adversely affected the hydrogen pick-up by the alloys. The percent pick-up of hydrogen varied from 6 to 8% for Zircaloy-4 cladding and was less than 1% for Zr-2.5Nb pressure tube specimens.

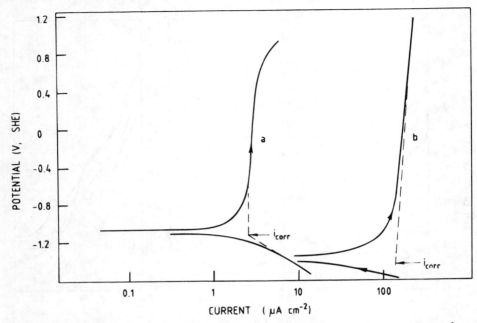

FIG. 7—*Polarization curves for Zr-2.5Nb alloy corroding in LiOH solutions: (a) two days at 315°C in 2.5 ppm Li solution and (b) 4 h at 315°C in 7000 ppm Li solution; 2 MPa hydrogen added to solutions; potentiodynamic scan at 1.2 V h^{-1}.*

Discussion

$LiOH-H_3BO_3$ Solution Chemistry and Corrosion

Corrosion of both Zircaloy-4 fuel cladding and Zr-2.5Nb pressure tube material bear the same type of relationship to the pH of the aqueous solutions and the concentrations of

TABLE 1—*Equilibrium constants for the ionization of water (k_w), dissociation of LiOH (k_{Li}) and acid ionization of H_3BO_3 (k_1 to k_4).*

Equilibrium Constant	Temperature		
	300°C	316°C	360°C
k_w	3.78 E-12	2.41 E-12	2.65 E-13
k_{Li}	7.88 E-3	4.53 E-3	5.74 E-4
k_1	62.1	56.7	48.5
k_2	86.5	75.2	54.2
k_3	74.4	53.9	24.2
k_4	8.8 E+3	6.93 E+3	4.42 E+3

IONIZATION OF H_3BO_3

$H_3BO_3 + OH^- = H_4BO_4^-$ (k_1)
$2H_3BO_3 + OH^- = B_2(OH)_7^-$ (k_2)
$3H_3BO_3 + OH^- = B_3(OH)_{10}^-$ (k_3)
$4H_3BO_3 + OH^- = B_4(OH)_{13}^-$ (k_4)

TABLE 2—*System chemistry of aqueous LiOH-H_3BO_3 solutions; concentrations of the alkali and the acid, total, and that remaining undissociated.*

LiOH (total) (ppm Li)	No Boron		H_3BO_3 Added								
			300 ppm B			600 ppm B			1200 ppm B		
	pH	LiOH (ppm Li)	pH	LiOH (ppm Li)	H_3BO_3 (ppm B)	pH	LiOH (ppm Li)	H_3BO_3 (ppm B)	pH	LiOH (ppm Li)	H_3BO_3 (ppm B)
					300°C						
1.4 E0	7.70	3.13 E-2	7.26	1.16 E-2	2.99 E+2	7.03	6.88 E-3	5.98 E+2	6.75	3.61 E-3	1.20 E+3
7.0 E0	8.34	6.20 E-1	7.93	2.52 E-1	2.93 E+2	7.71	1.52 E-1	5.91 E+2	7.43	8.08 E-2	1.19 E+3
3.5 E1	8.92	8.84 E0	8.57	4.5 E0	2.70 E+2	8.35	2.89 E0	5.59 E+2	8.08	1.6 E0	1.15 E+3
7.0 E1	9.14	2.44 E+1	8.83	1.42 E+1	2.47 E+2	8.63	9.55 E0	5.24 E+2	8.36	5.46 E0	1.10 E+3
3.5 E2	9.60	1.98 E+2	9.41	1.56 E+2	1.55 E+2	9.25	1.24 E+2	3.52 E+2	9.01	8.19 E+1	8.15 E+2
7.0 E2	9.78	4.51 E+2	9.64	3.90 E+2	1.10 E+2	9.51	3.35 E+2	2.51 E+2	9.31	2.47 E+2	6.04 E+2
3.5 E3	10.17	2.71 E+3	10.11	2.62 E+3	4.05 E+1	10.06	2.52 E+3	8.66 E+1	9.96	2.31 E+3	1.95 E+2
7.0 E3	10.33	5.70 E+3	10.29	5.59 E+3	2.50 E+1	10.26	5.48 E+3	5.22 E+1	10.20	5.25 E+3	1.13 E+2
					316°C						
1.4 E0	7.88	5.25 E-2	7.47	2.09 E-2	2.99 E+2	7.25	1.27 E-2	5.98 E+2	6.97	6.75 E-3	1.20 E+3
7.0 E0	8.51	9.24 E-1	8.13	4.31 E-1	2.94 E+2	7.92	2.07 E-1	5.91 E+2	7.65	1.47 E-1	1.19 E+3
3.5 E1	9.06	1.19 E+1	8.74	6.81 E+0	2.73 E+2	8.54	4.65 E+0	5.62 E+2	8.29	2.72 E+0	1.15 E+3
7.0 E1	9.26	3.07 E+1	8.98	1.99 E+1	2.54 E+2	8.80	1.43 E+1	5.31 E+2	8.55	8.80 E+0	1.11 E+3
3.5 E2	9.70	2.26 E+2	9.51	1.85 E+2	1.75 E+2	9.37	1.54 E+2	3.84 E+2	9.15	1.10 E+2	8.58 E+2
7.0 E2	9.87	4.98 E+2	9.72	4.40 E+2	1.32 E+2	9.61	3.87 E+2	2.92 E+2	9.42	3.01 E+2	6.71 E+2
3.5 E3	10.25	2.87 E+3	10.19	2.77 E+3	5.28 E+1	10.13	2.67 E+3	1.13 E+2	10.03	2.46 E+3	2.51 E+2
7.0 E3	10.41	5.95 E+3	10.37	5.83 E+3	3.30 E+1	10.33	5.72 E+3	6.87 E+1	10.26	5.48 E+3	1.48 E+2
					360°C						
1.4 E0	8.75	2.75 E-1	8.42	1.43 E-1	2.99 E+2	8.22	9.48 E-2	5.98 E+2	7.97	5.48 E-2	1.20 E+3
7.0 E0	9.27	3.03 E+0	8.99	1.99 E+0	2.95 E+2	8.92	1.47 E+0	5.93 E+2	8.59	9.48 E-1	1.19 E+3
3.5 E1	9.71	2.31 E+1	9.48	1.82 E+1	2.85 E+2	9.33	1.51 E+1	5.76 E+2	9.14	1.12 E+1	1.16 E+3
7.0 E1	9.8	5.12 E+1	9.67	4.28 E+1	2.76 E+2	9.53	3.70 E+1	5.61 E+2	9.34	2.89 E+1	1.14 E+3
3.5 E2	10.2	2.97 E+2	10.09	2.70 E+2	2.35 E+2	9.97	2.48 E+2	4.85 E+2	9.80	2.12 E+2	1.00 E+3
7.0 E2	10.42	6.15 E+2	10.26	5.74 E+2	2.04 E+2	10.16	5.37 E+2	4.25 E+2	10.00	4.73 E+2	8.89 E+2
3.5 E3	10.79	3.23 E+3	10.69	3.13 E+3	1.10 E+2	10.62	3.04 E+3	2.27 E+2	10.50	2.85 E+3	4.72 E+2
7.0 E3	10.94	6.52 E+3	10.87	6.39 E+3	7.10 E+1	10.82	6.27 E+3	1.45 E+2	10.73	6.01 E+3	2.95 E+2

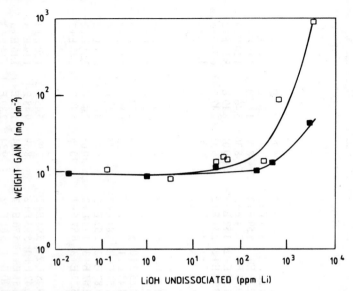

FIG. 8—*Weight gain, four-day exposures at 360°C, for Zircaloy-4 cladding in LiOH solutions; □ = 0 and ■ = 1200 ppm B added as H_3BO_3.*

undissociated LiOH. The weight gains show the same type of dependence of pH and on the concentration of undissociated LiOH. Therefore, the cause of accelerated corrosion at high lithium concentrations can be related to either pH or concentration of LiOH or to a combined effect of pH and LiOH [*3*]. For example, the surface of ZrO_2 is acidic compared to the pH of the solution, and an intrinsic pH effect will be the extent of acidic decomposition of the oxide surface. This is discussed in the subsection on ZrO_2 surface chemistry.

In the presence of added H_3BO_3, the concentration of LiOH (undissociated LiOH) in the solution decreases. However, the corrosion is found to be not directly related to only the concentration of LiOH. For example, examination of Fig. 8 shows that at the same LiOH concentration and at concentrations greater than 100 ppm lithium as LiOH, the weight gains are reduced in the presence of H_3BO_3. This inhibiting effect of H_3BO_3 can be related to the chemistry of the aqueous solution. The data for the system chemistry at 360°C with and without 1200 ppm boron (Table 2) are graphically illustrated in Figs. 9 and 10. It is seen that for comparable LiOH concentrations the pH of the solution is reduced when H_3BO_3 is present (Fig. 9) and that even at high concentrations of LiOH a considerable portion of H_3BO_3 remains nonionized as H_3BO_3. Inhibition by boron is thus simply not due its neutralizing action on LiOH in the bulk solution alone. Similar to acceleration of corrosion by lithium, the inhibition by boron can be related to a combined effect of pH and the concentration of H_3BO_3.

The corrosion behavior of Zr-2.5Nb alloy is similar to that of Zircaloy-4 up to a total lithium concentration of 50 ppm lithium, with and without added H_3BO_3. Major differences in behavior are observed at higher concentrations, of both lithium and boron. These are the occurrence of accelerated corrosion, in the absence of H_3BO_3, at a much lower lithium concentration of about 60 ppm compared to 350 ppm lithium in the case of Zircaloy-4 and the detrimental effect of added H_3BO_3 at high lithium and boron concentrations. In the absence of H_3BO_3 although the corrosion was accelerated at high lithium concentrations, the oxide films stayed adherent; but when H_3BO_3 was also present, oxide spallation and weight losses were observed. Therefore,

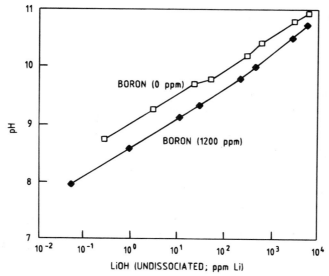

FIG. 9—*System chemistry of LiOH-H$_3$BO$_3$ solutions at 360°C, for total lithium concentration varying from 1.4 to 7000 ppm.*

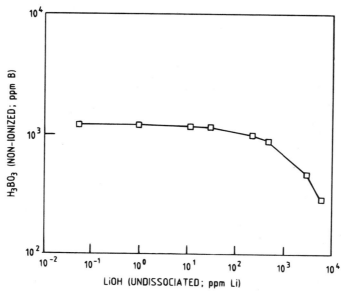

FIG. 10—*Variation in concentration of non-ionized H$_3$BO$_3$ with undissociated LiOH, at 360°C in solutions containing 1200 ppm boron as H$_3$BO$_3$ and 1.4 to 7000 ppm Li as LiOH.*

the combined effects of solution chemistry and oxide surface chemistry determine the corrosion behavior.

ZrO_2 Surface Chemistry

Zirconium oxide surfaces exposed to aqueous environments have been characterized by potentiometric titration and electrophoretic mobility measurements [13,14]. The surface is hydrated and the zero point of charge is quoted as 6 at room temperature. At elevated temperatures, the surface is expected to be even more acidic. The zero point of charge can be calculated to vary from 4.7 to 5.3 in the range of temperatures 300 to 360°C, taking into account the dissociation of water [15]. From the pH values listed in Table 2, it is seen that the LiOH-H_3BO_3 solutions are all quite alkaline to the zirconia surface at the corrosion conditions. The surface of the growing oxide crystallites exposed to the solution would thus undergo acidic decomposition and the oxide-solution interface can be expected to be similar to that schematically shown in Fig. 11b. The negative charge on the oxide surface is compensated by positively charged solvated lithium ions in the Stern layer (identified as inner and outer compact layer in the figure) [16]. This Stern layer is likely to be more diffuse at the corrosion temperatures than that of about two to three molecular diameters of water, the size usually taken to exist at room temperature. Undissociated LiOH and H_3BO_3 will be part of the hydrogen-bonded network of water molecules shown by the large circles in the figure.

Oxide Growth

Prior to corrosion, there is a thin (a few nanometres thick) oxide film on the specimens resulting from exposure to atmospheric air. The interface of this oxide surface with the aqueous solution is that schematically shown in Fig. 11b. Electrochemical reactions occurring at the alloy-oxide and oxide-solution interfaces provide the driving force for charge transport and gowth of the oxide. Anion vacancies generated at the ally-oxide interface transport the oxygen species by migrating to the oxide-solution interface and interacting with water molecules.

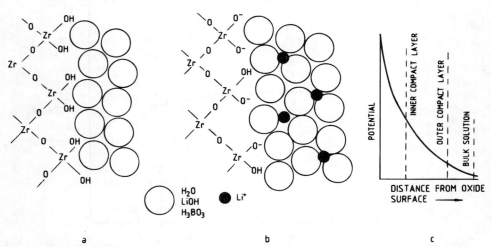

FIG. 11—Schematic representation of ZrO_2-aqueous solution interface: (a) oxide surface at zero point of charge (pH of solution same as ZPC), (b) oxide surface exposed to LiOH solution, and (c) potential decay away from oxide surface.

$$\text{>Zr} - \text{O}^- + 1 = \text{>Zr}^+ - 1 \quad (1a)$$
$$\text{>Zr}^+ - 1 + n\text{H}_2\text{O} = \text{>Zr} - \text{OH} + (n-1)\,\text{H}_2\text{O}\,\text{H}^+ \quad (1b)$$

where $\square$ is an anion vacancy and n represents hydrogen-bonded water molecules. The charge on the mobile oxygen species is assumed to be 2^-. Protons exist in the water as hydrated H_3O^+. Acidic decomposition of the surface hydroxyl produced regenerates the interface with the solution and the process gets repeated resulting in oxide growth.

$$\text{>Zr} - \text{OH} + \text{OH}^- + \text{Li}^+ = \text{>Zr} - \text{O}^- + \text{Li}^+(\text{H}_2\text{O}) \quad (1c)$$

Electron transport, occurring from the ally through the conduction band of the oxide, completes the cathodic reation.

$$\text{>Zr} - \boxed{e} + e + \text{H}_3\text{O}^+ = \text{>Zr} - \text{H} + \text{H}_2\text{O} \quad (1d)$$
$$2\,\text{>Zr} - \text{H} = 2\,\text{>Zr} - \boxed{e} + \text{H}_2 \quad (1e)$$

where $\boxed{e}$ represents an electron trapped at an anion vacancy. These reactions are considered to occur at the oxide-solution interface that is the outermost oxide layer for a nonporous pre-transition type corrosion film and is the oxide layer at the bottom of pores where solution incursion had occurred.

An important aspect of corrosion in general is structural development in the oxide film. For both Zircaloy-4 and Zr-2.5Nb, these structural characteristics are shown to be similar [17]. Nucleation followed by diametral grain growth occurs in the initial stages, during corrosion in water, resulting in a columnar structure of oxide grains [17]. In the bulk oxide, the atoms on the surface of the oxide grains, unlike the atoms inside the grains, have incomplete coordination neighbors and hence unfilled bonding electrons. It is proposed that the residual bonding forces lead to adsorption of water molecules on the oxide surface. The mismatch in orientation of the oxide grains may result in one or more layer of adsorbed water molecules on the surfaces of adjoining oxide grains. Experimental support for the presence of adsorbed water on the surface of oxide grains, in the form of hydrogen-bonded hydroxyls, is provided by the hydrogen-deuterium isotope exchange investigations [18]. Deuterium in the oxide after a number of years of exposure can be exchanged with hydrogen and re-exchanged with deuterium by short-term (a few days) exposures to light and heavy water vapor, respectively. A typical result for a 1-μm thick oxide is given in Fig. 12 that shows that deuterium can readily be exchanged for hydrogen and vice versa. Oxide film thickness, measured by infrared interferometry, had increased by 0.4 μm after an in-reactor exposure of 1.5 years at 270°C. No further increase in thickness, resulting from hydrogen-deuterium exchange experiments was detected. This confirms the observations reported by Woolsey and Morris that deuterium profiles in pre-transition oxides on Zircaloy-2 are essentially identical for the films grown solely in heavy water and those exposed only for a brief period to heavy water [19]. Combining the structural development of the oxide reported in the literature [17] with the surface chemistry of ZrO_2 discussed earlier, the pre-transition oxide growth in water and in dilute LiOH solutions with and without added H_3BO_3 can be represented by the schematic diagram, Fig. 13. The hydroxyl groups and adsorbed water are considered an integral part of the surface chemical structure of the oxide grains. Oxygen transport occurs by surface diffusion via anion vacancies that is much slower than proton transfer along the hydrogen-bonded hydroxyls on the surfaces of the oxide crystallites. The diameter of a water molecule is about 0.3 nm [20], comparable in size to the lattice spacing in monoclinic zirconia. However, for proton transfer along the hydrogen-bonded network to occur, steric requirements enabling the rotation of water molecules have to be met [20].

Post-transition oxide growth is structurally identified with development of porosity in the existing oxide and finer grain growth in the newly formed oxide [17]. According to the

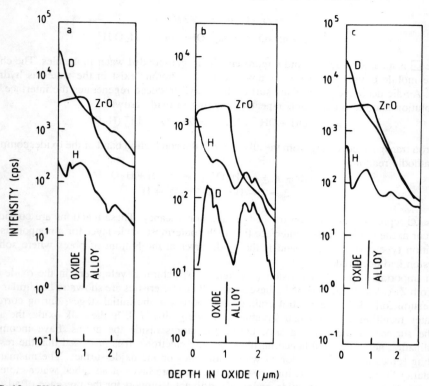

FIG. 12—*SIMS intensity profiles for hydrogen and deuterium in oxide film on the outside surface of Zr-2.5Nb alloy specimen, from a pressure tube removed from a CANDU reactor after an effective full power operation of 1.5 years: (a) as-received, (b) after exposure to 2.5 cm H_2O vapor at 300°C for seven days, and, (c) reexposed to 2.5 cm D_2O vapor at 300°C for seven days.*

proposed model, the oxide crystallites in post-transition films also would have a chemical structure similar to that in pre-transition films, namely, hydrogen-bonded hydroxyls and adsorbed water on their surfaces. However, the distinguishing feature of post-transition oxides is their porosity, because the size and distribution of pores (cylindrical and parallel-sided geometries) will determine the extent of solution incursion into the film. It is evident from Fig. 11c that on each surface of the adjoining oxide crystallites, forming the gap or pore, the charged layers exist in contact with the solution. The pore or gap between the surfaces of adjoining oxide grains has to be larger than 2 nm, about the size of eight molecular diameters of water, for the solution to have access in the oxide.

Therefore, the corroding solution may not penetrate all the pores. The transport of species during corrosion would follow the two paths, namely, that in solution inside pores >2 nm in size, which will be similar to that in bulk solution, and that along the surfaces of oxide crystallites, with intercrystalline gap <2 nm, which will be similar to that occurring in pre-transition films. Pore volume distribution measurements [8] in combination with high-resolution transmission electron microscopy of the oxide crystallite boundaries would provide an appropriate model for the physical structure of the oxide, to complement the chemical structure proposed here.

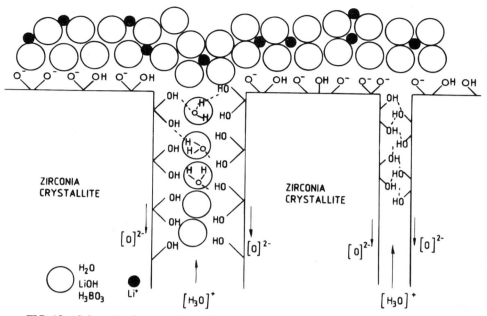

FIG. 13—*Schematic diagram of species transport during corrosion of zirconium alloys.*

During corrosion in concentrated LiOH solutions, the acidic decomposition of the oxide surface is enhanced by the increases in pH of the solution. A high concentration of LiOH existing as a part of the hydrogen-bonded water leads to the following reaction
The surface OLi groups, unlike the OH groups do not undergo acidic decomposition and their migration via the anion vacancies leads to surface OLi groups on the oxide grains. These OLi groups retard the diametral and columnar grain growth resulting in a fine-grained oxide and an enhanced corrosion rate [3]. Boric acid can function in two ways to inhibit the acceleration in corrosion caused by LiOH: (1) pH of the solution is lowered (Fig. 9) that reduces the enhanced acidic decomposition of the oxide surface and hence the extent of reaction with LiOH according to Eq 2 and (2) a considerable portion remains nonionized as H_3BO_3 that, forming part of the hydrogen-bonded water, reacts to remove surface OLi groups as

$$\mathrm{Zr}\begin{matrix}O^-\\ \\O^-\end{matrix} + 1 = \mathrm{Zr}\begin{matrix}\pm 1\\ \\O^-\end{matrix} \tag{2}$$

$$\mathrm{Zr}\begin{matrix}\pm 1\\ \\O^-\end{matrix} + \mathrm{LiOH} = \mathrm{Zr}\begin{matrix}\ \ \ \ \ H\\ \leftarrow\cdots O\ Li\\O^-\end{matrix}\quad \mathrm{Zr}\begin{matrix}OH\\ \\OLi\end{matrix}$$

$$\begin{array}{c}\diagdown\\ \diagup\end{array}\!\!Zr\!\!\begin{array}{c}\diagdown\\ \diagup\end{array}\!\!\begin{array}{c}\leftarrow\text{---}O\begin{array}{c}H\\ |\\ Li\end{array}\\ O^-\end{array} + H_3BO_3 = \begin{array}{c}\diagdown\\ \diagup\end{array}\!\!Zr\!\!\begin{array}{c}\pm 1\\ O^-\end{array} + H_4BO_4^- + Li^+ \quad (3)$$

Boric acid, as a Lewis acid, also reacts with ZrO_2 surface as

$$-\!\!\begin{array}{c}\diagdown\\ \diagup\end{array}\!\!Zr - O^- + H_3BO_3 = -\!\!\begin{array}{c}\diagdown\\ \diagup\end{array}\!\!Zr - O \text{-->} B(OH)_3^- \quad (4)$$

Spectroscopic and Electrochemical Implications

The model proposed for oxide growth is also in accordance with the spectroscopic and electrochemical observations. For example, thick oxides grown in concentrated LiOH solutions (>60 ppm for Zr-2.5Nb and >350 ppm lithium for Zircaloy-4 in our short-term tests) show two infrared characteristics that indicate access of the corroding solution in the bulk of the oxide. These are an absorption band at 3600 cm^{-1} that is characteristic of hydrogen-bonded hydroxyls and absorption peaks in the 1600 to 1400 cm^{-1} region that are characteristic of carbonate and borate species. For the carbonate to form, LiOH should to have been in the oxide that was converted to carbonate on exposure to air. Another possibility is carbonate present as an impurity in the solution becoming concentrated in the oxide [8]. Acid leaching dissolved the carbonate and washed out the lithium associated with it. This is also confirmed by the SIMS profiles that show a reduction only in lithium concentration throughout the film. From Fig. 11c, it is therefore inferred that pores in these oxide layers have to be 2 nm or larger in diameter for the solution to have gained access in the oxide.

The same conclusion, that in concentrated LiOH solutions the corroding solution has access through the bulk oxide, is arrived at from potentiodynamic polarization data. For films of comparable thicknesses (0.5 and 0.7 μm) grown and polarized in the dilute (2.5 ppm lithium) and concentrated (7000 ppm lithium) solutions, the anodic Tafel slopes are comparable but the corrosion current in the concentrated solution is about two orders of magnitude larger. Therefore, for the film growing in the concentrated solution, the polarization resistance has to be correspondingly less, meaning that only a small fraction of the total thickness controls the charge transport and the solution has access through the rest of the bulk of the oxide.

Depth profiles obtained in SIMS analyses may not be fully representative of the true distributions of the elements because of the possibility of sputter-induced changes. However, qualitative information may be inferred [8]. For example, the almost constant concentrations of lithium and boron isotopes in the bulk of thick oxides indicate that the free volume (integrated total volume between surfaces of adjoining oxide grains) and surface area (integrated total area of the surfaces of oxide grains) available per unit volume of the oxide are constant in the bulk oxide. The size of the pores and their distribution could be different, however. Similarly, the decreasing concentrations of lithium and boron near the oxide-alloy interfaces of thick oxides and in thin oxides (see Figs. 5 and 6) reflect a decrease in the available free volume and surface area with depth in the oxide. The steeper slopes in the case of thin pre-transition oxides grown in dilute lithium solutions suggest a more dense oxide film than that at the interface of thick oxides grown in concentrated lithium solutions. Acid leaching does not affect the lithium concentration profiles in the thin pre-transition oxides.

Thus from considerations of the surface chemistry of ZrO_2, the physical structure for the oxide revealed by SIMS depth profiles may be summarized as follows. In the case of films grown in concentrated lithium solutions, the corroding solution had access into the oxide. The oxide grain surfaces exposed to the solution are incorporated with Li$^+$ from ion-exchange (Eq 1) and -OLi from LiOH reaction (Eq 2) and with $H_3BO_4^-$ from H_3BO_3 reaction (Eq 4). These lithium and boron species are not leached by the acid. Similarly in the case of thin pre-transition films grown in dilute solutions (0.7 ppm lithium with 1200 ppm boron), non-leachable lithium and boron are found. However, only the outer surfaces (at the oxide-solution interface) of the oxide grains in these films had been exposed to the solution. Lithium and boron are present also in the bulk of these films. The sizes of both LiOH and H_3BO_3 are comparable to that of H_2O. It is proposed that LiOH and H_3BO_3 enter the free volume available between the surfaces of adjoining oxide grains. This is facilitated by the fact that LiOH and H_3BO_3 form a hydrogen-bonded network with water.

For the three species (Li, B, and OH) to exist on the surface of the oxide grains, an estimate can be made of the number of sites available and compared with the SIMS concentration profiles, for an idealized physical structure of the oxide film. Assuming the oxide is made up of columns that are 50 nm (base) and 0.5 µm (height) in dimension and that a surface site is circular with a radius of 0.4 nm, the total number of sites available (all these surface sites concentrations are expressed as atoms per unit volume of oxide) can be calculated as 1.7×10^{20}. In thin pre-transition films, the concentration of deuterium in a small volume near the surface is 10^{21} (Fig. 12, film grown in steam) and that of lithium in 10^{20} (Fig. 6, film grown in dilute lithium solution). The concentrations rapidly decrease with depth in the oxide. For thick films grown in concentrated lithium solutions (Fig. 5), the concentration of lithium in the bulk oxide is 4×10^{20} before leaching in acid and is 5×10^{19} after leaching and the boron concentration before and after leaching remains constant at 10^{20}. Thus, sites are available and the concentrations of the species are determined by the chemistry of the corroding solution and the accessibility of the oxide crystallite surfaces to the solution.

Oxide Dissolution

It has been suggested that preferential dissolution of cubic or tetragonal oxide crystallites could occur during corrosion in concentrated lithium solutions [21]. Dissolution of yttria (14%) stabilized cubic zirconia single crystals has been reported to occur but under conditions that are extremely severe compared to the corrosion environments referred to here. At 600°C and 100 MPa in 44 000 ppm lithium solution, a weight loss of about 0.5% is reported [22]. However, this is attributed to the greater solubility in the solution of yttrium components than zirconium components, because monoclinic zirconia is precipitated out [22]. Although oxide solubility as a means of pore generation is another mechanism to be considered for enhanced corrosion in concentrated lithium solutions, its role has to be confirmed by investigations on the formation and dissolution of cubic or tetragonal crystallites in corrosion films.

In the present investigation, the occurrence of oxide dissolution may be suspected in the case of Zr-2.5Nb alloy corroding in concentrated lithium solutions when H_3BO_3 was also present. The spalling of oxide as a fine powder resulting in weight losses is likely the result of a dissolution process. If dissolution was occurring, the resulting heterogeneity of the surface would scatter light and this may be the reason for obtaining featureless interference spectra (see Fig. 4a). Unlike Zircaloy-4 fuel cladding, the niobium alloy has a thin beta zirconium layer covering the alpha grains. The beta phase has about 20% niobium in solution compared to about 1% niobium in the alpha grains. The oxide on the beta phase and its surface chemistry are likely to be different from those on the alpha grains. In the absence of information on the type of oxide formed, it is speculative to propose a mechanism. Niobium has multiple valence states. For example, if niobium were in solid solution in a lower valence state in a zirconia-type oxide, then it might form a coordination-type complex with H_3BO_3 and dissolve. It is not clear, however, why a high lithium concentration is also needed for such a dissolution to occur. Similarly, if it is assumed that the zero point of charge for the oxide on Zr-2.5Nb alloy is less (more acidic) than that for the oxide on Zircaloy-4, then acceleration in corrosion could occur at a lower lithium concentration (Eq 2).

Conclusions

The surface chemistry of ZrO_2 in aqueous solutions and the aqueous chemistry of LiOH-H_3BO_3 play a dominant role in determining the corrosion behavior of Zircaloy-4 fuel cladding and Zr-2.5Nb pressure tube material.

Acceleration of corrosion is caused by, LiOH, undissociated LiOH. Normal oxide growth proceeds by the reaction of water with anion vacancies. When the concentration of LiOH is high, it participates in reaction with anion vacancies producing surface OLi groups. Unlike the OH groups, formed by reaction with water, the OLi groups do not decompose forming O^- on the surface. The diametral and columnar growth of oxide crystallites is impeded resulting in accelerated corrosion. Inhibition by boron occurs in two ways, namely, by lowering the pH of the solution at comparable LiOH concentrations and by interfering with the reaction between LiOH and anion vacancies to prevent the formation of surface OLi groups.

Lithium and boron in oxides formed in concentrated solutions are incorporated on the surfaces of oxide grains by ion exchange and by reactions with LiOH and H_3BO_3 in the solutions.

Lithium and boron in pre-transition films, grown in dilute solutions, exist in two different states. On the outer surfaces of the oxide grains, which have been exposed to the solution, lithium and boron are in the same state as in oxides formed in concentrated solutions. In the bulk of the oxide films, lithium and boron exist as adsorbed LiOH and H_3BO_3 on the surfaces of oxide grains. This is facilitated by their ability to form hydrogen-bond with surface hydroxyl groups.

Corrosion behavior of Zr-2.5Nb is different from that of Zircaloy-4. Inhibition by boron is only effective in dilute LiOH solutions. Boric acid addition is detrimental when LiOH concentrations are high.

Acknowledgments

The authors express their gratitude to the CANDU Owners Group (COG) for funding and supporting this project. V. C. Ling of Chalk River Laboratories provided technical assistance for the corrosion tests. FTIR spectroscopic measurements were done by L. Chee at Ontario Hydro Research. Surface Science Laboratories of the University of Western Ontario performed the SIMS analyses. The authors are grateful to C. G. Weisener and G. R. Mount for the SIMS measurements and data analyses.

References

[1] Ruvarac, A., "Group IV Hydrous Oxides—Synthetic Inorganic Ion Exchangers," *Inorganic Ion Exchange Materials,* A. Clearfield, Ed., CRC Press, Inc., Boca Raton, FL, 1982.
[2] McDonald, S. G., Sabol, G. P., and Sheppard, K. D. in *Zirconium in the Nuclear Industry: Sixth International Symposium, ASTM STP 824,* D. G. Franklin and R. B. Adamson, Eds., American Society for Testing and Materials, Philadelphia, 1984, pp. 519–530.
[3] Ramasubramanian, N., Preocanin, N., and Ling, V. C. in *Zirconium in the Nuclear Industry: Eighth International Symposium, ASTM STP 1023,* L. F. P. Van Swam and C. M. Eucken, Eds., American Society for Testing and Materials, Philadelphia, 1989, pp. 187–201.
[4] Perkins, R. A. and Busch, R. A. in *Zirconium in the Nuclear Industry: Ninth International Symposium, ASTM STP 1132,* C. M. Eucken and A. M. Garde, Eds., American Society for Testing and Materials, Philadelphia, 1991, pp. 595–611.
[5] Manolescu, A. V., Mayer, P., and Simpson, C. J., *Corrosion,* Vol. 38, 1982, pp. 23–31.
[6] Bramwell, I. L., Parsons, P. D., and Tice, D. R. in *Zirconium in the Nuclear Industry: Ninth International Symposium, ASTM STP 1132,* C. M. Eucken and A. M. Garde, Eds., American Society for Testing and Materials, Philadelphia, 1991, pp. 628–642.
[7] Ramasubramanian, N. and Schankula, M. H., *Proceedings,* Fifth International Symposium on Environmental Degradation of Materials in Nuclear Power Systems—Water Reactors, American Nuclear Society, Inc., Monterey, IL, 1991, pp. 150–155.

[8] Ramasubramanian, N. in *Zirconium in the Nuclear Industry: Ninth International Symposium, ASTM STP 1132*, C. M. Eucken and A. M. Garde, Eds., American Society for Testing and Materials, Philadelphia, 1991, pp. 613–626.
[9] McIntyre, N. S., Weisener, C. G., Davidson, R. D., Brennenstuhl, A., and Warr, B. D., *Journal of Nuclear Materials*, Vol. 178, 1991, pp. 80–92.
[10] Marshall, W. L. and Frank, E. U., "Ion Product of Water Substance, 0–1000°C, 1–10 000 Bars, New International Formulation and Its Background," *Journal of Physical Chemistry Reference Data*, Vol. 10, 1981, pp. 295–304.
[11] Wright, J. M., Lindsay, W. T., and Druga, T. R., "The Behaviour of Electrolytic Solutions at Elevated Temperatures as Derived From Conductance Measurements," Westinghouse Electric Power Corporation Report WAPD–TM–204, Bettis Atomic Power Laboratory, Pittsburgh, June 1961.
[12] Mesmer, R. E., Baes, C. F., and Sweeton, F. H., "Acidity Measurements at Elevated Temperatures, VI, Boric Acid Equilibria," *Inorganic Chemistry*, Vol. 2, 1972, pp. 537–543.
[13] Smith, G. W. and Salman, T., *Canadaian Metallurgical Quarterly*, Vol. 5, No. 2, 1966, pp. 93–107.
[14] Kagawa, M., Omori, M., and Syono, Y., *Journal, American Ceramic Society, Communications*, Vol. 69, No. 3, 1986, pp. C50–C51; Vol. 70, No. 9, 1987, pp. C212–C213.
[15] Tewari, P. H. and McLean, A. W., *Journal of Colloid and Interface Science*, Vol. 40, 1972, pp. 267–272.
[16] Adamson, A. W., *Physical Chemistry of Surfaces*, Wiley, New York, 1990, pp. 210–214 and 233–236.
[17] Sabol, G. P. and McDonald, S. G., "Stress Effects and the Oxidation of Metals," *Proceedings*, Metallurgical Society of the American Institute of Mining, Metallurgical and Petroleum Engineers, 1974, pp. 352–372.
[18] Ramasubramanian, N., unpublished results.
[19] Woolsey, I. S. and Morris, J. R., *Corrosion*, Vol. 37, 1981, pp. 575–585.
[20] Bockris, J. O'M. and Reddy, A. K. N., *Modern Electrochemistry*, Vol. 1, Plenum Press, New York, 1977, pp. 476–485.
[21] Cox, B. and Wu, C., *Journal of Nuclear Materials*, Vol. 199, 1993, pp. 272–284.
[22] Yoshimura, M., Hiuga, T., and Somiya, S., *Journal, American Ceramic Society*, Vol. 69, 1986, pp. 583–584.

DISCUSSION

G. P. Sabol[1] (written discussion)—(1) From data or your model, how much boron is required to mitigate the harmful effect of lithium?

(2) For your model, what is the porosity of the inner and outer oxide films?

N. Ramasubramanian and P. V. Balakrishnan (authors' closure)—(1) In our four-day exposures at 360°C, the harmful effect of LiOH is observed at concentrations >350 ppm lithium for Zircaloy-4 fuel cladding. The mitigation by H_3BO_3 seems to be a function of the lithium concentration. If it is expressed as a percentage reduction in weight gain, 100 (1 − ratio of weight gains with and without added H_3BO_3), then with increase in H_3BO_3 from 300 to 1200 ppm boron the reduction in weight gain increases from 83 to 85% at 700 ppm lithium and from 33 to 95% at 3500 ppm lithium.

The pH of the solution is determined by the dissociation of LiOH and the ionization of H_3BO_3. Corrosion (four-day weight gain) varies directly with undissociated LiOH above 300 ppm lithium and inversely with nonionized H_3BO_3. The database is too small to attempt even an empirical relation.

(2) In post-transition type films, the outer oxide layer has both a micro (pore diameter 0.5 to 2 nm) and a meso (pore diameter >2 nm) porosity and the inner layer has only the micro porosity.

B. Cox[2] (written discussion)—The question of when a pore is not a pore, but an expanded crystallite boundary is something that needs further study by electron microscopy. However, I would comment that we have continued to look at the direct effect of LiOH on oxide dissolution and the effect of H_3BO_3. We have not been able to observe any effect of H_3BO_3 on the dissolution process. Pore can be formed in nonporous oxides (formed in water) irrespective of whether H_3BO_3 is added to LiOH. The only discernible effect (barely outside experimental variation) may be that H_3BO_3 slightly inhibits reprecipitation of ZrO_2. Thus, any effect that H_3BO_3 is having on the corrosion rate must be through an effect on some other process, perhaps the rate of diffusion through the residual barrier oxide.

N. Ramasubramanian and P. V. Balakrishnan (authors' closure)—The mitigating effect of added H_3BO_3 on the corrosion of Zircaloy-4 fuel cladding at 360°C in LiOH solutions is observed at lithium concentrations >350 ppm in our four-day exposures. At lower lithium concentrations, the effect of H_3BO_3 is negligible. If H_3BO_3 is affecting the diffusion process through the residual barrier layer then such an effect does not seem to be operative at <350 ppm lithium.

During corrosion in solutions containing >350 ppm lithium, the mitigating effect of H_3BO_3 is evidenced both by weight gain and by infrared interferometric methods of oxide thickness measurements. Hydrogen pickup by the alloy is also commensurate with oxide thickness. Oxide dissolution, if any, was not indicated by these results.

In oxide films formed in high lithium solutions (>350 ppm lithium) with and without H_3BO_3 there is porosity. For example, SIMS profiles indicate almost the same near-surface concentration of lithium and boron isotopes. But in oxides grown in solutions without H_3BO_3, the porosity (allowing incursion by the aqueous solution) extends throughout the film, whereas in oxides grown in solutions with added H_3BO_3, a barrier layer behavior is observed.

Oxide dissolution as a means of porosity generation is an important mechanism to be considered. If the amount dissolved is small and if in addition there is reprecipitation then weight gain and interferometric methods may not provide the evidence.

---

[1] Westinghouse Electric Corporation, Pittsburgh, PA.
[2] University of Toronto, Ontario, Canada.

R. A. Ploc[3] *(written discussion)*—As an alternate explanation for 2-nm pores, is it possible that protons rather than OH moves from oxygen to oxygen at the crystallite boundaries? The proton would enter the electron cloud of the oxygen causing shrinkage and more rapid diffusion and a 2-nm diameter pore would not be required. Infrared spectroscopy would indicate an O-H bond length. The proton would hop from oxygen to oxygen along the crystallite boundary.

N. Ramasubramanian and P. V. Balakrishnan (authors' closure)—In our model, the 2-nm gap between the surfaces of adjoining oxide crystallites represents the minimum space required for the existence of "aqueous solution." If there are pores that allow incursion of the corroding solution, then such pores have necessarily to be >2 nm because of the surface charging effects arising from differences between the zero-point charge of the oxide and the pH of the solution. Proton transfer is the mechanism proposed for transport of hydrogen species. We are relating it to tunneling along the hydrogen-bonded network of hydroxyls and adsorbed water.

Y. S. Kim[4] *(written discussion)*—Please comment on the possibility of forming OLi^- on the oxide surface. I think that OLi^- is really a larger ion charge than O^{2-} so that it could not be fit into zirconia unit cell. Please comment on what would be the controlling mechanism in the oxide growth. Does your data say that the transport of oxygen or lithium and so on through pores is the main controlling factor?

N. Ramasubramanian and P. V. Balakrishnan (authors' closure)—In our model, we are not proposing the incorporation of lithium in the oxide lattice. All the lithium is on the surfaces of the oxide crystallites. In oxides grown on Zircaloy-4 fuel cladding, in concentrated lithium solutions (>350 ppm lithium), lithium is existing in the outer porous oxide (accessible to the corroding solution) as OLi. These are produced by ion-exchange and by reaction of anion vacancies with undissociated LiOH. In the inner nonporous oxide, surface reactions similar to those at the oxide-solution interface are not possible because the oxide crystallite surfaces are not accessible to the corroding solution. Lithium is existing as LiOH, hydrogen bonded to adsorbed water, on the oxide crystallite surface. Corrosion is controlled by oxygen transport through the inner nonporous layers whose thickness is determined by the lithium concentration in the solution.

---

[3] AECL Research, Chalk River, Ontario, Canada.
[4] KAERI, Daejeon, Korea.

Bo-Ching Cheng,[1] Richard M. Kruger,[1] and Ronald B. Adamson[1]

Corrosion Behavior of Irradiated Zircaloy

REFERENCE: Cheng, B.-C., Kruger, R. M., and Adamson, R. B., "**Corrosion Behavior of Irradiated Zircaloy,**" *Zirconium in the Nuclear Industry: Tenth International Symposium, ASTM STP 1245,* A. M. Garde and E. R. Bradley, Eds., American Society for Testing and Materials, Philadelphia, 1994, pp. 400–418.

ABSTRACT: There is ample evidence in the literature of the effects of reactor irradiation on the microstructure and corrosion behavior of zirconium alloys. Specifically, it has been shown that boiling water reactor (BWR) irradiation generally induces nodular corrosion and causes marked changes in precipitate structure and composition. The purpose of this study is to determine the effects of irradiation-induced microstructural changes on post-irradiation corrosion behavior and to gain insight into the operating in-reactor corrosion mechanisms.

Zircaloy-2 and Zircaloy-4 were irradiated in BWRs at a temperature near 561 K. Neutron fluences were at various values between 2 and 14×10^{25} n/m^2 ($E > 1$ MeV). Post-irradiation corrosion tests were conducted at 589, 673, and 793 K using standard techniques. Transmission electron microscopy (STEM) was conducted on unirradiated, as-irradiated, and corrosion-tested materials.

Post-irradiation test results include the following:

1. Nodular corrosion as indicated by 793 K steam testing was completely eliminated. Post-irradiation annealing at 923 K/2 h caused a return of nodules to specimen edges.
2. Uniform corrosion as indicated by 589 K water and 673 K steam tests was markedly increased relative to non-irradiated material.
   (*a*) The corrosion rate of welded Zircaloy-4 increased rapidly at short times and then achieved a lower steady rate for longer times.
   (*b*) The corrosion rate of alpha-annealed Zircaloy-2 and Zircaloy-4 was increased relative to unirradiated material and was linear with time. The corrosion rate was roughly proportional to neutron fluence. Post-irradiation annealing at 923 K/2 h reduced the rate substantially, but not to the unirradiated material value.

STEM studies of the Zircaloy-4 specimens indicate that the observed corrosion behaviors can be correlated to irradiation-induced increases in the matrix solute concentration. The effects of redistribution and re-precipitation of solute during the corrosion tests are also examined. These results are related to the basic mechanisms of both nodular and uniform corrosion in a BWR.

KEY WORDS: zirconium, nodular corrosion, uniform corrosion, solutes, precipitates, radiation effects, welds, microchemistry, zirconium alloys, nuclear materials, nuclear applications

Both Zircaloy-2 and Zircaloy-4 form a uniform protective oxide that thickens with time during residence in light water reactors. In boiling water reactors (BWRs), a localized form of corrosion resulting in the formation of a somewhat less protective oxide, called nodular oxide, has also been commonly seen. There have been instances where a combination of heavy nodular corrosion and copper-bearing crud deposition resulted in localized perforation of fuel

---
[1] Principal engineer, senior engineer, and manager, Fuel Materials Technology, respectively, GE Nuclear Energy, Vallecitos Nuclear Center, Pleasanton, CA 94566. Mr. Cheng is presently with the Electric Power Research Institute, Palo Alto, CA.

rods [1]. The nodular corrosion resistance of Zircaloy-2 and Zircaloy-4 has been found to be highly sensitive to the metallurgical conditions, as partially manifested by the size, composition, and distribution of the second-phase particles or precipitates [2].

The uniform oxide on Zircaloy fuel cladding has low growth rates in BWRs and normally remains as a thin black oxide to high exposures. Large patches of a light gray oxide of higher thicknesses have been reported to grow out of the black uniform oxide at high exposures in some BWRs [3]. After exposures up to 50 GWd/MT, local patches of oxide 30 to 60 μm thick were observed, whereas the predominant oxide thickness was <10 μm thick. Such a transition from a thin black oxide to a thicker, light gray one occurs much earlier, normally within the first or second cycle, in pressurized water reactors (PWRs). The higher growth rate and earlier transition of the uniform oxide in PWRs has been generally attributed to higher cladding surface temperatures. The uniform corrosion and the resulting hydriding are potential limiting factors for high exposure fuels in PWRs. Similar to the nodular corrosion in BWRs, the uniform corrosion resistance of Zircaloy has also been found to be sensitive to the alloy chemistry and the thermomechanical processing parameters.

Earlier mechanistic work [4] has suggested that, with the exception of certain high alpha annealing conditions, increased nodular corrosion resistance of Zircaloy correlates to increasing precipitate number density or decreasing particle size. Detailed studies led to conclusions that the nodular corrosion resistance of Zircaloy can be attributed to retention of a small quantity of solute elements including iron, chromium, and nickel in Zircaloy-2 or iron and chromium in Zircaloy-4. Other studies [5] confirm that elevated solute content in conjunction with small precipitate size does indeed result in high nodular corrosion resistance. The quantity of the solutes retained in the alloy matrix is influenced by the heat treatment and subsequent processing history of the Zircaloy products. In general, rapid quenching and subsequent low temperature annealing during processing favor the formation of smaller precipitates with higher number density and are believed to also favor higher solute concentration retained in the alloy matrix in a quasi-stable condition. Garzarolli et al. [6] have developed a cumulative annealing parameter approach to effectively integrate the annealing temperature and time during post-beta heat-treatment processing for correlation with the precipitate distribution in the alloy and the corrosion performance of Zircaloy.

A mechanism of how a variation in the concentration of solutes retained in the alloy matrix can influence the formation of nodular and uniform oxide on Zircaloy is schematically illustrated in Fig. 18 of Ref 4. It was proposed that a certain small amount of solutes in the alloy matrix is needed to dope the matrix of the growing uniform oxide to maintain stability of the oxide. Doping by the sub-valence solute ions, iron, chromium, and nickel (cationic valence of < +4) would create vacancies in the oxide matrix and allow it to grow by diffusion of oxygen through the oxide. At locations depleted in solute elements, the uniform zirconium oxide is near stoichiometry and cannot grow at the same rate as the adjacent substoichiometric oxide; thus, it is susceptible to breakdown leading to the formation of nodular oxide having a granular structure containing high porosity. Similar ideas have been presented by Taylor [7]. While doping of solutes in the growing oxide is needed to retard the formation of nodular corrosion, it may lead to higher growth rates for the uniform oxide. Thus, based on this hypothesis, an alloy matrix that is resistant to nodular corrosion can be susceptible to higher uniform oxide growth, and vice versa. Evidence of such effects of solute elements can be seen on irradiated Zircaloy components containing welds, where higher nodular corrosion resistance but thicker uniform oxide was found when compared with adjacent base metal, Fig. 16 of Ref 4. More recent experimental work by Garzarolli et al. [8] has demonstrated that a smaller cumulative annealing parameter favoring nodular corrosion resistance may increase the uniform corrosion rate.

The distribution of solute elements and precipitates in the alloy matrix can not only be influenced by thermomechanical processing parameters, but can also be altered by irradiation in light water reactors. Complete dissolution of very fine precipitates in Zircaloy rapidly quenched from the beta or alpha plus beta phase fields can occur at a fast neutron fluence as low as 1.2×10^{25} n/m^2 (>1 MeV), which is equivalent to less than one full-power year in BWRs [9,10]. Large spheroidal precipitates can sustain irradiation damages much longer, but are subjected to amorphization and loss of iron and nickel into the alloy matrix at high fast neutron dosages [11–13]. Such precipitate dissolution and amorphization might be anticipated to occur much more rapidly in PWRs, as the fast neutron flux in PWRs is typically 1.5 to 1.8 times of that in BWRs. However, more data are needed to confirm the effects of irradiation at the higher temperature of the PWRs. Nevertheless, when dissolution of precipitates occurs in reactors, the solutes originally tied up in the precipitates are dispersed into and retained in the alloy matrix. No re-precipitation has been reported in as-irradiated Zircaloy. Thus, Zircaloy matrix containing irradiation-induced defects is capable of retaining excess solutes, iron, chromium, and niobium, in the alloy matrix. Due to this irradiation-induced redistribution of solutes in irradiated Zircaloy, the uniform and nodular corrosion behavior of Zircaloy can be significantly altered by irradiation in light water reactors.

Questions can then be raised regarding whether irradiation-induced solute redistribution contributes to the transition from the black uniform oxide to the light gray oxide, as well as the rapid growth of the uniform oxide at high exposures in PWRs. Understanding these phenomena will shed light on methods for mitigation through metallurgical means.

The objective of this paper is to study the corrosion behavior of Zircaloy irradiated to various fluence levels in BWRs. Both the uniform and nodular corrosion behavior are studied in autoclave tests, and the results compared with the corrosion behavior in reactor. Microstructural changes are evaluated for correlation with the corrosion results.

Experimental Procedure and Materials

The specimens used in this experiment were obtained from standard non-heat-treated BWR components irradiated in two different BWRs. Details are given in Table 1. Channels and water rods are both non-fueled, non-heat-transfer components so their temperatures during irradiation are within 2°C (2 K) of the reactor water temperature of 288°C (561 K). Specimens

TABLE 1—*Material and irradiation histories.*

Material	Reactor	Component	Fluence, n/m^2 ($E > 1$ MeV)	In-Reactor Oxide, μm	Specimen
Zircaloy-4	BWR-A	channel, base metal; reference	15×10^{25} 0	100 0	A1–A4 A5, A6
Zircaloy-4	BWR-A	channel, weld; reference	15×10^{25} 0	25 0	W1–W8 W9, W10
Zircaloy-2	BWR-A	water rod	12×10^{25}	6	7, 8, 14
Zircaloy-2	BWR-B	water rod	2×10^{25} 4×10^{25} 6×10^{25} 0	10 16 21 0	1, 2 3, 4 5, 6, 11, 22, 23 20, 21

were obtained by cutting in the GE Vallecitos Nuclear Center hot laboratory. Channel specimens were rectangular coupons of 10 by 20 by 2 mm dimension. Oxide surface was removed by mechanical milling 0.18 mm from each surface, followed by 0.05 mm removal by etching. "Weld samples" contain a standard channel weld in about one-half of the specimen width. Water rod specimens were obtained by sectioning the 14.7-mm diameter by 0.76-mm wall thickness tubing into 12.5-mm-long segments. Oxide was removed by mechanical abrasion and etching to remove a total of 0.1 mm from both surfaces. Metallographic examination indicated hydride content of less than 100 ppm in all material.

Corrosion tests were conducted in standard refreshed autoclaves using ASTM procedures wherever possible. The tests conducted were 316°C (589 K)/1560 psig (10.8 MPa) saturated water; 400°C (673 K)/1500 psig (10.3 MPa) steam; and a modified GE two-step test of 410°C (683 K)/1750 psig (12.1 MPa) for 4 h followed by 520°C (793 K)/1750 psig (12.1 MPa) steam for 16 h.

Annealing of specimens for corrosion tests and electron microscopic examination was conducted under a vacuum of less than 10^{-5} torr.

Transmission electron microscopy and energy dispersive X-ray analysis (EDX) were conducted using a JEOL 100CX-II scanning transmission electron microscope (STEM). A double-tilt cryogenic specimen holder was used for the examination. The foil specimens were obtained by jet polishing in a 10% perchloric acid and 90% ethanol solution at 20 V and 243°K.

Results

Corrosion Behavior of Irradiated Zircaloy-4 Channels

Specimens prepared from the high-fluence Zircaloy-4 channels were tested in 316°C (589 K) saturated water and 400°C (673 K) 1500 psi steam for up to 160 days. The 316°C test data for the weld and base metal are shown in Fig. 1. The unirradiated reference base metal and weld specimens both had about the same low corrosion rate. The irradiated base metal specimens showed about the same low corrosion rate as the unirradiated specimens. The irradiated weld specimens showed increased corrosion. Two out of the four irradiated weld specimens showed a high linear corrosion rate of 3.0 mg/dm²/day, while the other two had only slight increases in the corrosion rate over the unirradiated reference specimens. The reason for the differences between the two sets of weld metal specimens is not known.

The 400°C data are shown in Fig. 2. The unirradiated base metal showed normal corrosion behavior with a transition to a linear growth rate of 0.25 to 0.30 mg/dm²/day after reaching a weight gain of about 40 mg/dm² or approximately 3 μm in thickness in about 40 days. The linear rate for unirradiated weld metal is slightly higher.[2] The irradiated base metal specimens rapidly achieved a linear rate of 2.1 mg/dm²/day. The three weld specimens all corroded rapidly in the beginning and reached a weight gain of about 200 mg/dm² or 14 μm in thickness in only 10 days. The corrosion rate of the weld specimens then reduced to a slower linear rate of 2.7 mg/dm²/day, which is nearly the same as the irradiated base metal.

Corrosion Behavior of Irradiated Zircaloy-2 Water Rods

The Zircaloy-2 water rod specimens having different levels of neutron fluence were tested in 400°C/1500 psig steam, and the weight gains are shown in Table 2. Figure 3 shows the effect of fast neutron fluence during the initial 160 days for selected specimens. It can be seen

---

[2] These specimens were tested out to 500 days, at which time the weight gain for weld specimens was 259 mg/dm² and for base metal specimens was 197 mg/dm².

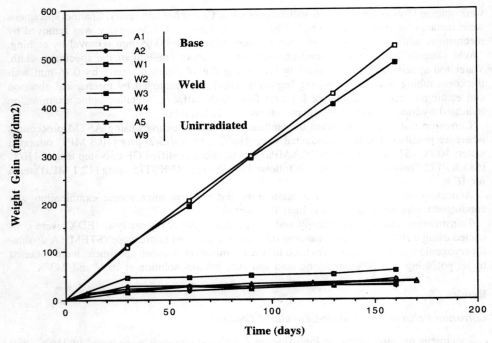

FIG. 1—*Corrosion of Zircaloy-4 in 316°C (589 K) saturated water. BWR channel material irradiated at 299°C (561 K), 15×10^{25} n/m² (E > 1 MeV).*

that the unirradiated reference, Specimen 21, and the low-fluence Specimen 2, tended to grow the initial oxide at slower, nonlinear rates, but eventually transitioned to linear kinetics. The higher fluence Specimens 6 and 8 established linear kinetics at the beginning. The linear growth rate increases with increasing the fast fluence. The Zircaloy-4 channel base metal had nearly the same growth kinetics, Fig. 2, as the comparable Zircaloy-2 water rod specimen at the same fluence level. The Zircaloy-4 channel weld specimen had a substantially higher oxide growth rate than any of the water rods.

The linear oxide growth rate of the irradiated specimens did not remain constant. Figure 4 shows that an increase in the linear kinetics occurred for all irradiated specimens at a weight gain of 120 to 150 mg/dm² or 8 to 10 μm thickness.

The effect of fast fluence on the 400°C corrosion is illustrated in Fig. 5 by comparing the weight gains after 271 days. It can be seen that the weight gain increases monotonically as the fast fluence, to increases to 12×10^{25} n/m².

Figure 4 also shows the corrosion behavior of two specimens that received special test treatments. Specimen 6, as well as Specimens 2 and 4 not shown, was annealed after 271 days of corrosion testing. The annealing induced a rapid rate increase followed by a rate characteristic of lower fluence specimens. Specimen 23 was tested without removing the reactor-formed oxide of about 20 μm in thickness. It can be seen that the specimen initially had a very low corrosion rate, but after 30 days achieved a linear rate similar to its siblings, Specimens 5 and 6, at similar testing times.

Visual inspection of the specimens indicates that the weight gain increases in the 316 and 400°C tests were due to thickening of the uniform oxide. After 300 days, the unirradiated

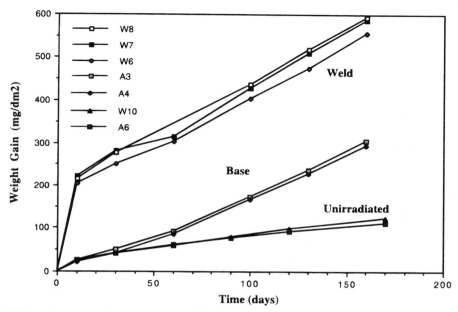

FIG. 2—*Corrosion of Zircaloy-4 in 400°C (673 K) steam. BWR channel material irradiated at 288°C (561 K), 15 × 10²⁵ n/m² (E > 1 MeV).*

specimens are shiny black, while the irradiated specimens range from dark to light gray. In the as-irradiated condition, the specimens are mottled black.

Nodular Corrosion Behavior

Irradiated and unirradiated Zircaloy-4 channel specimens and Zircaloy-2 water rod specimens were tested at 520°C (793K) for 16 h using a two-step procedure [14]. The results are shown in Fig. 6. In the unirradiated condition, weld areas of Zircaloy-4 specimens are nodule-free, whereas the base metal is completely covered with a thick sheet of coalesced nodular oxide. After irradiation to a high fluence, the entire specimen is shiny black, nodule-free. For the Zircaloy-2 water rod specimens, unirradiated material formed a few nodules on the outer tubing surface and on the specimen ends. After irradiation for either one, two, or three cycles, no nodules are observed. After annealing at 650°C (923 K) for 2 h, nodules returned only at the specimen ends.

Microstructure Studies Using Analytical Electron Microscopy

The irradiated Zircaloy-4 channel material used in this study has been extensively examined in earlier studies [10,11,15]. Results from those studies show the following for neutron irradiation at 288°C (561 K):

1. For the normal $Zr(Fe,Cr)_2$ precipitates in the base metal:
    (a) all became amorphous,
    (b) iron (and probably chromium) diffuse from the precipitates into the zirconium matrix,

TABLE 2—*Weight gains of irradiated water rod specimens tested in 400°C/1500 psig (673 K/10.3 MPa) steam, mg/dm^2.*

Days						Specimen								
	1	2	3	4	5	6	7	8	11	14	20	21	22	23
3	17	15	19	17	16	16	19	24	18	21	19	17	300	301
15	26	25	28	28	28	27	30	31	28	31	28	27	304	303
30	34	32	38	35	36	40	40	42	38	41	36	37	311	310
45	40	43	42	47	44	55	63	68	48	60	42	41	...	...
56	...	...	...	...	...	...	...	...	...	...	...	...	...	...
70	67	68	69	68	68	76	85	96	...	...	...	...	333	334
84	...	...	...	...	...	...	...	...	73	70	58	60	...	...
91	...	...	...	...	...	...	...	...	...	...	...	...	...	...
100	72	83	79	103	96	108	117	131	88	93	72	67	360	361
132	97	102	103	134	118	141	156	173	107	112	82	82	369	369
160	111	119	126	163	145	167	193	222	126	127	92	94	...	...
207	149	147	179	224	203	239	264	307	167	173	109	109	...	...
271	198	222[a]	247	312[a]	282	331[a]	367	433	226	244	127	130	...	...
286	212	...	264	...	295	404	...	...	...	...	...	...	...	...
301	216	301	276	394	312	419	...	...	...	...	...	...	...	...
327	244	313	315	429	349	450	...	...	...	...	...	...	...	...
355	272	...	346	448	381	473	...	...	...	...	...	...	...	...
362	283	...	356	460	392	488	...	...	...	...	...	...	...	...

Specimen	History
1	BWR-B, water rod, one cycle, etched
2	BWR-B, water rod, one cycle, etched
3	BWR-B, water rod, two cycle, etched
4	BWR-B, water rod, two cycle, etched
5	BWR-B, water rod, three cycle, etched
6	BWR-B, water rod, three cycle, etched
11	BWR-A, water rod, three cycle, annealed 650C/2 h, etched
7	BWR-A, water rod, seven cycle, etched
8	BWR-A, water rod, seven cycle, etched
14	BWR-A, water rod, seven cycle, annealed 650C/2 h, etched
20	BWR-B, water rod, unirradiated archive
21	BWR-B, water rod, unirradiated archive
22	BWR-B, water rod, three cycle, tested with in-reactor oxide intact
23	BWR-B, water rod, three cycle, tested with in-reactor oxide intact

[a] Specimens annealed in vaccum at 650C (923K) for 2 h.

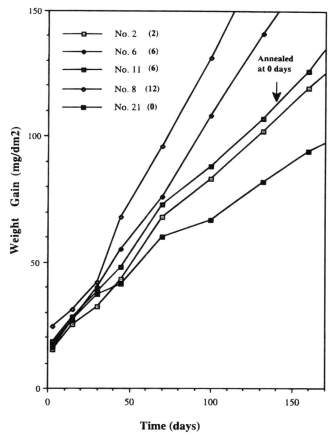

FIG. 3—*Corrosion of Zircaloy-2 in 400°C (673 K) steam. BWR water rod materials irradiated at 288°C (561 K). Fluences are given in parenthesis, 10^{25} n/m^2 (E > 1 MeV).*

(c) both of these effects increase with increasing fluence, being incomplete at 5×10^{25} n/m^2 and complete at 15×10^{25} n/m^2, and
(d) thermal annealing above 560°C (833 K) induces iron (and probably chromium) to diffuse back into the precipitates, and precipitation of iron on grain boundaries.

2. For the normal Zr_4(Fe, Cr) precipitates in the weld metal:
    (a) precipitates disappear (probably dissolve) even at low fluences,
    (b) iron and chromium form no visible precipitates, and
    (c) thermal annealing above 560°C (833 K) induces precipitation of iron on grain boundaries.

The Zircaloy-2 material designated as material from BWR-A has been previously examined by Kruger [16]. This material is similar to the other water rods tested in this program, except it was irradiated in another reactor to a higher fluence. Results from that study show the following for Zircaloy-2 irradiated at 288°C (561 K):

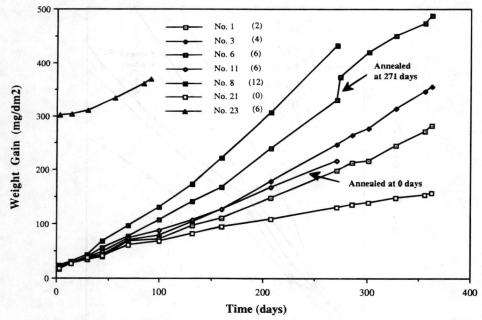

FIG. 4—*Corrosion of Zircaloy-2 in 400°C (673 K) steam. BWR water rod materials irradiated at 288°C (561 K). Fluences are given in parenthesis, 10^{25} n/m^2 (E > 1 MeV).*

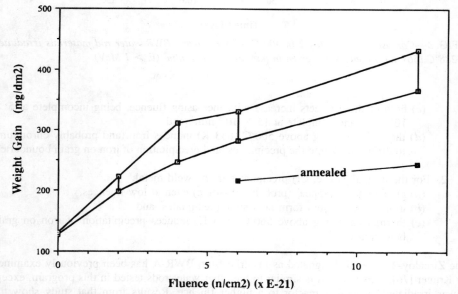

FIG. 5—*Effect of neutron fluence on corrosion of Zircaloy-2 in 400°C (673 K) steam at 271 days.*

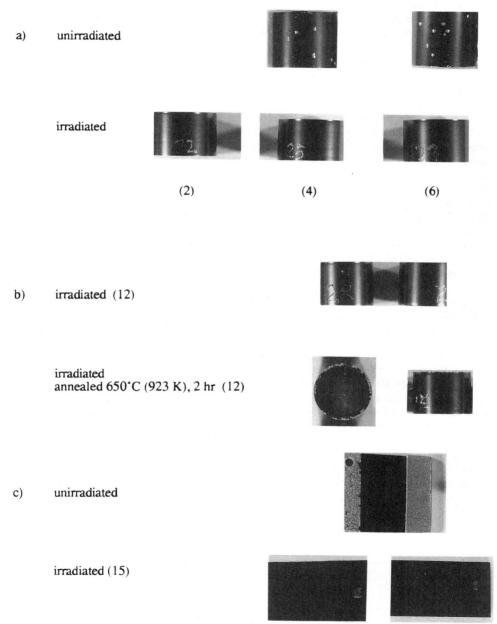

FIG. 6—*Corrosion in 520°C (793 K) steam. Numbers in ( ) are fluence, in 10^{25} n/m² (E > 1 MeV): (a) BWR-B water rods, (b) BWR-A water rods, and (c) BWR-A channel strip with weld in center.*

1. The behavior of the $Zr(Fe,Cr)_2$ precipitates is the same as in Zircaloy-4.
2. $Zr_2(Fe,Ni)$ precipitates do not become amorphous but iron and probably nickel diffuse into the base metal.
3. Thermal annealing above 575°C (848 K) results in precipitation of $Zr_2(Fe,Ni)$ and hexagonal $Zr(Fe,Cr)_2$ at grain boundaries.

In order to interpret the corrosion behavior of the irradiated specimens on the basis of solute distribution, the evolution of precipitates due to exposures to the corrosion test temperatures of 316°C (589 K) and 400°C (673 K) was investigated. Detailed work on both the Zircaloy-2 and Zircaloy-4 is in progress, and the results will be reported separately. The following is a summary of the results relevant to the present corrosion study.

Table 3 summarizes data on weld metal before and after corrosion testing. As previously shown, irradiation eliminates visible precipitates from the structure. Energy-dispersive X-ray analysis of the interlamellar boundaries, sites of precipitates before irradiation, indicates similar low concentrations of iron and chromium, while the matrix regions away from boundaries are slightly enriched in the fast diffusing iron. Thermal annealing at 400°C (673 K) for 10 days results in iron returning to the boundary, and after 250 days increased chromium is also detected there, with the matrix being depleted in both elements. After this annealing, a few zeta-phase precipitates can be observed at boundaries when imaged in dark field; after 250 days many precipitates are easily visible, Fig. 7a. The precipitates were not analyzed for composition or structure. At 316°C (589 K), the matrix stays enriched in iron even after 250 days, and the boundaries are only slightly enriched in iron and chromium relevant to the as-irradiated state, and no precipitates can be observed in either bright or dark field.

Table 4 summarizes data on Zircaloy-4 base metal before and after corrosion testing. The precipitates observed are intragranular, amorphous, and are characterized by the ratio χ_{Fe} = Fe/(Fe + Cr). Irradiation reduces χ_{Fe} from 0.6 to 0.2, as iron diffuses from the precipitates into the matrix. Other features include a high density of "black dot" a-component dislocation loops and long straight c-component dislocations having a density near $5 \times 10^9/cm^2$. Corrosion testing, through the effects of thermally induced annealing, causes a redistribution of iron.

For 316°C (589 K) testing, no effects are observed for 30 days, but after 200 days minor amounts of iron diffuse back to the precipitates. At 400°C (673 K), the process accelerates such that at 10 days substantial iron has diffused from the matrix to the precipitates and by 250 days the content of iron in precipitates doubled. Importantly, at 400°C (673 K), re-precipitation of iron occurs at grain boundaries after only a few days, resulting in a high density of small (<0.05 μm) zirconium-iron zeta-phase and Zr_3Fe precipitates, Fig. 7b. At 510°C (783 K), one day is sufficient to increase χ_{Fe} to 0.45 and causes precipitation of zeta-phase at the grain boundaries. In addition, at this temperature, the black dot damage is eliminated, and c-component dislocations become heavily jogged.

TABLE 3—*STEM analysis of irradiated Zircaloy-4 weld metal.*

Material	Boundary, % by weight		Matrix, % by weight	
	Cr	Fe	Cr	Fe
As-irradiated	0.21	0.28	0.03	0.19
673 K/10 day	0.21	0.63	0.04	0.14
673 K/250 day	0.67	1.91	0.02	0.07
589 K/10 day	0.30	0.39	0.03	0.12

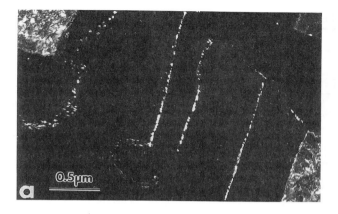

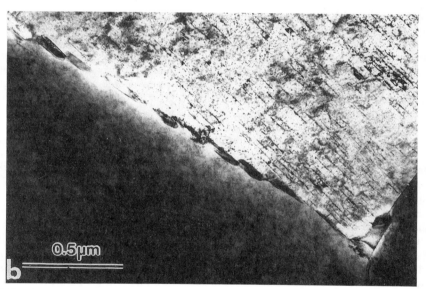

FIG. 7—*Intergranular precipitates in post-irradiation annealed Zircaloy-4:* (a) *weld, following 400°C (673 K) for 250 days and* (b) *base metal, following 400°C (673 K) for 3 days.*

Discussion

Since zirconium is a highly reactive metal, its stability in high temperature, high pressure water relies on the protective nature of the surface oxide, ZrO_2. The structure or morphology and chemistry of a growing ZrO_2, particularly the portion near the metal-to-oxide interface, can be controlled by the structure and chemistry of the underlying zirconium alloy. Earlier work [4] produced evidence that the distribution of solute elements, iron, nickel, and chromium in the alloy matrix of unirradiated Zircaloy can control the corrosion behavior. A Zircaloy with low solute concentration in the matrix is susceptible to nodular corrosion, while one supersaturated with solute elements is resistant to nodular corrosion. Increasing solute concentration in the alloy matrix, however, can increase the growth rate of the uniform oxide. The

TABLE 4—*STEM analysis of irradiated and unirradiated Zircaloy-4 base metal.*

Feature	Unirradiated Reference	Irradiated with Post-Irradiation Annealing						
		589 K/days			673 K/days			793 K/1 day
		None	30	200	3	10	250	
Fe/(Fe+Cr)	0.6	0.21	0.16	0.18 to 0.24	0.23	0.26 to 0.38	0.32 to 0.47	0.45
ppt size, μm	>0.3	0.3	0.3	...	0.24	...	...	0.29
Grain boundary precipitate	none	none	none	none	Zeta	Zeta, Zr$_3$Fe	Zeta, Zr$_3$Fe	Zeta

effects of solute elements in the alloy matrix on the behavior of a growing ZrO$_2$ is attributed to doping of the ZrO$_2$ by subvalent ions of the solutes. The results of the present study show that irradiation in BWRs, and most likely also in PWRs, will destabilize the precipitates in Zircaloy, resulting in an alloy matrix-enriched by the solutes. The degree of dissolution or amorphization of precipitates or both depends largely on the original size and distribution of precipitates as controlled by the fabrication procedure. The evolution of precipitates and matrix microchemistry under irradiation in BWRs is believed to result in two very important features of fuel cladding corrosion: (*a*) preventing nucleation of new nodules after a short length of irradiation; and (*b*) causing the uniform corrosion to grow at higher rates at high burnups. The effects on the uniform corrosion is expected to be the same as in PWRs. The present test results support these points.

The most clear evidence of how irradiation alters the corrosion and microstructure of Zircaloy can be learned from the behavior of the Zircaloy-4 weld specimens in this study. As can be seen in Fig. 1, the uniform oxide on the Zircaloy-4 channel weld metal grows linearly in 316°C (589 K) saturated water and reached a thickness of 34 μm (assuming that 14.7 mg/dm^2 equals 1 μm) after 160 days, as compared to 2 to 4 μm for unirradiated weld and base metal and irradiated base metal. In 400°C (673 K) superheated steam, the weld specimens also show a rapid oxide growth within the first 10 days, and subsequently reach a reduced, but still high, rate equivalent to that of the base metal. The laboratory test results are consistent with the corrosion behavior reported for irradiated Zircaloy-4 spacer welds found in two four-cycle discharged spacers retrieved from two BWRs at 288°C (561 K) [*4*]. The uniform oxides on those welds were found to be 3 to 4 times thicker than those on the adjacent base metal.

The rapid oxide growth on irradiated welds is striking. At 316°C (589 K), which is near the reactor operating temperature, vacancies are relatively immobile and, according to the results shown in Table 3, diffusion of iron and chromium is limited. If precipitates are present, they must be very thin or small or a combination in size as they cannot be observed in either bright or dark field. This is true even though the matrix concentrations are on the order of 300 ppm chromium and 1900 ppm iron (and grain boundary concentrations are even higher), which are well above recognized solubility limits [*17*]. However, conventional solubility does not consider the possible interaction of solute elements and irradiation produced defects peculiar to irradiated material. In addition, hypotheses have been presented [*18,19*] that propose that a nonequilibrium solute supersaturation can be produced by irradiation. This is likely to be the situation in the current BWR irradiations. At 316°C (589 K), re-precipitation of solutes is minor during the entire test, and the corrosion rate remains linear.

At 400°C (673 K), irradiation-produced defects are sluggishly mobile, and iron is shown to diffuse back to the interlamellar boundaries (Fig. 7*a*). The initial very high (>20 mg/dm^2/day) corrosion rate illustrates the powerful effects of solutes on oxidation. The process is very sensitive to iron distribution and form, as only a small decrease in matrix concentration and creation of boundary precipitates only barely visible in TEM dark field are sufficient to reduce the rate by a factor of 10, which, of course, is still high relative to unirradiated material.

The irradiated Zircaloy-4 channel base metal contains discrete amorphous precipitates that have lost about two thirds of their iron to the matrix. At 316°C, even this high iron concentration is not enough to accelerate corrosion. However, at 400°C (673 K), the irradiation-induced solutes in the matrix accelerate a transition in the oxide growth to a linear rate after only less than 10 days as opposed to 40 days observed for unirradiated Zircaloy-4. The oxide thickness of the irradiated Zircaloy-4 base metal is 3.2 times thicker than for the unirradiated specimens after 160 days. Although iron diffuses back to the original precipitates and reprecipitates at the grain boundaries during the entire test at 400°C (673 K), the corrosion rate is not affected. This can be interpreted either that there is enough iron in the matrix to sufficiently dope the growing oxide or that the tiny zeta-phase grain boundary precipitates are redissolved near the advancing ZrO_2 interface allowing sufficient doping of the oxide by iron, along the lines suggested in the literature [5,6,20,21]. These results on Zircaloy-4 indicate that acceleration of uniform corrosion occurs even in the absence of precipitates visible in TEM. It is therefore likely that peculiar characteristics of amorphous precipitates, such as the existence of local stress fields surrounding them, are not needed for altering the corrosion behavior.

The results of the Zircaloy-2 water rods are of interest because the fluence and thermal history are variable. The effects of fluence are quantified in Fig. 5 and Table 5. The monotonic increase in the 400°C (673 K) corrosion weight gain with increasing fluence is consistent with the known effect of fast neutron fluence on the amorphization of precipitates. The fluence at which all precipitates are "fully depleted" of iron and nickel depends on the initial precipitate size. For large precipitates (>0.2 μm), the process is complete only at high fluence, on the order of 10×10^{25} n/m² [11,12]. On the other hand, for small precipitates (<0.1 μm) the process can be complete at less than 1×10^{25} n/m² [22]. For the Zircaloy water rods in this

TABLE 5—*Post-transition linear corrosion rates of Zircaloy-2 tested in 673 K steam.*

Specimen	Fluence, $\times 10^{21}$/cm²	Rate, mg/dm²/day	Time Period
21	0	0.35	>160 days
1	2	0.83	>160 days
3	4	1.09	>160 days
4	4	1.05	<160 days
4	4	1.34	>160 days
4	4	1.05	>286 days, after annealing at 271 days
6	6	1.01	<160 days
6	6	1.48	>160 days
6	6	1.10	>286 days, after annealing at 271 days
11	6	0.81	>160 days after annealing at 0 days
23	6	1.03	56 to 91 days; reactor-formed oxide not removed
8	12	1.87	>130 days
8	12	1.15	<130 days
14	12	1.05	>160 days; annealed at 0 days

study, the precipitates fall in the "large" category, and the amount of iron and chromium depletion in $Zr(Fe,Cr)_2$ and $Zr_2(Fe,Ni)$ precipitates is expected to increase with fluence gradually at a rate similar to the fluence versus weight gain plot of Fig. 5.

The effectiveness of post-irradiation annealing in reducing the uniform oxide growth rate, as shown in Fig. 4, is another evidence that the uniform oxide growth kinetics is influenced by changes in the alloy matrix. High-fluence Specimens 11 and 14 (see Table 2) were annealed at 650°C (923 K) for 2 h before testing at 400°C (673 K) and compared to sibling Specimens 6 and 8. Such annealing eliminates all the "black dot" irradiation damage, reduces the metal hardness to the level of unirradiated material, and redistributes most of the matrix iron back to the original precipitates or to grain boundary precipitates, see the earlier section on results [16]. Figure 5 and Table 5 indicate that after annealing both specimens corrode at rates more typical of lower fluence material. Assuming that most of the irradiation-produced matrix iron and nickel is removed by the anneal (an unverified assumption), we then would conclude that solute iron and nickel are more effective in increasing corrosion rates than grain boundary-precipitated iron and nickel, but that the latter still do accelerate corrosion as they are incorporated into the oxide. Specimen 6 was annealed at 650°C (923 K) after having 271 days in 400°C (673 K) steam. Figure 4 and Table 5 show that after an initial large transient, the rate is lower than before annealing, but higher than sibling Specimen 11, which was annealed at zero time. (Similar behavior was observed in Specimens 2 and 4.) We attribute this behavior to a combination of effects on the existing oxide as well as in the metal. We cannot eliminate possible effects of hydrogen in the metal and oxide after this high weight gain, but that evaluation will have to await further material examination. It is likely, however, that annealing damaged or created porosity in the oxide, which was then reflected in the subsequent corrosion. It is possible that the transient behavior is caused by a high concentration of point defects existing after the anneal, but since it was not observed after annealing Specimen 11 at zero days, it seems unlikely.

With the in-reactor oxide intact on the corrosion test specimen, the oxide growth behavior in 400°C (673 K) steam is initially altered as shown in Fig. 4 for Specimen 23. The pre-existing oxide of 21 μm (300 mg/dm^2) reduced the uniform oxide growth to much lower rates for 30 days before returning it to a linear rate (1.03 μm/dm^2/day) nearly the same as its sibling Specimen 6, which was tested after the in-reactor oxide was removed. The results indicate that after a 30-day incubation time, the properties of the metal substrate, rather than the existing thick oxide, dominate the corrosion process. It appears then that the oxide growth is controlled by the thin advancing oxide near the metal-oxide interface, and the chemistry of the alloy matrix directly influences the chemistry of the advancing oxide.

Previous post-irradiation corrosion testing on neutron-irradiated and proton-irradiated Zircaloy, by Etoh et al. [23] and Kai et al. [24], respectively, has indicated that solute redistribution in the alloy matrix controls nodular corrosion behavior. Our results on Zircaloy-2 and Zircaloy-4 in the 520°C (793 K) two-step tests shown in Fig. 6 support this view. During the test, the microchemistry of the alloy matrix is evolving, such that, at the end of the one day test, solute iron has reprecipitated at the grain boundaries and at the original precipitates. It is probable that early in the test the high solute iron in Zircaloy-4 and iron and nickel in Zircaloy-2 is sufficient to inhibit nodule nucleation, and later in the test the small grain boundary precipitates have an effect. From earlier thermal annealing experiments [4] of a Zircaloy-zirconium couple, the amount of solute required to retard nodular nucleation is shown to be on the order of a few hundred ppm. Since over 2000 ppm of iron or 2500 ppm of iron and nickel are contained in Zircaloy-4 and Zircaloy-2, respectively, it is likely that an irradiated Zircaloy will have enough solutes in the matrix to retard nodular corrosion, even after some of the solutes are annealed out at 650°C (923 K). Thus, after annealing, only the ends of specimens, which are normally the most corrosion susceptible, developed nodular corrosion.

The results of this test program clearly support a very broad-based field experience that nodular corrosion on Zircaloy fuel cladding forms only early in reactor service. The nodular corrosion formation during the early irradiation stage can be influenced by the metallurgical condition and surface finish of the Zircaloy cladding. The apparent lack of new nodule nucleation on fuel rods after the first fuel cycle under normal operation can be attributed to the irradiation-induced precipitate dissolution or amorphization, or both, as found in this test program. A consequence of the irradiation-induced solute redistribution is an increase in the growth rate of the uniform oxide with irradiation time as is clearly seen throughout this test program. This is consistent with the observation of thicker greyish uniform oxide patches on some high burnup fuel rods in BWRs as discussed at the beginning of this paper. It is believed that this solute redistribution phenomenon also controls the uniform oxide growth behavior in PWRs. Due to the higher fast flux levels in PWRs, the transition from thin black oxide to thicker greyish oxide occurs rather early, mostly within the first or second fuel cycle. The irradiation-induced solute redistribution may also have led to accelerated oxide growth with increasing fuel burnup to high exposures in PWRs, as recently reported by Kilp et al. [25]. One final cautious note on the long-term uniform corrosion kinetics of Zircaloy is the potential effect of hydrogen. As hydrogen continues to accumulate with oxide growth, its potential effect on the oxide growth at high oxide thicknesses needs to be separated from that of the irradiation-induced microchemistry changes.

Conclusions

Zircaloy-2 and Zircaloy-4 have been irradiated in BWRs at 288°C (561 K). Post-irradiation corrosion testing and microstructure evaluations have led to the following conclusions:

1. Neutron irradiation through its effects on microchemistry strongly affects corrosion rates in water and steam.
2. Uniform corrosion rates are increased by pre-irradiation.
3. Initial microstructure influences the irradiation effect. Very small precipitates, such as are found in weld metal, disappear rapidly and completely. The resulting effects on the uniform oxide growth rates are very significant. Larger precipitates lose iron and nickel more slowly, so the effects are related to neutron fluence.
4. Matrix solute levels strongly affect corrosion processes. Small iron-rich precipitates induced by thermal annealing have a lesser secondary effect.
5. The dominant effect on corrosion is due to the underlying metal microchemistry. Oxide characteristics, however, can also affect corrosion rate. Changes in the microchemistry caused by thermal annealing induce corrosion rate transients that may not level out at rates characteristic of the metal microchemistry.
6. Nodular corrosion is eliminated in both Zircaloy-2 and Zircaloy-4 by pre-irradiation.
7. Corrosion mechanism hypotheses based on solute doping of the growing oxide are supported by the data. Small precipitates can also contribute to solute doping.

Acknowledgment

We gratefully acknowledge the assistance of R. E. Blood for his performance of the corrosion tests in this work.

References

[1] Marlowe, M. O., Armijo, J. S., Cheng, B., and Adamson, R. B., *Proceedings*, Topical Meeting on Light Water Reactor Fuel Performance, Orlando, FL, American Nuclear Society, 1985, pp. 3–73.

[2] *Zirconium in the Nuclear Industry: Seventh International Symposium, ASTM STP 939*, R. B. Adamson and L. F. P. Van Swam, Eds., American Society for Testing and Materials, Philadelphia, 1987, pp. 243–447.

[3] Baumgartner, J. A., "BWR Fuel Bundle Extended Burnup Program Final Report," DOE/ET/34031-18, UC-78, U.S. Department of Energy, Dec. 1984.

[4] Cheng, B. and Adamson, R. B. in *Zirconium in the Nuclear Industry: Seventh International Symposium, ASTM STP 939*, R. B. Adamson and L. F. P. Van Swam, Eds., American Society for Testing and Materials, Philadelphia, 1987, pp. 387–416.

[5] Kruger, R. M., Adamson, R. B., and Brenner, S. S., *Journal of Nuclear Materials*, Vol. 189, 1992, pp. 193–200.

[6] Garzarolli, F., Stehle, H., Steinberg, E., and Weidinger, H. in *Zirconium in the Nuclear Industry: Seventh International Symposium, ASTM STP 939*, R. B. Adamson and L. F. P. Van Swam, Eds., American Society for Testing and Materials, Philadelphia, 1987, pp. 417–430.

[7] Taylor, D. F., *Journal of Nuclear Materials*, Vol. 184, 1991, pp. 65–77.

[8] Garzarolli, F., Steinberg, E., Trapp-Pritsching, S., and Weidinger, H. G., *Atomwirstschaft*, Nov. 1989, p. 545.

[9] Cheng, B., Adamson, R. B., Bell, W. L., and Proebstle, R. A., *Proceedings*, International Symposium on Environmental Degradation of Materials in Nuclear Power Systems—Water Reactors, Myrtle Beach, SC, National Association of Corrosion Engineers, 1984, p. 274.

[10] Yang, W. J. S. and Adamson, R. B. in *Zirconium in the Nuclear Industry: Eighth International Symposium, ASTM STP 1023*, L. F. P. Van Swam and C. M. Eucken, Eds., American Society for Testing and Materials, Philadelphia, 1989, pp. 451–477.

[11] Yang, W. J. S., Tucker, R. P., Cheng, B., and Adamson, R. B., *Journal of Nuclear Materials*, Vol. 138, 1986, pp. 185–195.

[12] Griffiths, M., Gilbert, R. W., and Carpenter, G. J. C., *Journal of Nuclear Materials*, Vol. 150, 1987, pp. 53–66.

[13] Garzarolli, F., Dewes, P., Maussner, G., and Basso, H. H. in *Zirconium in the Nuclear Industry: Eighth International Symposium, ASTM STP 1023*, L. F. P. Van Swam and C. M. Eucken, Eds., American Society for Testing and Materials, Philadelpha, 1989, pp. 641–657.

[14] Cheng, B., Levin, H. A., Adamson, R. B., Marlowe, M. O., and Monroe, V. L. in *Zirconium in the Nuclear Industry: Seventh International Symposium, ASTM STP 939*, R. B. Adamson and L. F. P. Van Swam, Eds., American Society for Testing and Materials, Philadelphia, 1987, pp. 257–283.

[15] Yang, W. J. S. in *Effects of Radiation on Materials: 14th International Symposium, Vol. 1, ASTM STP 1046*, N. H. Packan, R. E. Stoller, and A. S. Kumar, Eds., American Society for Testing and Materials, Philadelphia, 1989, pp. 442–456.

[16] Kruger, R. M., "Precipitate Stability in Zircaloy-2," EPRI NP-6845-D, Electric Power Research Institute, Palo Alto, CA, May 1990.

[17] Charquet, D., Hahn, R., Ortlieb, E., Gros, J. P., and Wadier, J. F., *Zirconium in the Nuclear Industry: Eighth International Symposium, ASTM STP 1023*, L. F. P. Van Swam and C. M. Eucken, Eds., American Society for Testing and Materials, Philadelphia, 1989, p. 405.

[18] Nelson, R. S., Hudson, J. A., and Mazey, D. J., *Journal of Nuclear Materials*, Vol. 44, 1972, pp. 318–330.

[19] Frost, H. J. and Russell, K. C., *Acta Metallurgica*, Vol. 30, 1982, pp. 953–960.

[20] Garzarolli, F., Seidel, H., Tricot, R., and Gros, J. P. in *Zirconium in the Nuclear Industry: Ninth International Symposium, ASTM STP 1132*, C. M. Eucken and A. M. Garde, Eds., American Society for Testing and Materials, Philadelphia, 1991, p. 395.

[21] Kubo, T. and Uno, M. in *Zirconium in the Nuclear Industry: Ninth International Symposium, ASTM STP 1132*, C. M. Eucken and A. M. Garde, Eds., American Society for Testing and Materials, Philadelphia, 1991, pp. 476–498.

[22] Kruger, R. M. and Adamson, R. B., "Precipitate Behavior in Zirconium-based Alloys in BWRs," *Journal of Nuclear Materials*, Vol. 205, 1993, pp. 242–250.

[23] Etoh, Y., Kikuchi, K., Yasuda, T., Koizumi, S., and Oishi, M., *Proceedings*, International Topical Meeting on LWR Fuel Performance, Avignon, France, American Nuclear Society and European Nuclear Society, 1991, p. 691.

[24] Kai, J. J., Tsai, C. H., Shiao, J. J., Hsieh, W. F., Tu, C. S., Lee, Y. S., Lin, L. F., and Huang, K. Y., *Proceedings*, Fifth International Symposium on Environmental Degradation of Materials in Nuclear Power Systems—Water Reactors, American Nuclear Society, Monterey, CA, 1991, pp. 190–198.

[25] Kilp, G. P., Balfour, M. G., Stanutz, R. N., McAfee, K. R., Miller, R. S., Boman, L. H., Wolfhope, N. P., Ozer, O., and Yang, R. L., *Proceedings*, International Topical Meeting of LWR Fuel Performance, Avignon, France, American Nuclear Society and European Nuclear Society, 1991, p. 730.

DISCUSSION

Brian Cox[1] (written discussion)—If you take your post-transition linear oxidation rates for irradiated Zircaloy-4 and assume that fuel cladding in a PWR will oxidize at this rate as a function of burnup, this would give you an accelerating oxidation curve. How does such a curve compare with the accelerating oxidation curves usually seen for high burnup cladding?

B.-C. Cheng et al. (authors' closure)—We have presented the data in sufficient detail to produce such curves and would hope that those interested would do so. A general trend of increasing uniform corrosion rate at high burnups is indicated from our data. However, one must remember that the details of microstructure evolution may be different in PWRs and BWRs. For instance, the effect of irradiation temperature is significant.

G. P. Sabol[2] (written discussion)—If iron is going back into the precipitate during the corrosion tests, should not the corrosion rate decrease with increasing autoclave exposure time? Your data indicate otherwise. Please comment.

B.-C. Cheng et al. (authors' closure)—Some iron does go back into the original precipitates during corrosion testing at 400 or 520°C. In addition, new very small precipitates also form. In the text, we discuss the possibility that there are small precipitates that act in a way similar to solutes.

N. Ramasubramanian[3] (written discussion)—(1) The microchemistry of the oxide grown on irradiated Zircaloy material will be different if iron and chromium dissolved in the alloy matrix is also in solid solution in the oxide. Can the microchemistry of the oxide be equally important as the microchemistry of the underlying metal in affecting the corrosion?

(2) If iron, chromium, and niobium are dissolved out of the precipitation by irradiation, then nodule nucleation sites cannot be unambiguously identified with a precipitate or a matrix location that is free of precipitate. A point for clarification.

B.-C. Cheng et al. (authors' closure)—(1) Yes, we agree. Our point is that the solute in the metal is readily available to be taken into the oxide forming at the metal/oxide interface.

(2) That is correct. It does imply, however, that precipitates themselves are not required to prevent nodular corrosion. In irradiated weld material, the original precipitates are entirely dissolved and nodule initiation is entirely suppressed.

A. M Garde[4] (written discussion)—Did you investigate the effect of hydrides in the irradiated materials on its subsequent autoclave corrosion?

B.-C. Cheng et al. (authors' closure)—We did not investigate this directly. The hydride level in all of our as-irradiated materials was fairly low, less than about 100 ppm. We did do analysis for hydrogen in our irradiated Zircaloy-2 materials after 273 days in 400°C steam. The amount of hydrogen absorbed increased uniformly with increasing weight gain. The hydrogen pickup fraction, calculated in the standard way, was quite high and decreased with increasing weight gain. Pickup fraction for the highest fluence Material 8, which also had the highest weight gain, was 35%. For the lowest fluence Material 1, it was 53%. However, for unirradiated Material 20, it was 71%, indicating that the higher pickup fraction was not necessarily related to irradiation effects on the precipitates.

Y. S. Kim[5] (written discussion)—Your results show that the redistribution of iron into or out of the matrix is somehow related to the corrosion resistance. Please comment on the real cause for change in the corrosion of Zircaloy with the redistribution of iron?

---

[1] University of Toronto, Toronto, Ontario, Canada.
[2] Westinghouse NMD, Pittsburgh, PA.
[3] Ontario Hydro, Toronto, Ontario, Canada.
[4] ABB CE Nuclear Fuel, Windsor, CT.
[5] Korea Atomic Energy Research Institute, Daejom, Korea.

B.-C. Cheng et al. (authors' closure)—We do not claim to fully understand the mechanism, however, our thoughts are outlined in the discussion.

R. Holt[6] (written discussion)—Did you make any observations of the dislocation structure and its evolution with fast neutron fluence and, if so, was there any correlation between the rate of precipitate dissolution (for example, effect of precipitate size) with fluence and the appearance of c-component dislocations?

B.-C. Cheng et al. (authors' closure)—STEM work on high fluence Zircaloy-4 and Zircaloy-2 specimens is reported in the text. After high fluence, the dissolution process is well advanced and c-component dislocations are present in large numbers. We do not yet have data on the evolution of the microstructures as a function of fluence for the Zircaloy-2 materials in this study; that work is in progress. However, in an earlier work [22], we have shown that small precipitates lose almost all of their iron at very low fluence, less than 1×10^{25} n/m², but c-component dislocations do not appear until the fluence is considerably higher.

A. Motta[7] (written discussion)—(1) Have you any information on the state of the other alloying elements dissolved from the precipitates (chromium, niobium)? Do they also reprecipitate elsewhere in the matrix?

(2) Have you determined whether corrosion is enhanced or inhibited in samples where precipitates experience amorphization as compared to samples where the precipitates get dissolved without amorphizing?

B.-C. Cheng et al. (authors' closure)—(1) We have looked for but have not found any evidence of re-precipitation of any of the alloying elements in as-irradiated material. Post-irradiation annealing of material irradiated to 1.2×10^{26} n/m² ($E > 1$ Mev) results in intergranular precipitation of iron, nickel-rich precipitates at 575°C and of chromium, iron-rich precipitates at 625 to 750°C.

(2) We have no experience with material irradiated at temperatures in which amorphization does not occur. We have, however, published other work [5] on irradiated materials that indicates that small changes in solute content can have large effects on corrosion. Precipitates in that study remain crystalline.

---

[6] Chalk River Laboratories, Chalk River, Ontario, Canada.
[7] Pennsylvania State University, Department of Nuclear Engineering, University Park, PA.

Brett J. Herb,[1] Jack M. McCarthy,[2] Chun T. Wang,[1] and Heinz Ruhmann[3]

Correlation of Transmission Electron Microscopy (TEM) Microstructure Analysis and Texture with Nodular Corrosion Behavior for Zircaloy-2

REFERENCE: Herb, B. J., McCarthy, J. M., Wang, C. T., and Ruhmann, H., "**Correlation of Transmission Electron Microscopy (TEM) Microstructure Analysis and Texture with Nodular Corrosion Behavior for Zircaloy-2,**" *Zirconium in the Nuclear Industry: Tenth International Symposium, ASTM STP 1245,* A. M. Garde and E. R. Bradley, Eds., American Society for Testing and Materials, Philadelphia, 1994, pp. 419–436.

ABSTRACT: Zircaloy-2 controlled chemistry tubes were fabricated using different tubeshell annealing temperatures and tube reduction schedules. The results of this set of materials was previously presented in *ASTM STP 1132*. This paper presents new results from texture and transmission electron microscopy (TEM) microstructure investigations on the same set of materials.

Samples with various processing histories were examined by TEM microscopy to characterize the intermetallic precipitate distributions and matrix solute concentration. Matrix concentration profiles across a 2-μm length were obtained in scanning transmission electron microscopy (STEM) mode using an electron beam spot of approximately 0.07 μm on high and low corrosion resistant material. The influence of texture on corrosion was also examined by fabricating specimens with radial textures (f_r) values from approximately 0.3 to 0.6.

The results show that the tube reduction schedule has a very significant influence on nodular corrosion behavior. The accumulated reduction factor was established to define the influence of deformation during tube reducing on texture. High f_r on the final cladding suppressed the nodular corrosion behavior regardless of the precipitate size. This leads to conflicting evidence on the role of precipitate size and corrosion behavior. Micro-zones of solute segregation in the matrix were also detected above the solubility limits of α-zirconium on some final cladding specimens.

KEY WORDS: zirconium, zirconium alloys, nuclear materials, nuclear applications, nodular corrosion, texture (materials), deformation (materials), transmission electron microscopy, intermetallic precipitates

This paper is the continuation of the previous work [1] in which the effect of texture, annealing temperature, and process schedule on the nodular corrosion behavior of Zircaloy-2 cladding was studied. The previous paper concluded that only a poor correlation exists between the average particle sizes and nodular corrosion behavior. In addition, it concluded

---

[1] Process engineer, retired, respectively, Teledyne Wah Chang Albany, Albany, OR 97321.
[2] Senior research engineer, Oregon Graduate Institute of Science and Technology, Beaverton, OR 97006.
[3] Siemens AG, Power Generation Group (KWU) D-8520 Erlangen, Germany.

that the matrix chemistry is closely related to nodular corrosion. When a large area of the matrix became depleted in alloy elements, the corrosion resistance deteriorated.

The method of analysis used in the previous work had limitations since the particle size was determined by optical microscopy and the matrix composition was analyzed with an electron microprobe. Thus, further analysis on the tubing material was performed by transmission electron microscopy (TEM) and scanning transmission electron microscopy (STEM) techniques. The lateral resolution and sensitivity of the TEM/STEM allow for a more accurate characterization of the specimens to determine indicators or combinations of indicators for good nodular corrosion resistance. This paper reports the results of this examination.

Experimental Procedure

Processing History

The focus of this paper is on the barrier tubeshell from controlled chemistry Zircaloy-2 Heats 228293 and 227356 presented in the previous work [1]. Typical ingot chemical composition for both heats was Zr, 1.3% Sn, 0.18% Fe, 0.09% Cr, 0.07% Ni and 0.13% O. The hollow Zircaloy-2 billets were beta quenched, machined, and assembled with a zirconium liner. After heating to 650°C, the billets were extruded to 63.5 mm outside diameter (OD) by 10.9 mm wall (W) by length (L) and sawed into three sections to study the effect of annealing temperature on nodular corrosion behavior. The tubeshell sections identified as 1, 2, and 3 were then vacuum annealed for 2 h at 663, 732, and 800°C, respectively. The annealed tubeshell sections were sawed again into three sections and identified as A, B, and C to study the effect of reduction schedule on nodular corrosion. Thus, the tubeshell annealed at 663°C and tube reduced according to Schedule A was identified as A1; that annealed at 800°C and tube reduced according to Schedule C was identified as C3; and so forth. Schedule B was further divided into two different tube reducing schedules for the final pass that were identified as B.a or B.b. After each tube reducing pass, a 620°C for 2-h vacuum anneal was performed. The processing schedule is outlined in Fig. 1 and Table 1.

Autoclave Tests

Corrosion specimens were prepared as 12.7- to 38-mm-long rings after the zirconium liner was removed from the inside diameter (ID). The specimens were then pickled to achieve a surface removal of about 0.07 mm before testing.

High-pressure steam tests at 520°C, 10.3 MPa, and 24 h were conducted according to a modified procedure based on ASTM Practice for Aqueous Corrosion Testing of Samples in Zirconium and Zirconium Alloys (G 2). A 7.6-L static autoclave was used with a 1 MΩ-cm minimum resistance water. This test has been reported to rank in-pile corrosion resistance with a good correlation [2]. Several samples were also tested by a second laboratory to verify the initial weight gain results.

Precipitate Analysis Method

Tubing specimens were reduced to 125-μm thickness by machining and pickling in preparation for the punching of standard 3-mm diameter TEM disks. Care was taken so that each TEM disk represented the same mid-wall location. The disks were then electropolished in a 10% by volume perchloric acid in methanol solution at -40°C in a Fishione twin jet electropolisher. This produced TEM foils with large electron transparent regions that were examined in a Hitachi H-800 TEM/STEM analytical electron microscope with a 200-kV accelerating voltage or a Philips EM400T TEM with 100-kV accelerating voltage.

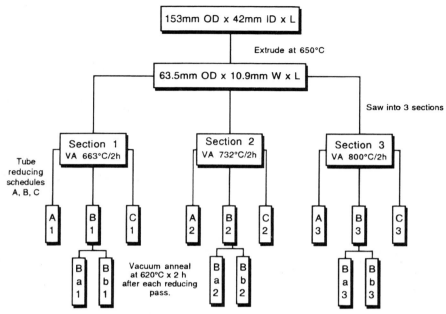

FIG. 1—*Process flow diagram for Zircaloy-2 cladding.*

Precipitate distributions were determined by different analysis techniques on the two electron microscopes. The more powerful 200-kV TEM analyzed thick foil sections at low magnifications to produce stereo pair images for the determination of foil thickness and precipitate diameter. Typical analyzed foil sections were 0.3 μm thick and imaged at a magnification of ×5000 to ×10 000. In contrast, the 100-kV TEM analyzed thin foil sections at high magnification. Images of 0.05 to 0.1 μm thick foil sections were typically obtained at a magnification of ×42 000. Once the foil thickness was determined, calculations were performed to produce results for the interparticle distance and number density. Approximately 200 precipitates per specimen were measured to enhance the reliability of the statistics.

Matrix Microanalysis Method

X-ray microanalysis was performed on the matrix using the Hitachi H-800 TEM/STEM with a beryllium window energy dispersive (EDS) X-ray detector and a multichannel analyzer. Matrix solute profiles were generated for 2-μm lengths of precipitate-free matrix. Approximately 0.07 μm converged beam spots were focused on the matrix at 0.2-μm intervals and held stationary for the 100-live time EDS X-ray spectrum acquisition. Matrix profiles were analyzed at the edge of the foil in regions of the same foil thickness to minimize beam spreading and X-ray absorption inaccuracies. Profiles were typically performed in regions of foil with a thickness of less than 0.05 μm at the edges of thickness fringes to reduce variation in the volumes of analysis. X-ray spectra were later processed to determine compositions using a standardless metallurgical thin film analysis program.

TABLE 1—*Reduction schedule for Zircaloy-2 cladding.*[a]

Schedule	OD, mm	Wall, mm	Q_ϵ[b]	%AR	f_r	ΣRF_i
A	63.5	10.9	...	...	0.32	...
	44.5	7.9	0.9	49.6	0.35	44.6
	31.8	5.9	0.8	47.2	0.31	82.4
	19.1	4.1	0.7	59.8	0.34	124.3
B	63.5	10.9	...	...	0.32	...
	44.5	7.9	0.9	49.6	0.35	44.6
	31.8	3.2	3.7	68.3	0.47	297.3
	19.1	2.1	0.8	61.0	0.45	346.1
B.a[c]	12.7	1.7	0.5	47.3	0.38	369.9
B.b[c]	12.7	0.8	2.7	73.3	0.59	544.0
C	63.5	10.9	...	...	0.32	...
	38.1	5.1	1.6	70.6	0.38	113.0
	19.1	1.3	2.2	86.2	0.54	302.6
	12.7	0.5	2.5	73.6	0.63	486.6

[a] All schedules after tube reducing were annealed 620°C/2 h.

[b] Natural strain ratio = $Q_\epsilon = \dfrac{\ln(t/T)}{\ln(d/D)}$

where

T = wall thickness of mother tube,
t = wall thickness of reduced tube,
D = mean diameter of mother tube, and
d = mean diameter of reduced tube.

[c] The last pass was reduced to two different wall thicknesses.

Results

520°C Steam Autoclave Tests

The weight gains and nodular ratings obtained from the previous work are shown in Table 2. The results are an average of at least three corrosion coupons. Samples from A1, B1, and C1 generally showed low weight gains with no nodules appearing on the OD surface. Samples from A2, B2, and C2 showed higher weight gains and some OD nodules, while samples from A3, B3, C3 had the highest weight gains and worst nodule ratings. Most specimens typically had more ID nodules than OD nodules, and the weight gains were lower at final size than at the intermediate stages.

Annealing Parameter and Reduction Schedule

Work by Steinberg [3] established the accumulated annealing parameter, ΣA_i, to describe the effect of heating temperatures and times on corrosion results in high pressure steam. The annealing parameter is derived from the relationship

$$\Sigma A_i = \Sigma t_i \exp(-Q/RT_i) \qquad (1)$$

where t_i is the time increment spent at temperature T_i and $Q/R = 40\,000$ K. In this study, however, two distinct types of corrosion performances were observed when the weight gains of final cladding specimens were plotted against ΣA_i in Fig. 2. The high weight gains correspond

TABLE 2—Modular corrosion data for Zircaloy-2 cladding, 520°C, 10.3 MPa, 24-h autoclave test.

Schedule	Wall, mm	ΣA_i, $\times 10^{-18}$	Average Weight Gain, mg/dm²	Nodular Rating[a] OD	Nodular Rating[a] ID
As-extruded	10.9	...	89	1	1
A1	10.9	0.6	61	1	1
	7.9	0.6	160	1	2
	5.9	0.7	204	1	3
	4.1	0.8	900	1	4
A2	10.9	10.4	60	1	1
	7.9	10.5	202	1	3
	5.9	10.5	435	3	4
	4.1	10.6	1258	4	4
A3	10.9	129	72	1	1
	7.9	129	3170	4	4
	5.9	129	1713	4	4
	4.1	129	2197	4	4
B1	10.9	0.6	61	1	1
	7.9	0.6	256	1	4
	3.2	0.7	145	1	3
	2.1	0.8	248	1	3
B1.a	1.7	0.8	149	1	...
B1.b	0.8	0.8	95	2	...
B2	10.9	10.4	60	1	1
	7.9	10.5	814	2	4
	3.2	10.5	454	3	4
	2.1	10.6	451	3	4
B2.a	1.7	10.7	235	3	...
B2.b	0.8	10.7	94	3	...
B3	10.9	129	72	1	1
	7.9	129	2060	4	4
	3.2	129	1240	4	4
	2.1	129	1358	4	4
B3.a	1.7	129	1390	2	...
B3.b	0.8	129	149	2	...
C1	10.9	0.6	61	1	1
	5.1	0.6	100	1	2
	1.3	0.7	183	1	4
	0.5	0.8	68	1	3
C2	10.9	10.4	60	1	1
	5.1	10.5	316	2	3
	1.3	10.5	367	1	4
	0.5	10.6	68	1	3
C3	10.9	129	72	1	1
	5.1	129	1822	4	4
	1.3	129	1056	4	4
	0.5	129	84	2	4

[a] Nodule rating: (1) = none, (2) = few, (3) = many, (4) = nodule coalescence, and (5) = full surface coverage (see Ref 12).

to Schedules A and B.a with low reduction schedule during tubing reducing, while the low weight gains correspond to Schedules B.b and C with a high reduction schedule. The low weight gain group shows very little dependence on the accumulated annealing parameter.

Other investigators have reported the relationship between the natural strain ratio (Q_ϵ) and area reduction (%AR) on radial texture (f_r) [4–6]. Generally, as the deformation increases due to diameter, wall thickness, or area reductions then a corresponding change in f_r also occurs.

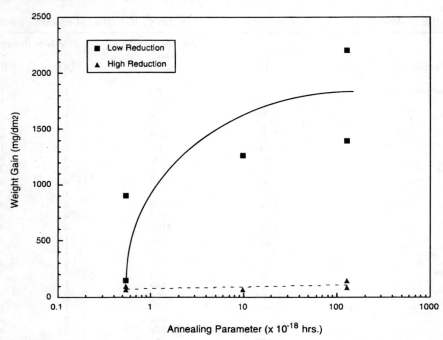

FIG. 2—*Effect of annealing parameter on the nodular corrosion resistance of Zircaloy-2 final cladding.*

Usually a cladding with a low Q_ϵ reduction schedule will have low f_r, while a cladding with a high Q_ϵ reduction schedule will have a high f_r. Figure 3 shows the correlation between radial texture and deformation in the form of a new unitless term call the accumulated reduction factor, ΣRF_i. The formula for the reduction factor is defined as

$$\Sigma RF_i = \Sigma(Q_{\epsilon i} \times \%AR_i) \qquad (2)$$

where i is the number of tube-reducing steps. Thus, a reduction factor of 45 is the result of one tube-reducing step from 63.5 mm × 10.9 mm to 44.5 mm × 7.9 mm while a reduction factor of 544 is the accumulation of four tube-reducing steps from 63.5 mm × 10.9 mm to the 12.7 mm × 0.8 mm final dimension. This is a logical relationship since f_r of the final cladding is the product of three or four tube-reducing steps. Linear regression analysis of the data from Schedules A, B.6, and C show the following relationship

$$f_r = 5.8 \times 10^{-4}\, \Sigma RF_i + 0.30 \qquad (R^2 = 0.88) \qquad (3)$$

The ΣRF_i term provides a better correlation with f_r than Q_ϵ, $\Sigma Q_{\epsilon i}$, %AR, or $\Sigma\%AR_i$.

The positive influence of high f_r on nodular corrosion behavior is clearly shown for the 732 and 800°C annealed material in Fig. 4. Plotted in the figure are the weight gain values and corresponding radial texture results of material throughout the tube-reducing sequence. Radial texture values at the far left of Fig. 4 reflect the as-extruded and annealed tubeshell while

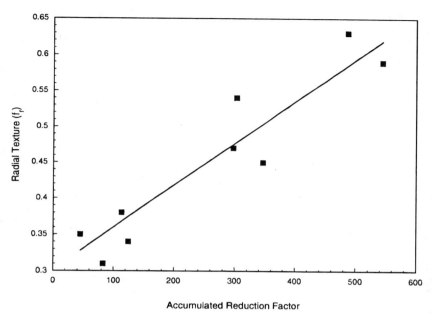

FIG. 3—*Effect of deformation during tube reducing on texture.*

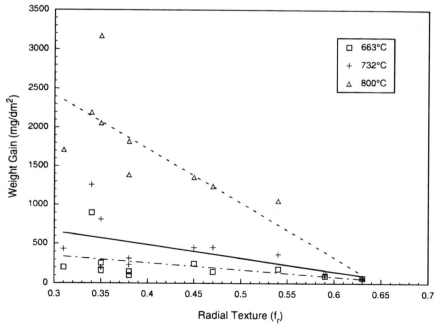

FIG. 4—*Effect of texture on the nodular corrosion resistance of Zircaloy-2 during tube reducing.*

increasing texture values toward the right of the x-axis represent the intermediate and final cladding. A high reduction schedule that results in high f_r can substantially improve the corrosion resistance of 800°C anneal material. A moderate improvement was also shown for the 732°C annealed material while only a minor improvement resulted for the 663°C annealed material. The 520°C weight gains converged for material from the 663, 732, and 800°C annealed lots provided that $f_r = 0.59$ to 0.65.

Precipitate Distribution

Table 3 lists the results of the TEM analysis of the intermetallic precipitate distributions using both the high and low magnification TEM techniques on the final cladding material. Measurements on the same foil specimens by the two different techniques produced equivalent results for mean precipitate diameter. The low magnification method measured a mean diameter of 0.16 µm on Specimen A2 and 0.06 µm on Specimen C1, while the high magnification method measured 0.17 µm and 0.07 µm, respectively. The number density results were also similar between the two techniques. The results show that there is little bias in combining data from the low- and high-magnification techniques.

The growth of the intermetallic precipitates due to the higher annealing temperatures and the resulting increase in accumulated annealing parameter, ΣA_i, is clearly evident. Typical bright-field TEM micrographs of final cladding show the precipitate growth for the 663, 732, and 800°C annealed material in Figs. 5a through c. Figure 6 shows the difference in mean precipitate diameter resulting from the three annealing temperatures with ΣA_i parameters of 0.8×10^{-18}, 11×10^{-18}, and 130×10^{-18}. The data is in agreement with work from Garzarolli et al. [7] on laboratory annealed specimens and production annealed specimens.

The amount of deformation during tube reducing is also influencing the correlation between the nodular corrosion behavior and the mean precipitate diameter. The results in Fig. 7 show that material with a large mean precipitate diameter can perform satisfactorily in the 520°C autoclave test if the tube is reduced with a high reduction schedule. However, if the tube is reduced with a low reduction schedule, then a large mean precipitate diameter is detrimental to the nodular corrosion resistance.

The related precipitate distribution parameters like the interparticle distance, number density, and maximum diameter are also listed in Table 3. Correlations similar to Fig. 7 can be made with the 520°C weight gain values and these attributes.

Matrix Composition

The results indicate that occasional micro-zones of solute enrichment can occur in the typically solute-depleted matrix. Figures 8 and 9 show matrix concentration profiles along a 2-µm distance on low and high weight gain final cladding specimens identified as B.b3 and H226411 from Ref *1*. In addition, Fig. 10 shows a profile from a low weight gain specimen identified as C1. Material from both B.b3 and H226411 were annealed at 800°C after extrusion, while material from C1 was annealed at 663°C. All the material was then tube reduced according to different reduction schedules with 620°C anneals after the passes to final cladding.

The specimen from H226411 had very little nickel (Ni) and chromium (Cr) and no iron (Fe) detected in the matrix while specimens from B.b3 and C1 had a slightly higher matrix concentration. Except for the peaks in the profiles of B.b3 and C1, the concentrations were typically Fe $\leq$ 0.02%, Cr $\leq$ 0.02%, and Ni $\leq$ 0.02% by weight that is assumed to be approximately equivalent to the solubility limit of α-zirconium. The peaks displayed in the profiles are of interest since they exceed the solubility limit and suggest that micro-segregation is occurring. Charquet et al. measured the solubility of iron, chromium, and iron plus chromium

TABLE 3—*TEM precipitate distributions of final cladding.*

Specimen	$\Sigma A_i \times 10^{-18}$	Mean Diameter, μm	Maximum Diameter, μm	Interparticle Spacing, μm[c]	Number Density, number/μm³	Number Counted
A1[a]	0.8	0.09	0.34	0.35	3.8	273
A2	10.7	0.16	0.46	0.44	1.9	252
A2[b]	10.7	0.17	0.90	0.31	5.9	452
A3	129	0.27	1.10	0.68	0.7	226
B.a1	0.8	0.09	0.36	0.30	6.3	315
B.b1	0.8	0.07	0.29	0.14	61	262
B.a3	129	0.22	0.84	0.65	0.6	254
B.b3	129	0.24	0.92	0.47	1.6	186
C1	0.8	0.06	0.30	0.19	25	274
C1[b]	0.8	0.07	0.60	0.19	24	530
C2	10.6	0.16	0.52	0.44	2.0	125
C3	129	0.25	0.76	0.69	0.5	156

[a] Schedule 1: ×1 = 663°C extrusion anneal where x = Reduction Schedule A, B, or C.
Schedule 2: ×2 = 732°C extrusion anneal where x = Reduction Schedule A, B, or C.
Schedule 3: ×3 = 800°C extrusion anneal where x = Reduction Schedule A, B, or C.
[b] Analyzed by high magnification technique.
[c] Interparticle spacing calculated from Ref 8.

FIG. 5—*Typical TEM micrographs of Zircaloy-2 final cladding: (a) 663°C annealed, (b) 732°C annealed, and (c) 800°C annealed.*

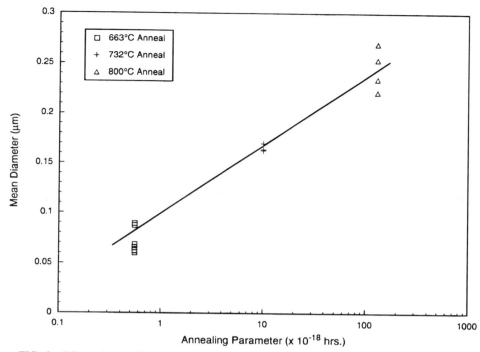

FIG. 6—*Effect of annealing parameter on the mean precipitate diameter of Zircaloy-2 final cladding.*

in a Zr-1.4Sn-0.1O alloy and found a maximum solubility of 0.012, 0.020, and 0.015% by weight, respectively [9]. Given the variability of the measuring method, the solubility limit was considered exceeded only when the concentration of iron, chromium, or nickel was ≥0.04% by weight. Figure 8, therefore, has five of the ten profile spots showing segregation effects while Fig. 10 has segregation at two of the ten spots. Figure 9 shows no segregation across the profile. The tin (Sn) content shows some variability across the profile but approaches nominal bulk ingot composition.

Additional analysis was performed to determine if secondary fluorescence of neighboring precipitates could be causing the peaks shown in the profiles. An approximately 0.07-μm beam spot was stepped toward a precipitate until elevated iron, chromium, and nickel was detected. The results indicate that secondary fluorescence does not occur until approximately 0.1 μm from the precipitate. Since no precipitates were typically observed closer than 5 μm from the profile spots, neighboring precipitates are probably not influencing the collected spectra. A 5-μm interparticle spacing is possible because instead of a three-dimensional space, the thin edge of the foil is essentially a two-dimensional plane. Although unsubstantiated, the solute element peaks could be the result of segregation at lattice defects like dislocations or loops. Dislocations may be visible at other tilt angles even though the region appeared defect free in the image. The segregation could be diffusion zones around precipitates just above or below the foil surface that were removed during electropolishing or submicroscopic clusters of solute atoms typically called Guinier-Preston zones (G.P. zones) that are not seen in the TEM image as coherent precipitates.

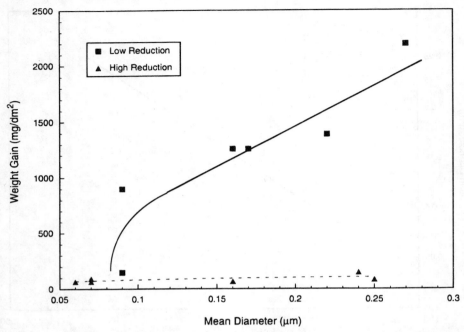

FIG. 7—*Effect of mean precipitate diameter on the nodular corrosion resistance of Zircaloy-2 final cladding.*

It must be pointed out that due to limitations in the X-ray collection process, the accuracy of the quantitative data can vary significantly for analysis on concentrations of less than 1% by weight. The matrix concentrations of iron, chromium, and nickel are near the minimum detection limits of the instrument. The statistical error for the points in the profiles are based on the number of X-rays collected for a given peak. The error ranged from approximately 12 to 30% of the measured compositions.

Discussion

Influence of Annealing Parameter and Reduction Schedule

The present work shows that nodular corrosion behavior in the final cladding can be acceptable even though the extruded tubeshell was annealed beyond typical ΣA_i ranges for optimized boiling water reactor (BWR) corrosion. Schedules B.b3 and C3 showed poor nodular corrosion behavior during the intermediate rocking steps. However, with each tube-reducing step and increase in accumulated reduction factor, the corrosion weight gains decreased. The tendency toward higher ID versus OD nodule ratings may be influenced by a slower ID cooling rate during beta quenching or a slightly different texture than OD material.

ΣRF_i was established to show the influence of deformation steps during tube reducing on radial texture. The reduction factor formula assumes that each tube-reduction step influences the texture. This assumptions follows the conclusion proposed by Cook et al. [10] in which the correlation between contractile strain ratio (CSR) and Q-ratio improved as more tube-

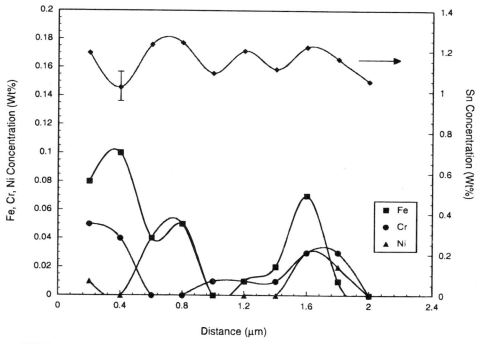

FIG. 8—*STEM microanalysis profile of matrix from B.b3 with low corrosion weight gain.*

reducing passes were included into the Q-ratio value. In fact, the intermediate reductions had a larger influence on texture than the final reduction in that work. Therefore, the development of texture can be described by the summation of deformation during the tube-reduction process.

The corrosion behavior of material with the different textures is remarkable. Charquet et al. [*11*] reported similar results on sheet that was rolled either parallel or perpendicular to the rolling direction. Material rolled in the parallel direction resulted in the (0002) poles inclined approximately 30°C to the transverse direction. The weight gain after a 500°C, 24-h autoclave test was 1300 mg/dm^2. The cross-rolled material, however, had the (0002) poles aligned perpendicular to the sheet and the weight gain decreased to 700 mg/dm^2. This finding is equivalent to high f_r for cladding tubes that also exhibit lower weight gains. The mechanistic effect of texture on nodular corrosion has not been reported in great detail.

The results indicate that when $\Sigma A_i \leq 10^{-18}$ then the nodular corrosion behavior of Zircaloy-2 is acceptable in 520°C steam autoclave tests. However, when $\Sigma A_i \geq 10^{-17}$, the tube-reducing schedule must be optimized for high reductions. Therefore, the tubeshell should be extruded to a larger wall thickness so the subsequent tube-reducing steps can instill the critical reduction factor and high f_r on the final cladding.

Influence of Precipitate Distribution

The precipitate distribution data provides conflicting evidence on understanding their influence on the nodular corrosion behavior of Zircaloy-2. The 663°C annealed material with its small precipitate size generally had excellent corrosion behavior regardless of the tube-reduction schedule. The 732 and 800°C annealed material had larger precipitates that were detrimental

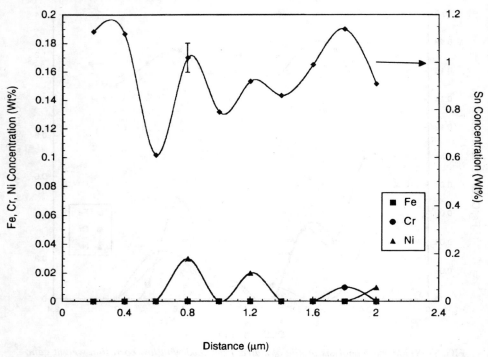

FIG. 9—*STEM microanalysis profile of matrix from H226411 with high corrosion weight gain.*

to the corrosion resistance when tube reduced with a low reduction schedule. This confirms the relationships reported between the precipitate size and corrosion behavior in the literature [13–15]. The Electrochemical-Mechanical-Passivity (EMP) model proposed by Weidinger et al. [16] describes corrosion behavior based on the influence of intermetallic precipitates. Rest potential measurements determined that the precipitates are more noble than the surrounding matrix, so they galvanically couple with the matrix to act as cathodes that reduce water to hydrogen while the matrix is the anode where zirconium is oxidized. The passive surface typical of nodular corrosion has very localized anodic sites that may develop into nodules. An increase in precipitate density will result in a more noble potential since the number of available cathodic sites increases the oxidation rate. An increase in the matrix area, however, results in more active potentials. It was also proposed that a microstructure with a low interparticle distance reduces the susceptibility of nodular corrosion by redistributing the surface oxide cracks at the nodular precursor sites.

Contrary to the proposed theories and models, the 732 and 800°C annealed material with large precipitates had excellent corrosion behavior provided it was tube reduced with a high reduction schedule. This agrees with the results presented by Bradley et al. [17] in which the primary influence on the 520°C autoclave weight gains was not the precipitate size. Texture appears to have larger influence than the precipitate size on the nodular corrosion behavior of cladding for material in this paper.

Matrix Composition

The previous paper [1] concluded that the depletion of solute elements in the matrix caused poor nodular corrosion behavior. This finding was based on electron microprobe data that

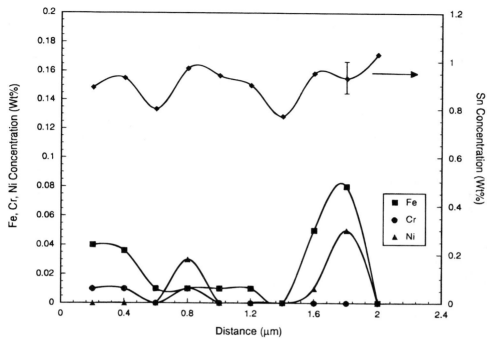

FIG. 10—*STEM microanalysis profile of matrix from C1 with low corrosion weight gain.*

showed fluctuations in the matrix composition. The microprobe, however, is best used as an instrument to rank the uniformity in precipitate density since the zone of beam excitation can be 1 to 2 μm while the interparticle distance is usually less than 1 μm. The STEM microanalysis profiles in precipitate-free matrix regions in the current work allows the matrix concentration to be estimated without the influence of precipitates.

The results of the present work indicate that a solute-depleted matrix exists with microzones of solute enrichment in the final cladding specimens. If the peaks along the profile length are excluded, then the results generally agree with the reported solubility limits. The STEM data also agrees with atom probe microanalysis on Zircaloy-2. Wadman and Andrén detected iron, chromium, and nickel at approximately 0.019%, 0.007%, and 0.018% by weight in the matrix of Zircaloy-2 [18]. Given a defect-free crystal lattice, supersaturation should not exist in the matrix due to the rapid diffusion rates of iron, chromium, and nickel at 663, 732, or 800°C extrusion annealing temperatures used in this study [19].

Following processing, zirconium alloys do not have an ideal lattice since intermetallic precipitates form by heterogeneous nucleation. Dislocations, loops, and grain boundaries can act as nucleation sites and also enhance diffusion of solutes during annealing. New precipitates have been shown to form at the dislocation networks during annealing [20]. In addition, the presence of submicroscopic tin clusters have been reported [21]. The possibility exists for the clustering of submicroscopic complexes (Zr-Sn-Fe, etc.) that are similar to G. P. zones but below the resolution of the TEM. Further analysis of matrix profiles may also determine if minor increases in matrix solute concentration will significantly increase the nodular corrosion resistance as proposed by Kruger et al. [22]. The paper claimed that increases in the matrix solute concentration improve the nodular corrosion behavior for specimens of the same precipitate diameter. This observation may be generalized as the reason for the different corrosion behavior between B.b3 and H226411 since both have a similar precipitate diameter but B.b3 is more corrosion resistant. However, given the significant influence of deformation on nodular corrosion behavior, this relationship is inconclusive.

Conclusions

1. High radial texture can substantially reduce the 520°C weight gains of Zircaloy-2 cladding that was annealed beyond the typical ΣA_i ranges for optimized BWR corrosion.
2. Deformation can be correlated to radial texture by the accumulated reduction factor that is the summation of deformation steps during tube reducing.
3. The role of precipitate diameter on corrosion behavior is unclear. Final cladding with both small and large mean precipitate diameters have acceptable nodular corrosion behavior.
4. Micro-zones of solute segregation exist in the matrix above the solubility limits of α-zirconium.

Acknowledgments

The authors are grateful for the funding of this work by the Oregon Metals Initiative and Teledyne, Inc. We also wish to thank both the Oregon Graduate Institute of Science and Technology (OGI) and Siemens KWU for performing the TEM measurements, and Sandvik Special Metals for testing some of the 520°C steam autoclave samples.

References

[1] Wang, C. T., Eucken, C. M., and Graham, R. A., "Investigation of Nodular Corrosion Mechanism for Zircaloy Products," *Zirconium in the Nuclear Industry: Ninth International Symposium, ASTM STP 1123*, C. M. Eucken and A. M. Garde, Eds., American Society for Testing and Materials, Philadelphia, 1992, pp. 319–345.
[2] Rudling, P. and Machiels, A. J., "Corrosion Performance Ranking of Zircaloy-2 for BWR Applications," *Zirconium in the Nuclear Industry: Eighth International Symposium, ASTM STP 1023*, L. F. P. Van Swam and C. M. Eucken, Eds., American Society for Testing and Materials, Philadelphia, 1989, pp. 315–333.
[3] Steinberg, E., Weidinger, H. G., and Schaa, A., "Analytical Approaches and Experimental Verification to Describe the Influence of Cold Work and Heat Treatment on the Mechanical Properties of Zircaloy Cladding Tubes," *Zirconium in the Nuclear Industry; Sixth International Symposium, ASTM STP 824*, D. G. Franklin and R. B. Adamson, Eds., American Society for Testing and Materials, Philadelphia, 1984, pp. 106–122.
[4] Tenckhoff, E., "A Review of Texture and Texture Formation in Zircaloy Tubing," *Zirconium in the Nuclear Industry: Fifth Conference, ASTM STP 754*, D. Franklin, Ed., American Society for Testing and Materials, Philadelphia, 1982, pp. 5–25.
[5] Konishi, T., Honji, M., Kojima, T., and Abe, H., "Effect of Cold Reduction on Anisotropy of Zircaloy Tubing," *Zirconium in the Nuclear Industry: Seventh International Symposium, ASTM STP 939*, R. B. Adamson and L. F. P. Van Swam, Eds., American Society for Testing and Materials, Philadelphia, 1987, pp. 653–662.
[6] Nagai, N., Kakuma, T., and Fujita, K., "Texture Control of Zircaloy Tubing During Tube Reduction," *Zirconium in the Nuclear Industry: Fifth Conference, ASTM STP 754*, D. Franklin, Ed., American Society for Testing and Materials, Philadelphia, 1982, pp. 26–38.
[7] Garzarolli, F., Steinberg, E., and Weidinger, G., "Microstructure and Corrosion Studies for Optimized PWR and BWR Zircaloy Cladding," *Zirconium in the Nuclear Industry: Eighth International Symposium, ASTM STP 1023*, L. F. P. Van Swam and C. M. Eucken, Eds., American Society for Testing and Materials, Philadelphia, 1989, pp. 202–212.
[8] DeHoff, R. T. and Rhines, F. N., *Quantitative Microscopy*, McGraw Hill, New York, 1968.
[9] Charquet, D., Hahn, R., Ortlieb, E., Gros, J.-P., and Wadier, J.-F., "Solubility Limits and Formation of Intermetallic Precipitates in ZrSnFeCr Alloys," *Zirconium in the Nuclear Industry: Eighth International Symposium, ASTM STP 1023*, L. F. P. Van Swam and C. M. Eucken, Eds., American Society for Testing and Materials, Philadelphia, 1988, pp. 405–422.

[10] Cook, C. S., Sabol, G. P., Sekera, K. R., and Randall, S. N., "Texture Control in Zircaloy Tubing Through Processing," *Zirconium in the Nuclear Industry: Ninth International Symposium, ASTM STP 1132*, C. M. Eucken and A. M. Garde, Eds., American Society for Testing and Materials, Philadelphia, 1991, pp. 80–95.

[11] Charquet, D., Tricot, R., and Wadier, J.-F., "Heterogeneous Scale Growth During Steam Corrosion of Zircaloy-4 and 500°C," *Zirconium in the Nuclear Industry: Eighth International Symposium, ASTM STP 1023*, L. F. P. Van Swam and C. M. Eucken, Eds., American Society for Testing and Materials, Philadelphia, 1989, pp. 374–391.

[12] Schemel, J. H., "Parametric Study of an Autoclave Test for Nodular Corrosion," *Zirconium in the Nuclear Industry: Seventh International Symposium, ASTM STP 939*, R. B. Adamson and L. F. P. Van Swam, Eds., American Society for Testing and Materials, Philadelphia, 1987, pp. 243–256.

[13] Andersson, T. and Vesterlund, G., "Beta-Quenching of Zircaloy Cladding Tubes in Intermediate or Final Size—Methods to Improve Corrosion and Mechanical Properties," *Zirconium in the Nuclear Industry: Fifth Conference, ASTM STP 754*, D. Franklin, Ed., American Society for Testing and Materials, Philadelphia, 1982, pp. 75–95.

[14] Maussner, G., Steinberg, E., and Tenckhoff, E., "Nucleation and Growth of Intermetallic Precipitates in Zircaloy-2 and Zircaloy-4 and Correlation to Nodular Corrosion Behavior," *Zirconium in the Nuclear Industry: Seventh International Symposium, ASTM STP 939*, R. B. Adamson and L. F. P. Van Swam, Eds., American Society for Testing and Materials, Philadelphia, 1987, pp. 307–320.

[15] Weidinger, H. G., Garzarolli, F., Eucken, C. M., and Baroch, E. F., "Effect of Chemistry on Elevated Temperature Nodular Corrosion," *Zirconium in the Nuclear Industry: Seventh International Symposium, ASTM STP 939*, R. B. Adamson, and L. F. P. Van Swam, Eds., American Society for Testing and Materials, Philadelphia, 1987, pp. 364–386.

[16] Weidinger, H. G., Ruhmann, H., Cheliotis, G., Maguire, M., and Yau, T.-L., "Corrosion-Electrochemical Properties of Zirconium Intermetallics," *Zirconium in the Nuclear Industry: Ninth International Symposium, ASTM STP 1132*, C. M. Eucken and A. M. Garde, Eds., American Society for Testing and Materials, Philadelphia, 1991, pp. 499–535.

[17] Bradley, E. R., Schemel, J. H., and Nyström, A-L., "Influence of Alloy Composition and Processing on the Nodular Corrosion Resistance of Zircaloy-2," *Zirconium in the Nuclear Industry: Ninth International Symposium, ASTM STP 1132*, C. M. Eucken and A. M. Garde, Eds., American Society for Testing and Materials, Philadelphia, 1991, pp. 304–318.

[18] Wadman, B. and Andrén, H.-O., "Microanalysis of the Matrix and the Oxide-Metal Interface of Uniformly Corroded Zircaloy," *Zirconium in the Nuclear Industry: Ninth International Symposium, ASTM STP 1132*, C. M. Eucken and A. M. Garde, Eds., American Society for Testing and Materials, Philadelphia, 1991, pp. 461–475.

[19] Hood, G. M. and Schultz, R. J., "Diffusion of 3D Transitional Elements in α-Zr and Zirconium Alloys," *Zirconium in the Nuclear Industry: Eighth International Symposium, ASTM STP 1023*, L. F. P. Van Swam and C. M. Eucken, Eds., American Society for Testing and Materials, Philadelphia, 1989, pp. 435–450.

[20] Yang, W. J. S. and Adamson, R. B., "Beta-Quenched Zircaloy-4: Effects of Thermal Aging and Neutron Irradiation," *Zirconium in the Nuclear Industry: Eighth International Symposium, ASTM STP 1023*, L. F. P. Van Swam and C. M. Eucken, Eds., American Society for Testing and Materials, Philadelphia, 1989, pp. 451–477.

[21] Hood, G. M. in *Zirconium in the Nuclear Industry: Eighth International Symposium, ASTM STP 1023*, L. F. P. Van Swam and C. M. Eucken Eds., American Society for Testing and Materials, Philadelphia, 1989, p. 749.

[22] Kruger, R. M., Adamson, R. B., and Brenner, S. S., "Effects of Microchemistry and Precipitate Size on Nodular Corrosion Resistance of Zircaloy-2," *Journal of Nuclear Materials*, Vol. 189, 1992, pp. 193–200.

DISCUSSION

H. G. Weidinger[1] (written discussion)—The work you presented is a very interesting example how one can override the bad nodular corrosion when large intermetallic precipitates are present by high radial textures. This offers the opportunity to use this combination of large precipitates with high radial texture to make Zircaloy that shows good nodular corrosion as well as good uniform corrosion. Did you perform tests to investigate the uniform corrosion also?

B. J. Herb et al. (authors' closure)—No uniform corrosion tests were performed during this experiment, however, a similar correlation is expected between radial texture and uniform corrosion. Graham et al.[2] showed a relationship between the orientation of the (0002) poles and both 360°C water and 420°C steam autoclave tests on Zircaloy-2 strip material. The highest weight gains were typically from Lot B, which had a texture of $f_N = 0.38$. For Lots A and C, the basal poles were orientated normal to the rolling direction ($f_N = 0.73$) and the uniform corrosion weight gains were consistently lower.

A. T. Motta[3] (written discussion)—Were there any microstructural features, such as dislocation loops, thicker regions of the foil, that would account for the segregation of iron and chromium that you observed? Have you eliminated the possibility that these might be small coherent precipitates that happened to be out of contrast?

J. M. McCarthy (authors' closure)—Care was taken to avoid regions with dislocations or other defects as seen in the TEM bright-field images. The specimens were also tilted to see if coherent precipitates and other low-contrast or no-contrast defects could be made visible. Due to tilting limitations imposed by the microscope and the foil, all possible orientations could not be achieved and visibility conditions for some defects might not have been satisfied. The measurement technique described in the Experimental Procedure section of the paper is important since thicker regions in the foil may suppress the iron, chromium, or nickel concentrations due to absorption. Conversely, thinner regions will increase the solute content. In the STEM examinations, we saw no evidence of dislocations, small coherent precipitates, diffusion zones, or G. P. zones but they may explain the observed segregation effects.

E. Tenckhoff[4] (written discussion)—With respect to the influence of texture on the nodular corrosion behavior: (1) with the change of the deformation parameters for the tube-reduction processes, surely you influenced the texture, especially the preferred orientation of the basal planes; (2) superimposed with this also are other changes in microstructure such as, grain size, dislocation density, and degree of recrystallization. Did you also look at these superimposed parameters and their possible effect on the results you have shown?

B. J. Herb et al. (authors' closure)—Grain size, dislocation density, and degree of recrystallization were not specifically examined. However, dislocation density and degree of recrystallization are not expected to influence the nodular corrosion behavior on the final cladding since the material was recrystallized at 620°C for 2 h. The microstructure of typical TEM foils was 100% recrystallized with few dislocations. Grain size could be a superimposed variable but the difference between lots is assumed to be small.

---

[1] The S. M. Stoller Corporation, Erlangen, Germany.
[2] Graham, R. A., Tosdale, J. P., and Finden, P. T., "Influence of Chemical Composition and Manufacturing Variables on Autoclave Corrosion of the Zircaloys," *Zirconium in the Nuclear Industry: Eighth International Symposium, ASTM STP 1023*, L. F. P. Van Swam and C. M. Eucken, Eds., American Society for Testing and Materials, Philadelphia, 1989, pp. 334–345.
[3] Pennsylvania State University, Department of Nuclear Engineering, University Park, PA 16802.
[4] Siemens, D-8520 Erlangen, Germany.

Takeshi Isobe,[1] Yutaka Matsuo,[1] and Yoshiharu Mae[1]

Micro-Characterization of Corrosion Resistant Zirconium-Based Alloys

REFERENCE: Isobe, T., Matsuo, Y., and Mae, Y., "**Micro-Characterization of Corrosion Resistant Zirconium-Based Alloys,**" *Zirconium in the Nuclear Industry: Tenth International Symposium, ASTM STP 1245*, A. M. Garde and E. R. Bradley, Eds., American Society for Testing and Materials, Philadelphia, 1994, pp. 437–449.

ABSTRACT: On the basis of experimental results using button ingots, three zirconium, tin, niobium, iron, and chromium (Zr-Sn-Nb-Fe-Cr) alloy compositions (VAZ series) with superior corrosion resistance and equivalent mechanical properties to those of conventional Zircaloy-4 were determined in a previous work. The alloys of these compositions were vacuum arc melted in 5-kg ingots and were fabricated into sheets with various annealing parameters. They were subjected to long-term corrosion tests in water at 633 K and tension tests at room and elevated (658 K) temperatures. Results showed that the corrosion weight gains and the hydrogen uptakes of VAZ alloys were about 50% and 20 to 60% of that of Zircaloy-4, respectively, and that the tensile strengths were equivalent to those of Zircaloy-4. Transmission electron microscopy/ energy dispersive X-ray (TEM/EDX) analysis showed that niobium precipitated in VAZ alloys containing only 0.2% by weight niobium.

In order to clarify the effects of niobium at low concentration on the corrosion property, zirconium-tin-niobium (Zr-Sn-Nb) alloys (iron/chromium precipitate-free) were prepared from a zirconium X-bar and subjected to corrosion test in 633 K water. Results showed that the corrosion resistance remarkably increased with increasing niobium content from zero up to 300 ppm and was approaching the level of VAZ alloys. It is thought that a very small amount of niobium, almost dissolved in the matrix, leads to the superior corrosion resistance of VAZ alloys in pure water.

KEY WORDS: zirconium, zirconium alloys, nuclear materials, nuclear applications, niobium, uniform corrosion, corrosion resistance, tensile properties, intermetallic precipitates, accumulated annealing parameter, zirconium X-bar

Experimental Procedures

Materials

The materials used in this study were melted into ~5 kg ingots. Double meltings, consisting of plasma arc and vacuum arc meltings, were conducted. As raw materials, Zr-xNb (x = 0.2 and 1.0% by weight) mother alloys, Zircaloy-4 sheets, tin sheets (99.99% purity) and iron (99.9%), chromium (99.9%), molybdenum (99.9%), and vanadium (99.9%) flakes were used. Table 1 summarizes the chemical analysis of the ingots. Three compositions are designated VAZ-01, VAZ-02, and VAZ-03 alloy, respectively. The ingots were converted to about 0.8-mm-thick sheets according to a standard zirconium-alloy processing sequence, shown in

---

[1] Senior engineer, manager, and manager, respectively, Central Research Institute, Mitsubishi Materials Corporation, 1-297 Kitabukuro-cho, Omiya, Saitama, Japan 330.

TABLE 1—*Ingot chemical analysis.*

Element	Alloy Composition, % by weight		
	VAZ-01	VAZ-02	VAZ-03
Sn	0.99	0.55	0.73
Fe	0.17	0.18	0.20
Cr	0.089	0.089	0.36
Nb	0.21	0.21	0.21
Mo	...	0.090	...
V	...	0.094	...
O	0.14	0.14	0.15

Fig. 1, consisting of beta-forging at 1289 K, beta-quenching, alpha-hot rolling at 873 K, cold rolling in three steps with intermediate alpha annealing in the range of 853 K to 1023 K for 2.5 h, and then four kinds of intermediate alpha-annealing precessing were done for each alloy (see Fig. 1). They were finally annealed in either stress-relieved (743 K for 2.5 h) or fully recrystallized (853 K for 2.5 h) condition. Sheets of VAZ-03 alloy intermediately annealed at lower temperatures (853 and 903 K) were not manufactured because of the limited workability.

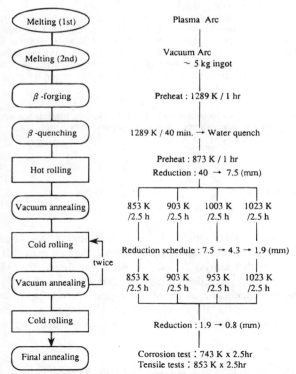

FIG. 1—*Processing sequence of sheet materials investigated.*

Moreover, standard tin (1.5Sn) Zircaloy-4 tubes manufactured by our company with about 1.5E-17 (h) of accumulated annealing parameter (ΣAi) [($Q/R = 40\,000$ K)] were used as a reference material on the mechanical and corrosion properties.

Experiments

Tension Test

Tension tests were conducted on the fully recrystallized materials, perpendicular to the rolling direction, at room and elevated (658 K) temperatures using an Instron-type testing machine, in accordance with ASTM Test Methods of Tension Testing of Metallic Materials (E 8-93).

Corrosion Test

Long-term corrosion tests were performed on the stress relieved materials in water at 633 K for 240 days at a pressure of 19 MPa in the static autoclaves, in accordance with the ASTM Practice for Aqueous Corrosion Testing of Samples of Zirconium and Zirconium Alloys (G 2-88).

Analytical Electron Microscopy

Transmission electron microscopy (TEM) and energy dispersive X-ray analysis (EDX) were conducted on the fully recrystallized materials at 200 kV accelerated voltage in a H-800/H-8010 (Hitachi, Ltd.) analytical electron microscope (AEM). Thin foils were prepared by chemical milling in an aqueous solution of 5% hydrogen fluoride/45% nitric acid and electropolishing in a solution of 10% perchloric acid/90% methanol at about 15 V and 233 K.

Results

Tensile Properties

Figures 2a, b, and c show the tensile properties at room temperature of VAZ-01, VAZ-02, and VAZ-03 alloys, respectively, as a function of intermediate annealing temperature. The 0.2% yield strength of fully recrystallized Zircaloy-4 tube (Zr-4/RX) is also shown as a reference. The ultimate tensile strength (UTS) and 0.2% yield strengths (YS) of VAZ alloys decreased with increasing intermediate annealing temperature, but the 0.2% yield strengths were equal to or higher than that of Zr-4/RX in the whole range studied in this work. In addition, there were no clear differences in tensile properties among the three VAZ alloys at all annealing conditions.

Figure 3 shows the tensile properties at elevated temperature (658 K) as a function of intermediate annealing temperature. VAZ-02 and VAZ-03 were tested only for the specimens intermediately annealed at 953 K. The strengths tended to decrease with increasing intermediate annealing temperature, as in the room temperature test. The 0.2% yield strengths of VAZ alloys intermediately annealed at temperatures lower than 953 K were almost equal to or higher than that of Zr-4/RX.

Corrosion and Hydrogen Absorption Properties

The corrosion weight gains as a function of exposure time for VAZ alloys intermediately annealed at 953 K [$\Sigma Ai = 1.5E-17$ (h), ($Q/R = 40\,000$ K)] are shown in Fig. 4, compared with that of stress-relieved Zircaloy-4 tube (Zr-4/SR). It shows that the corrosion resistance

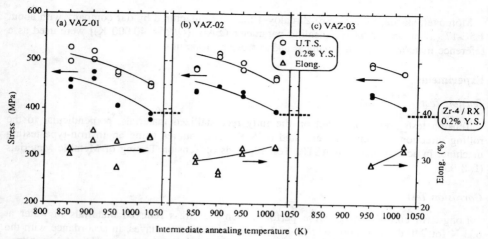

FIG. 2—*UTS, 0.2% YS, and elongation at room temperature of (a) VAZ-01, (b) VAZ-02, and (c) VAZ-03 alloys as a function of intermediate annealing temperature.*

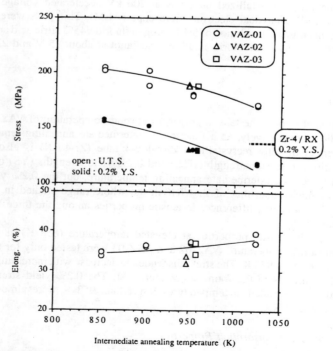

FIG. 3—*UTS, 0.2% YS, and elongation at elevated (658 K) temperature as a function of intermediate annealing temperature.*

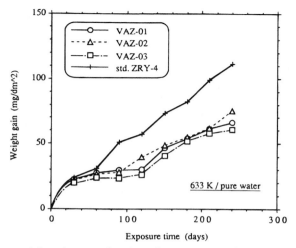

FIG. 4—*Corrosion weight gains as a function of exposure time for VAZ alloys (intermediately annealed at 953 K) and standard Zircaloy-4.*

of three VAZ alloys is significantly superior to that of Zr-4/SR and that the weight gains for 240 days were about 50% of that of Zr-4/SR.

Figure 5 shows the corrosion weight gains for 240 days as a function of the accumulated annealing parameter. The corrosion resistance of VAZ alloys decreases to some extent with increasing annealing parameter, but is significantly superior to that of Zr-4/SR in the whole range studied in this work.

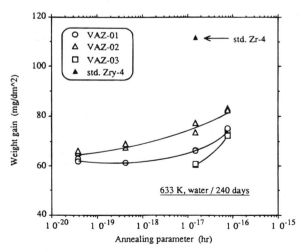

FIG. 5—*Corrosion weight gains as a function of accumulated annealing parameter at 633 K for 240 days for VAZ alloys.*

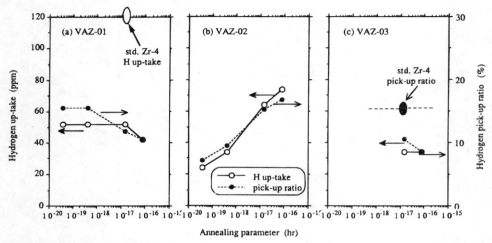

FIG. 6—*Hydrogen contents and pickup ratios as a function of accumulated annealing parameter after the corrosion testing (633 K, water/240 days): (a) VAZ-01, (b) VAZ-02, and (c) VAZ-03.*

Corrosion specimens after 240 days of testing were subjected to the hydrogen analysis. Hydrogen uptakes during the corrosion test and pickup ratios of VAZ alloys as a function of the annealing parameter are shown in Figs. 6a, b, and c, respectively, compared with those of Zr-4/SR. Hydrogen uptakes and pickup ratios of VAZ-01 and VAZ-03 alloys were not significantly affected by the annealing parameter, but those of VAZ-02 alloy increased with increasing annealing parameter. In general, hydrogen uptakes of three VAZ alloys were significantly smaller than that of Zr-4/SR (about 120 ppm).

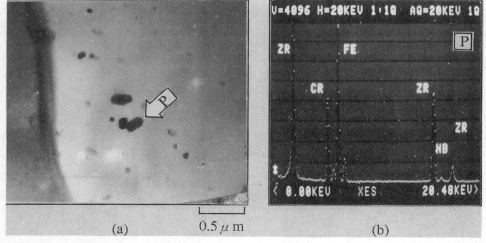

FIG. 7—(a) *TEM micrograph and* (b) *EDX spectrum of the precipitate (P) in VAZ-01 alloy intermediately annealed at 953 K [$\Sigma Ai = 1.5E\text{-}17(h)$].*

Microstructure Observations

TEM/EDX analysis was conducted on VAZ alloys with higher annealing parameters [$\Sigma Ai = 1.5E-17$ (h)]. In addition, the VAZ-01 alloy with the lowest parameter [$\Sigma Ai = 3.8E-20$ (h)] was examined to study the effect of the annealing parameter on the microstructure. Figures 7 to 10 show TEM micrographs and EDX spectra of the precipitates in VAZ alloys.

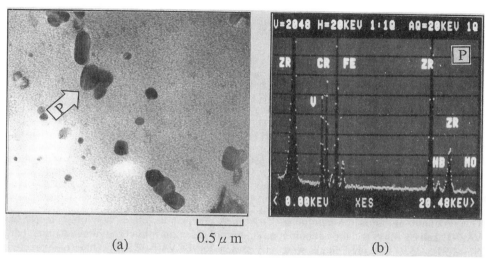

FIG. 8—(a) *TEM micrograph and (b) EDX spectrum of the precipitate (P) in VAZ-02 alloy intermediately annealed at 953 K [$\Sigma Ai = 1.5E-17(h)$].*

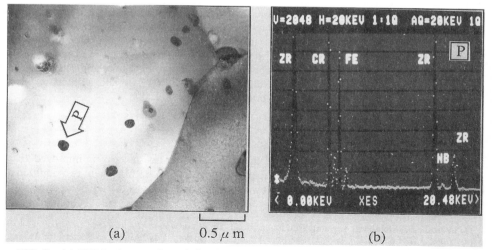

FIG. 9—(a) *TEM micrograph and (b) EDX spectrum of the precipitate (P) in VAZ-03 alloy intermediately annealed at 953 K [$\Sigma Ai = 1.5E-17(h)$].*

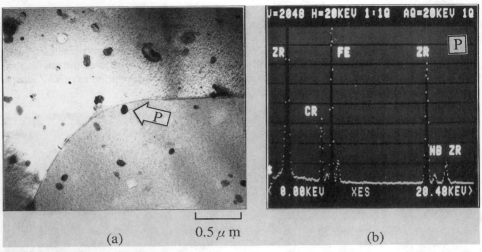

FIG. 10—(a) *TEM micrograph and* (b) *EDX spectrum of the precipitate (P) in VAZ-01 alloy intermediately annealed at 853 K* [$\Sigma Ai = 3.8E\text{-}20(h)$].

The zirconium-niobium-iron-chromium (Zr-Nb-Fe-Cr) type precipitates were only present for VAZ-01 and 03 alloys and the zirconium-niobium-molybdenum-iron-chromium-vanadium (Zr-Nb-Mo-Fe-Cr-V) type precipitates were only present for the VAZ-02 alloy. Other type precipitates (zirconium-niobium, zirconium-iron, zirconium-tin type, etc.) could not be identified in the observations of about 20 particles for each alloy including large and small ones. Moreover, EDX spectra of the VAZ-01 alloy with the lowest annealing parameter were similar to those of the higher parameter; that is, niobium usually precipitates in every VAZ alloy containing 0.2Nb regardless of orders of the annealing parameter.

Size distribution and average size of more than about 80 second-phase particles measured by TEM micrographs are shown in Figs. 11a to d. Precipitate sizes of VAZ-01 alloy with the lowest parameter (Fig. 11d) were smaller than those with a higher parameter (Fig. 11a). However, there were no clear differences in size distribution and average size among three kinds of VAZ alloys annealed at the same condition (Figs. 11a to c). The relationship between the average size and the accumulated annealing parameter of VAZ-01 alloys seems to be in good agreement with that of Zircaloys reported by Garzarolli et al.[2]

Discussion

Effects of Solute Niobium

In our previous paper,[3] it was reported that the corrosion resistance was improved to a significant extent by a small addition of niobium (about 0.05 to 0.2%). Then, two additional experiments were conducted on the very low niobium content alloys to clarify its effects.

---

[2] Garzarolli, F., Steinberg, E., and Weidinger, H. G. in *Zirconium in the Nuclear Industry: Eighth Symposium, ASTM STP 1023*, L. F. P. Van Swam and C. M. Eucken, Eds., American Society for Testing and Materials, Philadelphia, 1989, pp. 202–212.

[3] Isobe, T. and Matsuo, Y. in *Zirconium in the Nuclear Industry: Ninth International Symposium, ASTM STP 1132*, C. M. Eucken and A. M. Garde, Eds., American Society for Testing and Materials, Philadelphia, 1991, pp. 346–367.

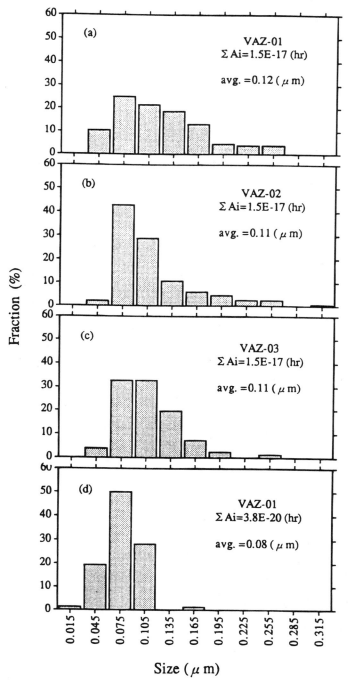

FIG. 11—*Size distributions and average sizes of second-phase particles in (a) VAZ-01, (b) VAZ-02, and (c) VAZ-03 alloys with higher annealing parameters [$\Sigma Ai = 1.5E\text{-}17(h)$] and (d) VAZ-01 alloy with the lowest parameter [$\Sigma Ai = 3.8E\text{-}20(h)$].*

At first, TEM/EDX analysis was performed on Zr-1.0Sn-0.2Fe-0.1Cr-xNb (x = 0.05 and 0.1% by weight) alloys fabricated in the previous work, with the same composition as the VAZ-01 alloy but with less niobium, to investigate the behavior of niobium for alpha-Zr matrix. Figure 12 shows the TEM micrograph and EDX spectra of both the precipitate and the matrix near the precipitate in the alloy containing 0.05Nb. It is apparent that the niobium can be detected in the precipitate by comparing the EDX spectra of the matrix near the precipitate. The Zr-Nb-Fe-Cr type precipitates were only observed for these lower niobium content alloys (0.05 and 0.1) as well as VAZ-01 alloy (0.2Nb), and then the intensity of

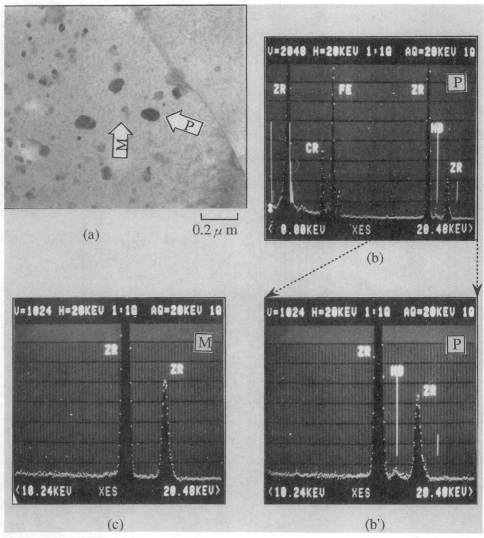

FIG. 12—(a) *TEM micrograph and EDX spectra of both (b)(b') the precipitate (P), and (c) the matrix (M) near the precipitate in the Zr-1.0Sn-0.2Fe-0.1Cr-0.05Nb alloy.*

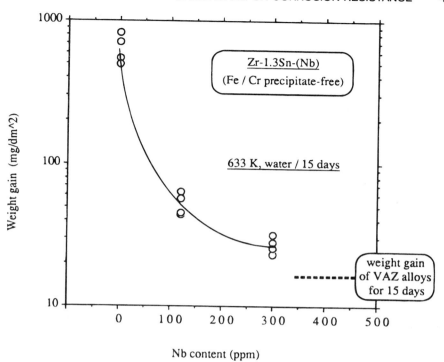

FIG. 13—*Effects of niobium content on corrosion weight gains in pure water for 15 days for Zr-1.3Sn matrix (iron/chromium precipitate-free materials).*

niobium on the EDX spectra decreased with decreasing niobium content in these alloys. Consequently, the concentration of niobium in the precipitates seems to change with the alloy composition in the range of 0.05 to 0.2% by weight.

Next, in order to clarify the effects of niobium in such a very low concentration on corrosion property, Zr-1.3Sn alloys containing niobium from zero up to 300 ppm were melted into about 200 g buttons from a zirconium X-bar and pure materials (niobium, tin, iron, and chromium), whose compositions were listed in Table 2. Charquet et al. reported that the highest total solubility of iron + chromium was 150 ppm at 1083 K in an alpha-phase of Zr-1.4Sn alloy with iron and chromium ratio similar to that in Zircaloy-4, and that the solubility decreased

TABLE 2—*Chemical analysis of iron/chromium precipitate-free materials.*[a]

	Alloy Composition		
Element	A	B	C
Sn, % by weight	1.32	1.34	1.32
Nb, ppm	...	120	300

[a]Fe = 30 ppm and Cr = 13 ppm.

with decreasing temperature.[4] To minimize the effects of iron and chromium, total iron + chromium content was controlled below the maximum solubility. These small ingots were converted to 0.8-mm-thick sheets, using a standard zirconium alloy processing sequence, consisting of hot rolling at 873 K, cold rolling in two steps with intermediate annealing at 903 K for 2 h, and final solution heat treatment in alpha-phase field (903 K for 72 h) followed by cooling in argon gas. The precipitates could not be observed by TEM analysis in these alloys. Figure 13 shows the corrosion weight gains in water at 633 K for 15 days as a function of niobium content. The corrosion resistance remarkably increases with increasing niobium content from zero up to 300 ppm in Zr-1.3Sn matrix and is approaching the level of VAZ alloys containing 0.2Nb and normal contents of iron and chromium. In addition, the corrosion weight gains of Zr-1.0Sn-0.2Fe-0.1Cr-xNb (x = 0.05 and 0.1% by weight) alloys for 15 days were almost same as those of VAZ alloys. Consequently, it is thought that a very small amount of niobium (about 300 to 500 ppm), almost dissolved in the matrix, leads to the superior corrosion resistance of Zr-Sn-Nb-Fe-Cr alloys like the VAZ series.

Conclusions

1. Mechanical properties and corrosion resistance of three kinds of zirconium-tin-niobium-iron-chromium (Zr-Sn-Nb-Fe-Cr) alloys (VAZ series) were evaluated.

2. Yield strengths (0.2% YS) at room and elevated (658 K) temperatures were almost equal to or higher than those of standard Zircaloy-4.

3. Corrosion weight gains were lower than that of standard Zircaloy-4 by about 50%. With increasing annealing parameters, the corrosion resistance decreased to some extent, but were superior to that of standard Zircaloy-4 in the whole range studied in this work.

4. Hydrogen uptakes were about 20 to 60% of that of standard Zircaloy-4.

5. Niobium precipitated in every VAZ alloy containing 0.2Nb. In addition, precipitates containing niobium were observed even in the very low niobium content alloy (0.05 Nb).

6. Corrosion resistance remarkably increased with increasing niobium content up to 300 ppm in Zr-1.3Sn matrix (iron/chromium precipitate-free material). It is thought that a very small amount of niobium, almost dissolved in matrix, leads to the superior corrosion resistance of the VAZ series.

[4] Charquet, D., Hahn, R., Ortlieb, Gros, J.-P., and Wadier, J.-F. in *Zirconium in the Nuclear Industry: Eighth Symposium, ASTM STP 1023*, L. F. P. Van Swam and C. M. Eucken, Eds., American Society for Testing and Materials, Philadelphia, 1989, pp. 405–422.

DISCUSSION

A. M. Garde[1] (written discussion)—Can you comment on the fabricability of the alloys you have investigated?

T. Isobe et al. (authors' closure)—Sheets of VAZ-03 alloy (containing high chromium) intermediately annealed at lower temperatures (853 and 903 K) were not manufactured because of their limited workability. But, there was no difficulty on the workability of the other sheets.

R. B. Adamson[2] (written discussion)—You state that the solubility of niobium in the zirconium-tin matrix is 0.05% by weight. What is the evidence for this? The normal phase diagram would predict a solubility of 0.5% by weight.

T. Isobe et al. (authors' closure)—As shown in Fig. 12, niobium can be detected in the precipitate in the alloy containing 0.05% by weight niobium by comparing with EDX spectra of the matrix near the precipitate. That is, the Zr-Nb-Fe-Cr type precipitates were observed for Zr-1.0Sn-0.2Fe-0.1Cr-xNb (x = 0.05 and 0.1% by weight) alloys as well as for VAZ-01 alloy (X = 0.2% by weight). Then, the intensity of niobium on the EDX spectra decreased with decreasing niobium content in these alloys. The concentration of niobium in the precipitates seems to change with the alloy composition in the range of 0.05 to 0.2% by weight (at the same time, the concentration of solute niobium for alpha zirconium matrix also changes with ones.)

The solubility limit is therefore not below 0.05% by weight for every alloy. Then, the statements concerning solubility limit were modified.

B. D. Warr[3] (written discussion)—We have found very low percent hydrogen uptakes in Zr-2.5Nb (<5%). Have you investigated the corrosion and hydrogen uptake of samples with higher niobium contents (to 1%)?

T. Isobe et al. (authors' closure)—Hydrogen uptakes in water corrosion testing decreased with increasing niobium content up to about 0.2% by weight, but increased with further additives of niobium (below 1.0% by weight)

---

[1] ABB-CE, Nuclear Fuel, Windsor, CT.
[2] GE Nuclear Energy, Pleasanton, CA.
[3] Ontario Hydro, Toronto, Ontario, Canada.

Ian L. Bramwell,[1] Timothy J. Haste,[2] David Worswick,[1] and
P. Donald Parsons[1]

An Experimental Investigation into the Oxidation of Zircaloy-4 at Elevated Pressures in the 750 to 1000°C Temperature Range

REFERENCE: Bramwell, I. L., Haste, T. J., Worswick, D., and Parsons, P. D., **"An Experimental Investigation into the Oxidation of Zircaloy-4 at Elevated Pressures in the 750 to 1000°C Temperature Range,"** *Zirconium in the Nuclear Industry: Tenth International Symposium, ASTM STP 1245,* A. M. Garde and E. R. Bradley, Eds., American Society for Testing and Materials, Philadelphia, 1994, pp. 450–465.

ABSTRACT: Most of the literature data on the oxidation of Zircaloy-4 in steam at high temperatures was generated at atmospheric pressure, but what little data does exist at elevated pressures indicates that oxidation may be enhanced. In view of the paucity of available data, a test program was undertaken to determine the oxidation characteristics of Zircaloy-4 in steam at high temperatures and pressures. Lengths of Zircaloy-4 were heated to temperatures between 750 and 1000°C while being exposed to flowing steam at pressures up to 18.6 MPa for durations up to 2500 s. Oxide thicknesses were measured by metallography that indicated that oxidation was enhanced by pressure over this temperature range. The deleterious effect increased with temperature up to 900°C where the oxide thicknesses were up to five times greater than expected from atmospheric pressure data. Where breakaway oxidation occurred, the local oxide thicknesses were up to a factor of seven greater than predicted. At 1000°C, the effect was less pronounced than at 900°C, with maximum enhancement factors of around two. At 800°C, the enhancement saturated around 15.2 MPa, when further increases in pressure to above normal pressurized water reactor (PWR) operating pressure had no further noticeable effect. Scatter in the data at 900°C prevented identification of any firm trends but the indications were that here the effect did not saturate in the pressure range considered. Thus, predictive models based on atmospheric pressure data are likely to substantially underestimate the extent of oxidation when Zircaloy-4 is heated in elevated pressure steam. These results, which will enable more confident estimates of oxidation under such conditions and form a good starting point for empirical modeling efforts, are reviewed in the context of existing literature data and computer models.

KEY WORDS: zirconium alloys, oxidation, high temperature, high pressure, steam, zirconium, nuclear materials, nuclear applications, radiation effects

There have been many studies of the oxidation performance of Zircaloy-4 in steam atmospheres at high temperatures, such that the kinetics are well quantified to 1600°C and beyond [1–4]. However, most of this oxidation data was generated at atmospheric pressure. Very little experimental work has been performed at elevated pressure, but such information as does

---

[1] Corrosion scientist, senior corrosion scientist, and department manager, respectively, Materials Science AEA Technology, Risley Laboratory, Warrington, Cheshire, UK.
[2] Senior scientist, reactor safety studies, AEA Technology, Winfrith Technology Centre, Dorset, UK.

exist in the literature indicates that oxidation may be enhanced in high pressure steam. In order to investigate this possibility and to fill in the gaps in the database, an experimental program was performed where lengths of Zircaloy-4 were heated to 750 to 1000°C in flowing steam at pressures up to 18.6 MPa. The results from these experiments are presented and compared with data generated under similar conditions at atmospheric pressure.

Background

Summary of the Zircaloy-Steam Reaction

Zirconium reacts with steam exothermically to form hydrogen and zirconium oxide (zirconia). The rate of reaction depends on temperature, following Arrhenius laws. At temperatures above about 1000°C in unlimited steam, and where the oxide film is adherent, the reaction is believed to be controlled by diffusion of oxygen anions through the anion-deficient zirconia structure [2]. At temperatures below the alpha to alpha + beta transus (about 820°C for Zircaloy-4), the oxidation results in a brittle outer oxide layer, with some diffusion of oxygen into the underlying metal. Above the transus temperature, a basically three-layer structure is formed; the outer layer of zirconia, an intermediate layer of alpha-phase zirconium that is stabilized by a high concentration of dissolved oxygen, and an inner layer of beta-phase zirconium into which a smaller concentration of oxygen has diffused.

Above 1050°C, the kinetics of oxide and oxygen-stabilized alpha layer growth are well described by the parabolic rate laws expected from diffusion theory [5–7]. Below 900°C, approximately cubic kinetics are observed for the oxide layer; stabilized alpha-layer growth remains parabolic. This behavior has been associated [5] with the formation of monoclinic rather than tetragonal oxide. At long times (for example, over 30 min at temperatures of 800°C), transition to "breakaway" oxidation, in which the outer oxide layer flakes off and is therefore no longer protective, is observed, leading to linear kinetics thereafter. Resistance to breakaway oxidation is observed in the phase change region (820 to 980°C) and in the beta phase region above 1050°C. Cubic/parabolic parametric models are available covering this range of temperatures; differences are noted among the rates of corrosion of Zircaloy from different sources that decrease with increasing temperature.

The literature on Zircaloy oxidation at atmospheric pressure is very extensive and will not be considered further here. Detailed reviews are available elsewhere, for example Refs 2 and 3.

Literature Review—Zircaloy Oxidation at High Pressures

In contrast to the database available for oxidation at atmospheric pressures, the data available for high pressure oxidation are relatively sparse. Early work [8–11] was primarily devoted to testing new alloys or establishing accelerated test procedures. It was mainly confined to lower temperatures (up to about 600°C) than those of interest here, and to pressures up to about 11.5 MPa (1700 psi). Enhancement of oxidation was observed in several cases, for example, by a factor of two for a zirconium-rich alloy in going from 3.4 to 11.5 MPa [10]; however, the results were not systematically organized and no general conclusions were drawn.

The first systematic and detailed study was that of Cox [12], who oxidized Zircaloy-2 in stagnant steam at pressures up to 34 MPa (5000 psi) at temperatures from 350 to 600°C. Strong enhancements of oxidation for pressures over 3.4 MPa (500 psi) were observed, of an order of magnitude or more, the influence of pressure increasing with temperature. Tests planned for 700°C were abandoned owing to operational problems with the apparatus.

A further detailed study was carried out by Pawel et al. [13,14] who oxidized Zircaloy-4 in flowing steam at temperatures of 905 and 1101°C at pressures of 3.4, 6.8, and 10.3 MPa (500, 1000, and 1500 psi, respectively). The measurements showed a progressive enhancement of oxidation with increasing pressure at 905°C, similar to that seen at 500 to 600°C in the work of Cox. However, no effect was seen at 1100°C. At higher pressures, the ratio of oxide to alpha thickness increased, suggesting a change of mechanism at high pressures.

The present test series was intended to extend the available database so that predictions of oxidation could be made without recourse to extrapolation in either temperature or pressure. Temperatures were chosen to fill in the gaps between 600 and 900°C and 900 and 1100°C so that the temperature dependence of the pressure effect could be more accurately determined. At the higher temperatures, the pressures were increased above those previously studied to determine whether the enhancement was progressive or whether saturation of the effect occurred.

Experimental Procedure

Lengths of stress-relief annealed Zircaloy-4 tubing were heated rapidly to high temperatures in pressurized steam in the ZORO rig, shown schematically in Fig. 1. The rig consisted of two interconnected pressure vessels, each heated by means of an external clam furnace. The lower vessel acted as a boiler, producing steam that was then fed to the upper reaction vessel where it passed over the specimen before flowing out to vent. The specimen was heated by direct resistance heating, one side being connected via a silver electrode to a power supply and control system, the other being earthed against the reaction vessel. The rig was operated in the manner described subsequently.

The boiler was filled with high-quality demineralized water and then the rig was flushed with helium. The valve on the vent line was closed to seal the rig that was then heated to a temperature sufficient to produce steam at the pressure required for the test. Once the necessary pressure had been achieved, a needle valve on the vent line was opened to create a flow of steam from the boiler, over the specimen, and out to vent while maintaining a constant pressure in the rig. The specimen was then heated at 5°C/s to the test temperature for the required period before rapid cooling and removal for examination.

A batch of Zircaloy-4 from a well-qualified source relevant to UK interests was used for these tests. The tubing was 9.5 mm in diameter with a wall thickness of 0.57 mm. This material has been used extensively in the past for studies of oxidation and creep rupture characteristics at similar temperatures to those of interest in this work (for example Ref 7). Roughly half the specimens tested were made from two 150-mm lengths of Zircaloy-4 with a solid bar of Zircaloy welded between them to form three sides of a rectangle. Later specimens were made by bending a single 550-mm length over a former into an inverted "U" shape. This technique had the obvious advantage that welding could be eliminated. Initial fears that the curvature of the upper part of the later specimens would enhance oxide spallation proved unfounded. Specimen temperature was controlled by spot welding two platinum-tipped, stainless-steel-sheathed thermocouples to one leg of the specimen. One of these was used to monitor and control the specimen temperature via a three-term control unit and temperature programmer, the other was used as an over-temperature trip as a precaution against the control thermocouple becoming detached during a test. This proved to be a particular problem when thick oxide films were generated during a test. The second thermocouple allowed the sudden temperature rise following detachment of the control thermocouple to be detected and the power to the specimen immediately cut off. In these circumstances, a result could still be obtained from a test even though the intended test duration had not been achieved.

During a test, the pressure could typically be maintained within 0.14 MPa (20 psi) of the intended value. The steam flow rate was not quantified, but adjustments were made during each test to maintain a good flow of steam in the vent line. The ability to control the specimen temperature accurately was found to be dependent on the test pressure such that the variation

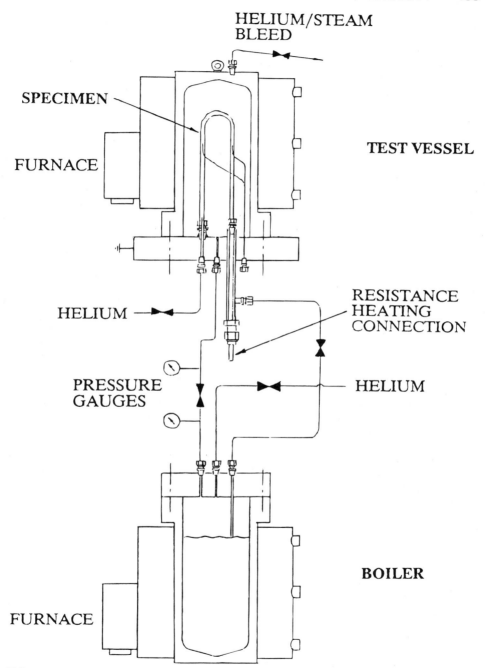

FIG. 1—*General arrangement of ZORO high-temperature and pressure steam oxidation rig.*

about the set point increased with steam pressure. For a test between atmospheric pressure and 11.7 MPa, the temperature could be controlled to ±2°C, but at higher pressures, control became more difficult such that at 18.6 MPa, the temperature was typically held within ±5°C of the target. The CANSWEL-2 computer code (a mechanistic clad ballooning modeling code developed by AEA Technology) has been used to assess any effect of these temperature variations on the expected oxidation by comparison of the oxide produced by a transient with typical temperature fluctuations and that from an idealized isothermal temperature profile. This showed any such errors to be negligible.

The CANSWEL-2 code has also been used to assess the likely amount of oxidation produced on each specimen during the heat-up and cooling cycles. This indicated that the following oxide thicknesses would be produced prior to and following the isothermal part of each test: 0.8 μm at 750°C, 1.2 μm at 800°C, 2.8 μm at 900°C, and 2.9 μm at 1000°C. The kinetics equations in Ref 6 have been used to assess the extra time at temperature that would be predicted to produce these values. This indicated that 14 s would be required at 750°C, 15 s at 900°C, and 4 s at 1000°C. These values are within the likely experimental errors and their significance decreases as the total exposure period increases. This of course assumes that oxidation was occurring at the predicted rate. Where enhancement has occurred, the errors due to the heating and cooling period are likely to be larger than the values quoted.

Following testing, each specimen was sectioned and the oxide and alpha layer thicknesses measured by conventional optical metallography. This involved taking approximately 15 measurements around the circumference of each specimen after sectioning transversely at the control thermocouple position. These measurements were then averaged to give a value for the oxide thickness and underlying alpha layer. The uniformity of the oxide around the circumference for the majority of specimens indicated that any temperature variations were negligible, as were any local cooling effects caused by the presence of the thermocouple. Unfortunately, the direct resistance heating method does not produce a uniform temperature along the length of the specimen so that the material above and below the control thermocouple experiences a different temperature that varies with time. This prevents further sections from being taken to allow, for example, analysis of hydrogen content. Analysis of test results is thus limited to the information that can be obtained from a single section at the control thermocouple position.

The same batch of Zircaloy-4 as used for these tests has been used previously [7] to model oxidation behavior between 700 and 1300°C in atmospheric pressure steam. Further unpublished data generated since Ref 7 has allowed this model to be refined, and this has been used to compare the following results at elevated pressure with atmospheric pressure prediction. The results of Leistikow et al. [6] that form a more extensive database are also shown.

Results

The test matrix involved oxidizing specimens at temperatures between 750 and 1000°C in flowing steam at pressures up to 18.6 MPa. The maximum test duration was 2500 s at the test temperature. The resultant oxide and alpha layers are shown in Table 1. The values predicted from the atmospheric pressure model [7] and from Ref 6 are also given, as is an "enhancement factor" of the measured oxide divided by that predicted using the model [7] and Leistikow et al. data [6]. Comparisons between prediction and metallography for each test temperature are shown graphically in Figs. 2 to 5.

At 750°C (Fig. 2), only one test was performed, of 300 s duration at 11.7 MPa; the oxide thickness was greater than predicted from Refs 6 and 7 but not significantly so when possible errors are considered. At 800°C, the test matrix was more extensive and in each case the amount of oxide present was consistently higher than predicted (Fig. 3), by factors of between

TABLE 1—Comparison of oxide and alpha layer thicknesses generated during tests at elevated pressure with prediction.

Temperature, °C	Pressure, MPa (g)	Time, s	This Work at Elevated Pressure		Predicted from Atmospheric Pressure Data [7]		Predicted from Atmospheric Pressure Data [6]		Enhancement Factor[a] Compared to Ref 7	Enhancement Factor[a] Compared to Ref 6	Effect on Polarized Light?[b]
			Oxide, μm	Alpha, μm	Oxide, μm	Alpha, μm	Oxide, μm	Alpha, μm			
750	11.7	300	7 ± 1	0	4 ± 2	2 ± 1	2	2	1.8	3.5	N
800	11.7	1000	30 ± 2	3 ± 1	8 ± 3	8 ± 3	5	6	3.8	6.0	N
800	15.2	300	20 ± 2	2 ± 1	5 ± 2	4 ± 2	3	3	4.0	6.6	N
800	15.2	1000	31 ± 2	4 ± 1	8 ± 3	8 ± 3	5	6	3.9	6.0	N
800	15.2	1500	42 ± 5	4 ± 1	9 ± 4	9 ± 4	5	7	4.7	8.4	N
800	18.6	300	16 ± 6	1 ± 1	5 ± 2	4 ± 2	3	3	3.2	5.3	N
800	18.6	1000	30 ± 3	1 ± 1	8 ± 3	8 ± 3	5	6	3.8	6.0	N
900	0	1000	15 ± 1	12 ± 1	14 ± 6	21 ± 7	11	18	1.1	1.4	N
900	0	1500	17 ± 1	15 ± 1	16 ± 6	25 ± 8	12	23	1.1	1.4	N
900	6.9	240	12 ± 3	1 ± 1	9 ± 4	10 ± 3	7	8	1.3	1.7	N
900	6.9	1500	28 ± 2	28 ± 2	16 ± 6	25 ± 8	12	23	1.8	2.3	Y
900	6.9	2500	49 ± 2	15 ± 1	18 ± 7	33 ± 9	15	28	2.7	3.3	Y
900	11.7	300	25 ± 5	9 ± 1	10 ± 4	11 ± 4	7	10	2.5	3.6	Y
900	11.7	1000	58 ± 4	15 ± 1	14 ± 6	21 ± 7	11	18	4.1	5.3	Y
900	11.7	1500	97 ± 3	38 ± 4	16 ± 6	25 ± 8	12	23	6.1	8.1	Y
900	15.2	300	...[c]	6 ± 1	10 ± 4	11 ± 4	7	10	...	...	Y
900	15.2	1000	98 ± 3	24 ± 2	14 ± 6	21 ± 7	11	18	7.0	8.9	Y
900	15.2	1500	48 ± 5	20 ± 2	16 ± 6	25 ± 8	13	21	3.0	3.7	Y
900	15.2	1500	55 ± 4	20 ± 2	16 ± 6	25 ± 8	13	21	3.4	4.2	Y
900	18.6	1000	66 ± 4	9 ± 1	14 ± 6	21 ± 7	11	18	4.7	6.0	Y
1000	6.9	300	24 ± 2	14 ± 2	19 ± 2	18 ± 5	16	20	1.3	1.5	Y
1000	6.9	615	19 ± 1	22 ± 1	25 ± 2	26 ± 8	23	29	0.8	0.8	N
1000	15.2	300	40 ± 3	15 ± 2	19 ± 2	18 ± 5	16	20	2.1	2.5	Y

[a] Enhancement factor = observed oxide/predicted oxide.
[b] N = no and Y = yes.
[c] Oxide spalled during sectioning.
NOTE—Errors shown are 1 standard deviation.

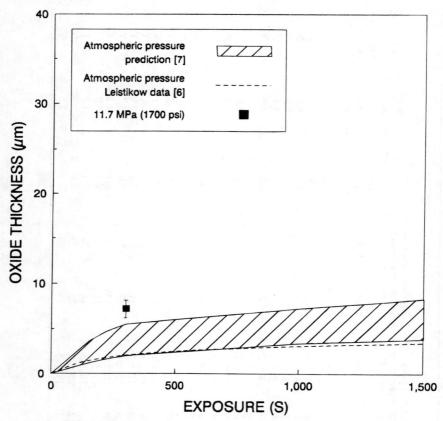

FIG. 2—*Oxide thickness as a function of test pressure at 750°C.*

three and five (five and eight versus the Leistikow et al. data). Another trend that was evident was that for those pressures used in the tests (11.7 MPa or greater), increasing the test pressure did not appear to alter significantly the amount of enhancement, indicating that the accelerating effect of pressure may have saturated. The oxidation process appeared to follow normal cubic kinetics, at least up to 1000 s. It is possible that the greater degree of enhancement in the 1500 s test may indicate the start of breakaway, but there is insufficient data to be certain. When sectioned, the appearance of the oxide on this specimen was identical to the others produced at this temperature.

In order to ensure that the oxidation enhancements were not caused by some unforeseen feature of the experimental system, such as large errors in temperature measurement, two control tests were performed at a pressure just sufficient to ensure a flow of steam over the specimen (0.2 MPa). The oxides generated during these tests, which were performed at 900°C, were within the prediction scatterband, indicating that the accelerated kinetics observed in the other tests did in fact result from the elevated pressure. Figure 4 shows these and the other results obtained at 900°C. Also shown are the results of Pawel et al. [*13*] at this temperature. At 6.9 MPa, oxidation was accelerated compared to prediction and, at least up to 1500 s, was in reasonable agreement with the Pawel data. However, the oxide thickness after 2500 s was greater than predicted by extrapolation of Pawel's results and may indicate a deviation from

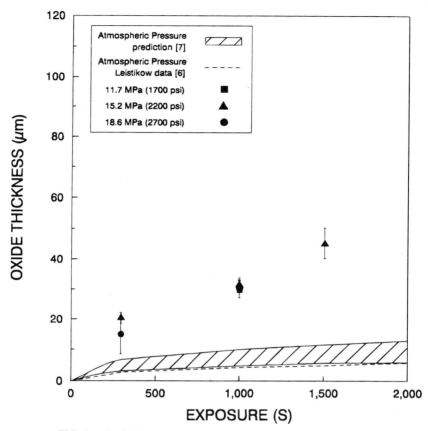

FIG. 3—*Oxide thickness as a function of test pressure at 800°C.*

cubic kinetics towards breakaway oxidation although, again, there is insufficient information to be certain.

At pressures of 11.7 MPa and above, it was more difficult to obtain a result from the tests because of oxide spallation on cooling specimens to room temperature that, when it occurred, prevented oxide thickness measurements from being taken. For those results that could be obtained, there was a greater degree of enhancement than at the lower pressures and more scatter in the data than at 800°C. Two specimens showed breakaway oxidation that resulted in an oxide almost 100 μm thick in each case. The other results appear to follow cubic kinetics but the scatter makes it difficult to ascertain any trends. It would appear that, for the time-pressure combinations studied, the accelerating effect of pressure did not saturate.

Three tests were performed at 1000°C where the oxide did not spall. These were naturally of short duration and the results are shown in Fig. 5. At 6.9 MPa, the oxide thicknesses were close to the predicted value and, at 15.2 MPa, the observed enhancement factor of 2.1 was smaller than seen in most cases at the other temperatures. These results indicate that the magnitude of the pressure effect at this temperature is lower than at the other temperatures in this study.

The structure of the oxides formed during these tests was examined to look for any differences that might explain the observed behavior. The films formed at 750 and 800°C were very

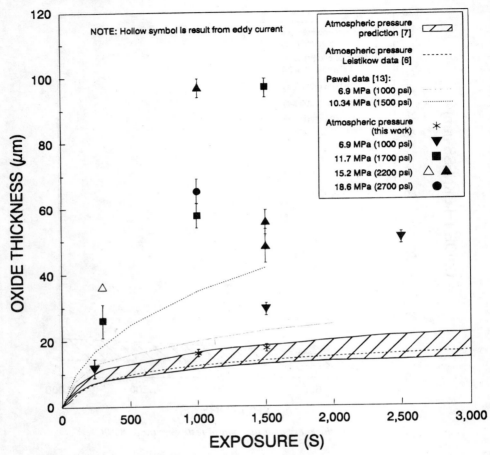

FIG. 4—*Oxide thickness as a function of test pressure at 900°C.*

uniform with little obvious porosity or other features, and polarized light had no effect on their appearance. The oxides formed at 900°C fell into two distinct groups. Those formed at atmospheric pressure and at 6.9 MPa after 240 s had uniform dense oxides similar in appearance to those formed at the lower temperatures. All the others had a nonuniform layered structure following initial polishing that, with the exception of the specimens where breakaway had occurred, was triplex in nature with a thin uniform outer oxide and inner duplex inhomogeneous layer (Fig. 6). This layer had many voids in it that, in the outer part, were aligned perpendicular to the oxide-environment interface, with more randomly oriented voids in the inner part. When examined under polarized light, the outer and middle layers appeared as a bright band and the duplex nature of the inhomogeneous layer could be clearly seen. These features were most unlikely to have been present during the tests but will be the result of the sectioning and polishing operations. However, the consistency in the nature of the observations for a large number of specimens indicated that they were illustrative of differences in the nature of the oxide crystallites in different parts of the oxide scale. Further careful polishing could remove these features, but the photographs in Figs. 6 and 7 were taken to illustrate the layering of

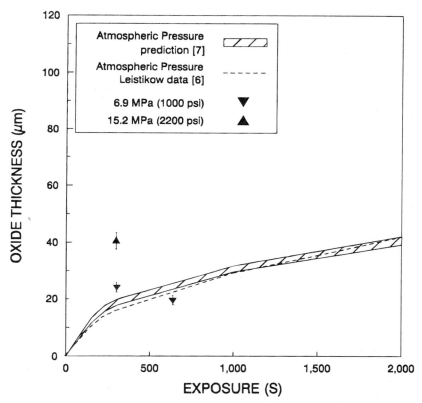

FIG. 5—*Oxide thickness as a function of test pressure at 1000°C.*

the oxides. On specimens with thicker oxides, there were regions around the circumference where the oxide had cracked during a test (Fig. 8) and increased the local oxide thicknesses above the averages in Table 1.

The two specimens with breakaway oxidation were found to have four distinct layers in the oxide (Fig. 7). In addition to the outer uniform and thick porous duplex layers, there was a thin layer adjacent to the oxide-metal interface that was uniform in parts (Fig. 7a), but there were regions with a striated appearance due to the removal of large crystallites perpendicular to the interface (Fig. 7b). On both specimens there were regions where the oxide had cracked and increased the local oxidation rate. The outer parts of the oxide films again appeared bright under polarized light.

At 1000°C, the two specimens with the thickest oxides had a triplex oxide structure and a bright outer band under polarized light. The other specimen had a uniform oxide and there was no effect of polarized light.

Discussion

The results presented here indicate that the oxidation of Zircaloy-4 in steam can be enhanced markedly at elevated pressure and that, under conditions where oxidation occurs at both high temperature and pressure, predictive models based on atmospheric pressure data will

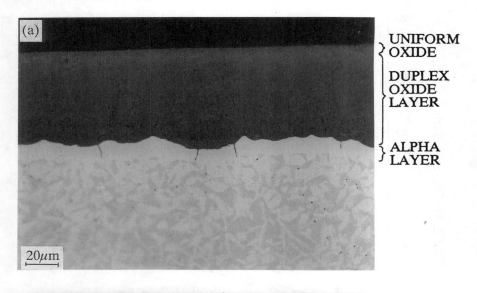
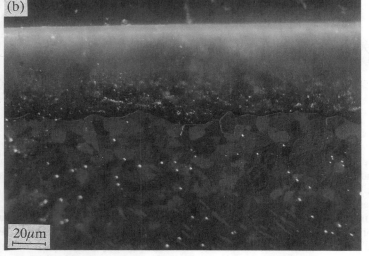

FIG. 6—*Oxide and oxygen-stabilized alpha layers on a specimen exposed at 900°C and 18.6 MPa for 1000 s, showing triplex nature of the oxide:* (a) *bright field illumination and* (b) *polarized light.*

underpredict the extent of oxidation. A number of specimens had oxide thicknesses over six times greater than predicted (over eight times compared to the Leistikow data [6]). Data at 800°C suggests that the effect may saturate at pressures above 11.7 MPa when further increases in pressure do not further significantly enhance the oxidation. Scatter in the data at 900°C makes such a conclusion at this temperature difficult. At 1000°C, the measured enhancements are smaller than at the other temperatures, indicating that the effect of pressure may be declining with further increases in temperature.

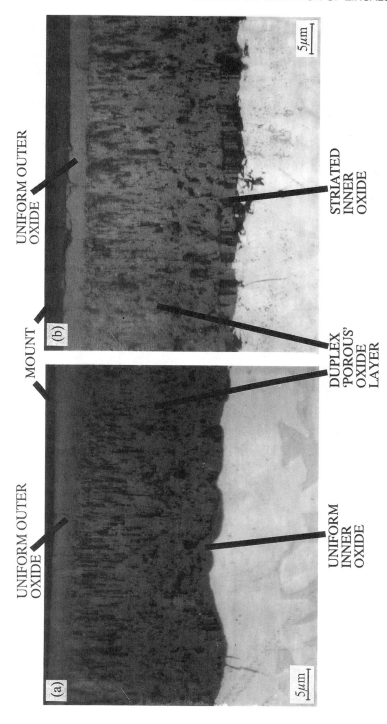

FIG. 7—*Oxide and oxygen-stabilized alpha layers on a specimen that exhibited breakaway oxidation, showing differences in the nature of the inner oxide layer:* (a) *uniform and* (b) *striated.*

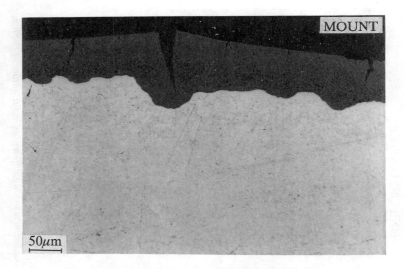

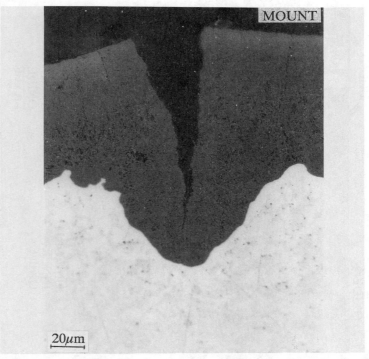

FIG. 8—*Cracks in oxide on a specimen exposed for 1000 s at 900°C and 18.6 MPa showing locally enhanced oxidation.*

In so far as a comparison can be made, these results are consistent with those of Pawel et al. [13,14] who performed similar tests at 905 and 1101°C with steam pressures up to 10.34 MPa. Pawel reported a progressive effect of pressure over the range of conditions studied at 905°C that was not evident at 1101°C. Their results at 905°C and 6.9 MPa for times up to 1800 s are in close agreement with our own data as shown in Fig. 4, and the observation of no pressure effect at 1101°C is consistent with the trend of reduced enhancements at 1000°C compared to lower temperatures. The data published by Cox [12] from tests at 500 and 600°C showed dramatic increases in post-transition weight gain with increasing pressure from 3.45 to 34.5 MPa without saturation of the effect.

The Leistikow data [6] do not predict the onset of breakaway oxidation at 900°C to times in excess of 10 000 s, but two specimens in these tests had breakaway after 1000 to 1500 s. The Leistikow data also indicate that the alpha layer thickness remains constant after the onset of breakaway. It was not possible to determine definitively whether this was the case in these tests, but the alpha layers were thicker than found on specimens exposed for equivalent times where breakaway had not occurred.

It has not proved possible to determine the mechanism for the deleterious effect of pressure on the kinetics of oxidation in steam. The temperature range over which enhancement has been observed in both these and literature experiments corresponds with the range over which alpha zirconium is present in the metal substrate, either as a single phase or together with beta phase. Enhancement has not been observed in the beta-phase region. However, this correlation is purely speculative. Other workers have suggested that porosity may play an important role. Cox suggested that porosity in the oxide develops more rapidly at high pressures that leads to an increase in oxidation rate. Pawel et al. found physical changes in the oxide that were correlated with enhanced oxidation in high pressure steam. When the oxide on their specimens was examined under polarized light, a bright band of material was observed in the outer part of the oxide on specimens that had experienced enhanced oxidation rates. The thickness of this band increased with steam pressure and was absent on specimens oxidized at atmospheric pressure. It was speculated that this effect could be produced by light reflected from microcracks developing along grain boundaries beneath the surface of the outer portion of the oxide, but no evidence of such a system of cracks could be found.

A bright band was observed in the oxides of some of the specimens from this program (Table 1). Those oxidized at 800°C had enhancement factors between 3.2 and 4.7 (5.3 and 8.4 versus [6]), but none had a bright band in the oxide under polarized light. In each case, the oxide was dense and uniform with little obvious porosity. At 900°C, only the specimens oxidized at atmospheric pressure and one exposed at 6.9 MPa for 240 s had a uniform oxide and no effect on polarized light. All the others had either a triplex or quadruplex structure and a bright band in the outer two layers. The thickness of this band varied between 45 and 65% of the oxide thickness but could not be correlated with the degree of oxidation enhancement or, in contrast to Pawel, to the test pressure. At 1000°C, two of the three specimens exhibited a bright band of material in the outer part of the oxide. The specimen exposed for the longest period did not show this effect, but it had the thinnest oxide of the three and was uniform in nature. Thus, the presence or absence of the bright band of material in the oxide was related to the oxide structure rather than the degree of enhancement or the oxidation temperature. The oxides with a layered structure all exhibited this band, but it has not proved possible to correlate the presence of this structure with the experimental conditions or a threshold oxide thickness. In general, the layered structure was found in the thicker oxides except for those formed at 800°C that were uniform.

The current database could form a starting point for the formulation of an empirical model for the pressure enhancement effect. More data would be desirable, to identify more precisely when the transition to breakaway oxidation with the concomitant change in time dependence occurs, no theory being available. The data are, however, insufficient for the formulation of a mechanistic treatment, since the mechanism responsible for the enhancement cannot yet be unequivocally identified.

Conclusions

The oxidation of Zircaloy-4 has been successfully measured under conditions of high temperature and pressure. The temperature range considered was 750 to 1000°C, with pressures up to 18.6 MPa and transient times up to 2500s. The data obtained cover a region not previously explored experimentally.

At 800 and 900°C, there was an accelerating effect of pressure on the oxidation of Zircaloy-4 that increased with both temperature and pressure. This led to the production of oxide films up to five times thicker than predicted using cubic/parabolic equations based on oxidation data from atmospheric pressure.

At 800°C, the effect of pressure seemed to saturate around 11.7 MPa (1700 psi) and further increases in pressure had no noticeable further influence. Scatter in the data at 900°C prevented identification of any firm trends at the higher test pressures, but the indications were that the effect did not saturate at this temperature. If the oxidation enters breakaway, local oxide films up to seven times greater than predicted may be formed.

At 1000°C, the magnitude of the effect was less than at 900°C, with a maximum observed acceleration factor around two compared with prediction, indicating that the influence of test pressure was declining with further increases in temperature. This is consistent with the null effect observed previously by other workers at 1100°C. The data obtained form a good basis for more reliable and confident estimates of cladding oxidation in high pressure steam than was previously possible, and a starting point for empirical modeling efforts. The generation of more data is desirable to enable further predictive improvements to be made.

Acknowledgments

This work was funded by the UK Health and Safety Executive.

References

[1] Viskanta, R. and Mohanty, A. K., "Heat-up of a Partially Uncovered PWR Fuel Rod in the Presence of Cladding Oxidation and Steam Dissociation," *Nuclear Engineering and Design,* Vol. 105, 1988, pp. 231–242.

[2] Parsons, P. D., Hindle, E. D., and Mann, C. A., "The Deformation, Oxidation and Embrittlement of PWR Fuel Cladding in a Loss-of-Coolant Accident," OECD/NEA State-of-the-Art Report, ND-R-1351(S), UK Atomic Energy Authority, Risley, Nov. 1985.

[3] Cox, B., "Oxidation of Zirconium and Its Alloys," *Advances in Corrosion Science and Technology,* Vol. 5, 1976, pp. 173–362.

[4] Kinnersly, S. R., et al. "In-vessel Core Degradation in LWR Severe Accidents," OECD/NEA State-of-the-Art Report, NEA/CSNI/R(91)12, Organization for Economic Cooperation and Development, Nov. 1991.

[5] Pawel, R. E. and Cathcart, J. V., "The Kinetics of Oxidation of Zircaloy-4 in Steam at High Temperatures," *Journal,* Electrochemical Society, Vol. 126, No. 7, July 1979, pp. 1105–1111.

[6] Leistikow, S., Schanz, G., versus Berg, H., and Aly, A. E., "Comprehensive Presentation of Extended Zircaloy-4 Steam Oxidation Results 600–1600°C," OECD-NEA-CSNI/IAEA meeting, Risø, Denmark, May 1983, IAEA Summary Report IWGFPT/16, International Atomic Energy Agency, Vienna.

[7] Haste, T. J., Harrison, W. R., and Hindle, E. D., "Zircaloy Oxidation Kinetics in the Temperature Range 700–1300°C," DCO 7861(S), IAEA-TC-657/4.7, presented at IAEA Technical Committee Meeting, Preston, UK, Sept. 1988, International Atomic Energy Agency, Vienna.

[8] Kneppel, D. S., "Effect of Temperature and Pressure on Steam Testing of Zirconium Alloys," NMI-1119, Oct. 1954.

[9] Wanklyn, J. N., et al., "The Corrosion of Zirconium and Its Alloys in High Temperature Steam," AERE-R-4130, UK Atomic Energy Authority, Harwell, Aug. 1962.

[10] Kass, S., "The Corrosion and Hydrogen Absorption Characteristics of Zirconium Alloys Containing Iron and Chromium in High Temperature Water and Steam," WAPD-TM-517, Westinghouse Electric Corporation, Feb. 1968.

[11] Johnson, A. B., Jr., "Corrosion and Failure Characteristics of Zirconium Alloys in High Pressure Steam in the Temperature Range 400–500°C," *Applications-Related Phenomena for Zirconium and Its Alloys, ASTM STP 458*, American Society for Testing and Materials, Philadelphia, 1969, pp. 271–285.

[12] Cox, B., "Accelerated Oxidation of Zircaloy-2 in Supercritical Steam," AECL-4448, Atomic Energy of Canada Limited, April 1973.

[13] Pawel, R. E., Cathcart, J. V., and Campbell, J. J., "The Oxidation of Zircaloy-4 at 900 and 1100°C in High Pressure Steam," *Journal of Nuclear Materials*, Vol. 82, 1979, pp. 129–139.

[14] Pawel, R. E., Cathcart, J. V., Campbell, J. J., and Jury, S. H., "Zirconium Metal-Water Oxidation Kinetics—V. Oxidation of Zircaloy in High Pressure Steam," ORNL/NUREG-31, Oak Ridge National Laboratory, Oak Ridge, TN, Dec. 1977.

DISCUSSION

Brian Cox[1] (written discussion)—You seem to have two very different types of oxides. Ones that apparently have large amounts of porosity (or oxide pull-out during polishing) but no radial cracks, and ones that have large radial cracks that increase the local oxidation rate but apparently little porosity. If the cracks have a large effect on the local oxidation rate, then clearly the oxide must present some sort of oxidation barrier. Is it possible to classify these two types of oxide according to the conditions under which they occur?

I. L. Bramwell et al. (authors' closure)—We do have two major types of oxide present, but the difference does not lie in whether the oxide contained radial cracks but in the presence or absence of a layered oxide structure. The cracks like those in Fig. 8 were generally found on the thicker oxides and were present irrespective of the observed oxide structure. The micrograph in the figure was taken after repolishing to remove the layered structure that had been present on that particular specimen after initial polishing, similar to the oxide in Fig. 6.

The presence of oxidation enhanced still further by the presence of a crack may suggest that the oxide was protective, but it is also possible that the crack penetrated the barrier layer at the oxide-metal interface, in which case, locally enhanced oxidation would be expected. We observed some smaller cracks that did not penetrate the thickness of the oxide and had no apparent effect on the local oxide thickness that indicates that the bulk of the oxide is non-protective and it is the barrier layer that is important. However, it is also feasible that these smaller cracks formed when cooling the specimens when no further oxidation was possible.

[1] University of Toronto, Toronto, Ontario, Canada.

Mechanical Properties

Richard A. Perkins[1] and Shih-Hsiung Shann[1]

Prediction of Creep Anisotropy in Zircaloy Cladding

REFERENCE: Perkins, R. A. and Shann, S.-H., **"Prediction of Creep Anisotropy in Zircaloy Cladding,"** *Zirconium in the Nuclear Industry: Tenth International Symposium, ASTM STP 1245*, A. M. Garde and E. R. Bradley, Eds., American Society for Testing and Materials, Philadelphia, 1994, pp. 469–482.

ABSTRACT: Due to the hexagonal crystal structure of zirconium and the radial orientation of the basal poles in Zircaloy cladding, the deformation of light water reactor Zircaloy fuel cladding is anisotropic. Plastic deformation of this cladding can be defined by the R and P factors that are the circumferential/radial and axial/radial contractile strain ratios under uniaxial deformation along the axial and circumferential direction, respectively. The in-reactor deformation performance of the cladding can be modeled with good accuracy if the R and P values of the irradiation-induced creep are known.

In a boiling water reactor (BWR) fuel assembly, most fuel rods have a hoop stress to axial stress ratio of about 2:1. The assembly also contains several fuel rods (tie rods) that connect the upper and lower tie plates. The tie rods have an additional stress component in the axial direction. BWR assemblies can, furthermore, contain water rods that are free of stress, and therefore, provide a measure of stress-free irradiation growth. Post-irradiation deformation measurements of these three types of rods are used to derive the R and P factors for BWR cladding during irradiation.

Laboratory tensile and creep tests at several temperatures have been performed on five types of cladding to determine if a short-term test could be used to obtain R values representative of the in-reactor values. For standard Zircaloy-2 cladding that was examined in-reactor, a good correlation with the in-reactor R value was obtained from both tensile and creep tests performed at 382°C. For another type of Zircaloy-2 cladding (late beta-quenched (LBQ) cladding), in-reactor deformation performance correlates better with the tensile test results than the creep test results. The other three types of cladding (Zircaloy-4) with differing textures and processing histories have exhibited significant differences in the R values in the laboratory tests.

KEY WORDS: zirconium alloys, contractile strain ratio, boiling water reactor, creep (materials), anisotropy, in-reactor deformation, zirconium, nuclear materials, nuclear applications, radiation effects

Zircaloy tubing used for nuclear fuel cladding in light water reactors is highly anisotropic due to the hexagonal crystal structure of zirconium at temperatures below approximately 800°C. During the normal fabrication process using pilgering and intermediate annealing, the crystallographic planes in the majority of grains become oriented so that the majority of the basal poles lie in the radial-circumferential plane of the cladding with the highest density of basal poles oriented at plus and minus 30° to 40° from the radial direction. Due to the preferred crystallographic orientations and the limited slip systems available in the hexagonal crystal structure, plastic deformation of the cladding is highly anisotropic. At a given temperature and neutron fluence, the in-reactor deformation of the cladding is determined by the cladding

---

[1] Senior engineers, Siemens Power Corporation, Nuclear Division, Richland, WA 99352.

texture and the applied stresses. The texture of the cladding can be measured by X-ray diffraction or indirectly determined by measuring the degree of anisotropic deformation during a tensile test with an axially applied load. The latter test (referred to as the CSR test) is easy to perform and provides a contractile strain ratio (designated as R to denote the circumferential/radial contractile strain ratio) that should be useful in predicting the in-reactor deformation behavior. The R values obtained in short-term room temperature tensile tests have been found to correlate well with the texture of the cladding [1], but the values obtained in this test do not appear to be representative of the in-reactor deformation. Testing has been performed at 382°C, to obtain R values under light loading for creep deformation and at higher loading consistent with the room temperature CSR test. These values have been compared with the R values obtained from measurements on irradiated fuel rods.

Creep Model

In-Reactor

The creep correlation for textured anisotropic Zircaloy cladding can be expressed by the following equations [2]

$$\sigma_g^2 = \frac{R(\sigma_r - \sigma_\theta)^2 + RP(\sigma_\theta - \sigma_z)^2 + P(\sigma_z - \sigma_r)^2}{P(R + 1)} \quad (1)$$

$$\begin{bmatrix} \dot{\epsilon}_r \\ \dot{\epsilon}_\theta \\ \dot{\epsilon}_z \end{bmatrix} = \frac{\dot{\epsilon}_g}{P(R + 1)\sigma_g} \cdot \begin{bmatrix} (R + P) & -R & -P \\ -R & R(P + 1) & -RP \\ -P & -RP & P(R + 1) \end{bmatrix} \begin{bmatrix} \sigma_r \\ \sigma_\theta \\ \sigma_z \end{bmatrix} \quad (2)$$

If the state of stress in the cladding does not change, the preceding strain rate equation can be integrated and written as

$$\begin{bmatrix} \epsilon_r \\ \epsilon_\theta \\ \epsilon_z \end{bmatrix} = \frac{\epsilon_g}{P(R + 1)\sigma_g} \cdot \begin{bmatrix} (R + P) & -R & -P \\ -R & R(P + 1) & -RP \\ -P & -RP & P(R + 1) \end{bmatrix} \begin{bmatrix} \sigma_r \\ \sigma_\theta \\ \sigma_z \end{bmatrix} \quad (3)$$

where

σ_r, σ_θ, and σ_z = radial, circumferential, and axial stress;
ϵ_r, ϵ_θ, and ϵ_z = stress-induced radial, circumferential, and axial strain;
σ_g and ϵ_g = generalized stress and strain; and
R and P = circumferential/radial and axial/radial contractile strain ratios.

For an engineering application of the model to predict strains in cladding under a given stress condition, the contractile strain ratios, R and P, need to be derived from post-irradiation deformation measurements.

It is generally assumed that the in-reactor stress-induced strain has two components, thermal and irradiation creep strain [3]. The correlation given by Gittus indicates that at typical reactor operating temperatures (280 to 320°C) the irradiation-induced creep strain is much larger than the thermal creep strain. The irradiation creep strain has been determined to have the following stress dependence [3,4]

$$\epsilon \propto \sigma^{1.23} \tag{4}$$

This relationship has been used up to a stress of 295 MPa [3].

In addition to the stress-induced strains, cold-worked stress-relieved Zircaloy cladding undergoes stress-free neutron irradiation-induced growth (length increase) in the axial direction. In single crystals of Zircaloy, the irradiation-induced stress-free growth is due to the formation of interstitial loops on prism planes and vacancy loops on basal planes [5–7]. In textured Zircaloy, with the majority of the basal poles at 30° to the radial direction, axial elongation and radial contraction take place with an insignificant change in the circumferential direction.

Ex-Reactor

The same model can be applied to determine the strain ratios for out-of-reactor test data (Eqs 1 and 2). For the 2:1 stress ratio (hoop to axial) test, the following equations are derived from Eq 2

$$\frac{\dot{\epsilon}_z}{\dot{\epsilon}_\theta} = \frac{1-R}{R}\frac{P}{2+P} \quad \text{or} \quad \frac{\epsilon_z}{\epsilon_\theta} = \frac{1-R}{R}\frac{P}{2+P} \tag{5}$$

And similarly for 1:1 stress ratio (hoop to axial) test

$$\frac{\dot{\epsilon}_z}{\dot{\epsilon}_\theta} = \frac{P}{R} \quad \text{or} \quad \frac{\epsilon_z}{\epsilon_\theta} = \frac{P}{R} \tag{6}$$

Depending on whether strain or strain rate is used, the R and P values for out-of-reactor test data can be calculated using the appropriate forms of Eqs 5 and 6.

For a uniaxial tensile or creep test along the tube axis (0:1 stress ratio), Eq 2 results in the following equations

$$\frac{\dot{\epsilon}_z}{\dot{\epsilon}_\theta} = \frac{R+1}{R} \quad \text{or} \quad \frac{\epsilon_z}{\epsilon_\theta} = \frac{R+1}{R} \tag{7}$$

The R value can be determined from the uniaxial test using the measured strain or strain rate ratio.

Boiling Water Reactor (BWR) Cladding Analysis

A BWR fuel assembly consists of a square array of fuel rods with upper and lower tieplates. The upper and lower endcaps of the fuel rods have long shanks that are captured in the upper and lower tieplates. Each BWR fuel assembly contains three different kinds of rods that are stressed differently.

1. Standard fuel rods have non-threaded upper and lower endcap shanks. Under the influence of the coolant and rod internal pressure, the cladding in a standard fuel rod is subjected to a biaxial state of stress with a ratio of hoop to axial stress of approximately 2:1.
2. Water rods do not contain fuel. Water flows through the rods, and the cladding in a water rod can be considered to be free of stress.
3. Tie rods are special fuel rods that have a lower endcap that is threaded into the lower tieplate and an upper endcap locking mechanism that holds the upper tieplate in place. The upper and lower tie plates are forced apart by spacer springs on standard fuel

rods and tie rods. Eight tie rods are typically used in a 9 by 9 fuel assembly, and these eight tie rods carry the force of 73 springs. Thus, the tie rods have an additional axial stress such that the hoop to axial stress ratio will be greater than 2:1.

Two types of post-irradiation measurements have been made: rod diameter and length. The fuel rod diameter was measured by the mechanical displacement of two rubbing shoes located on opposite sides of the fuel rod. The measured diameters were corrected for the oxide on the cladding surface. The rod length was measured by removing the upper tie plate and installing a reference plate with which the length of each rod can be measured directly. The rod diametral deformation is made up of two components: stress-induced thermal creep and stress-induced irradiation creep. Due to the texture of the cladding that has been measured, the effect of the stress-free irradiation growth on the diameter change is minor. All of the postirradiation data used in the analysis were measured before the onset of pellet-cladding interaction. There are three components in the measured rod length (axial deformation) changes: stress-induced thermal creep, stress-induced irradiation creep, and stress-free irradiation-induced growth.

To determine the values of R and P from the in-reactor measurements, the following assumptions were made [8]:

1. The thermal creep component in the post-irradiation measured deformation data can be neglected.
2. Contractile strain ratios of stress-induced irradiation creep do not change during irradiation.
3. Tie rods and standard fuel rods in a given fuel assembly experience an identical hoop stress.
4. The standard fuel rods have no axial stress other than from the system and internal gas pressures.
5. The thin wall approximation applies.

The measured length change of standard fuel rods minus the stress-free irradiation-induced component (determined through water rod growth measurements) gives the stress-induced creep component. Under thin-wall approximation, $\sigma_r = 0$ and $\sigma_\theta = 2\sigma_z$, and Eq 5 applies.

The stress-induced ratio, $\epsilon_z/\epsilon_\theta$, can be determined from the measured data. The values of R and P cannot be derived from Eq 5 alone. This equation only provides a correlation between R and P.

The stress-induced axial deformation of tie rods was obtained in the same way. Assuming a stress ratio, $x = \sigma_\theta/\sigma_z$, the stress-induced tie rod length change can be expressed as

$$\epsilon_z \propto \frac{\epsilon_g}{\sigma_g}\left(\frac{1-(x-1)R}{x(R+1)}\right)\sigma_\theta \tag{8}$$

The generalized stress in the tie rods can then be calculated from Eq 1 assuming that $\sigma_r = 0$

$$\sigma_g = \left[\frac{PR + R - \frac{2}{x}PR + \frac{PR}{x^2} + \frac{P}{x^2}}{P(R+1)}\sigma_\theta^2\right]^{1/2} \tag{9}$$

As indicated before, $\epsilon_g \propto \sigma_g^{1.23}$. With the provision that the standard fuel rods and the tie rods in a fuel assembly experience similar hoop stress early in life, it follows that

$$\frac{\epsilon_z \text{ tie rod}}{\epsilon_z \text{ fuel rod}} = \left[\frac{PR + R - \dfrac{2}{x} PR + \dfrac{PR}{x^2} + \dfrac{P}{x^2}}{R + \dfrac{PR}{4} + \dfrac{P}{4}} \right]^{-0.115} \frac{1 - (x - 1)R}{1 - R} \frac{2}{x} \qquad (10)$$

Using Eqs 5 and 10 and the measured fuel rod deformations, the R and P values for irradiation-induced creep of cold-worked stress-relieved Zircaloy-2 cladding were determined.

Laboratory Measurements

Five different types of cladding have been examined. Four types of standard production cladding, manufactured by different vendors, were made by a series of pilgering steps followed by recrystallization anneals. The fifth was LBQ cladding that received a beta-phase anneal and rapid quench before the next-to-last pilgering step. The final anneal for all cladding types was a stress relief anneal. Some of the general properties of the cladding types are given in Table 1.

The texture of the cladding has been measured by X-ray diffraction using the indirect pole figure method. The texture is defined by the Kearns factors that are obtained for the radial, circumferential, and axial directions. The Kearns factor gives the fraction of the basal poles oriented in each direction. Separate samples were made for each of the three directions. The sum of the Kearns factors should equal one.

In the standard CSR test, the cladding is strained to approximately 4% axial strain under axial loading. After removing the load, the changes in length and diameter of the cladding are measured and compared to the initial values. The measured length and diameter changes are used to calculate the circumferential and axial plastic strains. With the assumption of constant volume under plastic deformation, radial plastic strain can be determined. The circumferential strain of the cladding divided by the radial strain is defined as the R value. The CSR test is normally performed at room temperature. For this work, additional CSR tests were performed at 382°C.

Creep testing was performed at 340, 360, and 382°C. Samples were internally pressurized to obtain a hoop stress of 112 MPa (one test had a hoop stress of 120 MPa). The stress state in the cladding for this loading condition is very close to 2:1 with the axial stress being about half the hoop stress. In order to obtain other stress states, axial loads were applied to the samples with and without internal pressurization. To obtain a 1:1 stress state, an axial load equal to that exerted by the internal pressurization (56 MPa) was added to internally pressurized samples. To obtain a 0:1 stress state, an axial load of 56 MPa was applied to an unpressurized

TABLE 1—*General cladding characteristics.*

Cladding Type	Zircaloy Type	Tin Content, %	Cold Work in Last Pilger, %	Final Anneal Temperature, °C
A	2	1.5	80	500
B	4	1.5	80	500
C	4	1.5	30	500
D	4	1.3	80	470
LBQ	2	1.5	80	500

sample. Additional axial load tests were performed with a load of 112 MPa to obtain larger deformations. In order to distinguish between the various conditions and provide a comparison of the magnitude of the stresses, they are defined as 2:1, 2:2, 0:1, and 0:2, respectively. The tests were normally extended to 30 days, and the samples were removed at various time intervals (normally 3, 10, 20, and 30 days) to measure the change in length and diameter (with a digital vernier caliper and micrometer, respectively). The R and P values were calculated from the axial test data and by combining the data for the 2:1 and 2:2 stress ratio tests.

Results

In-Reactor

Rod length change data and rod creepdown data for Type A and LBQ fuel cladding in five reactors were used to determine the R and P values. Table 2 summarizes the number of data analyzed. The axial deformation data had one measured value per rod; multiple diametral deformation measurements were made along the rod at various axial locations.

Due to the somewhat different design, the spring force affected hoop to axial stress ratio in the tie rods varied from 2.3 to 2.8. Inserting these values and the measured circumferential (diametral) and axial strains of standard fuel and tie rods in Eqs 5 and 8, values for R and P were obtained. For the standard cold-worked, stress-relieved Zircaloy-2 (Type A) cladding at a stress-induced axial to circumferential strain ratio of 0.15, the values for R and P were 1.4 and 2.4, respectively. For the LBQ cladding, the R and P values were 2.5 and 2.0, respectively [9]. The uncertainties (one standard deviation) for the R and P values are about 5 and 15%, respectively.

Ex-Reactor

In order to compare the various cladding types, the texture and room temperature R-values are listed in Table 3. The Kearns factors for each cladding were not normalized to one. Significant differences are seen between the various cladding types. The texture of the Type B cladding is significantly less radially oriented than that for the other types. This is also seen in the fact that the R-value for this type is also lower than that for the others. No significant differences in texture seem to be present between the other standard cladding types. The circumferential Kearns factor for the LBQ cladding appears to be higher than for the other types. In general, the R-value that is measured at room temperature appears to correlate well with the amount of radial texture (F_R) developed in the cladding.

TABLE 2—*Database for BWR cladding analysis.*

	Assemblies		Rods		Data Points	
	Type A	LBQ	Type A	LBQ	Type A	LBQ
Fuel rod diametral deformation	14	7	219	86	1377	527
Fuel rod axial deformation	14	6	219	161	210	161
Water rod axial deformation	7	2	13	2	13	2
Tie rod axial deformation	7	2	28	16	28	16

TABLE 3—*Texture and room temperature CSR measurements for the cladding types.*

Cladding Type	Texture, Kearns factor			Contractile Strain Ratio
	Radial	Circumferential	Axial	
A	0.64	0.31	0.08	2.09
B	0.52	0.46	0.09	1.29
C	0.60	0.32	0.08	1.99
D	0.63	0.33	0.07	2.00
LBQ	0.61	0.37	0.08	1.98

The results of the creep tests are shown in Figs. 1a through e with the strain in the axial and circumferential directions plotted as a function of time. The creep strains for Types B and C are significantly lower than for the other types. For Type B, this is due to the less radially oriented texture, and, for Type C, because there was less cold work in the last pilger step. The lower creep rate for the Types B and C cladding has also been observed in the post-irradiation creepdown measurements (see Table 4). The higher creep rate for Type D is probably due to a slightly lower final annealing temperature.

As shown in Figs. 1a through e, the creep tests measured primary and secondary creep. Based on Eq 2-2 of Ref 10, the thermal creep strain for primary and secondary creep may be expressed by the following correlation

$$\epsilon_{th} \propto \sqrt{t}$$

Regression analyses of the creep test data were performed with this correlation, and the results were used to calculate the contractile strain ratios.

The results are given in Table 5. The R values obtained from the uniaxial creep tests for the standard cladding were generally much lower than values obtained from the biaxial creep tests. The differences were smaller at 340 and 360°C than at 382°C. Also, for the LBQ cladding, the R values obtained were higher than for the standard cladding. The 0:2 tests are more reliable than the 0:1 tests because larger strains were measured, which increases the accuracy.

The contractile strain ratio values obtained from the tensile tests at 382°C are given in Table 6. The R values are consistent with those obtained for the creep tests for the standard cladding run at the same temperature. The 0:1 and 0:2 tests (performed with an axial load) were not found to yield representative R values in the creep tests. Therefore, it appears that different slip systems must be active for the faster strain rate used in the tensile test than in the slower creep test. The performance of the LBQ cladding is significantly different from that of the standard cladding. For this material, similar values for R are obtained at room temperature and 382°C.

Discussion

In compression tests using zirconium single crystals, Tenckhoff [11] has observed that the deformation system changes with temperature. The predominant twinning system seen at room temperature differed from that observed at 250 and 500°C. Jensen and Backofen [12] have studied the deformation of zirconium and Zircaloy-4 sheets with strongly developed textures. For both materials, they observed that there is a change in deformation mechanisms at about 300°C. Below 300°C, the Zircaloy-4 deformed mainly by slip, but above 300°C twinning was also observed. Stehle et al. [13] observed that the mechanical anisotropy measured for Zircaloy

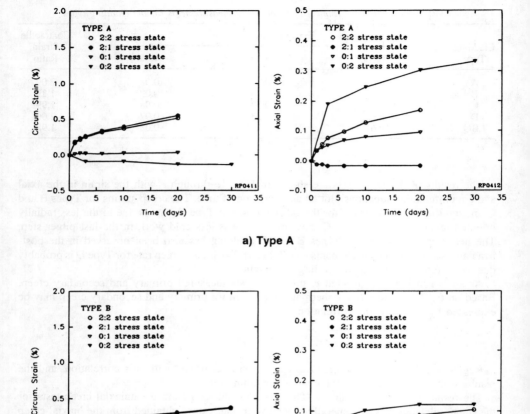

FIG. 1 (a and b)—*Circumferential and axial creep curves for cladding samples at 382°C. The ratio of the hoop stress to the axial stress is indicated with "2" representing 112 MPa.*

cladding at 400°C is low compared to that observed at room temperature. They concluded that additional slip systems are probably operative at the higher temperatures. Therefore, it appears that the deformation of Zircaloy cladding may occur by different mechanisms at room temperature and 382°C. On the other hand, Beauregard et al. [*14*] have reported that no change in R value for Zircaloy-4 cladding (annealed at 500°C) was seen over a test temperature range of 20 to 800°C. The R value for the cladding they evaluated was about 1.4, indicating that the cladding did not have a strong radial texture. We saw no significant change in the R values between room temperature and 382°C for Type B cladding, which also did not have a strongly radially oriented texture.

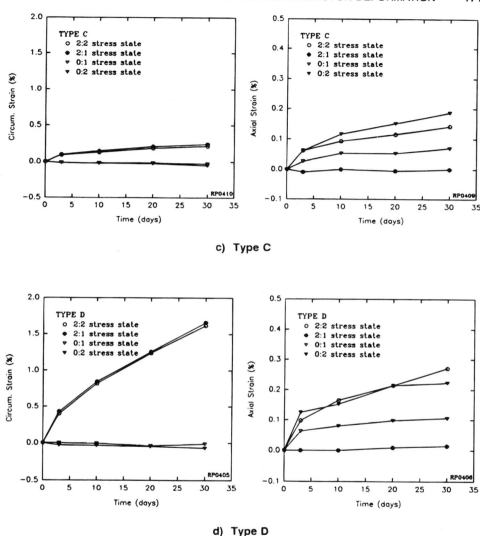

c) Type C

d) Type D

FIG. 1 (c and d)—*Continued*.

In our results, we have seen three different temperature effects. The highly textured standard cladding types that we have examined in this study have had lower R values at 382°C than at room temperature. The cladding that was not highly textured (Type B) had about the same low R value at both temperatures. The LBQ cladding had a high R value at both temperatures. It appears that the cladding with the low radial texture has no change in deformation mechanism which is consistent with the results of Beauregard et al. [*14*] with similarly textured cladding. However, the highly textured cladding undergoes a change in deformation mechanisms such that the R values at 382°C are significantly lower than that at room temperature. For LBQ cladding, which has a somewhat higher density of circumferentially oriented basal poles, there

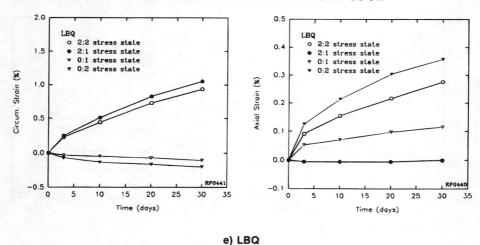

e) LBQ

FIG. 1 (e)—Continued.

TABLE 4—Comparison of in-reactor and out-of-reactor thermal creep.

Cladding Type	Out-of-Reactor Thermal Creep, %[a]	Relative In-Reactor Creepdown, %
A	49/0.60/0.07[b]	
B	4/0.27/0.02	0.81 (9)[c]
C	8/0.22/0.03	0.81 (36)
D	7/0.72/0.01	1.11 (24)
LBQ	12/0.65/0.10	1.0 (25)

[a]Ten days/382°C/120 MPa hoop stress at 2:1 hoop to axial stress ratio.
[b]Number of measurements/average circumferential strain/standard deviation.
[c]Relative creepdown with respect to cladding Type A at the same burnup (burnup in MWd/kgU).

does not appear to be a change in deformation mechanisms; the operative mechanism at room temperature appears to also be operative at the higher temperature.

The R value that has been obtained from the in-reactor measurements of the Type A cladding in BWRs is 1.4. This is much lower than the R value of about 2 that is obtained for the cladding in room temperature CSR testing normally performed on cladding. Although the result of the normal CSR test gives a good indication of the degree of radial texture development in the cladding, it does not yield a value representative of the in-reactor deformation. CSR tests and biaxial creep tests performed at 382°C have yielded R values of about 1.2 which is closer to the in-reactor result. The R value obtained for cladding with a poor radial texture is lower than the others at room temperature, but at 382°C, it is the same as that with the more developed radial texture. Therefore, the in-reactor deformation of this cladding would be expected to be similar to the others. Unfortunately, at this point, there is not sufficient in-reactor data to evaluate Types B, C, and D cladding. Currently, unstressed samples of these materials are being irradiated to determine the irradiation growth component to the deformation. The in-reactor R value for LBQ cladding is 2.5. The biaxial creep test results gave an R value similar to that for standard cladding, but the axially loaded creep tests and the high temperature CSR test gave a higher value for the LBQ cladding. Therefore, it may be, at least in a qualitative

TABLE 5—R and P values calculated from creep test results.

Cladding Type	Temperature, °C	Stress Ratio, σ_θ/σ_z	R and P	
A	340	2:1 and 2:2	$R = 1.5$	$P = 1.0$
	340	0:1	$R = 1.1$	
	360	2:1 and 2:2	$R = 1.1$	$P = 0.6$
	360	0:1	$R = 1.1$	
	382	2:1 and 2:2	$R = 1.4$	$P = 0.4$
	382	2:1 and 2:2	$R = 1.2$	$P = 0.4$
	382	0:1	$R = 1.1$	
	382	0:2	$R = 0.7$	
B	382	2:1 and 2:2	$R = 1.2$	$P = 0.3$
	382	0.1	$R = 0.2$	
	382	0:2	$R = 0.2$	
C	382	2:1 and 2:2	$R = 1.1$	$P = 0.7$
	382	0:1	$R = 0.6$	
	382	0:2	$R = 0.3$	
D	382	2:1 and 2:2	$R = 1.1$	$P = 0.2$
	382	0:1	$R = 0.2$	
	382	0:2	$R = 0.4$	
LBQ	382	2:1 and 2:2	$R = 1.1$	$P = 0.3$
	382	0:1	$R = 3.9$	
	382	0:2	$R = 1.3$	

TABLE 6—382°C CSR measurements for the cladding types.

Cladding Type	CSR Values		
	Samples	Average	Standard Deviation
A	10	1.29	0.29
B	4	1.27	0.15
C	2	1.21	0.30
D	6	1.16	0.31
LBQ	6	1.91	0.35

manner, possible to have a short-term laboratory test that can provide cladding R values that can be used to predict in-reactor deformation behavior.

The P values that were obtained in the 382°C creep tests were in the range of 0.2 to 0.8, which is very different from that obtained in-reactor. From similar creep testing of cold-worked, stress-relieved cladding at 400°C, Murty and Adams [15] obtained R and P values of 1.5 and 0.4, respectively. Therefore, it appears that these out-of-reactor tests are only useful for obtaining R values for comparison to the in-reactor measurements.

Conclusions

The deformation of standard highly radially textured, stress-relieved cladding in a BWR is described by R and P factors of about 1.4 and 2.4, respectively. For LBQ cladding, values of 2.5 and 2.0, respectively, were obtained.

Creep testing at 382°C with a hoop stress of 112 MPa and a biaxial stress state of hoop stress/axial stress of 2:1 and 1:1 yields R values consistent with the in-reactor results for standard cladding but not LBQ cladding.

CSR testing performed at 382°C yields R values that are representative of the in-reactor deformation for both types of cladding. The R values obtained for several types of cladding vary significantly between the room temperature CSR testing and the creep and CSR testing performed at 382°C.

References

[1] Van Swam, L. F., Knorr, D. B., Pelloux, R. M., and Shewbridge, J. F., "Relationship Between Contractile Strain Ratio R and Texture in Zirconium Alloy Tubing," *Metallurgical Transactions*, Vol. 10A, April 1979, p. 423.

[2] Murty, K. L. and Adams, B. L., "Biaxial Creep of Textured Zircaloy I: Experimental and Phenomenological Descriptions," *Materials Science and Engineering*, Vol. 70, 1985, pp. 169–180.

[3] Gittus, J. H., Howl, D. A., and Hughes, H., "Theoretical Analysis of Cladding Stress and Strains Produced by Expansion of Cracked Fuel Pellets," *Nuclear Applications and Technology*, 1970, pp. 40–46.

[4] Christensen, R., "SPEAR Fuel Reliability Code System, General Description," EPRI-NP-1378, Electric Power Research Institute, Palo Alto, CA, 1980, pp. 5–8, 5–9.

[5] Buckley, S. N., in *Irradiation Growth, in Properties of Reactor Materials and Effects of Radiation Damage*, D. J. Littler, Ed., Butterworths, London, 1962.

[6] *Creep of Zirconium Alloys in Nuclear Reactors, ASTM STP 815*, D. G. Franklin, G. E. Lucas, and A. L. Bement, Eds., American Society for Testing and Materials, Philadelphia, 1983.

[7] Fidleris, V., Tucker, R. P., and Adamson, R. B., "An Overview of Microstructure and Experimental Factors That Affects Growth Behavior of Zirconium Alloys," *Zirconium in the Nuclear Industry: Seventh International Symposium, ASTM STP 939*, R. B. Adamson and L. F. P. Van Swam, Eds., American Society for Testing and Materials, Philadelphia, 1987, pp. 49–85.

[8] Shann, S. H. and Van Swam, L. F., "Creep Anisotropy of Zircaloy-2 Cladding During Irradiation," *Transactions*, 11th International Conference on Structure Mechanics in Reactor Technology, Paper C 03/5, Vol. C, International Association for Structural Mechanics in Reactor Technology, 1991, pp. 67–72.

[9] Shann, S. H. and Van Swam, L. F., "Influence of Manufacturing Process on the In-Reactor Creep Anisotropy of Stress Relieved Zircaloy-2 Cladding," *Proceedings*, SMiRT-12 Conference, Paper C 02/1, International Association for Structural Mechanics in Reactor Technology, Stuttgart, Germany, 15–20 Aug. 1993.

[10] Lindquist, K., Henson, J., and Santucci, J., "Evaluation and Modification of COMETHE III-J," EPRI-NP-2911, Electric Power Research Institute, Palo Alto, CA, 1983.

[11] Tenckhoff, E., "{1071} Zwillingsbildung in Zirkonium-Einkristallen bei erhöhten Verformungstemperaturen," *Zeitschrift für Metallkunde*, Vol. 63, 1972, pp. 729–734.

[12] Jensen, J. A. and Backofen, W. A., "Deformation and Fracture of Alpha Zirconium Alloys," *Canadian Metallurgical Quarterly*, Vol. 11, 1972, pp. 39–51.

[13] Stehle, H., Steinberg, E., and Tenckhoff, E., "Mechanical Properties, Anisotropy, and Microstructure of Zircaloy Canning Tubes," *Zirconium in the Nuclear Industry (3rd Conference), ASTM STP 633*, A. L. Lowe, Jr., and G. W. Perry, Eds., American Society for Testing and Materials, Philadelphia, 1977, pp. 486–507.

[14] Beauregard, R. J., Clevinger, G. S., and Murty, K. L., "Effect of Annealing Temperature on the Mechanical Properties of Zircaloy-4 Cladding," *Proceedings*, SMiRT-IV, Paper C3/5, San Francisco, 1977.

[15] Murty, K. L. and Adams, B. L., "Multiaxial Creep of Textured Zircaloy-4," *Mechanical Testing for Deformation Model Development, ASTM STP 765*, Rhode and Swearengen, Eds., American Society for Testing and Materials, Philadelphia, 1982, pp. 382–396.

DISCUSSION

H. S. Rosenbaum[1] (written discussion)—To what extent could differences in the microstructures of the materials (degree of recrystallization, grain shape anisotropy) influence your results? Would you expect your data to be applicable to fully recrystallized materials?

R. A. Perkins and S. H. Shann (authors' closure)—All of our results are for stress-relieved cladding. As the cladding becomes recrystallized and rotation of the prism planes around the basal poles occurs, this would certainly affect our observations. For example, we have observed that the room temperature CSR value of the cladding increases as the degree of recrystallization increases although the radial texture (F_R) remains the same. Therefore, our results for stress-relieved cladding would not be applicable to recrystallized cladding.

R. J. Comstock[2] (written discussion)—Your laboratory tests were performed at much higher stresses than those experienced by the cladding during in-reactor exposure. Can you comment on the assumptions one must make in trying to simulate in-reactor deformation by a high stress, short-term laboratory test and whether or not the assumptions are valid for Zircaloy?

R. A. Perkins and S. H. Shann (authors' closure)—Our study was made to determine if a high strain rate test performed out-of-reactor could be used to predict R and P values that are representative of in-reactor performance. On this basis, no assumptions were made before the study. The lack of a fit for the P values shows that we have not duplicated the in-reactor behavior. However, the R value is more critical in that it shows more variation in-reactor.

R. Holt[3] (written discussion)—In out-of-reactor creep tests at high stress, the sensitivity of the creep rate to the magnitude of stress is large (for example, fifth power). For in-reactor creep at reactor operating stresses, the sensitivity of creep rate to stress is approximately linear. Our calculations based on the models described in our earlier paper show that creep anisotropy is different with different stress sensitivities even if the slip systems are the same. Further, the Hill formulation is not able to describe the anisotropy and high stress sensitivity.

R. A. Perkins and S. H. Shann (authors' closure)—Communications with Dr. K. L. Murty (North Carolina State University) and Dr. G. E. Lucas (University of California, Santa Barbara) were made to estimate the stress at which the dependency of strain rate on stress changes from a linear function to a fifth order power function. Based on the SMiRT-12 paper C02/4 and from Ref 6, the stress level used in our tests (112 MPa) is close to but still lower than the transition stress for the tested cladding (80% cold work in the last pilger step).

We have performed limited creep tests at 85 MPa, and the results are in agreement with the 112 MPa data. Therefore, it is likely that our laboratory creep tests are in the same stress dependence regime as experienced in-reactor.

D. Schrire[4] (written discussion)—Donaldson has shown that the growth of an oxide layer on a Zircaloy tube can cause an axial tensile stress that results in axial creep. Hydrides may also increase the dimensions of a Zircaloy component. Were there any measurements performed to ascertain the oxide and hydride levels in the cladding tubes and water rods, to ensure that these effects did not make a significant contribution to the measured in-pile elongation?

R. A. Perkins and S. H. Shann (authors' closure)—The oxide thickness on all the tubing has been measured. At the low uniform oxide thicknesses that are obtained in the boiling water reactors (usually less than 30 μm) and the normal wall thickness of about 750 μm, the contribution to the increase in length of the fuel rods due to oxidation is negligible. The water

---

[1] General Electric Nuclear, San Jose, CA 85125.
[2] Westinghouse Electric Corporation, Science & Technology Center, Pittsburgh, PA 15235.
[3] AECL, Chalk River Laboratories, Chalk River, Ontario, Canada.
[4] ABB ATOM, 5-72163 Västerås, Sweden.

rods would have oxide on both the inner and outer surface, but the oxide thickness is less than for fuel rods and the temperature of the water rods is lower. From Donaldson's results, a very minor contribution to elongation due to the stress from the oxide would be expected at 280°C. No direct measurement of the hydrogen content has been made. The hydrogen content of the tubing measured in this study should be less than 300 ppm based on the oxide measurements. This would not cause a significant change in the length.

E. Ross Bradley[1] *and Anna-Lena Nyström*[2]

Microstructure and Properties of Corrosion-Resistant Zirconium-Tin-Iron-Chromium-Nickel Alloys

REFERENCE: Bradley, E. R. and Nyström, A.-L., "**Microstructure and Properties of Corrosion-Resistant Zirconium-Tin-Iron-Chromium-Nickel Alloys,**" *Zirconium in the Nuclear Industry: Tenth International Symposium, ASTM STP 1245*, A. M. Garde and E. R. Bradley, Eds., American Society for Testing and Materials, Philadelphia, 1994, pp. 483–498.

ABSTRACT: Long-term autoclave tests in steam at 673 K and mechanical property (tensile and creep) tests have been conducted to evaluate the effects of chromium (0 to 0.3%) and nickel (0 to 0.08%) additions on the corrosion and mechanical properties of dilute zirconium-tin-iron alloys. The base alloys contain 0.4% iron and either 0.5% or 1.0% tin.

Multiple regression analysis showed tin concentration to have a statistically significant effect on corrosion, hydrogen pickup, creep, and tensile strength. The corrosion resistance of the alloys increased significantly when the tin concentration decreased from 1.0 to 0.5%. An additional improvement in corrosion resistance was observed with increasing chromium content. Nickel additions had no apparent effect on corrosion for the alloy compositions examined. Lower hydrogen pickup fractions were measured for alloys containing 0.5% tin compared to alloys containing 1.0% tin. No significant effects of chromium or nickel additions to hydrogen pickup were observed.

Creep rate decreased with increasing tin concentration and to a lesser extent with increasing chromium content. Nickel additions produced a large increase in creep rate for some alloys but not others. This suggests a possible interaction between nickel and the other alloy additions in determining creep strength.

Intermetallic particle type and size distribution varied with alloy composition. Chromium additions tended to decrease the particle size distribution of $Zr(Fe,Cr)_2$ intermetallic particles. Conversely, nickel additions produced large particles of the $Zr_2(Fe,Ni)$ type intermetallic. Changes in the intermetallic particle distributions and matrix compositions are discussed with regard to the observed changes in corrosion and mechanical properties of the alloys.

KEY WORDS: zirconium alloys, autoclave testing, uniform corrosion, tensile properties, creep properties, hydrogen pickup, intermetallic precipitates, zirconium, nuclear materials, nuclear applications, radiation effects

The current trend to extend discharge burnups for improved fuel utilization in pressurized water reactors (PWRs) imposes additional demands on the performance of Zircaloy-4 fuel cladding. Waterside corrosion is an area of particular concern and may limit the achievable burnup for existing alloys. Other concerns for implementing extended discharge burnups include hydrogen pickup and mechanical properties of the cladding.

To meet the challenge of extended burnups, an extensive effort has been made over the past ten years to optimize the corrosion performance of Zircaloy-4. Thermal history was

---

[1] Senior development metallurgist for zirconium, Sandvik Special Metals Corporation, Kennewick, WA 99336-0027.
[2] Research metallurgist, AB Sandvik Steel, S-811 81 Sandviken, Sweden.

identified as an important factor in corrosion resistance and was modeled by a cumulative annealing parameter [1,2]. Although differences exist in the various formulations of the "anneal parameter," it is generally agreed that high anneal parameters that produce large intermetallic particles improve the corrosion resistance of Zircaloy in PWR environments. However, Bradley [3] found that anneal temperature was more important than anneal parameter in determining corrosion behavior, and particle size did not correlate with corrosion behavior. Similar conclusions have recently been reported by Perkins [4].

Additional improvements in corrosion performance of the Zircaloys has been achieved by controlling the composition within the ASTM Specification for Zirconium and Zirconium Alloy Ingots for Nuclear Application (B 350-91) range for Zircaloy-4 and Zircaloy-2 [5,6]. The tin concentration is controlled near the lower limit for both Zircaloy-4 and Zircaloy-2, while the iron and nickel contents are controlled near the upper end of the range for Zircaloy-2. Although these modifications in composition and processing have improved the corrosion resistance of the Zircaloys, new alloys are required to meet the challenge of higher burnups in modern high-temperature reactors.

Decreasing the tin concentration below the current specification levels for the Zircaloys provides an attractive means for improving the corrosion resistance as can be seen by reviewing the early work on the development of Zircaloys [7]. Tin was originally added to zirconium to overcome the adverse corrosion effects of nitrogen, which was an inherent impurity in the early manufacture of zirconium sponge. Additions of iron, chromium, and nickel were found to improve the corrosion resistance of zirconium-tin alloys and led to the development of Zircaloy-2. Further improvements in corrosion resistance were obtained by decreasing the tin concentration, Zircaloy-3A, -3B, and -3C, but were not commercially pursued because of decreased strength and increased "stringer" formation. Zircaloy-4 was developed at a later date to reduce the hydrogen pickup caused by the nickel addition to Zircaloy-2. Because nitrogen and other impurities are not a problem in modern zirconium production, improvement of the corrosion resistance by reducing the tin concentration is possible if acceptable mechanical properties and hydrogen pickup fractions can be obtained.

This study examines the effects of chromium and nickel additions on the corrosion, hydrogen pickup, and mechanical properties of Zr-0.5Sn-0.4Fe (Zircaloy-3B) and Zr-1.0Sn-0.4Fe-based alloys. Microstructural examinations were included to aid in evaluating the observed behavior.

Experimental Procedures

Materials

Thirteen alloy slabs (35 by 15 by 110 mm) were prepared by tungsten electrode arc melting in a low partial pressure of ultra-high purity argon gas. To ensure homogeneity, each alloy was remelted at least three times with the previous ingot cut into small sections, inverted, and repositioned lengthwise in the water-cooled copper mold prior to remelting. Chemical analysis from each end of selected ingots verified the homogeneity of the melting practice. The composition of the thirteen alloys is given in Table 1.

The small ingots were processed into strip material by the combination of hot- and cold-working operations shown in Fig. 1. After beta-quenching at 8-mm thickness, the intermediate hot rolling and anneals were all conducted at 975 K. Final vacuum anneal conditions were 773 K for 3.5 h that produced a partly recrystallized structure in the alloys.

Testing

Long-term corrosion tests were conducted on alloy coupons (40 by 25 by 0.8 mm) in a static steam autoclave operating at 673 K and 10.4 MPa pressure. The coupons were etched

TABLE 1—*Chemical composition of alloys.*

Alloy	Composition, % by weight				
	Sn	Fe	Cr	Ni	O_2
S2-1	0.47	0.39	0.04	0.08	0.10
S2-2	0.50	0.40	0.09	0.00	0.11
S2-3	0.53	0.41	0.20	0.00	0.12
S2-4	0.52	0.41	0.00	0.05	0.12
S2-5	0.52	0.41	0.09	0.05	0.10
S2-6	0.47	0.39	0.19	0.05	0.10
S2-7	1.04	0.41	0.00	0.00	0.11
S2-8	1.00	0.39	0.10	0.00	0.12
S2-9	0.93	0.35	0.18	0.00	0.11
S2-10	0.92	0.35	0.00	0.05	0.11
S2-11	0.92	0.37	0.09	0.05	0.11
S2-12	0.91	0.38	0.18	0.05	0.12
S2-13	1.02	0.39	0.28	0.00	0.11

in an aqueous hydrofluoric-nitric acid mixture, rinsed in running water, and finally rinsed in hot distilled water prior to autoclaving. Weight gains were periodically measured at approximately 14-day intervals and the total exposure time was 518 days. The general test procedures were in accordance with ASTM Practice for Aqueous Corrosion Testing of Samples of Zirconium and Zirconium Alloys (G 2-88). After 518 days of exposure, the corrosion coupons were analyzed for hydrogen by the inert gas fusion technique using a LECO Model RH-1 gas analyzer.

Specimens for tension and creep testing were machined from the 0.8-mm-thick strip material with the tensile axis parallel to the rolling direction. Tension tests were conducted at room temperature using procedures in general agreement with ASTM Test Methods for Tension Testing of Metallic Materials (E 8-93).

Creep tests were conducted at 673 K with the axial stress being 120 MPa. The axial strain was intermittently measured; total creep time varied between 70 and 1200 h depending on the creep behavior of the specific alloy.

Specimens for optical metallography (OM) and scanning electron microscopy (SEM) were prepared by standard metallographic techniques. The chemical etch for OM specimens consisted of a 60-mL glycerine, 30-mL HNO_3, 40-mL HF solution followed by electro-etching in a 9% H_2SO_4 electrolyte at 30 V.

Specimens for SEM examinations were chemically polished and etched using a 60% glycerine, 10% HNO_3, 30% HF solution. This etch was followed by immersion in warm (50°C) HNO_3 solution to remove surface artifacts produced by the etch. The specimens were examined in a JEOL JSM-840 microscope, equipped with a LINK AN100000 energy dispersive X-ray spectrometer (EDS).

Thin foils for analytical transmission electron microscopy (ATEM) examinations were prepared by grinding strip material to approximately 0.1 mm thickness. Disks, 3 mm in diameter, were then punched out of the strip and electrochemically thinned to perforation in a 15% perchloric acid in methanol solution. The thin foils were examined in a JEOL 2000FX transmission electron microscope (TEM) containing a LINK AN100000 EDS analysis system.

The effect of composition on the corrosion and mechanical properties of the alloys was evaluated statistically by a standard multiple linear regression analysis program. Output of the program included the linear regression coefficients and level of significance for each variable. For the present work, an addition was considered to have a statistically significant effect if

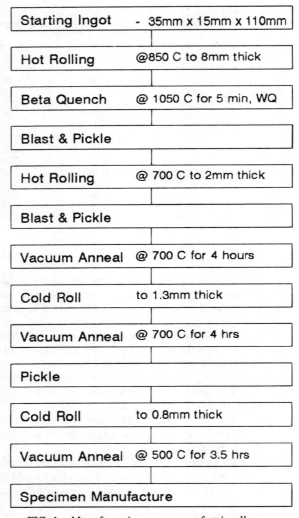

FIG. 1—*Manufacturing sequence of strip alloys.*

the probability of producing the same effect randomly was less than 5%. Interactions between variables or nonlinear effects were not considered in the analysis.

Results

Mechanical Properties

Tin concentration was the most important variable affecting the mechanical properties of the alloys as shown in Table 2. The room temperature yield and tensile strength increased by 40 and 30 MPa, respectively, when the nominal tin content increased from 0.5 to 1.0%. Multiple regression analysis of the yield strength data showed tin to be the only statistically

TABLE 2—*Summary of mechanical properties.*

Alloy	Yield Strength, MPa	Tensile Strength, MPa	Elongation, %	Steady-State Creep Rate, %/h
S2-1	362	464	21	...
S2-2	374[a]	474[a]	27[a]	...
S2-3	402	507	20	0.024
S2-4	379[a]	473[a]	20[a]	...
S2-5	340[a]	444[a]	34[a]	...
S2-6	370	475	20	0.092
S2-7	412	504	24	0.007
S2-8	426[a]	517[a]	26[a]	0.003
S2-9	405	489	23	0.004
S2-10	386[a]	474[a]	27[a]	0.016
S2-11	408[a]	504[a]	27[a]	0.004
S2-12	409[a]	514[a]	28[a]	0.004
S2-13	435	523	22	0.001
Zircaloy-4	504[a]	613[a]	21[a]	0.001
Zircaloy-2	490[a]	602[a]	22[a]	0.001

[a] Average of two samples.

significant composition variable. Variations in nickel or chromium did not affect yield or tensile strength.

The steady-state creep rate values listed in Table 2 were determined by regression analysis of the linear regions of the creep strain versus time measurements. Correlation coefficients were greater than 0.99, indicating the good fit of the data to the assumed linear steady-state creep regime. Alloys containing 0.5% tin and lower chromium concentrations exhibited extremely high creep rates and were not included in the analysis. The test temperature and stress level were too high for this class of alloys to make meaningful comparisons among the alloys. However, some general trends within the various alloys can be made from the existing creep data.

The effect of tin and nickel on creep is illustrated in Fig. 2. Increasing the tin concentration from 0.5 to 1.0% dramatically decreases creep strain. Creep rates for the 1.0Sn-0.4Fe-0.2Cr- (0 and 0.05% nickel) alloys approach those of Zircaloy-2 and Zircaloy-4, but the higher tin level of the Zircaloys provides additional creep strength.

Another interesting feature shown in Fig. 2 is the effect of nickel on creep rate. Nickel adversely affects the creep strength of the 0.5Sn-0.4Fe-0.2Cr alloys, but has no significant effect for similar alloys containing 1.0% tin or the Zircaloys. However, an adverse effect of nickel on creep strength was also observed for 1.0% tin alloys containing less than 0.2% chromium as shown in Fig. 3. It appears that nickel additions adversely affect creep strength under certain conditions, but there is insufficient information to quantify the potential interactions based on the present data.

Increasing the chromium concentration tended to increase the creep strength of the various alloy groups. This tendency is shown in Fig. 4 for the 1% tin alloys without nickel additions. Although a positive correlation is observed in Fig. 4, multiple regression analysis did not indicate a statistically significant effect of chromium on creep rate. The experimental variability in the creep data obtained from the thin strip materials was too large to give a high degree of significance to the linear correlation.

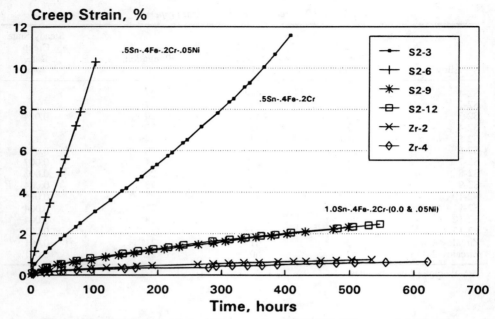

FIG. 2—*Effect of tin and nickel additions on the 673 K, 120 MPa creep behavior.*

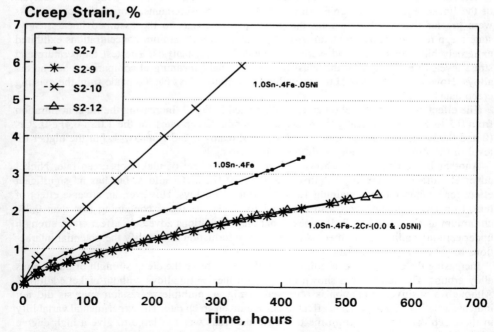

FIG. 3—*Effect of nickel and chromium additions on the 673 K, 120 MPa creep behavior.*

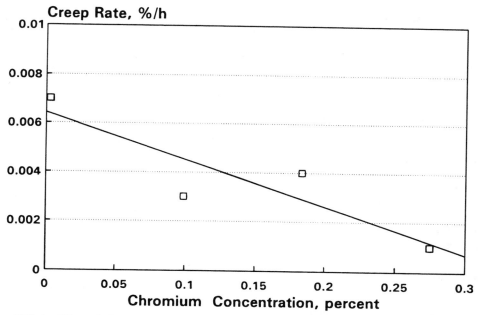

FIG. 4—*Effect of chromium concentration on the 673 K, 120 MPa steady-state creep rate of Zr-1.0Sn-0.4Fe-xCr alloys.*

Corrosion and Hydrogen Pickup

Results from weight gain and hydrogen pickup measurements after 518 days of exposure in the 673 K, 10.3 MPa steam autoclave environment are summarized in Table 3. Each data point represents the average of two measurements.

TABLE 3—*Summary of weight gain and hydrogen analysis (673°K steam autoclave test for 518 days).*

Alloy	Weight Gain, mg/dm^2	Hydrogen Concentration, ppm	Hydrogen Pickup Fraction, %
S2-1	169.1	490	54
S2-2	149.6	409	50
S2-3	146.5	399	51
S2-4	182.6	502	48
S2-5	157.2	460	51
S2-6	138.1	461	62
S2-7	204.9	708	65
S2-8	200.4	729	67
S2-9	177.3	633	69
S2-10	199.0	676	65
S2-11	192.2	752	68
S2-12	183.9	608	60
S2-13	180.5	634	68

Weight gain curves produced by 673 K steam autoclave testing are shown in Figs. 5 and 6, respectively, for the 0.5% and 1.0% tin alloys. The alloys show cyclic behavior throughout the 518-day exposure period. At 518 days of exposure, weight gains for the 0.5% tin alloys ranged from 137 to 186 mg/dm^2 compared to 179 to 209 mg/dm^2 for the 1.0% tin alloys.

Multiple regression analysis of the 518-day weight gain data identified tin and chromium as important factors at the 1% significance level. Nickel additions had no significant effect on the measurements. Weight gain increased with increasing tin concentration and decreased with increasing chromium concentration as shown in Fig. 7. Chromium was more effective in reducing corrosion in the 0.5% tin alloys as evidenced by the difference in slope of the two curves. For alloys containing 0.18 to 0.20% chromium, reducing the tin content from 1.0 to 0.5% produced approximately a 25% reduction in weight gain. A similar comparison for alloys containing no chromium shows about 10% reduction in weight gain for the low tin alloys.

Hydrogen pickup fraction as a function of the chromium content of the alloys is shown in Fig. 8. The data show tin concentration to have a significant effect at the 1% significance level. The average pickup fraction for the 1% tin alloys was 0.67 compared to 0.53 for the 0.5% tin alloys. Chromium or nickel additions had no significant effect on the hydrogen pickup in these alloys.

Microstructure

A partly recrystallized microstructure with various particle distributions was observed by the various examination techniques. Evidence of incomplete recrystallization was seen only by TEM examinations. The degree of recrystallization varied extensively in localized regions, making quantitative estimates extremely difficult. Detailed studies of selected variants indicated

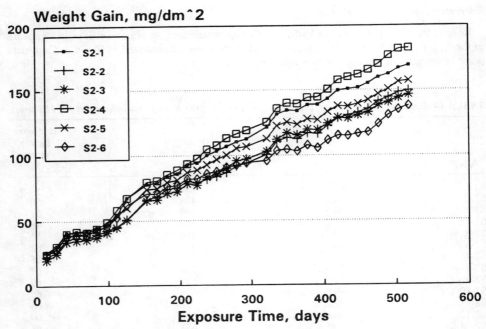

FIG. 5—*Corrosion weight gain as a function of exposure time for the 0.5% tin alloys.*

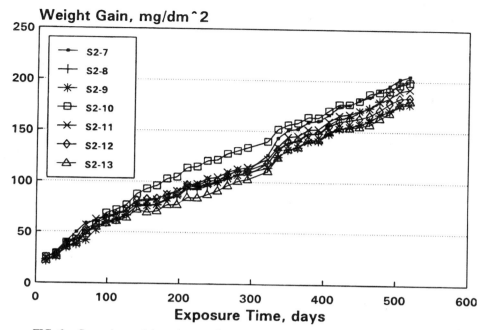

FIG. 6—*Corrosion weight gain as a function of exposure time for the 1.0% tin alloys.*

the area fraction of recrystallized material ranged from about 30 to 60% with no definite correlations between alloy composition or particle distribution being observed. However, the area of sample examined by the TEM technique was necessarily small, 0.003 mm^2, and the measured degree of recrystallization can only be considered as a qualitative estimate.

Three types of intermetallic particles were identified by the ATEM examinations. The base alloys without chromium or nickel additions contained a coarse distribution of zirconium-iron type precipitates. Adding chromium to the alloys produced a fine distribution of $Zr(Fe,Cr)_2$, while $Zr_2(Fe,Ni)$ and $Zr(Fe,Cr)_2$-type precipitates were found in alloys containing both chromium and nickel.

The effect of chromium additions on particle size distributions is shown in Fig. 9. A dramatic shift to smaller particles is seen in the distribution for Alloy S2-6 that contains 0.2% chromium. Average particle size for Alloy S2-6 was 0.28 μm compared to 0.49 μm for Alloy S2-4.

Figure 10 illustrates the independent effect of nickel on the particle size distributions. The two alloys have similar compositions except for the addition of 0.5% nickel in Alloy S2-6. The size distributions for the two alloys are very similar below 0.8 μm that primarily reflects the size distribution of the $Zr(Fe,Cr)_2$ precipitates. The larger particles in Alloy S2-6 can be attributed to the $Zr_2(Fe,Ni)$-type precipitates.

Discussion

Results of this study show the tin concentration has the predominant influence on both the mechanical properties and corrosion behavior of the alloys examined. Decreasing the tin concentration improves the corrosion resistance and hydrogen uptake but decreases the tensile strength and creep resistance. Chromium additions to the base Zr-Sn-0.4Fe alloys tended to

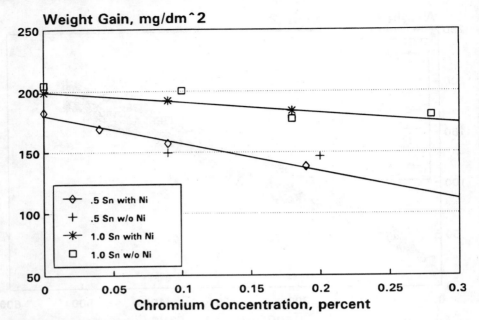

400 C Autoclave Data (518 days)

FIG. 7—*Corrosion weight gain as a function of chromium concentration for 0.5% and 1.0% tin alloys.*

improve both the corrosion resistance and creep resistance while nickel additions had no effect on corrosion resistance and a possible detrimental effect on creep behavior. These observations regarding the effect of alloy additions on the mechanical properties, corrosion, and hydrogen pickup are discussed separately later.

Mechanical Properties

Tin and oxygen are the primary strengthening elements in the Zircaloys and the current results reflect this predominance. The yield strength increased by about 40 MPa when the tin content increased from 0.5 to 1.0%. A similar increase in yield strength for comparable tin contents was reported by Isobe [8] for Zr-0.2Fe-0.1Cr-0.5Nb-Sn alloys. However, an increase in room temperature tensile strength with increasing (iron plus chromium) content was also reported for their alloys that was not seen in the present results. This difference may be related to the higher iron content of the present alloys. Alternatively, the present alloys were partly recrystallized and variations in degree of recrystallization could mask differences caused by alloy additions.

Creep strength of the alloys can be attributed to a combination of matrix hardening and precipitate hardening. The dramatic increase in creep strength with increasing tin concentration, Fig. 2, primarily reflects the contribution of tin to matrix hardening. The steady-state creep rate decreased from 0.092%/h for Alloy S2-3 (0.5% tin) to 0.004%/h for Alloy S2-9 (0.93% tin). Except for the tin concentration, the composition and particle distributions for both alloys were very similar, indicating that the improvement in creep strength is from matrix hardening.

The apparent increase in creep strength with increasing chromium concentration for the 1.0% tin alloys, Fig. 4, may be related to precipitate hardening. Chromium additions to the

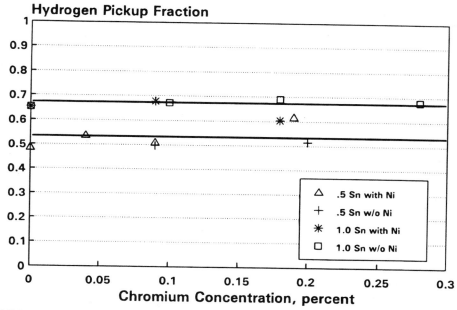

FIG. 8—*Hydrogen pickup as a function of chromium concentrations for 0.5% and 1.0% tin alloys.*

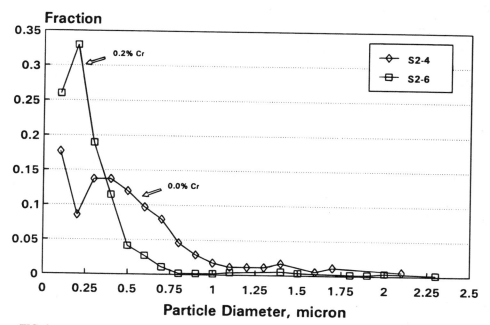

FIG. 9—*Effect of chromium on the particle size distribution of Zr-0.5Sn-0.4Fe-0.05Ni-xCr alloys.*

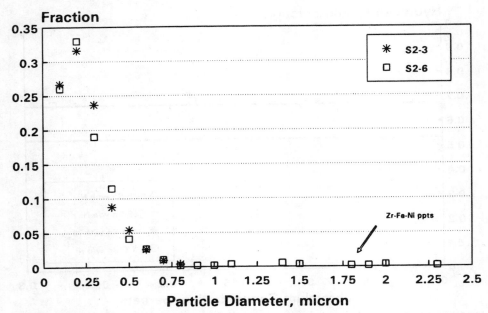

FIG. 10—*Effect of nickel on the particle size distributions of Zr-0.5Sn-0.4Fe-0.2Cr alloys.*

base alloys decreased the size and increased the density of intermetallic precipitates. This would increase precipitate strengthening, but the magnitude of the effect is not known.

An alternative explanation for the effect of chromium is additional matrix hardening due to a higher concentration of chromium in the matrix. However, Charquet et al. [9] reported that the solubilities of iron and chromium are very low in a Zr-1.4Sn matrix. Maximum solubility for iron plus chromium was only 150 ppm at about 1083 K and decreased at lower temperatures. Wadman [10] confirmed the low matrix concentrations of iron, chromium, and nickel in Zircaloy-2 and Zircaloy-4 by field ion microscopy. Consequently, additional matrix hardening would require a supersaturation that is not consistent with these previous studies.

Corrosion and Hydrogen Pickup

Corrosion of the Zircaloys and similar zirconium-based alloys is very complex, and the corrosion mechanisms are not clearly understood after more than 40 years of intensive investigations. Alloy composition [5,6], thermal history [1,2], and precipitate type and distributions [11,12] are all important factors in determining corrosion behavior. However, there are strong interactions between variables that prevent consistent correlations between the many variables and measured corrosion behavior [3].

The present study shows that improved corrosion behavior is obtained by decreasing the tin concentration and increasing the chromium content of the Zr-0.4Fe-Sn-Cr-Ni alloys, Fig. 7. All alloys received the same thermomechanical processing, and differences in corrosion behavior can be attributed to variations in matrix composition or precipitate type and distribution.

Comparing the particle size distributions and weight gains from Alloys S2-3 and S2-9 suggests that the detrimental effect of tin on corrosion is associated with matrix composition that is consistent with the model proposed by Thomas [13] for uniform corrosion of zirconium alloys. Particle type and diameter distributions for Alloys S2-3 and S2-6 are nearly identical and, therefore, cannot explain the difference in corrosion behavior.

The improved corrosion resistance with increasing chromium content, Fig. 7, could be associated with matrix composition or particle distributions or both. The total iron plus chromium content of the alloys increased with increasing chromium content and both elements are known to improve the corrosion resistance of zirconium-tin alloys [7]. However, as discussed previously, the solubility of iron and chromium are very low in zirconium alloys [9,10]. The iron/chromium ratio in the metal matrix would change with chromium additions, but large increases in the total iron plus chromium content would not be expected and the improved corrosion resistance may be attributed to the observed changes in type and distribution of zirconium-iron and zirconium-iron-chromium type precipitates. Nickel additions had no significant effect on weight gains of the alloys, although large $Zr_2(Fe,Ni)$ precipitates were in the microstructure.

Results of the hydrogen analysis were unexpected based on previous measurements of zirconium alloys. It is well documented that nickel additions increase the hydrogen pickup of zirconium and zirconium alloys [7,14]. The lack of a significant effect of nickel on hydrogen pickup in the present study may be related to the lower tin and high iron plus chromium levels present in these alloys. For binary zirconium alloys, Berry [15] reported hydrogen pickup fractions decreased with additions of iron and chromium, remained constant with tin additions, and increased with nickel additions. The total iron plus chromium content of the present alloys ranged from 0.4 to 0.6%, and it seams reasonable that the beneficial effect of these alloy elements could minimize the adverse effect of 0.05% nickel additions. Additional work on identifying the interaction effects of alloy additions on hydrogen pickup is needed to explain the present results.

Conclusions

1. Decreasing the tin concentration from 1.0 to 0.5% improves the 673 K autoclave corrosion resistance and hydrogen pickup of Zr-0.4Fe-xSn-yCr alloys but decreases the room temperature tensile strength and 673 K creep strength.

2. Chromium additions to Zr-0.4Fe-xSn alloys improve the 673 K autoclave corrosion resistance and 673 K creep strength. Chromium additions have no effect on hydrogen pickup in these alloys.

3. Nickel additions to Zr-0.4Fe-xSn-yCr alloys have no effect on 673 K autoclave corrosion or hydrogen pickup.

References

[1] Andersson, T., Thorvaldsson, T., Wilson, A., and Wardle, A. M., "Improvements in Water Reactor Fuel Technology and Utilization," *Proceedings,* IAEA Symposium, Stockholm, Sweden, 1987, International Atomic Energy Agency, Vienna, pp. 435–439.

[2] Garzarolli, F., Steinberg, E., and Weidinger, H. G., "Microstructure and Corrosion Studies for Optimizing PWR and BWR Zircaloy Cladding," *Zirconium in the Nuclear Industry: Eight International Symposium, ASTM STP 1023,* L. F. P. Van Swam and C. M. Eucken, Eds., American Society for Testing and Materials, Philadelphia, 1989, pp. 202–212.

[3] Bradley, E. R., Schemel, J. H., and Nyström, A.-L., "Influence of Alloy Composition and Processing on the Nodular Corrosion Resistance of Zircaloy-2," *Zirconium in the Nuclear Industry: Ninth International Symposium, ASTM STP 1132,* C. M. Eucken and A. M. Garde, Eds., American Society for Testing and Materials, Philadelphia, 1991, pp. 304–318.

[4] Perkins, R. A., Peterson, D. T., and Busch, R. A., "Effect of Annealing on the Uniform and Nodular Corrosion Resistance of Zircaloy-4," *Proceedings,* NACE Annual Conference and Corrosion Show, National Association of Corrosion Engineers, Houston, 1992.

[5] Eucken, C. M., Finden, P. T., Trapp-Pritsching, S., and Weidinger, H. G., "Influence of Chemical Composition on Uniform Corrosion of Zirconium-Base Alloys in Autoclave Tests," *Zirconium in the Nuclear Industry: Eighth International Symposium, ASTM STP 1023*, L. F. P. Van Swam and C. M. Eucken, Eds., American Society for Testing and Materials, Philadelphia, 1989, pp. 113–127.

[6] Graham, R. A., Tosdale, J. P., and Finden, P. T., "Influence of Chemical Composition and Manufacturing Variables on Autoclave Corrosion of the Zircaloys," *Zirconium in the Nuclear Industry: Eighth International Symposium, ASTM STP 1023*, L. F. P. Van Swam and C. M. Eucken, Eds., American Society for Testing and Materials, Philadelphia, 1989, pp. 334–345.

[7] Kass, S., "The Development of the Zircaloys," *Corrosion of Zirconium Alloys, ASTM STP 368*, American Society for Testing and Materials, Philadelphia, 1964, pp. 3–27.

[8] Isobe, T. and Matsuo, Y., "Development of Highly Corrosion Resistant Zirconium-Base Alloys," *Zirconium in the Nuclear Industry: Ninth International Symposium, ASTM STP 1132*, C. M. Eucken and A. M. Garde, Eds., American Society for Testing and Materials, Philadelphia, 1991, pp. 346–367.

[9] Charquet, D., Hahn, R., Ortlib, E., Gros, J.-P., and Wadier, J.-F., "Solubility Limits and Formation of Intermetallic Precipitates in Zr-Sn-Fe-Cr Alloys," *Zirconium in the Nuclear Industry: Eighth International Symposium, ASTM STP 1023*, L. F. P. Van Swam and C. M. Eucken, Eds., American Society for Testing and Materials, Philadelphia, 1989, pp. 405–422.

[10] Wadman, B. and Andren, H.-O., "Microanalysis of the Matrix and the Oxide-Metal Interface of Uniformly Corroded Zircaloy," *Zirconium in the Nuclear Industry: Ninth International Symposium, ASTM STP 1132*, L. F. P. Van Swam and C. M. Eucken, Eds., American Society for Testing and Materials, Philadelphia, 1991, pp. 461–475.

[11] Rudling, P., Vannesjo, K. L., Vesterlund, G., and Massih, A. R., "Influence of Second-Phase Particles on Zircaloy Corrosion in BWR Environment," *Zirconium in the Nuclear Industry: Seventh International Symposium, ASTM STP 939*, R. B. Adamson and L. F. P. Van Swam, Eds., American Society for Testing and Materials, Philadelphia, 1991, pp. 292–306.

[12] Andersson, T. and Thorvaldsson, T., "Influence of Microstructure on 400 and 500°C Steam Corrosion Behavior of Zircaloy-2 and Zircaloy-4 Tubing," *Fundamental Aspects of Corrosion on Zirconium Base Alloys in Water Reactor Environments*, IWGFPT/34, International Atomic Energy Agency, Vienna, 1990, pp. 237–248.

[13] Thomas, D. E., "Corrosion in Water and Steam," *Metallurgy of Zirconium*, B. Lustman and F. Kerze, Jr., Eds., McGraw Hill, New York, 1955, pp. 3–27.

[14] Harada, M., Kimpara, M., and Abe, K., "Effect of Alloying Elements on Uniform Corrosion Resistance of Zirconium-Based Alloys in 360°C Water and 400°C Steam," *Zirconium in the Nuclear Industry: Ninth International Symposium, ASTM STP 1132*, C. M. Eucken and A. M. Garde, Eds., American Society for Testing and Materials, Philadelphia, 1991, pp. 368–391.

[15] Berry, W. E., Vaughan, D. A., and White, E. L., "Hydrogen Pickup During Aqueous Corrosion of Zirconium Alloys," *Corrosion 17*, Vol. 17, 1961, p. 82.

DISCUSSION

G. P. Sabol[1] (written discussion)—(1) Was testing performed at any conditions other than 673 K (400°C)? If not, do you have any information on the activation energy for post-transition corrosion? Why was the 673 K (400°C) steam test chosen for screening of the alloys?

(2) It is widely known that decreasing tin content will improve the corrosion resistance in autoclave steam tests, and the best corrosion is obtained at zero tin, but with iron and chromium. Why were your low compositions limited to 0.5 and 1.0% tin?

E. R. Bradley and A.-L. Nyström (authors' closure)—(1) The autoclave tests of these alloys were conducted only at 673 K (400°C) and, therefore, we have no information on the activation energy for the post-transition corrosion. The 673 K (400°C) steam test was chosen for screening the alloys because we believe long-term exposure to 673 K (400°C) steam provides a good correlation with the in-reactor uniform corrosion behavior of Zircaloy materials.[2] In addition, our 673 K (400°C) steam autoclave data have shown a good correlation with the relative ranking produced by 633 K (360°C) water autoclave tests for Zircaloy-type materials. The primary advantage of the 673 K (400°C) steam test is the lower exposure time needed to separate the material variants. Autoclave testing for uniform corrosion resistance at temperatures above 673 K (400°C) is questionable because of the change in mechanism from uniform to nonuniform corrosion at higher temperatures.

(2) The primary objective of this study was to systematically evaluate the effect of chromium, nickel, and tin additions to the corrosion and mechanical properties of Zircaloy-3B (Zr-0.5 Sn-0.4Fe). Zircaloy-3B is known to have excellent corrosion resistance, but its mechanical properties are generally considered too low for reactor applications. Tin is a major strengthening element in zirconium alloys, and reducing the tin concentration below 0.5% did not seem appropriate for this study.

M. H. Koike[3] (written discussion)—For the in-reactor use, irradiation creep and growth will occur in addition to the thermal creep for your tests. In order to minimize the irradiation creep and growth with the corrosion resistance, what do you think is the ideal alloy composition for Zircaloy?

E. R. Bradley and A.-L. Nyström (authors' closure)—Irradiation creep, growth, and corrosion resistance are all important characteristics of fuel rod cladding. The "ideal composition" for Zircaloy would be a compromise based on the specific effects of alloy composition on each of these properties. Decreasing the tin concentration improves the corrosion resistance but decreases the tensile and thermal creep strength. It is very difficult to select specific alloy compositions based only on unirradiated test data because there is not a definite relationship between thermal and irradiation creep. Irradiation tests of promising low tin alloys are being conducted to identify an ideal composition for PWR applications.

D. Khatamian[4] (written discussion)—In your corrosion results, you showed variability from 140 to 180 mg/dm². Do you know what is the cause of the variability?

E. R. Bradley and A.-L. Nyström (authors' closure)—The weight gains for the 0.5% tin alloys varied from 138 to 183 mg/dm² after 512 days exposure to the 673 K (400°C) environment

---

[1] Westinghouse NMD, Pittsburgh, PA.
[2] Schemel, J. J., Charquet, D., and Wadier, J.-F., "Influence of the Manufacturing Process on the Corrosion Resistance of Zircaloy-4 Cladding," *Zirconium in the Nuclear Industry (Eighth International Symposium), ASTM STP 1023*, L. F. P. Van Swam and C. M. Euchen, Eds., American Society for Testing and Materials, Philadelphia, 1989, pp. 141–152.
[3] PNC, O-arai Engineering Center, Japan.
[4] AECL Research, Chalk River Laboratories, Chalk River, Ontario, Canada.

as shown in Fig. 5. The chromium content of the various alloys is the cause of this variation as seen in Fig. 7. Variation between duplicate samples of the same alloy was less than 10 mg/dm^2.

Nobuaki Yamashita[5] (written discussion)—Do you have an intention to expand your work to vary the content of iron?

E. R. Bradley and A.-L. Nyström (authors' closure)—At this time, we have no plans to evaluate the effect of iron concentration on the properties of these alloys.

C. D. Williams[6] (written discussion)—Your presentation included reference to partially recrystallized structures in the experimental materials. Did you generate quantitative information on the extent of recrystallization as a function of alloy chemistry and intermetallic particle distribution?

E. R. Bradley and A.-L. Nyström (authors' closure)—Measuring the degree of recrystallization in zirconium alloys is extremely difficult because the degree of recrystallization can be estimated only by transmission electron microscopy (TEM). Detailed TEM examinations of four selected alloys showed that the area fraction of recrystallized material from small areas (approximately 0.003 mm^2) ranged from about 30 to 60%. No definite correlations between the amount of recrystallization and alloy composition or particle distribution were observed. However, the degree of recrystallization varied extensively in local regions of each material, and these measurements can only be considered as qualitative estimates. Developing quantitative information on the degree of recrystallization would require an extensive TEM effort that was beyond the scope of the present work.

[5] GE Company, Wilmington, NC.
[6] GE Nuclear Energy, Wilmington, NC.

Steven B. Wisner,[1] Myron B. Reynolds,[1] and Ronald B. Adamson[1]

Fatigue Behavior of Irradiated and Unirradiated Zircaloy and Zirconium

REFERENCE: Wisner, S. B., Reynolds, M. B., and Adamson, R. B., "**Fatigue Behavior of Irradiated and Unirradiated Zircaloy and Zirconium,**" *Zirconium in the Nuclear Industry: Tenth International Symposium, ASTM STP 1245,* A. M. Garde and E. R. Bradley, Eds., American Society for Testing and Materials, Philadelphia, 1994, pp. 499–520.

ABSTRACT: As a normal part of reactor operation, Zircaloy components in the core of boiling water reactors (BWRs) are subjected to oscillating loads. It then becomes important to assess the fatigue behavior of core materials. These include Zircaloy-2 used as fuel cladding, zirconium used as liners in barrier fuel, and Zircaloy-4 used for fuel channels.
Fatigue testing was performed on unirradiated and irradiated materials. Fully reversed uniaxial fatigue tests were conducted at constant total strain amplitudes between 0.3 and 1.4% at 616 K in air. Fatigue crack growth testing was conducted on conventional compact tension (CT) specimens at 293 and 561 K. Unirradiated material was tested in air and water, and irradiated material was tested in air.
Crack growth rates, expressed as a function of ΔK (applied stress intensity range), were determined to be insensitive to neutron irradiation, but were increased by a factor of 2 to 4 in water, depending on oxygen content of the water.
Fatigue life is shown to be strongly dependent on the partitioning of plastic and elastic strain during the test. In general, the softer zirconiums have longer fatigue lives than Zircaloy. Irradiation reduces the fatigue life of all materials in the low cycle regime, particularly when applied plastic strain is used as the test variable.
The results obtained support the current design basis for fuel rods and channels and reflect the excellent observed performance of BWR core components.

KEY WORDS: zirconium, zirconium alloys, nuclear materials, nuclear applications, radiation effects, axial fatigue, zirconium alloy composition, fatigue crack growth, neutron fluence, boiling water reactor environment, compact tension specimen

Utilities often now operate their reactors in load following automatic frequency control modes. In addition, complex thermohydraulics within the reactor create conditions of oscillating pressures and loads. Therefore, it is important to understand the fatigue properties of materials used in the reactor core. These materials include zirconium, Zircaloy-2, and Zircaloy-4. Rowland et al. [1] have reported no negative effects on performance of zirconium barrier fuel rods given extensive power cycling in a test reactor. This confirms the observed general excellent performance of fuel components under whatever cyclic loading conditions are present in reactor.
Nevertheless, although data already exists for fatigue life [1–6] and crack growth rate [7–10] for Zircaloy, there does not appear to be a clear consensus on the effects of irradiation of testing mode on the fatigue behavior. Therefore, a comprehensive program has been established to provide understanding of some fundamental aspects of fatigue for zirconium alloys.

---

[1] Senior engineer, principal engineer (retired), and manager, respectively, GE Nuclear Energy, Fuel Materials Technology, Vallecitos Nuclear Center, Pleasanton, CA 94566.

In this paper, data is reported on the relative fatigue behavior of unirradiated and irradiated zirconium and Zircaloy sheet under cyclic loading conditions representative of boiling water reactor (BWR) fuel rod design. Results are presented on recrystallized zirconium of three purities and on recrystallized Zircaloy-2. These materials are used in BWR barrier fuel rods [11].

Data was also generated for the fatigue crack growth rates and crack growth thresholds for Zircaloy-4 under simulated environmental (water chemistry) conditions found in the core of a BWR. The applied cyclic stress was characterized by a minimum-to-maximum ratio (R) of approximately 0.8 and a frequency of 3 Hz, which are proposed to be typical for BWR channel service. The experimental difficulty of crack growth measurements in the actual thermal, chemical, and fast neutron environment of a nuclear reactor necessitated that a sequence of experiments be undertaken that would provide a basis for predicting crack growth behavior in the BWR.

These tests were:

1. determine crack growth thresholds at 50 Hz in air for different load ratios at 293 and 561 K,
2. compare crack growth rates at 3 and 50 Hz over the linear portion of the da/dN versus ΔK curve in air at 293 and at 561 K,
3. measure crack growth rates at 3 Hz in water at 561 K and 6.89 MPa at 0.1 to 0.2 ppm dissolved oxygen level and in air saturated water at 561 K and 6.89 MPa, and
4. compare crack growth rates at 3 Hz in irradiated and unirradiated Zircaloy-4 in air at 561 K.

Experimental

Fatigue Specimens

Fatigue specimens, Fig. 1a, were machined from blanks cut transverse to the rolling direction from nominally 1.27-mm-thick, 60% cold-worked unirradiated crystal bar zirconium, low-oxygen sponge zirconium, reactor grade sponge zirconium, and recrystallized Zircaloy-2 sheet. This specimen design is similar to one used by Martin [12], who found that buckling could be avoided during compression for up to 1.5% total axial strain amplitude. All specimens were recrystallized by annealing at 850 K for 2 h in vacuum that resulted in an equiaxed grain microstructure of ~10 μm in size. Table 1 gives the chemical composition for these materials and Table 2 lists the texture f-parameters for the (0002) plane.

Irradiated specimens were fabricated from 1.27-mm-thick sheet material. Zircaloy-2 was irradiated at ~644 K in a sodium environment in the EBR-II (Experimental Breeder Reactor) to fluences between 1.5 to 7.7 × 10^{25} n/m² ($E >$ 1 MeV). Crystal bar zirconium was irradiated at 600 K in an inert environment in a BWR to a fluence of 1.05 × 10^{25} n/m² ($E >$ 1 MeV). Zircaloy-4 was irradiated at 561 K as a BWR channel to a fluence of 15 × 10^{25} n/m² ($E >$ 1 MeV). The original channel thickness of 2.03 mm was reduced to 1.27 mm by machining both sides of the sheet. None of the irradiated specimens had appreciable oxide when fatigue tested, although Zircaloy-2 and zirconium were tested with as-irradiated surfaces.

Fatigue Crack Growth Specimens

All unirradiated specimens were fabricated from recrystallized Zircaloy-4 in the form of 2.54-mm-thick rolled sheet. The specimen configuration was the compact tension (CT) type with the W dimension equal to 50.8 mm and B thickness equal to the sheet thickness (Fig. 1b). It should be noted that this thickness was not great enough to satisfy the ASTM Test Method for Plane-Strain Fracture Toughness of Metallic Materials (E 399-90) criterion for K_{Ic}

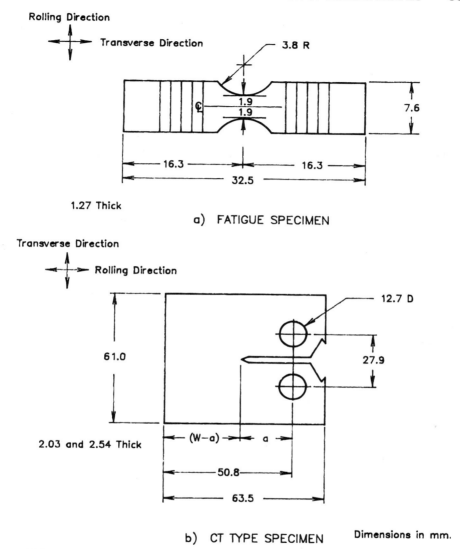

FIG. 1—(a) *Fatigue specimen and* (b) *fatigue crack growth (CT) specimen design.*

tests, but was adequate for low ΔK fatigue crack growth tests since there is experimental evidence that this thickness requirement can be relaxed severalfold without compromising the validity of the fatigue crack growth rate tests [13,14]. The term, ΔK, is defined as the applied stress intensity factor range. Specimens were oriented to give crack growth in the longitudinal direction in the strip. Chemical composition, texture f-parameters, and tensile properties for this material are given in Tables 1, 2, and 3. Irradiated CT specimens were fabricated from nominally 2.03-mm-thick fuel channels that had been exposed in a BWR at 561 K. The fluence for this material ranged from 0.3 to 7.4 × 10^{25} n/m² ($E > 1$ MeV) and there was a black oxide on the outer surfaces of these specimens.

TABLE 1—*Chemical analysis of Zircaloy and zirconium used for fatigue testing.*

Material	Element, % by weight				
	Sn	Fe	Cr	Ni	O
FATIGUE SPECIMEN MATERIALS					
Zircaloy-2	1.53	0.18	0.10	0.04	0.13
Zircaloy-4[a]	1.45	0.21	0.10	...	0.13
Low-oxygen sponge Zr	...	...	...	...	0.05
Crystal bar Zr	...	...	...	...	<0.005
Reactor-grade sponge Zr	...	...	...	...	0.10
FATIGUE CRACK GROWTH SPECIMEN MATERIAL					
Zircaloy-4	1.56	0.21	0.10	...	0.15

[a]Typical values.

TABLE 2—*Texture parameters for Zircaloy and zirconium for the (0002) plane.*

Material	f_n	f_l	f_t	Σ_f
FATIGUE SPECIMEN MATERIALS				
Zircaloy-2	0.718	0.105	0.178	1.001
Zircaloy-4[a]	0.749	0.120	0.139	1.008
Low-oxygen sponge Zr	0.730	0.091	0.183	1.004
Crystal bar Zr	0.659	0.083	0.259	1.001
Reactor-grade sponge Zr	0.649	0.085	0.217	1.005
FATIGUE CRACK GROWTH SPECIMEN MATERIAL				
Zircaloy-4	0.696	0.113	0.190	0.996

[a]Typical values.

TABLE 3—*Tensile properties of unirradiated Zircaloy-4 strip used for crack growth testing.*

Specimen	0.2% Yield Stress, MPa	Ultimate Stress, MPa	Uniform Elongation, %	Total Elongation, %	Test Temperature, K	Strain Rate, s^{-1}
T1 T[a]	411.6	477.8	10.5	30.8	297	0.000083
T2 T	413.7	479.2	10.2	31.9	297	0.000083
T1 L[a]	365.4	465.4	12.0	29.2	297	0.000083
T2 L	375.8	474.4	12.0	28.0	297	0.000083
T3 T	143.4[b]	189.6	16.4	47.3	561	0.000083
T4 L	142.0[b]	180.6	16.4	41.5	561	0.000083
T4 L	165.5[b]	215.8	15.3	45.9	561	0.00083

[a]T = transverse specimen and L = longitudinal specimen.
[b]Specimens exhibited yield drop behavior.

Test Procedure–Fatigue Tests

For all fatigue tests, the cyclic load range, ΔL, was applied to the specimen in the axial direction and the total strain amplitude was controlled across the minimum width of the specimen as shown in Fig. 2. Therefore, for these experimental conditions, the total strain amplitude in the axial direction for the tensile half-cycle has to be calculated and is given by

$$\epsilon_t = L/A_0 E + C\Delta W_p/W_0 \qquad (1)$$

where

- L = axial tensile load amplitude (N),
- A_0 = initial specimen cross-section area (mm^2),
- E = Young's modulus (MPa),
- C = the plastic contractile strain ratio [15] that is a measure of the anisotropy of the material
- ΔW_p = the plastic displacement amplitude of the specimen gage width for the tensile side of the cycle taken from the experimental hysteresis loop at midlife (mm), and
- W_0 = initial specimen gage width (mm).

$L/A_0 E$ in Eq 1 is the axial elastic strain amplitude and $C\Delta W_p/W_0$ is the axial plastic strain amplitude. The Young's modulus for all materials were determined as a function of temperature from X-ray diffraction texture analysis that generated elastic compliance coefficients [16]. The value is near 75 000 MPa for all materials. It is assumed that irradiation has no effect on these constants. Tests were controlled by cycling at a constant total diametral displacement amplitude that was converted to axial strain amplitude as described.

The value for C was determined experimentally from a series of uniaxial tension tests on the test materials that had been pulled to 5% total strain at 616 K. Based on pre- and post-test measurements of the specimen permanent deformations, C was calculated and is given by the ratio of ϵ_a/ϵ_w, where ϵ_a and ϵ_w are engineering plastic strains for the specimen axial

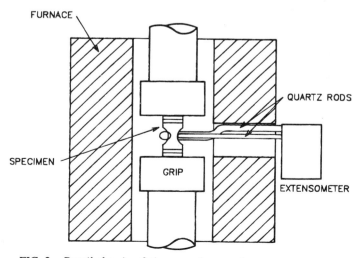

FIG. 2—*Detail showing fatigue specimen and extensometer setup.*

and width directions, respectively. Table 4 lists the values of C found for the unirradiated test materials. Contractile strain ratio tests were also conducted on irradiated Zircaloy-2, but the results of these tests were difficult to interpret due to the heterogeneous deformation of the material. The results gave C values that ranged from 1.1 to 1.6. Therefore, for consistency, the values found for the unirradiated materials were used for the irradiated materials.

All fatigue tests were conducted in air at a frequency of 0.5 Hz using a servohydraulic tension-compression load frame and specially designed pull bar and grip assembly. Heating was obtained from a resistance split furnace centered around the load train and specimen. Temperature was controlled by a Type K thermocouple that contacted the upper tab region of the specimen. This thermocouple provided a millivolt feedback signal to a proportional band controller. Temperature during the test was maintained at 616 or 561 K following a pre-test hold for 30 min to attain thermal equilibrium. All tests were conducted in strain control using a ± 0.1 mm range extensometer fixed to the specimen so that the displacement of the specimen minimum gage width was sensed (Fig. 2). With the exception of specimens that fractured and the Zircaloy-4 specimens, tests were stopped after steady-state tensile load (stabilized hysteresis loop) decreased 5%.

Test temperature, load range, and specimen gage width displacement were continuously monitored and recorded during each test. Also, hysteresis loops at 10, 50, and 100 cycles, selected intervals during test, and at "failure" were recorded on an X-Y recorder.

Test Procedure–Fatigue Crack Growth Tests

All crack growth tests were run on a closed-loop servohydraulic test machine. For the fatigue tests done in air at 293 K, specimen loads were determined using a load transducer and signal conditioner. Crack growth rates were determined by periodically removing the cyclic load and measuring the crack depth with an optical cathetometer with the specimen under a moderate stationary load.

In the elevated temperature tests (561 K) conducted in air on unirradiated specimens, temperature was maintained by a small low-voltage furnace designed to fit closely about the specimen and test yokes. Knife edges were machined into these yokes to accommodate a standard clip gage for determination of load-point deflection. Temperature was controlled by a proportional controller. Two slits were provided in the back wall of the furnace through which were inserted the arms of the clip gage. The clip gage was cooled by jets of air directed into the gage housing.

Tests on irradiated specimens were done with the test machine located in a small shielded cell equipped with manipulators for remotely loading and unloading specimens from the test machine. Test temperature was measured by a thermocouple pressed against the side of the specimen. In these tests, and subsequent tests where crack growth depth could not be measured

TABLE 4—*Experimentally determined contractile strain ratio for materials used for fatigue testing.*

Material	C
Zircaloy-2/4	1.14
Low-oxygen sponge Zr	1.11
Crystal bar Zr	1.21
Reactor-grade sponge Zr	1.19

in situ, crack depth was determined from changes in the compliance of the specimen. This was done by using either a crack depth-compliance relationship given by Hudak et al. [17] or the compliance values obtained from a number of specimens where crack depth had been determined by cathetometry and a plot of specimen compliance versus crack depth was constructed covering the crack depth range needed for the tests. These measurements were done at 293 K. In subsequent tests at elevated temperature, compliance was measured then converted to 293-K compliance by multiplying by the ratio of elastic moduli at 561 to 293 K, which is equal to 1.12. These moduli were determined from experimental load-deflection curves for unirradiated Zircaloy-4. Using the computed 293-K compliance value, crack depth was read from the compliance versus crack depth plot.

Elevated temperature tests in water were conducted in a chamber made from a stainless steel pipe cross that was mounted within the test machine load frame and fitted with flanges designed to permit entry of specimen pull bars through spring-loaded graphite-filled Teflon seals. Water temperature was measured by a sheathed thermocouple located near the test specimen. Pressure was measured by a pressure transducer. Load-point displacement amplitude was measured by a clip gage engaging a stationary knife edge and a movable knife edge clamped to the lower pull bar. The force exerted on the lower pull bar was measured by a load transducer. Water was pumped through the cross from a reservoir at high pressure at approximately 60 mL/min, and the desired test pressure was maintained by a gas-ballasted relief valve. The reservoir was equipped with a demineralizer recirculation loop using two mixed resin filters for cleanup. Desired oxygen level was obtained by bubbling nitrogen containing 0.5% oxygen and 1.6% hydrogen through the reservoir. Conductivity and oxygen content were monitored by sensors mounted in the recirculation lines at the relief valve exit. For the tests in air-saturated water, air was bubbled through the reservoir by an aquarium pump.

The procedure for determining specimen compliance in the pressure chamber differed somewhat from that used in tests in air because the friction forces exerted on the pull bars by the Teflon seals caused the specimen load-point displacement to lag behind the force applied by the test machine actuator. Therefore, the load values indicated by the load transducer did not represent the true loads on the specimen. It was possible, using a simple spring and dashpot model for the test system, to calculate specimen load and load-point deflection from load and displacement measurements made outside the test chamber and measured load train compliance and calculated pull-rod extrusion forces. This made it possible in later experiments to calculate all required crack growth parameters from transducer readings obtained with a computer data acquisition system connected to the test system. This system produced complete hard-copy graphic da/dN versus ΔK records for each test.

Results and Discussion

Low-Cycle Fatigue Properties

Table 5 presents the results on unirradiated materials for the fatigue behavior for Zircaloy-2 and three purities (ppm O_2) of zirconium. Table 6 lists the results for fatigue performance for irradiated Zircaloy-2, Zircaloy-4, and crystal bar zirconium. The data in the tables were obtained for each specimen from analysis of the experimental recordings of the tensile loads and hysteresis loops. In general, a stabilized hysteresis loop was attained in about 10% of life, which indicated that cyclic strain hardening for the unirradiated materials or softening for the irradiated materials had ceased (Figs. 3 and 4). The cyclic loads and loop widths for the tensile side of the reversal, used to calculate the total strain amplitude, were taken from the hysteresis loops at mid-life.

Except as noted, specimens were tested to failure, defined as the number of cycles at which tensile load decreased ~5% from a steady-state condition. This was the point in the test where

TABLE 5—*Results for fatigue tests on unirradiated Zircaloy-2 and zirconium in 616-K air at 0.5 Hz.*

Specimen	ΔW_p, mm[a]	L, N[a]	ϵ_t, %	ϵ_e, %	ϵ_p, %	Cycles to Fail	Comments
ZIRCALOY-2							
62	0.0343	1032	1.32	0.30	1.02	211	cracks
63	0.0249	979	1.01	0.28	0.73	463	crack
82	0.0117	876	0.59	0.25	0.34	1 775	cracks
80	0.0114	810	0.57	0.23	0.34	2 232	crack
81	0.0061	796	0.41	0.23	0.18	7 181	crack
139	0.0020	627	0.24	0.18	0.06	37 598	fracture
137	0.0012	565	0.20	0.16	0.04	95 183	fracture
141	0.0008	552	0.17	0.15	0.02	206 972	fracture
140	0.0003	489	0.14	0.14	<0.01	283 091	fracture
LOW-OXYGEN SPONGE ZR							
65	0.0386	605	1.27	0.16	1.11	408	cracks
64	0.0259	592	0.91	0.16	0.75	910	crack
74	0.0111	520	0.46	0.14	0.32	3 393	cracks
75	0.0058	436	0.29	0.12	0.17	13 512	cracks
125	0.0026	267	0.15	0.07	0.08	92 975	fracture
127	0.0027	276	0.15	0.08	0.07	104 575	fracture
124	0.0019	258	0.12	0.06	0.06	196 310	fracture
126	0.0015	236	0.11	0.06	0.05	295 295	fracture
123[b]	0.0013	245	0.10	0.07	0.03	345 773	fracture
CRYSTAL BAR ZR							
51	0.0386	543	1.37	0.14	1.23	398	cracks
59	0.0389	556	1.37	0.15	1.22	412	cracks
52	0.0254	467	0.93	0.12	0.81	998	cracks
57	0.0109	472	0.47	0.13	0.34	3 831	cracks
58	0.0061	445	0.31	0.12	0.19	10 854	cracks
60	0.0061	409	0.30	0.11	0.19	11 376	cracks
REACTOR-GRADE SPONGE ZR							
68	0.0381	930	1.41	0.24	1.17	283	cracks
67	0.0356	890	1.34	0.23	1.11	308	cracks
66	0.0241	805	0.96	0.21	0.75	600	cracks
69	0.0254	845	1.00	0.22	0.78	690	crack
70	0.0127	694	0.54	0.18	0.36	2 269	crack
71	0.0056	574	0.32	0.15	0.17	8 687	cracks

[a] See Eq 1.
[b] Test interrupted.

fatigue crack(s) had propagated to a depth such that the compliance of the system began to increase. Also, when sufficient crack depth and crack opening had been attained, a cusp in the hysteresis loop was observed to form near the tail of the compression reversal due to crack closure. Post-test visual examination revealed that for all specimens the cracks initiated in the regions of highest stress at the specimen edge on one or both sides near the minimum width of the gage region.

Tests were conducted between total strain amplitude limits. However, because the hysteresis loops stabilize early in life (Figs. 3 and 4), the tests are also conducted under nominal constant

TABLE 6—*Results for fatigue tests on irradiated Zircaloy-2, Zircaloy-4, and zirconium in 616 K air at 0.5 Hz.*

Specimen	Fluence, 10^{25} n/m²	ΔW_p, mm[a]	L, N¹	Calculated Axial Strain Amplitude			Cycles to Fail	Comments
				ϵ_t, %	ϵ_e, %	ϵ_p, %		
ZIRCALOY-2								
149	1.51	0.0145	1624	0.87	0.45	0.42	705	cracks
95	2.13	0.0109	1441	0.72	0.40	0.32	2216	crack
148	7.12	0.0051	1312	0.52	0.37	0.15	3263	fracture
91	3.15	0.0053	1223	0.52	0.36	0.16	3741	fracture
147	7.65	0.0046	1001	0.44	0.30	0.14	5584	cracks
ZIRCALOY-4								
309	15.0	0.0070	2263	0.85	0.65	0.20	386	crack
310	15.0	0.0040	2028	0.70	0.57	0.13	814	crack
308	15.0	0.0020	1822	0.56	0.50	0.06	2330	fracture
312	15.0	0.0015	1764	0.53	0.48	0.05	1820	fracture
311	15.0	0.0008	8427	0.44	0.42	0.02	8427	crack
CRYSTAL BAR ZIRCONIUM								
154	1.05	0.0251	1085	1.09	0.30	0.79	180	fracture
158	1.05	0.0170	1103	0.86	0.31	0.55	306	fracture
160	1.05	0.0198	890	0.87	0.24	0.63	350	cracks
155	1.05	0.0102	890	0.57	0.25	0.32	998	cracks
157	1.05	0.0114	747	0.57	0.21	0.36	1794	cracks
159	1.05	0.0038	756	0.33	0.21	0.12	4900	cracks

[a] See Eq 1.

plastic strain amplitude limits. Once the applied load reaches saturation and the hysteresis loops become stable, the applied plastic strain is related to applied total strain through the shape of the loop. Examples of stable loops are shown in Fig. 5.

Figure 6 shows the total axial tensile strain amplitude versus cycles to failure for unirradiated zirconium and Zircaloy-2 along with a best fit O'Donnell and Langer curve for unirradiated Zircaloy-2 tested under axial fatigue [2]. Curves for the Zircaloy-2 and low-oxygen sponge zirconium were constructed using linear regression analysis. Comparison of the curves for Zircaloy-2 fatigue life shows reasonably good agreement with O'Donnell and Langer in the low-cycle regime but reduced life in the high-cycle regime.

Figure 7 shows the fatigue data re-plotted for constant applied plastic strain. It is noted that a Coffin-Manson relationship [18,19] appears to hold, that is, the data adheres to the equation, $\Delta\epsilon_p N_f^\alpha = c$. For instance, for unirradiated Zircaloy-2, $\alpha = 0.55$ and $c = 0.22$. Compared to Hosbons [20], the slope is similar but the fatigue life is somewhat lower.

In the low cycle regime, fatigue life is a function of material purity that affects strength and ductility. Figure 5a and Table 5 illustrate that for a given applied total strain, the resulting plastic strains for Zircaloy and zirconium are about the same, but the applied loads are different. As is normal for unirradiated material that deforms homogeneously, fatigue life increases with the ability to accommodate plastic strain, that is, ductility that is inversely proportional to strength. Therefore, fatigue life increases in the following order: Zircaloy-2, reactor grade sponge zirconium, sponge zirconium, and crystal bar zirconium, as shown in Figs. 6 and 7.

In the high-cycle regime, there is a reversal in the order of fatigue life for constant applied total strain. This is a result of the partitioning of plastic and total strain. At small values of

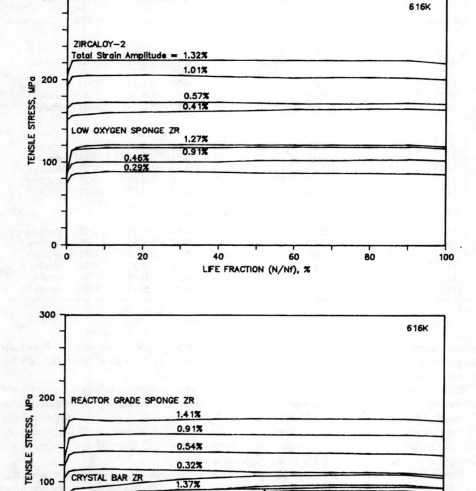

FIG. 3—*Cyclic strain hardening behavior of unirradiated Zircaloy-2 and zirconium at 616 K.*

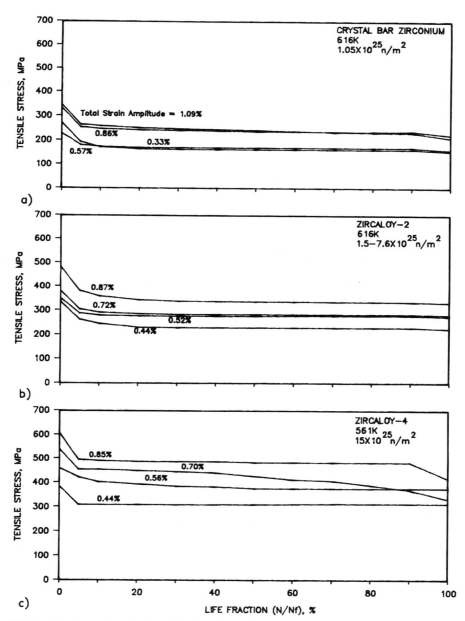

FIG. 4—*Cyclic strain softening behavior of irradiated Zircaloy-2 and crystal bar zirconium at 616 K and Zircaloy-4 at 561 K.*

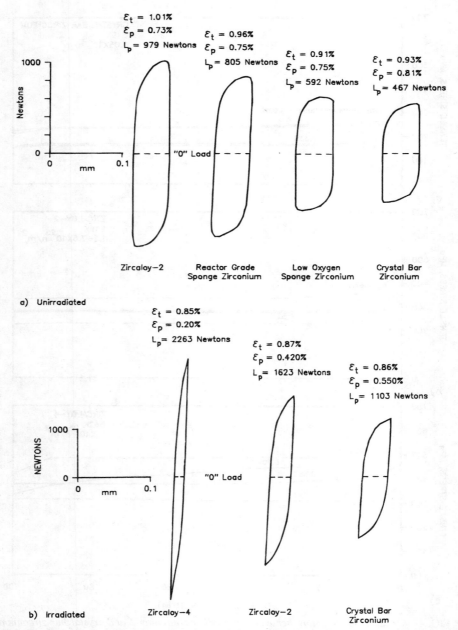

FIG. 5—*Mid-life stabilized hysteresis loops for* (a) *unirradiated and* (b) *irradiated Zircaloy and zirconium.*

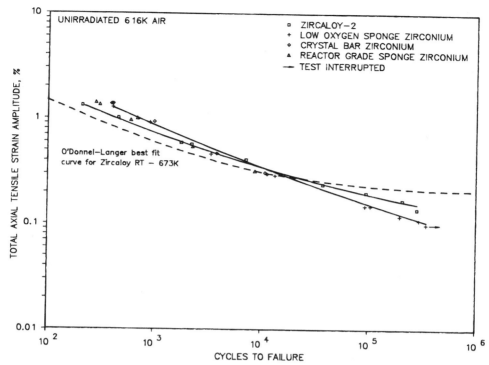

FIG. 6—*Total axial tensile strain amplitude versus cycles to failure for unirradiated Zircaloy-2 and zirconium.*

applied strain, the plastic component is much lower for the higher strength Zircaloy, so the fatigue life is higher relative to sponge zirconium. For reference, in Fig. 7, the data of Kubo et al. [*21*] is also shown. The origin of the discrepancy is not known, but it is noted that Kubo et al. tested in bending and converted applied strain to plastic strain through calculation.

Figure 8 shows the data for irradiated Zircaloy-2, Zircaloy-4, and crystal bar zirconium plotted for total and plastic tensile strain amplitudes versus cycles to failure. For constant total applied strain, there is not an unambiguous separation of the data, but crystal bar zirconium does appear to have a lower fatigue life than the Zircaloys. Figure 5b and Table 6 illustrate that for these irradiated materials, zirconium has the highest component of plastic strain for a given applied total strain. For irradiated materials, which deform in a very hetereogeneous manner, higher plastic strain will then result in lower fatigue life. Irradiated Zircaloy, like many other alloys, has been shown to deform by a dislocation channeling mechanism [*22*] whereby strain is concentrated in local swaths that are cleared of irradiation damage by avalanches of dislocations moving along slip planes. Details of the process have not been studied for fatigued Zircaloy, but earlier work in irradiated copper [*23*] showed that cyclic strain was accommodated within a limited number of dislocation channels, resulting in large slip steps (both "intrusions" and "extrusions") at the surface. These slip steps undoubtedly serve as fatigue crack initiation sites. The larger plastic strains in zirconium compared to the higher strength Zircaloys apparently result in lower fatigue life for the zirconium.

For constant applied plastic strain, Fig. 8b, the high-fluence Zircaloy-4 material has the lowest fatigue life. The maximum loads of the hysteresis loops in Fig. 5b indicate that the

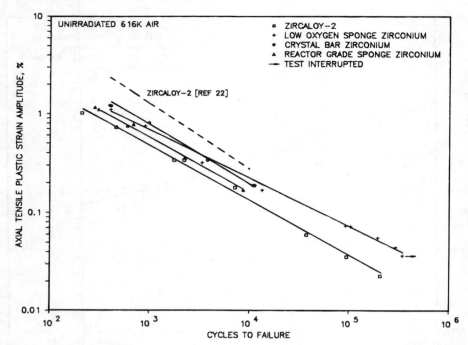

FIG. 7—*Axial tensile plastic strain amplitude versus cycles to failure for unirradiated Zircaloy-2 and zirconium.*

irradiation conditions (higher fluence, lower irradiation temperature) resulted in more irradiation hardening for Zircaloy-4 specimens than for Zircaloy-2 or zirconium. This results in a lower macroscopic ductility and increased tendency for localized deformation through the dislocation channeling mechanism and, thus, lower fatigue life for the Zircaloy-4.

Comparison of fatigue lives of unirradiated and irradiated material is given in Fig. 9. It is clear that irradiation reduces the fatigue life in the low-cycle regime, particularly for crystal bar zirconium. The trend agrees with O'Donnell and Langer [2], but is the opposite of Pettersson [3] and Nakatsuka et al. [4].

Crack Growth Behavior of Zircaloy-4

Figure 10 shows the crack growth threshold behavior of Zircaloy-4 in air at 293 and 561 K for a stress ratio of $R = 0.8$. An increase in temperature increases the crack growth rate. This agrees with the trend reported by James [8]. It was found in these crack growth threshold tests that if ΔK values were normalized by dividing them by the specimen elastic modulus at the test temperature, the relationship between crack growth rate, da/dN, and the normalized stress intensity factor range, $\Delta K/E$, was not temperature dependent within the temperature range covered by the experiments, although the load ratio, R, did influence the crack growth rate.

Normalized crack growth thresholds determined at a cyclic load frequency of 50 Hz are given in Table 7. These correspond to threshold ΔK values ranging from approximately 1.2 to 2.2 MPa$\sqrt{m}$, depending on temperature and load ratio. For the purposes of our tests, however, only $R = 0.8$ will be considered.

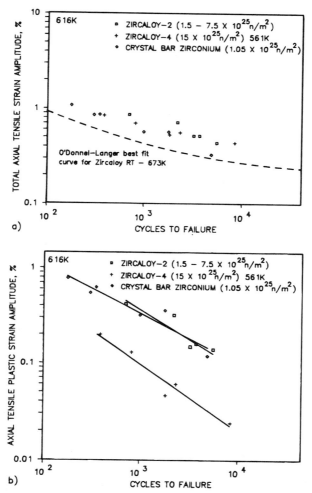

FIG. 8—*Tensile axial strain amplitude, (a) total and (b) plastic, versus cycles to failure for irradiated Zircaloy-2, Zircaloy-4, and crystal bar zirconium.*

Crack growth rates in air at 561 K for the unirradiated material are given in Fig. 11. For a load ratio, $R = 0.8$, the crack growth rate in air at 561 K is given to a good approximation by the relationship

$$da/dN = (5.47 \times 10^{-11})\Delta K^{3.13} \qquad (2)$$

where da/dN is in metres per cycle and ΔK is given in MPa$\sqrt{m}$. Between 3 and 50 Hz, frequency dependence was found to be negligible based on the linear portion of the da/dN versus ΔK log-log plot.

Crack growth rates measured in elevated temperature pressurized water are plotted, as a function of ΔK, in Fig. 12. It can be seen that water at 6.89 MPa and 561 K and containing

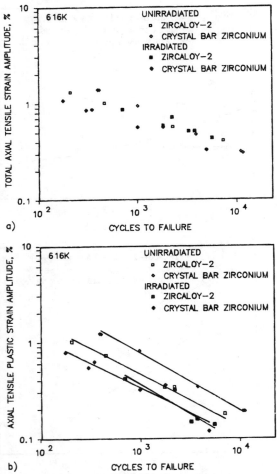

FIG. 9—*Comparison of tensile axial strain amplitude, (a) total and (b) plastic, versus cycles to failure for unirradiated Zircaloy-2 and crystal bar zirconium with irradiated Zircaloy-2 and crystal bar zirconium.*

0.1 to 0.2 ppm of dissolved oxygen approximately doubles the crack growth rate at 3 Hz in comparison to the rate in air at the same temperature and frequency. Increasing the oxygen content to 6 to 8 ppm further increases the growth rate three to four times the corresponding rate in air.

Crack growth rates measured in the irradiated specimens at 561 K are shown in Fig. 13. For neither the higher nor the lower fast neutron fluences do the measured crack growth rates in the irradiated material differ significantly from those measured in the unirradiated reference Zircaloy-4 at the same frequency and load ratio. In Figs. 12 and 13, the straight line represents Eq 2, the mean of the unirradiated test data in air.

Pickles and Picker [7,8] have reported on similar tests conducted at 533 or 555 K. They also observed similar trends; that is, no effect of irradiation and an increase in crack propagation

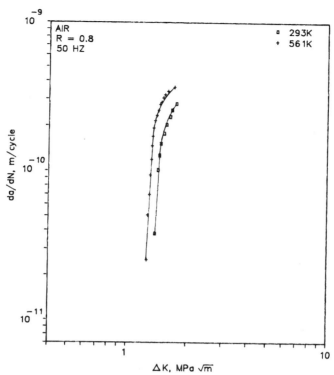

FIG. 10—*Crack growth threshold for unirradiated Zircaloy-4 in air at 293 and 561 K and a stress ratio of* R = 0.8.

rates in water. In addition, where their data overlaps with ours, around 10^{-8} m/cycle, the values of ΔK are similar. Walker and Kass [10] report an irradiation increased crack propagation rate, but comparison with our data is complicated because the Walker and Kass experiments were conducted at 293 K and at very high ΔK values.

Comparison of crack propagation between different experiments is made difficult because of possible significant effects of test variables. Crack growth thresholds, particularly at elevated temperature, can be affected by the load ratio, R, through the effects of creep on the crack extension process. Surface condition can affect fatigue life and crack growth rates. For instance, Mowbray [24] has demonstrated that fatigue life is decreased by the presence of thin oxide layers on the Zircaloy surface, such as were present on the irradiated specimens in our tests. Our observations on the increase in crack growth rates in high temperature water are apparently related to stress corrosion phenomena that are influenced by both cyclic frequency and load ratio.

TABLE 7—*Normalized crack growth thresholds.*

R-ratio	$\Delta K_{th}/E$, $\sqrt{m}$
Large (0.8)	1.6×10^{-5}
Small (<0.5)	2.5×10^{-5}

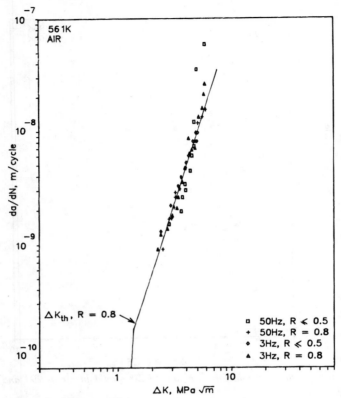

FIG. 11—*Effects of test parameters on 561-K fatigue crack growth rates in air in unirradiated Zircaloy-4 channel strip.*

Conclusions

Fatigue life and crack growth rate experiments were conducted on Zircaloy and zirconium specimens. Tests were conducted primarily in the temperature range of BWR interest, 561 to 616 K. Materials were irradiated to fluences between 1 to 15×10^{25} n/m² ($E > 1$ MeV). Conclusions based on this work are:

1. Fatigue life is dependent on the partitioning of plastic and elastic strain during the testing. Partitioning is influenced by material strength as affected by alloy content and irradiation and by testing technique.
2. Applied plastic strain more clearly defines fatigue life than does applied total (elastic plus plastic) strain. The conclusions given are based on applied plastic strain.
3. For unirradiated material, the softer zirconiums have a higher fatigue life than Zircaloy. The more pure the material, the higher the fatigue life.
4. Irradiation lowers the fatigue lives of Zircaloy and zirconium. The effect is larger for zirconium than for Zircaloy. High fluence appears to have a larger effect than low fluence. The effect appears to be related to strain localization induced by irradiation.

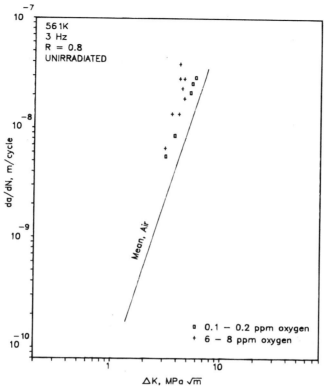

FIG. 12—*Effect of 68 bar, 561-K water on normalized fatigue crack growth rates in air for Zircaloy-4 channel strip.*

5. Fatigue crack growth rates increase slightly with temperature for a load ratio of $R = 0.8$ and can be normalized by dividing by the elastic moduli at the test temperature.
6. Fatigue crack growth rates at 3 Hz, $R = 0.8$, and ΔK in the range of 3 to 10 MPa$\sqrt{m}$ are
    (a) insensitive to neutron irradiation, and
    (b) increased by exposure to water, the increase being larger with increasing oxygen concentration in the water.
7. The results support the current design basis for fuel rods and channels in BWRs.

Acknowledgments

The authors wish to extend special credit to G. H. Henderson of GENE for conducting the fatigue tests. Also credit is due to W. R. Andrews and R. A. Williams formerly of GE CRD for determining the crack growth thresholds.

References

[1] Rowland, T., Iwasaki, R., Iwano, Y., Matsumoto, T., Rosenbaum, H., Nakatsuka, M., Maru, A., and Yoshizawa, A., "Power Cycling Operation Tests of Zr-Barrier Fuel," *Proceedings*, International

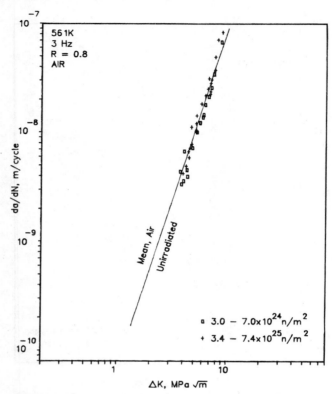

FIG. 13—*Effect of fast fluence on normalized fatigue crack growth rates in Zircaloy-4 channel strip.*

Topical Meeting on LWR Fuel Performance, American Nuclear Society and European Nuclear Society, Avingon, France, 1991, p. 808.
[2] O'Donnell, W. J. and Langer, B. F., *Nuclear Science and Engineering,* Vol. 20, 1964, pp. 1–12.
[3] Pettersson, K., *Journal of Nuclear Materials,* Vol. 56, 1975, pp. 91–102.
[4] Nakatsuka, M., Kubo, T., and Hayashi, Y. in *Zirconium in the Nuclear Industry: Ninth International Symposium, ASTM STP 1132,* C. M. Eucken and A. M. Garde, Eds., American Society of Testing and Materials, Philadelphia, 1991, pp. 230–245.
[5] Lee, D., *Metallurgical Transactions,* Vol. 3, 1972, pp. 315–322.
[6] Brun, G., Pelchat, J., Floze, J. C., and Galimberti, M., *Zirconium in the Nuclear Industry: Seventh International Symposium, ASTM STP 939,* R. B. Adamson and L. F. P. Swam, Eds., American Society of Testing and Materials, Philadelphia, 1987, pp. 597–616.
[7] Pickles, B. W. and Picker, C., "The Effects of Environment and Neutron Irradiation on the Fatigue Crack Growth Behavior of Zircaloy-2," *Proceedings,* International Conference on Dimensional Stability and Mechanical Behavior of Irradiated Metals and Alloys, Brighton, UK, 1983, pp. 179–182.
[8] Pickles, B. W. and Picker, C. in *Proceedings,* Sixth International Conference on Structural Mechanics in Reactor Technology, F6/4, Paris, France, North Holland, 1981, pp. 1–8.
[9] James, L. A., *Nuclear Application and Technology,* Vol. 9, 1970.
[10] Walker, T. J. and Kass, J. N., "Variation of Zircaloy Fracture Toughness with Irradiation," *Zirconium in Nuclear Applications, ASTM STP 551,* American Society for Testing and Materials, Philadelphia, 1974, pp. 328–359.
[11] Rosenbaum, H. S., Rand, R. A., Tucker R. P., Cheng, B., Adamson, R. B., Davies, J. H., Armijo, J. S., and Wisner, S. B., "Zirconium-Barrier Attributes," *Zirconium in the Nuclear Industry: Seventh*

International Symposium, ASTM STP 939, R. B. Adamson and L. F. P. Van Swam, Eds., American Society for Testing and Materials, Philadelphia, 1987, pp. 675–699.

[12] Martin, J. F., "Cyclic Stress-Strain and Fatigue Properties of Sheet Steel as Affected by Load Spectra," *Journal of Testing and Evaluation*, Vol. 11, No. 1, Jan. 1983, pp. 66–74.

[13] Sullivan, A. M. and Crooker, T. W., "The Effect of Specimen Thickness Upon the Fatigue Crack Growth Rate of A516-60 Pressure Vessel Steel," *Journal of Pressure Vessel Technology*, Vol. 99, No. 2, May 1977, pp. 248–252.

[14] Clarke, W. G., Jr., "Effect of Temperature and Section Size on Fracture Crack Growth in Pressure Vessel Steel," *Journal of Materials*, Vol. 6, No. 1, March 1971, pp. 134–149.

[15] Van Swam, L. F. P., Knorr, D. B., Pelloux, R. M., and Shewbridge, J. F., "Relationship Between Contractile Strain Ratio R and Texture in Zirconium Alloy Tubing," *Metallurgical Transactions*, Vol. 10A, April 1979, pp. 483–487.

[16] Rosenbaum, H. S. and Lewis, J. E., "Use of Pole Figure Data to Compute Elasticity Coefficients of Zirconium Sheet," *Journal of Nuclear Materials*, Vol. 67, 1977, pp. 237–283.

[17] Hudak, S. J., Jr., Saxena, A., Bucci, R. J., and Malcolm, R. C., "Development of Standard Methods of Testing and Analyzing Fatigue Crack Growth Rate Data," Technical Report AFML-TR-78-40, Air Force Materials Laboratory, Air Force Systems Command, Wright-Patterson Air Force Base, OH, May 1978.

[18] Coffin, L. F., Jr., *Transactions*, American Society of Mechanical Engineers, Vol. 76, 1954, pp. 931–949.

[19] Manson, S. S., NASA Technical Note 2933, NASA, Lewis Research Center, Cleveland, OH, 1954.

[20] Hosbons, R. R., *Fatigue at Elevated Temperatures, ASTM STP 520*, American Society for Testing and Materials, Philadelphia, 1973, pp. 482–490.

[21] Kubo, T., Motomyia, T., and Wakashima, Y., *Journal of Nuclear Materials*, Vol. 140, 1986, pp. 185–196.

[22] Adamson, R. B., Wisner, S. B., Tucker, R. P., and Rand, R. A., *The Use of Small-Scale Specimens for Testing Irradiated Materials, ASTM STP 888*, W. R. Corwin and L. E. Lucas, Eds., American Society for Testing and Materials, Philadelphia, 1986, pp. 171–185.

[23] Adamson, R. B., *Philosophical Magazine*, Vol. 17, 1968, p. 681.

[24] Mowbray, D. F., *Journal of Nuclear Applications*, Vol. 1, 1965, p. 39.

DISCUSSION

Clement Lenaignan[1] (written discussion)—Is the fatigue crack surface that you obtained in fatigue crack growth rate measurement somehow related to the grain orientation?

S. B. Wisner et al. (authors' closure)—Fatigue crack growth rate will be dependent on crystallographic texture and on grain shape. Our materials contain equiaxed, recrystallized grains so that should not be a factor. We measured crack growth rate in the sheet rolling direction in which the texture parameter is 0.11. It is likely that the crack growth rate would be different in different directions.

[1] CEA, Grenoble, France.

Didier Gilbon[1] *and Claude Simonot*[1]

Effect of Irradiation on the Microstructure of Zircaloy-4

REFERENCE: Gilbon, D. and Simonot, C., "**Effect of Irradiation on the Microstructure of Zircaloy-4,**" *Zirconium in the Nuclear Industry: Tenth International Symposium, ASTM STP 1245,* A. M. Garde and E. R. Bradley, Eds., American Society for Testing and Materials, Philadelphia, 1994, pp. 521–548.

ABSTRACT: This paper deals with the irradiation growth and the radiation-induced changes in the microstructure of Zircaloy-type materials. The experimental irradiations that were conducted at 400°C in Siloé (metallurgical test reactor) and Phénix (fast breeder reactor) show that the growth of both alpha-recrystallized and beta-quenched Zy-4FORT (high tin and oxygen contents) is accelerated after a dose of about 4 dpa (displacements per atom). In correlation with this acceleration, a high density of basal-plane c-component dislocations is revealed by transmission electron microscope (TEM) examinations. At this temperature, the structure of precipitates remains fully crystalline, but radiation-induced dissolution occurs and allows the formation of Zr_5Sn_3 particles evenly dispersed throughout the matrix. To allow comparisons with the microstructural evolutions produced in pressurized water reactor (PWR) conditions, TEM examinations were also performed on standard Zircaloy-4 irradiated as recrystallized guide tube and stress-relieved cladding. In recrystallized Zircaloy-4 irradiated to 10×10^{21} n/cm² ($T \approx 320°C$), a high density of basal-plane c-component dislocations is also observed, the $Zr(Fe,Cr)_2$ Laves phases undergo a partial crystalline-to-amorphous transformation, and their outer rim is severely depleted in iron. In stress-relieved Zircaloy-4 irradiated to the same neutron fluence, irradiation does not produce any significant recovery of the initial dislocation network. Although the Laves phases undergo an amorphous transformation only in the lower part ($T \approx 335°C$), a marked dissolution is observed all along the cladding (335 to 380°C).

KEY WORDS: zirconium alloys, zirconium, neutron irradiation, irradiation growth, dislocations, precipitates, amorphization, dissolution, nuclear materials, nuclear applications, radiation effects

The main limitation to reach high burnup levels in water-cooled nuclear reactors comes from the corrosion of zirconium-alloy core components. However, other phenomena, such as irradiation growth and creep, have to be considered for design purposes. When improvements in corrosion resistance allow higher burnup, or operation under more severe irradiation conditions in future reactors, the growth and creep properties will become increasingly important. In order to anticipate the behavior of zirconium-based alloys, it is necessary to perform experimental irradiations in a wide range of temperatures and neutron fluences, because a great number of data shows that accelerated growth [*1–13*] and creep [*14*] may occur at high doses, at least for the alpha-recrystallized and beta-quenched metallurgical states. Moreover, this accelerated regime seems coincident with the formation of basal-plane c-component loops and the progressive dissolution of intermetallic phases that are present in zirconium-based

---

[1] Research scientist, Transmission Electron Microscopy Group, and Ph.D. student, respectively, CEA, DTA/CEREM-DTM, Service de Recherches Métallurgiques Appliquées, C. E. Saclay, 91191 Gif sur Yvette cedex, France. Mr. Simonot is sponsored by CEA and FRAMATOME.

alloys [15–18]. These radiation-induced changes in microstructure might in turn affect the corrosion resistance of highly irradiated materials [19,20]. Thus, in addition to experimental irradiations, detailed microstructural investigations appear essential to obtain a better understanding of the damage production.

The aim of the current work is to summarize briefly the results of experimental irradiations that were conducted on recrystallized and beta-quenched Zircaloy-4FORT irradiated at 400°C in Phénix (fast breeder reactor or FBR) and Siloé (metallurgical test reactor or MTR), then to describe in detail the irradiation-induced changes in the microstructures. In a second step allowing comparisons with the microstructural evolutions produced in pressurized water reactor (PWR) conditions, transmission electron microscopy (TEM) examinations were also performed on standard Zircaloy-4 material irradiated as guide tube (recrystallized) and fuel cladding (stress-relieved) up to high burnup levels.

Experimental Procedure

Materials

The experimental irradiations were conducted on Zircaloy-4FORT tubing with an outer diameter of 9.5 mm and a wall thickness of 0.6 mm. This alloy (Zy-4F) is a tin- and oxygen-modified version of Zircaloy-4 developed by CEA for use in the alpha-recrystallized condition to obtain a better resistance to diametral creep-down than the standard stress-relieved (SR) Zircaloy-4 cladding material, and to reduce the internal pressure of the rod. In the current work, two metallurgical conditions were studied: recrystallized (RX) and beta-quenched (β-Q). The chemical compositions of Zy-4F and of standard Zircaloy-4, and the final heat treatments and the corresponding f parameters (volume fraction with 0002 poles oriented in axial directions) are listed in Table 1. The results of burst and creep control tests are summarized in Table 2.

Irradiation Conditions and Growth Results

RX and β-Q Zy-4F were irradiated as stress-free tubes (90 mm in length) in the two lower stages of an experimental rig in Phénix for 258 EFPD. A similar experiment was conducted in Siloé for 270 EFPD. Both irradiations were performed in NaK environment at 400°C, using copper and niobium detectors to measure the values of neutron fluences and damage productions in Siloé and in the two stages of Phénix. All these irradiation conditions are summarized in Table 3.

TABLE 1—*Materials and metallurgical conditions.*

Alloy	Chemical Composition, % by weight				Condition—Final Heat Treatment	f parameter
	Sn	Fe	Cr	O_2		
Zy-4F	1.73	0.20	0.115	0.164	RX 50% cold worked + 2 h 575°C	0.147
					β-Q 1050°C β-annealing + argon-cooled at 50°C/s	0.331
Zircaloy-4	1.2	0.18	0.07	0.09	RX cold worked + α-recrystallized	0.135
	1.7	0.24	0.13	0.15	SR cold worked + stress-relieved	0.103

TABLE 2—Results of burst and creep control tests.

Alloy		Burst Tests, 400°C, 10 MPa/min			Creep Test (400°C, 250 h, 127 N/mm^2) Deformation, %
		Yield Strength, N/mm^2	Ultimate Strength, N/mm^2	Uniform Strain, %	
Zy-4F	RX	200	290	10	0.39
	β-Q	233	289	8.3	0.02
Zircaloy-4	RX	190	265	10	0.80
	SR	350	450	2.2	1.51

TABLE 3—Irradiation conditions.

Reactor Material	Condition	T, °C	Neutron Flux, 10^{14} n/cm^2/s (E > 1 MeV)	Conversion, n/cm^2 (for 1 dpa)	Damage, dpa
Phénix	RX, β-Q	400	1.31 to 2.11 1.73^a	2.2 × 10^{20}	12.7 to 21.9 17.5^a
Zy-4F	RX, β-Q	400	0.53 to 1 0.77^a	2.2 × 10^{20}	5.24 to 10.3 7.6^a
Siloé	RX	400	2.34	7 × 10^{20}	7.84
Zy-4F	β-Q	400	2.06	7 × 10^{20}	6.95

^{a} Mean value.

Before irradiation, the length of tubes was measured along four generating lines. Similar measurements were performed repeatedly during irradiation in Siloé, and only at the end of irradiation in Phénix. The elongation deformations of both β-Q and RX Zy-4F are plotted versus dose (dpa NRT) in Fig. 1. For RX tubes irradiated in Siloé (see inset Fig. 1), the initial rapid growth is immediately followed by a slow steady state. However, after a critical damage of about 4 dpa (displacements per atom), an acceleration seems to occur. Apparently, the effects of neutron spectrum and damage rate can be neglected, because the Phénix results confirm this acceleration (Fig. 1) and show that above 6 dpa, growth increases linearly with increasing dose. The growth of the β-Q material is much lower than the RX one because of a less sharp texture (Table 1), but it undergoes a very similar acceleration after about 4 dpa.

Analytical Electron Microscopy

The radiation-induced changes in microstructure were studied by TEM. These examinations were first carried out on the lower parts of stress-free tubes irradiated in Phénix and Siloé (RX and β-Q Zy-4F, Table 3). Then, to allow comparison with the microstructural evolutions that are produced in PWR conditions, TEM examinations were also performed on a guide tube (RX Zy-4) irradiated for five cycles in Fessenheim 2, and on a fuel pin cladding (SR Zy-4) irradiated for four cycles in Gravelines 3. All of the specimens and their corresponding irradiation conditions are listed in Table 4. The temperature of the samples taken from the guide tube was approximately the temperature of water at the appropriate level. The values

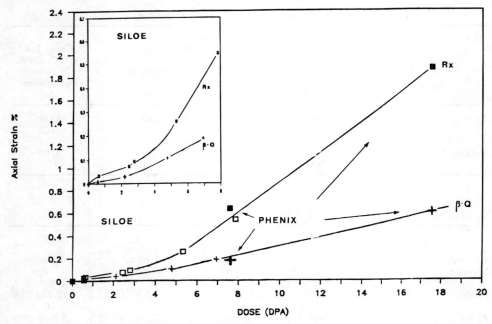

FIG. 1—*Axial strain of β-Q and RX Zy-4F irradiated at 400°C in Siloé and Phénix.*

of temperature indicated for the two samples taken from the cladding were computed taking into account the average linear power during the last cycle (170 W/cm), the local temperatures of water (295 and 322°C, respectively), the thicknesses of the oxide that were measured after irradiation (25 and 86 μm), and the position of thin foils in the thickness of tubing.

Thin foils for TEM examinations were prepared as follows. Small strips were first cut from the different tubes and mechanically thinned down to 100 μm. This allowed 3-mm disks to be punched out, a few of which (Table 4) were annealed under vacuum (1 10^{-7} torr, 1 h at 600°C). As-irradiated and annealed disks were then jet-polished in a 10% $HClO_3$, 20% C_4H_9-O-C_2H_4OH, and 70% C_2H_5OH solution at 30 V and 15°C. All of the TEM examinations were conducted in a Philips EM430 microscope equipped with a LaB6 filament and operated at 300 kV. The microstructural evolution (dislocation loops and lines, precipitates, and so on) was studied by the usual techniques of selected area diffraction (SAD), bright/dark field imaging, and weak beam imaging. The scanning and energy dispersive devices (STEM/X-EDS) also allowed the study of the radiation-induced microchemical changes in precipitates and matrix. For these experiments, an electron probe diameter of about 7 nm was used, giving an average count rate of 1000 cps (from 0.400 to 20.480 keV). The contents in zirconium, tin, iron, chromium, nickel, and silicon were computed on a TRACOR TN5500 system by processing the experimental X-ray spectra (150 s acquisition time) with the MICROQ software.

Results

Microstructure of RX and β-Q Zy-4F Irradiated in Phénix

Dislocation Microstructure—In RX and B-Q Zy-4F irradiated to 5 and 13 dpa, the dislocation microstructure consists mainly of a tangled network of *a*-type dislocation lines (Fig. 2), although

TABLE 4—*Materials examined by transmission electron microscopy.*

Reactor	Alloy	Temperature, °C	Neutron Dose, 10^{21} n/cm² ($E > 1$ MeV)	Damage, dpa
Phénix	RX Zy-4F	400	1.2	5
	RX Zy-4F	400	2.9	13
	β-Q Zy-4F	400	1.2	5
	β-Q Zy-4F	400	2.9	13
Siloé	RX Zy-4F	400	5.5	8
	β-Q Zy-4F	400	4.8	7
	β-Q Zy-4F	350	2.35	3.4
PWR	RX Zy-4[a]	320	9.5	13.6
	SR Zy-4[a]	335	9.6	13.7
	SR Zy-4[a]	380	9.7	14.2

[a] As-irradiated and annealed 1 h at 600°C.

a few alignments of loops ($b = 1/3\langle 11\bar{2}0\rangle$) can still be detected in β-Q samples irradiated to the lower dose (Fig. 2d). As shown in Table 5, the densities of these dislocation lines are quite similar for the two metallurgical conditions and increase very little with increasing damage.

In addition to these *a*-type dislocations, a high density of *c*-component defects can be observed. In prism plane foils, they appear as a uniform distribution of lines parallel with the *c*-axis when imaged with a 0002 diffracting vector (Fig. 2c), and exhibit stacking fault contrast when imaged with a $10\bar{1}0$ diffracting vector (Fig. 2b). This suggests that these *c*-component defects are the large, faulted, basal-plane dislocation loops ($b = 1/6\langle 20\bar{2}3\rangle$) that are generally considered as relevant to the onset of accelerated growth [21]. In β-Q samples (Figs. 2e and f), non-basal *c*-component dislocations are also seen, especially in the vicinity of grain- and lath-boundaries. Such defects could be a result of the stress generated by continuous anisotropic growth of adjacent grains and subgrains. The number of these non-basal defects increases from 5 to 13 dpa, so that in β-Q Zy-4F the total density of *c*-component dislocations becomes significantly higher than in RX Zy-4F (Table 5).

Irradiation Effects on Precipitates—In RX Zy-4F, the two intermetallics that were present prior to irradiation (mainly $Zr(Fe,Cr)_2$, and $Zr_2(Fe,Ni)$ to a much lesser extent) can still be detected after irradiation at 400°C (Fig. 3). The diffraction patterns show that their crystal structures remain hexagonal and tetragonal, respectively, with no sign of amorphization. However, the precipitate periphery (Fig. 3a) shows irregularities, suggesting that partial dissolution could have occurred upon irradiation. The particle size distributions (Fig. 4) confirm that their average diameter decreases from 310 nm in the control (unirradiated) specimen to 210 and 190 nm in specimens irradiated to 5 and 13 dpa, respectively. Moreover, the X-ray microanalysis profiles that were performed on $Zr(Fe,Cr)_2$ precipitates indicate that chemical re-solution was induced by irradiation. Whereas before irradiation (Fig. 5a), the iron/chromium (Fe/Cr) ratio remains constant and close to 2 throughout the particles, after irradiation (Fig. 5b), it drops continuously from the center to the periphery of precipitates and reaches a value of 1 at interface with the matrix. This microchemical change is mainly the result of preferential dissolution of iron, the chromium profile being roughly the same before and after irradiation. An asymmetry is observed on the profiles of Fig. 5b, but the interfaces between matrix and precipitates, and hence the matrix contribution, were not symmetrical. Four other profiles were

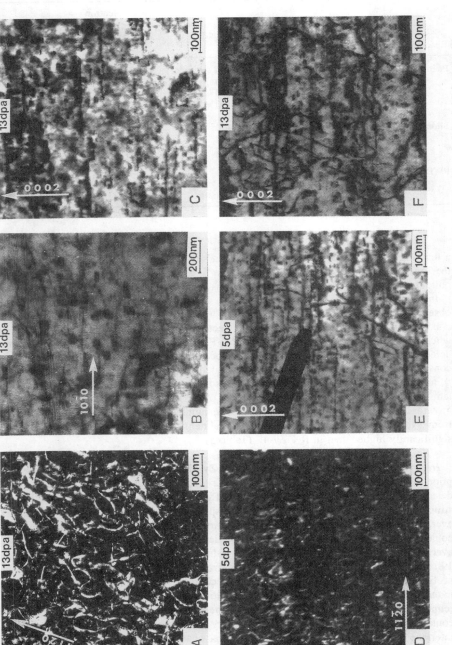

FIG. 2—*Dislocation microstructure in Zy-4F irradiated at 400°C in Phénix: (a, b, and c) RX Zy-4F and (d, e, and f) β-Q Zy-4F.*

TABLE 5—*Dislocation densities for RX and β-Q Zy-4F irradiated at 400°C in Phénix.*

Material	Dose, dpa	a-Type Dislocations, m^{-2}	c-Component Dislocations, m^{-2}
RX Zy-4F	5	7.5×10^{14}	9.1×10^{13}
	13	9.1×10^{14}	9.2×10^{13}
β-Q Zy-4F	5	7.7×10^{14}	9.5×10^{13}
	13	9.6×10^{14}	1.7×10^{14}

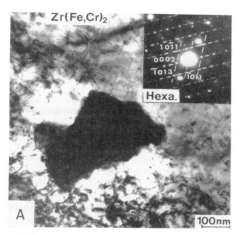 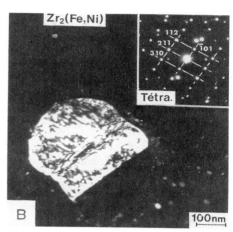

FIG. 3—*Morphology and crystal structure of Zr(Fe,Cr)$_2$ and Zr$_2$(Fe,Ni) precipitates observed in RX Zy-4F after irradiation to 13 dpa in Phénix (400°C).*

performed and gave similar informations. In addition, the local composition were measured at the center and edges of many irradiated particles. All of them confirmed that the Fe/Cr ratio drops from the center to the periphery of precipitates.

In unirradiated β-Q samples (Figs. 6a and b), finely sized Zr(Fe,Cr)$_2$ particles were present as alignments within the lath-boundaries (cubic structure), or were dispersed in the matrix close to some prior β-grain boundaries (cubic or hexagonal structure). After irradiation to 5 and 13 dpa (Figs. 6c and d), the lath structure tends to vanish, the precipitates dispersed in the matrix disappear, and the number density of particles aligned within the lath boundaries decreases with increasing damage production.

Radiation-Induced Phases and Voids—After irradiation to 5 and 13 dpa, a uniform distribution of new particles was detected in both RX and β-Q Zy-4F samples. In basal plane foils (Figs. 7a and b), these small precipitates have hexagonal shapes with two sets of facet orientations, and the SAD pattern can be analyzed as corresponding to a hexagonal crystal structure with a lattice parameter, $a = 0.846$ nm, and the orientation relationships, [0001] // [0001]α, ⟨10$\bar{1}$0⟩ // {21$\bar{3}$0}α. In prism plane foils (Figs. 7c and d), they appear as thin platelets parallel to the basal plane, and the diffraction pattern can be indexed using a lattice parameter, $c = 0.579$ nm, as deduced from the moiré fringe spacing (2.35 nm) observed in bright-field

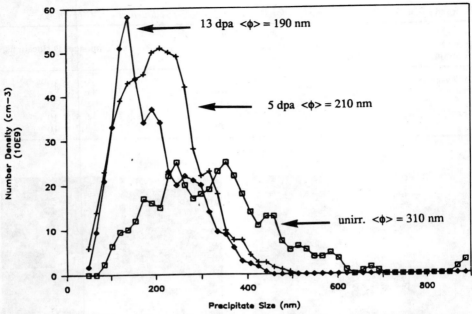

FIG. 4—*Effect of irradiation on precipitate size distribution in RX Zy-4F (Phénix, 400°C).*

images ($g = 0002\alpha$, Fig. 7e). Moreover, all these particles contain tin and a small amount of iron, as shown on Fig. 7f where the apparent solute contents are plotted versus zirconium percent to take into account large matrix contributions. Both these structural and chemical features suggest that the precipitates are Zr_5Sn_3 phases stabilized by iron. Their mean sizes, number densities, and volume fractions are listed in Table 6 for both materials and damage levels.

In addition to these Zr_5Sn_3 particles that are evenly dispersed throughout the matrix, rod-shaped precipitates with their long axis parallel to $\langle 11\bar{2}0\rangle\alpha$ directions were observed in RX samples, most often in the vicinity of Laves phases (Fig. 8a). In spite of very large matrix contributions, X-ray microanalysis revealed that these elongated particles were rich in iron and might contain some tin. In β-Q samples, a high density of rod-shaped precipitates with their long axis parallel to [0001]α were also observed adjacent to the lath boundaries (Fig. 8b). These precipitates, similar to the defects reported by Griffiths [22], could result from radiation-induced dissolution of $Zr(Fe,Cr)_2$ particles, and reprecipitation of solute.

Finally, voids were detected in β-Q Zy-4F irradiated to 5 and 13 dpa (Fig. 9), but not in RX Zy-4F. Most voids seem to have nucleated adjacent to basal c-component dislocations. At 5 dpa, the cavities appear as elongated platelets parallel with the basal plane, while at 13 dpa, the cavities are roughly equiaxed. This change in morphology with radiation damage seems to be the result of growth along the c-direction and shrinkage along the a-directions, as predicted by the diffusional anisotropy difference (DAD) theory [23,24].

Microstructure of β-Q and RX Zy-4F Irradiated in Siloé

The microstructures of RX and β-Q Zy-4F irradiated at 400°C in Siloé are very similar to the microstructures observed after irradiation in Phénix. Thus, basal-plane c-component dislocations were again observed and the micrographs in Fig. 10 give a clear illustration of

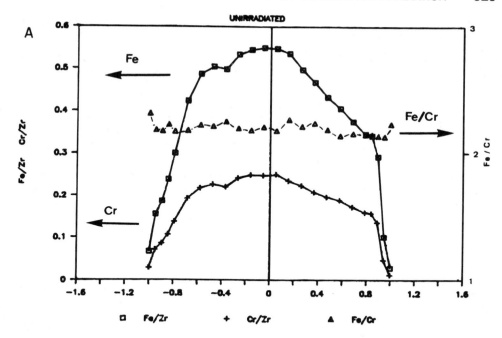
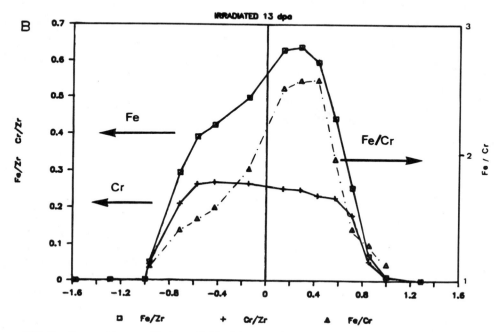

FIG. 5—*Composition profile for a $Zr(Fe,Cr)_2$ particle in RX Zy-4F:* (a) *unirradiated specimen and* (b) *irradiated at 400°C in Phénix.*

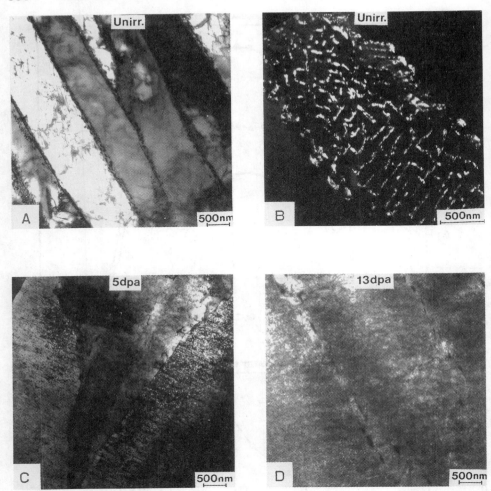

FIG. 6—*Evolution of Zr(Fe,Cr)₂ precipitates in β-Q Zy-4F:* (a *and* b) *unirradiated specimen and* (c *and* d) *irradiated at 400°C in Phénix.*

their faulted nature. Zr_5Sn_3 particles were also detected in both metallurgical states, while a few voids were only seen in β-Q samples. Consequently, the conversion factors between neutron fluence and damage level seem to render a correct account of the differences in neutron spectra. However, it is worth noting that for Siloé irradiations, both basal-plane c-component dislocations and Zr_5Sn_3 particles are less evenly dispersed throughout the matrix than for Phénix irradiations.

On the other hand, the micrographs in Fig. 11 allow comparison of the c-component dislocation microstructures for β-Q samples irradiated at 400°C up to a dose of 7 dpa corresponding to the accelerated growth regime, and at 350°C up to a dose of 3.5 dpa where this regime is far from being reached. The nonbasal dislocations that were present at some lath boundaries, have climbed or slipped onto basal plane at 400°C (Fig. 11a), and a high density of basal plane dislocations have nucleated within matrix (Fig. 11b). In contrast, the dislocations present

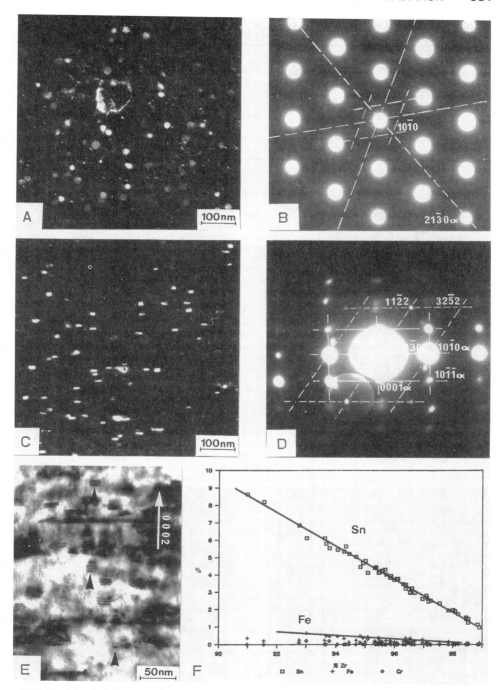

FIG. 7—*Morphology, crystal structure, and X-ray microanalysis of the Zr_5Sn_3 particles observed in RX and β-Q Zy-4F irradiated at 400°C (Phénix, 13 dpa).*

TABLE 6—*Size and number density of Zr_5Sn_3 particles in RX and β-Q Zy-4F irradiated at 400°C in Phénix.*

Material	Dose, dpa	Thickness, nm	Diameter, nm	Density, m^{-3}	Volume Fraction, %
RX Zy-4F	5	5.5	14	2.4×10^{20}	0.019
	13	6.5	19	2×10^{21}	0.36
β-Q Zy-4F	5	5.4	12.5	1×10^{21}	0.065
	13	5.8	18	2.3×10^{21}	0.32

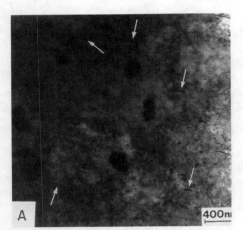 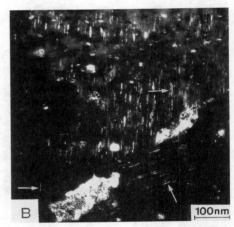

FIG. 8—*Rod-shaped precipitates in Zy-4F irradiated to 13 dpa in Phénix (400°C): (a) RX Zy-4F and (b) β-Q Zy-4F.*

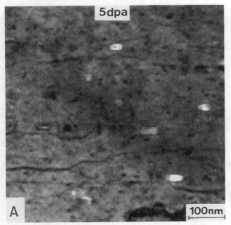 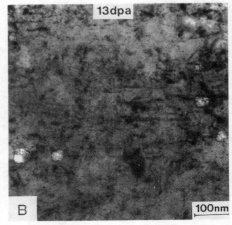

FIG. 9—*Radiation-induced voids in β-Q Zy-4F irradiated at 400°C in Phénix.*

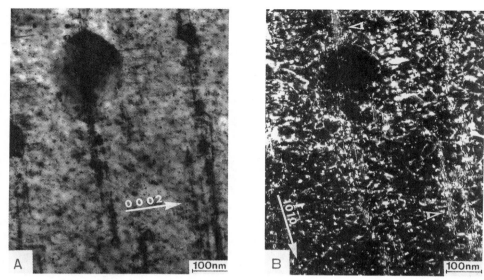

FIG. 10—*Dislocation microstructure in RX Zy-4F irradiated at 400°C in Siloé.*

at 350°C prior to irradiation mainly exhibit some signs of splitting into partials (arrowed in Fig. 11c), and only a residual contrast originating from the alignment of *a*-type loops can be detected in the matrix (Figs. 11c and d).

Microstructure of RX Zy-4 Irradiated in PWR (Guide Tube)

As-Irradiated Condition—After irradiation at about 320°C for 5 cycles, the dislocation microstructure of RX Zy-4 mainly consists of an inextricable network of *a*-type dislocation loops or lines. However, for prism plane foils, images using a 0002 diffracting vector allow a large number of basal-plane *c*-component defects to be distinguished (Fig. 12a) and the corresponding stacking faults can be seen again by using a 1010 diffracting vector (Fig. 12b). The density of these *c*-component dislocations is twice that for the previous irradiations at 400°C, but the main difference between these experimental and PWR irradiations is in the evolution of precipitates.

While the $Zr(Fe,Cr)_2$ precipitates remained crystalline up to high doses at 400°C, they have undergone a very significant crystalline-to-amorphous transformation at 320°C. This phenomenon is clearly illustrated by the difference in contrast between the center and the outer rim of large particles (Fig. 13a) and the presence of diffuse rings on the corresponding SAD pattern. The thickness of the amorphous layer (about 120 nm) is independent of the shape and size of particles and the smallest precipitates have undergone a complete transformation. Moreover, the composition profiles that were performed on duplex particles (Fig. 13b) show that the amorphous transformation is accompanied by a severe depletion in iron: its content drops sharply at the interface between crystalline and amorphous regions, decreases continuously across the amorphous zone, and finally drops again at the interface with matrix. At the same time, the chromium content seems to be roughly constant throughout the precipitate, so that the Fe/Cr ratio drops from 2 in the crystalline center down to less than 0.5 in the outer rim.

In contrast to $Zr(Fe,Cr)_2$ phases, the structure of $Zr_2(Fe,Ni)$ phases remained fully crystalline (Fig. 14). However, the periphery of the largest particles show saw-tooth-like irregularities,

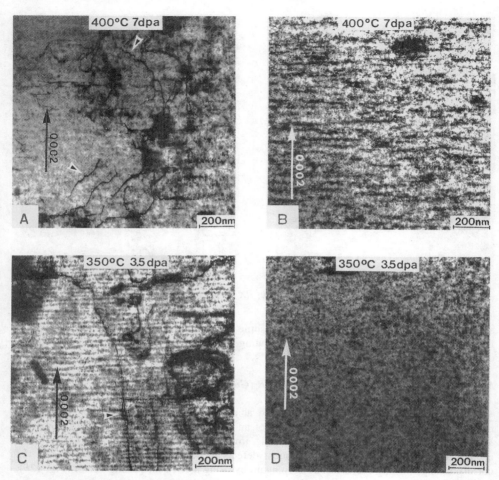

FIG. 11—*Dislocation microstructure in β-Q Zy-4F irradiated in Siloé at 400°C (a and b) and 350°C (c and d) (b) and (d) are areas of the matrix not containing dislocations prior to irradiation.*

and the smallest tend to have a lamellar morphology. These observations already suggest that amorphous transformation is not necessary to obtain a significant resolution during irradiation.

600°C Annealing—This post-irradiation annealing produced a considerable recovery of the dislocation microstructure but allowed retention of a loose network of *a*-type lines. Moreover, basal-plane *c*-component dislocations could still be detected, although unfaulting seems to have occurred during annealing and their density has been reduced by a factor of 2 (Figs. 12c and *d*). As for the partially amorphous $Zr(Fe,Cr)_2$ particles, their outer rim has undergone a slight microcrystalline transformation (Fig. 13c) and, at the same time, the local Fe/Cr ratio increased back to 1 (Fig. 13*d*), instead of 0.5 in the as-irradiated condition.

However, the most striking effect of annealing is to reveal a copious precipitation at grain boundaries (Fig. 15). The largest particles (about 350 by 150 nm) were identified as orthorhombic Zr_3Fe phases (Fig. 15*a*), and their number density was roughly estimated to be 3×10^{17}

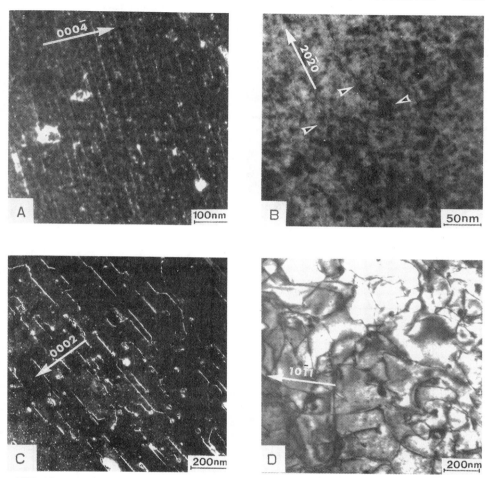

FIG. 12—*Dislocation microstructure in RX Zy-4 irradiated at 320°C in a PWR: (a and b) as-irradiated and (c and d) after annealing at 600°C.*

m^{-3}. The volume fraction of Zr$_3$Fe precipitates seems consistent with the amount of iron dissolved from the original precipitates. A more precise comparison cannot be made because the initial precipitate size distribution is not known and because Zr$_3$Fe particles are located at grain boundaries. Moreover, during annealing, part of dissolved iron comes back to the original precipitates.

In addition to these blocky precipitates, much smaller particles ($\approx$8 nm in diameter) are also observed at grain boundaries and in the vicinity of amorphized Zr(Fe,Cr)$_2$ precipitates (Fig. 15b). The X-ray microanalysis experiments (Fig. 15c) did confirm that two different phases were present: one is rich in both iron and chromium with a Fe/Cr ratio close to 2 and corresponds to the small particles mentioned earlier, the other contains only iron and corresponds to the intergranular Zr$_3$Fe precipitates. These results are in agreement with the deep depletion in iron observed in the amorphous regions of Laves phases, but they also show that a small amount of chromium has been dissolved during irradiation.

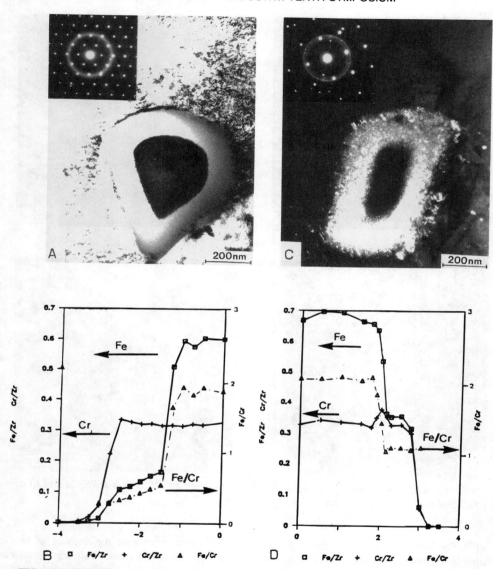

FIG. 13—*Crystalline to amorphous transformation of $Zr(Fe,Cr)_2$ particles in RX Zy-4 irradiated at 320°C in a PWR:* (a and b) *as-irradiated and* (c and d) *after annealing at 600°C.*

Microstructure of SR Zy-4 Irradiated in PWR (Cladding Material)

As-Irradiated Condition—In both samples taken from the cladding ($\approx$335 and 380°C, respectively), the irradiation up to 1×10^{22} n/cm² ($E > 1$ MeV) did not produce a significant recovery in the structure of cold-worked and stress-relieved Zircaloy-4 (Fig. 16). The local deformations and dislocation densities are so high that a detailed analysis seems impossible. However, for very thin prism-plane foils, weak-beam and bright-field images using a 0002 diffracting vector

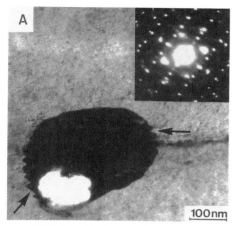 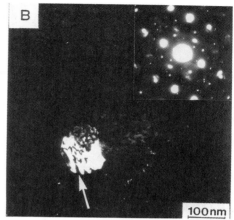

FIG. 14—*Evolution of $Zr_2(Fe,Ni)$ particles in RX Zy-4 irradiated at 320°C in a PWR.*

gave some information about the *c*-component dislocations. Thus, Fig. 16*b* shows a nonbasal dislocation wall close to a grain or subgrain boundary. Farther from this boundary, a lower density of similar dislocations and a few basal-plane defects can be detected. Moreover, Fig. 16*c* might indicate that the dislocations that were present prior to irradiation have climbed onto basal planes. This micrograph also confirms that basal-plane dislocations are present and the image using a 10$\bar{1}$0 diffracting vector shows the corresponding stacking faults (arrowed in Fig. 16*d*), in spite of intense moiré fringes that could be the result of superficial oxidation. These microstructural details suggest that a few faulted basal-plane *c*-component dislocation loops were produced by irradiation.

In both samples, a few $Zr_2(Fe,Ni)$ particles were found again, with a (bct) crystal structure and shape irregularities similar to those observed in the RX Zy-4 guide tube. In contrast to these Zintl phases, the evolution of Laves phases were found highly dependent on irradiation temperature. At 335°C, these particles experienced a complete amorphous transformation (Fig. 17*a*) and a severe depletion in iron affecting the whole amorphous volume (Fig. 17*b*). At about 380°C, the $Zr(Fe,Cr)_2$ particles keep their hcp crystal structure (Fig. 17*c*), and the composition profiles are similar to those observed for RX Zy-4F irradiated at 400°C in Phénix: the Fe/Cr ratio drops from 2 in the center of particles to 1 at their edges (Fig. 17*d*). Moreover, the very low number of precipitates that were detected at both levels, but especially at the lower one, suggest that a pronounced dissolution occurred during irradiation.

600°C Annealing—Whereas for the RX guide tube, this post-irradiation treatment retained part of the dislocation network, it induced full recrystallization of the samples taken from the SR cladding. Nevertheless, for the specimens irradiated at about 335°C (Fig. 18*a*), this annealing confirmed that most initial intermetallic phases experienced a complete dissolution, and revealed a copious precipitation of new small particles (22 nm in average diameter and 3×10^{20} m^{-3} in number density). These particles form alignments with the same direction all across thin foils and seem to reproduce the elongated grain and subgrain boundaries typical of the SR condition. Their morphology and striated contrast suggest that they could be tiny $Zr(Fe,Cr)_2$ phases, with a Fe/Cr ratio close to 2 as shown on Fig. 18*b* where iron and chromium apparent contents are plotted versus zirconium percent to account for the matrix contribution relative to each individual particle. As for the samples irradiated at about 380°C (Fig. 18*c*), the same

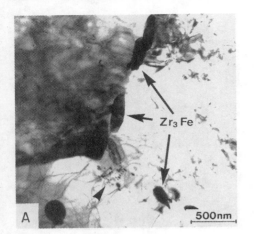
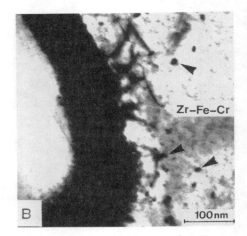
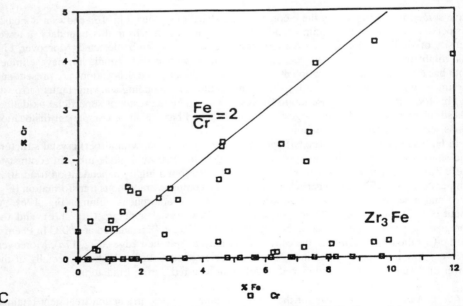

FIG. 15—*RX Zy-4 irradiated at 320°C in a PWR: Zr_3Fe and Zr-Fe-Cr particles revealed by annealing at 600°C.*

annealing revealed a similar distribution of smaller particles (17 nm in average diameter) having a slightly higher Fe/Cr ratio (Fig. 18d) and showed that a significant part of the initial phases were still present ($\approx$97 nm in diameter and 1.5×10^{18} m^{-3} in number density). Finally, it is worth noticing that in this last sample, X-ray microanalysis experiments allowed the detection of a few tin-rich particles that are very similar to the Zr_5Sn_3 phases observed in RX and β-Q Zy-4F irradiated at 400°C in Phénix and Siloé.

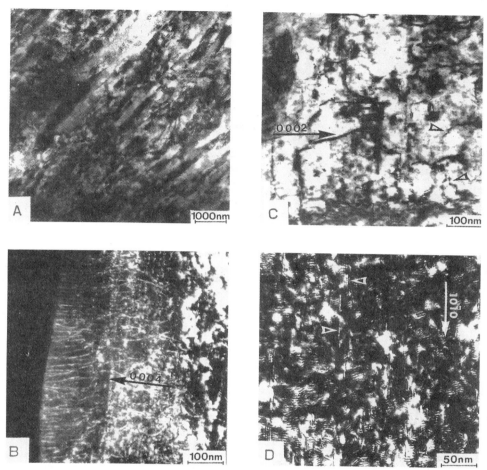

FIG. 16—*Dislocation microstructure in SR Zy-4 irradiated at 380°C in a PWR.*

Discussion

The TEM examinations reported earlier allowed us to study the radiation-induced changes in the microstructures of Zy-4F and Zy-4 for three different metallurgical states (β-Q, RX, and SR). Moreover, different irradiation conditions were used (FBR, MTR, and PWR) and the samples that were chosen for microstructural investigations allowed us to cover a significant range of damage level, from 3 to about 15 dpa, and to study precisely a range of irradiation temperatures comprised between 320 and 400°C. In the following sections, we will discuss the results concerning the effects of irradiation on precipitates and dislocations in relation with the observed in-reactor behaviors and we will compare these results with those published in the literature.

Effect of Irradiation on Precipitates

In recent years, the radiation-induced changes in structure, composition, and morphology of precipitates in Zircaloys have been the subject of increasing study [22,25–32] because they

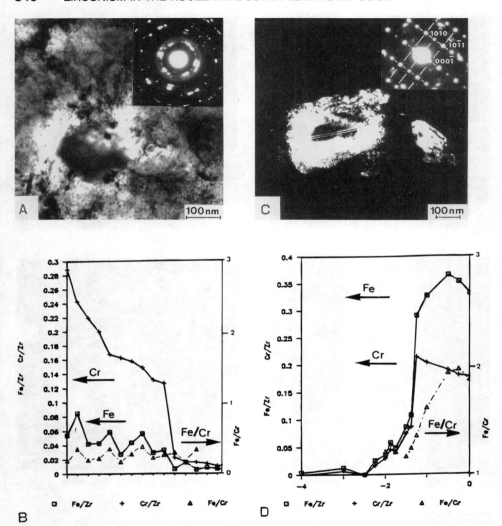

FIG. 17—*Radiation-induced change in structure and composition of $Zr(Fe,Cr)_2$ particles in SR Zy-4 irradiated at 335°C (a and b) and 380°C (c and d) in a PWR.*

are suspected to affect their in-reactor behavior, especially, irradiation growth, creep, and corrosion resistance. All of these investigations led to three separate temperature regimes.

At Low Temperatures (T ≈ 80°C)—Both $Zr_2(Ni,Fe)$ and $Zr(Fe,Cr)_2$ that are the predominant second phases in Zircaloy-2 and Zircaloy-4, respectively, undergo a rapid and complete crystalline-to-amorphous transformation. This transformation is the result of radiation-induced disordering of the crystal structure, and the irradiation temperature is too low to induce dynamic recrystallization or significant dissolution of the alloying elements.

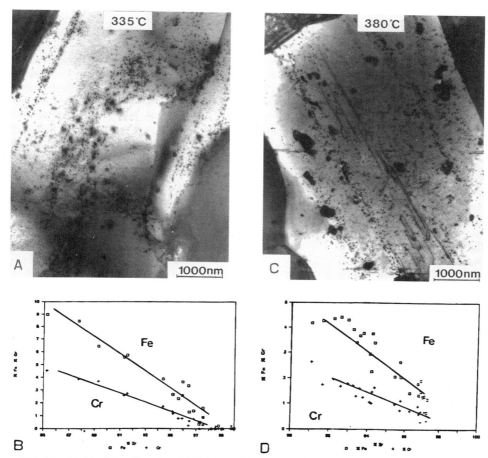

FIG. 18—*SR Zy-4 irradiated at 335°C (a and b) and 380°C (c and d) in a PWR: distribution and composition of precipitates after annealing at 600°C.*

At Intermediate Temperatures (T ≈ 300°C)—The Zr_2(Ni,Fe) particles remain crystalline, while the $Zr(Fe,Cr)_2$ particles undergo a partial amorphous transformation. This second temperature regime was by far the most extensively studied, since it is relevant to many zirconium-alloy core components in water-cooled nuclear reactors. Both experimental and theoretical works [26–33] gave a better understanding of the progressive amorphization of Laves phases and allowed the development of powerful models. The crystalline-to-amorphous transformation starts at the periphery of particles and seems mainly driven by the slight deviation from stoichiometry that occurs when iron diffuses out of $Zr(Fe,Cr)_2$ phases during irradiation. Then, the amorphous layer rejects a high quantity of iron to attain a new equilibrium and the reaction can progress inwards as the neutron fluence increases. Both chromium and iron seem to diffuse by interstitial mechanisms [34,35] but the much higher diffusion rate of iron explains its preferential dissolution from $Zr(Fe,Cr)_2$ particles while their chromium content is much less affected.

In the current work, this scenario was clearly encountered for the RX Zy-4 guide tube specimen irradiated at about 320°C in a PWR. The partial amorphous transformation of

In the current work, this scenario was clearly encountered for the RX Zy-4 guide tube specimen irradiated at about 320°C in a PWR. The partial amorphous transformation of $Zr(Fe,Cr)_2$ phases was effectively observed and the composition profiles showed that a very severe depletion in iron occurred within the amorphized outer rim. Moreover, the preferential dissolution of iron was confirmed by the post-irradiation annealing since a copious precipitation of blocky Zr_3Fe phases was observed at grain boundaries. By contrast, the structure of $Zr_2(Fe,Ni)$ particles (similar to those observed in Zircaloy-2 but having a much lower nickel content) remained fully crystalline, although a partial dissolution with a marked directional character [13,36] occurred during irradiation. All these results are in perfect agreement with the literature.

A rather similar situation was observed for SR Zy-4 specimens taken from the lower part ($T = 335°C$) of a cladding irradiated in PWR. However, the $Zr(Fe,Cr)_2$ particles were found fully amorphous, although the fluence level is the same as for the RX guide tube. This difference in amorphization (partial for RX Zy-4, complete for SR Zy-4) could be the result of differences in irradiation temperatures or initial precipitate size distributions. It is very unlikely that an increase in temperature (by 15°C) enhances the amorphization rate, and the second hypothesis has to be preferred. Nevertheless, the main difference between RX and SR materials is not the amorphization but the dissolution of second-phase particles: a post-irradiation annealing at 600°C revealed, or induced, an intense precipitation of tiny zirconium-iron-chromium particles (with a Fe/Cr ratio close to 2) and confirmed that most of $Zr(Fe,Cr)_2$ phases experienced a complete dissolution in SR samples. It has already been suggested that a single-phase solid solution may be achievable at very high fluence [37,38]. However, that such an event could occur in SR Zy-4 irradiated to 10×10^{21} n/cm^2 has not been realized before. Moreover, to explain the difference in dissolution between RX and SR materials irradiated to the same dose, a difference in initial precipitate size distribution is not sufficient, all the more, as a high initial dislocation density was not expected to enhance dissolution [39]. This would suggest that there is competition between dissolution and amorphization (as far as it can be separated from dissolution), and that at 335°C dissolution is close to becoming the dominant process.

At High Irradiation Temperatures (T > 370°C)—Amorphization was not detected, but dissolution of $Zr_2(Fe,Ni)$ and $Zr(Fe,Cr)_2$ particles, solute redistribution, and secondary phase formation throughout the matrix and at grain boundaries were observed by several authors [40–43]. In the current work, such a situation was clearly met for RX and β-Q Zy-4F irradiated at 400°C in Siloé and Phénix. Iron rich rod-shaped precipitates and a high density of Zr_5Sn_3 particles evenly dispersed throughout the matrix were seen for both metallurgical conditions. This Zr_5Sn_3 phase, which has already been observed by several authors in this high irradiation temperature range [39,42], is the result of radiation-enhanced diffusion of tin and is stabilized by iron to the detriment of the Zr_4Sn phase expected from the binary zirconium-tin phase diagram [44,45]. The volume fractions that have been measured in RX and β-Q Zy-4F irradiated to 13 dpa and the X-ray microanalysis results corrected for matrix contributions show that more than 200 ppm of iron had to be available. This estimation also confirms that a preferential dissolution of iron from intermetallic particles occurred during irradiation [39].

A similar situation was also observed for SR Zy-4 specimens taken from the upper part ($T = 380°C$) of the cladding irradiated in a PWR. In contrast with samples taken from the lower part ($T = 335°C$), the $Zr(Fe,Cr)_2$ particles detected after irradiation were found fully crystalline. This difference clearly shows that the highly irradiated part of a PWR fuel rod cladding cannot be considered as a monolith from a microstructural point of view, and this does not seem to have been realized before. Another very striking feature is the extent of dissolution that is achievable in the absence of crystalline-to-amorphous transformation. The post-irradiation annealing revealed very similar distributions of small zirconium-iron-chromium particles in both parts of cladding, and their corresponding volume fractions only differ by a factor of 2. These results clearly show that the amorphous transformation is not necessary to obtain very significant dissolution, even if the dissolution process does seem faster at intermediate temperatures.

Finally, several authors [13,36] have suggested a kind of competition between radiation-induced dissolution and radiation-enhanced diffusion, the former being the dominating process at lower temperatures whereas the latter could even be fast enough at higher temperatures to cause a growth of the precipitates. Our findings on RX and β-Q Zy-4F irradiated at 400°C, as on SR Zy-4 irradiated at 380°C, do not support this hypothesis: neither the precipitate size distributions, nor the intensity of secondary phase precipitation suggest any ripening of the precipitates.

In-Reactor Behavior and Radiation-Induced Changes in Microstructure

It is now well established that for recrystallized zirconium alloys, the first two stages of growth (initially rapid, then with a low steady state rate) are linked with the nucleation and development of a mixture of prism-plane a-type dislocation loops ($b = \frac{1}{3}\langle 11\bar{2}0\rangle$) with interstitial and vacancy characters. By contrast, the accelerated growth regime that occurs at high neutron fluences is highly correlated with the formation of large, faulted, basal-plane c-component dislocation loops ($b = \frac{1}{6}\langle 20\bar{2}3\rangle$) that are invariably vacancy in character. In the current work, the growth experiments that were performed at 400°C showed that an acceleration was produced in RX, as in β-Q, Zy-4F irradiated to about 4 dpa, and the TEM examinations revealed a high density of basal-plane c-component defects. By contrast, these defects were not detected in β-Q Zy-4F irradiated at 350°C to a dose of 3.5 dpa corresponding to the quasi-saturation regime. All these results are in perfect agreement with the literature.

In the SR Zy-4 cladding, the irradiation to 10×10^{21} n/cm^2 in a PWR did not produce a significant recovery of the initial dislocation network, as reported by other authors [18,39]. However, the TEM examinations confirmed the presence of a high proportion of c-component dislocations that are considered responsible for the high growth rate of SR materials [21]. In the RX Zy-4 guide tube irradiated to a very similar neutron fluence, the TEM examinations have shown that a high density of basal-plane c-component dislocations were produced by irradiation. This finding is in agreement with many published data and could suggest that the accelerated growth regime has been reached.

At present, the rapid development of basal-plane c-component loops and the high growth rate in the breakaway regime can be explained by the diffusional anisotropy difference theory [46]. The nucleation process of these loops and the onset of acceleration are more difficult to understand but have been shown highly dependent on the matrix content in alloying elements and impurities [18,47]. In the current work, the observation of basal-plane c-component defects was always coincident with marked amorphization or dissolution or both of the precipitates. Thus, a high quantity of iron (and chromium) becomes available in the matrix and could enhance diffusional anisotropy [48] by coupling with vacancies [49,50], and could segregate to modify the local c/a ratio and the basal-plane stacking fault energy [16,48]. Moreover, in Zy-4F irradiated at 400°C in Phénix and Siloé, the high O$_2$-content and the progressive lowering of the tin-content by Zr$_5$Sn$_3$ phase formation could also favor the nucleation of vacancy-type c-component defects [16].

Finally, as suggested by several authors [51,52], it is worth noticing that the radiation-induced dissolution of precipitates and the redistribution of solute elements by local segregation and secondary phase formation, could also affect the mechanical resistance of zirconium-based alloys, and especially their corrosion resistance since the initial precipitate size distribution and the matrix content in solute elements are considered important factors.

Conclusions

In this work, TEM examinations were performed on Zy-4F irradiated in Siloé and Phénix (RX and β-Q samples), then on standard Zircaloy-4 irradiated in a PWR (RX guide tube and SR fuel rod cladding). The main conclusions can be summarized as follows:

1. In RX and β-Q Zy-4F irradiated in Phénix and Siloé (400°C), as in RX Zy-4 irradiated in a PWR (≈320°C), the TEM examinations revealed a high density of basal-plane c-component dislocations that are typical of the accelerated growth regime.
2. In SR Zy-4 irradiated in a PWR (335 to 380°C), the irradiation does not produce a significant recovery of the initial dislocation network. The faster growth rate of SR materials is explained by the high proportion of c-component dislocations that are present prior to irradiation.
3. The effect of irradiation on precipitates is highly temperature dependent and various phenomena can be observed: crystalline-to-amorphous transformation, dissolution, and secondary phase transformation.
4. In RX and β-Q Zy-4F irradiated at 400°C, the structure of $Zr(Fe,Cr)_2$ and $Zr_2(Fe,Ni)$ particles remain crystalline but radiation-induced dissolution occurs and allows the formation of iron-rich rod-shaped precipitates and iron-stabilized Zr_5Sn_3 platelets, the latter being evenly dispersed throughout the matrix.
5. In RX Zy-4 irradiated at 320°C, the $Zr(Fe,Cr)_2$ Laves phases experience a partial amorphous transformation. The amorphous regions are severely depleted in iron, and a post-irradiation annealing at 600°C revealed a copious precipitation of blocky Zr_3Fe particles at grain boundaries.
6. In fuel rod cladding, two situations can be observed: in the lower part of cladding (≈335°C), the Laves phases experience a fully amorphous transformation, whereas they remain fully crystalline in the upper part (≈380°C). Although more precipitates are still present in the upper part, the common feature of both these parts is the marked radiation-induced dissolution, and a post-irradiation annealing at 600°C reveals similar distributions of small secondary phase particles that are rich in iron and chromium and have a Fe/Cr ratio close to 2.

Acknowledgments

The authors would like to thank Mm O. Rabouille and D. Nunes (CEA-DRN-SETIC) for the preparation of thin foils from irradiated materials, Mrs M. P. Hugon and Mm C. Rivera and Ph. Legendre for their assistance in the area of transmission electron microscopy.

References

[1] Northwood, D. O., Gilbert, R. W., Bahen, L. E., Kelly, P. M., Blake, R. G., Jonstons, A., Madden, P. K., Faulkner, D., Bell, W., and Adamson, R. B., "Characterization of Neutron Irradiated Damage in Zirconium Alloys. An International "Round Robin" Experiment," *Journal of Nuclear Materials*, Vol. 79, 1979, pp. 379–394.
[2] Murgatroyd, R. A. and Rogerson, A., "An Assessment of the Influence of Microstructure and Test Conditions on the Irradiation Growth Phenomenon in Zirconium Alloys," *Journal of Nuclear Materials*, Vol. 90, 1980, pp. 240–248.
[3] Rogerson, A. and Murgatroyd, R. A., " 'Breakaway' Growth in Annealed Zircaloy-2 at 353K and 553K," *Journal of Nuclear Materials*, Vol. 113, 1983, pp. 256–259.
[4] Williams, J., Darby, E. C., and Minty, D., "Irradiation Growth of Annealed Zircaloy-2," *Zirconium in the Nuclear Industry: Sixth International Symposium, ASTM STP 824*, D. G. Franklin and R. B. Adamson, Eds., American Society for Testing and Materials, Philadelphia, 1984, pp. 376–393.

[5] Rogerson A., "Irradiation-Induced Dimensional Changes in Polycrystalline Iodide Zirconium and Zircaloy-2," *Zirconium in the Nuclear Industry: Sixth International Symposium, ASTM STP 824,* D. G. Franklin and R. B. Adamson, Eds., American Society for Testing and Materials, Philadelphia, 1984, pp. 334–408.

[6] Tucker, R. P., Fidleris, V., and Adamson, R. B., "High Fluence Irradiation Growth of Zirconium Alloys at 644 to 725 K," *Zirconium in the Nuclear Industry: Sixth International Symposium, ASTM STP 824,* D. G. Franklin and R. B. Adamson, Eds., American Society for Testing and Materials, Philadelphia, 1984, pp. 427–451.

[7] Willard, H. J., "Irradiation Growth of Zircaloy (LWBR Development Program)," *Zirconium in the Nuclear Industry: Sixth International Symposium, ASTM STP 824,* D. G. Franklin and R. B. Adamson, Eds., American Society for Testing and Materials, Philadelphia, 1984, pp. 452–480.

[8] Fidleris, V., "The Irradiation Creep and Growth Phenomena," *Journal of Nuclear Materials,* Vol. 159, 1988, pp. 22–42.

[9] Carpenter, G. J. C., Zee, R. H., and Rogerson, A., "Irradiation Growth in Zirconium Single Crystals: A Review," *Journal of Nuclear Materials,* Vol. 159, 1988, pp. 86–100.

[10] Rogerson, A. and Zee, R. H., "Irradiation Growth of Zirconium-Tin Alloys at 353 and 553K," *Journal of Nuclear Materials,* Vol. 152, 1988, pp. 220–224.

[11] Rogerson, A., "Irradiation Growth in Annealed and 25% Cold-Worked Zircaloy-2 between 353–673K," *Journal of Nuclear Materials,* Vol. 154, 1988, pp. 226–285.

[12] Rogerson, A., "Irradiation Growth in Zirconium and Its Alloys," *Journal of Nuclear Materials,* Vol. 159, 1988, pp. 43–61.

[13] Garzarolli, F., Dewes, P., Maussner, G., and Basso, H., "Effects of High Neutron Fluences on Microstructure and Growth of Zircaloy-4," *Zirconium in the Nuclear Industry: Eighth International Symposium, ASTM STP 1023,* L. F. P. Van Swam and C. M. Eucken, Eds., American Society for Testing and Materials, Philadelphia, 1989, pp. 641–657.

[14] Causey, A. R., Fidleris, V., and Holt, R. A., "Acceleration of Creep and Growth of Annealed Zircaloy-4 by Preirradiation to High Fluences," *Journal of Nuclear Materials,* Vol. 139, 1986, pp. 277–278.

[15] Holt, R. A. and Gilbert, R. W., "⟨c⟩ Component Dislocations in Annealed Zircaloy Irradiated to about 570K," *Journal of Nuclear Materials,* Vol. 137, 1986, pp. 185–189.

[16] Griffiths, M. and Gilbert, R. W., "The Formation of c-Component Defects in Zirconium Alloys during Neutron Irradiation," *Journal of Nuclear Materials,* Vol. 150, 1987, pp. 169–181.

[17] Fidleris, V., Tucker, R. P., and Adamson, R. B., "An Overview of Microstructural and Experimental Factors that Affect the Irradiation Growth Behavior of Zirconium Alloys," *Zirconium in the Nuclear Industry: Seventh International Symposium, ASTM STP 939,* R. B. Adamson and L. F. P. Van Swam, Eds., American Society for Testing and Materials, Philadelphia, 1987, pp. 49–85.

[18] Griffiths, M., Gilbert, R. W., and Fidleris, V., "Accelerated Irradiation Growth of Zirconium Alloys," *Zirconium in the Nuclear Industry: Eighth International Symposium, ASTM STP 1023,* L. F. P. Van Swam and C. M. Eucken, Eds., American Society for Testing and Materials, Philadelphia, 1989, pp. 658–677.

[19] Garzarolli, F. and Stehle, H., "Behaviour of Core Structural Materials in Light Water-Cooled Power Reactors," *Proceedings,* IAEA International Symposium on Improvement in Water Reactor Fuel Technology and Utilization, Stockolm, Sweden, IAEA SM 288/24, International Atomic Energy Agency, Vienna, pp. 387–407.

[20] Kruger, R. M., Adamson, R. B., and Brenner, S. S., "Effects of Microchemistry and Precipitate Size on Nodular Corrosion Resistance of Zircaloy-2," *Journal of Nuclear Materials,* Vol. 189, 1992, pp. 193–200.

[21] Holt, R. A., "Mechanisms of Irradiation Growth of Alpha-Zirconium Alloys," *Journal of Nuclear Materials,* Vol. 159, 1988, pp. 310–338.

[22] Griffiths, M., "A Review of Microstructure Evolution in Zirconium Alloys during Irradiation," *Journal of Nuclear Materials,* Vol. 159, 1988, pp. 190–218.

[23] Woo, C. H., "Effects of Anisotropic Diffusion on Irradiation Deformation," *Radiation-Induced Changes in Microstructure: 13th International Symposium, ASTM STP 955,* F. A. Garner, A. S. Packan, and N. H. Kumar, Eds., American Society for Testing and Materials, Philadelphia, 1987, pp. 70–89.

[24] Griffiths, M., Styles, R. C., Phillip, F., Frank, W., and Woo, C. H., "Study of Point Defect Mobilities in Zirconium by the Direct Observation of Dislocation Climb and Cavity Growth during Electron Irradiation in a HVEM," *Journal of Nuclear Materials,* to be published.

[25] Yang, W. J. S. and Adamson, R. B., "Beta-Quenched Zircaloy-4: Effects of Thermal Aging and Neutron Irradiation," *Zirconium in the Nuclear Industry: Eighth International Symposium, ASTM*

STP 1023, L. F. P. Van Swam and C. M. Eucken, Eds., American Society for Testing and Materials, Philadelphia, 1989, pp. 451–477.

[26] Yang, W. J. S., "Precipitate Stability in Neutron-Irradiated Zircaloy-4," *Journal of Nuclear Materials,* Vol. 158, 1988, pp. 71–80.

[27] Gilbert, R. W., Griffiths, M., and Carpenter, G. J. C., "Amorphous Intermetallic in Neutron Irradiated Zircaloy after High Fluences," *Journal of Nuclear Materials,* Vol. 135, 1985, pp. 265–268.

[28] Harris, L. and Yang, W. J. S., "Amorphous Transformation of Laves Phases in Zircaloy and Austenitic Stainless Steel upon Irradiation," *Radiation-Induced Changes in Microstructure: 13th International Symposium, ASTM STP 955,* F. A. Garner, A. S. Packan, and N. H. Kumar, Eds., American Society for Testing and Materials, Philadelphia, 1987, pp. 661–675.

[29] Motta, A. T., Olander, D. R., and Machiels, A. J., "Electron Irradiation-Induced Amorphization of Precipitates in Zircaloy-2," *Effects of Radiation on Materials: 14th International Symposium, Vol. I, ASTM STP 1046,* N. H. Packan, R. E. Stoller, and A. S. Kumar, Eds., American Society for Testing and Materials, Philadelphia, 1989, pp. 457–469.

[30] Motta, A. T., Lefebvre, F., and Lemaignan, C., "Amorphization of Precipitates in Zircaloy under Neutron and Charged Particle Irradiation," *Zirconium in the Nuclear Industry: Ninth International Symposium, ASTM STP 1132,* C. M. Eucken and A. M. Garde, Eds., American Society for Testing and Materials, Philadelphia, 1991, pp. 718–739.

[31] Motta, A. T. and Lemaignan, C., "A Ballistic Mixing Model for the Amorphization of Precipitates in Zircaloy under Neutron Irradiation," *Journal of Nuclear Materials,* Vol. 195, 1992, pp. 277–285.

[32] Pecheur, D., Lefebvre, F., Motta, A. T., Lemaignan, C., and Charquet, D., "Effect of Irradiation on the Precipitate Stability in Zr Alloys," *Proceedings,* 3rd Conference on Evolution of Microstructure in Metals During Irradiation, 1992, Muskoka, Canada, to be published in *Journal of Nuclear Materials.*

[33] Griffiths, M., "Comments on Precipitate Stability in Neutron Irradiated Zircaloy-4," *Journal of Nuclear Materials,* Vol. 170, 1989, pp. 294–300.

[34] Hood, G. M., "Point Defect Diffusion in α-Zr," *Journal of Nuclear Materials,* Vol. 159, 1988, pp. 149–175.

[35] Hood, G. M. and Shultz, R. J., "Chromium Diffusion in α-Zirconium, Zircaloy-2 and Zr-2.5 Nb," *Journal of Nuclear Materials,* Vol. 200, 1993, pp. 141–143.

[36] Etoh, Y. and Shimada, S., "Neutron Irradiation Effects on Intermetallic Precipitate in Zircaloy as a Function of Fluence," *Journal of Nuclear Materials,* Vol. 200, 1993, pp. 59–69.

[37] Chung, H. M., "Phase Transformations in Neutron-Irradiated Zircaloys," *Radiation-Induced Changes in Microstructure: 13th International Symposium, ASTM STP 955,* F. A. Garner, A. S. Packan, and N. H. Kumar, Eds., American Society for Testing and Materials, Philadelphia, 1987, pp. 670–699.

[38] Yang, W. J. S., "Microstructural Development in Neutron Irradiated Zircaloy-4," *Effect of Radiation on Materials: 14th International Symposium, ASTM STP 1046,* N. H. Packan, R. E. Stoller, and A. S. Kumar, Eds., American Society for Testing and Materials, Philadelphia, 1989, pp. 442–456.

[39] Griffiths, M., Gilbert, R. W., and Carpenter, G. J. C., "Phase Instability, Decomposition and Redistribution of Intermetallic Precipitates in Zircaloy-2 and -4 during Neutron Irradiation," *Journal of Nuclear Materials,* Vol. 150, 1987, pp. 53–66.

[40] Griffiths, M., Gilbert, R. W., Fidleris, V., Tucker, R. P., and Adamson, R. B., "Neutron Damage in Zirconium Alloys Irradiated at 644 to 710K," *Journal of Nuclear Materials,* Vol. 150, 1987, pp. 159–168.

[41] Herring, R. A. and Northwood, D. O., "Microstructural Characterization of Neutron-Irradiated and Post-Irradiation Annealed Zircaloy-2," *Journal of Nuclear Materials,* Vol. 159, 1988, pp. 386–396.

[42] Woo, O. T. and Carpenter, G. J. C., "Radiation-Induced Precipitation in Zircaloy-2," *Journal of Nuclear Materials,* Vol. 159, 1988, pp. 397–404.

[43] Yang, W. J. S., Tucker, R. P., Cheng, B., and Adamson, R. B., "Precipitates in Zircaloy: Identification and the Effects of Irradiation and Thermal Treatment," *Journal of Nuclear Materials,* Vol. 138, 1986, pp. 185–195.

[44] Tanner, L. E. and Levinson, D. W., "The System Zirconium-Iron-Tin," *Transactions ASM 52,* American Society for Metals, Metals Park, OH, 1960, pp. 1115–1136.

[45] Abriata, J. P., Bolcich, J. C., and Arias, D., "Tin-Zirconium," *Bulletin on Alloy Phase Diagrams,* Vol. 4, No. 2, 1983, pp. 2087–2089.

[46] Woo, C. H., "Theory of Irradiation Deformation in Non-Cubic Metals: Effects of Anisotropic Diffusion," *Journal of Nuclear Materials,* Vol. 159, 1988, pp. 237–256.

[47] Griffiths, M., Gilbon, D., Regnard, C., and Lemaignan, C., "HVEM Study of the Effects of Alloying Elements and Impurities on Radiation Damage in Zr-Alloys," *Proceedings,* 3rd Conference on

Evolution of Microstructure in Metals During Irradiation, 1992, Muskoka, Canada, to be published in *Journal of Nuclear Materials*.

[48] Griffiths, M., "Microstructure Evolution in hcp Metals during Irradiation," *Philosophical Magazine A*, Vol. 63, No. 5, 1981, pp. 835–847.

[49] King, A. D., Hood, G. M., and Holt, R. A., "Fe-Enhancement of Self-Diffusion in α-Zr," *Journal of Nuclear Materials*, Vol. 185, 1991, pp. 174–181.

[50] Forlerer de Svarch, E. and Rodriguez, C., "On the Influence of Iron on the Zr-α (hcp) Self-Diffusion," *Journal of Nuclear Materials*, Vol. 185, 1991, pp. 167–173.

[51] Kai, J. J., Tsai, C. H., Shiao, J. J., Hsieh, W. F., and Tu, C. S., "Effects of Irradiation on the Microstructural Evolution and Corrosion Resistance of Zirconium Alloys," Poster Session, Ninth International Symposium on Zirconium in the Nuclear Industry, American Society for Testing and Materials, Kobe, Japan, 1990.

[52] Lee, Y. S., Huang, K. Y., Kai, J. J., and Hseih, W. F., "Effects of Proton Irradiation on the Microstructural Evolution and Uniform Corrosion Resistance of Zircaloys," *Proceedings*, 3rd Conference on Evolution of Microstructure in Metals During Irradiation, 1992, Muskoka, Canada, to be published in *Journal of Nuclear Materials*.

DISCUSSION

R. Holt[1] (written discussion)—Did you see diffusion of iron back into the original $Zr(Cr,Fe)_2$ precipitates during the post-irradiation annealing as observed by Adamson et al. in an earlier paper? It seems to be almost "conventional wisdom" now that the redistribution of iron from intermetallics into the α-matrix during irradiation causes the appearance of $\langle c \rangle$-component basal-plane loops and, hence, accelerated growth. However, as Adamson mentioned in the discussion in an earlier paper, there are cases where the iron is redistributed very rapidly (because of a very fine initial precipitate distribution) but the $\langle c \rangle$-component dislocations have not yet appeared. Hence, it seems that further work is required to confirm the "cause" of accelerated growth.

D. Gilbon and C. Simonot (authors' closure)—For the samples taken from the RX Zy-4 guide tube irradiated in a PWR, the main effect of annealing at 600°C was to reveal a copious precipitation of blocky Zr_3Fe particles at grain boundaries. However, the composition profiles that were performed on partially amorphous $Zr(Fe,Cr)_2$ particles, showed that a part of iron diffused back into the original precipitates during this post-irradiation annealing. The Fe/Cr ratio of their amorphous rim increased back to 1, instead of less than 0.5 in the as-irradiated condition (see Fig. 13).

Brian Cox[2] (written discussion)—Would you care to speculate on why you saw so many tin-rich precipitates on specimens irradiated in Siloé and Phénix yet apparently none in the fuel cladding, despite the rather small temperature differences between the irradiations?

D. Gilbon and C. Simonot (authors' closure)—Tin-rich precipitates were also detected in the upper part of the fuel cladding ($T \approx 380°C$), but their number density was much lower than in Zy-4F samples irradiated at 400°C. The higher tin content of Zy-4F could partly explain this difference. However, as reported in a previous paper [22], the formation of tin-rich precipitates is temperature and flux dependent, being observed at lower temperature ($\approx 370°C$) only in samples irradiated with high neutron fluxes to high fluences. Hence, in addition to the rather small temperature differences, the higher dose rates could explain why we saw so many Zr_5Sn_3 particles on specimens irradiated in Siloé and Phénix. This hypothesis is confirmed by the results obtained on samples irradiated in Phénix for the same time, at the same temperature, but with two different neutron fluxes (see Table 6).

---

[1] Chalk River Laboratories, Chalk River, Ontario, Canada.
[2] University of Toronto, Toronto, Ontario, Canada.

Annie Soniak,[1] Sylvie Lansiart,[1] Jacques Royer,[1] Jean-Paul Mardon,[2] and Nicolas Waeckel[3]

Irradiation Effect on Fatigue Behavior of Zircaloy-4 Cladding Tubes

REFERENCE: Soniak, A., Lansiart, S., Royer, J., Mardon, J.-P., and Waeckel, N., "**Irradiation Effect on Fatigue Behavior of Zircaloy-4 Cladding Tubes**," *Zirconium in the Nuclear Industry: Tenth International Symposium, ASTM STP 1245,* A. M. Garde and E. R. Bradley, Eds., American Society for Testing and Materials, Philadelphia, 1994, pp. 549–558.

ABSTRACT: Since nuclear electricity has a predominant share in French generating capacity, pressurized water reactors (PWRs) are required to fit grid load following and frequency control operating conditions. Consequently, cyclic stresses appear in the fuel element cladding. In order to characterize the possible resulting clad damage, fatigue tests were performed at 350°C on unirradiated material or irradiated stress relieved Zircaloy-4 tube portions, using a special device for tube fatigue by repeated pressurization.

It appears that, for high stress levels, the material fatigue life is not affected by irradiation. But the endurance fatigue limit undergoes a decrease from the 350 MPa value for unirradiated material to the 210 MPa value for the material irradiated for four cycles in a PWR. However, this effect seems to saturate with irradiation dose: no difference could be detected between the two cycles' results and the corresponding four cycles' results. The corrosion effect and the load following influence were also investigated: they do not appear to modify the fatigue behavior in our experimental conditions.

KEY WORDS: zirconium, zirconium alloys, fatigue behavior, cladding tubes, irradiation effects, repeated pressurization, power cycling effects, nuclear materials, nuclear applications, radiation effects

Since nuclear electricity has a predominant share (75%) in French generating capacity, pressurized water reactors (PWRs) are required to fit grid load following and frequency control operating conditions. Consequently, cyclic stresses appear in the fuel element cladding. In order to evaluate the possible resulting clad damage, Commissariat à l'Energie Atomique (CEA), FRAMATOME, and Electricité de France (EDF) have jointly undertaken a large-scale fatigue resistance investigation program concerning both unirradiated and irradiated materials.

The first aim of this program is to determine the pure fatigue behavior of stress-relieved Zircaloy-4 tubes under internal pressure and to give separately the orders of magnitude of different cumulative effects, namely, the irradiation dose effect, the load following and frequency control effect, and the corrosion effect.

This paper shows the results obtained on irradiated material compared to previous ones [1,2] obtained on unirradiated material. The tube portions have been cut from FRAGEMA standard fuel clad irradiated in EDF power plants or in an experimental CEA reactor called

---

[1] Research engineer, manager of materials studies, and laboratory head, respectively, Commissariat à l'Energie Atomique, Centre d'Etudes de Saclay, 91191 Gif-sur-Yvette Cédex, France.
[2] Consulting engineer, Framatome Nuclear Fuel Division, Fragema, 69456 Lyon cédex 06, France.
[3] Senior engineer, Electricité de France/Septen, 69628 Villeurbanne cédex, France.

CAP (Chaufferie Avancée Prototype) that is located in Cadarache, France. A special device for tube fatigue testing by repeated pressurization has been introduced in a hot cell that is quite similar to the one designed for the unirradiated material experiments.

Material Characteristics

Metallurgical Characteristics

The samples were taken from stress-relieved Zircaloy-4 cladding tubes. The chemical composition (% by weight) is as follows for the studied samples:

$$1.37 < Sn < 1.52 \quad 0.19 < Fe < 0.23$$
$$0.10 < Cr < 0.12 \quad 0.105 < O_2 < 0.133$$

The impurity contents are within the ranges given in the classical Zircaloy-4 specification. The initial microstructure and texture [1] are fully typical of the FRAGEMA specification. The effect of irradiation on microstructure is reported in Ref 3.

Irradiation Conditions

Pressurized Water Reactor (PWR)—Some of the tested clads come from rods that have been irradiated for two or four cycles in EDF nuclear plants without load following. Other standard clads have been irradiated for two, three, or five cycles in EDF nuclear plants operating on load following and frequency control (50 to 60% of irradiation time for the ones irradiated two or three cycles). The mean irradiation temperature, the average rod local burnups, and the fluences for specimens taken at 5 to 6 grid span are reported in Table 1. The inlet and outlet coolant temperatures range from 285 to 325°C.

Experimental CEA Reactor (CAP)—Tube portions from standard cladding irradiated in the CAP CEA experimental reactor have also been tested. The CAP reactor enables load following and frequency control operating conditions to be simulated in order to test the fuel rod behavior. The assemblies have 17 by 17 geometrical characteristics, except for rod length that is half that of typical PWR rods. The mean irradiation temperature is also somewhat different in the CAP reactor being about 30°C below the commercial PWRs. Indeed, the inlet and outlet

TABLE 1—*Irradiation conditions of the 5 to 6 grid span of the clads irradiated in EDF nuclear plant.*[a]

Material	T_{moy}, °C	Rod Local Burnup, MWd/tU	Fluence, 10^{25} n/m^2
Two cycles in PWR without power cycling	350	26 000	4.06
Four cycles in PWR without power cycling	360	52 000	8.58
Two cycles in PWR with power cycling	345	24 000	3.85
Three cycles in PWR with power cycling	355	41 000	6.54
Five cycles in PWR with power cycling	360	63 000	10.7

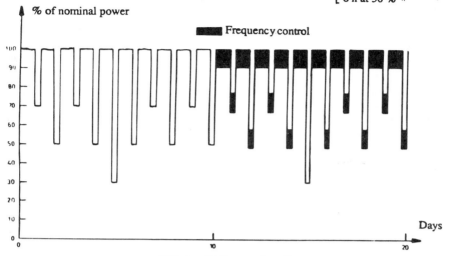

FIG. 1—*CAP operating diagram.*

coolant temperatures are equal to 267 and 287°C, respectively. So, the mean irradiation temperature in the CAP reactor is about 30°C below the PWR's one. The cyclic process consists essentially of the two successive periods (Fig. 1):

1. during ten days, there are daily power cut backs to 70, 50, and 30% of the nominal power, without frequency control;
2. during the ten following days, frequency control ranging within 5% of the nominal power is superimposed on the load following conditions.

The rod local burnup is around 40 000 MWd/tU, in each case. Fluences are not available for those claddings.

Mechanical Characteristics

The mechanical properties (conventional yield strength = $R_{p0.2}$, ultimate tensile strength = R_m, uniform elongation = A_r, and total elongation = A_t), were determined from burst tests at 350°C for a strain rate of $2.5 \ 10^{-4} \ s^{-1}$ for unirradiated and irradiated materials (Table 2).

TABLE 2—*Burst mechanical properties at 350°C of unirradiated and irradiated materials used in this study.*

Material	$R_{p0.2}$, MPa	R_m, MPa	A_r, %	A_t, %
Unirradiated	423	517	2	3.4
CAP	615	669	1.15	...
Four cycles in PWR	612	637	0.49	...

Standard out-reactor creep tests (400°C, 130 MPa, 240 h) were also performed on several tubes. The diametral strains induced by the internal pressure ranged from 1.24 to 1.80% for unirradiated material and from 0.16 to 0.21% for tubes irradiated for four cycles in an EDF nuclear plant.

Experimental Method

Fatigue tests were performed at 350°C using a device for tube fatigue testing by repeated internal oil pressurization. The pressure variations obey a periodic triangular signal between a very low pressure and a pre-defined maximum pressure.

The frequency influence in the range 0.5 to 1 Hz was first studied on unirradiated specimens [2]. The 0.5 Hz frequency was first chosen to minimize the influence of creep, and thus to be as close as possible to pure fatigue conditions, while remaining compatible with a correct strain recording. However, some tests were performed subsequently at 1 Hz. It appears that the influence of frequency on fatigue life does not occur for high stress levels. The most important effect observed essentially concerns the rise of the endurance fatigue limit with the frequency increase. To minimize the creep-fatigue interaction, the 1 Hz frequency was chosen for all the irradiated specimens. The investigated life ranges from 10^3 to 10^6 cycles.

It must be emphasized that the aim of this study was to characterize the pure fatigue behavior of Zircaloy-4 cladding tubes under repeated pressurization. So the test conditions were very different from the conditions undergone by the cladding during power cycling in an EDF nuclear plant. In the case of load following, particularly, the period is much higher than in an EDF nuclear plant (one day rather than 1 s) and the stress level is much lower, because of rapid stress relaxation.

Samples were 120 mm in length, 9.5 mm in outer diameter, and 0.57 mm in thickness. They were heated in a four-zone furnace with four thermocouples for regulation and another one for the sample temperature measurement.

The diametral strain was deduced by measuring the displacements of four LVDT (linear variable differential transformer) sensors scanning two perpendicular external diameters. The average of the two diameters was used for the mean strain value.

The hoop stress, σ, was calculated according to the following expression

$$\sigma = P \times \frac{\phi_i}{2e}$$

where P = the internal pressure, and ϕ_i and e = the average internal diameter and the average wall thickness of the specimen, respectively.

Fatigue Results

Irradiation Effect

A low cycle fatigue curve in the range of 10^3 to 10^6 cycles was plotted for unirradiated specimens at 1 Hz [2]. These data and those obtained on cladding tubes irradiated in the CEA CAP experimental reactor and on specimens taken at 5 to 6 space grid of cladding tubes irradiated two and four cycles in an EDF nuclear plant are shown in Fig. 2.

It appears, on one hand, that for high stresses ($\approx$500 MPa) the material fatigue life is not affected by irradiation. On the other hand, the endurance fatigue limit undergoes a decrease from the 350 MPa value for unirradiated material to 250 MPa for the specimens irradiated in the CAP reactor and to 210 MPa for specimens irradiated four cycles in an EDF nuclear plant.

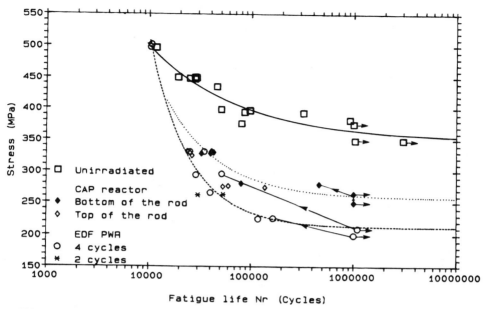

FIG. 2—*Irradiation effect on the S-N curve (maximum hoop stress versus the number of cycles to rupture) at 350°C and 1 Hz:* → *unfailed specimens and* ↘ *specimens tested at a σ_1 stress during 10^6 cycles then tested at a $\sigma_2 > \sigma_1$ stress up to rupture.*

The irradiation effect seems to saturate with the irradiation dose because the number of cycles to rupture for cladding tubes irradiated two or four cycles is approximately equal for a stress of 265 MPa. It is considered that this stress level is a clear indication of the different fatigue behaviors of specimens irradiated in various conditions, while keeping high accuracy on the number of cycles to rupture.

It was shown in Ref 2 for unirradiated specimens that the *S-N* curve, where the maximum stress is reported as a function of the number of cycles to rupture N_r, can be rationalized by a Langer [4] formulation

$$\sigma = \frac{E}{4\sqrt{N_r}} \log \frac{100}{100 - RA} + \sigma_0$$

where E is the mean elastic modulus measured at the first cycle while RA and σ_0 are fitted parameters. This formulation cannot be used for irradiated samples because of the form of the *S-N* curve. Stress versus the number of cycles to rupture can be rationalized by

$$\sigma = A + \frac{B}{N_r^n}$$

with the *n* coefficient of about 1.13. This formulation suggests endurance fatigue limits of 250 and 210 MPa for specimens irradiated in the CAP reactor or in an EDF nuclear plant, respectively, and is consistent with the large sensitivity of irradiated specimens to maximum stress. The investigated ratio, $\sigma_{max}/R_{p0.2}$, is however smaller for irradiated than for unirradiated material.

In addition, the differences observed in the fatigue behavior of specimens irradiated in the CAP reactor or in an EDF nuclear plant, respectively, may be attributed to differences in the irradiation conditions (for example, temperature) for tests all performed in the same conditions (temperature and frequency).

Environmental Effect

To highlight a possible environmental effect, that is, the influence of the oxide thickness or hydrogen concentration or both, specimens were taken from different levels of cladding tubes irradiated four cycles in an EDF nuclear plant. The maximum hoop stress is reported as a function of the number of cycles to rupture in Fig. 3 for specimens taken at 2 to 3, 5 to 6, and 6 to 7 grid spans. At spans 5 to 6 and 6 to 7, the irradiation temperature is higher and, consequently, the specimens are more oxidized and hydrided (Table 3).

In Table 3, we have reported oxide thickness and hydrogen concentrations. For the CAP specimens, the values are somewhat smaller than the ones observed on cladding tubes irradiated in an EDF nuclear plant because of the lower irradiation temperature.

No significant difference in terms of number of cycles to rupture can be seen between the different levels in spite of differences in oxide thickness and hydrogen concentrations (Fig. 3).

Power Cycling Effect

Pressurized Water Reactor (PWR)—Some fatigue tests were done for a stress of 265 MPa on cladding tubes irradiated two, three, or five cycles in an EDF PWR operating on load

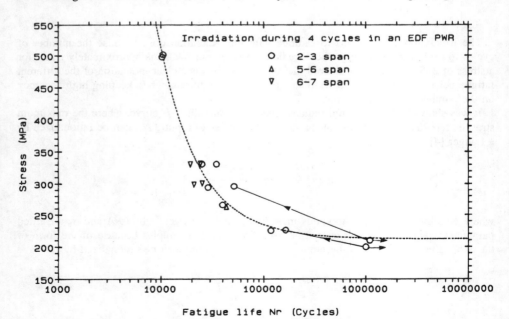

FIG. 3—*Environmental effect on the S-N curve at 350°C and 1 Hz for specimens irradiated four cycles in an EDF PWR:* → *unfailed specimens and* ↘ *specimens tested at a σ_1 stress during 10^6 cycles then tested at a $\sigma_2 > \sigma_1$ stress up to rupture.*

TABLE 3—*Zirconium oxide layer thickness and hydride contents of irradiated cladding tubes.*

Material	$T_{moy}{}^a$ of Cladding, °C	Oxide Thickness, μm	Hydrogen Concentration, ppm
CAP	...	12	60
PWR four cycles 2 to 3 span	335	29	280
PWR four cycles 5 to 6 span	360	54	520
PWR four cycles 6 to 7 span	365	65	600

$^a T_{moy}$ is the mean temperature of the cladding during irradiation.

following and frequency control. The results thus achieved are reported on the Wöhler curve of Fig. 4. The fatigue lives of these specimens are identical or even slightly higher than those obtained previously. Therefore, the load following and frequency control do not seem to modify the fatigue behavior of irradiated cladding tubes. Rowland et al. [5] also have not observed differences between the fatigue strength of Zircaloy-2 claddings irradiated without power cycling and Zircaloy-2 claddings that had been previously power cycled.

CEA Experimental Reactor (CAP)—In order to examine a possible influence of power cycling on fatigue behavior of specimens irradiated in the CAP experimental reactor, two different regions of the clad have been tested: (1) the bottom that is slightly influenced by the power variations and (2) the top zone (located around the level of the control rod lower end) that is much more influenced by these variations. The maximum hoop stress is shown

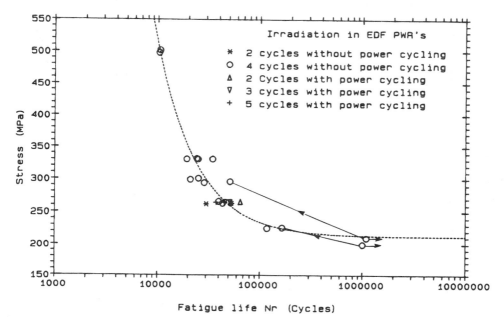

FIG. 4—*Power cycling effect on the S-N curve at 350°C and 1 Hz for specimens irradiated in EDF PWR:* → *unfailed specimens and* ↘ *specimens tested at a σ_1 stress during 10^6 cycles then tested at a $\sigma_2 > \sigma_1$ stress up to rupture.*

as a function of the number of cycles to rupture in Fig. 5. A small reduction of the number of cycles to rupture seems to appear for specimens that were taken at the top of the claddings. The slight effect may be due to power cycling but also to the influence of irradiation temperature on fatigue behavior. Indeed, the upper zone has been irradiated at a somewhat higher temperature than the lower one and, thus, at a temperature closer to the cladding irradiation temperature in the EDF nuclear plants. Therefore, it is not surprising to find, at the top of the clad, that fatigue life values approach the corresponding EDF nuclear plant results (Fig. 2). Consequently, no pure power cycling effect could be identified.

Discussion

The stabilized plastic strain is defined as the total deformation of the sample at rupture minus the plastic strain at the first cycle divided by the total number of cycles to rupture. It has been shown on unirradiated material in Ref 2 that the stabilized plastic strains are very small in comparison with the plastic strain observed at the first cycle. In fact, at the first cycle the plastic strain is equal to the monotonic one corresponding to the equivalent burst test, but, in the following cycles, the plastic strain becomes very small and rapid cyclic hardening can be observed. For irradiated specimens, because of irradiation hardening, the plastic strain at the first cycle is insignificant. In Fig. 6, the maximum stress is reported as a function of stabilized cyclic deformation. At higher stress levels (≈ 500 MPa), the number of cycles to rupture are quite similar but the stabilized plastic strain is three times smaller for irradiated specimens than for unirradiated ones.

In addition, the stabilized plastic strain is reported as a function of the number of cycles to rupture in Fig. 7 for cladding tubes unirradiated and irradiated four cycles in an EDF PWR.

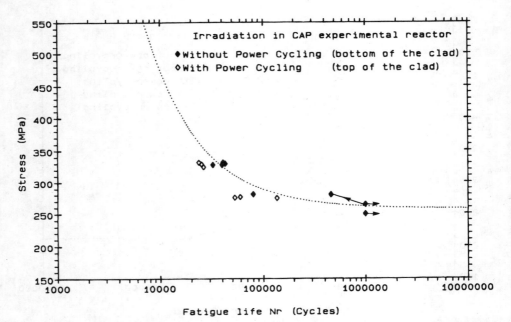

FIG. 5—*Power cycling effect on the S-N curve at 350°C and 1 Hz for specimens irradiated in the CAP CEA experimental reactor:* → *unfailed specimens and* ↘→ *specimens tested at a σ_1 stress during 10^6 cycles then tested at a $\sigma_2 > \sigma_1$ stress up to rupture.*

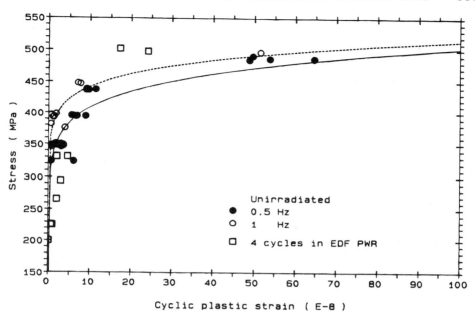

FIG. 6—*Maximum stress as a function of cyclic plastic strain for cladding tubes unirradiated and irradiated four cycles in an EDF PWR.*

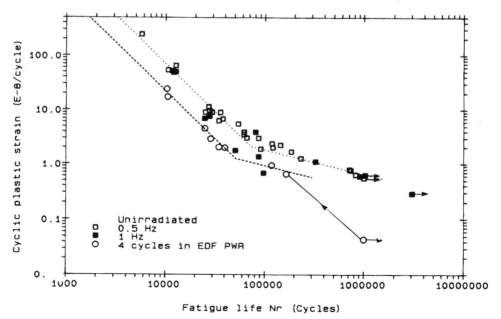

FIG. 7—*Cyclic plastic strain versus the number of cycles to rupture for cladding tubes unirradiated or irradiated four cycles in an EDF PWR:* → *unfailed specimens and* ↘ *specimens tested at a* σ_1 *stress during* 10^6 *cycles then tested at a* $\sigma_2 > \sigma_1$ *stress up to rupture.*

A frequency independent relationship between the total number of cycles and the stabilized plastic strain has been established on unirradiated cladding tubes in Ref 2. It appears that the results obtained on irradiated specimens can be rationalized by a type law Manson-Coffin [6], $\epsilon = AN_r^n$, where the n coefficient is the same as the one deduced on unirradiated specimens [2], at least for sufficiently high stresses when the cyclic plastic strain is strongly dependent on fatigue life. However, for the same number of cycles to rupture, N_r, the stabilized plastic strain is smaller for irradiated material than for unirradiated material by a factor of 3 for $6000 < N_r < 80\,000$.

Conclusion

Pure fatigue tests were performed at 350°C at 1 Hz for irradiated tube portions from FRAGEMA fuel standard claddings, using a special device for tube fatigue by repeated internal pressurization. The tube portions were irradiated in an EDF nuclear plant or in the CEA CAP experimental reactor.

It appears that, for high stress levels, the material fatigue life is not affected by irradiation. But the endurance fatigue limit undergoes a decrease from the 350 MPa value for unirradiated material to the 210 MPa value for four cycles power reactor irradiation. However, this effect seems to saturate with the irradiation dose.

Furthermore, the corrosion effect and the load following influence do not appear to modify the fatigue behavior in our experimental conditions.

Acknowledgments

We wish to thank Mr. Durocher and Mr. Pinte who have performed the hot cell tests.

References

[1] Lemoine, P., Darchis, L., Pelchat, J., Mardon, J. P., and Grosgeorge, M., "Sustained Fatigue of Zircaloy-4 Claddings—Nonirradiated Material," IAEA-TC-624/18, International Atomic Energy Agency, Lyon, France, 1987.

[2] Soniak, A., Pelchat, J., Mardon, J. P., and Permezel, P., "The Influence of Frequency on Fatigue Behaviour of Unirradiated Zircaloy-4 Cladding Tubes," *Transactions,* SMIRT 11, Vol. C, Tokyo, 1991.

[3] Gilbon, D. and Simonot, C., "Effect of Irradiation on the Microstructure of Zircaloy-4," in this volume.

[4] O'Donnel, W. J. and Langer, B. F., "Fatigue Design Basis for Zircaloy Components," *Nuclear Science and Engineering,* Vol. 20, 1964, pp. 1–12.

[5] Rowland, T. et al., "Power Cycling Operation Tests of Zr-Barrier Fuel," *Proceedings,* International Topical Meeting on LWR Fuel Performance, Fuel for the 90's, Avignon, France, 1991, pp. 808–817.

[6] Coffin, L. F., Jr., *Transactions,* American Society of Mechanical Engineers, Vol. 76, 1954, pp. 931–949.

Roberto E. Haddad[1] *and Alberto O. Dorado*[2]

Grain-by-Grain Study of the Mechanisms of Crack Propagation During Iodine Stress Corrosion Cracking of Zircaloy-4

REFERENCE: Haddad, R. E. and Dorado, A. O., **"Grain-by-Grain Study of the Mechanisms of Crack Propagation During Iodine Stress Corrosion Cracking of Zircaloy-4,"** *Zirconium in the Nuclear Industry: Tenth International Symposium, ASTM STP 1245,* A. M. Garde and E. R. Bradley, Eds., American Society for Testing and Materials, Philadelphia, 1994, pp. 559–575.

ABSTRACT: This paper describes the tests conducted to determine the conditions leading to cracking of a specified grain of metal, during the iodine stress corrosion cracking (SCC) of zirconium alloys, focusing on the crystallographic orientation of crack paths, the critical stress conditions, and the significance of the fractographic features encountered. In order to perform crystalline orientation of fracture surfaces, a specially heat-treated Zircaloy-4 having very large grains, grown up to the wall thickness, was used. Careful orientation work has proved that intracrystalline pseudo-cleavage occurs only along basal planes. The effects of anisotropy, plasticity, triaxiality, and residual stresses originated in thermal contraction have to be considered to account for the influence of the stress state. A grain-by-grain calculation led to the conclusion that transgranular cracking always takes place on those bearing the maximum resolved tensile stress perpendicular to basal planes. Propagation along twin boundaries has been identified among the different fracture modes encountered.

KEY WORDS: zirconium, zirconium alloys, nuclear materials, nuclear applications, stress corrosion, iodine, single crystals, crystallographic orientation, applied stress, twin boundaries

Stress corrosion cracking (SCC) induced by fission product has been the most probable cause of pellet-cladding interaction (PCI) failures in Zircaloy tubing used to contain nuclear fuel [1]. Many different mechanisms of crack initiation and propagation have been proposed, involving the effect of large inclusions or second-phase particles, chemical attack, hydriding, etc. However, in many circumstances, transgranular cleavage-like cracking has been produced in grains that show no metallurgical inhomogeneities at the initiation sites [2]. Instead, it has been found that the texture of the material had a strong influence, with batches of tubing with basal poles preferentially oriented in the radial direction being less susceptible. This is consistent with the idea of a cracking mechanism based on an energy balance, that is, SCC would happen every time the applied stress was high enough to break the zirconium-zirconium bond; then the separation should be easiest along basal planes, the theoretical cleavage (less bound) planes [3]. However, this cannot happen without the weakening help of the corrodant element, since the required energy is much higher than that needed to activate other deformation mechanisms. No cleavage of zirconium has been reported in fracture tests conducted in inert atmosphere at temperatures as low as $-195°C$ [4], while reductions in ductility in impact tests are mainly

---
[1] Staff researcher, CNEA (Gcia. de Desarrollo), Av. Libertador 8250, 1429 Buenos Aires, Argentina.
[2] Fellowship holder, CONICET, Rivadavia 1917, 1033 Buenos Aires, Argentina; now at ASTRA S. A., 9000 Chubut, Argentina.

related with the presence of hydrides [5]. SCC transgranular propagation cannot proceed by cleavage (at the speed of sound), since the formation of the right chemical environment at the crack tip is required. Nevertheless, the final result may not be distinguishable in terms of fractography and brittleness; for that reason, the process is referred to as "pseudo-cleavage."

If SCC in zirconium alloys depends on the aforementioned energy balance, there should be a definite relationship between the physical parameters of the crystal, the geometry of the samples, the applied stress system, and the chemistry of the environment. This implies, for a given chemistry, that the propagation plane should be the basal plane, the fractured grains should bear the maximum tensile stress resolved perpendicular to basal planes, and the stress system should allow this value to reach the minimum required to separate neighboring basal planes.

Various attempts to measure the orientation of transgranular iodine SCC in Zircaloy have given different results [6–11]. In tests conducted on tubes machined from thick highly textured plates [6], that is, with angular variations in grain orientation, the crack density distribution showed a maximum at about 10° to 30° from the direction of maximum density of tangential basal poles, as if propagation would preferentially proceed in planes with that inclination from basal planes. Also, optical microscopy determinations [7] indicated propagation at about 10° to 15° off the basal direction. The main phenomenologic correlation of these values are the orientation of transgranular cracking in titanium alloys and the habit planes for hydrides ($10\bar{1}7$) [8,9]. Although brittle zirconium hydrides can act as initiators for SCC, from fractographic studies, it can be concluded that they do not play any essential role during propagation [10]. The only reference of close to basal transgranular propagation [11] corresponds to tests conducted by means of an etch-pit technique that permits comparisons between traces of transgranular fractures and traces of basal planes in metallographic samples with up to 10% of plastic deformation. The authors claim that the angular error is about 3°, and in some of the tests the separation is higher, including that corresponding to the picture shown in the paper. Also, the coincidence of traces do not imply coincidence of planes, since there may be many different planes with the same trace.

The aim of this study is to shed light on the possible operating mechanism, corroborating the crystallographic orientation of transgranular propagation of SCC in zirconium alloys, and establishing dependences from the type and magnitude of the stresses involved.

Experimental

In order to establish the crystalline orientation of the SCC propagation, specially heat-treated Zircaloy-4 tubes having very large grains, disk-shaped, a few millimetres in diameter, and grown up to the wall thickness were cracked in an environment containing iodine. Once a crack was produced, careful orientation work on the fractured grains permitted the fracture planes to be determined. After that, it was possible to calculate the applied stress on that individual grain as a function of its orientation and the general stress system, taking also into account the effect of symmetry, material anisotropy, grain boundary interactions (specially thermal contraction stresses) between differently oriented neighboring grains, axiality of stress state, etc.

Materials

Tests were conducted on stress-relieved Zircaloy-4 tubing of the type used as cladding in the fuel of the nuclear power plant Atucha-I, Argentina. Their external diameter is 11.9 mm with 0.55 mm of wall thickness. The chemical composition of the alloy is given in Table 1.

TABLE 1—*Chemical composition of the alloy used.*

Element	Alloy Concentration, % by weight
Sn	1.55
Fe	0.21
Cr	0.105
Hydrogen content	<50 ppm, average 8 ± 3 ppm

The samples were cut 15 cm long, triply degreased in boiling trichloroethane, and rinsed in tridistilled water prior to annealing. Care was taken to avoid contamination with hydrogen-producing species.

In order to grow the grains, several different heat treatments were tried, based on earlier recrystallization work [12]. The two finally selected resulted in partial (PSR) and total (TSR) secondary recrystallization (Fig. 1). They included heating the central part of the samples up to 840°C for 10 min, then decreasing the temperature to 800 or 820°C, respectively, keeping them at these values for 24 h followed by furnace cooling. This procedure was conducted under vacuum, giving the following grain structures: (*a*) isolated disk-shaped single crystals a few millimetres in diameter and extending over the wall thickness, surrounded by small (about 50 μm in diameter) grains (PSR structure); and (*b*) single crystals only (TSR structure).

Test Method

Internally pressurized tube-type SCC tests [13] were conducted, with 99.99% pure argon as inert gas; in all but a few control cases, one gram of bisublimated iodine was added, giving a concentration well above the SCC threshold values. The working temperature was 400°C. The determination errors in geometrical variables, as well as in temperature and pressure, were less or equal to 0.5%.

Grain Orientations

Failures were detected when drops in pressure, monitored with an X-T recorder connected to a transducer, had been produced. After cooling down, the cracks were located by means of bubble production with the help of a slight gas overpressure. The samples were cut transversally close to the leaking place and polished until the crack had appeared in the metallography. All the necessary care was taken to avoid damage to the crystals. After a nitrofluoric acid pickling, the grain structure was visible under polarized light. This permitted the placement of a small piece of fluorescent material in each grain to be used as a target for an X-ray beam. A Laue diagram was obtained of each grain of the section to provide full orientation. On those samples containing a fissure, a second diagram was made with X-rays directed perpendicular to the inner surface as shown in Fig. 2; the traces of the crack on both surfaces were compared with the corresponding traces of the basal planes.

Stress Analysis

A full grain-by-grain stress analysis was conducted on four failed samples (140 grains), assuming linear and isotropic behavior, cylindrical symmetry, and absence of residual stresses. Given these conditions, the diagonal components of the stress tensor, σ_{rr}, σ_{tt}, and σ_{zz}, have a simple dependence on the radial position [14]. Tests and estimations to evaluate the validity

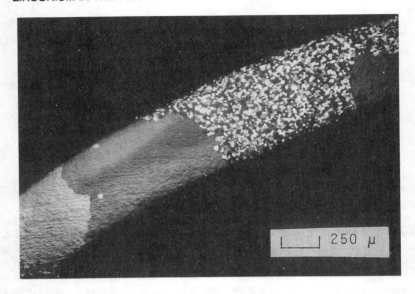

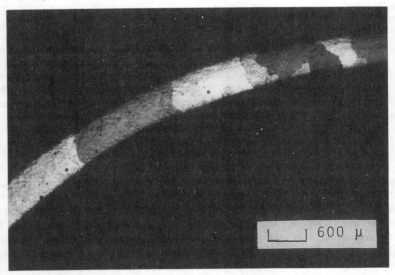

FIG. 1—*Metallographies of heat-treated cladding: (a) PSR and (b) TSR.*

of the approximations were made. It was concluded that the only factor that could introduce any significant error is the presence of anisotropic thermal contraction residual stresses. This could arise during cooling from the annealing temperature and generate grain-to-grain interactions.

Since the wall thickness is normally about one order of magnitude smaller than the radius, the stress component, σ_{rr}, calculated at the inner surface becomes negligible when compared with σ_{tt} and σ_{zz} and the test is considered to be a biaxial test (thin wall approximation); however, when working with single crystals, the resolved tensile stress perpendicular to the basal plane is

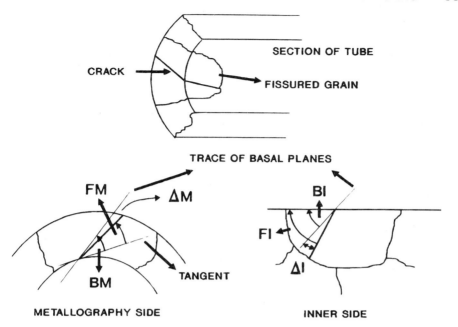

FIG. 2—*Angular reference for X-ray orientation (schematic).*

$$\sigma_{0002} = \sigma_{tt}(c \cdot t)^2 + \sigma_{rr}(c \cdot r)^2 + \sigma_{zz}(c \cdot z)^2 \qquad (1)$$

where

σ_{0002} = resolved tensile stress perpendicular to the basal plane,
c = unit vector normal to basal planes,
t = unit vector in the tangential direction,
r = unit vector in the radial direction, and
z = unit vector in the axial direction.

Owing to the original texture of the material used, the crystals tend to be oriented with their basal poles in the radial direction, making the term proportional to $(c \cdot r)^2$ no longer negligible.

Influence of the Stress State Axiality

When studying the way the stresses operate during the SCC process, it is important to consider the type of stress system applied. It has been easier to produce iodine SCC in zirconium alloys in tests where a biaxial stress state was imposed on the samples than in those in which it was uniaxial [13]. This will probably depend mostly upon the texture and orientation of the samples. The influence of the axial/tangential stress ratio has been studied [1] for a given tube texture in biaxial tests. However, in polycrystalline material, the individual grains probably never suffer a pure uniaxial or biaxial tension, but, are subjected instead to a complex interaction with neighboring grains; thus, from those results, it is not possible to establish easily the mechanism by which SCC is affected.

From the stress calculations [*14*], tubes with a reduced wall thickness will have a relative lower radial component of stress, meaning that the stress system will tend to be biaxial. In an attempt to study the influence of this effect, several tests were conducted in tubes in which the thickness had been reduced either by lathe machining or pickling in a 42% HNO_3 + 8% HF solution to produce different thickness reductions. After that the samples were rinsed in boiling tridistilled water according to the recommendations contained in the ASTM Practice for Aqueous Corrosion Testing of Samples of Zirconium and Zirconium Alloys (G 2-74), the rest followed the normal procedure described for the majority of the samples.

Results

Typical SCC evidence was found in the tests conducted with iodine: considerable loss of ductility, stress to failure lower than measured in an inert environment, and cleavage-like transgranular cracking.

Orientation of Fractures

Table 2 contains the data corresponding to ten determinations of angles between traces of basal planes and cracks, both on the inner surface of the tubes and on the metallography, together with reference indications. As it can be seen in all cases, the agreement is excellent, with differences less than 1°, indicating that SCC transgranular propagation by pseudo-cleavage occurs only along basal planes.

Resolved Stresses

The results of the full stress analysis conducted on four tube samples containing fissures are shown in Table 3; the values of stress components normal and tangential to the basal planes are included and the grains that suffered the SCC process are marked. Negative numbers indicate compressive stresses.

It can be clearly seen that in all four samples examined the fractured grains are those having the maximum value of tensile stress resolved perpendicular to the basal planes. In Sample 1, where two fissures were found, Grain 1 (the most highly stressed) cracked first before Grain 4, the second most highly stressed. In this sample, Grain 5, having the maximum resolved shear stress parallel to basal planes, has not cracked.

TABLE 2—*Angles between fissures and basal planes traces in cracked grains of failed samples, as defined in Fig. 2.*

Sample Number	Angles (°) as defined in Fig. 2					
	FM	BM	ΔM	FI	BI	ΔI
1	47	46	1	62	63	1
2	36	37	1	60	59	1
3	26.5	27	0.5	62	63	1
4	35.5	35	0.5	71	72	1
5	45	45	0	78	77	1
6	37.5	38	0.5	81	80.5	0.5
7	45.5	46	0.5	86	86	0
8	53.5	53	0.5	85	84.5	0.5
9	39	39	0	77	77	0
10	48	48.5	0.5	83	84	1

TABLE 3—Resolved stresses on basal planes (MPa).

Grain Number	Samples							
	1		2		3		4	
	Normal	Tangential	Normal	Tangential	Normal	Tangential	Normal	Tangential
1	52[a]	52	23[a]	68	45[a]	80	7	54
2	−5	22	0	48	−9	31	−3	44
3	−5	19	−7	28	2	36	−4	41
4	33[a]	52	−2	44	17	65	−14	15
5	29	53	−10	21	−14	11	12	61
6	5	35	10	57	−11	23	7	55
7	21	48	3	45	−13	16	4	39
8	6	39	5	55	3	53	−3	41
9	10	42	−4	42	36	70	−10	20
10	−2	26	14	61	14	63	14	63
11	−8	13	1	46	−10	29	−5	38
12	13	43	−4	30	26	68	0	47
13	−7	19	1	48	3	53	28	70
14	−6	20	1	42	−13	16	−11	19
15			−2	40	−1	46	−14	12
16			−7	32	−7	35	0	39
17			7	57	−9	22	35	75
18			13	63	−14	13	−7	25
19			−4	31	−11	19	−12	22
20			−1	45	−11	19	14	52
21			10	60	−11	19	−5	38
22			−15	8	−8	34	69[a]	84
23			−15	7	9	59	−12	17
24			−1	41	2	50	−6	38
25			17	66	31	72		
26					−5	30		

[a]Fissured grain.

The main inconsistency is related to the low stress value (23 MPa) associated to the fissured grain of Sample 2, because Grains 9, 12, and 25 of Sample 3 and Grains 13 and 17 of Sample 4 have higher resolved tensile stresses but did not suffer SCC.

Table 4 contains the time to failure data for nine tests, indicating the tensile component of stress resolved perpendicular to the basal planes of each of the fractured grains. It is not possible to establish a correlation between these variables, as it would be an asymptotic value for that stress component for the longer times (stress threshold).

TABLE 4—Tensile stresses resolved perpendicular to basal and failure times.

	Sample Number								
	1	2	3	4	5	6	7	8	9
σ_{0002}, Mpa	50	28	44	70	49	72	95	53	79
Time, h	38	59	59	57	95	6	60	10	38

Reduced Thickness Tests

Twenty samples with the central portion pickled to reduce their thicknesses were tested in iodine and analyzed metallographically. Table 5 shows a number of samples where incipient, through wall, or no cracks were encountered. It includes the results obtained with samples without reduction.

It is clearly more difficult to produce cracks in the tubes with the thinner wall, although many incipient cracks could be observed. Normally, there are many initiation sites in each affected grain.

TABLE 5—*Thinning effect on microcracks density.*

Thinning, %	0	10	17.5	24	26	27	40
Number of Tests	3	3	1	1	1	1	1
Density, number of cracks/cm	2.1	0	0.05	0	0.05	0.1	0
	1.5	0.15					
	2.8	0.1					

Analysis by Microscopy

Figure 3 shows a fissured single crystal in a PSR tube; the crack starts at the center of the grain, propagates transgranularly at a far-from-radial angle, and, after reaching the grain boundary, continues through the microstructure following the general radial direction.

On the surface of a through-wall fracture in a macrograin (Figs. 4a and b), typical pseudo-cleavage features can be seen: river patterns with the steps bearing the characteristic hexagonal fluting.

The habit planes of zirconium hydrides could be clearly seen in metallography on samples with a higher than normal hydrogen concentration (Fig. 5). In these cases, the angular difference between SCC propagation and hydride precipitation was evident (Fig. 6).

FIG. 3—*Transgraular crack in a macrocrystal of a PSR tube.*

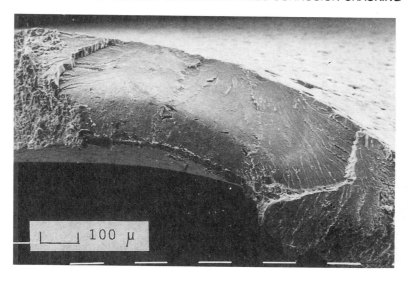

FIG. 4—*Transgranular fracture surface of a single crystal: (a) general view and (b) detail showing a step.*

A crack in a crystal that had been deformed prior to the test (Fig. 7) follows a path that changes directions while propagating transgranularly along different crystal planes. However, in a sample showing many deformation features, it is clear that the fissure leaves the basal plane only to follow the angle of twin boundaries. All the directions of the SCC propagation, twin boundaries, and hydride precipitates can be seen simultaneously. The corresponding fracture surface is shown in the stereo pair of Fig. 8, in which the common cleavage-like facets can be seen on the basal planes. A smooth curved surface corresponds to the intergranular portion (Fig. 9), and the most interesting feature (seen in Fig. 10) is the propagation along a

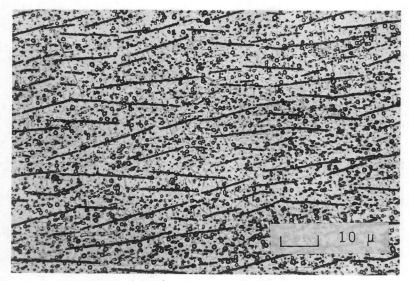

FIG. 5—*Hydride precipitation revealed on a macrograin.*

twin boundary (not fractographically identified before) consisting of a flat surface with very few details and no river patterns or steps.

Discussion

Single crystals grown in Zircaloy-4 tubes have been cracked in an environment containing iodine. The evidence leads to the conclusion that SCC transgranular fissures propagate along basal planes, only in those grains having the maximum tensile stress resolved perpendicular to them.

Since the basal planes are those with the least binding energy, these observations suggest that cracking occurs when the stress is high enough to break the zirconium-zirconium bond, reduced by the chemical effect of iodine.

The errors eventually introduced by the approximations involved in the stress analysis have not been high enough to obscure the interpretations of the results obtained in individual samples. Sample to sample inconsistency may be probably related to residual stresses induced by thermal contraction or from uncontrolled variables changing in each test, like differences in the chemical environment of each sample, variable oxide thickness, or hydrogen concentration originated in the previous manipulation, heat treatment, etc. The lack of correlation between resolved stresses and time to failure may be due to these effects, or it may indicate that a stress threshold does not exist. This could arise because at testing temperature the atoms have an energy distribution, and some of them might be able to diffuse in a redistributive movement that leads to a crack. A mechanism based on this surface mobility could well explain the iodine-SCC susceptibility in zirconium alloys [15].

Consistent results are only obtained when the radial stress component is taken into account, meaning that the stress system applied to pressurized tube-type samples must be considered triaxial. This component seems to play an essential role, since SCC diminishes when the stress system tends to a biaxial state, as follows from the reduced wall thickness results. From the point of view of stresses, the two main consequences of thinning the wall at constant pressure

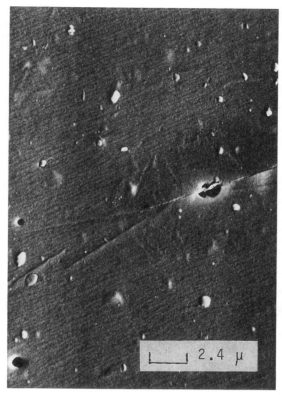

FIG. 6—*SEM image showing angular difference between an SCC fissure and hydride platelets.*

are the increase of the tangential and axial components while the radial component remains constant. This implies that while total stresses are increasing (including the component resolved perpendicular to basal planes), the whole system tends to be biaxial. The fact that, in thinned wall tubing, the cracks stop at depths of about 10 to 30 µm independently of the duration of the tests, indicates that, at that point, the axiality of the stress system is more important to the cracking than the value of the stress. In a single crystal, the result of uniaxial stressing will depend on the orientation; if the basal planes are parallel to the stresses, the stress will be relaxed by prismatic slip, plastic deformation will occur, and the component normal to the basal planes will be zero; if instead, the basal planes are perpendicular to the stresses, the only possible plastic relaxation will be by pyradimal slip (or twinning) at a higher critical resolved shear stress (CRSS) and it will be easier to increase the normal stresses to the necessary values. Still, they may not be high enough to promote SCC. In a biaxial stress state, there may be a higher constraint to deformation, but it does not seem to be sufficient. A minimum radial/tangential stress ratio may be needed to inhibit the activation of deformation modes; that is, taking into account the observation of small twins produced just at the very tip of the cracks partially propagated in thin wall cladding, it should be possible to estimate the critical stress ratio as a function of the final depths of the cracks and relate it with the threshold conditions to activate twinning at the crack tip.

FIG. 7—*Transgranular crack that changes direction in a deformed and hydrided single crystal.*

Another possibility to explain the effect of thinning the wall is related to the fact that, in heat-treated material, the grain boundaries are areas that go from the inner to the outer side; then, reducing the thickness implies making smaller areas of grain boundaries, that lessens the probability that a transgranular crack would intercept them. It may be possible that accumulation of dislocations due to the intersection of a crack path with a grain boundary would permit the intensification of stresses at the crack tip.

The fractographic analysis permits association with features seen in polycrystalline material. The basal propagation produces the same transgranular pseudo-cleavage features typical of iodine SCC tests conducted in as-received fuel cladding and present in fracture surfaces of in-reactor failed fuel pins: flat facets, river patterns, and the so called "fluting." Their crystallographic formation had been formerly proposed after a careful microscopy analysis [16], and this orientation work has more accurately confirmed it. The same analogies can be drawn for intergranular propagation and hydride precipitation (in this case, with a clear crystallographic distinction from transgranular SCC). Fracture surfaces corresponding to twin boundary propagation have been identified for the first time. These are flat and featureless, combining the characteristics of the grain boundary (absence of details) and the planar aspect that corresponds to the twin boundary shape that conserves the crystalline coherence. These may serve as a reference for the interpretation of the strange flat featureless details observed in high deformation rate SCC tests conducted in iodine with organic compounds [17] in which fast crack propagation was observed. For tubes with the normal cladding texture (basal poles mainly oriented in the radial direction), the crack advance could not be possible only by pseudo-cleavage on-basal planes, rather in a great proportion must proceed by different mechanisms such as intergranular separation or ductile rupture. The former mechanism would permit a slow crack growth, specially if the grains are elongated in shape, as normally occurs. The

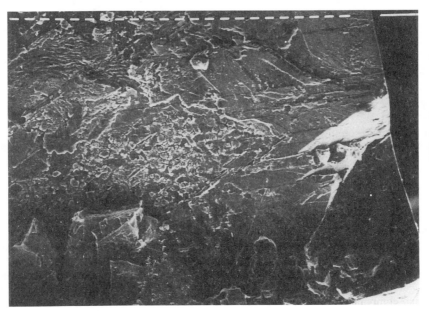

FIG. 8—*Fracture surface corresponding to the crack shown in Fig. 7.*

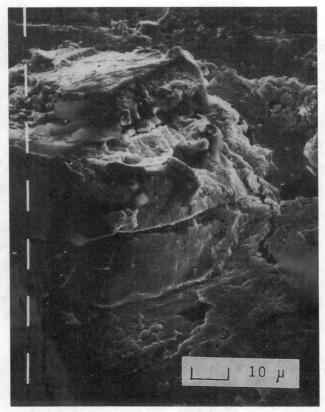

FIG. 9—*Surface corresponding to an intergranular portion of the crack of Fig. 7.*

latter mechanism implies a heavy mechanical interaction. The option of following a twin boundary gives the chance for a faster way, with higher resolved stresses perpendicular to the propagation surface.

Conclusions

1. Transgranular pseudo-cleavage of zirconium alloys produced during iodine SCC occurs along basal planes.
2. Cracking starts in those grains bearing the maximum tensile stress resolved perpendicular to the basal planes.
3. A triaxial stress state seems to be necessary for the process to start and continue.
4. For the calculation of the stresses produced in individual grains in pressurized tubes, the radial component should be included.
5. Fractographic features encountered in macrograins reproduce at a higher scale those known in polycrystalline material.
6. Twin boundaries have been identified as an alternate path for SCC propagation; their fractographic appearance has been studied.

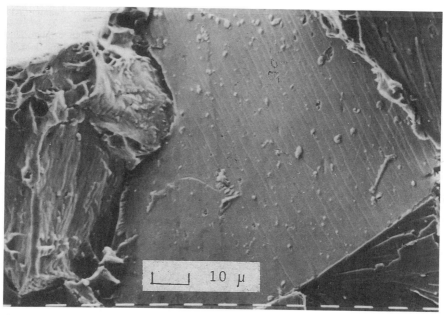

FIG. 10—*Fracture surface corresponding to propagation along twin boundary in the crack of Fig. 7.*

7. The results support a decohesion model in which mechanical, chemical, and thermal energies combine to exceed the binding energy of the zirconium atoms. This is a thermodynamic balance and the energy distribution statistics and the kinetics must be taken into account for a more detailed behavior prediction. The energy distribution statistics will perhaps make it impossible to find a threshold stress value for SCC; the kinetics will depend upon variables that control the movements of both the iodine and zirconium atoms at the tip of the crack, as described by the surface mobility based model for SCC [15].

References

[1] Cox, B., "Pellet-clad Interaction (PCI) Failures of Zirconium Alloy Fuel Cladding," *Journal of Nuclear Materials*, Vol. 172, No. 3, Aug. 1990, pp. 249–292.
[2] Haddad, R. and Cox, B., "On the Initiation of Cracks in Zircaloy Tubes by Iodine and Cs/Cd Vapours," *Journal of Nuclear Materials*, Vol. 138, 1986, pp. 81–88.
[3] Taylor, A., and Kagle, B. J. in *Crystallographic Data on Metal and Alloy Structures*, Dover Publications Inc., New York, 1963.
[4] Douglass, D. L., "The Metallurgy of Zirconium," *Atomic Energy Review, Supplement*, International Atomic Energy Agency, Vienna, 1971.
[5] Wallace, A., Shek, G., and Lepik, O., "Effect of Hydride Morphology on Zr-2.5%Nb Fracture Toughness," *Zirconium in the Nuclear Industry: Eighth International Symposium, ASTM STP 1023*, L. F. P. Van Swam and C. M. Eucken, Eds., American Society for Testing and Materials, Philadelphia, 1989, pp. 66–88.
[6] Peehs, M., Stehle, H., and Steinberg, E., "Out of Pile Testing of Iodine Stress Corrosion Cracking in Zircaloy Tubing in Relation with the PCI-Phenomenon," *Zirconium in the Nuclear Industry (4th Conference), ASTM STP 681*, American Society for Testing and Materials, Philadelphia, 1979, pp. 244–260.
[7] Pettersson, K. in Discussion of Peehs et al., *Zirconium in the Nuclear Industry (4th Conference), ASTM STP 681*, American Society for Testing and Materials, Philadelphia, 1979.

[8] Blackburn, M. J., Feeney, J. A., and Beck, T. R., "Stress Corrosion Cracking of Titanium Alloys," *Advances in Corrosion Science and Technology*, Vol. 3, 1973, p. 67.
[9] Vanhill, R. J. H., "Fractographic Interpretation of Subcritical Cracking in a High Strength Titanium Alloy IMI-550)," Report NLR-MP72025U, National Aerospace Laboratory, The Netherlands, 1972.
[10] Cox, B., "Hydride Cracks as Initiators of Stress-Corrosion Cracking of Zircaloys," *Zirconium in The Nuclear Industry (4th Conference), ASTM STP 681*, American Society for Testing and Materials, Philadelphia, 1979, p. 306.
[11] Kubo, T., Wakashima, Y., Amano, K., and Nagai, M., "Effects of Crystallographic Orientation on Plastic Deformation and SCC Initiation of Zirconium Alloys," *Journal of Nuclear Materials*, Vol. 132, 1985, pp. 1–9.
[12] Bush, S. H. and Kemper, R. S., "Recovery and Recrystallization of Zirconium and Its Alloys—Part 3: Annealing Effects in Zircaloy-2 and Zircaloy-3," HW-69680, AEC Research and Development Report, Richland, WA, May 1961.
[13] Cox, B. and Wood, J. C., "Iodine Induced Cracking of Zircaloy Fuel Cladding—A Review," *Proceedings*, Electrochemical Society Symposium on Corrosion Problems in Energy Conversion and Generation, New York, Oct. 1974, pp. 275–321.
[14] Timoshenko, S. P. and Wornonsky-Krieger, S., *Theory of Plates and Shells*, McGraw-Hill, New York, 1959.
[15] Galvele, J. R., "A Stress Corrosion Cracking Mechanism Based on Surface Mobility," *Corrosion Science*, Vol. 27, 1987, pp. 1–32.
[16] Aitchison, I. and Cox, B., "Interpretation of Fractographs of SCC in Hexagonal Metals," *Corrosion*, Vol. 28, 1972, pp. 83–87, 123.
[17] Cox, B. and Haddad, R., "Methyl Iodide as a Promoter of Iodine SCC of Zircaloys," *Journal of Nuclear Materials*, Vol. 137, 1986, p. 115.

DISCUSSION

Clement Lenaignan[1] (written discussion)—For samples tested in iodine, it is usually found that the crack starts as intergranular and later grows as transgranular with pseudo-cleavage. In your experiment, you have very large grains, and I would like to know if the crack starts at grain boundaries and grow after in the grain on the 0001 plane?

R. E. Haddad and A. O. Dorado (authors' closure)—Intergranular SCC initiation has often been seen in the polycrystalline part of tubes with a mixed structure (grains grown as large wall-to-wall single crystals surrounded by a microstructure composed by grains of about 20 µm diameter). However, transgranular cracks in our single crystals do not start at grain boundaries. On the contrary, all the through-wall cracks have initiated and propagated transgranularly in those macrograins, up to the point they have reached the grain boundary. This observation suggests that the crack propagation limiting step in polycrystalline material is the renucleation of the crack when passing from one grain to another.

E. Tenckhoff[2] (written discussion)—You described the phenomenon of SCC-propagation along the basal plane. What is the mechanism behind this phenomenon, for instance:

(1) is it an electrochemical process by anodic solution?
(2) is it a reduction of the banding/bonding forces by surface energy effects?
(3) is it a dislocation guided process, etc?

R. E. Haddad and A. O. Dorado (authors' closure)—Iodine SCC of Zircaloy may not be considered a typical electrochemical process by anodic dissolution since the aggressive environment is not a polar solvent and there is no charge transference through a metal-electrolite interface. Reduction of zirconium-zirconium bonds by adsorption of corrodant species at the crack tip is an accepted mechanism for liquid metal embrittlement; it is considered that it may apply for the SCC case too, given the phenomenological similarities (Ref *1* of this paper). A recently proposed surface mobility based SCC mechanism (Ref *15* of this paper) that serves to explain other types of environmental-assisted cracking as well allows for a better understanding of the kinetics of the process. The role of dislocations has not been clearly established yet. It seems easier to form a fissure if the line of the crack path intersects a grain boundary, indicating that, if the movement of dislocations is not permitted, the value of stress on the fracture plane could be increased to the values needed for rupture.

Brian Cox[3] (written discussion)—One reason for previous investigators apparently finding transgranular propagation planes that were 10° to 12° off basal planes may have been that they measured the orientations of crack segments that included steps caused by the edges of small "river patterns." Have you selected segments of the cracks that specifically avoid such steps in order to establish that the cracks are within ±1° of the basal plane?

R. E. Haddad and A. O. Dorado (authors' closure)—The orientation work was conducted on samples with a relatively low amount of deformation. The cracks were very fine lines with no branching, as seen in Fig. 3 of this paper. Branching and steps have been observed in samples with higher deformation values; in a few cases, lines apparently running at some angle of the basal traces were seen, but became stepped when looked at with some magnification.

---

[1] CEA, Granoble, France.
[2] Siemens, D-8520 Erlangen, Germany.
[3] University of Toronto, Ontario, Canada.

Oxide Characterization

Boel Wadman,[1] Zonghe Lai,[1] Hans-Olof Andrén,[1] Anna-Lena Nyström,[2] Peter Rudling,[3] and Håkan Pettersson[4]

Microstructure of Oxide Layers Formed During Autoclave Testing of Zirconium Alloys

REFERENCE: Wadman, B., Lai, Z., Andrén, H.-O., Nyström, A.-L., Rudling, P., and Pettersson, H., "**Microstructure of Oxide Layers Formed During Autoclave Testing of Zirconium Alloys,**" *Zirconium in the Nuclear Industry: Tenth International Symposium, ASTM STP 1245,* A. M. Garde and E. R. Bradley, Eds., American Society for Testing and Materials, Philadelphia, 1994, pp. 579–598.

ABSTRACT: The microstructure of oxide layers formed in steam in a 400°C, 10.3-MPa autoclave on different zirconium alloys was studied by transmission electron microscopy. Pre- and post-transition oxide layers on Zircaloy-4 with different heat treatments, and post-transition oxide layers on Zr-0.5Sn-0.53Nb were compared. Special attention was paid to the oxide-metal interface. In Zircaloy-4 with short annealing times and high post-transition corrosion rates, the interface had a disordered structure, and pores were found in the oxide very close to the interface. In Zircaloy-4 with low uniform corrosion rates, the interface consisted of highly ordered, columnar grains. The interface in Zr-0.5Sn-0.53Nb had a different appearance, with an intermediate phase of equiaxed grains between the columnar oxide and the metal. The hydrogen absorption of the zirconium alloys during oxidation was measured by the melt extraction technique on samples oxidized for 63, 147, and 343 days. The Zr-0.5Sn0.53Nb alloy had considerably lower hydrogen absorption than Zircaloy-4.

KEY WORDS: zirconium alloys, niobium, uniform corrosion, corrosion resistance, autoclave testing, oxide growth, oxide-metal interface, transmission electron microscopy, hydrogen absorption, zirconium, nuclear materials, nuclear applications, radiation effects

Zirconium alloys used today are satisfactory as pressurized water reactor (PWR) fuel cladding materials at the present reactor conditions. However, new and more corrosion-resistant materials are needed to permit future modifications of reactor chemistry and higher reactor temperatures. To improve the mechanistic understanding of the uniform corrosion behavior of different zirconium alloys, careful studies of the microstructure of the growing oxide layer may be of importance.

Recently, characterization of Zircaloy oxide layers produced without neutron irradiation has been performed, and the results point towards a tetragonal-to-monoclinic transformation in the oxide-metal interface at the transition from decreasing to approximately constant uniform corrosion rate [*1,2*]. In addition, there exists a variety of oxidation models based on transmission

---

[1] Research students and associate professor, respectively, Department of Physics, Chalmers University of Technology, S-412 96 Göteborg, Sweden.
[2] Research engineer, AB Sandvik Steel, R&D Centre, Physical Metallurgy, S-811 81 Sandviken, Sweden.
[3] Specialist, Nuclear Fuel Materials, Fuel Division, ABB Atom AB, S-721 63 Västerås, Sweden.
[4] Nuclear fuel engineer, Vattenfall Fuel, Nuclear Fuel Technology, S-162 87 Vällingby, Sweden.

electron microscopy studies of oxide layers on different zirconium alloys [3–7]. The oxidation process and irradiation-induced damage of the intermetallic precipitates in Zircaloys have also been thoroughly investigated [8–13]. The objective of the present work was to compare the microstructure of oxide layers before and after transition on different zirconium alloys. Oxidation in autoclave without neutron irradiation was used to produce oxide layers of comparable thickness from different materials. Special emphasis was given to the microstructure of the oxide-metal interface.

Experimental

Materials

The zirconium alloys studied in this report are Zircaloy-4 and Zr-0.5Sn-0.53Nb. The materials were produced as 0.8 mm sheet, and their chemical composition and manufacturing history is described in Table 1. The microstructure of the standard Zircaloy-4 Material 2 can be compared to that of nuclear fuel cladding tubes used in PWRs. From the same melt, Material 1 was fabricated as a low-annealed Zircaloy-4 reference material expected to have a high uniform corrosion rate. Figure 1 shows transmission electron micrographs of intermetallic precipitates in the three materials. In the Zr-0.5Sn-0.53Nb material, the precipitates were of the hexagonal $Zr(Fe,Cr)_2$ type also found in Zircaloy-4, but contained some niobium.

Autoclave Testing

To produce uniform oxide layers of different thickness, accelerated corrosion in an autoclave was used. Coupons were oxidized up to 343 days in steam in a static autoclave at 400°C, a test used to compare uniform corrosion resistance of different Zircaloy materials in PWRs. The pressure in the autoclave was 10.3 ± 0.7 MPa, and the water had a resistivity greater than 10^6 Ωcm, an oxygen concentration less than 45 ppb, and pH values between 5 and 8. The corrosion weight gain was measured every 20 days, and coupons were removed after 3, 21, 63, and 147 days to obtain specimens from different oxidation stages for hydrogen absorption measurements and electron microscopy.

Hydrogen Absorption

The hydrogen uptake after oxidation was measured on coupons after 63, 147, and 343 days in an autoclave. A melt extraction furnace was used, where the sample was fused at approximately 2100°C in a graphite crucible. The released hydrogen was transported by an argon flow to a thermistor cell where the amount of hydrogen was determined from the thermal conductivity of the gas. Two samples from each oxidized coupon were used, and the precision of the analysis was typically ±1.5 ppm when 100 ppm hydrogen was measured.

Electron Microscopy

Transmission electron microscopy (TEM) examinations of the oxide-metal interface were performed in a JEOL 2000 FX instrument equipped with a Link AN 10 000 system for energy dispersive X-ray spectrometry (EDX). The specimens were produced as cross sections of the oxide layer, the electron beam of the microscope being perpendicular to the oxide growth direction. This involved gluing two oxide surfaces of 2-mm-large specimens together, then cutting slices perpendicular to the oxide surfaces. The slices were mechanically polished and dimpled thin at the center, and the specimen was glued to a copper grid with a central hole. Further thinning of the free side of the cross section was performed until the thickness in the center was 20 to 30 μm. Dimpling without specimen rotation and with very low pressure on

TABLE 1—*Chemical composition and fabrication history of studied materials.*

Chemical Composition	Material 1 Low-Annealed Zircaloy-4	Material 2 Standard Zircaloy-4	Material 5 Zr-0.5Sn-0.53Nb
Sn, % by weight	1.38	1.38	0.50
Nb, % by weight	...	...	0.53
Fe, % by weight	0.20	0.20	0.22
Cr, % by weight	0.09	0.09	0.10
Ni, % by weight	0.003	0.003	0.003
Si, ppm	<30	<30	<30
Fabrication History	Log A^a = −15.7	Log A^a = −13.2	Log A^a = −14.9

Material 1	Material 2	Material 5
Beta-quenched 30×10 mm	Beta-quenched 30×10 mm	Beta-quenched 35×8 mm
Cold-rolled to 32×5 mm	Cold-rolled to 32×5 mm	Hot-rolled to 32×2 mm
Vac. annealed 565°C/1h	Vac. annealed 669°C/24h	Vac. annealed 580°C/8h
Cold-rolled to 27×2.5 mm	Cold-rolled to 27×2.5 mm	Cold-rolled to 35×1.3 mm
Vac. annealed 580°C/1h	Vac. annealed 580°C/1h	Vac. annealed 580°C/8h
Cold-rolled to 27×0.8 mm	Cold-rolled to 27×0.8 mm	Cold-rolled to 35×0.8 mm
Vac. annealed 580°C/1h	Vac. annealed 580°C/1h	Vac. annealed 500°C/3.5h
		Vac. annealed 580°C/1h

[a] Annealing parameter with activation energy, Q, given by Andersson et al. [14] Q/R = 31 700 K.

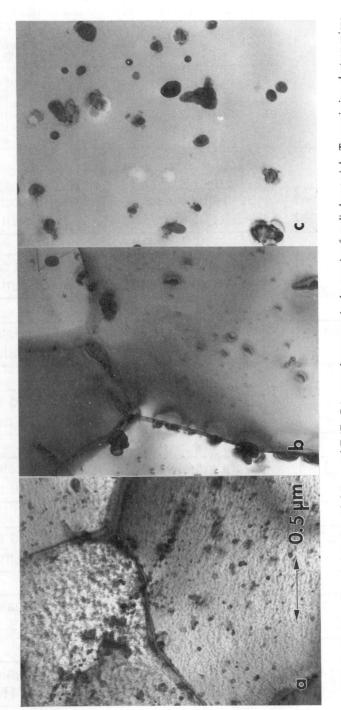

FIG. 1—*Intermetallic precipitates with hexagonal Zr(Fe,Cr)$_2$ crystal structure in the matrix of studied materials. Transmission electron micrographs from recrystallized specimens (annealed at 565°C for 1.5 h): (a) Material 1, low-annealed Zircaloy-4, (b) Material 2, standard Zircaloy-4, and (c) Material 5, Zr-0.5Sn-0.53Nb. Here, all studied precipitates also contained niobium.*

the specimen was found to decrease the mechanical damage during this step. To obtain electron transmission, the specimen was ion etched in a Gatan Dual Ion Mill with argon ions at 4.2 kV and a current of 0.3 mA. The specimen was then cooled with liquid nitrogen while it was supported between two copper grids to improve the thermal conductivity.

TEM examinations were performed on oxide layers before and after the corrosion rate transition in standard and low-annealed Zircaloy-4, and after transition in Zr-0.5Sn-0.53Nb, see Table 2. Several TEM specimens were prepared from each material variant, and up to three specimens from each oxidation stage were carefully investigated. The main interest was focused on the metal-oxide interface, but also outer parts of the oxide layer were studied. All specimens were studied in the as-prepared conditions, that is, no extra anneal was performed in order to facilitate the detection of precipitates in the oxide layer.

Scanning electron microscopy (SEM) and optical microscopy were used to observe the oxide thickness and the smoothness of the oxide front on carefully polished cross sections.

Results

Autoclave Testing

The oxidation weight gain of the coupons that had remained up to 343 days in the autoclave is shown in Fig. 2. For the standard Zircaloy-4 Material 2, the transition in oxidation rate seems to occur after 50 days in autoclave, when the weight gain is approximately 35 mg/dm^2. Eucken et al. [15] found transitions varying between 30 to 40 mg/dm^2 in similar tests of different Zircaloys. Material 5 showed a transition after 63 days in autoclave at similar weight gains, 35 to 50 mg/dm^2, which corresponds to observations by Isobe and Matsuo [16]. This material was tested at the same conditions as Zircaloy-4, because the main purpose of this autoclave test was to produce oxide layers of comparable thickness for TEM examination. In 400°C steam, it had a slightly higher post-transition corrosion rate than the standard Zircaloy-4. As expected, Zircaloy-4 with short annealing times had grown thick oxide layers already after 21 days and all remaining coupons were therefore removed from the autoclave at this time.

Hydrogen Absorption

The hydrogen absorption during oxidation is shown in Fig. 3 as a function of weight gain in the autoclave. After long times in the autoclave, Material 5 absorbed only about half the amount of hydrogen absorbed by Zircaloy-4. Figure 4 shows optical micrographs of the hydride formation through the thickness of the metal. The absence of heat gradient in the coupons oxidized in the autoclave gave rise to hydride bands more evenly distributed than after oxidation in the PWR, where the hydrides are concentrated to the metal part closest to the oxide interface [17]. In this study, 50-μm-thick hydride-free regions were observed close to the surface in Material 5.

Electron Microscopy

The results from TEM examinations of the oxide layers are summarized in Table 2. The length of metal-oxide interface studied was 1 to 3 μm, and between 4 and 30 μm^2 of outer oxide layer was investigated on all oxide variants.

Influence of Heat Treatment

Pre-Transition Oxide Layers—After three days in the autoclave, the oxide was 0.7 to 0.8 μm thick on the two Zircaloy-4 materials. The oxide-metal interface showed a similar appearance in both materials. Columnar grains, 10 to 20 nm wide and 50 to 100 nm long were growing in

TABLE 2—*Oxide layers studied by transmission electron microscopy.*

Coupon Number	Time in Autoclave, days	Weight Gain, mg/dm²	Investigation		Results: Gain Shape and Diameter, nm	
			Number of Thin Foils	Length of Studied Metal-Oxide Interface, μm	Metal-Oxide Interface	Outer Oxide Layer
Material 1: Low-Annealed Zircaloy-4						
14	3	15.4	3	2	columnar 10–30	columnar 20–30
15	21	511	1	...	...	mixed columnar and equiaxed 20–60
16	21	488	1	1	columnar 10–40 equiaxed 20	mixed columnar and equiaxed 20–60
Material 2: Zircaloy-4						
23	3	12.3	1	2	columnar 10–20	columnar 10–30
26	147	80	3	3	columnar 10–30	columnar 20 and equiaxed 20–40
Material 5: Zr-0.5Sn-0.53Nb						
59	147	93.3	1	1.5	suboxide 50–100 columnar 20–40	columnar 20–50

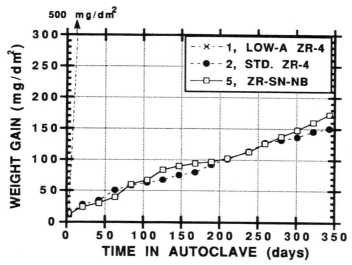

FIG. 2—*Corrosion weight gain during autoclave test (400°C, 10.3 MPa).*

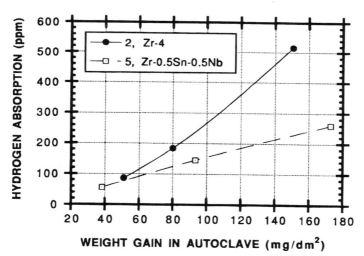

FIG. 3—*Hydrogen absorption as a function of weight gain in 400°C autoclave for standard Zircaloy-4 and Zr-0.5Sn-0.53Nb.*

direct contact with the metal in a direction perpendicular to the oxide-metal interface. No pores, cracks, or amorphous oxide layers were found in the interface or between oxide grains. The crystal structure of most oxide grains was monoclinic ZrO_2, but some tetragonal ZrO_2 was found in the interface of Coupon 23 of the standard Zircaloy-4 material, see Fig. 5a. As observed by SEM, the interface had an even appearance, but TEM revealed a fine-scaled irregularity, where the distance between growth fronts varied between 50 to 200 nm. One difference between the oxide-metal interfaces of the two materials was a higher dislocation density in the metal part of the interface of Coupon 14, the low-annealed Zircaloy-4.

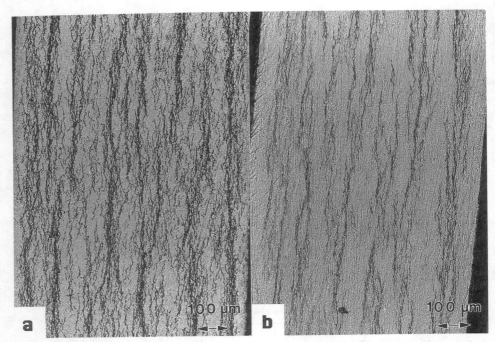

FIG. 4—*Optical micrographs of etched cross sections of samples oxidized 343 days in autoclave, showing the hydride distribution in the metal of* (a) *Standard Zircaloy-4 and* (b) *Zr-0.5Sn-0.53Nb.*

The thin oxide layers on both Zircaloy-4 materials consisted of dense layers of well-ordered columnar grains. No difference in grain size was observed at different distances from the interface. The oxide layers contained a fairly high amount of twin grain boundaries, and high-resolution microscopy frequently showed dislocations inside the oxide grains and in the grain boundaries, see Fig. 6a. As in all oxide layers studied on Zircaloy-4, the columns were oriented perpendicular to the oxide-metal interface.

Post-Transition Oxide Layers—After 147 days in the autoclave, Coupon 26 from standard Zircaloy-4 had an oxide thickness of approximately 5.5 μm. SEM examinations showed a smooth growth front even in this oxide-metal interface. In TEM, we observed columnar grains of approximately the same size and appearance as in the interface after three days, that is, no porosity between grains and no amorphous layer was present. Regions with a high dislocation density in the metal were observed at the interface.

In the low-annealed Zircaloy-4, the weight gains after 21 days were already so high (approximately 500 mg/dm^2) that a direct comparison between equally thick oxide layers could not be made. Here, the interface had a more disordered structure, with short columnar and equiaxed grains growing from the metal, see Fig. 5b. Intergranular pores and cracks were observed very close to the interface. No tetragonal oxide was found in the oxide-metal interface of the two materials. Also here, the low-annealed Zircaloy-4 contained regions with a high dislocation density in the metal close to the interface. In this material, the growth front was highly uneven as observed by SEM and optical microscope, and large differences in oxide thickness were found between different areas of the coupons.

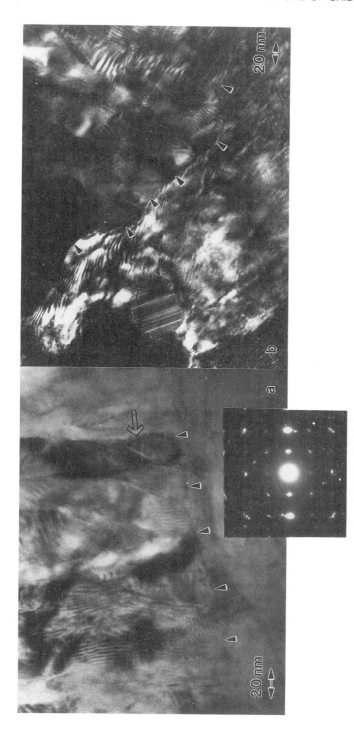

FIG. 5—*Oxide-metal interface in* (a) *standard and* (b) *low-annealed Zircaloy-4. Small arrows indicate the interface from the metal side.* (a) *Coupon 23 after three days in autoclave. Bright-field electron micrograph where the grain with dark contrast (large arrow) was tetragonal* ZrO_2. (b) *Coupon 16 after 21 days in autoclave, dark-field electron micrograph.*

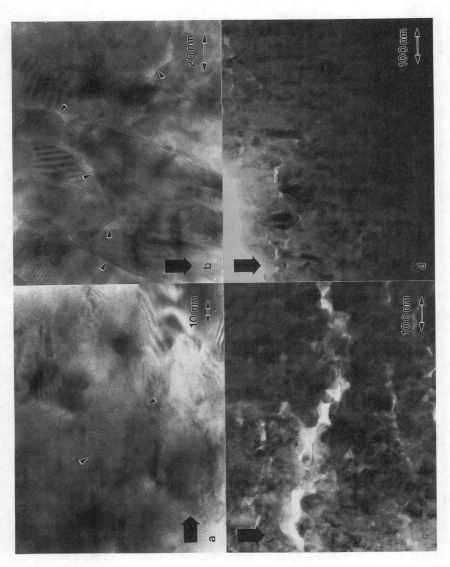

FIG. 6—*Outer oxide layer of standard Zircaloy-4, before and after transition (a,b), and low-annealed Zircaloy-4 after transition (c,d). Dark bold arrows indicate direction to the metal. A high dislocation density (see small arrows) was observed by high resolution electron microscopy. (a) Coupon 23, 0.8-em thick oxide. (b) Coupon 26, 5.5-em thick oxide. Small intergranular microcracks and pores are indicated by small arrows. (c) Porous and (d) dense region about 10-em from the interface of Coupon 16, 42-em thick oxide.*

In the outer part of the oxide layer on standard Zircaloy-4 (Coupon 26, Fig. 6b), the majority of the grains were columnar and monoclinic, but a few columnar tetragonal grains and equiaxed monoclinic grains were detected. The oxide was generally dense, with only very fine pores and microcracks between the grains. In some areas of the outer oxide, larger cracks parallel to the surface were found between equiaxed grains 20 to 40 nm in diameter. A partly oxidized precipitate 100 nm in diameter was found close to the oxide-metal interface. The outer oxidized part of the precipitate was microcrystalline, which was in accordance with earlier reports [1,9].

In one sample from low-annealed Zircaloy-4, a 42-µm-thick oxide layer could be studied all the way from the interface to the outer oxide surface. As expected, the layer was highly porous and cracked. The cracks and pores were all found in the grain boundaries, and lateral cracks were observed at a distance of 0.2 to 1 µm. The grain shapes were a mixture of equiaxed and columnar grains, 20 to 50 nm in diameter. Apart from the oxide-metal interface where the grains were small, no regular pattern of different grain sizes could be observed throughout the oxide layer, as opposed to observations by Bradley and Perkins [7]. The different porosity between different parts of the oxide layer could generally not be correlated to grain shapes or sizes. Only in one area of the oxide layer, 10 µm from the oxide-metal interface, did we observe a larger concentration of columnar grains in the dense oxide than in the adjacent porous oxide, see Figs. 6c and d. Another interesting observation was that a vast majority of the columnar grains remained elongated perpendicular to the outer oxide surface despite the otherwise disordered mixture of grain shapes and sizes in the oxide layer.

About 12 µm from the interface, a large area (3 by 3 µm) of columnar growth was observed, see Fig. 7. In this area, the columnar grains were all oriented with the (100) planes perpendicular to the oxide-metal interface.

Influence of Composition

Post-transition oxide layers of the standard Zircaloy-4 material with a weight gain of 80 mg/dm^2 have been compared with oxide layers of Zr-0.5Sn-0.53Nb with a weight gain of 93.3 mg/dm^2. In the oxide-metal interface, the most apparent difference between the materials was an intermediate crystalline phase between the metal and the oxide in the Zr-0.5Sn-0.53Nb material, see Fig. 8a. The phase consisted of equiaxed grains 50 to 100 nm in diameter in direct contact with both metal and oxide. A semi-quantitative EDX analysis of this intermediate phase showed that it contained mostly zirconium and oxygen, but we were not able to identify its diffraction pattern with any published zirconium-oxygen phase. Columnar monoclinic oxide grains, somewhat wider and shorter than in the standard Zircaloy-4 material, grew from this region. In this material, the columns were not always perpendicular to the interface as they were observed to grow in a 45° orientation from two different metal grains.

In the oxide layer of Zr-0.5Sn-0.53Nb, several precipitates were found. About 200 nm from the oxide-metal interface, two crystalline precipitates 50 nm in diameter and one amorphous precipitate 130 nm in diameter were studied, Figs. 8b and c. The diffuse ring in the diffraction pattern from the amorphous precipitate corresponds to a distance of 2.8 Å, similar to what was found by Pecheur et al. in Zircaloy-4 [9]. The iron/chromium (Fe/Cr) ratio of the precipitates was measured by EDX and was found to be approximately 3 in the oxide, somewhat higher in the outer rim of the amorphous precipitate compared to its inner part and the crystalline precipitates. EDX analysis showed that the Fe/Cr ratio of the precipitates in the unoxidized metal was 2.0 to 2.5, which agrees with the bulk material ratio of 2.2. The Fe/(Cr + Nb) ratio was 1.6 in the amorphous precipitate, and 1.1 in precipitates in the unoxidized material. Seven precipitates were analyzed, all of which contained approximately 9 atomic percent niobium. This can be compared with high-resolution EDX measurements on a nickel-iron-niobium alloy

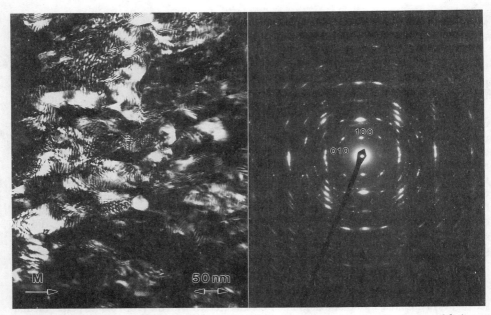

FIG. 7—*Part of the large columnar region found 12 μm from the interface in Coupon 16. Arrow indicate direction to the metal. Most of the monoclinic grains were oriented with the (100) plane perpendicular to the oxide-metal interface, see the corresponding $[001]_c$ selected area diffraction pattern.*

containing 0.09% niobium by weight, where some precipitates contained between 1 and 2.3 atomic percent niobium [18].

The outer part of the approximately 6-μm-thick oxide layer on Zr-0.5Sn-0.53Nb was also studied. Figure 8d shows the surface of the initial oxide layer, with the glue layer from specimen preparation in the upper part of the figure. Columnar grains, 20 to 50 nm wide, were observed, as well as some equiaxed grains 20 nm in diameter. Cracks in the oxide layer were found close to the crystalline precipitates, but otherwise only small microcracks were observed. The larger cracks between equiaxed grains that were found in some oxide layers of standard Zircaloy-4 were not observed in the niobium containing alloy.

Discussion

Influence of Heat Treatment

In Fig. 9, the oxide-metal interfaces of pre- and post-transition oxides in the different Zircaloy-4 materials can be compared schematically. It is clear that slow oxidation is connected to undisturbed growth of columnar oxide grains in the interface, irrespective of the original crystal structure of the oxide grains. This type of oxide morphology was found in the pre-transition oxide layers of both low-annealed and standard Zircaloy-4 that had similar pre-transition but very different post-transition oxidation rates. Surprisingly, the post-transition oxides of standard Zircaloy-4 also showed the same morphology in the oxide-metal interface. As the outer oxide layer of these specimens contained only a small amount of equiaxed grains,

FIG. 8—*Oxide layer on Coupon 59, Zr-0.5Sn-0.53Nb. (a) Dark-field transmission electron micrographs of the oxide-metal (O,M) interface. The arrowed grains consisted of an intermediate suboxide (S). From this region, columnar monoclinic ZrO_2 grains were growing. (b) Crystalline and (c) amorphous intermetallic precipitate 200 nm from the interface. (d) Initially formed oxide 6 em from the interface.*

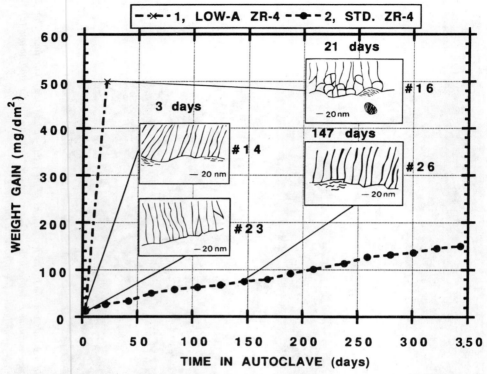

FIG. 9—*Schematic representations of oxide-metal interfaces of the two Zircaloy-4 variants at different weight gains in autoclave.*

the length of studied oxide-metal interface (3 µm) was not sufficient to verify our assumption that this grain morphology originates from the interface. However, in the low-annealed Material 1, equiaxed grains were observed in the interface of post-transition oxide that was less resistant to cracking and porosity. Similar structures were observed by Garzarolli et al. [1] in the oxide-metal interface of nodular oxide. As in this study, they found intergranular cracks very close to the interface.

Also, in the outer oxide layer, the transition to high corrosion rates can be related to oxide microstructure. The rate transition is generally connected with a formation of pores and cracks in the oxide layer. In the present study, no transcrystalline fracture was observed in columnar grains. Instead, all cracks formed intercrystalline, primarily at equiaxed grains. We found in standard Zircaloy-4 that cracks occurred mainly between equiaxed oxide grains. In the thick oxide layer of the low-annealed Zircaloy-4 material, the mixture of different grain shapes made it difficult to make the same general correlation, although some indications were found (Figs. 6c and d). The cracks parallel to the oxide surface may have been caused by specimen preparation [19]. However, the observation that no cracks were found in TEM specimens of pre-transition oxide layers and the periodicity of the cracks in post-transition oxides on Zircaloy-4 indicate that either cracks or a weakness in the oxide structure were present before specimen preparation.

An oxide growth mechanism proposed by Cox [20] is that the oxide layer initially grows as tetragonal or cubic ZrO_2 stabilized by the compressive stresses in the interface. At an oxide thickness where the stress has decreased sufficiently, a martensitic transformation to monoclinic zirconia occurs that could produce a twinned grain structure with pores at the columnar ends as in partially stabilized zirconia [21–23]. In this study, we have found some twin grain boundaries that could confirm a martensitic transformation. However, these grains are not in the majority throughout the oxide and also existed in pre-transition oxides. The pre-transition twinning would suggest that the pore formation starts at this early stage, something not observed in this study, and the existence of pre-transition porosity is disputed [24,2]. No large amount of tetragonal phase in the interface of pre-transition oxides could be found, as would be expected from the Raman spectroscopy studies by Godlewski et al. [2]. The low amount of tetragonal oxide was also reported by Garzarolli et al. [1], and could be explained by stress relief in thin TEM specimens resulting in a tetragonal-to-monoclinic transformation.

The strain situation in the oxide layer is such that grain boundaries perpendicular to the oxide-metal interface are in compression. This compression can be released if cracks parallel to the interface are formed, and the outer oxide layer expands radially from the crack. Areas with columnar grains are expected to have a high resistivity against intergranular fracture since they only have small areas of unconnected grain boundaries parallel to the metal-oxide interface.

In areas with equiaxed grains, however, continuous paths of grain boundaries parallel to the interface exist. When oxide growth has transported these areas some distance away from the interface, cracks may therefore develop that facilitate contact with the outer surface, under the assumption that porous grain boundaries perpendicular to the interface are also present. Water entering the cracks reduces the oxide diffusion length drastically, and a transition in oxide growth rate is observed.

Thus, the critical step is the formation of equiaxed grains at the interface. The orientation relationship between columnar grains and metal means that increasing tensile stresses parallel to the interface are built up in the metal until semi-coherent growth is no longer possible. Equiaxed incoherent oxide grains are then formed at the interface, possibly relaxing the stress so that new semi-coherent columnar grains may nucleate.

Influence of Composition

Material 5 and the standard Zircaloy-4 differ in composition primarily with respect to the niobium and tin concentrations. The difference in heat treatments that might also influence the oxide growth will not be treated here. TEM examinations of the oxide layers of the two materials showed different microstructures both in the oxide-metal interface, where an intermediate phase existed only in the niobium-containing alloy, and in the outer oxide layer, where a larger amount of equiaxed grains was found in the standard Zircaloy-4 material. As the interface studied in the niobium-containing alloy originates from only one specimen, this intermediate phase has to be further examined.

It is interesting to compare our results from Material 5 with our recent TEM studies of oxide layers on Zr-0.23Fe-0.2V produced by steam autoclave testing at 400°C. Also in this alloy, an intermediate phase was found in the oxide-metal interface of all specimens, where the approximately 3-μm interface was examined. Figure 10 shows the oxide-metal interface consisting of 20 to 100 nm large equiaxed grains, again with a diffraction pattern not corresponding to any phase known to us. The oxygen profile through the interface measured by EDX is shown in the picture, indicating a sub-oxide with no large amounts of alloying elements present. Above the new phase columnar monoclinic oxide grains, 10 to 30 nm wide and 50 to 200 nm long were observed. In the outer parts of the oxide, the grains were columnar and somewhat larger, up to 40 nm wide.

It is well known [25] that zirconium can dissolve a considerable amount of oxygen interstitially, and a diffusion profile of oxygen into the metal exists during oxidation of zirconium

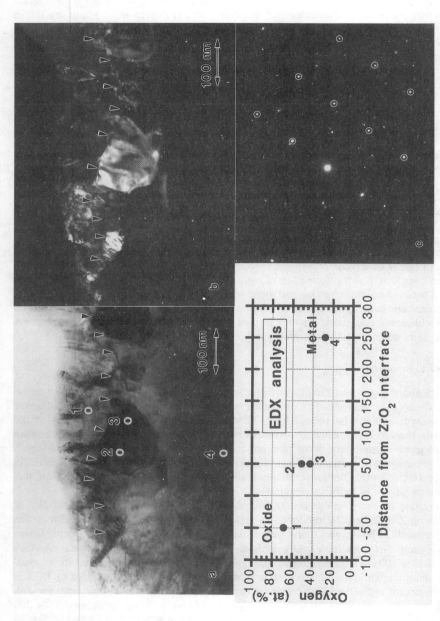

FIG. 10—*Oxide-metal interface in Zr-0.23Fe0.2V. (a) The composition of the intermediate grains (arrowed) was analyzed by EDX, see diagram. Numbers indicate position of analysis (beam size <20 nm, beam broadening 10 nm). (b) Dark-field electron micrograph, arrows again show the interface between ZrO_2 and the unidentified intermediate phase, see selected area diffraction pattern in (c), where spots corresponding to intermediate phase are enclosed in circles.*

(see analysis Number 4 in Fig. 10). In the niobium and vanadium containing alloys there seems to exist a separate crystalline sub-oxide phase between the oxide and the metal, possibly creating a barrier for hydrogen migration. The low hydrogen absorption of the Zr-0.5Sn-0.53Nb could also be favorable to in-pile corrosion resistance, as large concentrations of hydrides at the interface means lost coherence between oxide and metal [17]. Compared to Zircaloy-4, our hydrogen absorption measurements suggest that the oxide layer on Zr-0.5Sn-0.53Nb serves as a barrier against hydrogen diffusion, similar to what was observed on Zr-2.5Nb [3].

The amorphous precipitate observed may have undergone amorphization during specimen preparation, as it was found close to the specimen edge. The result is presented because of the interesting increase in atomic distance from 2.3 to 2.8 Å, also seen in the work by Pecheur et al. [9] where it was suggested to result from hydrogen inclusion. The amount of precipitates studied in this paper is not large enough for any definite conclusions, but our findings are in accordance with the observed behavior of similar precipitates in Zircaloy-4 [1,9,11]. The cracks noticed close to unoxidized precipitates clearly show that the retarded oxidation of the intermetallics disturb the symmetry and epitaxy of the growing oxide layers.

As a complement to the detailed work by Pecheur et al. [9], it would be interesting to study annealed precipitates in cross-section TEM specimens. In this study, we wanted to compare materials that were heat treated similar to PWR fuel claddings, and the small size of the precipitates made it difficult to observe them, whether oxidized or not, in the oxide layers.

Conclusions

1. In standard and low-annealed Zircaloy-4, the oxide-metal interface of less than 1-μm-thick pre-transition oxide layers had a similar appearance. Columnar grains 10, 20, 50, 100 nm grew in direct contact with the metal grains, and no amorphous layers were observed. The outer parts of these oxides were dense, without cracks or pores between the grains.

2. Post-transition oxide layers on standard Zircaloy-4 were similar to the pre-transition microstructure in the oxide-metal interface. In the outer oxide layer, most grains were columnar, but in some areas equiaxed grains with intergranular cracks were observed.

3. In low-annealed Zircaloy-4 with high post-transition corrosion rates, the oxide-metal interface was disordered, with short columnar and equiaxed grains. Cracks and pores and a mixture of equiaxed and columnar oxide grains characterized the outer oxide layer. Even in these thick oxide layers, most columnar grains were oriented in the oxide growth direction.

4. In post-transition oxide layers of Zr-0.5Sn-0.53Nb, an intermediate crystalline sub-oxide was found in a 50 to 100 nm-thick region in the oxide-metal interface. No amorphous phase was found in the interface. In oxide layers of similar thickness (5.5 μm) on standard Zircaloy-4, no intermediate phase was found. The intermediate layer might be a barrier against hydrogen migration, as the material absorbed only half the amount absorbed by Zircaloy-4 at equal weight gains. Unoxidized precipitates were found close to the interface in the oxide layer of Zr-0.5Sn0.53Nb.

5. Cracks in the oxide layers in all materials were associated with the presence of equiaxed oxide grains.

References

[1] Garzarolli, F., Seidel, H., Tricot, R., and Gros, J. P. in *Zirconium in the Nuclear Industry: Ninth International Symposium, ASTM STP 1132*, C. M. Eucken and A. M. Garde, Eds. American Society for Testing and Materials, Philadelphia, 1991, pp. 395–415.

[2] Godlewski, J., Gros, J. P., Lambertin, M., Wadier, J. F., and Weidinger, H. in *Zirconium in the Nuclear Industry: Ninth International Symposium, ASTM STP 1132*, C. M. Eucken and A. M. Garde, Eds., American Society for Testing and Materials, Philadelphia, 1991, pp. 566–594.

[3] Warr, B. D., Elmoselhi, M. B., Newcomb, S. B., McIntyre, N. S., Brennerstuhl, A. M., and Lichtenberger, P. C. in *Zirconium in the Nuclear Industry: Ninth International Symposium, ASTM STP 1132*, C. M. Eucken and A. M. Garde, Eds., American Society for Testing and Materials, Philadelphia, 1991, pp. 740–757.

[4] Harada, M., Kimpara, M., and Abe, K. in *Zirconium in the Nuclear Industry: Ninth International Symposium, ASTM STP 1132*, C. M. Eucken and A. M. Garde, Eds., American Society for Testing and Materials, Philadelphia, 1991, pp. 368–391.

[5] Kubo, T. and Uno, M. in *Zirconium in the Nuclear Industry: Ninth International Symposium, ASTM STP 1132*, C. M. Eucken and A. M. Garde, Eds., American Society for Testing and Materials, Philadelphia, 1991, pp. 476–498.

[6] Zhou, B.-X. in *Zirconium in the Nuclear Industry: Eighth International Symposium, ASTM STP 1023*, L. F. P. Van Swan and C. M. Eucken, Eds., American Society for Testing and Materials, Philadelphia, 1989, pp. 360–373.

[7] Bradley, E. R. and Perkins, R. A. in *Proceedings*, IAEA Technical Committee Meeting on Fundamental Aspects of Corrosion of Zirconium-Base Alloys in Water Reactor Environments, International Atomic Energy Agency, ISSN 1011–2766, Paper 9, 1990.

[8] Ploc, R. in *Zirconium in the Nuclear Industry: Eighth International Symposium, ASTM STP 1023*, L. F. P. Van Swan and C. M. Eucken, Eds., American Society for Testing and Materials, Philadelphia, 1989, pp. 498–514.

[9] Pecheur, D., Lefebvre, F., Motta, A. T., and Lemaignan, C., *Journal of Nuclear Materials*, Vol. 189, 1992, pp. 318–332.

[10] Motta, A. T., Lefebvre, F., and Lemaignan, C. in *Zirconium in the Nuclear Industry: Ninth International Symposium, ASTM STP 1132*, C. M. Eucken and A. M. Garde, Eds., American Society for Testing and Materials, Philadelphia, 1991, pp. 718–739.

[11] Griffiths, M., Gilbert, R. W., and Carpenter. G. J. C., *Journal of Nuclear Materials*, Vol. 150, 1987, pp. 53–66.

[12] Yang, W. J. S., *Journal of Nuclear Materials*, Vol. 158, 1988, pp. 71–80.

[13] Weidinger, H. G., Ruhmann, H., Cheliotis, G., Maguire, M., and Yau, T.-L. in *Zirconium in the Nuclear Industry: Ninth International Symposium, ASTM STP 1132*, C. M. Eucken and A. M. Garde, Eds., American Society for Testing and Materials, Philadelphia, 1991, pp. 499–535.

[14] Andersson, T., Thorvaldsson, T., Wilson, A., and Wardle, A. M. in *Proceedings*, IAEA Technical Committee Meeting on Fundamental Aspects of Corrosion of Zirconium-Base Alloys in Water Reactor Environments, International Atomic Energy Agency, Vienna, ISSN 1011–2766, Paper 9, 1990.

[15] Eucken, C. M., Finden, P. T., Trapp-Pritsching, S., and Weidinger, H. G. in *Zirconium in the Nuclear Industry: Eighth International Symposium, ASTM STP 1023*, L. F. P. Van Swam and C. M. Eucken, Eds., American Society for Testing and Materials, Philadelphia, 1989, pp. 113–127.

[16] Isobe, T. and Matsuo, Y. in *Zirconium in the Nuclear Industry: Ninth International Symposium, ASTM STP 1132*, C. M. Eucken and A. M. Garde, Eds., American Society for Testing and Materials, Philadelphia, 1991, pp. 346–367.

[17] Garde, A. M. in *Zirconium in the Nuclear Industry: Ninth International Symposium, ASTM STP 1132*, C. M. Eucken and A. M. Garde, Eds., American Society for Testing and Materials, Philadelphia, 1991, pp. 566–594.

[18] Rudling, P., Mikes-Lindbäck, M., Lehtinen, B., Andrén, H.-O., and Stiller, K., "Corrosion Performance of New Zircaloy-2 Based Alloys," in this volume.

[19] Cox, B., "Oxidation of Zirconium and Its Alloys," *Advances in Corrosion Science and Technology*, Vol. V, M. G. Fontana and R. W. Staehle, Eds, Plenum Press, New York, 1976, p. 173.

[20] Cox, B., "Are Zirconia Corrosion Films a Form of Partially Stabilised Zirconia (PSZ)?" Report AECL 9382, Atomic Energy of Canada Ltd, Chalk River Nuclear Laboratories, Ontario, Canada, 1987.

[21] Heuer, A. H., Claussen, N., Kriven, W. M., and Rühle, M., *Journal*, American Ceramic Society, Vol. 65, 1982, pp. 642–650.

[22] Porter, D. L. and Heuer, A. H., *Journal*, American Ceramic Society, Vol. 62, 1975, pp. 298–305.

[23] Muddle, B. C. and Hannink, R. H. J., *Journal*, American Ceramic Society, Vol. 69, 1986, pp. 547–555.

[24] Cox, B., *Journal of Nuclear Materials*, Vol. 148, 1987, pp. 332–343.

[25] Komarek, K. L. and Silver, M. in *Proceedings*, Symposium on Thermodynamics of Nuclear Materials, International Atomic Energy Agency, Vienna, 1962, pp. 749–774.

DISCUSSION

Brian Cox[1] *(written discussion)*—Have you oxidized these materials in water at around 350°C and examined the oxides? I ask this for two reasons. First, because alloys like your 0.5Sn-0.5Nb alloy may behave differently under such conditions; and second, because these alloys are usually exposed in water rather than steam, and we need to know that the evolution of the oxide microstructure is the same as in 400°C steam if we are to be able to interpret the large body of 400°C steam data that is being accumulated.

B. Wadman et al. (authors' closure)—No, we have not oxidized the materials other than in 400°C steam, as we were interested in comparing the new alloys with Zircaloy-4 materials. For these, we have found the used autoclave conditions to predict the corrosion performance in PWR environment. However, we are aware that this is not generally the case for niobium-containing alloys, at least not for alloys with higher than 1% niobium by weight. In a longer time perspective, we are very interested in examining oxidation in water at the temperatures you mentioned.

D. Khatamian[2] *(written discussion)*—(1) Have you studied the effects of alloy texture on the oxide grain morphology?

(2) What is the composition of the unidentified phase you spoke about?

B. Wadman et al. (authors' closure)—(1) We have not measured the texture in the base material, but we have studied the orientation of some metal grains close to the oxide layer. No correlation between metal grain orientation and oxide grain morphology was found.

(2) Only semi-quantitative EDX analysis was performed on the intermediate phase in Material 5, resulting in a sub-oxide with lower oxygen concentration than in ZrO_2, about 50 atomic percent.

B. Warr[3] *(written discussion)*—Recent TEM examinations of thin steam-formed oxides on Zr-2.5Nb alloys at Ontario Hydro Research Division Dr. Y.-P. Lin have shown similar edge-on microstructures as your observations on Zircaloy-4; that is, columnar grains extending to the interface with no amorphous interlayer and few pores. Based on your observations, do you feel it is likely that hydrogen uptake and acceleration in corrosion kinetics occur largely by cracks forming in the oxide extending directly to the metal-oxide interface where access of solution results in new oxide formation and rupture of the adjacent film?

B. Wadman et al. (authors' closure)—Based on our observations on Zircaloy-4 and similar niobium-containing alloys, we suggest that a nonporous oxide layer exists close to the metal-oxide interface in slowly growing oxide layers. In this zone, the oxygen is probably transported by grain boundary diffusion. In thicker oxide layers, cracks can form in oxide grain boundaries close to the interface if equiaxed grains are present. The distance between the nearest crack and the metal-oxide interface may be as small as 150 nm. We believe that the major direction of the cracks is parallel to the interface and that these cracks are interconnected by pores or perpendicular cracks. However, our results indicate that the transport of oxygen from the nearest parallel crack to the metal is again grain boundary diffusion.

R. A. Ploc[4] *(written discussion)*—Over the last few days, we have been told that changing the element content of silicon and tin may reduce the corrosion rate but the hydrogen pickup data is directly predictable from the corrosion kinetics. One author showed that changing the chromium content does not change the hydrogen pickup rate but reduced the corrosion rate. These contradictory data indicate that what we need to be shown is the hydrogen pickup data

---

[1] University of Toronto, Toronto, Ontario, Canada.
[2] AECL Research, Chalk River Laboratories, Chalk River, Ontario, Canada.
[3] Ontario Hydro, Toronto, Ontario, Canada M8Z 5S4.
[4] AECL Research, Chalk River, Ontario, Canada.

rather than concentrating on the corrosion rate. This is important from a mechanistic point of view since you have shown that some of your alloy changes affect an interface layer at the oxide-metal interface.

B. Wadman et al. (authors' closure)—We fully agree that the hydrogen uptake is important in order to characterize the oxidation mechanics of these materials, which is also why we made these measurements.

Peter Rudling,[1] Mirka Mikes-Lindbäck,[1] Börje Lethinen,[2] Hans-Olof Andrén,[3] and Krystyna Stiller[3]

Corrosion Performance of New Zircaloy-2-Based Alloys

REFERENCE: Rudling P., Mikes-Lindbäck, M., Lethinen, B., Andrén, H.-O., and Stiller, K., "**Corrosion Performance of New Zircaloy-2-Based Alloys,**" *Zirconium in the Nuclear Industry: Tenth International Symposium, ASTM STP 1245,* A. M. Garde and E. R. Bradley, Eds., American Society for Testing and Materials, Philadelphia, 1994, pp. 599–614.

ABSTRACT: A material development project was initiated to develop a new zirconium alloy, outside the ASTM specifications for Zircaloy-2 and Zircaloy-4, with optimized hydriding and corrosion properties for both boiling water reactors and pressurized water reactors. A number of different alloys were manufactured. These alloys were long-term corrosion tested in autoclaves at 400°C in steam. Also, a 520°C/24 h steam test was carried out. The zirconium metal microstructure and the chemistry of precipitates were characterized by analytical electron microscopy. The metal matrix chemistry was determined by atom probe analysis. The paper describes the correlations between corrosion material performance and zirconium alloy microstructure.

KEY WORDS: zirconium alloys, zirconium, nuclear materials, nuclear applications, radiation effects, zirconium alloy corrosion, zirconium alloy matrix, second-phase particles

Today, the lifetime of Zircaloy products such as fuel cladding tubes in boiling water reactors (BWRs) and pressurized water reactors (PWRs), and fuel outer channels in BWRs is limited by Zircaloy corrosion and hydriding performance in some reactors. The current trend to increase burn-up and to uprate the reactor power will result in even heavier Zircaloy corrosion and hydriding. Thus, there exists an incentive to improve corrosion and hydriding performance of the standard Zircaloy materials of today, that is, Zircaloy-2 (Zr-2) and Zircaloy-4 (Zr-4).

From a literature search, it appeared that increasing nickel, iron and decreasing tin would improve corrosion performance and increasing silicon content and adding niobium would improve hydriding performance. With these ideas, Teledyne Wah Chang, Albany (TWCA), manufactured a number of materials that subsequently were rolled into sheet by AB Sandvik Steel (ABSS). These materials were then tested for corrosion performance in 400 and 520°C steam. Materials with different 400 and 520°C corrosion performance were subsequently subjected to characterization of the microstructures of the different materials to assess what impact microstructural parameters such as second-phase particle (SPP), size distribution, SPP chemical composition, matrix composition etc. had on corrosion performance.

The rationale for characterizing the metal microstructure is to obtain a mechanistic understanding of the Zircaloy corrosion properties of a new Zr-2-based alloy.

---

[1] Specialist and manager, respectively, ABB Atom, S-721 63 Västerås, Sweden.
[2] Research metallurgist, Institute of Metals Research, Drottning Kristinas Väg 48, S-114 28 Stockholm, Sweden.
[3] Associate professors, Department of Physics, Chalmers University of Technology, S-412 96 Göteborg, Sweden.

Materials and Experimental Procedure

Material

Strips of seven new Zr-2-based alloys and a standard Zr-2 material were manufactured by TWCA (ingot manufacturing) and ABSS (strip rolling).

TWCA melted small ingots, approximately 2 kg in weight, by the induction-slag melting process. The ingots were pre-heated to about 1020°C for 20 min prior to forging. The materials were forged to a dimension of about 25 by 110 by 140 mm. The ingot chemical composition is tabulated in Table 1. The alloy designations relate to the element content that has been changed compared to the content in standard Zr-2. During the manufacturing of the ingots, it became clear that the aluminum content, that is regarded as an impurity, was too high. Therefore, the two alloys that we believed would be the most corrosion and hydriding resistant were re-melted, that is, NiFeNb and NiFeSiSn alloys. The re-melts were designated as NiFeNb' and NeFeSiSn' alloys.

At the dimension of 25 by 110 by 140 mm, the material was hot-rolled, beta-quenched, and subsequently three cold rolling steps with intermediate anneals followed. All materials were manufactured in the same way and the accumulated annealing parameter, log A, was calculated to be -15.9 by using Eq 1 [1]

$$A = \Sigma\, t_i \exp(-Q/RT) \tag{1}$$

where

i = ith heat treatment,
t_i = annealing time (h),
T_i = temperature (K),
Q = activation energy (cal/mole) = 63 000 cal/mole, and
R = the molar gas constant (cal/mole K).

All strips were delivered in re-crystallized annealed condition.

TABLE 1—*Ingot chemical composition of different strip materials.*

Alloy	Sn, % by weight	Ni, % by weight	Fe, % by weight	Cr, % by weight	Si, ppm	O, ppm	Nb, ppm	Al, ppm
Ref. Zr-2[a]	1.39	0.04	0.09	0.27	59	2100	...	88
NiFeNb	1.55	0.21	0.26	0.11	120	2000	1280	140
NiFeNb'[a,b]	1.48	0.15	0.20	0.10	105	1270	930	47
NiFeSi[a]	1.60	0.24	0.25	0.12	560	1750	...	140
NiFeSiSn[a]	1.03	0.15	0.21	0.07	510	1820	...	340
NiFeSiSn'[a]	0.97	0.12	0.18	0.09	225	1120	...	29
Nb	1.51	0.05	0.11	0.10	53	1420	1440	75
Si[a]	1.55	0.06	0.13	0.11	455	1460	...	105

[a]Materials examined in SEM and ATEM.
[b]Material examined in APFIM.

Autoclave Tests

A long-term corrosion test at 400°C/10.3 MPa was performed. Eight final strip samples of each material type were included, four in degreased condition and four in pickled condition.

A short-term corrosion test at 520°C/10.3 MPa for 24 h was also done. Four final strip samples of each material type were included, two in degreased and two in pickled condition.

Second-Phase Particle Characterization

To determine the frequency and size distribution of the SPPs, a scanning electron microscope, Jeol Model JSM6400, was utilized. The instrument is equipped with a high-performance backscattered electron detector (BSE) Link Tetra model. Micrographs obtained at a magnification of seven thousand times were processed and classified in histograms by an image analyzer. Analytical electron microscopy, Vacuum Generators BH501 instrument equipped with a field emission electron gun, was performed to assess the chemical composition of the SPPs. The system includes an energy dispersive spectrometer (EDS) and a microcomputer (Link QX 2000) for standardless quantification of X-ray spectra.

Sheet in the final stage of manufacturing was used for specimen preparation. Two different kinds of samples, one from each material type, were prepared as follows:

1. The SPP size distribution was evaluated using scanning electron microscopy of cross sections of the sheet. After careful mechanical grinding with 0.25-μm diamond paste, and a very short, 5-s, final electrochemical polishing (see 2) the specimen surface condition was good enough to differentiate between the SPPs and the matrix. This differentiation was done by pure Z-contrast when the microscope was working in the BSE mode with a low accelerating voltage.
2. Thin foils for investigation in the analytical transmission microscope were prepared by using the jet electropolishing technique. Disk specimens, 3 mm in diameter, were mechanically polished to 0.05 mm and finally thinned by electropolishing at a voltage of 20 V and in a mixture of 80% ethanol and 20% perchloric acid. To minimize preferential etching of either the SPPs or the surrounding matrix, a very cold electrolyte (-30°C) was used.

More than 40 different particles per material type were analyzed to assess their chemical compositions.

Matrix Analyses

Analysis of the matrix with atom probe field ion microscopy (APFIM) was performed on specimens from one material type. The equipment has been described elsewhere [2]. To prepare the needle-shape atom probe specimens, a two-step procedure was used:

1. electropolishing to produce a neck. This was done in a 5% by volume perchloric acid-35% isobutanol-60% methanol at a temperature of -20°C and at a voltage of 20 V; and
2. polishing off the neck to produce a sharp needle in a solution of 11-g LiCl and 22-g $Mg(ClO_4)_2$ in 900 mL butyl cellusolve and 500 mL methanol at a temperature of -20°C and at a voltage of 12 V.

Prior to analysis, all specimens were examined in a Philips EM 300 transmission electron microscope to study the tip radius and surface smoothness. The microstructure close to the specimen tip was also examined to determine whether the analyzed volumes were too close to grain boundaries or SPP.

The analysis was done at a specimen temperature of 80 K, and the pulse amplitude was kept at 15% of the standing voltage.

Results

400°C Corrosion Results

The corrosion results are presented in Table 2 and Figs. 1 and 2.

In Fig. 1, the oxide weight gain of pickled samples after 3, 14, 30, 60, and 101 days of exposure are the arithmetic mean values of 4, 3, 3, 2, 2, and 1 samples, respectively. It is

TABLE 2—*400°C corrosion data of new Zircaloy-2 alloys.*

Alloy	Surface Treatment	Weight Gain, mg/dm²				
		3 days	14 days	30 days	60 days	101 days
NiFeNb'	pickled	20	28	35	49	60
NiFeSiSn	pickled	21	31	50	91	168
NiFeSi	pickled	21	30	37	60	101
NiFeNb	pickled	20	30	37	57	82
Nb	pickled	22	32	41	61	84
Ref. Zr-2	pickled	23	45	41	82	
NiFeSiSn'	pickled	20	29	36	44	62
Si	pickled	21	35	47	63	136
NiFeNb'	degreased	16	25	31	51	71
NiFeSiSn	degreased	17	26	51	91	175
NiFeSi	degreased	15	23	31	60	97
NiFeNb	degreased	16	25	30	59	93
Nb	degreased	17	25	30	68	137
Ref. Zr-2	degreased	16	27	39	125	233
NiFeSiSn'	degreased	14	22	28	36	63
Si	degreased	16	26	36	109	166

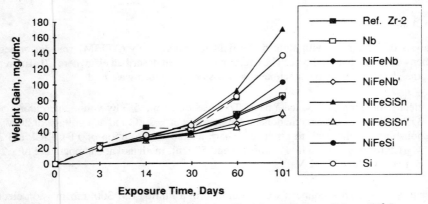

FIG. 1—*Corrosion performance of different alloys in pickled condition, 400°C steam.*

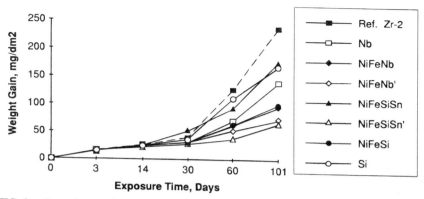

FIG. 2—*Corrosion performance of different alloys in degreased condition, 400°C steam.*

obvious that the NiFeNb′ and NiFeSiSn′ alloys have the lowest corrosion rate. The Ref Zr-2 sample was lost after the weighing after 60 days exposure and was consequently not inserted into the autoclave for the last 41 days of exposure.

In Fig. 2, the oxide weight gain values for degreased samples after 3, 14, 30, 60, and 101 days of exposure are the arithmetic mean values of 4, 3, 3, 2, and 2 samples. Also here, the NiFeNb′ and NiFeSiSn′ have the lowest corrosion rates. It is interesting to note that these alloys have four to five times lower oxide weight gains compared to that of the Ref Zr-2 alloy.

520°C Corrosion Results

The corrosion data at 520°C are shown in Table 3.

Also at 520°C, the NiFeNb′ and NiFeSiSn′ alloys in both pickled and degreased condition show significantly better corrosion performance than the Ref Zr-2 alloy.

TABLE 3—*520°C corrosion data of new Zr-2 base alloys.*

Alloy	Surface Treatment	Average Oxide Weight Gain (mg/dm²)	Standard Deviation (mg/dm²)
NiFeNb′	pickled	46	0
NiFeSiSn′	pickled	48	0
NiFeSi	pickled	54	4
NiFeNb	pickled	60	2
Nb	pickled	74	0
Ref Zr-2	pickled	89	4
NiFeSiSn	pickled	90	1
Si	pickled	129	115
NiFeSiSn′	degreased	49	4
NiFeNb′	degreased	51	1
NiFeNb	degreased	52	6
NiFeSi	degreased	53	1
Ref Zr-2	degreased	55	7
NiFeSiSn	degreased	57	33
Nb	degreased	60	11
Si	degreased	320	11

Second-Phase Particle Size Distribution

Typical SPP morphologies in the different material are shown in Fig. 3. The second-phase particle distributions for the different materials are shown in Fig. 4, where it seems that the NiFeNb' alloy has smaller SPPs than the rest of the materials.

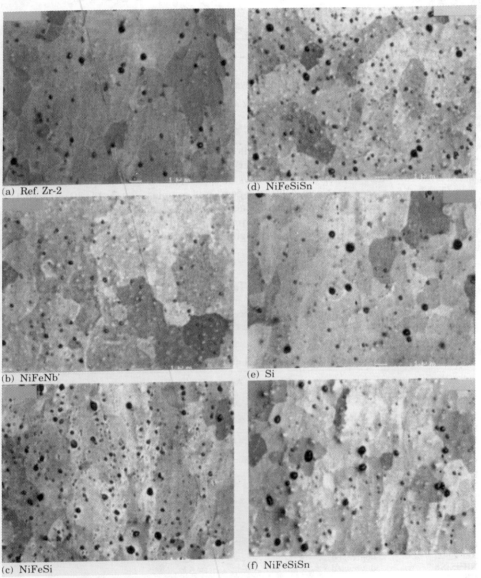

FIG. 3—*Typical scanning electron micrographs of different alloys, obtained with back-scattered electrons.*

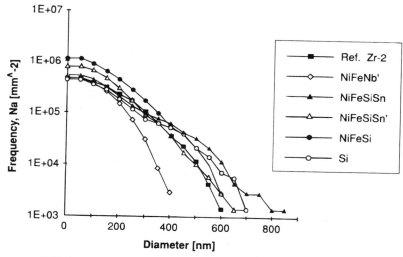

FIG. 4—*Second-phase particle size distribution in different alloys.*

Second-Phase Particle Chemical Composition

The second-phase particles found in the different materials were of three kinds: (1) Zr_2(Ni,Fe) tetragonal type, (2) $Zr(Cr,Fe)_2$ hexagonal type, and (3) Zr-Si. Figure 5 shows typical examples of the different kinds of SPPs.

Figure 6 presents the Cr/Fe and Ni/Fe ratio in the SPPs as a function of size of all analyzed particles. The sizes of the detected silicides (Zr-Si) are also plotted in this figure. It is clear that the chromium-bearing particles are the smallest and that the silicides are the largest. The nickel-bearing particles have an intermediate size. Since the plots are based on an analytical transmission electron microscopy ATEM of thin foils, the small SPPs are over-represented in these figures. It is strange that not even one nickel-bearing particle was found in the Ref. Zr-2 material with a nickel content of 0.04%. Between 1 to 2.3 atom percent of niobium was detected in seven SPPs in the NiFeNb' alloy. These particles were all smaller than 80 nm and were of the $Zr(Cr,Fe)_2$ type. The silicon content in the NiFeNb' alloy was 105 ppm (Table 1) and about 1.5 atomic% of silicon was detected in four SPPs. These SPPs were all of the Zr_2(Ni,Fe) type. No zirconium-silicon precipitates were observed in the NiFeNb' material.

The silicon element was found in four and one nickel-bearing SPP in the NiFeSiSn and Si alloys, respectively. The Si alloy also had two chromium-bearing SPPs containing silicon. The silicon content in both nickel- and chromium-bearing SPPs ranged from 1 to 2 atom percent.

Matrix Analysis

Results from atom probe analysis of the matrix of the NiFeNb' material is tabulated in Table 4.

There is no difficulty to detect niobium in the form of Nb^{3+}, however, it is impossible to quantify Nb^{2+}, due to overlap with ZrH^{2+}. Therefore, the amount of detected Nb^{3+} ions must be corrected by the measured ratio $(Zr^{2+} + Zr^{3+})/Zr^{3+}$ (100/44.35), which should be similar to the corresponding ratio for niobium. No chromium has been detected, indicating a chromium content of less than 100 ppm. The results for iron and nickel are for the major isotopes Fe-56 (91.66% of total iron) and Ni-58 (67.88% of total nickel) and have been isotope corrected.

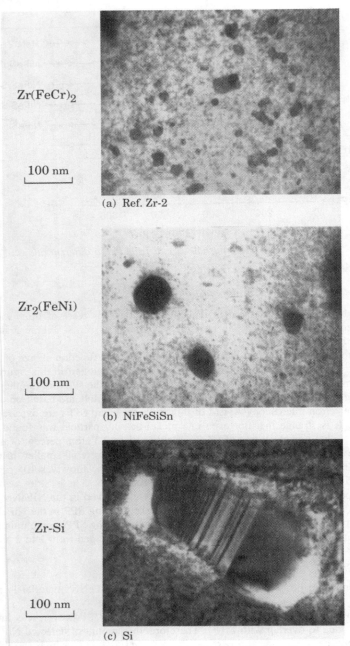

FIG. 5—*Three types of second-phase particles found in the different alloys, ATEM micrographs.*

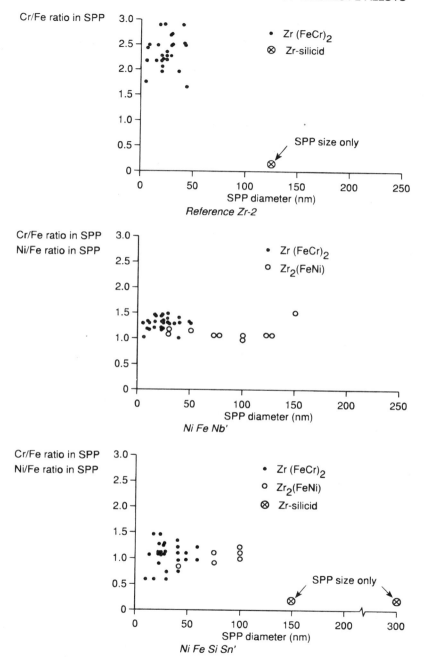

FIG. 6—*Cr/Fe and Ni/Fe ratios in SPP. Only the size of the silicides are shown, ATEM analyses.*

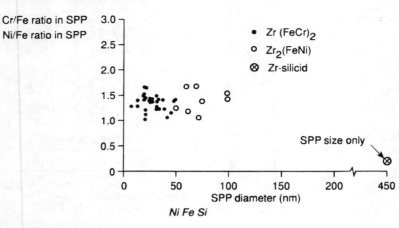

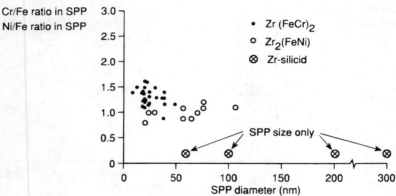

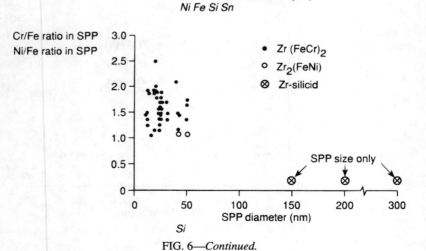

FIG. 6—*Continued.*

TABLE 4—*Atom probe analysis of Zircaloy matrix of NiFeNb' Material.*

Element	Ions Detected	Ions Corrected	Atom Percent	% by weight	Ingot Analysis, % by weight
C	4	4	0.0267 ± 0.0134	0.0035 ± 0.0018	0.014
O	118	118	0.7882 ± 0.0723	0.1386 ± 0.0018	0.127
Si+N	4	4	0.0267 ± 0.0134	...	(0.064 atom percent)
Cr	0	0	...	<0.01	0.095
Fe	5	5.45	0.0364 ± 0.0155	0.0223 ± 0.0100	0.199
Ni	2	2.95	0.0197 ± 0.0115	0.0126 ± 0.0089	0.147
Zr	14 677	14 677	98.0269 ± 0.2126	98.4353 ± 0.2135	(97.83)
Nb	4	9.02	0.0602 ± 0.0200	0.0616 ± 0.0204	0.093
Sn	152	152	1.0152 ± 0.0823	1.3261 ± 0.1071	1.48
Total	14 966	14 972.42	100.0000	100.000	

It is seen that the analyses of tin an oxygen correlates well with the ingot analyses. This is expected since these elements are found in solid solution. The concentrations of nickel, iron, and chromium in the matrix are low, 130, 220, and <100 ppm, respectively. This correlates well with earlier results that indicate that the solubility for these elements in the Zircaloy matrix is low, 100 to 200 ppm [*3*]. Almost all of the iron, chromium, and nickel are precipitated with zirconium in second-phase particles. The niobium addition of 930 ppm seems mostly to be in the matrix, about 600 ppm.

Discussion

Two plots are shown based upon the results from the 400°C steam test with degreased samples. However, the same trends as for these samples could also be seen for: (1) samples tested in pickled condition in 400°C steam and, (2) pickled and degreased samples in the 520°C steam test.

In Fig. 7, the oxide weight gain of degreased samples after 101 days of exposure in 400°C is correlated to the Fe/Cr ratio in the SPPs and the aluminum content in the ingot. It appears that the oxide weight gain increases with decreasing Fe/Cr ratio in the SPPs. Charquet et al. came to the same conclusion for Zr-4 type alloys with Fe/Cr ratios higher than 2.0 [*4*]. In Fig. 7, it is also indicated that increasing aluminum content increases the corrosion rate.

According to Kass [*5*], increasing the silicon content from 15 to 90 to 600 to 700 ppm in Zr-2 increases the post-transition corrosion rate in 400°C steam. This may be the reason that the silicon alloy with 455 ppm of silicon has such poor corrosion resistance, see Table 1. The oxide weight gains after 101 days of exposure in the 400°C steam test were 170 and 60 mg/dm^2 for the NiFeSiSn and NiFeSiSn' alloys, respectively, see Fig. 2. The former alloy has silicon and aluminum contents of 510 and 340 ppm, respectively. The corresponding contents in the latter alloy were 225 ppm of silicon and 29 ppm of aluminum, see Table 1. There exists also a significant difference in oxygen content between the NiFeSiSn and NiFeSiSn' alloys, that is, 1820 and 1120 ppm, respectively. However, according to Kass, increasing the oxygen content in Zr-2 from 800 to 1700 ppm did not affect the post-transition corrosion rate in 400°C steam [*5*]. Thus, very high levels of aluminum and silicon seem to accelerate corrosion rate.

In Fig. 8, the oxide weight gain for degreased samples after 101 days of exposure in the 400°C steam test is correlated to the ingot nickel and oxygen contents. Since the atom probe analysis yielded a matrix solubility of nickel in Zr-2-based alloys of about 130 ppm (see Table

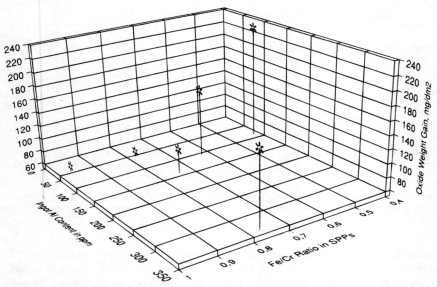

FIG. 7—*Oxide weight gain, degreased samples after 101 days of exposure in 400°C steam, correlated to average Fe/Cr ratio in SPP and ingot aluminum content in different materials.*

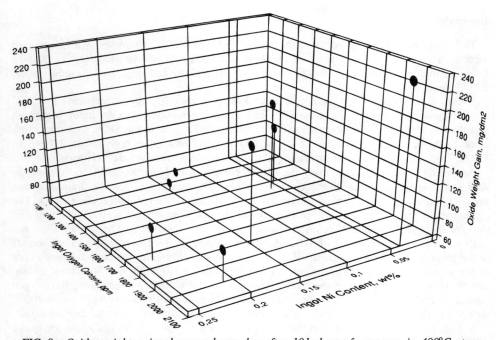

FIG. 8—*Oxide weight gain, degreased samples after 101 days of exposure in 400°C steam, correlated to ingot nickel and ingot oxygen content in different alloys.*

3) and the nickel content in the investigated alloys is ranging from 400 to 2400 ppm (Table 1) most of the nickel will be precipitated in the SPPs. Thus, increasing the nickel content in the ingot should also increase the volume fraction of nickel-bearing SPPs. It is indicated from Fig. 8, that increasing nickel content also decreases the oxide weight gain. Harada et al. found that adding nickel to Zr-4 dramatically improved the corrosion resistance. However, he argued that adding nickel over 0.04% by weight did not further improve corrosion resistance [6]. In Fig. 8, it seems that a significant reduction in corrosion rate by adding nickel beyond 0.05% by weight is obtained. Figure 8 also indicates that increasing the ingot oxygen content may increase the corrosion rate.

From the ATEM analysis, it appears that some of the silicon has a preference to concentrate in the $Zr_2(Ni,Fe)$ particles, about 1 to 2 atom percent. This situation that silicon is found in the SPPs may be due to the fact that silicon facilitates SPP nucleation. From the ATEM and atom probe analysis, it appears that one third of the niobium additions is concentrated in the chromium-bearing SPPs, that is, 1 to 2.3% by weight, while the rest, two-thirds, is located in the matrix. It also appears that adding niobium decreases the SPP sizes, see Fig. 4. It may be that niobium facilitates SPP nucleation, thus resulting in more and smaller SPPs.

Conclusions

Based upon the results on the manufactured test Zr-2-based alloys, the following conclusions can be drawn:

1. In Zr-2 base alloy with about 0.1% by weight niobium, about two thirds of the niobium seem to be in the matrix while one third is concentrated in the chromium-bearing SPPs. The niobium content in the $Zr(Cr,Fe)_2$ type SPP is about 1 to 2 atom percent. Additions of niobium appears to reduce the SPP size perhaps by facilitating SPP nucleation.
2. In Zr-2-based alloys with silicon additions, about 1 to 2 atomic percent of silicon is found in some of the SPPs. These SPPs are mostly of the $Zr_2(Ni,Fe)$ type. The element silicon seems to facilitate SPP nucleation.
3. The best of the new Zr-2-based alloys showed significantly improved corrosion performance compared to that of Zr-2 type material with high chromium content in the 400 and 520°C tests.
4. Increasing aluminum and oxygen contents seem to increase corrosion rate in the 400 and 520°C tests.
5. Increasing silicon content from 60 to 400 to 500 ppm increases 400 and 520°C corrosion rate.
6. Increasing the Fe/Cr ratio in the SPPs and increasing the volume fraction of nickel-bearing SPPs is correlated with higher 400 and 520°C corrosion resistance.

Acknowledgments

The authors thank Craig Eucken of Teledyne Wah Chang Albany and Thomas Andersson of Sandvik Steel for manufacturing the test alloys free of charge. The authors also gratefully acknowledge their comments and input to the paper.

References

[1] Andersson, T., Thorvaldsson, T., Wilson, A., and Wardle, A. M. in *Proceedings*, Symposium on the Influence of Thermal Processing and Microstructure on the Corrosion Behavior of Zircaloy-4 Tubing, International Atomic Energy Agency, (IAEA), Vienna, 1987, pp. 435–449.

[2] Wadman, B. and Andrén, H.-O, "Direct Measurements of Matrix Composition in Zircaloy-4 by Atom Probe Micro Analysis," *Zirconium in the Nuclear Industry: Eighth Symposium, ASTM STP 1023*, L. F. P. Van Swam and C. M. Euken, Eds., American Society for Testing and Materials, Philadelphia, 1989, pp. 423–434.

[3] Charquet, D., Hahn, R., Ortlieb, E., Gros, J. P., and Wadier, J. F., "Solubility Limits and Formation of Intermetallic Precipitates in ZrSnFeCr Alloys," *Zirconium in the Nuclear Industry: Eighth Symposium, ASTM STP 1023*, L. F. P. Van Swam and C. M. Eucken, Eds., American Society for Testing and Materials, Philadelphia, 1989, pp. 405–422.

[4] Charquet, D., Tricot, R., and Wadier, J.-F, "Heterogeneous Scale Growth During Steam Corrosion of Zircaloy-4 and 500°C," *Zirconium in the Nuclear Industry, Eighth Symposium, ASTM STP 1023*, L. F. P. Van Swam and C. M. Eucken, Eds., American Society of Testing and Materials, Philadelphia, 1989, pp. 374–391.

[5] Kass et al., "Effect of Silicon, Nitrogen and Oxygen on the Corrosion and Hydriding Absorption of Zircaloy-2," *Corrosion*, Vol. 20, 1964, 350t and U. S. Report, WAPDD-283, Nov. 1963.

[6] Harada, M., Kimpara, M., and Abe, K., "Effect of Alloying Elements on Uniform Corrosion Resistance of Zirconium-Based Alloys in 360°C Water and 400°C Steam," *Zirconium in the Nuclear Industry: Ninth International Symposium, ASTM STP 1132*, C. M. Eucken and A. M. Garde, Eds., American Society for Testing and Materials, Philadelphia, 1991, pp. 368–391.

DISCUSSION

R. B. Adamson[1] (*written discussion*)—The accumulated annealing parameter, A, appears to provide some guidance in determining the relationship between fabrication schedules and the corrosion performance. However, it is inherently sensitive to other parameters such as quenching rate, alloy composition, cold working schedule, etc. Yet, it sometimes appears that the number (A) is used as a "magic number" to predict corrosion resistance. Can you provide any insight from the experiments you conducted on the usefulness of the A parameter in predicting the corrosion behavior of your new alloys?

P. Rudling et al. (*authors' closure*)—The log A parameter is only a very rough number, indicating how the material was manufactured. The number by itself is not enough to predict the corrosion performance of the material. The exact manufacturing history is however proprietary information of the material manufacturer. The best data the material manufacturer can give, indicating the manufacturing history, is the log A parameter.

Nobuoki Yomashira[2] (*written discussion*)—Did you see any significance of carbon in the new alloy?

P. Rudling et al. (*authors' closure*)—We did not see any significant effect of carbon on corrosion performance.

Bo Cheng[3] (*written discussion*)—The fact that precipitates are subjected to dissolution or amorphization, or both, under irradiation in power reactors, as presented by many researchers, and that this might lead to accelerated uniform corrosion rates at high burnups, do you feel that increasing iron and nickel content is in the right direction for developing new alloys for high burning applications

P. Rudling et al. (*authors' closure*)—In these new alloys, we wanted predominantly to increase the nickel content to form more stable SPPs. Data in the open literature have shown that nickel-bearing SPPs are more stable than chromium-bearing SPPs. The latter has a larger tendency to dissolve during neutron irradiation.

R. A. Ploc[4] (*written discussion*)—Zr-2.5Nb and the Zircaloys show opposite trends in out-reactor compared to in-reactor corrosion rates. The corrosion rate for Zr-2.5Nb decreases in-reactor and the Zircaloys increase. The decrease for Zr-2.5Nb is often related to precipitation of excess Nb in the x-Zr Matrix, stimulated by the irradiation. Do you have any in-reactor data for your new alloys? Secondly, if you have no such data, have you examined these materials in a high voltage electron microscope that will cause excess niobium precipitation?

P. Rudling et al. (*authors' closure*)—We have not examined the materials in a high-voltage electron microscope. However, this summer we will get in-pile corrosion data after one reactor cycle.

Arthur Motta[5] (*written discussion*)—(1) In determining the niobium partition between the matrix and second-phase particles in your alloys, did you deduct the matrix content from the overall value or did you conduct a full material balance by evaluating the precipitation volume fraction.

(2) Is the niobium you fine in the $Zr(Cr,Fe)_2$ a solid solution of the form $(Zr,Nb)(Cr,Fe)_2$?

---

[1] General Electric Nuclear Energy, Pleasanton, CA 94566.
[2] General Electric, Wilmington, NC 28402.
[3] Electric Power Research Institute (EPRI), Palo Alto, CA.
[4] AECL-Research, Chalk River, Ontario, Canada.
[5] Pennsylvania State University, Department of Nuclear Engineering, University Park, PA 16803.

P. Rudling (authors' closure)—(1) The matrix niobium content was determined to be about 600 ppm by atom probe analysis, and the remaining 300 ppm of niobium that was added seemed to correspond to the TEM analysis of the second-phase particles (SPPs). Some of the chromium-bearing SPPs contained about 1 to 2 atomic percent niobium. This amount seems to correlate with the remaining 300 ppm of niobium.

(2) This was not determined, but most probably the niobium was in solid solution in the SPPs.

Hans-Jürgen Beie,[1] Alexander Mitwalsky,[2] Friedrich Garzarolli,[3]
Heinz Ruhmann,[3] and Hans-Jürgen Sell[3]

Examinations of the Corrosion Mechanism of Zirconium Alloys

REFERENCE: Beie, H.-J., Mitwalsky, A., Garzarolli, F., Ruhmann, H., and Sell, H.-J., "**Examinations of the Corrosion Mechanism of Zirconium Alloys,**" *Zirconium in the Nuclear Industry: Tenth International Symposium, ASTM STP 1245,* A. M. Garde and E. R. Bradley, Eds., American Society for Testing and Materials, Philadelphia, 1994, pp. 615–643.

ABSTRACT: Several mechanism-related aspects of the corrosion of zirconium alloys have been investigated using different examination techniques. The microstructure of different types of oxide layers was analyzed by transmission electron microscopy (TEM). Uniform oxide mainly consists of m-ZrO_2 and a smaller fraction of t-ZrO_2 with columnar grains and some amount of equiaxed crystallites. Nodular oxides show a high open porosity and the grain shape tends to the equiaxed type. A fine network of pores along grain boundaries was found in oxides grown in water containing lithium. An enrichment of lithium within such oxides could be found by glow discharge optical spectroscopy (GD-OES) depth profiling. In all oxides, a compact, void-free oxide layer was observed at the metal/oxide interface. Compressive stresses within the oxide layer measured by an X-ray diffraction technique were significantly higher compared to previously published values. Electrical potential measurements on oxide scales showed the influence of the intermetallic precipitates on the potential drop across the oxide. In long-time corrosion tests of Zircaloy with varying temperatures, memory effects caused by the cyclic formation of barrier layers could be observed. It was concluded that the corrosion mechanism of zirconium-based alloys is a barrier-layer controlled process. The protective properties of this barrier layer determine the overall corrosion resistance of zirconium alloys.

KEY WORDS: zirconium, zirconium alloys, corrosion, oxide layers, corrosion mechanism, barrier layer, nuclear materials, nuclear applications, radiation effects

Insight into the corrosion mechanism of zirconium-based alloys is still not well developed, despite numerous investigators who have put forward theories and models during the past 30 years. The complex interaction of parameters governing the corrosion of Zircaloy in aqueous environments makes it very difficult to predict its corrosion behavior. Even if the number of unknown parameters is reduced to a minimum and no radiation is present, as in out-of-pile experiments, the results may sometimes be surprising and unexpected. Studies that have been focused on the properties of the oxide scales formed during corrosion often lead to results that seem to contradict each other. Some of the most important oxide properties in this context are the microstructure, the electrical properties, the role of alloying elements, the compressive stress of the oxide film, and the influence of hydrogen and lithium.

In a previous publication [*1*], the results of an out-of-pile test program for characterization oxide layers of different types by various examination methods have been described. One of

---

[1] Siemens AG, Corporate Research and Development, D-91052 Erlangen, Germany.
[2] Siemens AG, Corporate Research and Development, D-81730 Munich, Germany.
[3] Siemens AG, Power Generation Group (KWU), D-91052 Erlangen, Germany.

the conclusions drawn from these investigations led to the formulation of a barrier layer concept to explain the uniform corrosion behavior of zirconium alloys. In this study, we present some further results of cross-sectional transmission electron microscopy (TEM) investigations on samples with uniform oxide, nodular oxide, and oxide grown in lithiated water that support this model. We also report on X-ray diffraction (XRD) measurements to determine internal compressive stresses in oxide films, on glow discharge optical spectroscopy (GD-OES) depth profiling of hydrogen and lithium within oxide scales, and on measurements of electrical potentials across oxide films in oxygen and water. Two types of zirconium alloys with different tin contents (1.2% and 0.5% tin) were used for these investigations. Both alloys contained 0.2% iron and 0.1% chromium. The results of these examinations together with findings of other workers are discussed, and some consequences for the understanding of the corrosion mechanism of Zircaloy will be shown.

TEM Investigations

An important aspect of the corrosion is the microstructure of the oxide layer that can be seen in detail only with near-atomic resolution. Therefore, different types of oxides were examined by cross-sectional TEM. Special emphasis was laid on the examination of the oxide layer in the vicinity of the metal/oxide interface. Pieces of cladding tubes were oxidized in water or steam to produce uniform and nodular oxides. One sample analyzed was corroded in water containing 70 ppm LiOH. The preparation of the cross-sectional TEM samples from these tubes was similar to the procedure described in Ref *1*.

Pre-Transition Uniform Oxide

Figure 1 shows cross sections of a uniform oxide in the pre-transition region (corroded in steam at 420°C/10.5 MPa/6 days). Columnar grains with a long-axis orientation perpendicular to the metal oxide interface are found in the entire oxide scale. In many cases, the outermost oxide contained equiaxed grains that were formed in the early stage of corrosion. Electron diffraction patterns showed textured monoclinic ZrO_2 (m-ZrO_2) and some small amount of tetragonal ZrO_2 (t-ZrO_2). The monoclinic and tetragonal phases had an orientational relationship ($\{-111\}$ and $\{111\}$ m-ZrO_2 parallel to $\{111\}$ t-ZrO_2) as would be expected, if t-ZrO_2 transforms to m-ZrO_2 via a diffusionless process. Very fine horizontal pores can be seen between the end faces of the columnar grains. At several locations, some larger horizontal pores or cracks can be observed. The density and size of pores in such pre-transition oxides varies from sample to sample and also on the same sample from one location to another. But the pores, especially those near the metal/oxide interface, are always closed and had no connection to the oxide surface.

Larger cavities or cracks have been found in all samples analyzed by TEM, the largest ones are visible in cross-sectional SEM images too. One may argue that these cracks are artifacts produced during sample preparation and do not represent a feature of the original oxide. This may be true, but in either case these cavities indicate locations of high stress that has been relieved by buckling. Whether this stress relief has occurred during the corrosion process, during cooling of the sample, or during the sample preparation is not known.

Post-Transition Uniform Oxide

The columnar grain structure with m-ZrO_2 as the predominant phase and a minor fraction of t-ZrO_2 was also found in post-transition oxides (see the example in Fig. 2, corrosion in steam at 420°C/10.5 MPa/318 days). But in the outermost layer and at mid-thickness, there exist many large pores and cracks parallel to the oxide surface with a varying density at

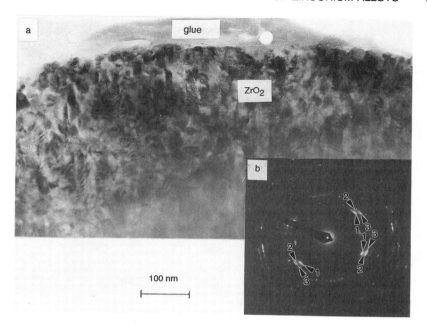

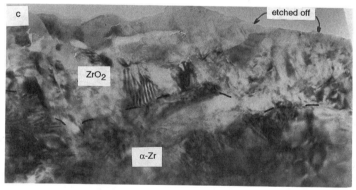

FIG. 1—*Cross-sectional TEM micrographs of an uniform pre-transition oxide grown on Zircaloy in steam at 420°C/10.5 MPa/6 days: (a) section near to the oxide surface, (b) electron diffraction pattern from the surface section in (a), and (c) section at the metal/oxide interface.*

different locations. Additionally, many equiaxed pores with sizes between 10 and 20 nm are visible. Adjacent to the metal/oxide interface, the number of pores is much smaller and they tend to be isolated within a dense matrix. Figure 3 shows a high-resolution TEM image (HRTEM) of one of ten locations examined at the metal/oxide interface in the same sample. A comparison of the patterns in the monoclinic ZrO_2 and the α-Zr grains shows that the (002) planes of m-ZrO_2 are parallel to the (002) planes of α-Zr at this location, indicating an epitaxial growth of m-ZrO_2 on the zirconium grain.

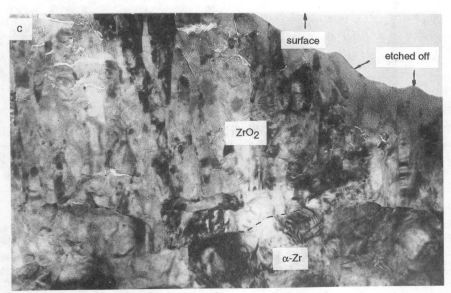

FIG. 2—*Cross-sectional TEM micrographs of a uniform post-transition oxide grown on Zircaloy in steam at 420°C/10.5 MPa/318 days: (a) section near to the oxide surface, (b) electron diffraction pattern from the surface section in (a), and (c) section at the metal/oxide interface.*

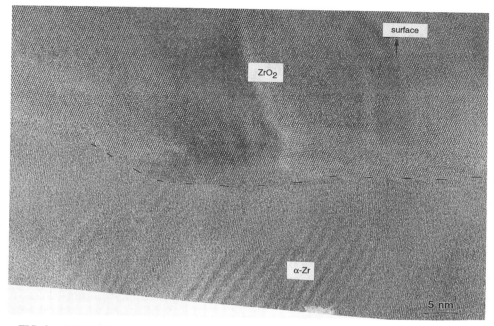

FIG. 3—*HRTEM image of the same oxide as in Fig. 2 at the interface between* m-ZrO_2 *and* α-Zr *grains showing fringes of* m-ZrO_2 *planes (002) parallel to* α-Zr *planes (002).*

Nodular Oxide

The TEM micrographs in Fig. 4 show a nodular oxide (grown in steam at 520°C/12.5 MPa/16 h) at mid-thickness and near the metal/oxide interface. Typically, we found equiaxed grains with a size of 20 to 50 nm, globular pores with diameters of 5 to 20 nm, and inhomogeneously distributed large horizontal cavities. Generally, the oxide has a very loose structure. Only adjacent to the metal/oxide interface a thin denser columnar structure was found. Electron diffraction patterns revealed pure monoclinic oxide.

Oxide Grown in Lithium-Containing Water

The oxide in Fig. 5 was formed in water containing 70 ppm LiOH at 350°C and 16.5 MPa for 294 days. It has a similar microstructure as found in other post-transition oxides. The fraction of equiaxed grains seemed to be larger compared to oxides grown without LiOH additions. A significant difference however is that over nearly all the entire oxide scale the oxide grains were surrounded by zones of low medium atomic number (width about 2 nm) that might be attributed to pores (open grain boundaries). This fact indicates that a lithium-containing environment attacks the oxide grain boundaries and supports the hypothesis proposed by Cox [2] that LiOH causes a preferential dissolution of ZrO_2 at the grain boundaries. The dense oxide layer without a porous network at the metal/oxide interface is thinner in oxide scales formed in an environment containing LiOH compared to that formed without LiOH. This explains why the corrosion rate is increased under the influence of LiOH.

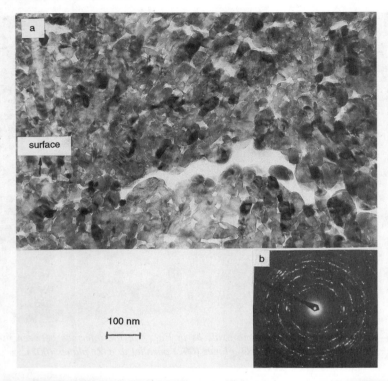

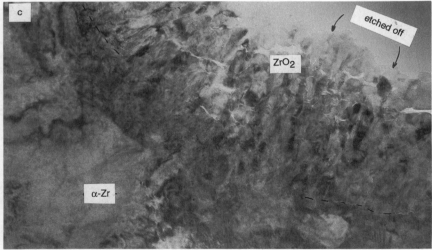

FIG. 4—*Cross-sectional TEM micrographs of a nodular oxide grown on Zircaloy in steam at 520°C/12.5 MPa/16 h:* (a) *section at mid-thickness,* (b) *electron diffraction pattern from the section at mid-thickness in* (a), *and* (c) *section at the metal/oxide interface.*

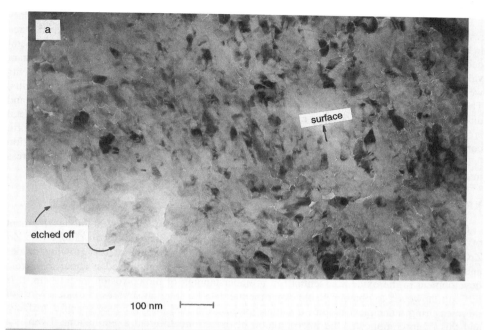

FIG. 5—*Cross-sectional TEM micrographs of an oxide grown on Zircaloy in water containing 70 ppm LiOH at 350°C/16.5 MPa/294 days:* (a) *section at mid-thickness and* (b) *section at the metal/oxide interface.*

Other Results from TEM Investigations

Samples with a tin content of 1.2% showed no significant differences compared to the results from the 0.5% tin samples described earlier. A variation of the tin content near the metal/oxide boundary and within the oxide layer could not be detected by X-ray microanalysis (XMA) in TEM (spatial resolution of XMA is about 40 nm).

XMA analysis of the oxygen/zirconium (O/Zr) atom ratio at the metal/oxide phase boundary revealed a sudden decrease of the oxygen content within the metal directly beyond the interface in a sample with nodular oxide. On a sample with uniform oxide, the O/Zr ratio in the metal decreased continuously within a distance of 1 μm from the phase boundary towards a constant value that was due to the surface oxidation of the TEM sample (see Fig. 6). These measurements were reproduced several times at different locations on both samples with the same results. The different oxygen profiles can be explained considering the different rates of nodular and uniform corrosion. In the nodular mode, the movement of the phase boundary into the metal is faster than the oxygen diffusion in zirconium. The behavior is reversed in the case of uniform corrosion. However, the thickness of the oxygen diffusion zone found after uniform corrosion is larger than calculated from published diffusion data, probably due to the large tensile stress in the metal underneath the oxide.

XRD Examinations

One of the important aspects for the corrosion mechanism of Zircaloy is the development of compressive stresses within the oxide, due to the large Pilling-Bedworth ratio of the Zr/ZrO_2 system. These stresses strongly influence the microstructure of the growing oxides and their resistance against mechanical failure. Therefore, the knowledge about the stress development in the oxide is essential for an understanding of the corrosion mechanism. It is difficult, however, to obtain reliable experimental data for internal stresses in oxides. Most of the measuring techniques require a sample preparation that may produce specific artifacts, and this often complicates the interpretation of data from different examinations. Even if a relative comparison of stress data would be sufficient (for example, for studying the internal oxide stress as a function of time during an autoclave test loop), so that the absolute precision

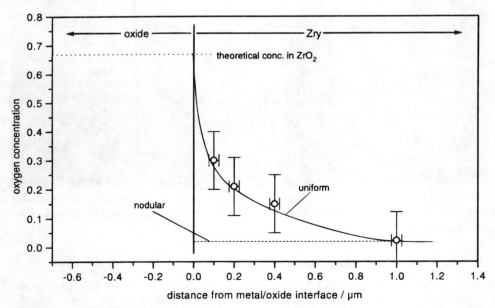

FIG. 6—*Oxygen profiles in the metal near the metal/oxide interface of a sample with uniform oxide (same as in Fig. 2) and of a sample with nodular oxide (same as in Fig. 4).*

of a certain method is of minor interest, the time-consuming preparation of samples requires a large effort. Another point in this context is the explanation for the presence of metastable tetragonal or cubic ZrO_2 in oxide scales grown on Zircaloy. According to pressure-temperature phase diagrams of ZrO_2, tetragonal or other phases of higher density are stable at pressures above 3 GPa [3]. However, most of the published experimental data are far below this value, so that the contribution of compressive stresses to the stabilization of t-ZrO_2 remains in question.

If planar sheet samples are to be examined, XRD offers the advantage that it does not require any sample preparation and it is a nondestructive method. We used the so-called $\sin^2\Psi$ technique for the determination of the internal oxide stress of some samples to compare these values with data obtained by a bending test method [1]. The method is based on the evaluation of peak shifts in the diffraction diagram that can be observed when a sample having internal stresses parallel to the surface is tilted for some amount with respect to the incident X-ray beam. The present study used a Siemens D500 diffractometer equipped with sample holder on a 2Θ mount that permits to turn the sample against the incident X-ray beam by angels of $-90° \leq \Psi \leq 55°$. The basic equation used for evaluation of the peak shift, $\Delta\Theta$, from the peak angle, Θ_0, of the stress-free sample as a function of the tilting angle, Ψ, is

$$\epsilon(\Psi) = -\cot\Theta_0 \Delta\Theta = \epsilon_0 \, \sigma \, \frac{1+\nu}{E} \sin^2\Psi \qquad (1)$$

where ϵ, ϵ_0, σ, ν, and E denote the lattice strain, the strain at $\Psi = 0$, the stress parallel to the sample surface, the Poisson's ratio, and the Young's modulus of the oxide, respectively. More about this method and further details about its application to stress measurements on zirconium oxide films may be found in Refs 4 through 6.

From Eq 1, it can be seen that the observed lattice strain, ϵ, increases with the diffraction angle, Θ_0. Thus, the error at small angles, Θ_0, where the highest peak intensities of m-ZrO_2 are found, is too high to give reliable values for ϵ. At larger Θ_0 angles, however, the X-ray diffraction patterns of ZrO_2 and the underlying Zircaloy coincide at most peak locations. In this region solely, the $\{-223\}$ peak of m-ZrO_2 (at $2\Theta_0 = 119.6°$ using $Cr_{K\alpha}$ radiation) was found not to be influenced by zirconium. The values used for ν and E were 0.23 and 180 GPa, respectively. In Fig. 7, the measured compressive stress values determined in two sets of samples corroded in steam at 420°C and 10.5 MPa are plotted versus the exposure time. The error bars on the stress curve represent the statistical error that results from fitting Eq 1 to the measured peak shifts. The corresponding oxide thickness and the fraction of the t-ZrO_2 is also shown. The fraction of t-ZrO_2 was estimated using the ratio of areas of the $\{-111\}$ and $\{111\}$ peaks of m-ZrO_2 and t-ZrO_2, respectively. Generally, the compressive stresses (1500 to 2200 MPa) found in this study are two to three times larger than previously published values determined with the bending-strip method [1]. They are in the same order of magnitude as the stress values reported by Roy and David [7] (compressive stress of 1200 MPa for a 1.9 μm oxide on Zircaloy-2) that were also calculated from the displacement of m-ZrO_2 peaks in XRD patterns relative to m-ZrO_2 powder (but peak angles were measured without tilting the samples, $\Psi = 0$). With all methods, the average compressive stresses are determined. Because the stress level is expected to be higher at the metal/oxide interface compared to the outer oxide region, it may be possible that the threshold for the existence of t-ZrO_2 of about 3 GPa is exceeded near the metal/oxide interface. Therefore, the high average values determined by XRD support the hypothesis that t-ZrO_2 is partially stabilized by stress more than those values measured before. But as long as the stress profile in the oxide is unknown, this will remain an open question. A correlation between the fraction of t-ZrO_2 and the stress level has not yet been found. As can be seen in Fig. 7, the fraction of tetragonal oxide in the samples

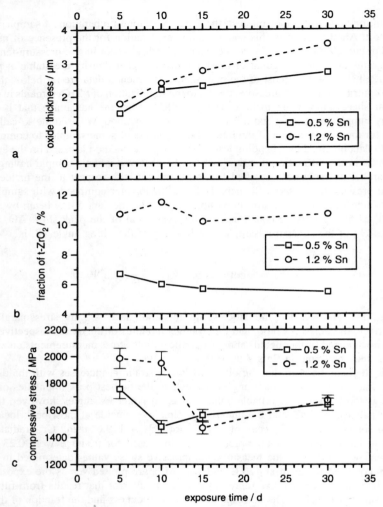

FIG. 7—*Oxide thickness* (a), *fraction of tetragonal* ZrO_2 (b), *and compressive stress of oxide layers* (c) *grown on Zircaloy in steam at 420°C/10.5 MPa.*

containing 1.2% tin was higher than in the low-tin samples. But the stress curves of both sets of samples do not show a certain trend. To come to a clearer decision, it is necessary to investigate more samples at higher corrosion times and to compare the results with those of similar experiments using other techniques for stress determination.

The tetragonal oxide found in the thin TEM samples with electron diffraction is certainly not stabilized by stress since an almost complete stress relaxation of the thin transparent TEM specimens ($\leq$100 nm) is assumed. So, the existence of other oxide phases than monoclinic ZrO_2 is at least partially an effect of the extremely small grain size or of doping the oxide with alloying elements, or both.

GD-OES Depth Profiles

For a better understanding of the complex corrosion mechanism, a knowledge of the distribution of corrosion-affecting species like hydrogen and lithium within the oxide is demanded. GD-OES is an versatile technique for in-depth profiling of the composition of surface layers [8,9]. The Grimm-type lamp [10] is the most commonly used source for depth profiling with OES. With this type of source, the material to be analyzed is sputtered from a planar sample surface (typical dimensions of the analyzed spot are in the range of some millimetres in diameter) in a glow discharge process (usually in an argon plasma). The excitation by the glow discharge causes the elements to emit light at specific wavelengths that are detected with an optical spectrometer separated from the source by an gas-tight optical window. If the sputtering rate is constant, the emission intensity at the specific wavelengths versus time is a measure for the concentration profiles of the corresponding elements in the sample. In principle, a quantification of the element distribution is also possible. But this would require suitable calibration standards to determine the sputter rate of the matrix material and the emission yields of the elements of interest. The depth resolution of GD-OES depends on the ratio of the sputter rate to the data acquisition time, on the initial smoothness of the sample surface, and on the homogeneity of the sputtering process. The former one was not the limiting factor in the case of the oxides to be analyzed, because the theoretical depth resolution of GD-OES is in the nanometre range. But generally, the surface roughness and especially an inhomogeneous removal of material from the surface flattens the element profiles. The examinations were performed with a Leco GDS-750 glow discharge spectrometer that has a 750 mm holographic grating in Paschen-Runge mount and allows the simultaneous measurement of up to about 40 elements by means of photomultipliers positioned at the corresponding wavelengths.

Sheets of Zircaloy-4 (15 by 30 by 1 mm) were pickled and oxidized in water in an static autoclave at 350°C/10.5 MPa. One set of samples was corroded in water containing 70 ppm of LiOH at the same temperature and pressure. Samples from both sets with comparable weight gains in the pre-transition region were used for the GD-OES analysis. To achieve a good electrical conductivity at the oxide surface that is necessary to start the glow discharge, the samples were coated with a gold film (thickness ≈ 600 nm) by d-c magnetron sputtering from a gold target prior to the analysis. The examined spots on the sample surfaces were 4 mm in diameter

Figures 8 and 9 show the intensity versus time plots of zirconium, oxygen, hydrogen, and lithium emission lines of two samples with pre-transition oxide scales grown in water and in water plus LiOH, respectively. To indicate the original surface of the oxide layer, the gold emission lines are also plotted. The two diagrams show uncalibrated raw data. That means, the intensity ratios of the elements do not represent their real fractions within the samples. But as described earlier, the variation of the signals with time is a sensitive indicator for the relative distribution of the elements within the samples.

The sample oxidized in the LiOH environment (Fig. 9) exhibits a distinct enrichment of lithium in the oxide scale with a maximum within the upper third of the layer. The height of the lithium emission in the sample corroded without LiOH (Fig. 8) represents the background or impurity level of this material. Additionally, a comparison of the hydrogen and oxygen profiles near the oxide surface of the two samples indicates that the hydrogen and the oxygen content of the LiOH-treated sample is increased within the uppermost oxide layer. The TEM investigations of thick corrosion layers described earlier have shown that the corrosion in the LiOH environment causes "open grain boundaries" within the oxide that propagate into the oxide scale close to the metal/oxide interface. Thus, the high hydrogen and oxygen levels in the outermost layer of this pre-transition oxide (Fig. 9) may be attributed to strongly bonded H_2O or OH^- at these grain boundaries. In both corrosion films, a hydrogen peak near the

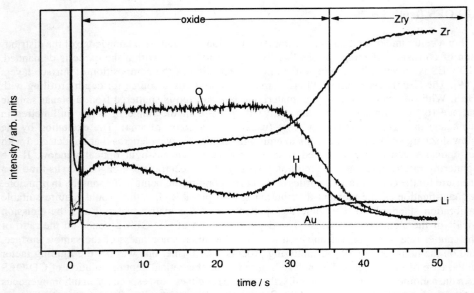

FIG. 8—*GD-OES depth profiles of a pre-transition oxide layer (thickness = 1.7 µm) on Zircaloy corroded in water (350°C/10.5 MPa/90 days). Sputtering rates for oxide = 51 nm/s and for metal = 117 nm/s.*

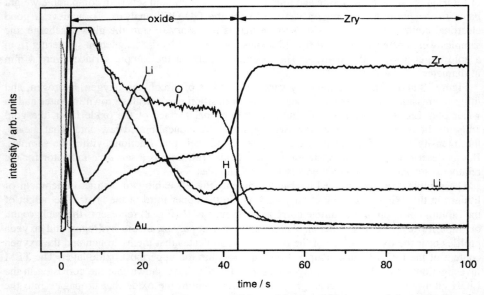

FIG. 9—*GD-OES depth profiles of a pre-transition oxide layer (thickness = 1.8 µm) on Zircaloy corroded in water plus 70 ppm LiOH (350°C/10.5 MPa/90 days). Sputtering rates for oxide = 44 nm/s and for metal = 100 nm/s.*

oxide/metal interface could be observed. A possible explanation for this phenomenon may be a kind of space charge effect that will be discussed later. It should be mentioned, however, that the interpretation of details in GD-OES profiles at boundaries between two different matrices is difficult (especially, if the profiles are flattened), because the sputter rate (and as a consequence the emission, as can be seen on the lithium background level in Fig. 8) changes or the glow discharge process may show fluctuations or both at such an interface if not properly controlled [8]. On the other hand, the sputter rate of the metal is larger than that of the oxide, so if hydrogen would be enriched at the metal side of the interface, then one should expect an increasing hydrogen emission as soon as the phase boundary is crossed (the location of the metal/oxide phase boundary in the two examples is assumed to be the turning point of the zirconium line).

Electrical Potential Measurements

The influence of the electrical potential across oxide scales on the corrosion of zirconium alloys has been discussed frequently. We performed potential measurement in dry oxygen and in high pressure water.

Measurements in Dry Oxygen

Electrical potential differences across oxide scales grown on Zircaloy have been measured in dry oxygen at temperatures up to 500°C. Due to the separation of ionic charges (O^{2-}) the electrical potential difference measured with metallic electrodes across an oxide layer is given by

$$E = E_0 - R_e I_e = \frac{kT}{4e} \ln \frac{p^i_{O_2}}{p^o_{O_2}} - R_e I_e \qquad (2)$$

where k, T, and e denote Boltzmann's constant, the absolute temperature, and the elementary charge, respectively. The oxygen partial pressure, pO^i_2, represents the decomposition pressure of ZrO_2 in equilibrium with the metal at the metal/oxide interface, and pO^o_2 is the oxygen partial pressure at the outer oxide surface. E_0 is that voltage that would be observed in the case of pure ionic conduction in the oxide. Generally, this voltage is decreased by the electronic resistance, R_e (contribution of electrons and holes), of the oxide and the corresponding current, I_e. With pure oxygen at atmospheric pressure present at the oxide surface ($pO^o_2 = 1$ atm), the potential difference E_0 is determined by the Gibb's free energy of formation of ZrO_2 ΔG_{ZrO_2}

$$E_0 = \frac{kT}{4e} \ln \frac{p^i_{O_2}}{p^0_{O_2}} = \frac{kT}{4e} e^{\frac{\Delta G_{ZrO_2}}{RT}} \qquad (3)$$

For temperatures between 300 and 500°C, values of around -2.5 V can be calculated from Eq 3.

Samples of 20 by 20 by 1 mm in size cut from Zircaloy-4 sheet material were chemically polished with a mixture of HNO_3/HF (45/10 vol. parts) in glycerol and rinsed in water. After oxidizing in pure oxygen (0.1 MPa) at 400 or 500°C, they were coated with a platinum film (area ≈ 10 by 10 mm, thickness ≈ 50 to 100 nm) by d-c magnetron sputtering from a platinum target. Prior to the platinum deposition the surface was cleaned by subsequent reference resputtering in an Ar-O_2 mixture and pure argon to achieve a good adhesion of the metallic film. Platinum wires (⌀ 65 μm) were spot welded to the Zircaloy sheet at the edges where the oxide scale was scraped off. The platinum surface electrodes were also contacted with thin platinum wires that were attached using a gold paste. The samples were placed between two flat resistance heaters ($\oslash$ 50 by 1 mm quartz disks with a platinum thin film resistor at one side) in a water-cooled gas-tight stainless steel chamber. The voltage between the metal sheet and the platinum electrode was measured with a digital voltmeter (input impedance > 10^{14} Ω) at temperatures up to 500°C in a constant flow of oxygen. Errors caused by thermoelectric voltages were less than 100 μV. Many oxide scales in the range between 0.5 and 2 μm had very low resistances at room temperature after deposition of the platinum films. This was also observed when the sputtering parameters were changed or evaporated gold films were used as electrodes. Examination by scanning electron microscopy (SEM) revealed small spots ($\leq$100 nm) on the oxide surface. XMA-SEM analysis at these locations showed increased amounts of iron and chromium compared to the surrounding oxide. From this, it was concluded that large intermetallic precipitates have caused an electrical short circuit. The mean size of the precipitates in the alloy was about 500 nm, and the largest particles were in the range of 1.5 μm. At larger oxide thicknesses (2 to 3 μm) or using an alloy with smaller precipitate sizes (intermetallics not larger than 1 μm), the probability of short circuits was clearly lower and the room temperature resistance of the oxides without short circuits reached up to some megaohms (MΩ). The potential differences measured at temperatures between 300 and 500°C across such high resistive oxides (metal negative with respect to the oxide surface) were in the range of some 100 mV (depending on the temperature) some time after the final temperature was reached, but diminished within minutes (at high temperatures) or at latest after 1 or 2 h (at lower temperatures). Figure 10 shows the potential differences of two Zircaloy-4 samples with pre-transition oxides recorded versus time at 350 and 400°C, respectively. This time behavior was attributed to an increasing number of local electric breakdowns at the sites where precipitates (probably covered only by a very thin oxide layer) were located. Even during long heat treatments for more than 100 h at 500°C, no re-establishing of a potential could be observed, although one would expect this, because the oxide should have grown to a thickness that cannot be bridged by large intermetallics. Additionally, some experiments were carried out with samples that have not been oxidized prior to the deposition of the electrodes. No significant potential differences could be detected during oxidation of these samples in oxygen. This seems to be inconsistent with the results of Urquhart et al. [11] who performed similar experiments using platinum thin film electrodes in high-pressure steam and oxygen at 500°C and always found an increase of the potential difference (without breakdown effects) as soon as the oxide thickness was larger than the precipitate size. On post-transition oxides, however, this breakdown was not observed. Figure 11 shows the potential difference recorded at 350°C across an oxide that had an initial thickness of about 5.4 μm. After the temperature setpoint was reached, the voltage increased up to more than 300 mV. The reason for the stepwise voltage increase is not quite clear. One may suppose that this is caused by generation of micro cracks within the oxide layer that increased the resistance of the scale discontinuously. The sample was cooled down to room temperature after about 5 h in the first run. In a second run, the experiment was repeated by again heating up to 350°C. In this case, the potential difference was established immediately after the final temperature was reached. This indicates, that the delayed potential increase at the beginning of the first run was involved by a kind of defect equilibration after the previous bombardment of the oxide surface during the sputter deposition of the electrode. If such a equilibration process takes place at the first heat-up, then the time behavior of the pre-transition oxides in Fig. 10 may be interpreted as a consequence of an increasing potential difference by equilibration of the oxide with the environment that in turn leads to local oxide breakdown caused by high electrostatic fields at the intermetallic particles.

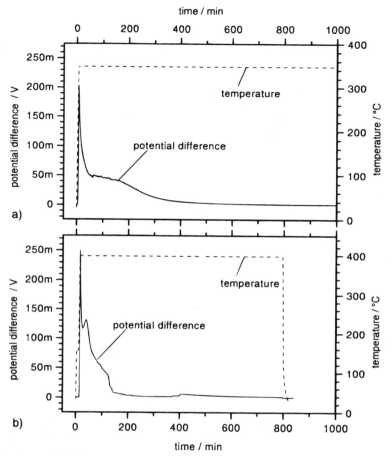

FIG. 10—*Potential difference (metal negative with respect to the oxide surface) across pretransition oxides (2.4 μm) in oxygen at 0.1 MPa versus time, recorded at 350°C (a) and 400°C (b).*

Measurements in High-Pressure Water

Electrical long-time in-situ measurements at 350°C in high-pressure water (with additions of 500 ppm boron, 300 ppm potassium, and 1.5 ppm lithium) were performed in an autoclave with a set of Zircaloy sheets. During this experiment, open-circuit potentials between the Zircaloy metal and a platinum electrode in the water were determined during the corrosion. Figure 12 shows the open-circuit potentials (metal-negative) of two samples for a period of 160 days and the corresponding weight gains that were determined at various points under the same conditions separately. The potentials decrease within the first few days from some hundreds of millivolts to a minimum at about five days of exposure. Then the potentials stabilize around 70 to 130 mV. At the transition (110 and 150 days of exposure for the 1.6% and 0.9% tin sample, respectively), the potentials increase to 160 to 200 mV. Obviously, the level of the potential is correlated to the growth rate of the oxide.

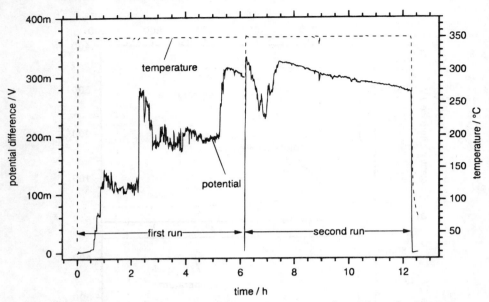

FIG. 11—*Potential difference (metal negative with respect to the oxide surface) across post-transition oxide (5.4 μm) in oxygen at 350°C/0.1 MPa versus time.*

This experiment will not be discussed in detail here (it is still in progress, and further results will be published elsewhere [*12*]), but it should demonstrate that the effects observed in dry oxygen are partially caused by the measuring conditions.

It should be noted that the absolute values of potentials determined in this experiment in water may not be directly comparable to the voltage drops across oxides measured with deposited platinum electrodes (see Eq 1), because the oxides are electrically coupled by an electrolyte and the influence of additional interface reactions and corresponding potentials must be considered. Nevertheless, it is obvious that in particular the electrical breakdown of thin pre-transition films is observed only when metallic electrodes were used to contact the oxide surfaces. Large electrostatic fields that may locally develop across thin oxide films covering the intermetallics can cause an electrical breakdown at these sites only if a sufficient quantity of charges is available at the oxide surface within very short times to provide the high breakdown currents. This requirement is fulfilled with a metal film at the oxide surface, whereas the electronic surface conductivity without a metal film may be too low to carry a very localized breakdown current. In thicker post-transition oxide scales, the intermetallic precipitates are isolated and partially dissolved within the oxide and thus a breakdown is not possible. Although this interpretation basically seems to be plausible, some points remain in question. One of these points is the inconsistencies of the experiments with platinum surface electrodes with that of Urquhart et al. [*11*]. Another aspect mentioned earlier is the observation that potential differences across pre-transition films were first observed when the oxide thickness was about twice the size of the largest precipitates. This point will be discussed later.

Long-Time Corrosion Tests with Varying Temperatures

In long-time corrosion tests, a series of Zircaloy samples was corroded in water at 105 MPa and varying temperatures (see Fig. 13). In the first step that corresponds to the negative times

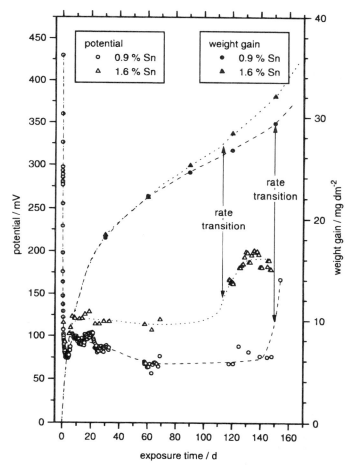

FIG. 12—*Open-circuit potential (metal negative) and corresponding weight gain of Zircaloy in high-pressure water at 350°C versus exposure time.*

in Fig. 13, the samples were treated at 350°C with exposure times up to 500 days. At exposure time zero (end of pre-treatment at 350°C), the specimens had reached different weight gains. The cyclic corrosion behavior up to this point is clearly visible. Then the corrosion was continued at a temperature of 300°C for a period of 960 days. At 300°C, the overall corrosion rate was lower, as was to be expected. But shortly after temperature reduction, some samples exhibit an increased rate for a certain period, whereas others abruptly change to a lower growth rate. The cyclic weight gain curves at 350°C indicate, that the corrosion is mainly controlled by barrier layers that grow with a decreasing rate and loose their protective character at certain thickness by development of cracks or pores. After this breakdown (transition), the next sequence starts, and a new compact layer is formed at the metal/oxide interface, grows, fails, and so on. The thickness, at which a compact layer fails increases with the temperature, whereas the time to transition decreases. So, if the temperature is decreased when the actual barrier layer is thick, then a transition is initiated, and the rate shortly after the temperature reduction increases. This happens, because a new compact layer builds up, and the growth

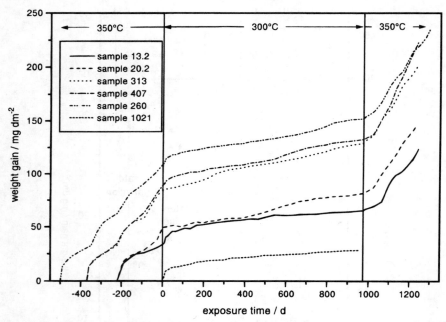

FIG. 13—*Weight gain curves of Zircaloy-4 with the same composition and processing history corroded in high-pressure water at varying temperatures.*

rate immediately after a transition at 300°C is higher than the rate immediately before a transition at 350°C. On the other hand, if there exists a thin barrier layer at the moment the temperature is reduced, it remains protective at 300°C, and the corrosion rate abruptly decreases to the corresponding lower pre-transition rate. A similar behavior was found with another set of samples at a temperature change from 400 to 300°C. Another kind of "memory effect" could be observed when the temperature was raised again to 350°C at an exposure time of 960 days. Immediately after this point, the increase of the corrosion rate is lower than expected at 350°C. With the barrier layer concept, this can be explained as follows. Because the corrosion rate of the barrier layer is larger at higher temperatures, the oxide first continues to grow somewhat faster. As soon as it has reached its critical thickness, which is higher than the critical thickness at 300°C, a new sequence of breakdown and generation of new compact layers is started, and the overall corrosion rate tends to the "normal" value one expects at 350°C.

Discussion

As a consequence of the observed cyclic long-time corrosion behavior of the TEM examinations of previously published results [1] and several observations made by other authors, it is concluded that the corrosion of Zircaloy is a barrier-layer controlled process. It should be noted here, that most of the explanations that will be used subsequently for a qualitative description of the corrosion are not new. Some of them are found in books, papers, and reviews published 20 or 30 years ago.

Pre-Transition Corrosion

Initially, by reaction of zirconium with water or oxygen or both different sub-oxides are formed sequentially at the metal surface. Further growth of the initial oxide film is accompanied by formation of substoichiometric ZrO_2 (for more details about the initial steps of the corrosion see Ref *13*). This oxide (which may remain amorphous for a certain time [*1*]) crystallizes to small equiaxed tetragonal grains. When the size of these grains reaches some critical value, they transform to the monoclinic modification. Some amount of t-ZrO_2 will survive, because it is stabilized due to its small crystallite size by high compressive stress at the metal/oxide interface or by dissolved alloying elements or both. A large hydrogen concentration at the metal/oxide interface facilitates the transformation from tetragonal to monoclinic [*1*]. The further oxide formation at the metal/oxide interface leads to a preferential growth of the monoclinic crystals. In uniform corrosion, this process leads to a dense layer with a columnar structure consisting mainly of monoclinic zirconium dioxide. In nodular oxides, equiaxed grains are formed at the metal/oxide interface and no, or probably only a thin compact layer, exists (see TEM results). The mobile ionic species in zirconium dioxide are oxygen ions. Therefore, further growth of the oxide film takes place by incorporation of oxygen at the oxide surface, diffusion of oxygen ions to the metal/oxide interface, transport of electrons in the opposite direction, and formation of oxide at the metal/oxide interface.

Intermetallic precipitates in Zircaloy are known to have a high oxidation resistance compared to the zirconium matrix [*14*]. These particles remain in metallic state for quite long times [*15*], and as long as the oxide thickness is smaller than the size of the large precipitates, they can serve as easy paths for electrons migrating form the metal to the oxide surface.

Zirconium and its alloys take up hydrogen when corroding in aqueous environments. If there exists a dense oxide layer on the metal, hydrogen reaches the metal either by migration of protons within the oxide matrix (the mechanism of proton migration in ZrO_2 is still unknown) or by diffusion of neutral hydrogen via intermetallic precipitates as long as they are not isolated within the oxide matrix.

As the diffusion distance of mobile species increases as the oxide thickens, the corrosion rate decreases with time. Because of the large Pilling-Bedworth ratio of Zr/ZrO_2, the formation of an adherent oxide film on the zirconium metal results in compressive stresses within the oxide. This stress may be relieved partially by plastic deformation and at later stages by buckling of the oxide that leads to some porosity, mainly isolated fine voids.

Transition

At some critical oxide thickness, a growth rate transition occurs. The transition starts at single points and spreads quite fast over the whole surface. This has been seen very clearly in pressurized water reactors (PWRs), less pronounced in lithium-containing water, and sometimes it can also be observed out-of-pile in high-pressure water. Figure 14 shows a sample that was exposed to water containing 70 ppm LiOH at 350°C for 140 days. At the light grey spots, the transition has already occurred, as was confirmed by cross-sectional SEM. Two layers separated by micro-cracks were found at these locations.

Finally, an open network of pores that permits direct access O_2 or H_2O to the metal surface is formed. Therefore, the corrosion rate increases again after this transition. Some amount of the tetragonal grains, which are still present in the dense layer before the transition occurs, transforms to m-ZrO_2 at the transition. This may have two reasons. As discussed in the XRD section, the tetragonal phase within the dense layer may be stabilized by the large compressive stress that is relieved at the transition. The second reason may again be the increased concentration of hydrogen that favors the monoclinic ZrO_2 [*1*].

Post-Transition Corrosion

After the transition, a new dense layer is formed at the metal/oxide interface. One has to consider, however, that the environment at the boundary between this compact barrier layer

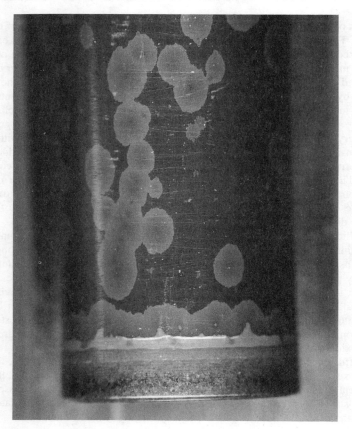

FIG. 14—*Appearance of corroded Zircaloy tube exposed to high-pressure water containing 70 ppm LiOH at 350°C for 140 days.*

and the porous scale will not remain the same as in the pre-transition region. After the first transition, the decomposition of water into hydrogen and oxygen takes places at the barrier layer surface. After diffusion through the dense layer, oxygen is consumed by the oxidation of zirconium at the metal/oxide interface but only a small fraction of hydrogen is incorporated into the metal. As a consequence, the hydrogen partial pressure at the boundary between the dense layer and the porous scale increases and a hydrogen concentration gradient within the porous oxide layer builds up. The higher hydrogen partial pressure may now increase the hydrogen uptake of the metal. Another possible effect that also depends to some degree on the properties of the porous layer is an enrichment of other species present in the coolant. One of the most important candidates is lithium that has a harmful influence on the corrosion resistance of Zircaloy (see the results of TEM examinations).

The sequence of formation of a barrier layer and cracking at some critical thickness may repeat several times and, finally, the corrosion rate seems to approach a nearly constant value. The cyclic nature of Zircaloy corrosion after the first transition can be more or less significant. Generally, one can observe that if a cyclic behavior is visible, it changes smoothly to a linear corrosion mode. The observed post-transition weight gain is the average of corrosion cycles, which appears to be linear on a macroscopic scale, because the growth rate will never be exactly the same at every location on a sample surface. At longer corrosion times, the small phase shift of cycles at different locations accumulates, and what we finally observe is a superposition of an increasing number of cycles at different stages.

Barrier Layer Mechanisms

The overall corrosion resistance of Zircaloy will mainly be governed by the quality and the kinetics of the barrier layer at the metal/oxide interface, because the porous part of the post-transition oxide can easily be penetrated by the corroding species compared to the much slower solid-state diffusion within the dense barrier layer. The migration of charged species in the growing compact oxide layer along grain boundaries and via lattice diffusion depends on several parameters interacting with each other. The classic oxidation theory of Wagner based on lattice diffusion of charge carriers gives a parabolic rate constant. Because the corrosion of Zircaloy tends to cubic time behavior, different models have been proposed that postulate grain boundary diffusion as the rate limiting factor. It will be shown later that probably both diffusion mechanisms are rate limiting depending on the location within the barrier layer.

Despite the complex behavior in the first approach, it may be helpful to describe the problem qualitatively in a simplified manner in terms of point defect equilibrium with fixed boundary conditions and a constant oxide thickness [16–18]. For this purpose, consider a dense zirconium oxide layer in contact with Zircaloy metal at one side and exposed to an gaseous or liquid environment of O_2 or H_2O at the other side (see Fig. 15). Assuming that oxygen vacancies are the predominant ionic defects in zirconium oxide, the nonstoichiometry of ZrO_{2-x} may be written as

$$ZrO_2 = ZrO_{2-x} + \frac{x}{2} O_2(\text{env}) \tag{4}$$

This equation means that nonstoichiometry increases with decreasing oxygen partial pressure. The formation of oxygen vacancies involves the transfer of an oxygen atom from its regular lattice site to the environment. If it is assumed that the resultant oxygen vacancies are doubly charged, this reaction may be written using the Kröger-Vink notation

$$O_O = V_O^{\cdot\cdot} + 2e' + \frac{1}{2} O_2(\text{env}) \tag{5}$$

where O_O, $V_O^{\cdot\cdot}$, and e' denote an oxygen ion, O^{2-}, at its regular lattice site, a doubly charged oxygen vacancy, and an electron, respectively. Applying the law of mass action, the equilibrium is given by

$$K_{V_O^{\cdot\cdot}} = 2[V_O^{\cdot\cdot}]n^2 p_{O_2}^{1/2} \tag{6}$$

with the concentration of electrons, n, and the equilibrium constant, $K_{V_O^{\cdot\cdot}}$. Neglecting any other point defects electroneutrality requires that

$$2[V_O^{\cdot\cdot}] = n = (2K_{V_O^{\cdot\cdot}})^{1/3} p_{O_2}^{-1/6} \tag{7}$$

In oxides growing on zirconium alloys, a certain amount of alloying elements (tin, iron, chromium, and nickel) is dissolved substitutionally on regular zirconium sites in the ZrO_2

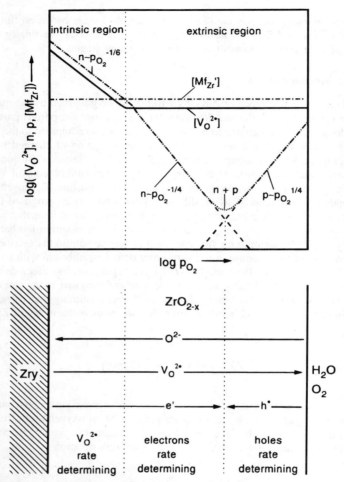

FIG. 15—*Concentration of charge carriers in ZrO_{2-x} as a function of oxygen partial pressure* (top), *and corresponding barrier layer on Zircaloy* (bottom).

matrix. If the valence of these foreign cations is lower compared to the parent cation, Zr^{4+}, the electroneutrality condition is

$$2[V_O^{2\bullet}] = [Mf'_{Zr}] + n \qquad (8)$$

where Mf'_{Zr} denotes a foreign cation at a zirconium site that has one effective negative charge (for example, Fe^{3+}, Cr^{3+}). For this "doped" oxide, two limiting conditions can be derived:

1. If $n = 2[V_O^{2\bullet}] \gg [Mf'_{Zr}]$, the foreign cations do not affect the native defect equilibrium (Eq 5) and the oxygen vacancy concentration is given by Eq 7 (intrinsic region). This will occur at the metal/oxide interface where a very low pO_2 (the oxygen dissociation pressure of ZrO_2) is fixed and a large deviation, x, from stoichiometry is to be expected.

2. If $[Mf'_{Zr}] \approx 2[V_O^{\bullet\bullet}] \gg n$, the oxygen vacancy concentration is determined by the content of dissolved foreign cations (doping level). In this extrinsic region at the outer part of the oxide layer, the electron concentration is given by

$$n = \frac{(2K_{V_O^{\bullet\bullet}})^{1/2}}{[Mf'_{Zr}]} p_{O_2}^{-1/4} \qquad (9)$$

At high oxygen partial pressures, the electron concentration may become sufficiently low so that electron holes, $h^{\bullet}$, may be generated

$$np = K_{np} \qquad (10)$$

Thus, the concentration of holes, p, has to be included in the electroneutrality condition

$$2[V_O^{\bullet\bullet}] + p = [Mf'_{Zr}] + n \qquad (11)$$

indicating that p increases with the oxygen partial pressure according to

$$p = \frac{K_{np}[Mf'_{Zr}]}{(2K_{V_O^{\bullet\bullet}})^{1/2}} p_{O_2}^{1/4} \qquad (12)$$

Depending on the amount of dissolved lower valent foreign cations, p may even exceed $[V_O^{\bullet\bullet}]$ at high oxygen partial pressures [19].

Figure 15 shows schematically how the concentrations of $V_O^{\bullet\bullet}$, n, p, and Mf'_{Zr} depend on the oxygen partial pressure. The x-axis (Fig. 14a) does not correspond directly to the pO_2 scale in Fig. 15b, but the comparison of the two figures should demonstrate the distribution of the different species within the oxide more generally. Oxygen vacancies are the predominant defect species in the outer oxide layer (assumed that $p \ll [V_O^{\bullet\bullet}]$), and, therefore, the transport of oxygen ions from the surface to the metal (or the migration of oxygen vacancies in the opposite direction) will be determined by the electronic conductivity in the extrinsic domain. This difference in oxygen mobility across the oxide was made responsible by Ellof et al. [20] for a space charge effect that they evaluated from weight gain curves of Zircaloy-4. In the outermost oxide layer space, charge effects caused by the p-n junction may also play a role [21]. Near the metal/oxide boundary, the concentrations of vacancies and of electrons are very high. Because the oxygen vacancies are expected to be less mobile than the electrons, the rate-determining species for diffusion near the metal will be the oxygen vacancies. The location of the boundary between the intrinsic and the extrinsic domain and of the p-n junction in the oxide and thus the relative contribution of ionic and electronic conductivities to the migration of the charge carriers will depend on the concentration of dissolved lower valent cations, on the oxygen partial pressure at the barrier layer surface, on the temperature, and on the layer thickness. Even if the temperature and the boundary conditions are held constant, the increase in oxide thickness will itself require a continuous readjustment of the defect concentrations at any point in the oxide. Methods for a computational treatment of these very complex problems can be found in Refs 22 and 23. But based on the simplified description of defect distributions, some general conclusions can be drawn that may help to interpret some phenomena. TEM examinations have shown that the barrier layer has a relatively fine grain structure. This has to be considered for a further refinement of the simple model described earlier.

At the grain boundaries, for example, the density of defects (especially oxygen vacancies) is certainly much higher than in the grain interior, and oxygen will diffuse via these grain boundaries to a large extent, at least in the outer part of the barrier layer. Close to the metal/oxide interface, however, the oxygen vacancy concentration and the electron density within the grains are so large that no significant difference between grain boundary and bulk diffusion is to be expected. This explains why we could never detect a preferential attack of the metal at the oxide grain boundaries by TEM examinations that should be seen if the migration of oxygen via grain boundaries is faster than bulk diffusion. That there really exists a high conductivity region at the metal/oxide interface can be seen in the cross-sectional SEM micrographs in Fig. 16. Figure 16a is a backscattered electron image of a pre-transition oxide on Zircaloy corroded in water at 350°C (oxide dark, metal light grey). The same location is shown in Fig. 16b as a secondary electron image. Due to charge accumulation effects, the oxide exhibits a two-layer structure in this picture. A dark band (width of 0.5 to 1 μm) within the oxide adjacent to the metal surface can be seen that has a higher conductivity compared to the uppermost oxide region (white).

An electrical potential difference across the oxide will be established under open circuit conditions with the metal being negative with respect to the oxide surface (see Eq 2). An increased hydrogen uptake of the metal by diffusion of protons in the oxide under electrical potential gradients has been discussed frequently in the literature. The potential difference between the metal and the layer close to the metal/oxide interface will be only small because of the high electronic conductivity of the oxide in this region. Therefore, it is possible that the proton migration enhanced by the electric field is slowed down at this location and a hydrogen accumulation results. This would explain the local hydrogen maxima observed in the GD-OES depth profiles.

Intermetallic precipitates locally serve as easy paths for electrons in thin oxide scales, as was shown earlier. At this stage, the electronic conductivity of the oxide matrix has only an influence at those locations where no precipitates are present and if the electronic surface conductivity is sufficiently low to prevent an electronic short-circuit effect over the total oxide area. With increasing oxide thickness, these particles become isolated in the oxide matrix and a metal-negative potential gradient will be established across the oxide scale. The high conductivity layer at the metal/oxide interface may be responsible for the observed breakdown of potential differences across pre-transition oxide scales in those cases where the oxides were thicker than the largest intermetallic particles. The intermetallics are slowly oxidized near the oxide surface [15], so the amount of iron and chromium dissolved within the ZrO_2 matrix rises, thereby increasing the p conduction at the surface (see Eq 12). Models for the oxide growth on Zircaloys based on space-charge accumulation at a p-n junction were already proposed by Hauffe [24] and Shirvington and Cox [21].

Additional parameters influencing the properties of the barrier layer are to be considered in the case of in-pile corrosion. For example, radiation will increase the overall density of defects, and thermal gradients will affect the defect distribution within the layer. Thus, the contribution of the different charge carriers to the oxide growth rate may be different compared to out-of-pile corrosion.

Conclusions

From the examinations on out-of-pile corroded samples of two zirconium alloys containing 0.2% iron, 0.1% chromium, and different tin amounts of 1.2 and 0.5%, the following can be concluded.

The microstructure at the metal/oxide interface after uniform corrosion of thin pre-transition oxides and of thicker post-transition oxides is similar (mainly columnar grains of monoclinic

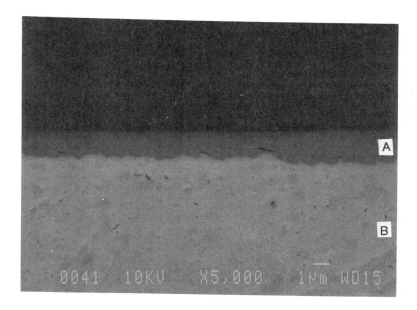

FIG. 16—*Cross-sectional SEM micrographs of pre-transition oxide on Zircaloy:* (top) *backscattered electron image (oxide dark (A), metal light grey (B)) and* (bottom) *secondary electron image at the same location as in* (top) *showing high conductivity layer (C) at the metal/oxide interface (dark band) due to charge accumulation effects.*

ZrO_2 with a smaller fraction of t–ZrO_2, epitaxial growth on α-Zr grains occurs, fine isolated pores). The number and size of pores increases in the direction of the oxide surface. Equiaxed oxide grains are found preferentially at mid-thickness and at the oxide surface. The uppermost layer on post-transition oxides shows open porosity that permits the molecular penetration by water and other corroding species. A network of fine pores spreading deeply into the oxide scale along grain boundaries was found when lithium was added to the corrosion medium. Element depth-profiling revealed an enrichment of lithium within the oxide scale.

The compressive stresses measured by means of a X-ray diffraction technique in pre-transition oxide films were significantly higher than values previously published. The fraction of t-ZrO_2 determined with XRD seems to be larger in oxides grown on Zr-1.2Sn compared to Zr-0.5Sn, but a correlation with the compressive stresses could not be detected.

Electrical measurement in dry oxygen and high-pressure water showed that intermetallic precipitates serve as easy paths for electrons in thin pre-transition oxides. The large electrical potential gradient between the metal/oxide interface and the oxide surface resulted in an electrical breakdown of thin oxide films covering the precipitates, when the electronic conductivity at the oxide surface was sufficiently high.

"Memory effects" were found in long-time corrosion tests with varying temperatures that could be explained by the existence of a dense barrier layer at the metal/oxide interface. The cyclic behavior of uniform corrosion often observed in the post-transition region supports this interpretation.

Except for the fraction of tetragonal oxide determined with XRD, no significant differences with respect to the tin contents of the alloys could be detected by the investigations performed in this study.

A barrier layer concept is proposed to describe the uniform corrosion of zirconium alloys. In this model, it is assumed that the corrosion resistance is governed by a protective layer at the metal/oxide interface. The properties of this layer were discussed in terms of defect distributions and migration of charge carriers within zirconium dioxide.

Acknowledgments

The authors gratefully acknowledge the help of O. Förster (Leco Instruments GmbH) for GD-OES analysis, S. Höfler and D. Keilholz for XRD measurements, H. Seidel for SEM and XMA examinations, and J. Schaller (Universität Erlangen-Nürnberg) for supplying data from his electrical in situ experiments.

References

[1] Garzarolli, F., Seidel, H., Tricot, R., and Gros, J. P. in *Zirconium in the Nuclear Industry: Ninth International Symposium, ASTM STP 1132*, C. M. Eucken and A. M. Garde, Eds., American Society for Testing and Materials, Philadelphia, 1991, pp. 395–415.
[2] Cox, B. and Wu, C., *Journal of Nuclear Materials*, Vol. 199, 1993, pp. 272–284.
[3] Kaliba, C., Ph. D. thesis, University of Erlangen-Nürnberg, Germany, 1989.
[4] Wofstieg, U., *Hüttentechnische Mitteilungen*, Vol. 29, 1974, pp. 175–184.
[5] Faninger, G., *Hüttentechnische Mitteilungen*, Vol. 31, 1976, pp. 16–18.
[6] Höfler, S., thesis, University of Erlangen-Nürnberg, Germany, 1992.
[7] Roy, C. and David, G., *Journal of Nuclear Materials*, Vol. 37, 1970, pp. 71–81.
[8] Bengtson, A., Eklund, A., Lundholm, M., and Saric, A., *Journal of Analytical and Atomic Spectrometry*, Vol. 5, 1990, pp. 563–567.
[9] Koch, K. H., Sommer, D., and Grunenberg, D., *Microchimica Acta* (Wien), Supplement 11, 1985, pp. 137–144.
[10] Grimm, W., *Spectrochim. Acta*, Vol. 23B, 1968, pp. 443–454.
[11] Urquhart, A. W., Vermilyea, D. A., and Rocco, W. A., *Journal*, Electrochemical Society, Vol. 125, 1978, pp. 199–204.

[12] Schaller, J., Ph. D. thesis, University of Erlangen-Nürnberg, Germany, to be published.
[13] Döbler, U., Knop, A., Ruhmann, H., and Beie, H.-J., "On the Initial Corrosion Mechanism of Zirconium Alloy: Interaction of Oxygen and Water with Zircaloy at Room Temperature and 450°C Evaluated by X-Ray Absorption Spectroscopy and Photoelectron Spectroscopy," in this volume.
[14] Weidinger, H. G., Ruhmann, H., Cheliotis, G., Maguire, M., and Yau, T. L., *Zirconium in the Nuclear Industry: Ninth International Symposium, ASTM STP 1132*, 1991, pp. 499–537.
[15] Pêcheur, D., Lefebvre, F., Motta, A. T., Lemaignan, C., and Wadier, J. F., *Journal of Nuclear Materials*, Vol. 189, No. 3, 1992, pp. 318–332.
[16] Kofstad, P., *Nonstoichiometry, Diffusion, and Electrical Conductivity in Binary Metal Oxides*, Krieger Publishing Company, Malbar, FL, 1983.
[17] Kofstad, P., *High Temperature Corrosion*, Elsevier Applied Science Publishers, London, 1988.
[18] Tuller, H. L., "Mixed Conduction in Nonstoichiometric Oxides," *Nonstoichiometric Oxides*, O. T. Sørensen, Ed., Academic Press, New York, 1981, pp. 271–335.
[19] Nasrallah, M. M. and Douglass, D. L., *Journal*, Electrochemical Society, Vol. 121, 1974, pp. 255–262.
[20] Eloff, G. A., Greyling, C. J., and Viljoen, P. E., *Journal of Nuclear Materials*, Vol. 199, 1993, pp. 285–288.
[21] Shirvington, P. J. and Cox, B., *Journal of Nuclear Materials*, Vol. 35, 1969, pp. 211–222.
[22] Fromhold, A. T., *Theory of Metal Oxidation, Vol. I—Fundamentals*, North-Holland Publishing Company, Amsterdam, 1976.
[23] Fromhold, A. T., *Theory of Metal Oxidation, Vol. II—Space Charge*, North-Holland Publishing Company, Amsterdam, 1980.
[24] Hauffe, K., *Werkstoffe und Korrosion*, Vol. 22, 1971, pp. 604–612.

DISCUSSION

B. Warr[1] (written discussion)—It is interesting to see that you have found peaks in hydrogen content at (just before) the metal-oxide interface in some samples, since we have observed similar effects with secondary ion mass spectrometry (SIMS). However, interpretation of such data is complicated by the effects of roughness of the oxide surface, sputter rates, and metal-oxide interface. Roughness would tend to "blur" out any real peaks that were present, and the peaks that you observe are very narrow. Could you comment please?

H.-J. Beie et al. (authors' closure)—It is true that the interpretation of such depth profiles is complicated by the roughness of the sample surface and the different sputtering rates of the oxide and the metal. But the simultaneous analysis of several elements (as in our experiments) gives additional information. The zirconium and oxygen profiles of the GD-OES analysis indicate the position of the metal/oxide interface. By examining the profiles of several other elements, which were also recorded (not shown in Figs. 8 and 9), we could exclude that the observed hydrogen peaks very close to the interface are artifacts caused by fluctuations of the sputtering process at the phase boundary. Therefore, we are quite sure that these hydrogen peaks in the oxide are real.

D. Khatamian[2] (written discussion)—In the hydrogen profile, you showed a peak near the oxide/metal interface and mentioned that the position of the peak is in the oxide very close to the interface. We have reported the existence of hydrogen peaks in the interface region and shown that the peak is in the alloy region just beneath the oxide.[3] Also we have shown that such peaks exist in hydrogen profiles obtained from anodic oxide.[4] We believe that under the influence of the stress field due to the growth of oxide (in the oxide front) the hydrogen moves to the region from the bulk of the alloy and not from the outside. Your explanation, as to the source of the hydrogen and under what influenced it to move to the region, does not seem to agree with the references just mentioned. Could you please comment on this?

H.-J. Beie et al. (authors' closure)—As explained earlier, we are sure that the hydrogen peaks observed in our measurements are in the oxide. A significant hydrogen enrichment in the metal beneath the oxide could not be detected. Therefore, we concluded that the hydrogen originates from the oxide surface and not from the alloy. The reason for the different explanations is that we observed peaks in the oxide and you found peaks in the metal. The depth of the hydrogen profiles reported in Ref *1* was calculated using some estimates. So is it possible that we all have seen the same peaks?

N. Ramasubramanian[5] (written discussion)—(1) If microstructures of the oxides formed in steam and lithiated aqueous solutions are different (columnar versus equiaxed) and corrosion is dependent on the oxide microstructure, then what is the rationale in using steam tests to rank alloys for in-reactor use?

(2) How significant is the change in rest potential in its relationships to kinetic transition? The potential is nearly steady, after a steep initial drop towards less cathodic values, during the pre-transiton growth when the corrosion rate is decreasing with oxide thickness. Then why and how should a small charge in rest potential towards more cathodic values correspond to an increase in corrosion rate and hence the rate transition?

---

[1] Ontario Hydro, Toronto, Ontario, Canada M8Z 5S4.
[2] AECL Research, Chalk River Laboratories, Chalk River, Ontario, Canada.
[3] Stern, A., et al., *Journal of Nuclear Materials,* Vol. 148, 1987, p. 257.
[4] Khatamian, D., *Proceedings,* International Electrochemical Society, Toronto, Oct. 1992.
[5] Ontario Hydro Research, Toronto, Ontario, Canada.

H.-J. Beie et al. (authors' closure)—(1) Indeed, this question is obvious. But we can not give a final answer, because these correlations between the microstructures of the oxides and the corrosion conditions should be proven by further examinations. To learn more about this, we are performing tests in steam, water, and lithiated water on the same materials more intensively. Despite these open questions, it has been shown, at least empirically, that steam tests are useful to rank alloys for in-reactor use.

(2) As mentioned in our paper, this experiment is still in progress, therefore, we cannot give a consistent interpretation at this stage. The potential may be interpreted as a measure for the resistance of the oxide, mainly that part of the oxide that is not penetrated by the electrolyte. After the steep drop at the beginning, the potential increases slowly or remains nearly constant, reflecting the formation of a dense barrier layer. At the rate transition, this layer becomes porous, and we expect to observe a decrease of the potential again. The main reason to present these potential curves was to show that electrical breakdown effects observed in dry oxygen were not observed in water.

U. Döbler,[1] A. Knop[1,2] H. Ruhmann,[3] and Hans-Jurgen Beie[4]

On the Initial Corrosion Mechanism of Zirconium Alloy: Interaction of Oxygen and Water with Zircaloy at Room Temperature and 450°C Evaluated by X-Ray Absorption Spectroscopy and Photoelectron Spectroscopy

REFERENCE: Döbler, U., Knop, A., Ruhmann, H., and Beie, H.-J., **"On the Initial Corrosion Mechanism of Zirconium Alloy: Interaction of Oxygen and Water with Zircaloy at Room Temperature and 450°C Evaluated by X-Ray Absorption Spectroscopy and Photoelectron Spectroscopy,"** *Zirconium in the Nuclear Industry: Tenth International Symposium, ASTM STP 1245*, A. M. Garde and E. R. Bradley, Eds., American Society for Testing and Materials, Philadelphia, 1994, pp. 644–662.

ABSTRACT: The initial stages of zirconium oxide formation on Zircaloy after water (H_2O) and oxygen (O_2) exposures have been investigated in situ using photoelectron spectroscopy and X-ray-absorption spectroscopy. The reactivity of the zirconium alloy with O_2 at room temperature is about 1000 times higher than for H_2O. Up to 100 L (1 L = 1 Langmuir unit = $1 \cdot 10^{-6}$ mbar $\cdot$ s) H_2O exposure, the reactivity of the zirconium alloy at 450°C is comparable to the room temperature reaction. At higher H_2O exposure, a sharp increase in the reaction rate for the high-temperature oxidation is observed. From the energy position of the Zr 3d photo emission line and their oxygen-induced chemical shifts, one can directly follow the formation of the oxide films. Two different substoichiometric oxides were found during reaction with water. Suboxide (1) is located at the zirconium/zirconium-oxide interface. Subsequently, a Suboxide (2) is concluded from the chemical shift of the zirconium photoelectrons. After an oxide thickness of 2 nm, the stoichiometric ZrO_2 phase is not yet developed.

KEY WORDS: zirconium alloys, corrosion, water, oxygen reactivity, X-ray photoelectron spectroscopy, X-ray absorption spectroscopy, corrosion mechanisms, zirconium, nuclear materials, nuclear applications

Because of their low thermal neutron capture cross section and excellent corrosion resistance, zirconium-based alloys are used as cladding and structural materials in nuclear power reactors. For a further improvement of the corrosion resistance under extended operating conditions, an understanding of the corrosion mechanisms is necessary. Extensive work has been carried out in order to gain insight into the formation of the corrosion layer by applying a wide spectrum of investigation methods. These methods include different length scales from which

---

[1] Sietec GmbH, Research Laboratory at BESSY, Lentzeallee 101, D-14195 Berlin, Germany.
[2] Institut für Festkörperphysik, TU-Berlin, C-10623 Berlin, Germany.
[3] Siemens AG, Power Generation Group (KWU), D-91058 Erlangen, Germany.
[4] Siemens AG, Corporate Research and Development, D-91508 Erlangen, Germany.

information is gained: that is, ranking from macroscopic (weight gain measurements) to near atomic distances using high-resolution transmission electron microscopy (HRTEM). Results found for the same material used in this study are reported by Beie et al. [1]. This work addresses to the initial state of oxide formation by the reaction with water and oxygen on Zircaloy samples. The experiments are performed so that the oxidation is followed directly in the system (in situ) without any kind of handling or sample preparation of the specimen that is usually necessary for ex situ-prepared corrosion layers. This in situ preparation gives a more direct insight into the surface processes for every sample that was later investigated. We have tried to cover both specimens.

The appropriate means for obtaining insight into the initial corrosion mechanisms that govern the first steps of oxide scale formation, even for the thickest corrosion layer, are surface-sensitive techniques such as low-energy electron diffraction (LEED), auger electron spectroscopy (AES), X-ray photoelectron spectroscopy (XPS), or other recent techniques.

Less work is reported in the literature concerning XPS that has the advantage of obtaining direct information about the chemical composition of the formed species. Surface composition of natural dense oxide buildup on zirconium has been investigated by different authors [2–7]. A summary of stable and metastable W-O-compounds are compiled in Ref 8. The formation of suboxides of very different composition has been reported [9–12]. Room temperature adsorption experiments with oxygen have been reported [13–18] as well as two studies performed at high temperature [18,19].

To our knowledge, only one in situ experiment was reported using water as the reactive medium at low temperature and low exposures [20]. The objective of this work was therefore to carry out in situ experiments on Zircaloys with water as the reactant and to compare the formation of oxide with oxygen treatment under similar conditions. The formation of the reaction products as well as the characterization of the native oxides is followed, applying XPS and X-ray absorption spectroscopy (XAS), by taking advantage of having access to a radiation source with continuous variable photon energy at the BESSY Synchrotron in Berlin.

Investigation Methods

Photoelectron Spectroscopy

In XPS, soft X-rays were used to excite the core electrons of the material under investigation. The X-ray photons have sufficiently high energy so that core electrons leave the solid and can be detected in an energy-dispersive electron analyzer. Typically, the electrons are analyzed with respect to their number (intensity) and their kinetic energy. When the kinetic energy and the photon energy are known, one can determine the binding energy of the electrons with respect to the vacuum level. Changes in the local chemical environment of an atom induce a variation in the binding energy of the core electron. The difference in the chemical environment can be, for example, the replacement of a zirconium neighbor atom by an oxygen atom. Oxygen is more strongly electronegative than zirconium and therefore induces a charge transfer toward the oxygen atom. This results in a higher binding energy of the zirconium core electron, which is detected in the photoelectron spectrum as an additional chemically shifted line to the lower kinetic energy of the released photoelectrons relative to the zirconium metal bulk peak.

X-Ray Absorption Spectroscopy

The chemical environment can be determined as well, using absorption spectroscopy. When sweeping the photon energy, the absorption coefficient of the irradiated material shows a step-like increase, the absorption edge, because the photons have the proper energy to excite a new electron shell. In a metal, the onset corresponds to the binding energy with respect to the Fermi level.

The near-edge region defined by near-edge X-ray absorption spectroscopy (NEXAFS), the first 50-eV above the absorption edge, is influenced and characterized by the chemical structure of the local environment. In simple cases, the NEXAFS spectrum can theoretically be calculated, but in complex situations it is usually being used as a fingerprint method for relative comparisons. If absorbing atomic species are localized in a thin film on top of a substrate, the height of the edge jump is proportional to the number of absorbing atoms and therefore proportional to the thickness of the film.

Experimental

The 10 by 10 mm samples were cut from Zircaloy sheet materials containing 1.3% tin, 0.2% iron, 0.1% chromium and 0,13% oxygen. They were mechanically polished using a 0.5-μm fine grain diamond polish in the final step. Electropolishing or pickling has not been used in order to avoid possible changes in an extended surface stoichiometry by contamination. This has been concluded by the findings for different surface-treated samples in pre-test investigations. Also ion bombardment of the samples after oxidation (autoclave oxidized or in situ oxidized) has been avoided. Initial precleaning of samples before in situ oxidation was done by 5-keV argon ion bombardment under normal or grazing incidence at room temperature or at 450°C. In all cases, even after repeated bombardment cycles (typical duration of the bombardment cycle was 60 min), a trace of oxygen was left on the surface that is attributed to the oxygen content of the Zircaloys.

Two different kinds of oxidized samples have been investigated. First, ex situ oxidized material was prepared in a static stainless steel autoclave under steam at 400°C and 105 MPa. The autoclave was purged with argon prior to closeup. Second, in situ oxidation was performed in the ultra-high vacuum system introducing oxygen or water by opening a leak valve to an amount that established a constant pressure of $1 \cdot 10^{-6}$ mbar in the system. Different amounts of reactive gases are defined by different time intervals that the sample is exposed to the gas pressure. (The product of gas pressure and exposure time defines the exposure of the sample measured in Langmuir.) This Langmuir unit is an average for the number of gas molecules that hit 1 cm^2 of the surface in one second and is used as a standard measure in adsorption experiments under high vacuum conditions. The proportionality between exposure and actual coverage frequently has been proven experimentally using different surface-sensitive investigation methods. After a defined exposure, the so-formed intermediate states of oxidation are investigated by X-ray absorption and photoelectron spectroscopy without any additional change of an experimental condition such as temperature or removal from the vacuum system. The applied experimental pressure has nothing to do with equilibrium decomposition pressures of zirconia, which is orders or magnitude lower.

The photo emission measurements were performed at the Synchrotron radiation light source in the BESSY using the SX-700 monochromator. The mounted samples were held under ultra-high vacuum conditions ($<2 \cdot 10^{-10}$ mbar) during the measurements and during the heat-treatment experiments. Resistant heating of the samples was performed by an electrical current via copper wires attached to the Zircaloy samples. Sample temperature was controlled applying a pyrometer and regulated by adjusting the electric current. Photo emission spectra were taken using a hemispherical mirror analyzer collecting the photoelectrons in normal emission. The sample-to-analyzer geometry and the photon energy was kept fixed at a constant information depth of 0.6 nm for the emitted photoelectrons.

The X-ray absorption measurements were performed by using the total electron yield of emitted electrons or by the detecting the fluorescence photons from the decay of the core hole.

Absorption spectra were taken above the oxygen (O 1s), iron (Fe 2p), chromium (Cr 2p), and several zirconium (Zr) lines. The two registration techniques have different information depths of about 20 nm (using electron yield) and 5 μm (fluorescence), respectively. Therefore, it is possible to discriminate between surface and bulk effects.

Results and Discussion

Characterization of the Clean Surface

After the initial sputter cleaning, the untreated surface of the samples are characterized by photoelectron spectroscopy. Independent of all cleaning procedures, there was a trace of oxygen remaining detectable. It corresponds to an amount of less than 0.1 monolayer of zirconium oxide and is probably due to the bulk oxygen of the alloys. This could be proved by X-ray absorption measurements above the O 1s edge in the fluorescence mode (bulk sensitive) that showed the existence of oxygen in the volume.

The clean surface was investigated by XPS. As an example, a typically measured spectrum and a least square fit of the spectrum are shown in Fig. 1. Because of its asymmetry, the Zr 3d photo emission spectrum of the clean surface cannot be explained by a simple $Zr\ 3d_{3/2}$ and $Zr\ 3d_{5/2}$ doublet (see Fig. 1). For sufficient accurate simulations (see the deviation noise below the spectrum), an additional line at higher kinetic energy has to be introduced. This line is attributed to zirconium atoms at the metal/vacuum interface. Zirconium atoms that are localized at the surface do not have their full coordination as they would have in the bulk metal and are identified by a chemical shift of their photoelectrons. The intensity of the surface component is rather high. It is greatly reduced because the metal surface is oxidized, and the missing bonds are saturated by oxygen atoms. This surface-correlated line has not yet been reported. The detection is made possible due to the good resolution of the soft X-ray monochromator and the high surface sensitivity of the experimental setup. This high surface sensitivity can only be achieved by using synchrotron radiation as an X-ray source, where one can easily tune the monochromator to an energy where the photoelectrons of this kinetic energy have its least mean free path. A good proof for the attribution of this signal is really surface induced is the observed variation of the intensity as a function of the oxidation process. On the clean sample using X-ray photo emission only, an iron signal (Fe 2p) can be detected with a poor signal-to-noise ratio. Chromium and tin were not observed on the surface. In the absorption spectra in fluorescence mode, iron and oxygen were clearly seen, but similar to the XPS investigation, no chromium or tin signal was detected. The absence of tin in the spectra is unexpected because the crosssection of Sn 3d is about 50% higher than of Fe 2p, and the concentration of tin is a factor of 7.5 higher than the iron concentration in the Zircaloy samples investigated. There possibly exists a tin depleted region near the surface. The origin of this region is unknown, but it is certainly not artificially induced by preferential sputtering during the cleaning procedure.

Heat Treatment Experiments at the Clean Surface—Heating of the clean zirconium alloy for 60 min to 450°C in ultra-high vacuum (UHV) results in the visual appearance of small shiny spots about 0.5 mm in diameter on the surface. Absorption measurements on these spots show an iron enrichment that is demonstrated by the increase of the Fe 2p edge jump compared to the normal surrounding surface in Fig. 2. Unfortunately, a quantitative determination of this iron concentration increase is not possible because of the unknown focus diameter of the monochromator and the exact beam position. During heat treatment, iron segregates onto the surface. In the shiny spot on the sample, no chromium is detected. Segregation of iron to the surface of a single zirconium crystal after 52 h of heat treatment to 820°C has been found by

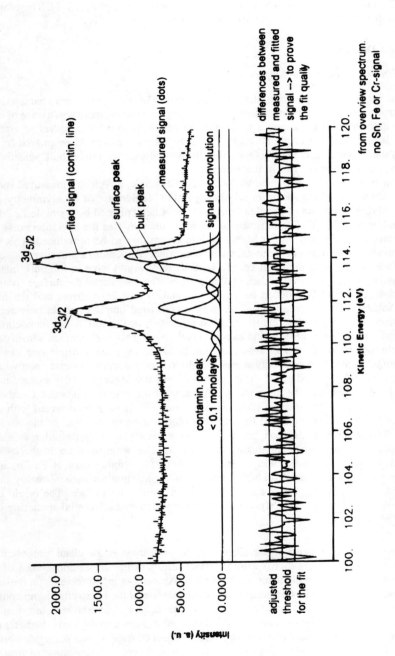

FIG. 1—*Photo emission spectrum of the Zr 3d doublet for the clean Zircaloy surface.*

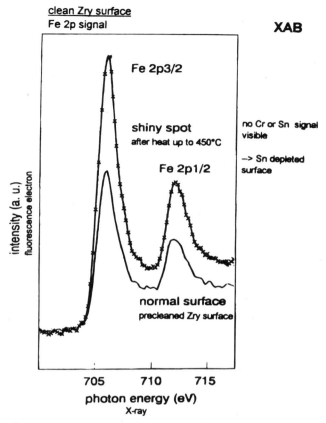

FIG. 2—*X-ray absorption spectrum that shows the locally segregated iron enrichment spots after heat treatment at 450°C due to the increase of the Fe 2p absorption edge.*

Hood and Schultz [21] from results of different surface- and bulk-sensitive concentration measurements. It is remarkable that even at significantly lower temperatures and shorter annealing times that locally uneven distribution of segregated iron could be observed on polycrystalline material.

In Autoclave Ex Situ Oxidized Samples and Native Oxide

Two samples of the same alloy have been oxidized ex-situ in a stainless steel autoclave under steam at 400°C, 105 MPa. Prior to oxidation, the surface was mechanically polished similar to the samples used in the in-situ oxidation experiments. The thickness of the pre-transition oxide obtained after three-day autoclave treatment was 0.5 µm. The post-transition oxide formed after 30 days exhibited a 3.5-µm thickness as determined by infrared spectroscopy (FTIR).

In Fig. 3 (upper diagram), absorption spectra (surface sensitive mode using electron yield) of the ex-situ oxidized samples above the O 1s threshold are compared to the native oxide and to a 2-nm-thick in-situ oxide. The onset of the absorption edge coincides for all the oxides. In the energy regime between 535 and 560 eV, three peaks labeled A to C are dominating the

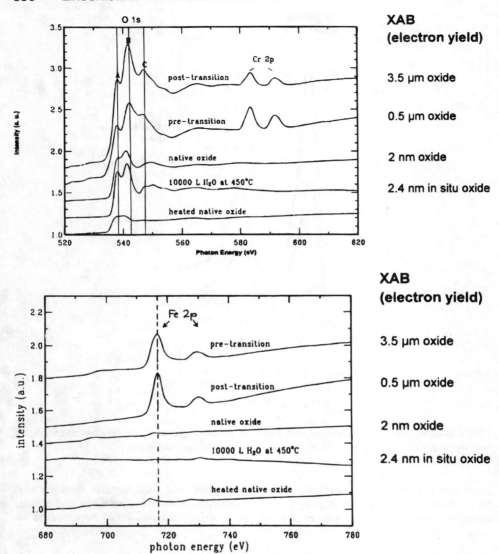

FIG. 3—*Absorption spectra above the O 1s and the Fe 2p threshold of various zirconium oxides. The differences in the NEXAFS structure are correlated with structural differences. They can be interpreted using a fingerprint method (see text). The origin of the high chromium signal of the autoclave oxides is an open question.*

spectrum. Whereas Peak A is at the same energy for all oxides, Peak B is found at higher energies for the autoclave samples compared to the native oxide. Between the pre-transition oxide and the post-transition oxide, the intensities of Peaks B and C are different. In the pre-transition oxide, Peaks B and C are not as well defined as in the post-transition oxide. One can argue that this corresponds to the not as well-defined local structural environment. The in-situ oxide, prepared at 450°C, has not reached the full ZrO_2 structure as can be seen from the different energetic position and intensity of Peaks B and C compared to the post-transition oxide. The presence of the metal-oxide interface is obviously still an influence. The shape of the spectrum of the native oxide is similar to that of the thin in-situ grown oxide. Here, as well, the structures are not as sharply defined. This can probably be understood by the presence of residual H_2O or OH groups in the native oxide. This is additionally supported by a multiple peak structure of the O 1s signal measured with XPS on the native oxide reported by Ref 4.

Heating of the native oxide results in the disappearance of nearly 70% of the oxide film (Fig. 3). This is not due to thermal desorption of the oxide, but to the diffusion of oxygen atoms into the bulk material as shown by Kaufmann [4]. Nethertheless, a residual oxide film is maintained as can be seen in the spectrum plotted in the bottom of the upper diagram of Fig. 3.

Oxides grown under autoclave oxidation showed a considerable amount of iron and chromium at the surface. The Cr $2p_{3/2}$ and Cr $2p_{1/2}$ respectively Fe $2p_{3/2}$ and Fe$2p_{1/2}$ doublets can be seen in the two top spectra of the upper and lower diagrams of Fig. 3 around 580 eV respectively 718 eV. On the clean Zircaloy surface, chromium and iron were below the detection limit. After in-situ oxidation, they were not observed as well. The intensity of the chromium and iron line is too high to be induced by improper handling, such as using a chromium-containing pair of tweezers. The reason must be either strong element selective segregation during long-term hot oxidation in the autoclave or an unknown path of contamination from the autoclave itself. (Nickel is not observed after the autoclave oxidation).

Due to the inhomogeneous charging effects of the insulating zirconium oxide, it was not possible to obtain reliable and reproducible photoelectron spectra from the autoclave samples under our experimental conditions.

In Situ Oxidation Experiments

In Situ Oxidation with Water at Room Temperature—The samples precleaned under UHV were exposed to water, and the oxidation reaction was followed by XPS. Figure 4 shows photoemission data from a Zircaloy sample that was oxidized using controlled exposures of H_2O at room temperature. The exposure range was from 1 L H_2O (1 L = 1 Langmuir = 1 · 10^{-6} mbar · s) to 1 · 10^5 L H_2O. The photon energy for excitation of the Zr 3d doublet was chosen to be 295 eV, so that the kinetic energy of the photo electrons is around 110 eV. Therefore, the mean free path of the electrons is about 0.6 nm. The spectra are analyzed in detail, as shown in Fig. 5, using a least square fitting procedure.

Up to four doublets were fitted to the data. The spin-orbit splitting (the difference between the two corresponding peaks of a doublet) of 2.37 eV was determined by the evaluation of the spectra from clean Zircaloy and of the native zirconium oxide. The difference of the two values was less than 0.03 eV. For the following analysis, it was kept fixed for all the lines. The line width of the oxide peak and of the bulk metal was similarly fixed. On the clean surface, the additional surface-correlated line was narrower than the oxide and the bulk metal line, which nearly had the same width.

In Fig. 5 typical spectra for 1 L, 100 L, and 1 · 10^5 H_2O are shown. For the simulation of the 1 L H_2O spectrum, the bulk metal line, the surface component, and a first substoichiometric oxide are needed. The existence of the suboxide has been reported earlier [10]. Using these three doublets, the fit is excellent. No higher oxidation states are visible at this low coverage. At the 100 L H_2O exposure, a fourth line, the second substoichiometric oxide, Suboxide (2), is well pronounced. The bulk metal and the surface lines are reduced in intensity and Suboxide (1) is still growing. At the 1 · 10^5 L H_2O exposure, the zirconium oxide layer is strongly

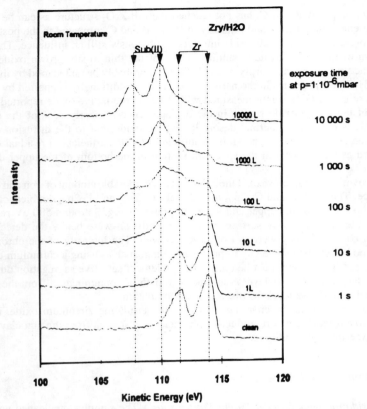

FIG. 4—*A set of photo emission spectra for the room temperature H_2O exposure using the flow of intensity from the metal peak of the clean sample to the oxide peak with increasing exposure times.*

dominated by the second oxide peak. It has changed its energy position toward the ZrO_2 value and the native oxide, but has not reached it yet.

The detailed analysis of the spectra is summarized in Fig. 6. On the upper side of the diagram, the intensities of the various lines normalized to the Zr 3d bulk metal peak are shown as a function of the H_2O dosage. While Suboxide (1) is increasing slowly with exposure, suboxide (2) is increasing more rapidly from coverage corresponding to an exposure of 100 to 1000 L of H_2O. From this, we conclude that suboxide (1) is located at the interface, whereas Suboxide (2) extends to the layers on top of it. While the energy for Suboxide (1) (lower diagram of Fig. 6) relative to the bulk metal position is fairly constant at 1.5 ± 0.2 eV, the energy for Suboxide (2) is decreasing continuously from -2.7 eV down to -3.4 eV at $1 \cdot 10^5$ L H_2O. The binding energy difference between the metal and the ZrO_2 peak was measured in Ref *11* to be about 4.6 eV. Since the surface peak and the bulk peak were interpreted as a single line in the reference, the energy difference to the ZrO_2 peak must be compared to the average energy of the surface and the bulk peak. This would reduce the value to about 4 eV, which is still larger than the position of Oxide (2). The existence of two types of suboxide is in agreement with the findings for the O_2 reaction in Ref *11*, but the values for their energetic position are different. Suboxide (1) was reported to be at -1.2 eV and Suboxide (2) at -3.3 eV. Both suboxides were found at fixed energies.

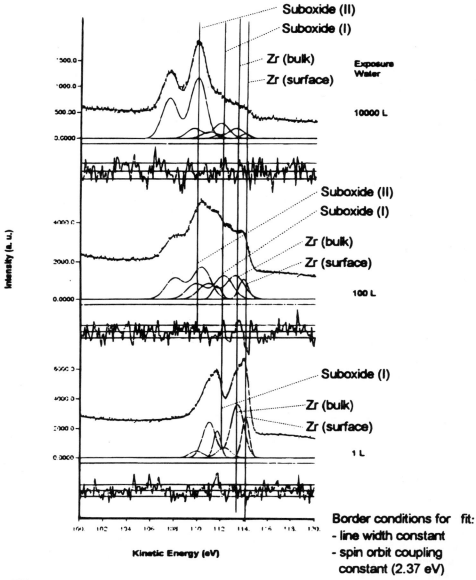

FIG. 5—*Representative least square fit procedures of the Zr 3d photo emission spectra for 1 L, 100 L, and $1 \cdot 10^5$ L H_2O exposures at room temperature on the Zircaloy. The curves in the lower part of the plots show the differences between experimental and calculated spectra.*

In situ Oxidation at 450°C—The in situ reaction of the zirconium with H_2O at 450°C is shown in Fig. 7. The data analysis was done analogously to the room temperature spectra. Only very small differences appear in the experimental spectra up to the 10 L H_2O exposure. An exposure of 100 L was necessary to produce the intensity of the oxide peaks clearly visible

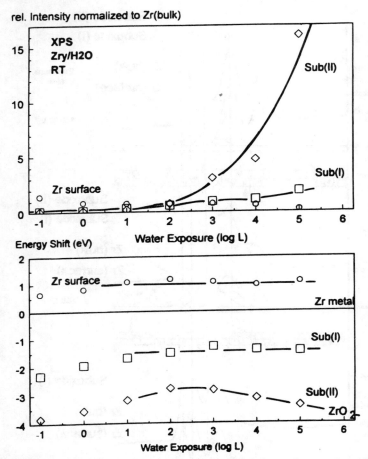

FIG. 6—*Detailed analysis of the spectra of Fig. 4 using least square fit procedures. Energy shifts are measured against the metal bulk line. The intensities are normalized to the bulk metal peak.*

in the raw spectra, and at 1000 and 10 000 L exposures, a dramatic decrease of the zirconium metal lines is observed according to the well-established oxide-correlated signals. The detailed analysis is summarized in Fig. 8. In the upper diagram, similar to Fig. 6 for the room temperature measurement, the intensity as a function of the H_2O exposure is plotted. Suboxide (1) is slowly increasing, whereas Suboxide (2) increases rapidly from 100 L on. At the same exposures, the surface peak is greatly reduced. It will be shown later that, after the 100 L H_2O exposure, a monolayer of oxide is developed.

A comparison between room temperature and 450°C kinetics of Suboxide (2) formation is presented in Fig. 9. The sharp increase for high temperature reaction can be clearly seen. The dashed line indicates the expected behavior, starting at low exposures, if only temperature is responsible for the increased growth of the oxide layer (Arrhenius-like behavior).

Comparison of Reactivity to Water and Oxygen—Similar in situ oxidation experiments were performed using clean Zircaloy samples and oxygen as the reactive gas. From recorded XPS-

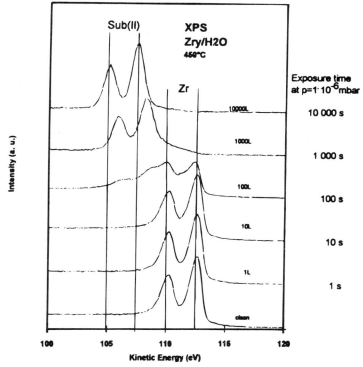

FIG. 7—*Photo emission spectra as a function of H_2O exposure at 450°C showing the shift of the Zr 3d doublet with oxide formation.*

spectra, significant differences in the reactivity between water and oxygen to Zircaloy can be concluded, as shown in Fig. 10. For comparison, typical results of the H_2O and O_2 reactions with the Zircaloy sample are shown. In the left part of the diagram, a set of spectra for the room temperature water reaction are plotted. A spectrum for the very large 5 mbar · min (10^8 L) exposure has been added. Even at this high exposure, the bulk metal peak did not totally vanish in contrast to the situation for the oxygen oxidation shown in the spectra of the right diagram. The onset of the clean spectrum at the high kinetic energy site is marked as a dashed line in the diagram. The origin of the shift in the onset observed at higher coverages is induced by the decrease of the surface line. Higher reactivity of O_2 can be seen in the right part of the figure. In contrast to the H_2O case, after 5 mbar · min O_2 exposure, the substrate signal vanished completely. A detailed comparison of the spectra showed similar intensities in the spectrum for an oxide formed by 111 L O_2 to one of $1 \cdot 10^5$ L H_2O. This is a factor of ~1000 in the exposure time and therefore reactivity. This means that if the same number of gas molecules is offered to the Zircaloy surface and defined by identical exposures, significant differences in the tendency exists between water and oxygen to form an oxide layer at the Zircaloy surface. In the case of gas mixtures containing water and oxygen in a combating surface adsorption process, the higher reactivity of oxygen favors the formation of O adsorption layers. This is a situation, for example, in non-purged autoclave experiments.

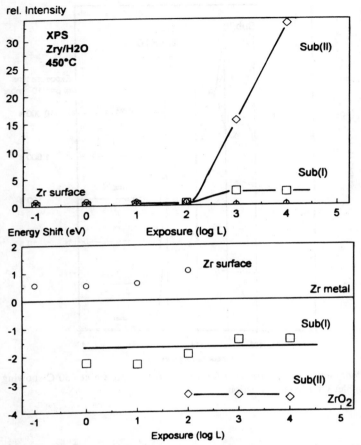

FIG. 8—*Detailed analysis of the spectra of the high temperature water adsorption experiment (Fig. 7) using least square fit procedures. Energy shifts are measured against the metal bulk line.*

Determination of Scale Thickness from XPS Measurements—The oxide film thickness on top of the zirconium alloy has been determined to be a function of the water exposure, by assuming an exponential attenuation of the escaping photoelectrons [17,22] and using the following relationship

$$I_1 e^{-\frac{d}{\lambda \cdot \sin \theta}} = I_2 C (1 - e^{-\frac{d}{\lambda \cdot \sin \theta}}) \tag{1}$$

Here, I_1 is the intensity of the Zr 3d line from the top oxide layer and I_2 from the underlying Zircaloy substrate, d is the thickness of the oxide film, and θ is the takeoff angle of the photoelectrons from the surface. At our detection geometry, θ equals 90° and is kept constant. The mean free path of the photoelectrons, λ, at kinetic energies ~120 eV is 0.6 nm and is thought to have comparable values in the metal and in the zirconium oxide. The difference in the density is corrected by the factor, C, which is the ratio of the number densities of zirconium atoms in the zirconium alloy and in ZrO_2. This can be expressed in terms of the

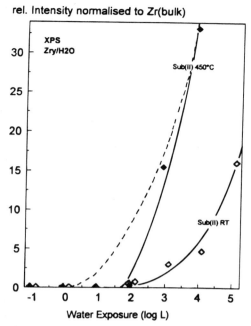

FIG. 9—*Comparison of reactivity for the oxidation of Zircaloy by water at room temperature and 450°C.*

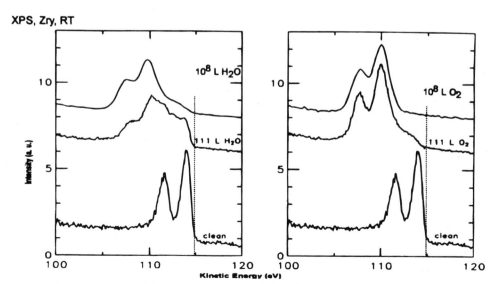

FIG. 10—*Comparison of the room temperature H_2O and O_2 reaction with the zirconium alloy. The spectrum of $1 \cdot 10^5$ L H_2O compares well to the spectrum of 111 L O_2. For the same coverage, a factor of nearly 1000 higher H_2O exposure is needed. Note the shift in the onset due to the decrease of the surface component at higher coverages.*

respective mass densities, ρ_{Zr} and ρ_{ZrO_2}, the atomic weight A_{Zr}, and the molecular weight M_{ZrO_2} as

$$C = \frac{\rho_{ZrO_2}}{\rho_{Zr}} \cdot \frac{A_{Zr}}{M_{ZrO_2}} \quad (2)$$

In the near interface region, the stoichiometry of the oxide film varies, so that Eq 1, which describes the case of a homogeneous thin oxide film on the metal, is not strictly fulfilled. Nevertheless, for the ease of calculation, the oxide film is considered to consist of ZrO_2 only, although the intensity, I_1, is taken to be the sum of all the oxygen-induced Zr 3d lines. I_2 is the sum of the surface-induced line and the zirconium bulk line. The oxide thickness can now be calculated from Eq 1 to be

$$d = \lambda \cdot \sin\theta \cdot \ln\frac{I_1 + CI_2}{CI_2} \quad (3)$$

Using this equation, the oxide thickness of the zirconium alloy that reacted with water at room temperature and at 450°C is shown in Fig. 11 as a function of the H$_2$O exposure in logarithmic and linear scale. The appropriate intensities are taken from Figs. 6 and 8, respectively. The oxide thickness increases sharply after an exposure of 100 L for the high temperature reaction. The base oxide thickness from where the rapid increase starts is 0.3 to 0.4 nm, which corresponds approximately to the oxide monolayer thickness. The linear (lower) plot expresses the nonlinear growth of the oxides, even at the onset of oxidation.

From the ZrO$_2$ bulk densities, one can estimate the thickness, d_0, corresponding to a monolayer to be

$$d_0 = \sqrt[3]{\frac{M_{ZrO_2}}{\rho_{ZrO_2} N_{AV}}} \quad (4)$$

where N_{AV} is the Avogadro number. For ZrO$_2$, d_0 equals 0.33 nm, which is approximately the base oxide thickness. The highest oxide thickness of 2.4 nm at 10 000 L is already close to the detection limit of the method, because it is necessary to have a non-zero signal from the oxide and the bulk metal (internal reference), simultaneously. At this thickness, the metal peak intensity is only 2% of the oxide intensity, and the signal-to-noise ratio is becoming significant. For the room temperature reaction, a continuous increase of oxide thickness can be observed.

For comparison, we would like to note that the thickness of a native oxide on a one-year laboratory-stored sample can be estimated from the fact that even using conventional laboratory XPS of Mg K$_\alpha$ of 1253.6 eV radiation, it was not possible to observe the bulk metal line. In that case, the photo electrons have a kinetic energy of ~1080 eV and a mean free path, λ, of 1.8 nm. Using Eq 3 and taking into account the minimal detectable intensity, I_2, determined by the signal-to-noise level of the experimental device, the thickness of the native oxide can be calculated to be at least 7 nm.

Interpretation of the Findings

The low reactivity of the Zircaloy at 450°C in the initial exposure region is surprising. In comparison to the room temperature behavior (Fig. 11), formation of an oxide layer initially develops only after the formation of a monolayer oxide. At room temperature, a more continuous increase to thicker oxide is observed. High surface temperatures can only promote the oxidation

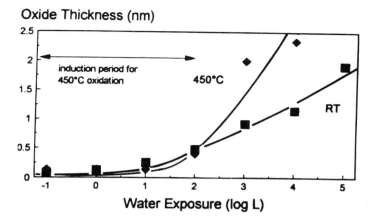
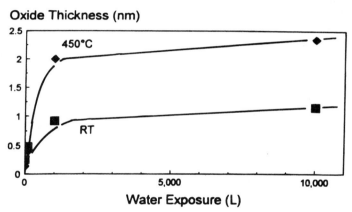

FIG. 11—*Comparison of the oxide thickness for the H_2O reaction at room temperature and at 450°C. In the logarithmic scale, no saturation takes place up to the highest measured exposures. In the linear scale, the high reactivity at low exposure is demonstrated.*

reaction after this initial oxide layer has been established. Temperature dependent activated oxidation is observed only after this induction period on top of a monolayer of zirconium oxide. This can possibly be explained by the following model.

The first step of the room temperature reaction is the adsorption of water molecules in the form of dipole layers as shown schematically in Fig. 12. Even at room temperature, a significant fraction of adsorbed molecules will dissociate into OH- and H-radicals. This would explain as well the nature of the native oxide, which consists of an oxide containing residual H_2O or OH groups as deduced from X-ray absorption spectra and the multiple oxygen structure in the XPS spectra. The adsorbed radicals are transformed into ionic oxygen species using electrons from the zirconium. In that way, oxide formation is performed. XPS reveals the development of this first oxide as Suboxide (1). Our experiment cannot distinguish between the surface and subsurface position of the incorporated oxygen ion in comparison to the dimension of a Zirconium atom layer big information depth of 0.6 nm for the photoelectrons. Subsurface inclusion has been proposed by Ref 23 as an explanation for the unexpected change

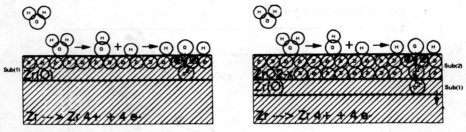

FIG. 12—*Formation of zirconium oxides with water at room temperature schematically showing the different steps that led to the formation of an oxide layer.*

of the measured work function. But this subsurface oxygen inclusion, in principle, would be not in contradiction with the formation of suboxide (1) found in our investigation. At higher coverages Suboxide (2) is developed.

At elevated temperatures the increased fraction of OH species is converted to oxygen and hydrogen, as shown in Fig. 13. This enhanced dissociation should lead to a significantly faster formation of the zirconium oxide system just described. The 450°C temperature is not high enough for desorption of the oxygen atoms, but a significant diffusion of oxygen into the bulk material is possible as shown by Anges Electron Spectroscopy (AES) depth profile measurements [4]. The diffusion of adsorbed oxygen reduces the amount of oxygen atoms that are available for oxidation. In this way, higher exposures are necessary (this is equivalent to the longer times that water molecules are available for adsorption) to form the suboxide (1) layer and further corrosion layers. Oxygen diffusion and the development of suboxide defect structures govern the kinetics of the corrosion reaction. We assume that the oxide system formed on top of the clean Zircaloy is responsible for the corrosion rate also on thick oxide layers. The rate determining step for every corrosion is the transportation of oxygen ions through the oxide layer systems. Suboxide (1) plays the role of a rate determining active layer

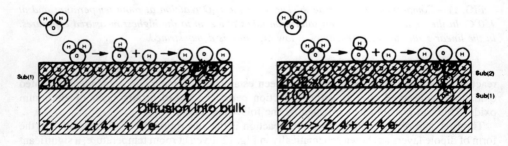

At high temperature: Two combatting reactions
Oxidation to O2-
Diffusion into bulk

---> Induction period for the formation of Sub(1) Zr(O)

FIG. 13—*Formation of zirconium oxides with water at 450°C schematically showing the different steps that led to the formation of an oxide layer.*

that is responsible for the corrosion process. This study shows that before the formation of corrosion layers can start a thin sub-monolayer structure will be developed. This sub-layer with high nonstoichiometry plays the role of a kinetic determining interlayer. The mechanistic aspects that describe the role of charge transfer through a zirconium oxide barrier layer are given in Ref 1. The extension where then the Suboxide (2) structures are converted to stoichiometric ZrO_2 is more than 2 nm, which is significantly thicker than in the case of O_2 oxidation [10].

Conclusion

The main conclusions drawn from the present study are summarized as follows:

1. Evaluation of the spectra found on clean Zircaloy showed the existence of a surface-induced Zr 3d core level shift that was observed for the first time.
2. It has been demonstrated that the oxide formation of the zirconium alloy near the surface results in two different suboxides: Suboxide (1) that is located near the surface and Suboxide (2) that dominates with increasing layer thickness.
3. The energy position of the chemically-shifted Zr 3d line moves toward the value of the stable ZrO_2, but does not reach it within the first 2 nm during water reaction. This implies that the intermediate region, where Suboxide (2) is converted into ZrO_2, is not yet finished within this dimension.
4. In situ oxidation experiments with water and oxygen were performed. A kinetic evaluation of the spectra obtained revealed significant differences in the reactivity. An 100 L O_2 exposure results in the same coverage as 100 000 L H_2O.
5. An induction period for the corrosion reaction is observed. The thickness of the oxide layer after reaction with water at 450°C is initially low, but sharply increases as a function of the exposure. The base thickness, from where the increase starts, corresponds to approximately one monolayer.

References

[1] Beie, H.-J., Mitwalsky, A., Garzarolli, F., Ruhmann, H., and Sell, H.-J., "Examinations of the Corrosion Mechanism of Zirconium Alloys," in this volume.
[2] West, P. E. and George, P. M., *Journal of Vacuum Science and Technology,* Vol. A5, No. 4, 1987, p. 1124.
[3] Steiner, P., Sander, I., Siegwart, B., Hüfner, S., and Fresenius, Z., *Analytische Chemie,* Vol. 329, 1987, pp. 272–277.
[4] Kaufmann, R., Klewe-Nebenius, H., Moers, H., Pfenning, G., Lennet, H., and Ache, H., *Surface and Interface Analysis,* Vol. 11, 1988, pp. 502–509.
[5] Tapping, R. L., *Journal of Nuclear Materials,* Vol. 107, 1982, pp. 151–158.
[6] Valyukhov, D. P., Golubin, M. A., Grebenschichokov, D. M., and Shestopalovu, V. I., *Sov. Phys. Solid State,* Vol. 24, No. 9, 1982, p. 1594.
[7] Rao, C. N. R. and Subba, G. V. in *Transition Metal Oxides NSRDS,* Monograph 49, National Bureau of Standards, Washington, DC, 1974.
[8] Steep, S. and Riekert, A., *Journal of Less Common Metals,* Vol. 17, 1969, p. 429.
[9] Ackermann, R. J., Garg, S. P., and Rauh, E. C., *Journal, American Ceramic Society,* Vol. 60, 1977, p. 341.
[10] Barr, T. L., *Journal of Vacuum Science and Technology,* Vol. 14, 1977, p. 660.
[11] Kumar, L., Sarma, D. D., and Krummacher, S., *Applied Surface Science,* Vol. 32, 1988, p. 309.
[12] Jungbluth, B., Siccking, G., and Papachristos, T., *Surface Interface Analysis,* Vol. 13, 1988, p. 135.
[13] Fromm, E. and Mayer, O., *Surface Science,* Vol. 74, 1978, p. 259.
[14] Foord, J. S., Goddard, P. J., and Lambert, R. M., *Surface Science,* Vol. 94, 1980, p. 339.
[15] Hoflund, G. B. and Cox, D. F., *Journal of Vacuum Science and Technology,* Vol. A1, 1983, p. 1837.
[16] Veal, W., Lam, D. J., and Westlake, D. W., *Physics Review B,* Vol. 19, 1979, p. 2856.

[17] Sen, P., Sarma, D.-D., Budhani, R. C., Chopra, K. L., and Rao, C. N. R., *Journal of Physics F (Metal Physics)*, Vol. 14, 1984, p. 565.
[18] Yashonath, S., Sen, P., Hegde, M. S., and Rao, C. N. R., *Journal of the Chemical Society, Faraday Transactions I*, Vol. 79, 1983.
[19] Schneider, H., *Mikrochimica Acta*, Supplement 8, 1978, p. 144.
[20] Zehringer, R., Hauert, R., Oelhafen, P., and Güntherodt, H.-J., *Surface Science*, Vol. 215, 1989, p. 501.
[21] Hood, G. M. and Schultz, R. J. in *Zirconium in the Nuclear Industry: Eighth International Symposium, ASTM STP 1023*, L. F. P. Van Swam and C. M. Eucken, Eds., American Society for Testing and Materials, Philadelphia, 1989, p. 435.
[22] Krishnan, G. N., Wood, B. J., and Cubicciotti, D., *Journal*, Electrochemical Society, Vol. 128, 1981, p. 191.
[23] Zhang, C.-S., Flinn, B. J., Mitchell, I. V., and Norton, P. R., *Surface Science*, Vol. 245, 1991, p. 373.

Joel Godlewski[1]

How the Tetragonal Zirconia is Stabilized in the Oxide Scale that is Formed on a Zirconium Alloy Corroded at 400°C in Steam

REFERENCE: Godlewski, J., "**How the Tetragonal Zirconia is Stabilized in the Oxide Scale that is Formed on a Zirconium Alloy Corroded at 400°C in Steam**," *Zirconium in the Nuclear Industry: Tenth International Symposium, ASTM STP 1245*, A. M. Garde and E. R. Bradley, Eds., American Society for Testing and Materials, Philadelphia, 1994, pp. 663–686.

ABSTRACT: Zircaloy-4, in three different metallurgical forms (stress relieved, recrystallized, and β-quenched), was oxidized at 400°C, in steam, up to 95 days. For each sample, the fraction of tetragonal zirconia was measured by X-ray diffraction and Raman spectroscopy. These two techniques show the presence of several zones containing tetragonal zirconia: a zone rich in the oxide near the metal-oxide interface and an other zones with lower concentrations in the rest of the pre-transition layers. For the post-transition samples, the external sublayer contains only a small amount of tetragonal zirconia.

Measurements of residual stresses by X-ray diffraction in the metal underlying the oxide show that the metal is under tensile stress state and that the stress values vary with oxidation duration. The level of the stress depends on the metallurgical form of the initial metal.

The low penetration of X-rays in the material also made it possible to show the presence of a very high stress gradient near the metal-oxide interface that can explain the high proportion of tetragonal zirconia near the interface.

The study of the incorporation of intermetallic precipitates in the oxide and their chemical changes was carried out by electron microprobe analysis on taper cross sections of the oxide. This technique makes it possible to perform a large number of point analyses that yield satisfactory statistics for the variation in the iron/chromium (Fe/Cr) atom ratio of the precipitates in the oxide.

The intermetallic precipitates are incorporated into the oxide layer and then undergo a chemical change starting at a particular distance from the metal/oxide interface. The characteristic values of the Fe/Cr ratio before oxidation (1.6 for stress relieved and recrystallized conditions and 0.8 for β-quenched samples) are progressively spread out during oxidation. This change could correspond to an oxidation of intermetallic precipitates with segregations of iron at the precipitate-oxide interface, as shown in the literature.

The oxidation of the precipitates is accompanied by a volume change that should lead to the formation of a stress field around the precipitates and could stabilize the neighboring tetragonal phase. When the precipitates are completely oxidized, the stress field disappears and there is a transformation of the tetragonal phase to a monoclinic form, leading to the kinetic transition.

Stress relaxation is shown by a decrease of the tensile stresses in the metal underlying the oxide that is undergoing kinetic transition.

KEY WORDS: zirconium alloys, annealed state, recrystallized state, beta-quenched phase, tetragonal zirconia, residual stresses, stress gradients, intermetallic precipitates, X-ray diffraction,

---

[1] Research engineer, Commissariat à l'Energie Atomique (CEA), DTA-CEREM-DECM-SRMA Centre d'Etudes de Saclay, 91191 Gif sur Yvette, France.

stress measurements, Raman spectroscopy, electron microprobe analysis, zirconium, nuclear materials, nuclear applications, radiation effects

The composition and the metallurgical form of Zircaloy-4 makes it possible to obtain satisfactory corrosion resistance under present in-reactor conditions. To increase the burnup up to 60 000 MWd/t, it is important to improve the corrosion resistance of the zirconium cladding while retaining good mechanical properties. This has led to additional studies on the oxidation mechanisms of zirconium alloys in order to better understand the effects of the metallurgical state and the chemical composition on these mechanisms.

A recent study [1] showed that tetragonal zirconia is present in the oxide layer formed at 400°C and that it is transformed into a monoclinic phase in the kinetic transition. Tetragonal zirconia is stable at temperatures above 1150°C, but it can be stabilized at lower temperatures in the presence of high compressive stresses or, if elements like yttrium or calcium are introduced, into the zirconia lattice.

Several studies [2–7] have shown the presence of high compressive stresses in the oxide layer. However, the results obtained, summarized in Table 1, are scattered and there are no data on the level of the stresses at the metal-oxide interface, a region where there is an appreciable amount of tetragonal zirconia. The additive elements also have a significant effect on the corrosion resistance. Iron and chromium, which form intermetallic precipitates, affect the corrosion resistance either by their size or their composition, but the mechanism for their effects is not clearly established.

The objective of this work is to study the effect of the stresses at the metal-oxide interface in the metal underlying the oxide, as well as the role of the iron and chromium additives on the stabilization of tetragonal zirconia. The study focused on the two following aspects:

1. measurement of the stress levels at the metal-oxide interface as a function of oxidation time, and
2. modification of the composition of the intermetallic precipitates on their incorporation into the oxide layer.

TABLE 1—*Maximum stress values in zirconia determined by different authors.*

References	Technique Used	Zirconium-Based Alloy[a]	Oxidizing Medium	Temperature, °C	Stress, GPa
David et al. [2]	X-ray diffraction	Zr-Zy-2	O_2	500	1.17
Roy et al. [3]	deflection method	Zy-2	O_2	500	0.27 to 2.6
Bradhurst et al. [4]	curvature method	Zr-Zy-2	O_2	500 to 700	0.62 (500°C)
Gadiyar et al. [5]	curvature method	Zr-Zy-2 Zr-1 Nb	steam	400 550	0.5 to 1.3 0.2 to 0.6
Antoni [6]	deflection method	Zy-4	steam	450	3
Garzarolli et al. [7]	curvature method	Zy-4	steam	400	0.2 to 0.9

[a] Zr = zirconium, Zy-2 = Zircaloy-2, and Zy-4 = Zircaloy-4.

Experimental Materials and Oxidation Kinetics

The study was carried out using Zircaloy-4 from one ingot that was produced as tubing in three different metallurgical states: stress relieved (at 440°C), recrystallized (at 580°C), and β-quenched phase (at 1050°C). The alloys are described in Table 2. The precipitates are mainly the $Zr(Fe,Cr)_2$ type with a hexagonal structure. They are uniformly distributed in the alloys except for the β-quenched phase, which has segregated precipitates along the inter-lamellar boundaries produced in the β-quenching. For the stress relieved and recrystallized alloys, the size of the precipitates is 0.1 to 0.2 μm, while the range for the β-quenched phase alloy is 0.04 to 0.15 μm. Autoclave corrosion tests were conducted at 400°C in steam at a pressure of 10.3 MPa for a duration of 1 to 95 days.

Quantitative Determination of the Tetragonal Zirconia in the Oxide Layers

X-ray diffraction and Raman spectroscopy were used for the localization and quantitative measurements of the tetragonal zirconia in the oxide layers as a function of the oxidation time. The Raman spectroscopy was carried out on a DILOR XY 800 confocal Raman spectrometer. Radiation excitation was the 514.5 nm. The analyses were made either at the oxide cladding surface or with the oxide taper cross sections following the distribution of the tetragonal zirconia in the oxide layers.

The quantitative determination of the tetragonal zirconia in the oxide layers was performed using equations proposed by Gravie et al. [8] for X-ray diffraction measurements and by Clarke et al. [9] for those carried out with Raman spectroscopy.

The Internal Stresses at the Metal-Oxide Interface

In our study, we used X-ray diffraction to measure the stresses in the metal underlying the oxide. One advantage of this technique is that it is nondestructive and it makes it possible to describe the mean stress of the metal underlying the oxide and its distribution as a function of the distance from the metal-oxide interface. In previous works [2–7], the internal stress measurements were made either by the radius of curvature of the sample or by deflection. These two techniques give the average stress over the whole oxide layer thickness.

The Stress Measurement at the Metal-Oxide Interface by X-Ray Diffraction

The residual stress determinations by X-ray diffraction use the interlattice distance, d_{hkl}, of the $\{hkl\}$ planes as a strain gage [10,11]. In a material under stress, d_{hkl} varies with the direction of the $\{hkl\}$ planes with respect to the stress. In conventional X-ray diffraction, d_{hkl} planes

TABLE 2—Chemical composition of materials.

Material	Chemical Composition					Annealing Parameters,[a] ΣAi (h)
	Sn, percent by weight	Fe, percent by weight	Cr, percent by weight	O, ppm		
Zy4R	1.3	0.2	0.1	1200	recrystallized	$>1 \times 10^{-17}$
Zy4B	1.3	0.2	0.1	1200	beta-quenched	0
Zy4D	1.3	0.2	0.1	1200	stress relieved	$>1 \times 10^{-17}$
Zy4	1.45	0.2	0.1	1200	recrystallized	$>1 \times 10^{-17}$

[a] The cumulative annealing parameter was calculated using a $Q/R = 40\ 000$ K.

parallel to the surface of the sample are measured. The principle of the method is to rotate the sample by angle ψ (Fig. 1) and to position other grains of the same $\{hkl\}$ index so that they diffract the X-rays.

The interlattice distance is related to the position of the $2\theta_{\phi\psi}$ peaks using Bragg's law. The deformation, $\epsilon_{\phi\psi}$ of the diffraction volume is given by the following equation for the ϕ direction of the sample

$$\epsilon_{\phi\psi} = (d_{\phi\psi} - d_0)/d_0 \tag{1}$$

where d_0 is the interlattice distance of the $\{hkl\}$ planes without stresses.

For an isotropic continuous material and uniform elastic stress state, the measured local deformation is related to the components of the stress tensor by the equation

$$\epsilon_{\phi\psi} = (1 + v)/E)(\sigma_{11}\cos^2\phi + \sigma_{12}\sin2\phi + \sigma_{22}\sin^2\phi)\sin^2\psi - v/E(\sigma_{11} + \sigma_{22}) \tag{2}$$

E and v are the Young's modulus and the Poisson's ratio, respectively.

For a tubular sample, the tangential and axial stresses are obtained, respectively, when the axis of the tube is parallel ($\phi = 0°$) and when it is perpendicular ($\phi = 90°$) to the goniometer axis (Fig. 1).

The internal stress measurements by X-ray diffraction are made in the metal underlying the oxide through this oxide layer. It was not possible to directly measure the stresses in the oxide by this technique, since the zirconia formed is made up of very small grains (with a size of less than 50 nm) and has a very marked texture. In addition, the wide-angle peaks of zirconia are superimposed on those of the metal.

A Siemens Type F Ω goniometer was used to measure the internal stresses. The position and the intensity of the diffraction peaks were measured with an ELPHYSE linear localization detector. The measurements were made for the (10.4) family of planes that correspond to a 2θ angular position = 156.69° for the chromium K_α radiation.

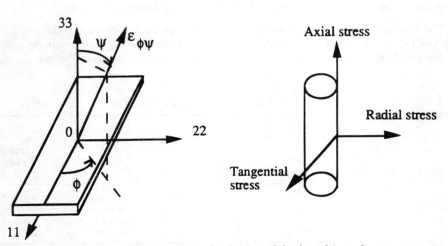

FIG. 1—*Coordinate system and definition of the ϕ and ψ angles.*

Stress Distributions at the Metal-Oxide Interface

In the X-ray diffraction internal stress measurements, if the material has a stress gradient at the surface, the deformation measured as a function of $\sin^2\psi$ is no longer linear. The expression for the depth of X-ray penetration in the case of a Ω-type mounting is given by the following equation as a function of ψ

$$\tau_{\{hkl\}} = \frac{\sin^2(\theta_{\phi\psi\{hkl\}}) - \sin^2\psi}{2\mu\sin(\theta_{\phi\psi\{hkl\}})\cos\psi} \tag{3}$$

where μ is the linear absorption coefficient of the material.

For a material with a stress gradient, the deformation, $\langle\epsilon_{\phi\psi}\rangle$, is expressed as follows

$$\langle\epsilon_{\phi\psi}\rangle = \int_0^e \epsilon_{\phi\psi}(z)\exp(-z/\tau)dz / \int_0^e \exp(-z/\tau)dz \tag{4}$$

where e is the total thickness analyzed.

By substituting Eq 2 in Eq 3, we obtain

$$\langle\epsilon_{\phi\psi}\rangle = (1 + \upsilon)/E)(\langle\sigma_{11}\rangle\cos^2\phi + \langle\sigma_{12}\rangle\sin 2\phi + \langle\sigma_{22}\rangle\sin^2\phi)\sin^2\psi \tag{5}$$
$$-\upsilon/E(\langle\sigma_{11}\rangle + \langle\sigma_{22}\rangle)$$

with

$$\langle\sigma_{ij}\rangle = \int_0^e \sigma_{ij}(z)\exp(-z/\tau)dz / \int_0^e \exp(-z/\tau)dz \tag{6}$$

or

$$\langle\sigma_{ij}\rangle = \sigma_{ij}(z = 0) + \int_0^e \exp(-z/\tau)g_{ij}(z)dz \tag{7}$$

where $g_{ij}(z)$ is the stress gradient

$$g_{ij}(z) = \frac{d}{dz}\sigma_{ij}(z) \tag{8}$$

We have calculated the depth that the X-rays penetrate as a function of $\sin^2\psi$ for the chromium K_α radiation and for the (10.4) family of planes. The results are shown in Fig. 2.

If we wish to measure local stresses close to the interface, we must use high values of ψ. But if a tubular sample is used, when the value of ψ is high, the errors due to defocusing are very large. To overcome this problem, the measurement of a stress gradient close to the metal-oxide interface was obtained using a Zircaloy-4 sheet oxidized in steam at 400°C to 2 μm oxide thickness.

The Incorporation of Intermetallic Precipitates in the Oxide Layers

Recent work [7,12–14] has shown that the intermetallic precipitates are incorporated in the oxide layer where they undergo microstructural modifications. The characterization of these precipitates is generally carried out either by SIMS (secondary ion mass spectrometry) or

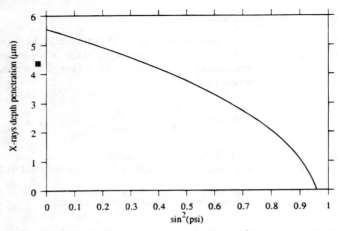

FIG. 2—*Penetration of X-rays as a function of $sin^2\psi$ in the Zy4R alloy.*

more generally by transmission electron microscopy (TEM). In this work, electron microprobe analysis was used to study the composition of the precipitates at different oxidation stages. The samples analyzed are listed in Table 3. For each of these, we prepared a taper cross section of the oxide at an angle of 1°. Over the apparent oxide thickness (which was multiplied by a factor of 60 due to the angle of the cut), we distinguished several analysis zones: that is, near the metal-oxide interface, near the external oxide surface, and at intermediate distances between these. In each zone, analyses were made at 400 points with a step of 1 μm, which made it possible to obtain satisfactory statistics for the various compositions of the precipitates. The volume analyzed by this technique was of the order of 1 μm³ for each analysis point.

Results

Tetragonal Zirconia Oxidation Kinetics and Distribution in the Oxide Layer

The kinetics of the oxidation at 400°C in steam for the three alloys are shown in Fig. 3. They are compared to the oxidation kinetics of an annealed Zircaloy-4 alloy oxidized under

TABLE 3—*Samples analyzed by an electron microprobe.*

Alloy	$\Delta m/S$, mg/dm²	Distance to Metal/Oxide Interface Where the Analysis is Performed, μm				
Zy4R	29	0.1	1	2.1	...	...
	36	0.1	1	2.3	...	...
	44	0.1	0.5	1.6	2.8	...
	50.5	0.1	0.4	0.9	2.1	3.5
Zy4B	30	0.1	0.4	0.9	1.7	...
	36.5	0.1	1	2.3	...	...
	46.5	0.1	0.5	1.6	2.9	...
	52	0.1	0.7	2.4	3.3	...
Zy4D	32.5	0.1	1	2.2	...	...
	39.5	0.1	0.8	1.7	2.6	...
	58	0.1	0.4	0.9	2.1	3.5
	73	0.1	0.5	0.9	2.6	4.8

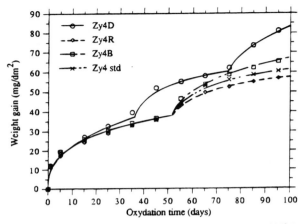

FIG. 3—*Oxidation kinetics of the alloys in steam at 400°C.*

the same conditions. Under these oxidation conditions, and for a given percentage of tin, the final, recrystallized metallurgical form shows the best corrosion resistance [15,16]. In the pretransition period, the Zy4B alloy has a corrosion resistance comparable to the Zy4R alloy; on the other hand, it is oxidized more rapidly after kinetic transition, in good agreement with previous results [17,18].

The volume fraction of tetragonal zirconia integrated over the total oxide thickness obtained by X-ray diffraction is plotted against the oxidation time in Fig. 4 for the three alloys. The variation is comparable to the results obtained previously [1] for Zircaloy-4 alloys with variable sizes of precipitates or for a Zr-1Nb alloy that was oxidized under identical conditions. An appreciable percentage of tetragonal zirconia is observed for a short oxidation time, then it decreases up to kinetic transition. Above this, there is a drop in the volume fraction of tetragonal zirconia. In the post-transition region, the volume fraction of tetragonal zirconia again increases, to a lesser extent, up to a new kinetic transition as for the Zy4D alloy.

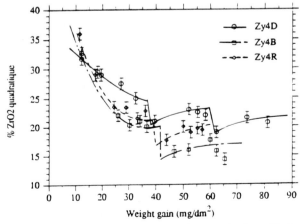

FIG. 4—*Variation of the tetragonal zirconia volume fraction in the oxide layer as a function of oxidation time.*

The Raman spectra obtained at the Zy4R alloy surface that were oxidized for different times are given in Fig. 5 and show the concentration variations for the tetragonal phase in the external zone of the oxide. When the oxidation time is short, the oxide surface is made up of about 40% tetragonal zirconia and 60% monoclinic zirconia. The concentration of the former decreases to about 20 to 25% up to kinetic transition. For the post-transition samples, the external layer has a low concentration of the tetragonal phase of about 5 to 7%. The Raman spectra obtained at different levels in the oxide layers for the Zy4R sample, oxidized for 65

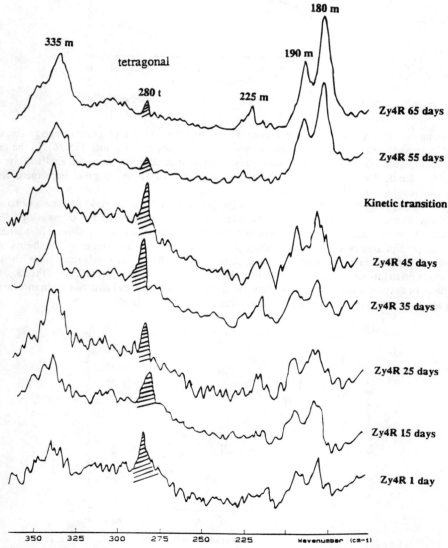

FIG. 5—*Variation of the Raman spectra obtained at the Zy4R alloy surface oxidized in steam at 400°C, as a function of oxidation time.*

days, are shown in Fig. 6. In this post-transition sample, the total thickness of the oxide can be divided into three sublayers with varying tetragonal zirconia compositions as was observed in a previous work [1].

For each of the alloys, we determined the average concentration of the tetragonal phase in each oxide sublayer for pre- and post-transition samples. The values are given in Table 4 where, for any alloy, the average volume fraction of the tetragonal zirconia in the sublayer near the metal-oxide interface does not vary between the pre- and post-transition samples. The external sublayer of the pre-transition samples has an average tetragonal zirconia volume fraction equivalent to the proportion obtained in the intermediate oxide sublayer observed for the post-transition samples for the same alloy. The average tetragonal zirconia volume fraction

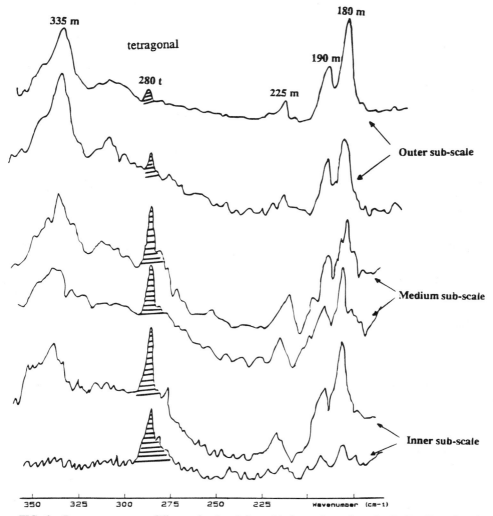

FIG. 6—*Raman spectra at different depths of the oxide layer formed on the Zy4R alloy after 65 days in steam at 400°C.*

TABLE 4—*Tetragonal zirconia volume fraction in different oxide scales formed during pre- and post-transition corrosion kinetics measured by Raman spectroscopy.*

	Tetragonal Zirconia Percent in Oxide Layers Formed during Pre-Transition Period	
Alloy	Inner Subscale	Outer Subscale
Zy4R	38 to 42	23 to 25
Zy4D	31 to 34	14 to 16
Zy4B	33 to 36	21 to 24

	Tetragonal Zirconia Percent in Oxide Layers Formed during Post-Transition Period		
Alloy	Inner Subscale	Medium Subscale	Outer Subscale
Zy4R	38 to 42	23 to 25	7 to 8
Zy4D	31 to 34	14 to 16	5 to 6
Zy4B	33 to 36	21 to 24	5 to 9

variation in the external sublayer of the oxide, as a function of oxidation time, shows that the kinetic transition observed is related to the transformation of the tetragonal phase to a monoclinic phase.

Stress Variations and Distributions in the Metal Beneath the Oxide

Before following the stress variations in the metal as a function of oxidation time, we measured the level of stresses at the surface of the starting alloys. To do this, we measured the stress at the surface of the tubes after the successive removal of the metal by chemical dissolution (Fig. 7). A significant level of compressive stresses, which decreases rapidly when going away from the surface of the tube, is observed at the surface of the as-received tubes. Starting from some 30 μm, the stresses are tensile and have a constant value (10 to 50 MPa). Where the stresses are high, the layer is removed in the sample preparation by chemical pickling before oxidation.

Using X-ray diffraction, we can measure the stresses in the metal beneath the oxide. Two types of information can be obtained: (1) the value of the average stress and (2) information on the stress distribution in the metal beneath the oxide (indicating a stress gradient).

The general shape of the average stress curves in the metal beneath the oxide is correlated with oxidation kinetics, as is shown in Fig. 8. These can be described as follows:

1. There is an increase of the different stress tensor components in the pre-transition period.
2. The kinetic transition is characterized by a drop in the stress level.
3. In the post-transition period, a new increase in the stress tensor components is shown simultaneously with the growth of a new oxide layer described by a parabolic kinetic law.

The average tangential and axial stresses for the three metallurgical forms are plotted versus the oxidation time in Fig. 9.

Garzarolli et al. [7] have measured the stress level in the oxide layer for Zircaloy-4, recrystallized and oxidized in steam at 400°C, by the radius of curvature method for different oxidation times. They found a compressive stress of -400 MPa in the oxide in a 2-μm-thick layer. In our case, for the Zy4R covered with a 2-μm-oxide layer, we obtained an average tensile stress of 150 MPa over a distance of 4 μm.

GODLEWSKI ON TETRAGONAL ZIRCONIA 673

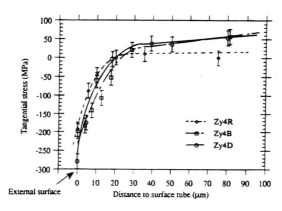

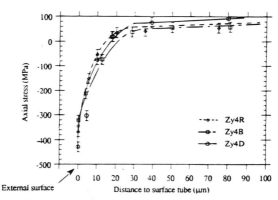

FIG. 7—*The axial and tangential stress profiles at the as-received tube surfaces.*

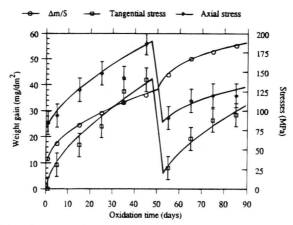

FIG. 8—*The axial and tangential stress variations and the oxidation kinetics, as a function of oxidation time, for the Zy4R alloy oxidized in steam at 400°C.*

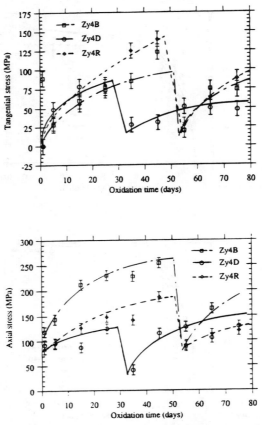

FIG. 9—*The axial and tangential stress variations in the metal beneath the oxide, as function of oxidation time.*

By using the low-penetration X-rays in the material and knowing that this varies with the angle ψ, it is possible to show that there is a stress gradient near the metal-oxide interface. The measurement was made using a sheet sample of recrystallized Zircaloy-4 covered with an oxide layer of about 2 μm. This type of measurement cannot be performed on tube samples because the very high defocalization of the X-ray beam induces inaccurate stress values for high value ψ. The expression for the stress gradient that gives the best fit to the deformation versus the $\sin^2\psi$ curve is a second-order polynomial function. The distribution of stresses in the metal beneath the oxide is shown in Fig. 10.

For our experimental conditions ($\psi_{max} \pm 60°$) and our sample, we were able to measure to a distance of 1 μm from the metal-oxide interface. The stress level values at the metal-oxide interface in Fig. 10 were obtained by extrapolation.

The shape of the stress gradient shows that:

1. There is a large stress level in the underlying metal near the metal-oxide interface. It would appear to be of the order of 1 to 1.3 GPa.
2. The stress gradient does not extend over an appreciable distance from the metal-oxide

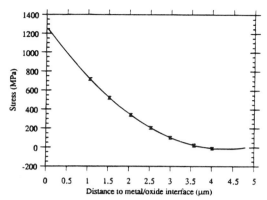

FIG. 10—*The tangential stress gradient at the metal-oxide interface on a Zircaloy-4 plate oxidized in steam at 400°C for 25 days.*

interface. For this sample, covered with a 2-µm oxide layer, the stress varies from 1.3 GPa to 200 MPa over a distance of 2.5 µm.
3. For distances greater than 3.5 µm from the metal-oxide interface, the stress level becomes equal to that measured in the metal before oxidation.

The stress values obtained near the metal-oxide interface are high and exceed the proof yield strength of the bulk metal, but extend over a relatively short distance. The mechanical properties and the composition of this underlying metal are not very well known. One of the important parameters is the amount of oxygen dissolved in the metal, because it is known that oxygen in solution in the zirconium increases its mechanical strength. In addition, the geometry of the metal-oxide interface that has an appreciable "roughness" must also play an important role in the level of acceptable stress for the material.

The tetragonal zirconia volume fraction reported in the Table 4 is the amount of tetragonal zirconia measured by Raman spectroscopy in each sublayer for all pre- and post-transition samples of the alloys. The tetragonal zirconia volume fraction near the metal-oxide interface shows that for a given alloy this does not vary appreciably, whatever the oxidation time. This suggests that the stress level at the metal-oxide interface is probably the same for any oxidation time. Consequently, the difference in the average stress measured at different oxidation times could be accounted for by the distance over which the stress gradient extends. This distance increases with the oxidation time in the pre-transition period.

When the kinetic transition occurs, the tetragonal zirconia in the outer part of the oxide is changed into monoclinic zirconia. This brings about stress relaxation in the outer sub-scale, which is also shown by a decrease in the average stress values for the underlying metal. In fact, as the oxidation process continues during kinetic transition, the stress levels at the metal-oxide interface are the same as those in the pre-transition region.

In post-transition kinetics, a new increase is observed for the average stresses in the metal. This can be related to the medium sub-scale growth that has the same tetragonal phase ratio as that the outer sub-scale formed during the pre-transition period.

Kinetics of the Incorporation of the Precipitates in the Oxide

For each of the metallurgical forms studied, we have measured the iron and chromium atom percent in the matrix using electron microprobe analysis. The size of the electron beam used is the order of 1 μm^3. Because the average diameter of the precipitates is less than 0.3 μm, the concentrations of iron and chromium are only apparent concentrations due to the large contribution of the matrix. We cannot, therefore, determine the absolute iron and chromium contents in the precipitates. However, due to the low solubility of these elements in α-zirconium, the ratio corresponds to that of the precipitates. This can be visualized in the plot of the iron atom percent as a function of chromium determined for the Zy4R alloy (Fig. 11). The points are distributed on both sides of a straight line with little scatter. The slope of this line corresponds to the ratio of iron to chromium in the precipitates, that is, 1.6 in this case.

We have measured the iron to chromium atomic ratio for different regions of each sample: that is, in the metal and at differents depths of the oxide layer. The distributions obtained for the initial metallurgical forms are shown in Fig. 12. The Zy4R and Zy4D alloys before oxidation have precipitates with an iron/chromium (Fe/Cr) ratio close to 1.6. Conversely, for the β-quenched alloy (the β quenching was the last step in the manufacturing operation), this ratio is lower and is about 0.8. The analyses carried out for the oxide layers show that the precipitates undergo composition changes when they are incorporated into these layers. In Fig. 13, we have plotted the Fe/Cr ratio values measured at different depths in the oxide layer that formed on the Zy4R alloy during 65 days. No matter the type of alloy or the oxidation time, judging from the analysis in the oxide near the metal-oxide interface, the precipitates do not appear to have undergone appreciable chemical changes. In effect, the distribution of the Fe/Cr atom ratios is close to observations made in the metal (Fig. 12).

The composition of the precipitates changes when going away from the metal-oxide interface. The Fe/Cr atom ratio distribution is spread over both sides of the average ratio. This result can be explained by the precipitates oxidation into the oxide scale. When the precipitates are incorporated into the oxide, they should be progressively oxidized, but probably with slower kinetics than that of the surrounding matrix. The zirconium contained in the precipitates is oxidized first [13]. Iron and chromium have a low solubility in zirconia, and they are rejected near the precipitate-oxide interface. In the precipitate region in the oxide layer, there zones where the Fe/Cr ratio is higher than for the initial precipitate, and others where it is lower, as shown in the work of Kubo et al. [12], Pecheur et al. [13,14], and Weidinger et al. [19].

The analyses for the outside regions of the post-transition oxides show that the composition of the precipitates no longer changes. This probably corresponds with the fact that the precipitates are completely oxidized in this oxide region.

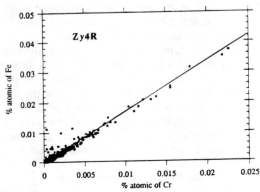

FIG. 11—*The iron atomic percent as a function of the chromium atom percent in the precipitates in the Zy4R matrix.*

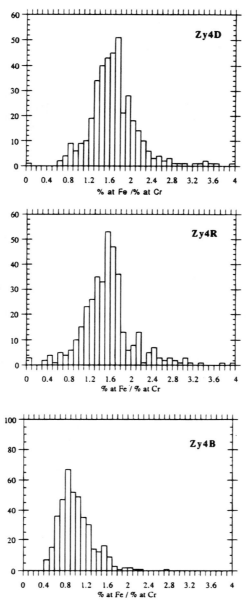

FIG. 12—*The Fe/Cr atomic ratio distributions for the intermetallic precipitates in the three initial alloys.*

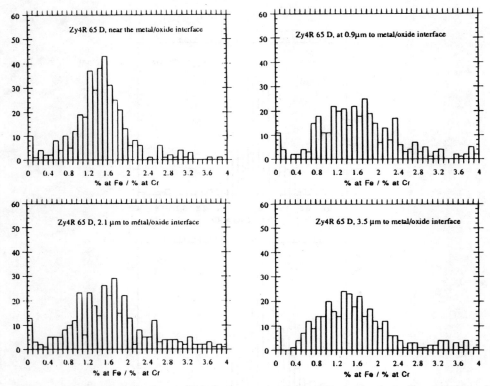

FIG. 13—*The Fe/Cr atomic ratio distributions measured at different depths in the oxide formed on the Zy4R alloy after 65 days at 400°C.*

For the Zy4D and Zy4R alloys that have almost the same Fe/Cr ratios distribution in the matrix, the precipitates experience the same changes in chemical composition during oxidation. For these two alloys, the modification of the distribution of iron and chromium atom ratios shows a marked variation starting at a distance of 0.5 μm from the metal-oxide interface. Conversely, the Zy4B alloy, whose distribution is centered around a value of 1 before oxidation, shows changes that are a little different compared to those of the other alloys. The variation in the chemical composition of the precipitates appears to be faster. The analyses made at a distance of 0.1 μm to the metal-oxide interface show that the precipitates have an extended Fe/Cr ratio distribution that is equivalent to those observed in the two other alloys for distances greater than 0.5 μm from the metal-oxide interface.

The results are in good agreement with those obtained by higher resolution techniques like transmission electron microscopy [*12–14*], but the advantage of the electron microprobe analysis technique used in this work is that the analysis is made for a large number of precipitates (more than 1000 precipitates) making it possible to obtain satisfactory statistics.

Discussion

This study shows function of stress distribution when examining the region of the metal-oxide interface and the variation of the composition of the precipitates in the oxide layer in the presence of the tetragonal zirconia.

The tetragonal zirconia distribution in the oxide layers determined by Raman spectroscopy is as follows: near the metal-oxide interface there is a tetragonal zirconia-rich zone (40% of tetragonal zirconia) that extends a short distance, then a less rich zone that corresponds to the remainder of the oxide layer thickness for the pre-transition sample. For post-transition samples, the tetragonal phase distribution can be similarly described for the medium sublayer formed after the transition; on the other hand, the outside region of the oxide includes only a small proportion of tetragonal zirconia. The stabilization of the tetragonal zirconia in each of the sublayers can be explained by the presence of internal stresses induced at the metal-oxide interface and by oxidation of the precipitates that can induce a stress field between the oxide and the precipitate.

X-ray diffraction makes it possible to measure the stress level in the metal beneath the oxide and shows the presence of a stress gradient near the metal-oxide interface. This gradient suggests that the value of the stress in the metal is very high at the metal-oxide interface. The metal and the oxide are in mechanical equilibrium, then in the oxide there is a zone near the interface where the compressive stresses are very high. This provides an explanation for the presence of appreciable amounts of tetragonal zirconia near the metal-oxide interface.

If it is assumed that the material is continuous between the oxide and the metal, there must be a transition zone between the tensile stresses in the metal and the compressive stresses in the oxide. This zone must be able to support very high stresses. The topography of the metal-oxide interface that is not planar, the dissolving of a small amount of oxygen in the metal beneath the oxide, and the probable existence of an amorphous or microcrystallized zone in the first oxide layers can explain the absence of spallation between the metal and the oxide in spite of the high stress level.

From the results, we can try to propose a mechanism for the oxidation of these alloys. The first oxide layers that are formed are going to create a high level of stress due to the difference in volume between the oxide and the metal (the ratio given by Pilling and Bedworth is 1.56 for the zirconia/zirconium system). This will give compressive stresses in the oxide and tensile stresses in the underlying metal. The microstructural studies of the oxide layers show that the preferential growth of certain crystallite orientations minimize the stress and give a columnar oxide. The oxide grains grow along the axis perpendicular to the interface because the compressive stress state in the oxide opposes the growth in the other directions.

In addition, the precipitates play a role in the generation of additional stresses in the oxide layer. When the matrix oxidation occurs, the precipitates are not oxidized at the same time as the surrounding metal. They are incorporated in the oxide as intermetallic precipitates. Starting at some distance from the metal-oxide interface, they are going to be progressively oxidized. The zirconium that has the highest reactivity towards oxygen is oxidized first. Because the iron and chromium, which make up the precipitate, have low solubility in zirconia, only small amounts are incorporated into the zirconia. This makes it possible to stabilize the cubic or tetragonal phase by forming vacancies in their crystal lattice, as Pecheur et al. [13,14] observed. The remaining iron and chromium are rejected towards the precipitate-oxide interface where they will segregate.

When the zirconium in the precipitate is oxidized to form zirconia, stresses are caused by the volume differences between the initial precipitate and the newly-formed zirconia (Fig. 14). The differential oxidation of the precipitates inside the oxide layers provides a new source of stresses that can keep the surrounding tetragonal phase stable. This is shown by stress increases in the oxide layers, and also tensile stress increase in the metal beneath the oxide, especially in the pre-transition period (Fig. 9).

The kinetic transition corresponds to the transformation from tetragonal to monoclinic zirconia starting from the external interface [1]. This transformation occurs when the stresses are no longer high enough to stabilize the tetragonal phase. This corresponds to the end of

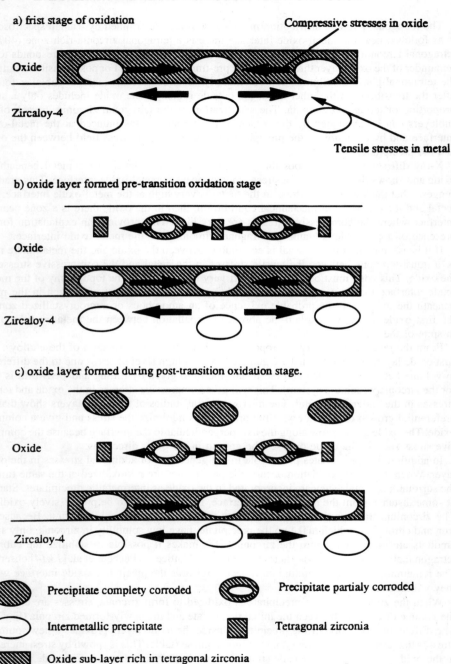

FIG. 14—*Schematic representation of the progressive oxidation for the intermetallic precipitates incorporated in the oxide layer.*

oxidation for some of the precipitates that would no longer locally produce additional stresses in the oxide.

There is a decrease in the overall tetragonal zirconia volume fraction in the oxide and a stress relaxation in the oxide thickness and in the underlying metal corresponding to the kinetic transition. After transition, the medium oxide sublayer that is formed has the same properties as those formed in the pre-transition phase; therefore, the stress level again increases in the oxide and in the underlying metal.

During kinetic transition, in the outside region of the oxide, some of the tetragonal zirconia does not undergo the phase transformation to monoclinic structure. This tetragonal phase probably results from the zirconium oxidation that makes up the precipitates and that is chemically stabilized by the presence of iron and chromium in its crystal lattice. This is confirmed by the Raman spectroscopy analyses of the external layers for the oxidized samples for the post-transition period, where only a low concentration of tetragonal zirconia stabilized by the additive elements in the precipitate is observed (Fig. 5). This is in good agreement with the recent work of Cox [20] who has also characterized the oxide layer by Raman spectroscopy.

The thermomechanical treatments during fabrication of the cladding tube also affect the composition and the size of the precipitates. Precipitates that are too small will be rapidly oxidized, decreasing the pre-transition period. On the other hand, large precipitates take more time to oxidize, but are less effective in stabilizing tetragonal zirconia in their environment. A compromise in precipitate size will optimize the corrosion resistance of the zirconium alloys [15–16].

Conclusions

The objective of this work was to obtain increased knowledge of the corrosion mechanisms for Zircaloy-4 alloys by studying the factors affecting tetragonal zirconia stabilization observed into oxide layers. This was done using three Zircaloy-4 alloys with different final metallurgical forms that were oxidized in steam at 400°C.

We were interested in the stresses that developed in the metal near the metal-oxide interface and in the oxidation mechanism of the precipitates in the oxide layer. This work has shown that there is a tensile stress gradient in the underlying metal near the metal-oxide interface. The high stress values at the interface can explain the presence of an appreciable proportion of tetragonal zirconia stabilized by compressive stresses in the oxide near the interface.

The second part of this study, concerning the incorporation of the precipitates in the oxide scales, shows that the variation of the composition of this precipitate could be correlated to delayed oxidation, as shown in previous works [13,14,19]. This oxidation leads to the formation of cubic or tetragonal zirconia inside the precipitates and a stress field around them. This surrounding stress field stabilizes the adjacent tetragonal zirconia up to the kinetic transition.

The stresses around the precipitates are caused by the volume difference between the initial precipitate and the newly-formed zirconia. When the precipitates are completely oxidized, the volume variations stop and the stress fields surrounding them decreases, and the metallurgical form of the starting material influences the corrosion resistance because the mechanical properties of the alloy, like creep, can modify the stress induced by the oxidation process at the metal/oxide interface. The tetragonal zirconia is transformed to the monoclinic phase.

The fraction of tetragonal zirconia that remains in the external region of the post-transition oxide layers is that which could be formed by oxidation of the precipitates and should be stabilized by the iron and chromium.

The metallurgical form of the starting material influences the corrosion resistance because of different creep behaviors of the alloys accommodation capability of the stresses induced by the oxidation process.

Acknowledgments

The author gratefully acknowledges the help of the DILOR society and Mr. Majoube (DSM-DRECAM-SPAM) for the Raman spectroscopy.

References

[1] Godlewski, J., Gros, J. P., Lambertin, M., Wadier, J. F., and Weidinger, H., "Raman Spectroscopy Study of the Tetragonal-to-Monoclinic Transition in Zirconium Oxide Scales and Determination of Overall Oxygen Diffusion by Nuclear Microanalysis of O^{18}," *Proceedings, Zirconium in the Nuclear Industry: Ninth International Symposium ASTM STP 1132*, C. M. Eucken and A. M. Garde, Eds., American Society for Testing and Materials, Philadelphia, 1991, p. 416.
[2] Roy, C. and David, G., "X-ray Diffraction Analysis of Zirconia Films on Zirconium and Zircaloy-2," *Journal of Nuclear Materials*, Vol. 37, 1970, p. 71.
[3] Roy, C. and Burgess, B., "A Study of the Stresses Generated in Zirconia Films during Oxidation of Zirconium Alloys," *Oxide Metals*, Vol. 2, 1970, p. 235.
[4] Bradhurst, D. H. and Heuer, P. M., "The Influence of Oxide Stress on Break Away Oxidation of Zircaloy 2," *Journal of Nuclear Materials*, Vol. 37, 1970, p. 35.
[5] Gadiyar, H. S. and Balachandra, J., "Stress Measurements and Structural Studies during Oxidation of Zirconium Base Alloys," *Transactions, The Indian Institute of Metals*, Vol. 28, No. 5, 1975, p. 391.
[6] Antoni-le Guyadec, F., "Approche mécano-chimique de l'oxydation du Zircaloy-4," Thesis, Institut National Polytechnique de Grenoble, 1990.
[7] Garzarolli, F., Seidel, H., Tricot, R., and Gros, J. P., "Oxide Growth Mechanism on Zirconium Alloys," *Zirconium in the Nuclear Industry: Ninth International Symposium, ASTM STP 1132*, C. M. Eucken and A. M. Garde, Eds., American Society for Testing and Materials, Philadelphia, 1991, p. 395.
[8] Gravie, R. C. and Nicholson, P. S., "Phase Analysis in Zirconia Systems," *Journal, American Ceramics Society*, Vol. 55–6, 1972, p. 303.
[9] Clarke, D. R., et al., "Measurement of the Crystallographically Transformed Zone Produced by Fracture in Ceramics Containing Tetragonal Zirconia," *Journal of American Ceramics Society*, Vol. 65–6, 1982, p. 284.
[10] Maeder, G., et al., "Present Possibilities for the X-Ray Diffraction Method of Stress Measurement," *Non Destructive Testing International*, Vol. 10, 1981, p. 235.
[11] Noyan, I. C., et al., "Residual Stress Measurement by X-Ray Diffraction and Interpretation," *Materials Research and Engineering*, Springer-Verlag, 1987.
[12] Kubo, T. and Uno, M., "Precipitate Behavior in Zircaloy-2 Oxide Films and Its Relevance to Corrosion Resistance," *Zirconium in the Nuclear Industry: Ninth International Symposium, ASTM STP 1132*, C. M. Eucken and A. M. Garde, Eds., American Society for Testing and Materials, Philadelphia, 1991, p. 476.
[13] Pecheur, D., et al., "Precipitate Evolution in the Zircaloy-4 Oxide Layer," *Journal of Nuclear Materials*, Vol. 189, 1992, p. 318.
[14] Pecheur, D., Lefebrere, F., Motta, A. T., Lemaignan, C., and Charquet, D., "Oxidation of Intermetallic Precipitates in Zircaloy-4" in this volume.
[15] Charquet, D., "Improvement of the Uniform Corrosion Resistance of Zircaloy-4 in the Absence of Irradiation," *Journal of Nuclear Materials*, Vol. 160, 1988, p. 186.
[16] Foster, J. P., et al., "Influence of Final Recrystallization Heat Treatment on Zircaloy-4 Strip Corrosion," *Journal of Nuclear Materials*, Vol. 173, 1990, p. 164.
[17] Andersson, T., et al., "Beta-quenching of Zircaloy Cladding Tubes in Intermediate or Final Size—Methods to Improve Corrosion and Mechanical Properties," *Zirconium in the Nuclear Industry: Fifth Symposium, ASTM STP 754*, D. Franklin, Ed., American Society for Testing and Materials, Philadelphia, 1982, p. 75.
[18] Bangaru, N. V., et al., "Effect of Beta Quenching on the Microstructure and Corrosion of Zircaloys," *Zirconium in the Nuclear Industry: Seventh International Symposium, ASTM STP 939*, R. B. Adamson and L. F. P. Van Swam, Eds., American Society for Testing and Materials, Philadelphia, 1985, p. 341.
[19] Weidinger, H. G., Ruhmann, H., Cheliotis, G., Maguire, M., and Yau, T.-L., "Corrosion-Electrochemical Properties of Zirconium Intermetallics," *Zirconium in the Nuclear Industry: Ninth International Symposium, ASTM STP 1132*, C. M. Eucken and A. M. Garde, Eds., American Society for Testing and Materials, Philadelphia, 1991, p. 499.
[20] Cox, B., et al., "Dissolution of Zirconium Oxide Films in 300°C LiOH," *Journal of Nuclear Materials*, Vol. 199, 1993, p. 272.

DISCUSSION

R. A. Ploc[1] (written discussion)—You seem to have shown a correlation between corrosion rate and the tetrogonal zirconia. What might also be important is the continuity of the tetragonal layer. Do you have any TEM micrographs to support your hypothesis?

J. Godlewski (author's closure)—In the case of pre-transition samples, the distribution of tetragonal zirconia in the oxide thickness is continuous. This continuity has been shown by Raman spectroscopy on taper cross sections. We do not have any TEM micrographs to support this, because tetragonal zirconia is essentially stabilized by compressive stresses. When we prepare samples for TEM examinations, the stresses were relaxed and a large amount of tetragonal zirconia was transformed to the monoclinic phase.

A. T. Motta[2] (written discussion)—(1) You say that the oxidation of the precipitates is accompanied by a volume change that leads to a stress field to help stabilize tetragonal zirconia. If the onset of oxidation creates stresses, what is the theoretical basis for your hypothesis that they then disappear upon complete oxidation?

(2) Did you independently confirm by TEM that the state of precipitate oxidation is dependent on its location in the oxide layer?

J. Godlewski (author's closure)—(1) The corrosion of intermetallic precipitates inside the oxide scale is accompanied by a volume change. This precipitate volume expansion contradicts the stress decrease in the oxide layer by the preferential growth of certain crystallite orientations. This creates a stress field around the precipitates that stabilizes the surrounding tetragonal zirconia. When the intermetallic precipitates are completely corroded, their volumes stop increasing and their stress fields decrease progressively, because the oxide layers relieve this stress.

(2) We did not confirm by TEM that the state of precipitate oxidation is dependent on its location in the oxide layer. This hypothesis is supported by the work of Pecheur et al. [*13,14*].

---

[1] AECL Research, Chalk River, Ontario, Canada.
[2] Pennsylvania State University, Department of Nuclear Engineering, University Park, PA.

John Schemel Award Paper

Dominique Pêcheur,[1] *Florence Lefebvre*,[1] *Arthur T. Motta*,[2] *Clément Lemaignan*,[1] *and Daniel Charquet*[3]

Oxidation of Intermetallic Precipitates in Zircaloy-4: Impact of Irradiation

REFERENCE: Pêcheur, D., Lefebvre, F., Motta, A. T., Lemaignan, C., and Charquet, D., "**Oxidation of Intermetallic Precipitates in Zircaloy-4: Impact of Irradiation,**" *Zirconium in the Nuclear Industry: Tenth International Symposium, ASTM STP 1245,* A. M. Garde and E. R. Bradley, Eds., American Society for Testing and Materials, Philadelphia, 1994, pp. 687–708.

ABSTRACT: Intermetallic precipitates are known to play a critical role in the oxidation process of Zircaloys. Since under irradiation they undergo structural changes, a specific study was conducted to analyze whether these transformations modify the oxidation behavior of the Zircaloy-4. Oxidation kinetics in autoclave were measured on reference, ion irradiated, and neutron irradiated materials. In the case of ion-irradiated samples, the oxidation kinetics are changed, while in the case of neutron-irradiated cladding, no significant change is observed after 60 days of oxidation. The behavior of reference and irradiated precipitates during the growth of these oxide layers was analyzed using analytical scanning transmission electron microscopy. Close to the metal-oxide interface, precipitates are incorporated unoxidized in the oxide layer. Then, when oxidized, at a few hundreds of nanometers from this interface, they undergo two major evolutions: their structure becomes either nanocrystalline or occasionally amorphous and an iron redistribution and depletion is observed. In the case of precipitates previously made amorphous by irradiation, a similar behavior is observed. The role of precipitates on the oxidation of the Zircaloy-4 is discussed in terms of interaction of the precipitates with the zirconia layer (stability of the dense oxide layer) and oxidation kinetics.

KEY WORDS: zirconium, precipitates, oxidation, oxide layer, oxidation kinetics, zirconia, irradiation, amorphization, analytical scanning transmission electron microscopy, zirconium alloys, nuclear materials, nuclear applications, radiation effects

In light water reactor (LWR) conditions, the oxidation of Zircaloy cladding is one of the main limitations to the extension of fuel rod burnups.

Many studies were conducted to evaluate the oxidation process in autoclave environments. The oxidation kinetics are generally divided in two main regions separated by a transition [1]. During the first one, called the pre-transition region, the oxide that is formed is dense, is composed of monoclinic and tetragonal ZrO_2 crystallites, and the growth rate decreases continuously. After the transition, the oxidation rate increases and remains approximately constant in the second stage. Most of the outer part of the oxide layer is now porous [2,3] and composed exclusively of monoclinic ZrO_2 crystallites [4]. Several metallurgical parameters greatly influence the oxidation kinetics. Among them, the influence of the nature and the

---

[1] Research engineers and head of Laboratory, respectively, CEA-Grenoble, DTP/SECC, 85X, 38041 Grenoble Cedex, France.
[2] Assistant professor, Pennsylvania State University, Department of Nuclear Engineering, University Park, PA 16802.
[3] Research engineer, Cézus, Centre de Recherches d'Ugine, 74340 Ugine, France.

distribution of intermetallic precipitates on oxidation is clearly established, but the origin of such an effect is not well understood [5–7].

However, these results concerning the oxidation kinetics of Zircaloys in autoclave environments cannot be transferred to reactor environments where a significant increase of the oxidation rate is measured when the oxide thickness exceeds a few microns [8,9]. To understand such an effect, various approaches are proposed based on the differences of the chemical environment in-reactor and in-autoclave [10] or on the radiolysis phenomena within pores [11,12]. Since it is known that precipitates in Zircaloys play a significant role in the oxidation process and because precipitates evolve under irradiation [13–18], it is important to evaluate the effect of the irradiation-induced amorphization of the precipitates on the oxidation kinetics.

The oxidation rates of reference, ion-irradiated, and neutron-irradiated materials were therefore measured in the autoclave to separate the effect of chemical reactor environments from those of radiation damage in the cladding. Several oxide layers grown on reference and ion-irradiated materials have been characterized by transmission electron microscopy (TEM). Some have already been described [19]. In addition, new observations on thicker oxides are briefly summarized to introduce local oxygen analysis. The oxide layer formed in-reactor on a three-cycle irradiated cladding sample was described. These observations were discussed, and particular attention are paid to determine the role of intermetallic precipitates on oxidation kinetics.

Experimental Methods

Materials

The reference material used in this work is standard Zircaloy-4 (Zy4 STD) furnished by Cézus of Ugine, France. Samples were machined out of a 1-mm-thick plate (Zr-1.48Sn-0.21Fe-0.11Cr-0.10O) and fully recrystallized at 948 K for 3 h and 30 min after cold rolling ($\Sigma Ai = 4 \cdot 10^{-19}$ h using $Q/R = 40\,000$ K). To increase precipitate size and thereby facilitate their observation, some specimens of the standard Zircaloy-4 were subjected to an additional coarsening heat treatment at 1055 K for 50 h. This material with large precipitates is referred to as Zy4 HT. Some samples (20 by 30 mm) have been irradiated on both faces with 1.5 MeV He$^+$ ion irradiations in the Van de Graaff ion accelerator of the Centre d'Etudes Nucléaires de Grenoble, at 77 K and to a dose of 0.4 displacements per atom (dpa), to simulate neutron irradiation and produce, at the surface, amorphous precipitates. Details of the irradiation conditions are described elsewhere [19,20].

Reference and irradiated recrystallized Zircaloy-4 cladding materials have also been used in this work. The chemical compositions of these materials are Zr-1.75Sn-0.23Fe-0.10Cr-0.11O and Zr-1.71Sn-0.21Fe-0.12Cr-0.15O, respectively. The irradiated cladding samples (25-mm-long tubes) were exposed to two and three irradiation cycles in the experimental BR3 reactor at about 590 K to fluences of 4.0 and 5.4 10^{25} n · m^{-2}, respectively. The oxide thickness produced during three cycles has been estimated, when preparing thin foils, to be of the order of 10 μm.

Oxidation and Examinations

Reference and ion-irradiated samples were oxidized in autoclave (10.3 MPa steam, 673 K). The neutron-irradiated samples, before being oxidized, were mechanically polished to remove the oxide layer grown in-reactor whereas their internal surface remained as-received. The neutron-irradiated and reference cladding samples were oxidized together in a shielded autoclave (15 MPa steam, 673 K). After 60 days of oxidation, three samples were removed for metallographic examinations while the autoclave testing of other specimens continued.

Oxide layers formed in-autoclave and in-reactor were examined using JEOL 1200 EX and a TOPCON 002B 200 kV scanning transmission electron microscopes. Compositions were determined by energy dispersive X-ray spectroscopy. This foils were prepared from the specimen regions at the metal-oxide interface, in the middle of the oxide layer and at the oxide-water interface. The experimental method used for preparing these thin foils has been described in Refs *19* and *20*. In the case of neutron-irradiated materials, the same method was used within a shielded cell.

In order to follow the evolution of precipitates from the very beginning of their incorporation into the dense oxide layer to their complete stabilization in the porous oxide layer, different thicknesses of oxide were grown in-autoclave and characterized as 1, 2, 4, and 14 µm (that is, 3, 20, 40, and 420 days of oxidation, respectively) in the case of reference materials and 1, 2, and 10 µm (that is, 3, 20, and 120 days of oxidation, respectively) in the case of ion-irradiated materials. In the case of the neutron-irradiated materials, thin foils have been prepared in the middle of the oxide layer grown in-reactor for three cycles.

Experimental results

Oxidation of Reference and Irradiated Zircaloy-4

Oxidation kinetics obtained on reference and ion-irradiated Zircaloy-4 are presented in Fig. 1. The ion-irradiated material exhibits a higher oxidation rate than the reference one. This increase is particularly significant for Zy4 HT irradiated to a dose of 0.4 dpa once the oxide thickness is greater than 3 µm. Surprisingly, this increase of the oxidation rate is maintained even after the irradiation-damaged surface layer is consumed by the autoclave oxidation process. In the case of standard material irradiated at 0.6 dpa, a spalling of the oxide occurs after 60 to 80 days of oxidation.

In the case of neutron-irradiated cladding, after 60 days of autoclave oxidation, different behaviors are observed on the inside and outside tube surfaces. No significant differences in oxide thicknesses exist between the external surfaces of the irradiated material and the reference unirradiated specimen. In both cases, the thickness of the external oxide layer ranges between 2 and 3 µm. On the other hand, the internal surface of the cladding samples irradiated for 2 and 3 cycles exhibited a much thicker oxide layer ($\approx$15 µm) than the reference specimens ($\approx$2 µm) (Fig. 2). Moreover, this thick oxide layer presented cracks with a tendency to spalling.

Oxidation of Reference $Zr(Fe,Cr)_2$ precipitates

The precipitates present in the Zircaloy-4 sheet used in this work are mainly $Zr(Fe,Cr)_2$ Laves phases with both C14 hcp and C15 fcc crystalline structure and an iron/chromium ratio close to 1.7. In the case of Zy4 STD, their average diameter is equal to about 130 nm while in Zy4 HT, it is close to 330 nm.

In the Dense Oxide Layer—At a few hundred nanometers away from the metal-oxide interface, unoxidized and oxidized precipitates are found. The unoxidized ones have the same crystalline structure (hcp or fcc) and the same iron/chromium (Fe/Cr) ratio (1.7) as in the precipitates in the metallic matrix, and no oxygen is detected (Fig. 3). The oxidized precipitates are composed of nanocrystallites, a few nanometers in size. Their structure, identified by

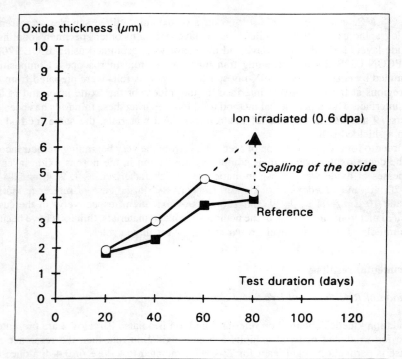
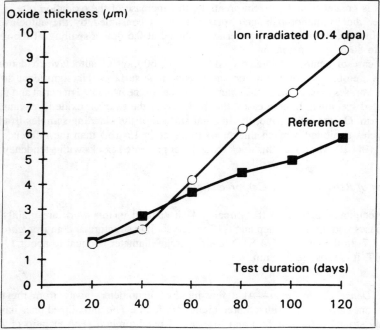

FIG. 1—*Oxidation kinetics in a 15 MPa steam test autoclave of reference and ion-irradiated Zircaloy-4 at 400°C:* (top) *Zy4 STD with normal precipitates, and* (bottom) *Zy4 HT with large precipitates.*

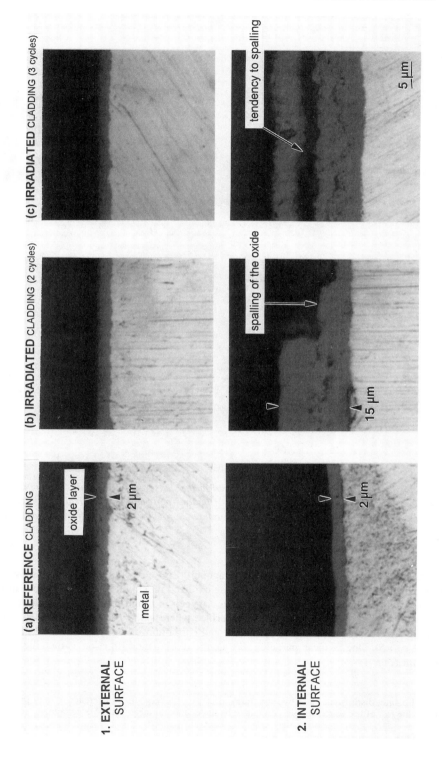

FIG. 2—Oxide layers formed, in shielded autoclave after 60 days of oxidation, (1) on the external surface and (2) on the internal surface of: (a) a reference cladding, (b) two cycles of neutron-irradiated cladding, and (c) three cycles of neutron-irradiated cladding.

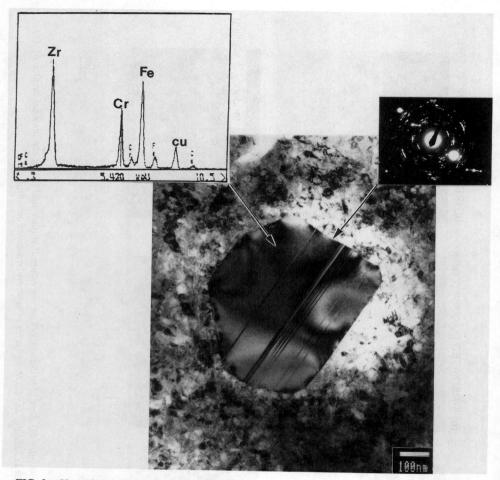

FIG. 3—*Unoxidized $Zr(Fe,Cr)_2$ precipitate observed in the dense oxide layer at a few hundred nanometers from the metal-oxide interface. Both its structure (hcp) and its chemical composition (Fe/Cr ≈ 1.7) are similar to those observed for the reference precipitates present in the metallic matrix.*

electronic diffraction, corresponds to cubic (or tetragonal)[4] ZrO_2 with small amounts of monoclinic ZrO_2. Chemical composition analysis of these oxidized precipitates shows a Fe/Cr ratio smaller than 1.7 indicating a slight iron depletion. Oxygen is found throughout precipitates except, occasionally, at the precipitate-matrix interface where small particles, a few tens of nanometers in size, composed exclusively of iron and chromium in the metallic state can be found (Fe/Cr ≈ 1.5). One of these particles was identified by nanodiffraction as metallic bcc iron-chromium (Fig. 4).

---

[4] Since TEM cannot distinguish these two structures, because of a very small difference in their lattice parameters, in the rest of the paper, "cubic" ZrO_2 is to be understood to mean "cubic or tetragonal" ZrO_2.

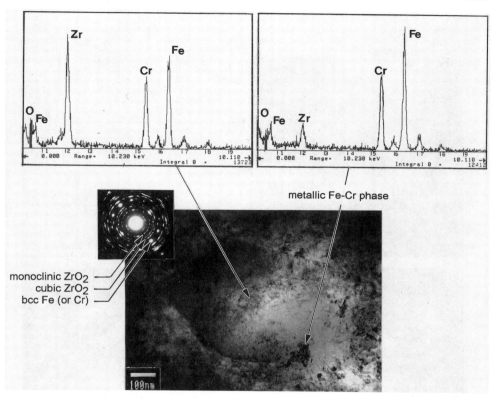

FIG. 4—*Oxidized precipitate observed in the dense oxide layer at a few hundred nanometers from the metal-oxide interface. It is composed of nanocrystallites of cubic and monoclinic ZrO_2 and, at the edge of the precipitate, of a bcc iron-chromium phase with no trace of oxygen. Only minor evolution of the Fe/Cr ratio is observed with respect to the reference precipitates (Fe/Cr > 1).*

At about 1 μm away from the metal-oxide interface, in the dense oxide layer, all precipitates are observed to be nanocristallized and oxidized except a few amorphous ones [*19*]. Nanocrystallized precipitates have the same structure as described earlier, but their chemical composition has considerably changed (Fig. 5). Their Fe/Cr ratio is now decreased to less than 1 and some iron is segregated at the edges of the precipitates. This segregation occasionally corresponds to metallic bcc iron. Only a small percentage of chromium is found in it (Fe/Cr > 3).

In the Porous Oxide Layer—At a few microns or a few tens of microns away from the metal-oxide interface, all the precipitates observed are nanocrystallized and oxidized. Their crystalline structure has not changed nor has their chemical composition (cubic ZrO_2, Fe/Cr < 1). Nanocrystallites within precipitates remain small (a few nanometers in size) compared to those observed in the surrounding matrix (a few tens of nanometers). However, the metallic iron-rich bcc phase is no longer observed, indicating an iron dissolution into the matrix.

These observations are summarized in Fig. 6. Close to the metal-oxide interface, the precipitates are incorporated, unoxidized, in the oxide layer implying that they oxidize later than the

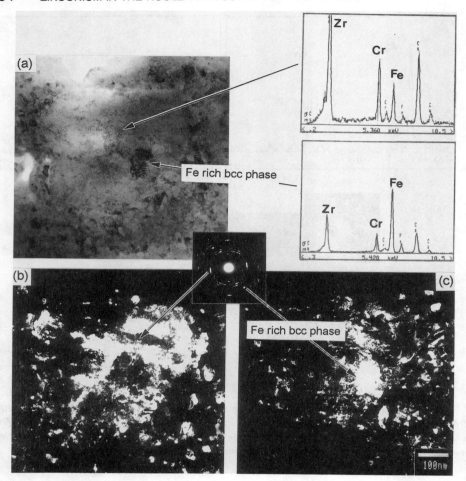

FIG. 5—(a) *Oxidized precipitate observed in the dense oxide layer, at about one micron from the metal-oxide interface, (b) a dark field from the oxide rings shows it is still composed of nanocrystallites of cubic and monoclinic ZrO_2, and (c) a bcc iron and chromium phase is also detected. The Fe/Cr ratio of the precipitate is now decreased to less than one and the bcc phase is now richer in iron.*

zirconium matrix. When oxidized, they are composed essentially of nanocrystallites of cubic ZrO_2. During their oxidation, they undergo a drastic iron redistribution leading to an iron depletion that is more pronounced in the middle of the precipitates and occasionally to a metallic iron precipitation at the edges of the precipitates. Finally, the iron partially dissolves in the ZrO_2 matrix. These results are valid for both Zy4 STD and HT.

Oxidation of Irradiated $Zr(Fe,Cr)_2$ Precipitates

Ion-Irradiated Precipitates—The 1.5 MeV He, ion-irradiated Zircaloy-4 samples present amorphous $Zr(Fe,Cr)_2$ precipitates within a few microns of the surface. Their nearest neighbor

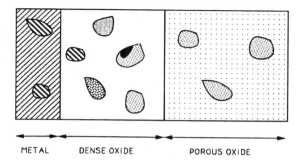

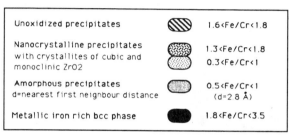

FIG. 6—*Schematic view of the structural and chemical composition evolutions experienced by the reference $Zr(Fe,Cr)_2$ precipitates during their incorporation in the Zircaloy-4 oxide layer grown in-reactor.*

spacing is equal to 2.2 Å. As previously reported [*21*], the chemistry of these precipitates amorphized by ion irradiation has not changed (Fe/Cr ≈ 1.7).

In the dense oxide layer—At a few hundred nanometers away from the metal-oxide interface, both amorphous and nanocrystallized precipitates are observed. Amorphous precipitates have a nearest neighbor spacing ranging between 2.2 and 2.8 Å and nanocrystallized precipitates are composed of ZrO_2 cubic and monoclinic crystallites. Both have a Fe/Cr ratio close to 1.5, indicating a slight iron depletion. At about 1 µm away from the metal-oxide interface, in the dense oxide layer, both amorphous (Fig. 7) and nanocrystallized precipitates (Fig. 8) are still observed. The amorphous precipitates, usually the smallest (<450 nm), are characterized by a nearest neighbor spacing close to 2.8 Å (for 2.2 Å before oxidation). Concurrently, oxygen is detected in these amorphous precipitates and their Fe/Cr ratio is decreased to about 0.5. Metallic bcc phases composed of iron (≈90%) and chromium (≈10%) are observed at the edges of some of these precipitates.

The nanocrystallized oxidized precipitates, composed of cubic and monoclinic ZrO_2 crystallites, have a higher Fe/Cr ratio (1.5) and are usually larger than amorphous ones. They also occasionally present a metallic bcc iron-chromium external layer.

In the porous oxide layer—At a few microns away from the metal-oxide interface, amorphous and nanocrystallized precipitates are present. Their structures are similar to those just described, and their Fe/Cr ratio is close to 0.3. No iron-rich metallic bcc phases are found in this outer layer.

The results are summarized in Fig. 9: when incorporated in the oxide layer, the irradiated precipitates behave similar to the reference ones. They experience structural and chemical

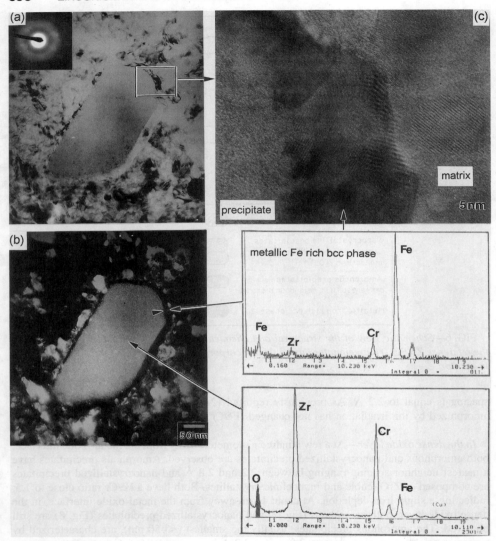

FIG. 7—*Ion-irradiated precipitate observed in the dense oxide layer at about one micron from the metal-oxide interface. (a) It is still amorphous but its first neighbor distance has increased during the oxidation process from 2.2 to 2.8 Å and its Fe/Cr ratio has decreased to less than 1, (b) a metallic iron-rich bcc phase with no trace of oxygen is observed surrounding the precipitate, and (c) high-resolution image of the precipitate-matrix interface.*

changes. Small precipitates (<450 nm) remain amorphous, but their nearest neighbor spacing increases during oxidation (from 2.2 to 2.8 Å) while incorporating oxygen. The largest precipitates become nanocrystallized with cubic ZrO_2 and a small amount of monoclinic ZrO_2. Both undergo a large amount of iron redistribution and depletion, occasionally with a transient metallic iron precipitation, and finally the iron partially dissolves in the ZrO_2 matrix.

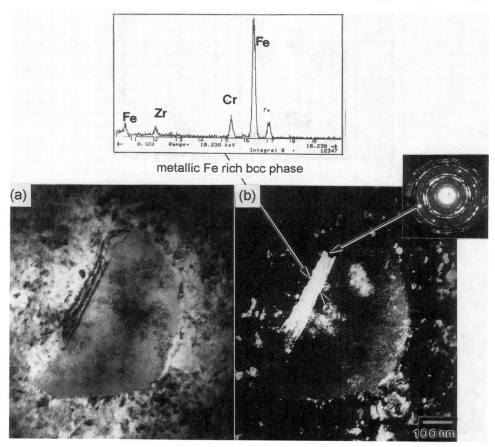

FIG. 8—*Ion-irradiated precipitate observed in the dense oxide layer at about one micron from the metal-oxide interface. (a) It is oxidized and composed of nanocrystallites of cubic and monoclinic ZrO_2 and (b) an iron-rich bcc phase, with no trace of oxygen, is detected at the edge of this precipitate.*

Neutron-Irradiated Precipitates—The state of the precipitates that were present in the cladding materials irradiated for two and three cycles were extensively described in Ref *17*. In the metallic matrix, the $Zr(Fe,Cr)_2$ precipitates experience an amorphous transformation and an iron depletion leading to a progressive dissolution of these precipitates into the matrix. After two cycles, all the precipitates are either amorphous or partially amorphous. Their size is significantly decreased compared to the reference cladding. After three cycles, all are amorphous and their sizes decreased further. The nearest neighbor spacing of these amorphous precipitates is equal to 2.2 Å.

To analyze these precipitates in the oxide layer, thin foils were prepared in the middle of the oxide layer grown in reactor during three cycles of irradiation. All the precipitates observed are completely amorphous with a nearest neighbor spacing equal to 2.7 Å and a Fe/Cr ratio close to 0.6 (Fig. 10). The initial average precipitate size has not changed, indicating that they were incorporated into the oxide before being dissolved into the metallic matrix. However, since those precipitates were incorporated in the oxide layer before the third irradiation cycle,

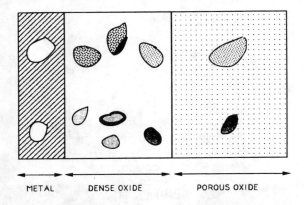

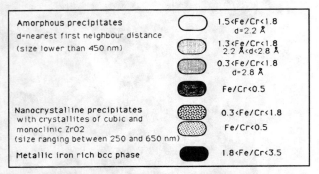

FIG. 9—*Schematic view of the structural and chemical composition evolutions experienced by the ion-irradiated amorphous $Zr(Fe,Cr)_2$ precipitates during their incorporation in the Zircaloy-4 oxide layer formed in-autoclave.*

at a time when they may have been only partially amorphous, it is impossible to determine if they became fully amorphous before being oxidized or during the oxidation process. The observation of thin foils prepared at the metal-oxide interface and at the water-oxide interface, currently in progress, will give us critical information with respect to this point.

Discussion

Oxidation Mechanism of Intermetallic Precipitates

During the precipitate oxidation process, three different steps can be discussed:

1. In the porous oxide layer and in the outermost part of the dense oxide layer, all precipitates are oxidized; whereas, close to the metal-oxide interface, a large number of them remain in their original crystalline structure. The oxidation of intermetallic precipitates is delayed compared to the zirconium matrix. This observation is in good agreement with those previously reported in Zircaloy-2 and -4 [22,23]. Moreover, at the beginning of precipitate oxidation, some bcc iron (and chromium) is found indicating

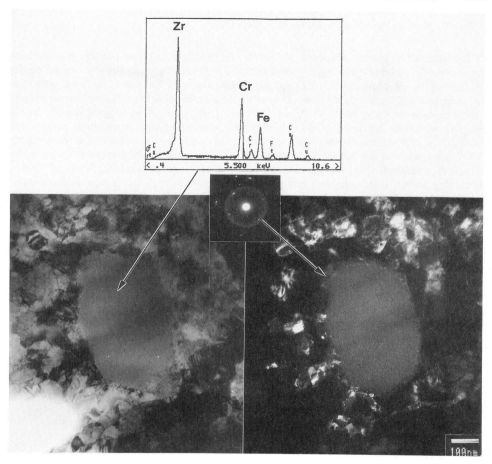

FIG. 10—*Neutron-irradiated precipitate observed in the middle of the oxide layer grown in-reactor after three cycles of irradiation at about 590 K. It is amorphous with a first neighbor distance equal to 2.7 Å and the Fe/Cr ratio decreased to less than one. No metallic iron-rich bcc phase is detected.*

a delayed oxidation of iron and chromium with respect to zirconium in precipitates.
2. As reported in Refs 22 and 24, when oxidized, all the precipitates are essentially composed of nanocrystallites of cubic ZrO_2. Since cubic zirconia is not supposed to be stable at 673 K [25], some additional stabilization factor must operate within the precipitates.
3. Moreover, the Fe/Cr ratio of oxidized precipitates progressively decreases as the metal-oxide interface moves off, from the nominal value 1.7 to an average value of 0.5 in the porous oxide. In the dense oxide layer, some iron segregates at the matrix-precipitate interface and occasionally precipitates as metallic bcc iron, while most of the chromium remains homogeneously distributed within precipitates. In the porous oxide layer, these local iron enrichments are no longer detected. Similar evolutions have already been reported [23,24], but the existence of a metallic iron-rich bcc phase, free of oxygen,

appears to be a new experimental observation.

The selected oxidation observed within the $Zr(Fe,Cr)_2$ precipitate can be discussed by comparing the value of the oxygen partial pressure in the oxide layer with the oxygen partial pressure of the equilibria Zr/ZrO_2, Cr/Cr_2O_3, and Fe/Fe_3O_4. At 673 K, according to an isothermal cut through the Ellingham's diagram, these equilibria are respectively achieved for an oxygen partial pressure equal to 10^{-75}, 10^{-50}, and 10^{-35} atm [26]. An evaluation of the oxygen partial pressure within the oxide layer can be performed considering that, at the inner metal-oxide interface, it corresponds to the value for which zirconium and zirconia are in equilibrium (10^{-75} atm), while at the outer oxide-vapor interface, it is equal to the oxygen pressure in the autoclave (estimated from the dissociation reaction of water, between 10^{-4} and 10^{-10} atm, depending on the concentration of oxygen dissolved in the water). In the porous oxide layer, the oxygen partial pressure can be considered as constant while, in the dense oxide, the oxygen chemical potential (μ_{O_2}), that is, log P_{O_2}, would decrease linearly from the outer oxide surface to the metal-oxide interface as proposed in Ref 27. Thus, as illustrated in Fig. 11, in the first third of the dense oxide layer, close to the metal-oxide interface, the oxygen partial pressure would increase from 10^{-75} atm to 10^{-50} atm allowing only zirconium to be oxidized. In the second third of the oxide layer, it would increase from 10^{-50} atm to 10^{-35} atm allowing chromium to now be oxidized while iron would still remain in its metallic state. Then, when the oxygen partial pressure would reach 10^{-35} atm, the oxidation of iron could take place. This simplified thermodynamic approach considering zirconium, chromium and iron as pure elements is in good agreement with secondary ion mass spectrometry (SIMS) analysis of

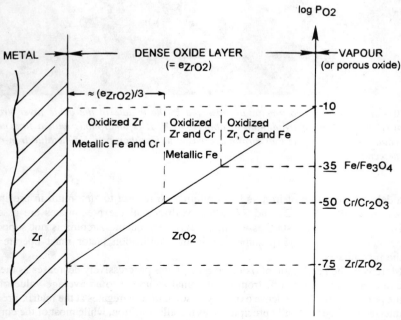

FIG. 11—*Stability domain of zirconium, iron, and chromium at the metallic state in the dense oxide layer versus the distance from the metal-oxide interface and corresponding value of the oxygen partial pressure.*

Zircaloy-4 oxide layers showing different oxidation states for zirconium, iron, and chromium depending on the location of the area analyzed, with respect to the metal-oxide interface [28].

While these thermodynamic considerations give a satisfactory explanation to the delayed oxidation of chromium and iron with respect to zirconium within precipitates, they are not sufficient to understand the delay in precipitate oxidation compared to the matrix. Since no oxygen is detected in these unoxidized monocrystalline precipitates imbedded in the oxide layer, a need to add kinetic considerations to the previous thermodynamic arguments is clear. This has not been incorporated in this work due to the lack of diffusion data for the various species involved (oxygen, zirconium, iron, and chromium) in different phases and across different interfaces.

To understand the existence of cubic ZrO_2 within oxidized precipitates, some stabilizing factors such as compressive stresses, crystallite sizes, or chemical effects should be proposed:

1. The intense compressive stresses produced by the oxide growth are known to stabilize cubic zirconia [29]. However, the large proportion of cubic zirconia observed in the oxidized precipitates is maintained in the porous oxide, where the remaining compressive stresses are weak. Moreover, in thin foils, the stresses are supposed to be relaxed in the oxide layer because of the dissolution of the metal [4]. If these intense compressive stresses were the only stabilizing factor, cubic ZrO_2 could not be observed. Thus, the compressive stresses do not appear to be the major stabilizing mechanism of cubic ZrO_2 in the precipitates.
2. According to Ref 30, cubic zirconia can be stabilized by an ultra-fine microstructure. The small size of the nanocrystallites formed in the precipitates (a few nanometers in size) is then in good agreement with the existence of cubic zirconia. So, a size effect could be involved in the stabilization of this kind of zirconia in the precipitates.
3. During the oxidation process, the Fe/Cr ratio of the oxidized precipitates decreases but never falls under 0.5. This indicates that even in the porous oxide layer there is still a significant amount of iron and chromium in the oxidized precipitates. Because almost no iron oxide nor chromium oxide has been detected in the precipitates, we can suppose that the remaining amount of iron and chromium is dissolved in the zirconia of the oxidized precipitates. In other respects, a slight contribution of iron to the stabilization of tetragonal ZrO_2 has been shown in Ref 31, that is, when Fe_3O_4 is introduced into tetragonal zirconia, the temperature of the tetragonal to monoclinic transformation is significantly decreased (such an effect is not observed with Cr_2O_3). Since the formation of Fe_3O_4 was observed during the oxidation of Zr_3Fe precipitates [32,33] and, since we have also observed traces of Fe_3O_4 in oxidized precipitates, it is therefore reasonable to assume that metallic iron is oxidized as Fe_3O_4 and is then dissolved in zirconia contributing to the stabilization of tetragonal (or cubic) zirconia within the precipitates. Similar stabilization effects of chromium can be excluded because the oxidation of bulk $ZrCr_2$ only produces monoclinic zirconia [34].

The driving force behind the iron depletion in the precipitate, behind the metallic bcc iron precipitation, and behind the partial iron dissolution into the matrix is not clearly understood. The metallic iron-rich precipitation can be interpreted in terms of selective oxidation of zirconium, iron, and chromium, but the partial iron dissolution in the ZrO_2 matrix remains unexplained especially because of a lack of data concerning the iron solubility and diffusion in zirconia. However, the oxidation of the matrix disrupts the thermodynamic equilibrium between precipitates and matrix. Some chemical changes between these two phases (ZrO_2 and $ZrCr_2$) are not unexpected.

Influence of the Precipitates on the Oxidation Kinetics

At the very beginning of their oxidation, zirconium and Zircaloy-4 have the same behavior. They both form tetragonal zirconia close to the metal-oxide interface, within a few tens of nanometers [35,36]. But when the oxide layer becomes thicker, the oxidation behavior of pure zirconium diverges; that is, tetragonal ZrO_2 is observed in Zircaloy-4 [4] but not in zirconium [37]. This specific behavior of Zircaloy-4 with respect to pure zirconium can be associated to the existence of $Zr(Fe,Cr)_2$ precipitates.

Once incorporated in the oxide layer, at a few hundred nanometers away from the metal-oxide interface, they undergo a large volume expansion (estimated at a few tens of a percent [20]) due to their oxidation into ZrO_2 nanocrystallites, and some iron is simultaneously dissolved into the ZrO_2 matrix. Since both compressive stresses and adequate alloying elements are known to stabilize tetragonal ZrO_2, the precipitates, through their structural evolution (volume expansion) and their chemical evolution (iron dissolution into the oxide layer), would contribute to extend the stabilization of the tetragonal ZrO_2 in the oxide layer further away from the metal-oxide interface. Since the oxidation kinetics are correlated to the destabilization of tetragonal ZrO_2, by the mechanisms described earlier, the $Zr(Fe,Cr)_2$ precipitates would contribute to postpone the kinetic transition and to the formation of a thicker dense oxide layer. This could explain the slower oxidation rate of Zircaloy-4 with respect to pure zirconium.

Irradiation Effect on the Oxidation of the Zircaloy-4

During the oxidation of ion-irradiated Zircaloy-4, the amorphous precipitates behave like reference ones. When incorporated into the oxide layer, their structure evolves during oxygen incorporation and some iron, initially distributed in the precipitate homogeneously, segregates at the edges, occasionally precipitates as bcc metallic iron, and then dissolves into the matrix. Similar results are also obtained in the case of oxide layers formed in-reactor where the observed precipitates are amorphous with an increased first neighbor spacing and a significant iron depletion. Since these structural and chemical evolutions of precipitates during oxidation are similar for reference and irradiated materials, the amorphization of the precipitates without chemical change induced by irradiation cannot alone explain the acceleration of the oxidation process observed in the reactor environment.

However special attention should be paid to the dissolution of amorphous precipitates in reactor conditions. As shown in neutron-irradiated cladding, the precipitate size, which is known to greatly influence the oxidation kinetics, decreases considerably under irradiation. On the other hand, the iron, being transferred before oxidation from the amorphous precipitates into the metallic matrix, will not be available to dissolve in the zirconia layer during precipitate oxidation. Therefore, amorphization, by accelerating precipitate dissolution, will modify the oxidation kinetics of neutron-irradiated claddings.

The comparison between the oxidation rates measured in-autoclave on the external surfaces of reference cladding samples and of neutron-irradiated cladding samples (for two and three cycles) gives an answer for short oxidation times. After 60 days of oxidation, they both formed the same oxide thickness. So, during the first steps of the oxidation process, as long as the oxide layer remains thin (lower than 3 μm), neither the presence of radiation defects, the amorphization, and the dissolution of precipitates nor the iron enrichment of the metallic matrix appear to have any major effect on the oxidation process. These oxidation tests have continued to verify if, after 200 days, this result is still valid or if neutron-irradiated cladding samples behave differently compared to reference samples for large oxide thicknesses. Since the large acceleration reported in-reactor occurs when the oxide thickness exceeds a few microns, these new results will be of great interest.

When comparing the oxidation kinetics of reference and ion-irradiated materials, a significant increase in the oxidation rate is observed in the latter case. Such an effect on ion-irradiated materials was also reported during the oxidation of lithium and hydrogen ion irradiated materials [10,38]. Three kinds of ion irradiation induced damage can be involved in this phenomenon: precipitate amorphization, point defects in the matrix, or chemical changes due to ion implantation. Considering the previous discussions and since the acceleration of the oxidation rate is observed even when the oxide-metal interface has grown far behind the thickness of the area damaged by the ion beam, it is reasonable to exclude the first two phenomena. The higher oxidation rate of ion irradiated materials seems to be correlated to the presence of implanted ions.

This higher oxidation rate is parallel to the high oxidation rate of the internal surfaces of neutron-irradiated cladding samples where a significant amount of fission product is implanted by recoil (about 3% atom over 10 μm). Thus, an enhancement of oxidation can be correlated to different cases of ion irradiation (including fission recoils), where chemical doping is significant.

Generally, according to both the microstructural observations and the comparison between the oxidation kinetics of reference and irradiated materials (with ions or neutrons), it can be concluded that the radiation defects and the microstructural evolutions produced by irradiation in the material cannot fully account for the increase of the oxidation rate measured for the ion-irradiated samples.

Conclusions

This study, dealing with the oxidation of intermetallic precipitates in reference, ion-irradiated, and neutron-irradiated materials, has given the following results:

1. During the growth of the oxidation layer, reference and irradiated $Zr(Fe,Cr)_2$ precipitates are first embedded unoxidized in the zirconia. Later, they experience a structural evolution (formation of cubic and monoclinic ZrO_2 nanocrystallites) and a chemical evolution (iron redistribution and depletion). These evolutions were described in detail and some interpretations were proposed. Moreover, they are correlated to the oxidation kinetics of zirconium alloys in terms of tetragonal ZrO_2 stabilization leading to a better understanding of the role of the precipitates on the oxidation process of these alloys.

2. Because results are similar for both reference and irradiated precipitates, the irradiation-induced amorphization of the precipitates alone does not appear to be able to modify the oxidation rate of the irradiated zirconium alloys. And more generally, after 60 days of oxidation in-autoclave, the microstructural evolution induced by neutron irradiation in the cladding material (amorphization and dissolution of precipitates, iron enrichment of the metallic matrix, creation of defects) does not change the oxidation rate of neutron-irradiated Zircaloy-4. However, the oxidation rate of helium ion-irradiated materials is higher than those of reference materials. This phenomenon should be interpreted in terms of chemical changes caused by ion implantation.

Acknowledgments

Special thanks are given to C. Albaric, M. Muscillo, R. Salot, and D. Venet for preparing, oxidizing, and characterizing the neutron-irradiated claddings; to M. Dupuy for his expert assistance in TEM; and to L. Coudurier for stimulating discussions. The work of D. Pêcheur is partially supported by Fragema-Framatome Combustible and Cézus in the framework of a joint thesis research contract with CEA.

References

[1] Hillner, E. in *Zirconium in the Nuclear Industry, ASTM STP 633*, A. L. Lowe, Jr., and G. W. Parry, Eds., American Society for Testing and Materials, Philadelphia, 1977, p. 211.

[2] Cox, B., *Journal of Nuclear Materials*, Vol. 29, 1969, p. 50.
[3] Urquhart, A. W. and Vermilyea, D. A., *Zirconium in Nuclear Applications, ASTM STP 551*, American Society for Testing and Materials, Philadelphia, 1974, p. 463.
[4] Godlewski, J., Gros, J. P., Lambertin, M., Wadier, J. F., and Weidinger, H. in *Zirconium in the Nuclear Industry, Ninth International Symposium, ASTM STP 1132*, C. M. Eucken and A. M. Garde, Eds., American Society for Testing and Materials, Philadelphia, 1990, p. 416.
[5] Garzarolli, F. and Stehle, H., *Proceedings*, IAEA International Symposium on Improvements on Water Reactor Fuel Technology and Utilization, IAEA, Stockholm, International Atomic Energy Agency, Vienna, Sept. 1986, p. 387.
[6] Eucken, C. M., Finden, P. T., Trapp-Pritsching, S., and Weidinger, H. G. in *Zirconium in the Nuclear Industry, Eighth International Symposium, ASTM STP 1023*, L. F. P. Van Swam and C. M. Eucken, Eds., American Society for Testing and Materials, Philadelphia, 1989, p. 113.
[7] Charquet, D., *Journal of Nuclear Materials*, Vol. 160, 1988, p. 186.
[8] Billot, P., Beslu, P., Giordano, A., and Thomazet, J. in *Zirconium in the Nuclear Industry, Eighth International Symposium, ASTM STP 1023*, L. F. P. Van Swam and C. M. Eucken, Eds., American Society for Testing and Materials, Philadelphia, 1989, p. 165.
[9] Garzarolli, F., Boomer, R. P., Stehle, H., and Trapp-Pritsching, S., *Proceedings*, International Topical Meeting on LWR Fuel Performance, American Nuclear Society–European Nuclear Society (ANS-ENS), Orlando, April 1985, p. 3.
[10] Billot, P., Beslu, P., and Robin, J. C., *Proceedings*, International Topical Meeting on LWR Fuel Performance, American Nuclear Society—European Nuclear Society (ANS-ENS), Avignon, April 1991, p. 770.
[11] Johnson, A. B., "Zirconium Alloy Oxidation and Hydriding under Irradiation," EPRI NP 5132, Electric Power Research Institute, Palo Alto, 1987.
[12] Lemaignan, C, *Journal of Nuclear Materials*, Vol. 187, 1992, p. 122.
[13] Gilbert, R. W., Griffiths, M., and Carpenter, G. J. C., *Journal of Nuclear Materials*, Vol. 135, 1985, p. 265.
[14] Griffiths, M., Gilbert, R. W., and Carpenter, G. J. C., *Journal of Nuclear Materials*, Vol. 150, 1987, p. 53.
[15] Griffiths, M., *Journal of Nuclear Materials*, Vol. 159, 1988, p. 190.
[16] Yang, W. J. S., Tucker, R. P., Cheng, B., and Adamson, R. B., *Journal of Nuclear Materials*, Vol. 138, 1986, p. 185.
[17] Lefebvre, F, Ph.D. thesis, Institut National des Sciences Appliquées, Lyon, France, Jan. 1989.
[18] Etoh, Y. and Shimada, S., *Journal of Nuclear Materials*, Vol. 200, 1993, p. 59.
[19] Pêcheur, D., Lefebvre, F., Motta, A. T., Lemaignan, C., and Wadier, J. F., *Journal of Nuclear Materials*, Vol. 189, 1992, p. 318.
[20] Pêcheur, D., Ph.D. thesis, Institut National Polytechnique, Grenoble, France, Jan. 1993.
[21] Lefebvre, F. and Lemaignan, C., *Journal of Nuclear Materials*, Vol. 171, 1990, p. 223.
[22] Bradley, E. R. and Perkins, R. A., *Proceedings*, IAEA Technical Committee Meeting on Fundamental Aspects of Corrosion of Zirconium Base Alloys in Water Reactor Environments, Portland, 11–15 Sept. 1989, IWGFPT/34, International Atomic Energy Agency, Vienna, 1990, p. 101.
[23] Kubo, T. and Uno, M. in *Zirconium in the Nuclear Industry, Ninth International Symposium, ASTM STP 1132*, C. M. Eucken and A. M. Garde, Eds., American Society for Testing and Materials, Philadelphia, 1990, p. 476.
[24] Garzarolli, F., Seidel, M., Tricot, R. and Gros, J. P. in *Zirconium in the Nuclear Industry, Ninth International Symposium, ASTM STP 1132*, C. M. Eucken and A. M. Garde, Eds., American Society for Testing and Materials, Philadelphia, 1990, p. 395.
[25] Levin, E. M. and Mc Murdie, H. F., *Phase Diagramms for Ceramists*, Vol. 3, National Bureau of Standards (NBS), American Ceramic Society, Inc, 1975, p. 76.
[26] Coudurier, L., Hopkins, D. W., and Wilkomirsky, I., *Fundamentals of Metallurgical Processes*, 2nd ed., Pergamon Press Ltd., Headington Hill Hall, Oxford Ox3, OBW, England, Press, 1985.
[27] Ramasubramanian, N., *Proceedings*, IAEA Technical Committee Meeting on Fundamental Aspects of Corrosion of Zirconium Base Alloys in Water Reactor Environments, Portland, 11–15 Sept. 1989, IWGFPT/34, International Atomic Energy Agency, Vienna, 1990.
[28] Cadalbert, R., Boulanger, L., Brun, G., Lansiart, S., Silvestre, G., and Juliet, P., *Proceedings*, International Topical Meeting on LWR Fuel Performance, American Nuclear Society–European Nuclear Society (ANS-ENS), Avignon, April 1991, p. 795.
[29] Arashi, H. and Ishigame, J. M., *Physica Status Solidi*, VoPA 71, 1982, p. 313.
[30] Garvie, R. C., *Journal of Physical Chemistry*, Vol. 82, No. 2, 1978, p. 218.
[31] Davidson, S., Kershaw, R., Dwight, K., and Wold, A., *Journal of Solid State Chemistry*, Vol. 73, 1988, p. 47.

[32] Ploc, R. and Cox, B., *Proceedings*, K. T. G. Specialist's *Workshop on Second Phase Particles and Matrix Properties on Zircaloys,* ErPungen, 1–2 July 1985, Bonn: Kerntechnische Geseffschaft, 1987, pp. 67–83.
[33] Ploc, R. in *Zirconium in the Nuclear Industry, Eighth International Symposium, ASTM STP 1023,* L. F. P. Van Swam and C. M. Eucken, Eds., American Society for Testing and Materials, Philadelphia, 1989, p. 498.
[34] De Gelas, B., Beranger, G., and Lacombe, P., *Journal of Nuclear Materials,* Vol. 28, 1968, p. 185.
[35] Beranger, G., Ph.D. thesis, Faculté des Sciences de l'Université de Paris, Paris, France, 1967.
[36] Denoux, M., Ph.D. thesis, Faculté des Sciences de l'Université de Paris, Paris, France, 1965.
[37] Del Pietro, H., private communication, Centre de Recherches de Cezus, Ugine, France.
[38] Kai, J. J., private communication, Department of Nuclear Engineering, National Tsing Hua University, Taiwan.

In-Reactor Corrosion

Friedrich Garzarolli,[1] *Ranier Schumann,*[1] *and Eckard Steinberg*[1]

Corrosion Optimized Zircaloy for Boiling Water Reactor (BWR) Fuel Elements

REFERENCE: Garzarolli, F., Schumann, R., and Steinberg, E., "**Corrosion Optimized Zircaloy for Boiling Water Reactor (BWR) Fuel Elements**," *Zirconium in the Nuclear Industry: Tenth International Symposium, ASTM STP 1245,* A. M. Garde and E. R. Bradley, Eds., American Society for Testing and Materials, Philadelphia, 1994, pp. 709–723.

ABSTRACT: A corrosion optimized Zircaloy has to be based primarily on in-boiling water reactor (in-BWR) results. Therefore, the material parameters affecting corrosion were deduced from results of experimental fuel rod irradiation with systematic variations and from a large variety of material coupons exposed in water rods up to four cycles.

The major material effect is the size and distribution of precipitates. For optimizing both early and late corrosion, the size has to stay in a small range. In the case of material quenched in the final stage, the quenching rate appears to be an important parameter. As far as material chemistry is concerned, the in-BWR results indicate that corrosion in BWRs is influenced by the alloying elements tin, chromium, and the impurity silicon.

In addition to corrosion optimization, hydriding is also considered. A large variation from lot to lot under identical coolant condition has been found. The available data indicate that the chromium content is the most important material parameter for hydrogen pickup.

KEY WORDS: zirconium, zirconium alloys, nuclear materials, nuclear applications, corrosion, nodular corrosion, boiling water reactors, in-reactor behavior, fuel rod cladding, material samples, microstructure effect, material chemistry effect, radiation effects

In boiling water reactors (BWRs), Zircaloy-2 is now used for the cladding of fuel rods and Zircaloy-4 or Zircaloy-2 for the structural components of the fuel assemblies. The corrosion environment in the core (boiling water and saturated steam at 288°C and 70 bars) is, from the standpoint of the corrosion resistance, at rather low temperatures but contains a rather high oxygen content formed by radiolysis and traces of oxidative radicals. In this environment, a thin uniform oxide layer is formed on Zircaloy-components that generally looks black. Also, white spots often arise in this black film, and appear in metallographic cross sections as lens-shaped oxide spots along with a local significant increase of oxide layer thickness [1]. At long exposure times, these nodules can grow together to form an almost uniform thick oxide layer. This type of corrosion is termed, "nodular corrosion."

The extent of nodular corrosion varies significantly from reactor to reactor and even from cycle to cycle and also depends on certain Zircaloy material properties [1]. The reasons for the variations between reactors are still not well understood. Conductivity transients at the beginning of the fuel element life can initiate severe nodular corrosion [2]. The nodular corrosion can be simulated out of pile in high pressure (>60 bar) steam tests at high temperatures (450 to 550°C). Such tests as well as analysis of BWR results have shown that the sensitivity

---

[1] Deputy director, Nuclear Fuel Cycle Joint Technology and Materials Issues; research engineer, irradiation experiments; and deputy director, Zircaloy Technology, respectively, Siemens AG, Power Generation Group (KWU) D-91050, Erlangen, Germany.

to nodular corrosion can be correlated to the precipitate number density or to the precipitate size or to the accumulated annealing parameter (ΣAi), which considers all annealing temperatures and times after the last β-quenching [1]. Today, generally Zircaloy with improved resistance to nodular corrosion is used.

Recently, it was pointed out that at high burnups an accelerated corrosion occurs in Zircaloy with very fine precipitates and very high resistance against nodular corrosion. This accelerated corrosion is probably similar to the accelerated uniform corrosion observed in pressurized water reactor fuel rods with Zircaloy-4 cladding [3]. In the future, however, the anticipated batch-burnups will be increased. Therefore, it will be necessary to further optimize the material for cladding and structure for higher burnups and longer exposure times.

Some fraction of the corrosion hydrogen is absorbed by the Zircaloy during exposure and can embrittle the metal. This hydrogen pickup is especially important for thin structural parts, like the spacers, where corrosion attacks the sheet from both sides. The data for hydrogen pickup of BWR fuel rods, components, and material samples [1] reveals quite a large scatter. Little information is available on the particularities of the water chemistry and the material responsible for this large scatter in hydrogen pickup. However, for future higher burnups and longer exposure times, it will be important to evaluate the factors affecting the hydrogen pickup to reduce the maximum hydriding rate.

Database for Material Optimization

A large out-of-pile database exists worldwide today from tests in 500 to 520°C high pressure steam on the effects of material condition and chemical composition. However, the validity of these results for the behavior in BWRs has to be considered carefully. Both the large difference in temperature as well as the effect of neutron flux may change the influence of some material parameters. Different temperature dependencies for certain material parameters can even invert the ranking with respect to nodular corrosion, as is obvious from Fig. 1. In out-of-pile corrosion tests in water at 300°C, nodular corrosion originates in zirconium alloys

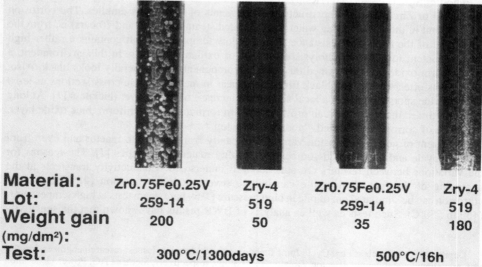

FIG. 1—*Effect of the test temperature on ranking to nodular corrosion.*

containing only transition elements (iron, chromium, and vanadium) but not tin after extended exposure times (several hundred days). This can be seen in Fig. 1 for ZrO-75FeO-25V (Zircaloy-4 is shown for comparison). These alloys, on the other hand, behave extremely well in 500 to 520°C corrosion tests. Several studies on the effect of chemical composition applying high-temperature high-pressure steam tests (for example, Refs 4 and 5) propose to keep tin as low as possible to minimize nodular corrosion. The data from BWR experiments do not confirm this conclusion, as will be shown later, probably due to the much lower operation temperature. Therefore, the basis for optimization of zirconium material for BWR application must be mainly the experience from tests performed in BWR.

A rather large database for fuel element corrosion behavior in BWRs has been collected over the years. However, for the analysis of these data, the following aspects have to be considered. The existing database indicates that the ranking with respect to corrosion can change with exposure time. At short exposure times, materials with large precipitate sizes (>0.15 μm) generally have thicker oxide layers than materials with fine precipitates. At long exposure times or at high burnups, the ranking may be the opposite, as can be seen in Fig. 2. Because of this influence of exposure time on corrosion ranking of different materials, conclusions relevant for high burnups must be based on high burnup (long time) experiments. Another problem for the analysis of the data from BWRs is the large differences of corrosion sensitivity between different reactors and even different exposure periods. Figure 3 shows the oxidation behavior of certain lots in different BWRs and reveals that both the corrosion rate as well as the time dependency can vary in different BWRs for the same material. From Fig. 3, one has to conclude that only results originating from the same reactor and from the same exposure period can be reliably analyzed.

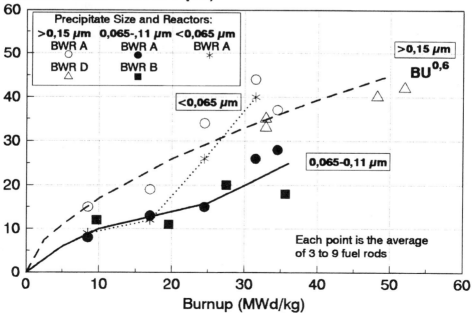

FIG. 2—*Corrosion behavior of different Zircaloy-2 BWR fuel claddings.*

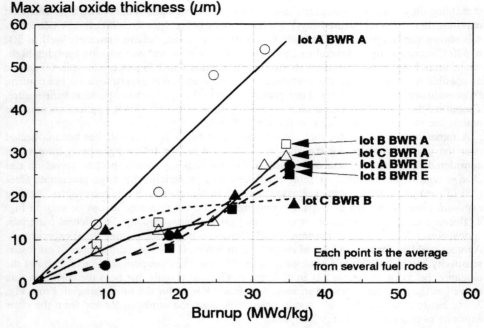

FIG. 3—*Corrosion behavior of certain Zircaloy-2 lots in different BWRs.*

If all these aspects are considered, the database, which can be used for a high burnup material optimization, becomes more limited. Highly reliable data can be derived from experimental fuel elements, where the claddings of the different fuel rods have been varied with respect to material condition and chemical composition after an exposure of more than 1000 effective full power days (EFPD) corresponding to a burnup of more than 30 MWd/kg. This type of fuel element is called a pathfinder element. Another valid data set available at Siemens are the results from corrosion coupons exposed to the BWR environment within segmented water test rods (WTRs) replacing fuel rods. The materials studied in these tests are from many different Zircaloy-2 tubing lots from several Zircaloy-4 strip lots and contain these materials that were studied earlier out of pile in 500°C high-pressure steam for nodular corrosion [4] and in steam at 400°C and water at 350°C for uniform corrosion [6]. Figure 4 shows one segment and the arrangement of the corrosion coupons. Stainless steel pins were used to fix

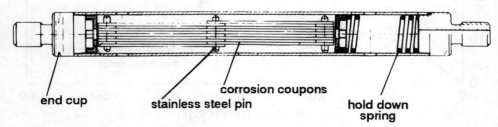

FIG. 4—*Water rod segment with corrosion coupons.*

the individual coupons to simulate the effect of austenitic springs, which often cause locally enhanced corrosion. The interior of these WTRs is open to the coolant. The WTRs were exposed up to four cycles to an equivalent burnup of 30.2 MWd/kg. Measurements of the weight gain of the individual samples were performed after one, two, and four cycles in hot cells. The reactor selected for this test unfortunately changed its sensitivity with respect to nodular corrosion and became insensitive during the test period, but the stainless steel pins sustained a high sensitivity, at least locally. The known corrosion increase at positions opposite the stainless steel or nickel-base alloys obviously is not affected much by the water chemistry. Material coupons with a high tendency to nodular corrosion showed nodular corrosion on the whole surface and high weight gains, material coupons with a medium sensitivity to nodular corrosion revealed white oxide only around the holes for the stainless steel pins and low to medium weight gain, whereas samples with a low corrosion sensitivity showed only uniform black oxide and moderate weight gain. The typical appearance of samples with different corrosion sensitivities after an exposure time of two cycles is shown in Fig. 5.

Material Parameters Affecting Corrosion in BWR

The effect of the "size of precipitates" on corrosion in a BWR has been found to be quite complex, as can be seen from Fig. 6. The results shown in this diagram were obtained from one pathfinder fuel element with different Zircaloy-2 fuel rod claddings. (The data contain different melts with similar chemistry. Some of these melts were tested with different precipitate sizes.) An examination of the various claddings for oxide layers was performed after one, two, three, and four irradiation cycles. The size of the precipitates was determined as the geometrical mean from high-magnification (>40.000) transmission electron microscopy. At short exposure times (one and two cycles), all materials with precipitate sizes up to 0.12 μm

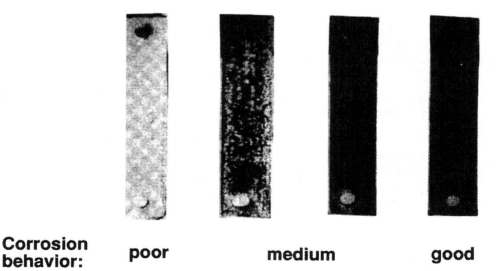

FIG. 5—*Typical appearance of corrosion coupons exposed in water rods in a BWR.*

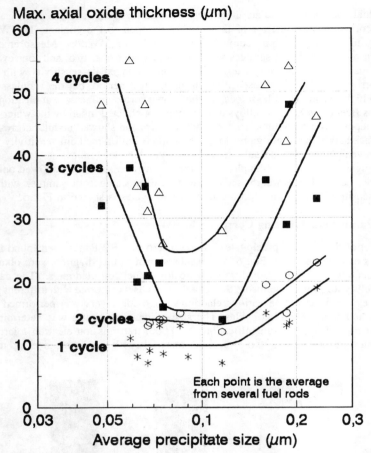

FIG. 6—*Effect of the precipitate size on corrosion of experimental fuel rods with Zircaloy-2 cladding in one test assembly in a BWR.*

exhibit low corrosion, whereas those with larger precipitate size show increasing corrosion with increasing precipitate size. At long exposure times (three and four cycles), increased corrosion is not only seen for materials with large precipitates but also in materials with extreme fine precipitates. Lowest corrosion is now observed for materials with a medium precipitate size (0.075 to 0.12 μm). This precipitate size can be achieved by a fabrication routine that leads to an accumulated annealing parameter of $2 \cdot 10^{-19}$ to $1.5 \cdot 10^{-18}$ h (for $Q/R = 40\,000$ K). This finding is in aggreement with out-of-pile corrosion tests, which have shown that nodular corrosion (500°C tests) can be minimized if the accumulated annealing parameter is below $1 \cdot 10^{-18}$ h, whereas low uniform corrosion (tests at 350 to 420°C) needs an accumulated annealing parameter above several 10^{-19} h (for example, Ref [1]).

The effect of the quench rate was studied in the WTRs with samples quenched from the β-temperature with different controlled quenching rates. Figure 7 reveals that at high exposure times (four cycles), both the very high as well as the low quenching rates, result in high corrosion. At lower exposure times (one and two cycles), high corrosion was only seen with

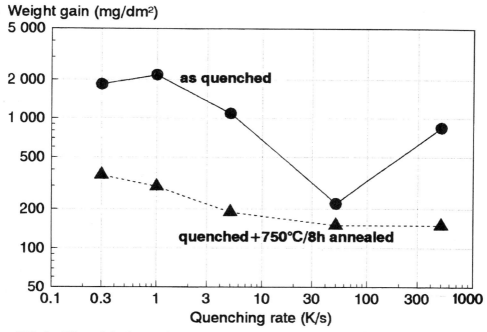

FIG. 7—*Effect of the β-quenching rate of corrosion of Zircaloy-4 in BWR after four cycles of exposure in water test rods.*

slowly quenched samples. The late accelerated corrosion of the fast quenched samples is similar to the observed late accelerated corrosion of fuel rods with very fine precipitates mentioned before. Figure 7 also shows that annealing of the quenched samples at 750°C for 8 h improves the corrosion behavior and that low corrosion of annealed samples needs high quenching rates (≥ 50 K/s). These results are similar to the findings in a high-temperature high-pressure out-of-pile test [4]. However, the response of such tests on the nodular corrosion sensitivity of β- or (α + β)-quenched samples varies quite drastically. The majority of tests do not form nodular corrosion in such material, independent of quenching rate [5].

A pronounced effect of the "tin content" on corrosion in BWR is obvious from Fig. 8. Here, the weight gains measured on corrosion coupons exposed in a WTR at the same axial position for four cycles is shown versus the tin content of these samples. The figure indicates that nodular corrosion increases significantly with decreasing tin content below the ASTM range for Zircaloy-2 and -4 and less pronounced within the ASTM range. As outlined earlier, this finding is in contradiction to out-of-pile tests in high pressure steam at 500°C (for nodular corrosion) (for example, Refs 5 and 6) and in pressurized water and steam at 350 to 420°C (for uniform corrosion) (for example, Ref 7) but seems to be in agreement with long time tests in water at 300°C as shown earlier. Tin obviously plays a dual role, it shows a beneficial behavior at the lower temperatures (≤ 300°C) but is detrimental at higher temperatures (350 to 520°C).

In addition, the WTR data also reveal an effect of the "silicon content" as shown by Fig. 9. Here the four-cycle weight gain data are plotted versus the silicon content. Lowest weight gains are observed with highest silicon contents. A beneficial effect of silicon corrosion was

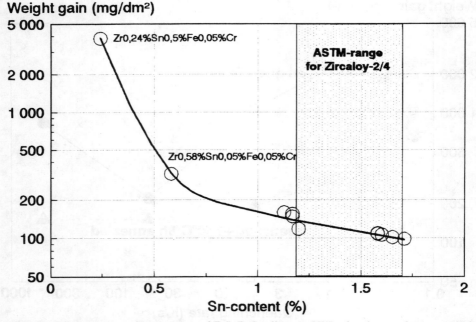

FIG. 8—*Effect of tin on the corrosion of ZrSnFeCr-alloys in BWR after four cycles exposure in water test rods. (All samples have an equivalent heat treatment and are Zircaloy-4 with a low silicon content if not specifically noticed.)*

found also in out-of-pile tests at 500°C (nodular corrosion) as well as at 350 to 400°C (uniform corrosion) [5,7].

Furthermore, according to the WTR data, chromium has a beneficial effect on in BWR corrosion (Fig. 10). There are some indications that iron may also have a positive effect on corrosion in BWR, but probably less than chromium. The conclusions with respect to the effect of nickel are not definitive. In the WTR samples, nickel had the lowest influence of the three transition elements (chromium, iron, and nickel). This ranking of the transition elements with respect to nodular corrosion derived from tests in BWR is inverse to the ranking found in high-temperature high-pressure steam tests. In the latter test, nickel is by far the most effective element to eliminate nodular corrosion, whereas chromium was found to have the smallest positive effect (for example, Refs 5 and 6). Long time tests for uniform corrosion at 400°C (for example, Ref 7) or at 370°C have shown a pronounced beneficial effect for iron and smaller effects for nickel and chromium (for example, Ref 7).

With respect to the other alloying elements and impurities studied with the WTR-corrosion coupons, it was concluded that oxygen has a small negative effect, whereas carbon and phosporus have no effect. Again, the conclusions from in-BWR tests with respect to oxygen, carbon, and phosphorus are not in agreement with the conclusions from out-of-pile testing of the same materials [5,7].

Material Parameters Affecting Hydrogen Pickup

Figure 11 shows the hydrogen pickup of the various Zircaloy-2 and -4 material coupons exposed in a WTR at a certain axial position for two cycles. It reveals quite a large scatter

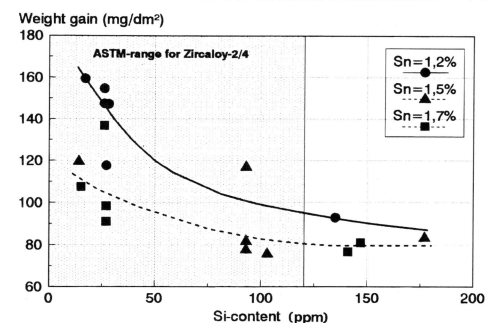

FIG. 9—*Effect of silicon and tin on the corrosion of Zircaloy-4 in BWR after four cycles exposure in water test rods. (All samples are Zircaloy-4 and have the same heat treatment.)*

especially for Zircaloy-2 and on average the same pickup fraction for Zircaloy-2 and -4. Comparing the scatter observed here with published data from different BWRs, it must be concluded that by far most of the large scatter observed with Zircaloy-2 and -4 in BWRs is caused by material variations.

Evaluation of the data from out-of-pile long time testing at 350°C measured over the years has shown that the pickup fraction is not a constant but is very low before transition, increases drastically at the transition and saturates at weight gains of about 100 mg/dm^2 ($\approx$7 μm). This behavior means that samples that corrode faster show a lower pickup fraction if evaluated after the same exposure time at rather large weight gains. The WTR-data and the other Siemens BWR-data on hydrogen pickup indicate that the pickup fraction versus weight gain relationship in BWR is very similar as found in 350°C long-term out-of-pile tests. Therefore, in the BWR data, only results from equivalent exposure times and equivalent weight gains can be simply correlated to the pickup fraction.

Analyses of the out-of-pile data at 350°C considering the previously mentioned conclusions indicate that the pickup fraction of Zircaloy-4 depends to a large extent on the chromium content (Fig. 12). Pickup fraction scatters between the extremes within the allowable ASTM band are low if the chromium content is above 0.15% and up to three times higher if below 0.1%. An evaluation of the BWR-MTR data after two cycles of irradiation for the effect of chromium indicates, in principle, a similar trend (Fig. 13).

BWR Corrosion Optimized Zircaloy for High Burnups

For high burnup applications, the fabrication routine should contain a fast β-quenching step ($\geq$50 K/s) and should lead to a size of precipitates between 0.075 and 0.12 μm. Tin should

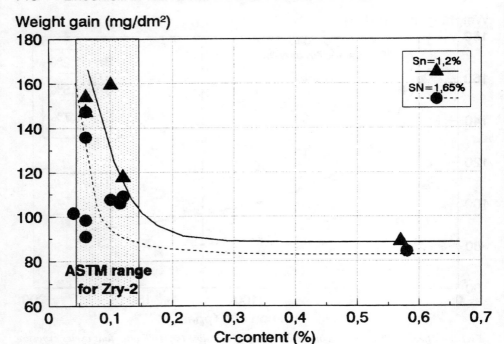

FIG. 10—*Effect of chromium and tin on the corrosion of Zircaloy-4 in BWR after four cycles exposure in water test rods. (All samples have an equivalent heat treatment.)*

be in the middle or upper range of the ASTM band of Zircaloy-2 and -4. Chromium and probably also iron should be high, preferably somewhat above the ASTM range of Zircaloy-2 and -4. Silicon should be added as an alloying element.

References

[1] Garzarolli, F. and Holzer, R., "Waterside Corrosion Performance of Light Water Power Reactor Fuel," *Nuclear Energy,* Vol. 31, No. 1, 1922, pp. 65–86.
[2] Jang, R. L. et al., "Fuel Performance Evaluation for EPRI Program Planning," *Proceedings,* ANS-ENS International Topical Meeting on Light Water Reactor Fuel Performance, Avignon, France, 1991, p. 258.
[3] Brümmer, G. et al., "Optimierung der Zircaloy Korrosion für Siedewasser-Reaktoren," *Proceedings,* Jahrestagung Kerntechnik 92, Karlsruhe, Germany, 1992, p. 321.
[4] Garzarolli, F. et al., "Progress in the Knowledge of Nodular Corrosion," *Zirconium in the Nuclear Industry: Seventh International Symposium, ASTM STP 939,* R. B, Adamson and L. F. P. Von Swam, Eds., American Society for Testing and Materials, Philadelphia, 1987, pp. 364–386.
[5] Weidinger, H. G. et al., "Effect of Chemistry on Elevated Temperature Nodular Corrosion," *Zirconium in the Nuclear Industry: Seventh International Symposium, ASTM STP 939,* R. B, Adamson and L. F. P. Von Swam, Eds., American Society for Testing and Materials, Philadelphia, 1987, pp. 364–386.
[6] Graham, R. A. et al., "Influence of Chemical Composition and Manufacturing Variables on Autoclave Corrosion of the Zircaloys," *Zirconium in the Nuclear Industry: Eighth International Symposium, ASTM STP 1023,* L. F. P. Von Swam and C. M. Eucken, Eds., American Society for Testing and Materials, Philadelphia, 1989, pp. 334–345.
[7] Eucken, C. M. "Influence of Chemical Composition on Uniform Corrosion of Zirconium Base Alloys in Autoclave Tests," *Zirconium in the Nuclear Industry: Eighth International Symposium, ASTM STP 1023,* L. F. P. Von Swam and C. M. Eucken, Eds., American Society for Testing and Materials, Philadelphia, 1989, pp. 113–127.

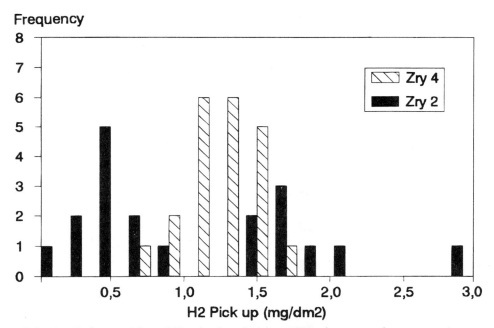

FIG. 11—*Hydrogen pickup of Zircaloy-2 and -4 in a BWR after two cycles exposure in water test rods.*

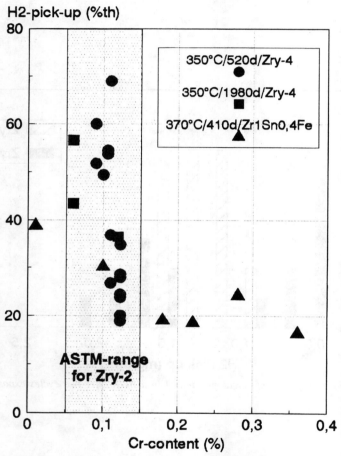

FIG. 12—*Hydrogen pickup of ZrSnFeCr-alloys out of pile in pressurized water.*

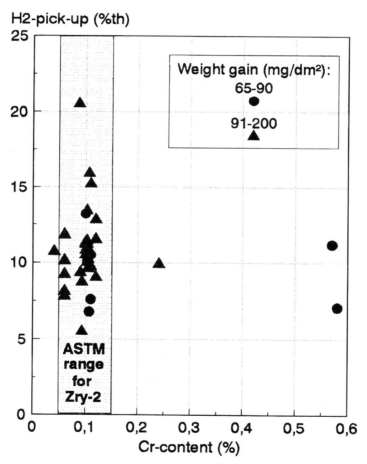

FIG. 13—*Hydrogen pickup of ZrSnFeCr-alloys in BWR after two cycles exposure in water test rods.*

DISCUSSION

P. Billot[1] (written discussion)—What is the origin of the material sensitivity to the power reactors?

F. Garzarolli et al. (authors' closure)—The cause of the different tendencies to nodular corrosion in different boiling water reactors is not fully known. Analysis of cladding oxide from two reactors with a high sensitivity has shown surprisingly high nitrogen concentrations in the oxide (about 1000 ppm) as well as somewhat increased nitrogen in the coolant (several 100 ppb). The suggestion that nitrogen in the coolant may influence nodular corrosion is also validated by in-pile loop tests performed in Halden where increased nodular corrosion was observed when a few 100 ppb $Zn(NO_3)_2$ was added [*1*]. However, nitrogen is not the only species that affects nodular corrosion. Increased nodular corrosion can also arise from conductivity transients (for example, due to a condenser leak). Such conditions affect only almost fresh Zry surfaces (with ≤ 1 month exposure) at the position of maximum power during the conductivity transient [*2,3*]. A certain correlation was also found between the tendency to nodular corrosion and effectivness of the clean up system [*2*]. The chemical species responsible for the later observations are not known. More than likely, there are several.

Bo Cheng[2] (written discussion)—You are proposing a higher tin as opposed to the industrial trend of reducing tin toward the lower end of the ASTM specification. Yet, the fuel components with lower tin has clearly had excellent corrosion resistance within the last six to seven years. Also, the corrosion weight gain and hydrogen pickup fraction appears to be rather high as compared to the broad database in existence. Would you please tell the processing history of the specimens? Could the results be affected by the processing history?

F. Garzarolli et al. (authors' closure)—The beneficial effect of reducing tin toward the lower end of the ASTM specification is in reactor only confirmed for the PWR environment, that is, at higher temperatures and at low oxygen contents. The conclusion that low Sn might be also beneficial for BWR results only from out of pile high pressure steam tests at 500 to 530°C. It clearly has to be mentioned again that the BWR irradiations showed larger effects only below the ASTM-range. The results reported here are from coupons that have seen rather high process-temperatures during fabrication. The Ai-parameter of these samples were above 1×10^{-17} h. More details of the fabrication history of these materials are reported in Ref *4* of the paper. Thus, the material studied should be more sensitive to nodular corrosion than a modern low temperature treated Zry. Therefore, it could be expected that a modern low tin Zircaloy-2 or -4 will not show the efffect. However, one must mention that the effect of Sn became more pronounced with increasing fluence. Therefore the Sn effect may become of more importance at the anticipated high burnups.

D. Schrire[3] (written discussion)—Observations of enhanced corrosion due to hydride concentrations at the underlying metal surface have been reported. Do you think that this effect could impact the validity of metal coupon corrosion tests for predicting cladding corrosion?

F. Garzarolli et al. (authors' closure)—Hydrogen pick up of BWR fuel rod cladding is, according to our data, rather low. Very seldom have we seen hydrogen concentrations above the thermal solubility at operation temperatures not even at burnups of 45 to 48 MWd/kg. For reactor/material conditions where hydrogen pick-up is accelerated, the formation of solid hydride rims at small gaps in the pellet column could occur. At such local positions corrosion may be increased in a similar way as has been seen occasionally on PWR fuel rods with thick

---

[1] CEA, CE Cadarache, St. Paul-lez Durana, France.
[2] Electric Power Research Institute (EPRI), Palo Alto, CA.
[3] ABB-ATOM, Vasteras, Sweden.

oxide layers. The goal of this paper was to find a material with a high corrosion resistance and a low tendency to pick up hydrogen. This material should also have the highest margin against a postulated hydride accelerated corrosion.

Brian Cox[4] (written discussion)—In your plot of weight gain after four cycles versus tin content, did you adjust the fabrication schedule in order to obtain similar structures? If you didn't, how do you rate the relative importance of tin content and structural differences in producing the observed results?

F. Garazolli et al. (authors' closure)—The materials listed in Fig. 8 have equivalent fabrication schedules. As far as it can be deduced from metallographic examination, all materials also have a closely similar microstructure.

Reference

[1] Shimada, S., et al., "Parametric Tests of the Effect of Water Chemistries on Corrosion of Zr-Alloys under Simulated BWR Condition," *Proceedings*, IAEA Technical Committee Meeting on Influence of Water Chemistry on Fuel Cladding Behavior, Rez, Chech Republic, 4th-8th Oct., 1993.
[2] Garzarolli, F. and Stehle, H., "Behavior of Structural Materials for Fuel and Control Elements in Light Water Cooled Power Reactors," *Proceedings*, IAEA International Symposium on Improvements in Water Reactor Fuel Technology and Utilization, Stockholm, Sweden, 10th-15th Sept. 1986, p. 387.
[3] Jung, R. L., "Fuel Performance Evaluation for EPRI Program Planning," *Proceedings,* ANS-ENS International Topical Meeting on Light Water Reactor Fuel Performance, Avignon, France, 1991, p. 258.

[4] University of Toronto, Ontario, Canada.

George P. Sabol,[1] *Robert J. Comstock,*[2] *Robert A. Weiner,*[1] *Paula Larouere,*[3] *and Robert N. Stanutz*[1]

In-Reactor Corrosion Performance of ZIRLO™[4] and Zircaloy-4

REFERENCE: Sabol, G. P., Comstock, R. J., Weiner, R. A., Larouere, P., and Stanutz, R. N., "**In-Reactor Corrosion Performance of ZIRLO and Zircaloy-4,**" *Zirconium in the Nuclear Industry: Tenth International Symposium, ASTM STP 1245,* A. M. Garde and E. R. Bradley, Eds., American Society for Testing and Materials, Philadelphia, 1994, pp. 724–744.

ABSTRACT: In-reactor and long-term autoclave corrosion data have been obtained on ZIRLO and three variants of Zircaloy-4: conventional (1.5% tin), low-tin, and beta-treated. In-reactor data from demonstration assemblies irradiated in the Virginia Power Company's North Anna Unit 1 reactor demonstrate the superiority of ZIRLO and, to a lesser extent, low-tin Zircaloy-4 over conventional Zircaloy-4. After two cycles of irradiation to an assembly burnup of 37.8 GWD/MTU, the average axial peak corrosion of ZIRLO was 32% that of conventional Zircaloy-4. Low-tin and beta-treated materials displayed average peak oxides 76% and 150% of that formed on conventional Zircaloy-4, respectively.

Autoclave corrosion tests of archive tubing have been performed in 633 K water, 672 K, 700 K, and 727 K steam, and in 633 K water containing 70 and 210 ppm lithium as the hydroxide. Correlation of the in-reactor data with the autoclave data indicates that the 633 K pure water test is the best qualitative indicator of in-reactor corrosion performance, and the 672 K steam test the poorest. Differences in in-reactor corrosion between ZIRLO and the Zircaloy-4 materials are consistent with the relative behavior of these materials in lithium hydroxide solutions. The relationships among the in-reactor and autoclave corrosion data, the microstructures, and the processing are discussed.

In addition to improved corrosion resistance, ZIRLO exhibits improved dimensional stability over Zircaloy-4. The in-reactor creep of ZIRLO is confirmed to be about 80% of that of Zircaloy-4, and irradiation growth is observed to be about 50% that of Zircaloy-4. These data are also presented and discussed.

KEY WORDS: corrosion, zirconium, zirconium alloys, autoclave, water, steam, oxide lithium hydroxide, fuel cladding, processing creep, growth, microstructure, nuclear materials, nuclear applications, radiation effects

More demanding PWR fuel duties have necessitated improvements in the corrosion resistance of fuel cladding. This need has been addressed by two parallel approaches; improvements to Zircaloy-4 and the development of new alloy compositions. The improvements to Zircaloy-4 have generally been to reduce tin content to the lower end of the specification range [1,2], to exercise greater control of impurities [1], and to optimize processing by application and control

---

[1] Consulting engineer, principal engineers, respectively, Westinghouse Electric Corporation, Nuclear Manufacturing Division (NMD), Pittsburgh, PA 15230.
[2] Program manager, Westinghouse Electric Corporation, Science and Technology Center (STC), Pittsburgh, PA 15235.
[3] Staff engineer, Virginia Power, Innsbrook Technical Center, Glen Allen, VA 23060.
[4] ZIRLO™ is a registered trademark of Westinghouse Electric Corporation.

of the so-called A-parameter [1–4]. Although there is increased activity in the development of alternate alloys [5–7], in-reactor data at high burnup have been reported only for the ZIRLO alloy [5]. Assemblies containing ZIRLO clad fuel were exposed in the BR-3 reactor and rod average burnups as great as 68 GWD/MTU and residence times up to 66 months were attained. Post-irradiation examination revealed ZIRLO with a corrosion improvement of up to 50% relative to Zircaloy-4. Additionally, ZIRLO also displayed lower irradiation growth and creep than Zircaloy-4.

Further data on the in-reactor performance of ZIRLO were provided from a demonstration program that commenced in 1987 in the North Anna Unit 1 reactor. Fuel rods in this program included ZIRLO and several types of Zircaloy-4: conventional (1.5% by weight tin), improved (low tin), and conventional Zircaloy-4 with a late stage beta treatment. At the end of the first cycle of exposure, the assembly burnup was 21.2 GWD/MTU and the benefits of both improved Zircaloy-4 and ZIRLO have been reported [8–10]. Specifically, improved Zircaloy-4 represented a corrosion benefit of about 22% over conventional Zircaloy-4, and ZIRLO provided a 39% benefit. In addition, the lower creep and irradiation growth observed for ZIRLO in BR-3 were verified by the one-cycle North Anna data [8,10].

Subsequently, two-cycle data have been obtained from the North Anna demonstration program. An assembly burnup of 37.8 GWD/MTU was obtained and the qualitative ranking of the two Zircaloys and ZIRLO continued [11], but with the benefit of ZIRLO being much more pronounced. The objective of this paper is to provide a complete characterization of the materials in the North Anna demonstration assemblies, to update the in-reactor data obtained on these materials, and to compare the in-reactor corrosion behavior of the various cladding materials with corrosion data obtained from autoclave tests.

Materials, Test Conditions, Fuel, and Reactor Parameters

Processing of Tubing

Tubing was produced from three triple-melted ingots of the following compositions: Zircaloy-4 with tin near the middle of the composition range (1.20 to 1.70% by weight), Zircaloy-4 with tin in the lower half of the composition range, and ZIRLO (Zr-1.0Nb-1.0Sn-0.1Fe). The ingots were produced from zirconium sponge and the major constituents are listed in Table 1. The tubing produced from these ingots will be referred to as conventional Zircaloy-4 (CON Zr-4), improved Zircaloy-4 (IMP Zr-4), and ZIRLO, respectively.

The ingots were beta-forged and then sawed into billets that were beta-annealed and water-quenched. The billets were bored, machined, and then extruded. The Zircaloy-4 extrusions

TABLE 1—*Chemical analysis of ingots.*

	Percent by Weight Unless Specified		
Element	CON Zr-4	IMP Zr-4	ZIRLO
Niobium	<50 ppm	<50 ppm	1.02 to 1.04
Tin	1.48 to 1.52	1.31 to 1.39	0.96 to 0.98
Iron	0.22 to 0.23	0.18 to 0.22	0.094 to 0.105
Chromium	0.11 to 0.12	0.10 to 0.12	79 to 83 ppm
Oxygen, ppm	1100 to 1300	1200	900 to 1200
Carbon, ppm	180 to 200	110 to 140	60 to 80
Silicon, ppm	87 to 96	82 to 112	<40
Nitrogen, ppm	<20 to 25	<32	22 to 30
Zirconium	balance	balance	balance

were alpha-annealed in vacuum prior to cold pilgering to tube reduced extrusions (TREXs). ZIRLO extrusions were beta-annealed, air-cooled, conditioned to remove the oxide film, and then pilgered to TREXs. All TREXs were alpha-annealed to induce recrystallization of the material prior to further working.

TREXs from the three ingot chemistries were reduced to final size tubing (9.500 mm outside diameter/0.584 mm wall) by an alternating sequence of cold pilgering and vacuum annealing. The same four-pass reduction schedule was utilized for the three chemistries. Each vacuum anneal was performed in the alpha range to recrystallize the tubing with the exception of the final stress relief anneal. The recrystallization anneals of Zircaloy-4 were characterized by an A-parameter using a value for Q/R of 40 000 K [4]. A-parameters for the IMP Zr-4 and CON Zr-4 tubes were 17.7×10^{-18} h and 6.2×10^{-18} h, respectively. The higher value for the IMP Zr-4 tubing was a result of an increased annealing temperature at the TREX stage of the processing. A lower recrystallization temperature was used for ZIRLO to achieve a distribution of fine particles in the final size tubing. Each recrystallization anneal was in the temperature range of 853 to 873 K for approximately 4 h. This processing approach for ZIRLO was consistent with that previously presented [5,12].

In addition to the processing of the three ingots as just described, some tubes from the conventional Zircaloy-4 ingot were beta-quenched following the second cold pilger reduction. This was performed by translating the intermediate size tube through an induction coil to heat it into the beta range. Upon exiting the coil, the tube was cooled to a temperature high in the alpha range by spraying with a water mist at a cooling rate of about 25 K/s. The tube was subsequently cooled rapidly to room temperature by a water quench further downstream from the exit of the coil. After removal of the oxide film, the tube was reduced in two passes to final size utilizing an intermediate recrystallization and a final stress relief anneal. The A-parameter for this tubing was 1.1×10^{-18} h. This material is subsequently referred to as βQ Zr-4.

Final conditioning of all of the tubes included straightening, pickling, polishing of the tube outside diameter using 400 grit silicon carbide belts, and grit blasting the inside diameter surface. After ultrasonic inspection, the tubes were cut to length and cleaned for fabrication into fuel rods. Archive tubes were retained for laboratory testing and characterization.

Autoclave Tests

Long-term corrosion testing of samples from the four tubing lots was performed in several different autoclave environments. The testing was performed to completely define the corrosion properties and to identify correlations between the out-reactor and in-reactor corrosion behavior. Specimens, typically 2-cm long, were tested in the as-fabricated condition. Specimens were cleaned in Alconox detergent, sequentially rinsed in tap water, deionized water, and ethanol, and then blow-dried with warm air.

The autoclave tests were performed in a manner consistent with ASTM Practice for Aqueous Corrosion Testing of Samples of Zirconium and Zirconium Alloys (G 2-88). Tests in 633 K water were performed at saturation pressure while tests in high temperature steam (672 K, 700 K, and 727 K) were performed at 10.3 MPa. Testing at 633 K included both pure and lithiated water with lithium additions made in the form of LiOH at concentrations of 70 ppm lithium (0.01 molal) and 210 ppm lithium (0.03 molal). The interest in testing in lithiated solutions stems from the use of LiOH as the pH control agent in pressurized water reactors (PWRs), and the previously observed benefit of ZIRLO over Zircaloy-4 in autoclave tests and in the BR-3 reactor [5].

Multiple specimens (three to five) from each lot were corrosion tested in each test environment. Weight gains during testing were monitored by periodically interrupting the tests and weighing the samples. The reported weight gains are averages of the samples tested.

In-Reactor Demonstration

Fuel Rod and Assembly Design—Two special, removable-rod assemblies were fabricated in the spring of 1987 for irradiation in Virginia Power's North Anna Unit 1 reactor, beginning with Cycle 7 in June of 1987. The assemblies contain a 17 by 17 array of fuel rods clad with conventional Zircaloy-4, ZIRLO, and the modified Zircaloy-4 materials previously described. North Anna Unit 1 was chosen for the demonstration of improved waterside corrosion resistance because its high coolant temperatures (greater than 600 K outlet) and long (18 month) fuel cycles provide an aggressive environment suitable for delineating even subtle differences in in-reactor corrosion performance. Such conditions are typical of modern, high-efficiency PWRs.

North Anna Unit 1 contains 157 fuel assemblies that are of essentially identical mechanical design, but use several different enrichment levels. Each fuel assembly contains 264 fuel rods supported at eight intervals along their lengths by Inconel 718 grid assemblies. The fuel rods were pressurized with helium and seal-welded at the top end plug. The demonstration fuel rods were part of a region with a 3.60% (uniform) enrichment and internal pressure of 2.0 MPa. The fuel stack length for these rods was nominally 366 cm (144.0 in.). Each demonstration assembly contained 102 CON Zr-4, 10 IMP Zr-4, 12 βQ Zr-4, and 24 ZIRLO clad fuel rods, all of which were highly characterized prior to irradiation. The remaining rods in each assembly were clad with CON Zr-4.

Reactor Operating Conditions—The two demonstration assemblies began operation in Cycle 7. One assembly was removed after Cycle 7 for examination with an assembly burnup of 21.2 GWD/MTU. The second assembly continued operation in Cycle 8 and accumulated a burnup of 37.8 GWD/MTU. This second assembly was examined at the end of Cycle 8. The average coolant core outlet temperature at full power in Cycle 7 was 602 K. Cycle 8 operated at this temperature for the first four months of the cycle at which time a temperature reduction was instituted. The core outlet temperature was decreased to 599 K for approximately 1.5 months, and was then increased to 600 K for the remainder of the cycle. Plots of reactor power versus time for Cycles 7 and 8 are shown in Figs. 1 and 2, respectively. The fuel rod average linear

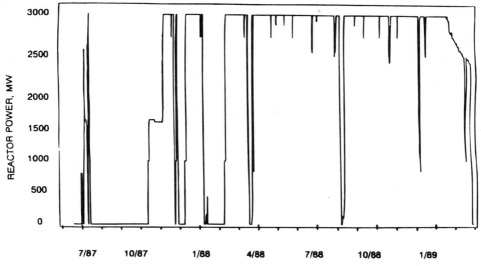

FIG. 1—*Reactor power versus time for North Anna Unit 1, Cycle 7 (full power = 2893 MW).*

power was 18.6 kW/m. The coolant was maintained at a pH of 6.9 at the core average temperature, and the lithium concentration near the start of the cycle was approximately 2.5 ppm.

PIE Data Acquisition

Fuel Rod Corrosion Measuring System and Methods—The oxide thickness measuring system (OTMS) nondestructively measures waterside corrosion film thickness on fuel rod cladding. The system utilized was developed by Westinghouse and is based upon an eddy current (EC) probe system manufactured by Fischer Tech Corp. of Windsor, Connecticut. Rods were removed from the assemblies for testing. Each rod was scanned in a straight line over most of its length at four orthogonal orientations. A digital sampling system recorded eddy current responses every 0.05 cm along the length of a rod. The digitized data were reduced by averaging the individual oxide measurements axially over 2.5-cm axial increments and all four azimuthal orientations, for example, the data reported at the 25-cm elevation are the average of all data points between 23.75 and 26.25 cm.

Standards were used to calibrate the OTMS system. A statistical analysis of the calibration standard data combined with the on-site system precision yielded an uncertainty at 1 σ of ± 0.8 μm for ZIRLO and ± 2.2 μm for Zircaloy-4. Past hot cell data [13] from Zion fuel rods indicate that eddy current corrosion measurements show a small positive bias when compared with metallographic corrosion data taken on the same rods. This same correction factor [13] was applied to the eddy current data obtained from the measurements of 15 one-cycle and 14 two-cycle rods.

Fuel Rod In-Reactor Growth Measurements—The lengths of peripheral fuel rods and internal fuel rods were measured by two different techniques. Peripheral rod lengths were inferred by an indirect measurement where rod lengths were calculated from measured rod/nozzle gaps and assembly length measurements. The axial gaps between fuel rods and assembly nozzle were measured from low-magnification TV tapes. All 64 peripheral rods lengths were determined in

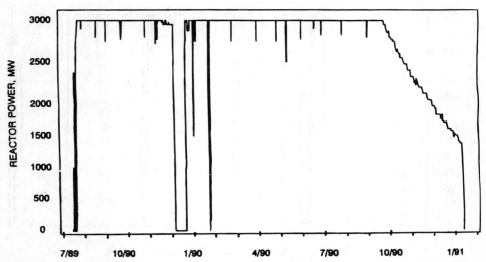

FIG. 2—*Reactor power versus time for North Anna Unit 1, Cycle 8 (full power = 2893 MW).*

this manner. The total uncertainty based on the calibration measurements and rod-to-nozzle gap measurement uncertainty is 5%.

Individual internal fuel rods were measured directly using the TV monitor to locate the top and bottom of the fuel rod, and the digital encoder of the motorized fuel rod handling tool to measure the rod length. The accuracy of the rod length measurements was ±0.025 cm. These measurements were performed on 6 one-cycle and 15 two-cycle internal rods.

Fuel Rod Profilometry System—The profilometry system nondestructively measures fuel rod diameters as a function of length. A constant-speed, motorized, fuel rod handling tool linearly translates the fuel rod past two stationary linear variable differential transformers (LVDTs), 45° apart. A digital sampling system records the LVDT response every 0.25 cm. Each rod is scanned two times to provide data at four orientations; fuel rod plenum areas are scanned two additional times to obtain data at a total of eight orientations.

The standards used to calibrate the profilometry system are stainless steel rods ground to precise diameters that span the range of expected measurements. A standard is measured prior to and after each fuel rod measurement. Precision and accuracy of the data are ±0.0005 cm. Data are typically displayed as diameters as a function of length. Each diameter listed at a particular elevation is the average of all data at a particular elevation over a 2.5-cm interval. Seventeen one-cycle rods were measured.

Results and Discussion

Properties of Tubing

Microstructure—Transmission electron microscopy (TEM) was used to characterize the microstructure of the four tubing lots. Of interest were the second-phase particles that were resolvable following a low-temperature recrystallization anneal at 823 K for 0.5 to 1 h. Average particle sizes of the four lots of tubing are provided in Table 2.

Scanning transmission electron microscopy-energy dispersive spectroscopy (STEM-EDS) quantitative microanalysis techniques were used to measure second-phase particle chemistries. In general, absolute particle chemistries were not obtained due to the presence of the zirconium matrix. Particle types were identified by analyzing a number of particles and then examining the relative concentrations of the alloying elements. Results from larger particles in which contributions from the zirconium matrix were minimized provided a means of estimating the absolute particle chemistries.

Particle analysis for the three types of Zircaloy-4 tubing revealed similar results. The alloying elements detected in the second-phase particles were iron and chromium with the predominant type of particle being zirconium-iron-chromium (Zr-Fe-Cr). The ratio of iron to chromium (atom percent) in the particles was comparable to the bulk ingot chemistry. The iron plus chromium concentration in the larger particles approached 60 atom percent, suggesting that the particles were $Zr(Fe,Cr)_2$ as typically observed in Zircaloy-4 [14,15]. In addition to the

TABLE 2—*Average particle size.*

Material	Average Size, nm
CON Zr-4	160
IMP Zr-4	183
βQ Zr-4	93
ZIRLO	79

Zr(Fe,Cr)$_2$ particles, there were a few Zr-Fe particles. The iron content was low due to the small size of the particles, and conclusive identification of the particles was not possible.

The main alloy additions present in particles analyzed in ZIRLO were niobium (Nb) and iron. A plot of niobium versus iron concentrations in Fig. 3 revealed that the particles were of two types. One type of particle was niobium-rich and the second type was Zr-Nb-Fe. The niobium-rich particles tended to be the smaller particles in the TEM micrographs shown in Fig. 4. Though not included in Fig. 3, low concentrations (1 to 2 atom percent) of chromium were detected in the Zr-Nb-Fe particles. Its presence was not unexpected as chromium is present in ZIRLO as a residual element.

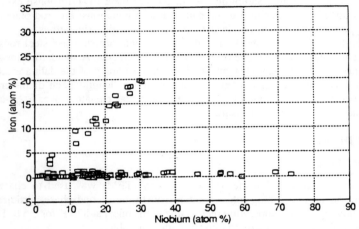

FIG. 3—*Iron and niobium concentrations in ZIRLO particles.*

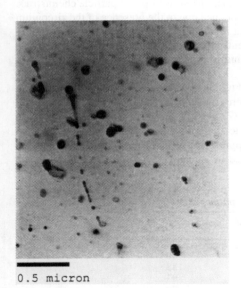 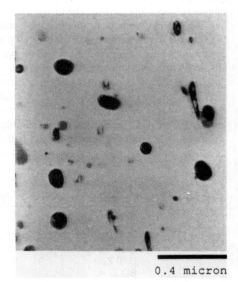

FIG. 4—*TEM micrographs of ZIRLO.*

The structure of the niobium-rich phase was determined to be body centered cubic with a lattice parameter of ~0.33 nm. These particles were identified as β-niobium as the structure was similar to that reported by Williams and Gilbert [16] for the β-niobium phase in Zr-2.5Nb. The Zr-Nb-Fe phase was hexagonal with lattice parameters of a = ~0.54 nm and c = ~0.87 nm. The maximum niobium and iron detected in the larger particles was 30% and 20%, respectively, suggesting the composition of the phase to be $Zr_{0.5}Nb_{0.3}Fe_{0.2}$ or possibly $(Zr_{0.625}Nb_{0.375})_4Fe$. Additional work is required to better define this precipitate.

The Zr-Nb-Fe particles observed in ZIRLO can be compared with those observed in other zirconium alloy systems containing niobium and iron. Miyake and Gotoh [17] suggested the presence of $(Zr_{1-x}Nb_x)(Fe_{1-y}Cr_y)_2$ intermetallic particles in Zircaloy-4 with niobium additions. While one might expect the ZIRLO particles to be similar, the STEM-EDS results did not support the presence of such particles. The iron levels in the ZIRLO particles were significantly lower than the 67 atom percent in the preceding intermetallic. In addition, the niobium-to-iron ratio in the ZIRLO particles was 1.5 compared to the ratio of 0.0 to 0.5 in the suggested intermetallic. Meng and Northwood [18] reported the presence of Zr-Fe-Cr particles in Zr-1Nb containing iron plus chromium impurities of 0.065%. This was contrary to the observations in ZIRLO in which niobium was always detected in the iron-containing particles. Zr-Fe-Cr particles have not been observed in ZIRLO. Glazkov et al. [19] reported on an alloy (Zr-1.0Nb-1.0Sn-0.5Fe) similar to ZIRLO. They identified two types of particles, namely, a β-niobium phase and a Zr_2Fe-type intermetallic containing niobium. The limited information on the Zr-Nb-Fe particle prevents direct comparison to the ZIRLO observations. However, the two particle types that they report (β-niobium and Zr-Nb-Fe) are identical to those detected in ZIRLO.

Mechanical Properties—Mechanical properties as determined from uniaxial tensile tests of the tubing are summarized in Table 3. The tensile strength of the βQ Zr-4 tubing shows a slight increase in strength relative to the conventional tubing, and the IMP Zr-4 tubing shows a slight decrease in strength due to a reduction in tin. The properties of ZIRLO are similar to those of the Zircaloys. In all cases, the tubing met the design requirements for the fuel rods.

Crystallographic texture parameters were measured on two of the four tubing lots. ZIRLO and IMP Zr-4 tubing have essentially equivalent texture values, although the contractile strain ratio (CSR) for the ZIRLO tubing was slightly lower.

TABLE 3—*Tubing properties.*

Property	CON Zr-4	IMP Zr-4	βQ Zr-4	ZIRLO
Room Temperature Tensile				
0.2% Offset, MPa	65.3	59.0	66.7	61.0
Ultimate, MPa	83.9	79.1	87.2	81.2
Elongation, %	15.0	18.3	15.0	16.5
658 K Tensile				
0.2% Offset, MPa	38.9	35.3	40.3	40.6
Ultimate, MPa	46.8	44.0	49.7	49.4
Elongation, %	18.5	18.5	17.0	15.0
CSR	1.53	1.49	1.30	1.29
Texture Parameters				
f_{radial}	...	0.492	...	0.496
$f_{circum.}$	...	0.393	...	0.397
f_{axial}	...	0.048	...	0.042

Autoclave Corrosion Tests—Corrosion weight gain data are plotted in Figs. 5 to 10. Post-transition corrosion rates for the tubing in each of the autoclave environments are summarized in Table 4. In the pure water and steam tests, ZIRLO samples were characterized by a single rate as there was no indication of a second breakaway during testing. The Zircaloy-4 samples exhibited accelerated rates or a second breakaway during long-term testing in steam environments at 672 K and 700 K. Spalling on the ends of the CON Zr-4 samples was observed after the final exposure in 700 K steam. Rates were calculated both before and after the second breakaway as indicated in the table. Testing of Zircaloy-4 in 727 K steam was terminated after six days. Localized corrosion of sample ends and identification numbers and nodular corrosion appeared on the IMP Zr-4 samples, thereby preventing a meaningful determination

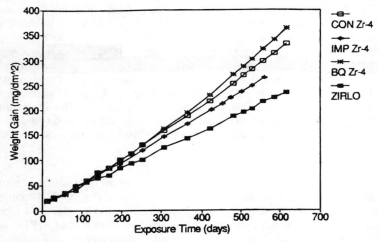

FIG. 5—*Autoclave corrosion results from testing in 633 K pure water.*

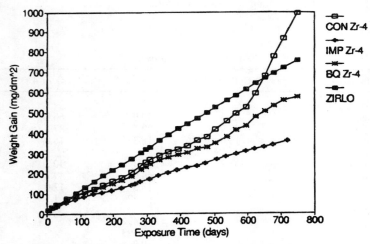

FIG. 6—*Autoclave corrosion results from testing in 672 K steam.*

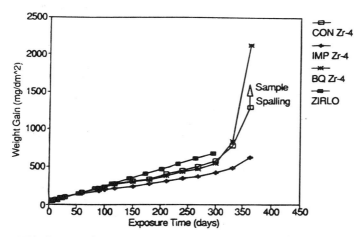

FIG. 7—*Autoclave corrosion results from testing in 700 K steam.*

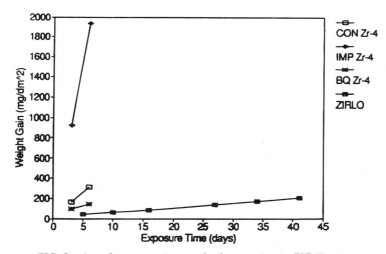

FIG. 8—*Autoclave corrosion results from testing in 727 K steam.*

of corrosion rate. The reported rate for CON Zr-4 was somewhat enhanced by localized corrosion in 727 K steam. In the lithiated water tests, ZIRLO exhibited the best behavior in both lithium concentrations. Zircaloy-4 exhibited accelerated rates in both environments while ZIRLO showed an acceleration only in 210 ppm lithium after the 28-day exposure.

A comparison of the corrosion rates in each of the test environments is shown in Table 5 in which the rates were normalized with respect to CON Zr-4. The comparison in the steam environments was based upon the rates observed in Zircaloy-4 prior to the acceleration observed late in the tests. It is noted that testing in 672 K and 700 K steam gave similar ranking of the materials prior to the onset of the exceedingly high corrosion rates. The higher kinetics at 700 K provide an advantage over testing at 672 K as the trends can be established in about one-half of the exposure time.

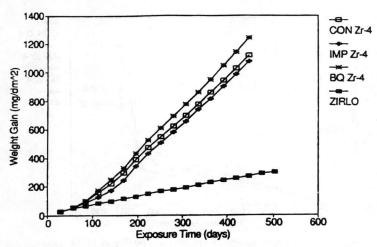

FIG. 9—*Autoclave corrosion results from testing in 633 K water containing 70 ppm lithium.*

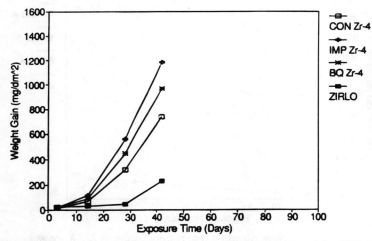

FIG. 10—*Autoclave corrosion results from testing in 633 K water containing 210 ppm lithium.*

In the pure water and steam tests (excluding 727 K), IMP Zr-4 exhibited the best behavior among the Zircaloy-4 lots. In addition to the lowest rates, IMP Zr-4 has shown no accelerated corrosion through 719 days of testing at 672 K and only the beginning of a second breakaway after 330 days of testing at 700 K. The appearance of nodules and localized corrosion on IMP Zr-4 and CON Zr-4 samples tested at 727 K indicates the difficulty in using high temperature ($\geq$727 K) steam tests to compare the uniform corrosion behavior of Zircaloy-4.

Both CON Zr-4 and βQ Zr-4 were processed from the same ingot but given significantly different thermal treatments. CON Zr-4 performed better at 633 K while βQ Zr-4 performed better in the high temperature (727 K) steam environments. This is consistent with previous observations of higher corrosion at low temperatures and lower corrosion at high tempera-

TABLE 4—*Post-transition autoclave corrosion rates from North Anna tubing.*

Material	Post-Transition Corrosion Rates, mg/dm²/day					
	633 K	672 K	700 K	727 K	633 K, 70 ppm Li	633 K, 210 ppm Li
CON Zr-4	0.584	0.870 (3.11)[a]	1.62 (>16.7)[a,b]	49.7[c]	2.88	29.7
IMP Zr-4	0.472	0.469	1.09 (4.56)[a]	ND[d]	2.89	44.6
βQ Zr-4	0.672	0.766	1.47 (41.7)[a]	14.3	3.18	37.3
ZIRLO	0.371	0.978	2.18	5.07	0.545	13.3

[a] Rate after second breakaway.
[b] Spalling on ends of the samples during final exposure.
[c] Localized corrosion on sample ends and identification number.
[d] Not determined because of localized and nodular corrosion.

tures for material processed with low A-parameters [1,3,20]. However, it is surprising that βQ Zr-4 performed as well as it did in 633 K water and 672 K steam. Late-stage-processed β-treated Zircaloy with a particle size less than 100 nm is usually reported to display very poor corrosion behavior in 623 K water [3]. Similar results have also been reported for late-stage β-treated Zircaloy in 672 K steam environments [1,3,21]. Presumably, the short time the βQ Zr-4 tubing was in the high alpha temperature range prior to quenching to room temperature resulted in a significant improvement in the autoclave behavior of this material.

ZIRLO exhibited the best autoclave behavior in 633 K water and the poorest in 672 K and 700 K steam prior to the onset of accelerated corrosion in Zircaloy-4. It is likely that continued testing in the steam environments will result in accelerated corrosion for all three Zircaloy-4 lots. This has already occurred for CON Zr-4 and βQ Zr-4 at 700 K after 300 days, and appears to have begun in IMP Zr-4 after about 330 days.

In-Reactor Corrosion Performance

The data from the oxide film eddy current traces for each material type were averaged at incremental positions of elevation and the data are shown in Figs. 11 and 12 for the one-cycle and two-cycle data, respectively. The figures show the relative ranking of the four types of materials to be consistent after both cycles, ZIRLO being the most corrosion resistant, IMP Zr-4 the next best, and βQ Zr-4 the poorest. Several features of the traces deserve attention. The reduced oxide thickness at the grid locations is expected because of enhanced heat transfer

TABLE 5—*Relative autoclave corrosion rates from North Anna tubing.*

Material	Post-Transition Corrosion Rates Normalized to CON Zr-4					
	633 K	672 K	700 K	727 K	633 K, 70 ppm Li	633 K, 210 ppm Li
CON Zr-4	100	100	100	100	100	100
IMP Zr-4	81	54	67	...	100	150
βQ Zr-4	115	88	91	28	110	126
ZIRLO	64	112	135	10	19	45

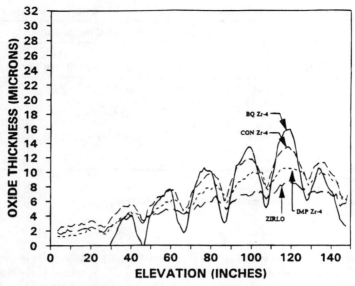

FIG. 11—*One-cycle oxide film eddy current traces.*

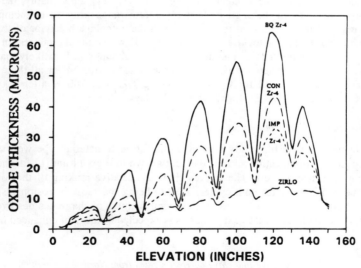

FIG. 12—*Two-cycle oxide film eddy current traces.*

from coolant mixing and flux suppression, but the severe reduction in the one-cycle data for the βQ Zr-4 suggests a strong flux dependence for this material. The observation of high oxide in the mid-spans and pronounced effect of the grids on the βQ Zr-4 is consistent with similar data reported by Garzarolli et al. [*3*] for tubing processed with an A-parameter of 3 × 10^{-19} h. Furthermore, Garzarolli et al. indicated a strong flux dependence on corrosion for Zircaloy-4 processed with A-parameters less than about 3 × 10^{-18} hour, corresponding to particle sizes less than about 100 nm. Therefore, the oxide trace for βQ Zr-4 confirms the Garzarolli et al. data for such material.

Discounting the one-cycle data for the βQ Zr-4, it is seen that the relative ranking of the materials is consistent along the length of the rods. It is also noticed that the benefit of IMP Zr-4 over CON Zr-4 is evident at the more severe conditions of the mid-span regions, but at the grids the corrosion films of these two Zircaloys are very similar. The ZIRLO benefit is displayed both at the mid-span regions and at the upper grid locations. Finally, and most importantly, it is noted that the behavior of ZIRLO relative to the three types of Zircaloy-4 improves with increasing burnup from one cycle to two cycles of irradiation.

The data in Figs. 11 and 12 represent averages of the oxides from the rods of the various materials. However, the burnups (BUs) vary somewhat for individual rods in an assembly and the peak oxide film thicknesses also vary among the rods within a material group. Table 6 is presented to allow some assessment of rod-to-rod variability. The range of the oxide film thicknesses observed for CON Zr-4 and IMP Zr-4 is almost a factor of two. In contrast, the range of thicknesses for ZIRLO is much less and the two values obtained on βQ Zr-4 are in close agreement. It is noted that within each material group that the rod with the thickest oxide is the rod with the highest burnup, suggesting a strong power dependence. The trend of increasing oxide thickness with increasing burnup is particularly evident for CON Zr-4 and IMP Zr-4. Finally, the average burnups of the rods in each material group are very close and are essentially equivalent. Thus, comparison of the different materials based on the average peak oxide thickness of each group is valid.

Table 7 gives a quantitative comparison of the behavior of the four materials based upon the oxide thicknesses depicted in Figs. 11 and 12. The average peak oxide film thicknesses are listed for each cycle, and the comparisons are based on a normalization to CON Zr-4. The

TABLE 6—*North Anna two-cycle individual rod corrosion and burnup (BU) data.*

Rod Number	Average Peak Oxide, μm	Rod Average BU, GWD/MTU	Average BU/ Material, GWD/MTU
		CON ZR-4	
E09	33.2	36.8	37.9
G12	32.8	37.1	
H07	48.8	38.2	
L10	58.7	39.3	
		IMP ZR-4	
D13	19.7	35.6	38.2
E07	31.1	37.4	
N11	35.3	39.8	
N07	45.7	40.1	
		βQ ZR-4	
F05	65.7	38.5	38.8
L13	63.5	39.1	
		ZIRLO	
E12	14.2	36.3	38.2
F08	13.4	37.4	
K11	13.1	38.8	
M06	17.0	40.3	

TABLE 7—*Corrosion data summary of North Anna demonstration fuel rods.*

Material	Average Peak Oxide, one-cycle		Average Peak Oxide, two-cycle[a]		Average Peak Oxide, two-cycle[b]	
	μm	% of CON Zr-4	μm	% of CON Zr-4	μm	% of CON Zr-4
CON Zr-4	13.5	100	43.0	100	53.8	100
IMP Zr-4	10.5	78	32.9	76	40.5	75
βQ Zr-4	16.0	118	64.5	150	64.6	120
ZIRLO	8.6	64	13.8	32	15.1	28

[a]Based on oxide thicknesses in Fig. 12.
[b]Based on rods in Table 6 with burnups >38 GWD/MTU.

peak oxide thickness of the IMP Zr-4 rods is 78% that of CON Zr-4 after one cycle and 76% after two cycles, a consistent benefit. The oxide on ZIRLO rods is 64% that of CON Zr-4 after one cycle, but improves to being only 32% that of CON Zr-4 after two cycles. In contrast, βQ Zr-4 becomes relatively worse from the first to the second cycle.

Due to the strong dependence of oxide thickness with burnup for CON Zr-4 and IMP Zr-4, comparisons were also made using data in Table 6 for which rod burnups were greater than 38 GWD/MTU. These comparisons are listed in the last column of Table 7. This is perhaps a more meaningful comparison as it compares the limiting rods from each group, that is, the rods with the thickest oxides. Interestingly, the relative oxide thicknesses for the Zircaloys are similar to those following one cycle of testing. However, the ZIRLO rods continue to show a relative improvement following the second cycle of irradiation.

In-Reactor Dimensional Stability

ZIRLO fuel rod cladding has improved in-reactor dimensional stability compared with CON Zr-4 and IMP Zr-4 cladding. The improved dimensional stability is seen in both the cladding growth and creep data obtained from the demonstration assembly fuel rods. Figure 13 shows a comparison of the ZIRLO, IMP Zr-4, and CON Zr-4 growth data, after one and two cycles of irradiation, with the range of previous experience with CON Zr-4. Previously reported [5] rod growth data for ZIRLO rods irradiated in BR-3 are also shown. These data show that ZIRLO cladding growth is significantly less than Zircaloy-4 growth, approximately 50% of the average Zircaloy-4 growth, and that the growth of the CON Zr-4 and IMP Zr-4 cladding is consistent with previous experience for Zircaloy-4. The reduced ZIRLO growth is slightly lower than the 60% of Zircaloy-4 growth observed in the BR-3 ZIRLO rods [5].

The profilometry data obtained after one cycle of irradiation has been used to evaluate the differences in the alloys' in-reactor creep behavior. Fuel-clad contact over most of the rod length occurred during the second cycle, and the profilometry data obtained on the two-cycle rods are controlled by the fuel pellet swelling rather than the cladding creep. Therefore, the two-cycle data cannot be used for creep analysis.

The relative creep behavior of ZIRLO and IMP Zr-4, compared to CON Zr-4, was assessed by taking the ratio of the average creepdown for all rods of each cladding type to the average creepdown of all the CON Zr-4 rods as a function of elevation along the rod. Figure 14 shows this comparison. The creepdown was relatively constant along most of the rods' length with the ZIRLO in-reactor creep being 83% of CON Zr-4 and IMP Zr-4 within 5% of CON Zr-4. The reduced ZIRLO creep was in good agreement with individual rod calculations that were 80% of the creepdown predicted by the Westinghouse Zircaloy-4 creep model. The creepdown

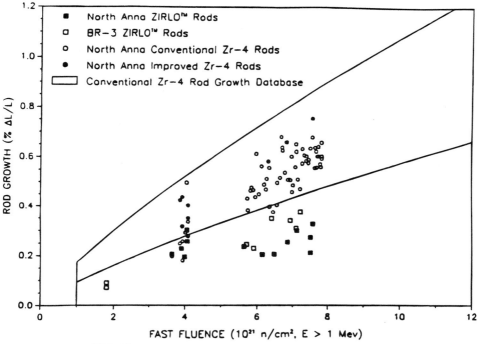

FIG. 13—*Rod growth of CON Zr-4, IMP Zr-4, and ZIRLO.*

decreased rapidly at the very top and bottom of the rods, where the fast flux decreased very rapidly. The apparent change in the creepdown ratios at the top and bottom of the ZIRLO and IMP Zr-4 rods was due to amplification of the effects of the slight offsets of the relative rod positions rather than an axial dependence of the relative creep rates.

The reasons for the improved in-reactor creep and growth of ZIRLO relative to Zircaloy-4 are not known. However, because ZIRLO and the Zr-2.5Nb alloy contain similar, very fine precipitates of β-niobium, it may be expected that these two alloys also share properties that are sensitive to the precipitate distribution. The in-reactor creep of heat-treated [22] and cold-worked [23] Zr-2.5Nb is reported to be superior to that of Zircaloy-2. TEM studies of precipitation and defect damage in irradiated Zr-2.5Nb led Williams and Ells [24,25] to suggest that irradiation strengthening occurred through niobium stabilization of small (<3 nm) interstitial aggregates, that is, a precipitation of niobium at the interstitial clusters or loops. Although the small defects were most prevalent in quenched material with excess niobium in solution, they were also observed in quenched and aged material and in the alpha grains of two-phase, α + α′, material. It is postulated that a similar mechanism of niobium stabilization of the irradiation-produced defects is operative in ZIRLO. At PWR temperatures and fluences, one might expect larger defects than those observed by Williams and Ells. These niobium-stabilized defects would be stronger obstacles to dislocation motion than niobium-free defects and could also serve as sinks for additional vacancy and interstitial defects, thus positively affecting both creep and growth.

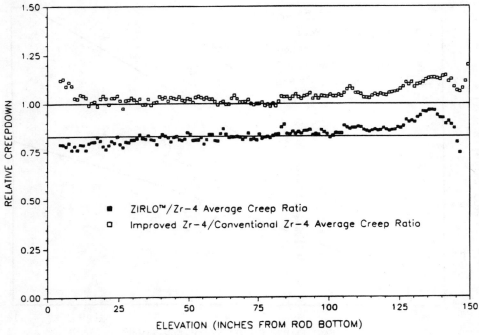

FIG. 14—*Creepdown of IMP Zr-4 and ZIRLO cladding relative to CON Zr-4.*

Correlation of In-Reactor and Out-Reactor Corrosion

A comparison of autoclave corrosion data with the in-reactor data reveals some interesting results. Tables 5 and 7 present the corrosion data normalized to CON Zr-4. It is immediately evident that the tests in 672 K and 700 K steam are completely remiss in identifying the good in-reactor performance of ZIRLO relative to the three Zircaloys. The steam tests also fail to identify the exceedingly poor in-reactor corrosion behavior of βQ Zr-4 relative to CON Zr-4. The only consistency between the two steam tests and the in-reactor data is the better performance of IMP Zr-4 relative to CON Zr-4 and βQ Zr-4. The data in 727 K steam correctly identifies ZIRLO as the best material in-reactor, but does not correctly rank the Zircaloy materials, possibly due to the confounding appearance of localized nodular corrosion on IMP Zr-4 and, to a lesser extent, on CON Zr-4.

The pure-water 633 K autoclave test gave the best indication of in-reactor behavior. This test and the one-cycle data were in very good quantitative agreement. After two cycles, the 633 K water test still gave the correct qualitative ranking of the materials, but in-reactor ZIRLO became progressively better. The relative performance of the peak Zircaloy-4 rods (BU > 38 GWD/MTU) remained similar to the one-cycle behavior and remained in quantitative agreement with the 633 K autoclave test.

One explanation of the separation of the ZIRLO data from those of the three Zircaloys with increased exposure from the first to the second cycles is thermal feedback caused by the growing oxide films. The lower corrosion rate of ZIRLO coupled with thermal feedback effects would explain the improvement relative to CON Zr-4. However, if the thermal feedback effects were large, one would also expect to see a greater separation among CON Zr-4, IMP Zr-4, and βQ Zr-4 with increased in-reactor exposure. This was not observed, suggesting other phenomena dominating the in-reactor corrosion behavior.

An alternate consideration is a lithium effect on corrosion. The data in the lithiated solutions, particularly at 70 ppm Li, indicate much better corrosion resistance of ZIRLO than the Zircaloys. Previous experience with binary Zr-Nb alloy tubing [5] also showed a marked sensitivity to accelerated corrosion in lithiated water environments as well as locally thicker oxides than either Zircaloy-4 or ZIRLO following testing in BR-3. The correlation between good performance in both lithiated water and in-reactor as well as poor performance in both lithiated water and in-reactor suggests a possible connection between lithium and corrosion in PWR environments. Corrosion rates of CON Zr-4, IMP Zr-4, and βQ Zr-4 in 70 ppm Li increased by a factor of about 5 to 6 over that in pure water compared to a rate increase for ZIRLO of about 1.5 times. While autoclave testing in pure water provided an initial quantitative ranking of the materials, testing in lithiated water highlighted the increased separation between ZIRLO and the Zircaloys observed from the first to the second cycle. This suggests that a low thermal rate in pure water as well as resistance to accelerated corrosion in moderately high lithium concentrations are reasonable out-reactor corrosion characteristics to screen potential new zirconium alloy compositions for improved in-reactor performance.

Conclusions

1. Demonstration fuel rods of ZIRLO and three variants of Zircaloy-4 cladding have been irradiated in North Anna Unit 1 reactor for two cycles to an assembly average burnup of 37.8 GWD/MTU. ZIRLO exhibited the best in-reactor corrosion performance with an average axial peak oxide thickness 32% of that formed on conventional Zircaloy-4.

2. Low tin Zircaloy-4 exhibited the second best performance and beta-treated Zircaloy-4 was the poorest with average peak oxides 76% and 150% of conventional material, respectively.

3. ZIRLO cladding exhibited increased in-reactor dimensional stability with about 80% of the creep and 50% of the growth of Zircaloy-4.

4. Autoclave corrosion testing in 633 K water correctly ranked the in-reactor performance of ZIRLO and the three variants of Zircaloy-4. Testing in steam at 672 K and 700 K failed to demonstrate the better performance of ZIRLO relative to Zircaloy-4.

5. The in-reactor corrosion resistance of ZIRLO improved relative to the Zircaloy-4 materials with increasing time. This separation of ZIRLO was consistent with reduced sensitivity of ZIRLO to accelerated corrosion in lithiated water.

References

[1] Kilp, G. R., Thornburg, D. R., and Comstock, R. J. "Improvements in Zirconium Alloy Corrosion Resistance," *Proceedings,* IAEA Technical Committee Meeting on Fundamental Aspects of Corrosion on Zirconium Base Alloys in Water Reactor Environments, International Atomic Energy Agency, Vienna, IWGFPT/34, 1990, pp. 145–157.

[2] Fuchs, H. P., Garzarolli, F., Weidinger, H. G., Bodmer, R. P., Meier, G., Besch, O. A., and Lisdat, R., "Cladding and Structural Material Development for the Advanced Siemens PWR Fuel 'Focus'," *Proceedings,* American Nuclear Society, European Nuclear Society (ANS, ENS) International Topical Meeting on LWR Fuel Performance, Avignon, 1991, p. 682.

[3] Garzarolli F., Steinberg, E., and Weidinger, H. G., "Microstructure and Corrosion Studies for Optimized PWR and BWR Zircaloy Cladding," *Zirconium in the Nuclear Industry: Eighth International Symposium, ASTM STP 1023,* L. F. P. Van Swam and C. M. Eucken, Eds., American Society for Testing and Materials, Philadelphia, 1989, pp. 202–212.

[4] Steinberg, E., Weidinger, H. G., and Schaa, A., "Analytical Approaches and Experimental Verification to Describe the Influence of Cold Work and Heat Treatment on the Mechanical Properties of Zircaloy Cladding Tubes," *Zirconium in the Nuclear Industry: Sixth International Symposium,*

ASTM STP 824, D. G. Franklin and R. B. Adamson, Eds., American Society for Testing and Materials, Philadelphia, 1984, pp. 106–122.

[5] Sabol, G. P., Kilp, G. R., Balfour, M. G., and Roberts, E., "Development of a Cladding Alloy for High Burnup," *Zirconium in the Nuclear Industry: Eighth International Symposium, ASTM STP 1023*, L. F. P. Van Swam and C. M. Eucken, Eds., American Society for Testing and Materials, Philadelphia, 1989, pp. 227–244.

[6] Isobe, T. and Matsuo, Y., "Development of Highly Corrosion Resistant Zirconium-Base Alloys," *Zirconium in the Nuclear Industry: Ninth International Symposium, ASTM STP 1132*, C. M. Eucken and A. M. Garde, Eds., American Society for Testing and Materials, Philadelphia, 1991, pp. 346–367.

[7] Harada, M., Kimpara, M., and Abe, K., "Effect of Alloying Elements on Uniform Corrosion Resistance of Zirconium-Based Alloys in 360°C Water and 400°C Steam," *Zirconium in the Nuclear Industry: Ninth International Symposium, ASTM STP 1132*, C. M. Eucken and A. M. Garde, Eds., American Society for Testing and Materials, Philadelphia, 1991, pp. 368–391.

[8] Balfour, M. G., Burman, D. L., Miller, R. S., Moon, J. E., Pritchett, P. A., Stanutz, R. N., Weiner, R. A., and Wilson, H. W., "Westinghouse Fuel Operating Experience at High Burnup with Advanced Fuel Features," *Proceedings*, American Nuclear Society, European Nuclear Society, ANS/ENS International Topical Meeting on LWR Fuel Performance, Avignon, 1991, p. 104.

[9] Kilp, G. R., Balfour, M. G., Stanutz, R. N., McAtee, K. R., Miller, R. S., Boman, L. H., Wolfhope, N. P., Ozer, O., and Yang, R. L., "Corrosion Experience with Zircaloy and ZIRLO™ in Operating PWRs," *Proceedings*, American Nuclear Society, European Nuclear Society, ANS/ENS International Topical Meeting on LWR Fuel Performance, Avignon, 1991, p. 730.

[10] Sabol, G. P., Schoenberger, G., and Balfour, M. G., "Improved PWR Fuel Cladding," *Proceedings*, IAEA Technical Committee Meeting on Materials for Advanced Water Cooled Reactors, International Atomic Energy Agency, Vienna, IAEA-TECDOC-665, 1992, p. 122.

[11] Miller, R. S., "Steps Toward High Burnup in PWRs: a U.S. Perspective," *Nuclear Energy*, Vol. 31, 1992, pp. 41–47.

[12] Sabol, G. P. and McDonald, S. G., "Process for Fabricating a Zirconium-Niobium Alloy and Articles Resulting Therefrom," U. S. Patent 4,649,023, 10 March 1987.

[13] Nayak, U. P., Kunishi, H., and Smalley, W. R., "Hot Cell Examination of Zion Fuel Cycle 5," Final Research Report EP80-16, Empire State Electric Energy Research Corporation, June 1985.

[14] Vander Sande, J. B. and Bement, A. L., "An Investigation of Second Phase Particles in Zircaloy-4 Alloys," *Journal of Nuclear Materials*, Vol. 52, 1974, pp. 115–118.

[15] Versaci, R. A. and Ipohorski, M., *Journal of Nuclear Materials*, Vol. 80, 1990, pp. 180–183.

[16] Williams, C. D. and Gilbert, R. W., "Tempered Structures of a Zr-2.5 wt % Nb Alloy," *Journal of Nuclear Materials*, Vol. 18, 1966, pp. 161–166.

[17] Miyake, C. and Gotoh, K., "Magnetic Study of Zircaloy, II Zircaloy with Additive of Niobium," *Journal of Nuclear Materials*, Vol. 184, 1991, pp. 212–220.

[18] Meng, X. and Northwood, D. O., "Intermetallic Precipitates in Zirconium-Niobium Alloys," *Zirconium in the Nuclear Industry: Eighth International Symposium, ASTM STP 1023*, L. F. P. Van Swam and C. M. Eucken, Eds., American Society for Testing and Materials, Philadelphia, 1989, pp. 478–486.

[19] Glazkov, A. G., Grigor'ev, V. M., Kon'kov, V. F., Moinov, A. S., Nikulina, A. V., and Sidorenko, V. I., "Corrosion Behavior of Zr-Nb-Sn-Fe Alloy," *Proceedings*, IAEA Technical Committee Meeting on Fundamental Aspects of Corrosion on Zirconium Base Alloys in Water Reactor Environments, International Atomic Energy Agency, Vienna, IWGFP/34, 1990, pp. 158–164.

[20] Charquet, D., Steinberg, E., and Millet, Y., "Influences of Variations in Early Fabrication Steps on Corrosion, Mechanical Properites, and Structure of Zircaloy-4 Products," *Zirconium in the Nuclear Industry: Seventh International Symposium, ASTM STP 939*, R. B. Adamson and L. F. P. Van Swam, Eds., American Society for Testing and Materials, Philadelphia, 1987, pp. 431–447.

[21] Andersson, T. and Thorvaldsson, T., "Influence of Microstructure on 400 and 500°C Steam Corrosion Behavior of Zircaloy-2 and Zircaloy-4 Tubing," *Proceedings*, IAEA Technical Committee Meeting on Fundamental Aspects of Corrosion on Zirconium Base Alloys in Water Reactor, Environments, International Atomic Energy Agency, Vienna, IWGFP/34, 1990, pp. 237–248.

[22] Gilbert, E. R., "In-Reactor Creep of Zr-2.5 at % Nb," *Journal of Nuclear Materials*, Vol. 26, 1968, p. 105.

[23] Ross-Ross, P. A. and Hunt, C. E. L., "The In-Reactor Creep of Cold-Worked Zircaloy-2 and Zirconium-2.5 wt % Niobium Pressure Tubes," *Journal of Nuclear Materials*, Vol. 26, 1968, p. 2.

[24] Williams, C. D. and Ells, C. E., "The Influence of Niobium in Irradiation Strengthening of Dilute Zr-Nb Alloys," *Philosophical Magazine*, Vol. 18, 1968, pp. 763–772.

[25] Ells, C. E. and Williams, C. D., "The Effect of Temperature during Neutron Irradiation or Subsequent Deformation in the Zr-2.5 wt % Nb Alloy," *Transactions, Journal*, Japan Institute of Metals, Vol. 9, Supplement 214, 1968, pp. 214–219.

DISCUSSION

P. Billot[1] (written discussion)—In Session C on oxide characterization, we saw many parameters that can have an influence on corrosion of zirconium-based alloys. For you, in connection with these works, what are the main parameters that could play a role to improve the corrosion behavior of ZIRLO (porosity of films formed, intermediate particles, inner barrier layer, etc.)?

G. P. Sabol et al. (authors' closure)—Although we do not know the mechanism by which ZIRLO is superior to Zircaloy-4, because the corrosion resistance in pure water is less than that of Zircaloy-4, we may expect that the barrier layer is more stable, and, therefore, thicker. The stable distribution of small precipitates may be easily incorporated into the growing oxide without significantly increasing the stresses in the oxide. Thus, the barrier oxide can attain a greater thickness before the formation of porosity and microcracks as present in the outer layers of the oxide.

The in-reactor superiority of ZIRLO over Zircaloy-4 ZIRLO is consistent with the greater resistance of ZIRLO to lithium-hydroxide solutions.

Brian Cox[2] (written discussion)—The apparently zero temperature dependence you saw in some of these tests reminded me of some experiments that we did many years ago where we saw apparently negative temperature coefficients over some ranges of temperature. The cause of this was the variation in the pressure dependence of the oxidation at different temperatures. Thus, to get an accurate temperature coefficient, you may need to measure the pressure dependence at each temperature and normalize the data to equal pressure.

G. P. Sabol et al. (authors' closure)—This corrosion response is indeed puzzling, and others are reporting similar behavior of Zircaloy-4 in the 673 K test. These tests were performed in the same autoclave laboratory in which we observed more normal behavior in tests performed earlier. Additionally, the behavior of this ZIRLO material is essentially the same as that tested earlier and as described in the ASTM meeting in San Diego. The Zircaloy-4, however, appears to behave differently.

M. Lindback[3] (written discussion)—(1) The corrosion resistance of ZIRLO is better than that of Zircaloy-4, but how does ZIRLO perform compared to Zircaloy-4 when regarding hydrogen pickup?

(2) Have you evaluated the pellet cladding mechanical interaction (PCMI) performance of ZIRLO? If yes, is the PCMI resistance similar to that of Zircaloys?

(3) The behavior of ZIRLO at elevated temperatures differs from that of the Zircaloys. How does this fact affect the fuel rod behavior under a postulated loss of coolant accident (LOCA)?

G. P. Sabol et al. (authors' closure)—(1) The hydrogen pickup behavior of ZIRLO has been determined from the BR-3 irradiation program and has been presented at the ASTM conference in San Diego (Ref 5). In the BR-3 data, the in-reactor corrosion resistance of ZIRLO was about 60% that of Zircaloy-4. Since the fractional hydrogen pickup was the same in both alloys, the in-reactor hydriding rate of ZIRLO was about 60% that of Zircaloy-4. In out-of-reactor autoclave tests, ZIRLO showed both lower corrosion and a lower fractional hydrogen pickup than did Zircaloy-4.

(2) The PCI performance has been evaluated by use of both SIMFEX tests with iodine and internal pressurization with Ar/I_2. These simulation tests indicate no difference between ZIRLO and Zircaloy-4.

(3) The temperatures for the initiation and completion of the formation of the beta phase are slightly lower for ZIRLO than for Zircaloy-4. All testing to support licensing of ZIRLO,

---

[1] CEA, CE Cadarache, St. Paul-lez Durana, France.
[2] University of Toronto, Toronto, Ontario, Canada.
[3] ABB Atom, Vasteras, Sweden.

including LOCA, has been performed, and full region licensing was completed for the first application in 1991.

A. M. Garde[4] *(written discussion)*—The corrosion properties of niobium-containing zirconium alloys are strongly dependent on the processing history. What can you say about the fabrication history of ZIRLO that you have investigated?

G. P. Sabol et al. (authors' closure)—The approach to the fabrication of ZIRLO was given in the ASTM San Diego conference (Ref 5). The processing is essentially as described in Ref 5 and in U. S. Patent No. 4,649,023 dated 10 March 1987 [*12*].

---

[4] ABB-CENF, Windsor, CT.

Young S. Kim,[1] *Karp S. Rheem,*[1] *and Duck K. Min*[1]

Phenomenological Study of In-Reactor Corrosion of Zircaloy-4 in Pressurized Water Reactors

REFERENCE: Kim, Y. S., Rheem, K. S., and Min, D. K., **"Phenomenological Study of In-Reactor Corrosion of Zircaloy-4 in Pressurized Water Reactors,"** *Zirconium in the Nuclear Industry: Tenth International Symposium, ASTM STP 1245,* A. M. Garde and E. R. Bradley, Eds., American Society for Testing and Materials, Philadelphia, 1994, pp. 745–759.

ABSTRACT: Uniform in-reactor corrosion has been examined on pressurized water reactor (PWR) Zircaloy-4 fuel claddings that were irradiated up to four cycles in the KORI Unit 1. The oxide layer that was formed on the Zircaloy-4 fuel claddings changes from a uniform black oxide into the nonuniform white granular oxides when the oxide layer reaches about 5 to 10 μm. With irradiation exposure, enough of the white granular oxides have nucleated above the black oxide layer to entirely cover the fuel cladding, resulting in a gray or white cladding surface. The change in the oxide layer growth pattern from the uniform black oxide layer into the nonuniform granular oxides resulted in the formation of macro-pores or cracks in the oxide layer, causing the corrosion rate to change into a linear rate from the cubic rate, especially at the beginning of fuel rod life. Consequently, the nucleation of white granular oxides is a phenomenological indicator of the transition of the in-reactor corrosion rate.

To explain the effect of hydride precipitates on the in-reactor corrosion, the corrosion behavior for a defective fuel rod with a through-hole and an intact fuel rod adjacent to it (both were irradiated for two cycles in the KORI Unit 1) has been investigated. Even though both of them were found to have almost identical operating power histories, their corrosion behavior was quite different: the maximum oxide layer thickness for the intact fuel rod was less than 10 μm, while the defective fuel rod had an abnormally thick oxide of 50 μm, especially at an elevation where a large extent of hydrides (corresponding to 1295 to 1520 ppm hydrogen) were precipitated at the cladding outer surface. Therefore, the hydrides that were locally precipitated on the cladding outer surface initiated the in-reactor corrosion enhancement to a degree that was dependent upon the extent of the hydrides. Furthermore, the corrosion enhancement by hydrides is likely to effectively occur late in the fuel life time or after more than three or four cycles of irradiation, thus leading to the enhanced post-transition corrosion rate.

KEY WORDS: in-reactor uniform corrosion, zirconium alloys, claddings, granular oxides, corrosion enhancement, hydrides, defective rods, zirconium, nuclear materials, nuclear applications, radiation effects

In-reactor corrosion of Zircaloy-4 pressurized water reactor (PWR) fuel claddings becomes a limiting factor for the fuel lifetime as fuel rod burnup tends to reach 60 000 MWd/MTU or more. Therefore, to assure the improved reliability of fuel cladding, a better understanding of the aqueous in-reactor corrosion of Zircaloy-4 was attempted.

Much effort was made to explain in-reactor corrosion of Zircaloy-4 fuel claddings based on the out-pile corrosion behavior. Comparisons of the in-reactor corrosion results with

---

[1] Project manager of Zirconium Alloy Development, manager of Nuclear Engineering Test & Evaluation Division, and principal researcher of PIE Engineering Department, Korea Atomic Energy Research Institute, PO Box 105, Yusung, Taejon, 305-600, Korea.

the out-pile corrosion data suggested some enhancement, especially in the post-transition period rather than in the pre-transition period [1,2]. In order to explain this phenomenon, several hypotheses such as the concentrated LiOH effect [3], the neutron irradiation effect [2,4], the thick film effect [5], and the hydride effect [6] were proposed. However, it is not clear as of now which of these hypotheses are responsible for the corrosion enhancement in-reactor. Therefore, the purpose of this study is to phenomenologically study the in-reactor corrosion of PWR Zircaloy-4 fuel claddings and to explain the corrosion enhancement mechanism. For the phenomenological study on the in-reactor corrosion in PWRs, the Zircaloy-4 fuel rods irradiated up to four cycles in the KORI Unit-1 were subjected to detailed examinations in the post-irradiation examination facility PIE of KAERI.

Experimental Details

Zircaloy-4 fuel rods, 10.72 mm outside diameter, were subjected to visual examination as well as destructive tests at the PIE facility of KAERI. A periscope was used to observe the growth pattern of the oxide layer on Zircaloy-4 fuel rods, while the oxide layer thickness and the hydride distribution were measured metallographically on the transverse sections of a fuel rod. The metallographic specimens were prepared by impregnating with resin, cutting, mounting, and then grinding/polishing. The quantity of hydrogen included in a given cladding tube specimen was determined through the vacuum extraction method by using the LECO RH404 analyzer. The precision of the measurement is within ± 5% relative to the calibration standard with known hydrogen levels.

The fuel rods were irradiated up to four reactor cycles in the KORI Unit-1. Table 1 summarizes their irradiation histories, and the average rod burnup ranged from 15 600 to 40 200 MWd/MTU maximum. Herein, the burnup represents the best estimated value calculated by using the reactor physics computer code, whereas the burnups of the D1 and D2 fuel rods represent the burnup taken from the calibration curve of the measured burnup versus the intensity of Cs-137 [7].

TABLE 1—*Irradiation histories of the examined fuel rods.*

Fuel Rod	Irradiation Cycles	Burnup, MWd/MTU	
		Measured	Calculated
A1	1	...	18 200
A2	1	...	15 600
D1[a]	2	16 100	...
D2[b]	2	17 900	...
B1	2	...	26 900
B2	2	...	25 400
C1	3	...	34 200
C2	3	...	29 200
C3	3	...	36 500
C4	3	...	34 700
G1	4	...	40 200
G2	4	...	37 600
G3	4	36 300	...

[a] The defective rod.
[b] The intact rod adjacent to Fuel Rod D1.

Results and Discussion

Phenomenological In-Reactor Corrosion

Figure 1 shows various features of the oxide layer formed on the Zircaloy-4 fuel rod (herein referred to as A1) irradiated to 18 200 MWd/MTU. The bottom part of the fuel rod experiencing the lowest cladding surface temperature exhibited lustrous black oxide as shown in Fig. 1d. With a distance from the bottom end, however, the white (or gray) granular oxide started to appear as shown in Fig. 1c. Around the middle part of the fuel rod, a lot of white granular oxides nucleated enough to partially mottle the pre-existing black oxide layer with patches of white granular oxides, as exhibited in Fig. 1b. Eventually, the whole surface at the top part of Fuel Rod A1 was fully covered with the white oxide granules and turned into the white oxide layer as shown in Fig. 1a where the surface temperature was the highest.

The emerging white granular oxide pattern changed with burnup. Figure 2 shows the surface appearances along the longitudinal direction of Fuel Rod G1 that was irradiated up to four cycles with a burnup of 40 200 MWd/MTU. As shown in Fig. 2f, the bottom end of Fuel Rod G1, subjected to the lowest cladding surface temperature, still exhibited lustrous black oxide. With an increasing distance from the bottom end (Fig. 2e), a few white granular oxides started to appear, as was observed on the low burnup fuel rod (Fig. 1c). Around the mid-plane, between the lowest spacer grid and the second one, a lot of white and large oxide granules nucleated, so that the oxide layer turned gray or white gradually, as shown in Figs. 2d and c. Compared to the white granular oxides observed on Fuel Rod A1 with lower burnup, their size and shape were larger and more spherical. It is interesting to note that the uncovered black oxide still is able to be seen at the places where the white granular oxides were not fully developed enough to cover the black oxide (as indicated in Fig. 2 at the arrow). At an increasing distance (corresponding to the middle part of the fuel rod), the whole surface of the fuel rod was fully covered by white oxide granules, thus exhibiting the white oxide layer as shown in Fig. 2a. However, at the grid position, corresponding to the middle part of the fuel rod, the shiny black oxide along with some white nucleated granular oxides turned up again, as shown in Fig. 2b. This indicates that the corrosion at the grid position proceeded more slowly than at other regions without a grid due to lower cladding surface temperature, and, to the low neutron flux effect, if any. In essence, the results shown in Figs. 1 and 2 suggest that the formation of white granular oxides is one of the in-reactor corrosion processes of a Zircaloy-4 cladding that is independent of burnup.

To compare the formation of the white granular oxide and the oxide layer thickness, the destructive test to microscopically measure the layer thickness on a transverse section was performed for the two-cycle-irradiated Fuel Rod B1 with the 26 900 MWd/MTU burnup. Figure 3 exhibits a typical example of how the appearance of the oxide layer changes with the layer thickness. As long as the oxide layer remained lustrous black, the layer thickness was found to be around 5 μm, irrespective of the axial elevation from the bottom end (Figs. 3c and d). However, when the oxide layer thickness was metallographically measured at the transverse section with both the black oxide layer and the white granular oxides coexisting as shown in Fig. 3b, it was found that the oxide layer thickness at the white oxide-nucleated region grew to about 15 μm and, by contrast, the black oxide layer next to it did not show any oxide layer growth, the thickness of which remained to be as low as 4 to 5 μm. Further, as the oxide layer completely turned white as shown in Fig. 3a, the layer thickness thickened to 25 μm. These results, therefore, indicate that white granular oxides begin forming as one of the oxide layer growth processes whenever the layer thickness reaches 5 to 10 μm. The trend could also be confirmed by correlating the surface appearances with the oxide layer

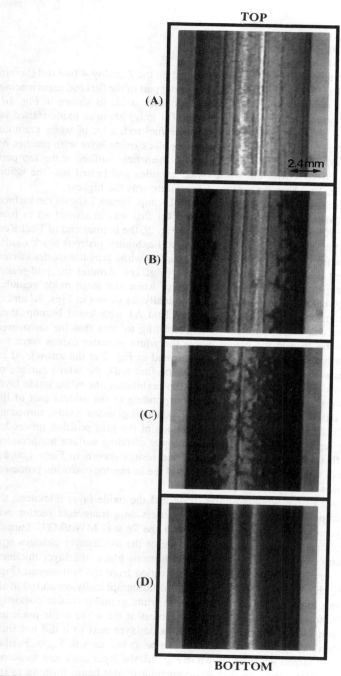

FIG. 1—*The appearances of the oxide film along the elevation on Fuel Rod A1 irradiated for one cycle in the KORI Unit-1: (a) the oxide film mostly covered with white granular oxides; (b) and (c) nonuniform nucleation of white (or gray) granular oxides above the black oxide at the upper half and at the middle, respectively; and (d) lustrous black oxide at the bottom.*

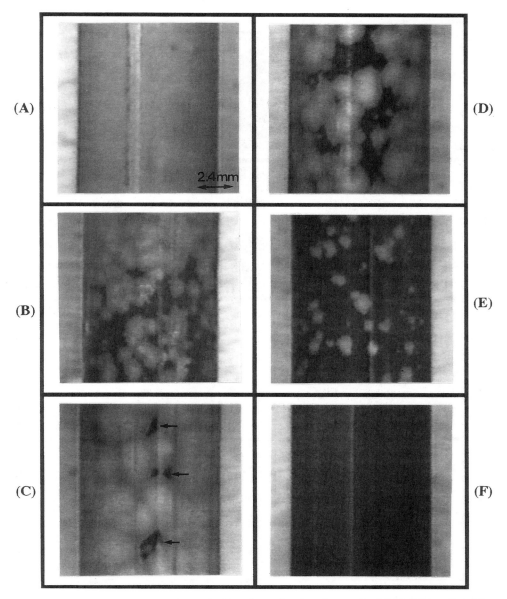

FIG. 2—*The appearances of the oxide film along the elevation on Fuel Rod G1 irradiated for four cycles in the KORI Unit-1:* (a) *the white granular oxide-covered surface at the top,* (b) *the oxide film at the grid position,* (c) *and* (d) *are the partial coverage by large white granular oxides at the lower half of the fuel rod,* (e) *the nonuniform nucleation of small white granular oxides above the black oxide, and* (f) *lustrous black oxide near the bottom.*

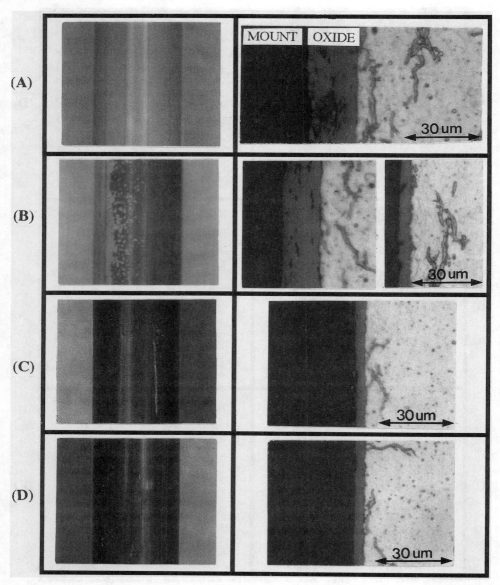

FIG. 3—*Evolution of the appearances and thicknesses of the oxide layer along the length of Fuel Rod B1 irradiated for two cycles in the KORI Unit-1:* (a) 2702 mm, (b) 1694 mm, (c) 1028 mm, and (d) 49 mm from the bottom.

thickness measured on the corresponding cross sections for the twelve fuel rods irradiated for one to four cycles. Table 2 summarizes the relationship of the surface appearance and the layer thickness. As already seen in Fig. 3, the white granular oxide is found to be formed whenever the layer thickness is in the 5 to 10 μm range.

When aqueous corrosion of a Zircaloy-4 fuel cladding occurs in-reactor, the oxide layer formed first is black and tight (Figs. 3c and d), the composition of which would be hypostoichiometric [8]. As the corrosion proceeds, however, white granular oxides are nonuniformly formed. Because the stoichiometric zirconia is white, creating the nonuniform formation of white granular oxides, it seems to suggest that the oxide layer turns to white stoichiometric from black hypostoichiometric. As the white oxide is formed by turning the top of the underlying nonstoichiometric zirconia into the stoichiometric zirconia, with the oxide advancing into the metal, the oxide layer would have two-layered zirconia, that is, hypostoichiometric zirconia underneath and the stoichiometric one above. There is supporting evidence that the oxide film of the irradiated Zircaloy-4 rod is kept nonstoichiometric up to the 5 μm thickness from the the oxide/metal interface and, beyond that, turns stoichiometric [9]. Therefore, the white granular oxide is likely to be nucleated above the black oxide. Moreover, the edges of white granular oxides appeared darker than their centers as shown in Fig. 2, thus leading to the suggestion that white granular oxides arise from the upper part of the underlying black oxide layer. Conversely, if the nucleation of white granular oxides started from the inside of the black oxide layer, that is, the oxide/metal interface, there should not have been any contrast between the edge and the center of a white granular oxide.

The results shown in Figs. 1 to 3 indicate that the oxide layer grows uniformly up to 5 μm thickness and then follows a nonuniform growth pattern with nucleation of the white granular oxides. Further, the oxide layer microstructures, in Fig. 3, show that the white granular oxide-covered layer does involve the formation of macro-pores or cracks and by contrast the uniform black oxide layer is very dense with little macro-pores. Therefore, the macro-pores or cracks observed only on the white granular oxide-covered layer are likely to suggest that the pores are associated with the nonuniform nucleation of white granular oxides. If this is the case, thermal conductivity of the oxide layer would be reduced linearly in proportion to the porosity. This may explain why the thermal conductivity of the irradiated oxide layer of 21 to 33 μm is 30% less than that of the unirradiated oxide layer with no white granular oxides nucleated [2]. As a result, the formation of white granular oxides will have an effect

TABLE 2—*Change of oxide layer thickness with the surface appearance of the oxide layer.*

Fuel Rod	Oxide Layer Thickness, μm		
	Black Oxide	Granular Oxide	White Oxide
A1	4 to 4.2	8.3	12 to 15.4
A2	2.6 to 3.7	7.7 to 8.0	13.3
B1	4.4 to 4.8	7.6 to 15	18.5 to 25
B2	3.3 to 4.2	...	23.3
B3	6	9.6	13.8
C1	5.5	...	13
C2	5.2 to 5.6	...	12.8
C3	5.5 to 6.5	8	27.4
C4	5.7 to 6.5	7.4	25.8
G1	6.4	...	13.0 to 26.0
G2	7.6	11.0 to 12.3	17.6 to 29
G3	5.8	11.0	14.6 to 22

on the increase in the oxide/metal interfacial temperature, thus leading to the increase in the corrosion rate. Thus, it is expected that the initial transition of the corrosion rate from the cubic rate to the linear rate occurs as a result of nonuniform nucleation of white granular oxides after the layer thickness reaches more than 5 to 10 μm. This suggestion is quite consistent with Garzarolli's results [2] where in-reactor corrosion becomes enhanced when the oxide thickness reaches about 5 μm together with the transition of the oxide layer from black to gray. Figure 4 shows the evolution of oxide layer thicknesses with local rod burnup. Here, the oxide layer thickness data are the average of three or four measured data at the circumference of a cross section. Due to the large scattering of the oxide layer thickness data even at a fixed burnup, probably because of the difference in cladding surface temperature, the upper and lower envelope curves are drawn in order to see a general trend in the change of the layer thickness data with burnup. Based on this figure and the published results by others [2,10,11], the corrosion rate seems to be enhanced after the 10 μm thickness and the transition corresponds to 5 to 10 μm. Therefore, we can conclude that the nucleation of the white granular oxides is the phenomenological indicator of the initial transition of the corrosion rate.

The Effect of Hydrides on the In-Reactor Corrosion Enhancement

To identify the effect of hydride precipitates on the corrosion enhancement, the corrosion behavior and the hydride distribution were correlated by examining the defective rod with a through-hole near the bottom and an intact rod next to it. Both the intact and defective Zircaloy-4 rods were irradiated for two cycles with a burnup of 17 900 and 16 100 MWd/MTU, respectively. The defective rod was found to have a round through-hole of about 2 mm in diameter at a distance of 675 mm from the bottom; this position was consistent with the grid spring location. Consequently, the through-hole is likely to result from fretting wear between a grid spring and the Zircaloy-4 cladding while operated in the reactor.

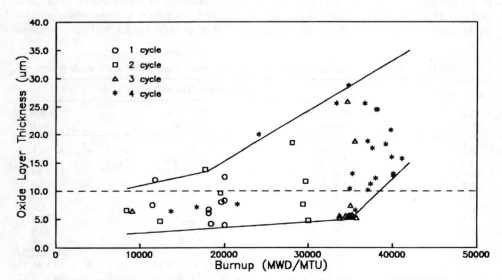

FIG. 4—*Evolution of oxide layer thicknesses with the local burnup (MWd/MTU).*

Figure 5 shows the abnormal amount and distribution of hydride precipitates along the defective cladding. Near the through-hole as shown in Fig. 5c, the amount of hydride precipitate was as low as that observed on the three-fourth elevation of the intact rod (see Figs. 6a and b). However, at the distance of 2227 to 2518 mm from the through-hole (2902 to 3193 mm from the bottom end), a large amount of hydrides precipitated on the outer cladding surface as shown in Fig. 5b. The extent of hydrogen at this elevation (2902 to 3193 mm from the bottom) was measured by using the vacuum extraction method at 1295 to 1520 ppm hydrogen (H). At a farther distance, 2917 mm, from the through-hole (3592 mm from the bottom), a much less amount of hydride precipitate, 345 ppm H, was observed. On the other hand, the intact cladding next to the defective one exhibited a small extent of hydride precipitate as shown in Fig. 6, even at the corresponding position where the abnormally large extent of hydride precipitates occurred in case of the defective fuel rod. Therefore, an abnormally large amount of hydride precipitate only on the defective Zircaloy-4 cladding is attributed to water

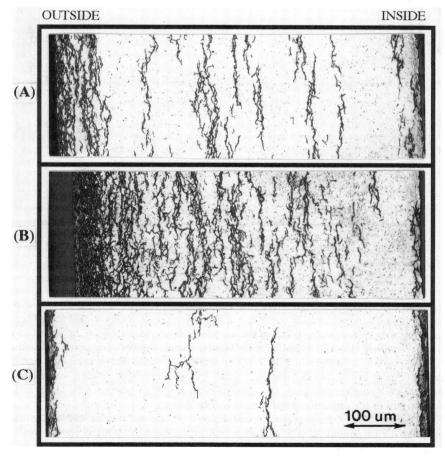

FIG. 5—*The abnormal distribution of hydride precipitates on defective Fuel Rod D1 irradiated for two cycles in the KORI Unit-1 at various elevations from the bottom: (a) 3592 mm, (b) 2902 mm, and (c) 655 mm.*

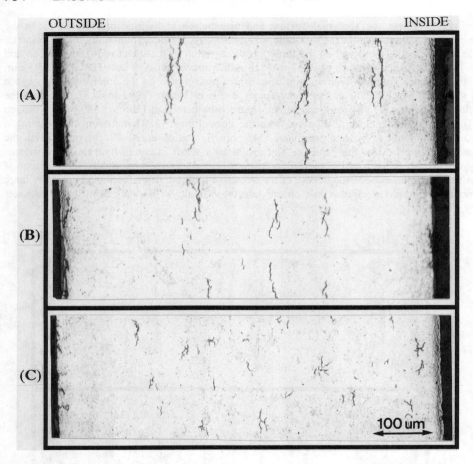

FIG. 6—*The normal distribution of hydride precipitates on the intact Fuel Rod D2 adjacent to Rod D1 at the following elevations:* (a) *3500 mm,* (b) *3200 mm, and* (c) *350 mm.*

ingress into the inside of the defective rod. For the intact rod, however, the main source of hydrogen comes from water-side corrosion of the outer-side Zircaloy-4 cladding. Therefore, the amount of hydrides precipitated on the cladding, as shown in Fig. 6, appeared to differ little with elevation, due to the similar oxide layer thickness between the lower part and upper part of the cladding as shown in Fig. 7. It is interesting to note that at the lower part of the intact rod (shown in Fig. 6c) the relatively small hydrides appeared uniform across the cross section and at the higher elevation, and the hydride precipitates became a bit larger and were located mainly in the circumferential direction (Fig. 6a).

When looking at the corrosion pattern on the defective and the intact cladding, the thickest abnormal oxide of around 50 μm was observed on the defective rod, as shown in Fig. 8b, especially at the corresponding location (2902 to 3193 mm from the bottom end) where the highest amount of hydrides (1295 to 1520 ppm H) precipitated locally at the outer side of the cladding. However, at the farther elevation of 3592 mm from the bottom, where far less amount of hydrides of 345 ppm H precipitated, the corrosion process proceeded slower resulting in

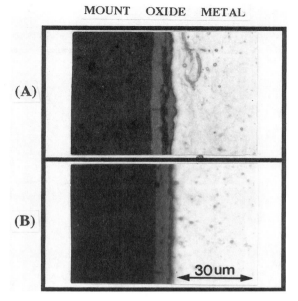

FIG. 7—*The oxide film formed on the intact Fuel Rod D2 adjacent to the defective Fuel Rod D1 at the elevations of* (a) *3174 mm and* (b) *2104 mm.*

the reduced oxide layer to around 11 μm as shown in Fig. 8a. On the contrary, the intact fuel rod, with a considerably small amount of hydrides as a result of little hydrogen pick-up from the inside of the cladding, exhibited a very thin oxide layer of less than 10 μm across the whole length, as shown in Fig. 7. Considering that the thickest oxide layer measured on the three- or four-cycle irradiated fuel rods was in the less than 30 μm range at the maximum, as shown in Table 2, the oxide layer thickness of as thick as 50 μm on the defective rod irradiated for only two cycles indicates that the corrosion enhancement must have occurred.

Because the defective rod and the intact rod adjacent to it underwent the same operating power histories, the effect of neutron flux on the enhanced corrosion can be disregarded in this case. Furthermore, due to the same rod power, the outer surface temperatures of both the fuel rods (one of the key factors in determining corrosion of Zircaloy-4 claddings) are expected to differ little among each other. This is because the outer surface temperature of the cladding is determined only by the coolant/clad heat transfer coefficient and the rod power, independent of the difference in the clad/pellet gap thermal conductivity. Consequently, the enhanced corrosion of the defective cladding can not be explained by the temperature effect. In addition, the oxide layer thicknesses of the irradiated rods obtained at the burnups of 16 000 to 18 000 MWd/MTU, equivalent to the burnups of the defective rod and the intact rod, respectively, ranged from 4 to 15 μm as shown in Fig. 4 despite the difference in cladding surface temperatures caused by different rod power histories. Therefore, the abnormally thick oxide layer of 50 μm only on the defective cladding is likely to be associated with a large amount of hydrides precipitated on the cladding, not with temperature effect.

Because the amount of hydrogen absorbed by an intact Zircaloy-4 cladding when corroded to the layer thickness of 52 to 58 μm was several hundreds ppm H (278 to 424 ppm) [*12*], the hydrogen pickup (as much as 1295 to 1520 ppm on the defective cladding) would be larger than expected, assuming that a part of the hydrogen released by the cladding outer

756 ZIRCONIUM IN THE NUCLEAR INDUSTRY: TENTH SYMPOSIUM

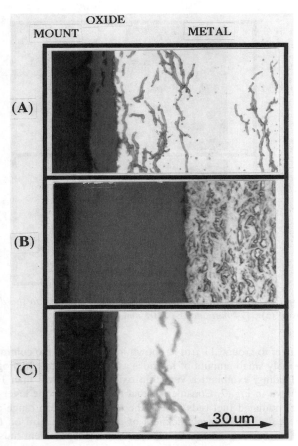

FIG. 8—*The abnormally thick oxide film on the defective Fuel Rod D1 at the corresponding elevations of:* (a) *3592 mm* (b) *2902 mm, and* (c) *655 mm.*

surface corrosion is picked up. Consequently, the hydrogen absorbed is believed to result mainly from the cladding inside because a lot of the hydrogen is generated from the reaction of steam with the cladding inner surface and also UO_2 pellets.

As water enters the defective rod through the hole, it flashes into steam. As a result of the reaction of this steam with the cladding inside and the UO_2 pellets, hydrogen starts to build up inside the defective cladding. Nevertheless, near the through-hole of the defective cladding, steam partial pressure is sufficiently large compared to hydrogen partial pressure, so that the break-down of the intact oxide will not occur. This inhibits diffusion of hydrogen into the cladding inside because the intact oxide acts as a barrier to hydrogen diffusion. Thus, only a small amount of hydride precipitate appears as shown in Fig. 5c. Since no gap was observed near the through-hole between the Zircaloy-4 cladding and UO_2 pellet due to Zircaloy-4 cladding and UO_2 pellet bonding, the availability of steam at the upper cladding gap 20 to 50 μm, will decrease with time compared to hydrogen because of the faster diffusion of hydrogen. When the hydrogen partial pressure in the gap at the upper cladding becomes higher than the critical value above which the intact barrier oxide undergoes breakdown, therefore, hydrogen is free to enter into the Zircaloy-4 cladding [13,14]. However, the breakdown process of the intact oxide layer under the high hydrogen partial pressure condition will be dependent upon temperature so that the largest amount of hydrogen will be picked up at the hottest part of the cladding. Moreover, the largest temperature gradient will enhance the movement of absorbed hydrogen into the outer surface of the cladding. As a result, the largest amount of hydrides is precipitated on the outer cladding surface of the hottest section, as shown in Fig. 5b.

Once hydrides build up at the outer surface of the defective cladding, the corrosion of the defective and the intact rods behaves differently. The defective cladding corrodes faster than the intact rod as a result of the corrosion enhancement by hydrides. This may explain why the defective cladding with highly concentrated hydrides exhibits the thickest oxide while the intact rod with comparable burnup has the oxide layer of less than 10 μm at the maximum. Further, the extent of the corrosion enhancement is quite dependent on the extent of concentrated hydrides: the oxide layer thicknesses were 11 μm on the cladding part with hydrides of 345 ppm H (Fig. 5a), 47 μm on the cladding part with hydrides of 1295 ppm H, and 56 μm at the maximum on the cladding part with 1520 ppm (Fig. 5b), respectively. Other supporting evidence for corrosion enhancement by hydrides can be found from the published data by Asher: the heated Zircaloy-2 with the concentrated hydrides locally at the tube outside had the abnormally thick oxide of 50 to 80 μm, while the unheated control tubes with uniformly distributed hydrides had the oxide layer of around 20 μm [15]. However, in order for hydride to effect the intact cladding, it needs enough time for the absorbed hydrogen to reach the temperature gradient and be concentrated as hydride precipitates. As shown in Fig. 6, even after the two-cycle irradiation, the hydrides were found to be more or less uniform across the cross section, therefore heavy build-up of hydrides near the oxide/metal interface is expected to happen only after the three- or four-cycle irradiation, as was observed on a highly burned Zircaloy-4 rod [6]. Consequently, the corrosion enhancement by hydrides will be put into effect only later in the fuel life time, thus leading to the increase in the post-transition corrosion rate. If so, the burnup-related corrosion enhancement suggested by Kilp [16], whereby the oxide layer in the fourth cycle grew to two or three times that formed at the third cycle, is likely to be associated with concentrated hydrides near the oxide/metal interface rather than with the neutron effect.

At this time, it is not clearly understood how the hydride precipitates accelerate the in-reactor corrosion. We may speculate that two hypotheses are possible. One hypothesis states that hydrides themselves corrode faster than the Zircaloy-4 matrix, as suggested by Asher [15]. Another states that hydrides facilitate transformation of tetragonal zirconia to a monoclinic one. Hydrides concentrated near the interface relieve the tensile stress on the metal substrate [6]. This, however, can lead to the relaxation of compressive stress on the oxide layer in mechanical equilibrium with the underlying metal. The relaxation of compressive stress that is a prerequisite to stabilize tetragonal zirconia as suggested by Godlewski [17], however, would cause either an enhanced transformation of tetragonal zirconia to a monoclinic one or that only monoclinic zirconia would be formed when the new oxide is developed. Because the pores are generated by the transformation of tetragonal zirconia to a monoclinic one, due to the accompanied volume expansion, there will be no intact oxide layer formation bringing on the enhanced corrosion rate. Further, because the extent of stress relaxation in the oxide is proportional to the amount of hydride precipitate at the interface, the corrosion enhancement is affected by the nonuniform distribution of hydrides as well as the extent of the concentrated hydrides, which is likely to increase with burnup.

Conclusions

1. As one of the in-reactor corrosion processes of Zircaloy-4 claddings, white granular oxides nucleate above the black oxide layer when the layer thickness grows to 5 to 10 μm, resulting in the formation of macro-pores in the oxide layer. Consequently, the transition of the corrosion rate that occurs when the oxide layer growth pattern turns to nonuniform granular oxides from the uniform black oxide layer is likely to be associated with the pores formed in the oxide layer as a result of nonuniform white granular oxides that were nucleated. Therefore, the nucleation of white granular oxides is a phenomenological indicator of the transition of the in-reactor corrosion rate.

2. Comparison of the oxide layer thickness on the defective fuel rod and the intact rod shows that enhanced corrosion occurs on the hottest part of the defective cladding where the heavily concentrated hydrides precipitate at the cold outer surface. Therefore, it is suggested that the hydrides that are locally precipitated on the cladding outer surface initiate in-reactor corrosion enhancement, depending on the extent of hydrides. Further, the corrosion enhancement by hydrides is likely to occur late in the fuel life time or after more than three or four cycles of irradiation, leading to the post-transition enhanced corrosion rate.

References

[1] Cox, B., *Proceedings,* IAEA Technical Committee on Fundamental Aspects of Corrosion of Zirconium Base Alloys in Water Reactor Environments, Portland, 11–15 Sept. 1989, International Atomic Energy Agency, Vienna, p. 167.

[2] Garzarolli, F., Bodmer, R. P., Stehle, H., and Trapp-Pritsching, S., "Progress in Understanding PWR Fuel Rod Water Side Corrosion," *Proceedings,* ANS International Conference on Light Water Reactor Fuel Performance, Orlando, 1985.

[3] Garzarolli, F., Pohlmeyer, J., Trapp-Pritsching, S., and Weidinger, H. G., *Proceedings,* IAEA Technical Committee on Fundamental Aspects of Corrosion of Zirconium Base Alloys in Water Reactor Environments, Portland, 11–15 Sept. 1989, International Atomic Energy Agency, Vienna, p. 65.

[4] Cox, B. and Fidleris, V. in *Zirconium in the Nuclear Industry: Eighth International Symposium, ASTM STP 1023,* L. F. P. Van Swam and C. M. Eucken, Eds., American Society for Testing and Materials, Philadelphia, 1989, p. 245.

[5] Johnson, A. B., Jr., *Proceedings,* IAEA Technical Committee on Fundamental Aspects of Corrosion of Zirconium Base Alloys in Water Reactor Environments, Portland, Sept. 1989, International Atomic Energy Agency, Vienna, p. 107.

[6] Garde, A. M. in *Zirconium in the Nuclear Industry: Ninth International Symposium, ASTM STP 1132,* C. M. Eucken and A. M. Garde, Eds., American Society for Testing and Materials, Philadelphia, 1991, p. 566.

[7] Noh, S. G., "Development of Post-Irradiation Examination and Evaluation Techniques for Nuclear Reactor Fuel (II)," Report KAERI/RR-1024/91, Korea Atomic Energy Research Institute, Taejeon, Korea.

[8] Douglass, D. L. and Wagner, C., *Journal of the Electrochemical Society,* Vol. 113, 1967, p. 671.

[9] Garzarolli, F., Schoenfeld, H., Scott, D. B., and Smerd, P. G., "Characterization of Zirconium Oxide Corrosion Film on Irradiated Zircaloy Clad PWR Fuel Rods," Report EPRI-NPSD-192, Electric Power Research Institute, Palo Alto, 1982.

[10] Thomazet, J., Mardon, J. P., Charquet, D., Senevet, J., and Billot, P., *Proceedings,* IAEA Technical Committee on Fundamental Aspects of Corrosion of Zirconium Base Alloys in Water Reactor Environments, Portland, 11–15 Sept. 1989, International Atomic Energy Agency, Vienna, p. 255.

[11] Van Swam, L. F. P. and Sham, S. H. in *Zirconium in the Nuclear Industry: Ninth International Symposium, ASTM STP 1132,* C. M. Eucken and A. M. Garde, Eds., American Society for Testing and Materials, Philadelphia, 1991, p. 758.

[12] Garde, A. M., "Hot Cell Examination of Extended Burnup Fuel from Fort Calhoun," Report DOE/ET/34030-11, 1986.

[13] Shannon, D. W., *Corrosion,* Vol. 67, 1963, p. 414.

[14] Une, K., *Journal of Less-Common Metals,* Vol. 57, 1978, p. 93.

[15] Asher, R. C. and Trowse, F. C., *Journal of Nuclear Materials,* Vol. 35, 1970, p. 115.

[16] Kilp, G. R., Balfour, M. G., Stanutz, R. N., McAtee, K. R., Miller, R. S., Bohman, L. H., Wolfhope, N. P., Ozer, O., and Yang, R. L., "Corrosion Experience with Zircaloy and ZIRLO in Operating PWRs," *Proceedings,* International Topical Meeting on LWR Fuel Performance, Avignon, France, 1991, p. 730.

[17] Godlewski, J., Gros, J. P., Lambertin, M., Wadier, J. F., and Weidinger, H. in *Zirconium in the Nuclear Industry: Ninth International Symposium, ASTM STP 1132,* C. M. Eucken and A. M. Garde, Eds., American Society for Testing and Materials, Philadelphia, 1991, p. 416.

Anand M. Garde,[1] *Satya R. Pati,*[1] *Michael A. Krammen,*[1] *George P. Smith,*[1] *and Robert K. Endter*[2]

Corrosion Behavior of Zircaloy-4 Cladding with Varying Tin Content in High-Temperature Pressurized Water Reactors

REFERENCE: Garde, A. M., Pati, S. R., Krammen, M. A., Smith, G. P., and Endter, R. K., "**Corrosion Behavior of Zircaloy-4 Cladding with Varying Tin Content in High-Temperature Pressurized Water Reactors,**" *Zirconium in the Nuclear Industry: Tenth International Symposium, ASTM STP 1245,* A. M. Garde and E. R. Bradley, Eds., American Society for Testing and Materials, Philadelphia, 1994, pp. 760–778.

ABSTRACT: Fuel rods clad with Zircaloy-4 with varying tin contents (1.33 to 1.58% Sn) and annealing parameters (1.0 to 4.1×10^{-17} h with $Q/R = 40\,000°K$) were irradiated in demonstration fuel assemblies in a high-temperature pressurized water reactor (PWR) to burnups in excess of 35 giga watt days per metric ton of uranium (GWd/MTU). The same cladding variants were subjected to long-term static water autoclave tests at 633°K of duration greater than 1100 days. Production fuel rods fabricated with low-tin (1.33 Sn) and high-tin (1.55 Sn) Zircaloy-4 cladding were also irradiated in regular fuel assemblies in two high-temperature PWRs to burnups up to 48 GWd/MTU. Poolside cladding oxide thickness measurements were conducted on 167 high-tin rods and 67 low-tin rods during refueling outages. The measured, circumferentially averaged, peak cladding oxide thickness values ranged from 3 to 113 µm. At high burnups, the oxide thickness on low-tin cladding was 30 to 40% lower than that on high-tin cladding. The long-term autoclave results also showed the beneficial effect of lower tin level on the corrosion rate, although to a lower degree than in PWRs. The 633 K water autoclave test appears to rank the corrosion resistance of the investigated Zircaloy-4 variants in the same order as in PWRs. Hydrogen analysis results indicate that tin level does not influence the hydrogen uptake of autoclave-tested samples. The observed in-PWR influence of tin level on the Zircaloy-4 cladding corrosion rate was incorporated in the ESCORE clad corrosion model by adjusting the pre-exponential term in the post-transition corrosion rate equation.

A corrosion rate acceleration at high burnups may be related to either hydride precipitation at the metal oxide interface or to degradation of the oxide thermal conductivity. The observed effect of tin on the uniform corrosion resistance of Zircaloy-4 in high temperature water is consistent with the corrosion mechanism that assumes the migration of O^{2-} anion vacancies through the oxygen deficient zirconia to be the rate-controlling process. It is postulated that lower tin level in Zircaloy-4 decreases the vacancy concentration in the oxide and, thereby, increases the corrosion resistance. It may be possible to further improve the corrosion resistance of low-tin Zircaloy-4 by optimizing the nitrogen and tin concentrations in Zircaloy.

KEY WORDS: corrosion, autoclave corrosion, tin content, corrosion modeling, hydrogen uptake, annealing effect, anion vacancy migration, zirconium, zirconium alloys, nuclear materials, nuclear applications, radiation effects

---

[1] Consulting engineer, supervisor, consulting engineer, and consulting engineer, respectively, ABB Combustion Engineering, Nuclear Fuel, Windsor, CT 06095.
[2] Principal engineer, ABB Combustion Engineering, Nuclear Services, Windsor, CT.

Zircaloy-4 with compositional limits as specified for the zirconium-tin alloy grade UNS Number R60804 in the ASTM Specification for Zirconium and Zirconium Alloy Ingots for Nuclear Application (ASTM B 350-91) has been traditionally used as the cladding and structural material for pressurized water reactor (PWR) fuel assemblies. To improve uranium utilization and reduce nuclear fuel cycle cost, there has been a continuous demand to increase the discharge burnup of PWR fuel assemblies since the late 1970s. It was recognized fairly early that the corrosion performance of Zircaloy-4 cladding would be a life-limiting phenomenon for higher burnup operation [1], particularly in reactors with high coolant temperatures. With that realization, investigations were undertaken by various researchers [2–26] to improve the corrosion performance of fuel rod cladding.

One approach in this regard was to improve the corrosion resistance of Zircaloy-4 by controlling the composition within the ASTM-specified range and also optimize the thermomechanical processing for the specific in-reactor environment. In the category of controlling composition within the ASTM B 350-91 specification limit, studies related to the effect of tin content received the first attention since, from the period of early development of zirconium-base alloys [27–29], it has been known that the corrosion resistance of zirconium-tin binary alloys decreased with increasing tin content. However, a minimum level of tin was found to be necessary in Zircaloy-4 to counteract the deleterious effect of nitrogen [27,28] and other impurities [29,30] that were normally present in sponge zirconium used in producing Zircaloy-4 ingots of that period. In contrast, the nitrogen content of modern commercial Zircaloy-4 ingots is lower, and as a result, the opportunity exists to improve corrosion resistance by reoptimizing the tin content of Zircaloy-4 [27,31–34]. Several investigators [2,3,12,14] reported autoclave test results on the effect of varying tin content, and the data confirmed the expected trend of lower weight gain with a reduction in tin content. Limited data have also recently become available to confirm that trend in-reactor [18,20]. However, sufficient details of the data, including companion autoclave test results on weight gain and hydrogen uptake, are not available to make a quantitative evaluation of the effect of tin content on the in-reactor corrosion and to provide insight into the mechanism for the effect of tin on the uniform corrosion resistance of Zircaloy-4 in a high-temperature PWR environment.

In this paper, in-reactor cladding oxide thickness data (measured on fuel rods from fuel assemblies irradiated in two PWRs, Reactors A and B) are presented for Zircaloy-4 cladding having three ranges of tin content: from 1.50 to 1.58; 1.44 to 1.50; and 1.32 to 1.44% by weight. Otherwise, the cladding tubes were manufactured to the same specification. These tin concentration ranges are referred to in the text as high-tin, intermediate-tin, and low-tin, respectively. While high-tin content represented the industry standard prior to mid-1980s, the low-tin content is the current industry standard. Also, a limited number of fuel rods with variations in the cladding fabrication annealing process were irradiated in Reactor A. Representative oxide thickness data from measurements of more than 200 fuel rods are reported. These data cover a fuel rod axial average burnup range of 6 to 49 giga watt days per metric ton of uranium (GWd/MTU) and maximum circumferentially averaged oxide thicknesses ranging from 3 to 113 μm. Long-term 633 K water autoclave corrosion data and post-corrosion-test hydrogen concentration measurements are presented. The autoclave data are used to evaluate the relationship between autoclave weight gain and in-PWR oxide thickness and to estimate the effect of tin content on the hydrogen pickup fraction. In-reactor oxide thickness data are analyzed by a fuel rod cladding corrosion model, ESCORE [35], accounting for the detailed power history and thermal hydraulic environments. Modifications to the cladding corrosion model are proposed to account for the observed effect of tin on Zircaloy-4 cladding corrosion.

Zircaloy-4 Characteristics

The characteristics of the different Zircaloy-4 cladding tube variants that are discussed in this paper are listed in Table 1. Table 1 also includes the Electric Power Research Institute (EPRI) standard material [36] that was tested in the long-term autoclave test program to facilitate comparison of corrosion test results from different autoclaves. The tin level varied from 1.33 to 1.55%. Silicon was used as an alloying element, and silicon concentration levels were about 100 ppm as shown in Table 1. The annealing parameter varied from 1.0×10^{-17} to 4.1×10^{-17} h. These values of integrated annealing parameter were calculated using a Q/R value of 40 000 K. The annealing parameter values near the low end of the investigated range represent commercial cladding tubes manufactured from 1986 to 1988. Cladding variants with the high annealing parameters were fabricated to study the effect of annealing process variations on the in-PWR corrosion of Zircaloy-4.

The final tube heat treatment (for all the variants listed in Table 1) was stress relief annealing (SRA) resulting in an elongated, partially recrystallized grain structure in the as-received tubing. The inside tube surface was pickled and the outside tube surface was belt-abraded for all the variants.

Experimental Procedure

Autoclave Testing

All of the cladding variants listed in Table 1 were corrosion tested in PWR fuel cladding geometry in two static autoclaves. The corrosion tests were conducted in deionized water at 633 K according to the procedure described in Ref 36. The total autoclave exposure time was up to 1331 days.

A single specimen of each type was subjected to hydrogen analysis after known autoclave test exposure. The selected samples were sectioned into ring segments and were subjected to hydrogen analysis at two different laboratories. Since zirconium calibration standards containing high hydrogen levels (>30 ppm) are not available, the hydrogen concentration was measured on adjacent pieces of the same specimen in two different laboratories to confirm the hydrogen concentration levels. The average value from all the replicate measurements was used to calculate the hydrogen uptake fraction results presented.

The oxide layers on both the inside and outside tube surface were not removed before the hydrogen analysis. To avoid loss of hydrogen from the metal due to heating, the specimens were not dried prior to hydrogen measurement. The hydrogen content of the cladding was measured at Laboratory A using a LECO Model RH1EN hydrogen analyzer in which a small

TABLE 1—*Zircaloy-4 cladding material characteristics used in Reactor A demonstration fuel assemblies and long-term autoclave testing.*

Variant	Sn, %	Fe, %	Cr, %	N, ppm	Si, ppm	ΣAi, 10^{-17} h
1[a]	1.54	0.21	0.10	24	98	1.2
2	1.54	0.21	0.10	17	98	4.1
3[b]	1.33	0.21	0.11	27	104	1.2
4	1.33	0.21	0.11	27	104	4.1
5	1.50	0.21	0.10	14	105	1.2
EPRI Standard[c]	1.55	0.20	0.11	24	83	1.0

[a]Representative of Reactor A high-tin production cladding.
[b]Representative of low-tin cladding used in Reactors A and B.
[c]Autoclave tested only—not irradiated in Reactor A.

piece of specimen (about 0.2 g weight) was fused in a graphite crucible by impulse furnace heating. The evolved gases were collected by a flow of pure argon. The hydrogen concentration was deduced from the change in the thermal conductivity of the gas mixture. National Bureau of Standards (NBS) calibration standards with known low hydrogen levels and α-titanium specimens containing higher hydrogen levels were used for the instrument calibration. The same technique for hydrogen measurement was used by Laboratory B, except for the use of a different hydrogen analyzer (Strohlein H-mat 251).

Poolside Oxide Thickness Measurement Technique

A nondestructive eddy-current technique was used to measure the zirconium oxide film thickness on irradiated fuel rods. The measurements were performed during refueling outages in the reactor spent fuel pool. The primary components used to perform the measurements are an eddy-current probe that contacts the surface of the rod and a Fischerscope® that measures variations in the impedance of the electrical field that is established between the probe and a fuel rod [37]. In addition, the measurement system includes a computerized data acquisition system and mechanical systems that align the probe in contact with the surface of the rod and translate the rod past the probe. Calibration of the eddy-current probe was frequently checked in the pool using autoclaved Zircaloy-4 tube specimens (same geometry as the fuel cladding) having oxide thicknesses encompassing the expected oxide thickness range on the fuel rods. The thickness of oxide on each of the autoclaved specimen was determined out of water using mylar, films of known thickness. It was determined that the variation of tin level in the autoclaved standards within the composition range from 1.3 to 1.6% Sn did not have a significant effect on the measurements.

Fuel rods were measured by one of two methods: either the fuel rods were removed from the assembly and then measured, or the rods were measured while residing in the assembly (that is, peripheral rods). Both methods generate a linear trace at a particular azimuthal orientation of the rod. The position along the rod of each thickness measurement is determined by an axial position transducer, and these data are mated with thickness data in a two-dimensional array of oxide thickness and location. For rods measured after removal from the assembly, multiple scans at selected azimuthal orientations (that is, 0°, 90°, 180°, 270°) result in an axial and circumferential map of oxide thickness for each rod. An axial profile at a single azimuthal orientation is obtained for rods measured in an assembly.

The oxide thickness traces were analyzed to determine the maximum oxide thickness for each rod. This thickness was computed by averaging all thickness measurements 1.25 cm above and below the specific elevation. The maximum oxide thickness observed along the length of a fuel rod is generally taken as a key parameter in determining its operational capability relative to cladding corrosion. Therefore, the maximum oxide thickness data on the measured fuel rods have been used in this study.

Results

The long-term autoclave corrosion weight gain results for the cladding variants are presented in Fig. 1. A comparison of corrosion weight gain results of Variants 1, 2, and 5 shows that for high tin Zircaloy-4 (~1.55% Sn), for the investigated range of annealing parameter, there was no significant difference in the 633 K water autoclave test results. These results are in agreement with the long-term 673 K steam autoclave test results of Schemel et al. [38]. The EPRI standard exhibited the highest weight gain. Although its tin level corresponds to the other high-tin variants, it was fabricated to a different specification requiring a higher as-fabricated strength level and at an earlier time frame compared to the rest of the variants listed

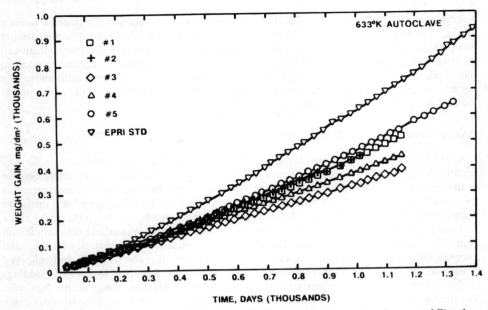

FIG. 1—*Corrosion weight gain as a function of autoclave test exposure time for several Zircaloy-4 tube variants.*

in Table 1. In recent years, the chemical homogeneity, changes in annealing practice and uniformity of Zircaloy-4 tube production lots have improved resulting in an improved corrosion resistance of Zircaloy-4. The low-tin Variants 3 and 4 (1.33% Sn) show lower weight gain values than the high-tin variants. The best corrosion resistance (that is, least weight gain) is exhibited by Variant 3, and the weight gain values are about 20% lower than those for the high-tin variants (1, 2, and 5). In comparison to the EPRI standard, after 1100 days of 633 K water autoclave exposure, Variant 3 shows a 45% lower weight gain.

The post-autoclave corrosion test cladding hydrogen concentration measurements and the calculated pickup fractions are presented in Table 2. A comparison of pickup fraction values for the high-tin Variants 1, 2, and 5 and the EPRI standard shows that the annealing parameter variations associated with the different annealing practices do not appear to affect the pickup fraction. The hydrogen pickup fraction for high-tin cladding is in the range of about 26 to 30%. The pickup fractions for the low-tin Variants 3 and 4 are in the same range, implying that the tin level variation within the investigated range has a negligible effect on the hydrogen pickup fraction. These latter measurements also show that, with respect to the annealing process variations within the investigated range for the low-tin Zircaloy-4, there is no significant effect on the hydrogen pickup fraction.

The hydrogen concentration values for specimens measured at Laboratory B are about 12% lower than the values on adjacent pieces of the same specimen that were measured at Laboratory A. This difference may be attributed to the differences in measuring equipment used between the two laboratories. The relative ranking of hydrogen concentrations among the variants, however, is the same for the measurements done at the two laboratories.

The results of the poolside measurements of cladding oxide thickness at Reactor A (on production fuel rods fabricated with cladding with high, intermediate, and low-tin Zircaloy-4) during successive refueling outages are presented in Fig. 2. The cladding tubes included in

TABLE 2—*Post-autoclave corrosion test (633 K water) hydrogen concentration measurements.*

Cladding Variant	Duration, days	Weight Gain after Corrosion Test, mg/dm^2	Measured Hydrogen,[a] ppm		Hydrogen Pickup Fraction, % (based on average H concentration)
			Laboratory A	Laboratory B	
1	808	335	609, 606, 602	516, 508	26.2
2	808	325	616, 581, 612	508, 539	26.8
3	1331	457	913, 822, 917	757, 757	26.5
4	1331	522	1006, 971, 999	846, 816, 842	26.3
5	524	217	469, 465	338, 324	29.9
EPRI Standard	661	371	775, 782	589, 573	25.8

[a]Oxide layer on specimen was not removed.

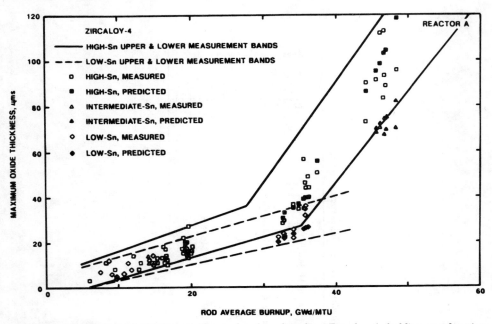

FIG. 2—*Maximum oxide thickness (measured and predicted) on Zircaloy-4 cladding as a function of rod average burnup for fuel rods irradiated in PWR-A.*

this figure, were fabricated over a range of time periods. The high burnup (37 to 48 GWd/MTU) fuel rods used early production batches (EPB) that were fabricated from 1981 to 1982. At that time, chemical homogeneity and annealing process uniformity was not as rigorously enforced as in the later years. This contributed to a wide scatterband in oxide thickness values. Cladding for the low burnup (<37 GWd/MTU) fuel rods were fabricated in 1987 with a fabrication history similar to Variants 1 and 3 listed in Table 1. Cladding oxide thicknesses were measured on a large number of fuel rods. The upper and lower bounds for the high-tin cladding shown in Fig. 2 represent oxide measurements on 138 fuel rods fabricated with high-tin cladding. Similarly, the scatterband for the low-tin cladding oxide thickness represents measurements on 32 individual rods with low-tin cladding (Variant 3). Individual oxide thickness values shown in Fig. 2 represent specific fuel rods selected (from a much larger database) for corrosion modeling. Details of corrosion modeling including other relevant information regarding Reactor A are discussed in a later section. Although an oxide thickness difference between low-tin and high-tin cladding at rod averaged burnup less than 20 GWd/MTU is not apparent, at higher burnups, there is a consistent trend of lower oxide thickness for low-tin cladding compared to the high-tin cladding. At a burnup of about 35 GWd/MTU, the average oxide thickness for low-tin cladding is about 40% lower than the average for the high-tin cladding.

Similar measurements for fuel rods fabricated with cladding with high-tin and low-tin Zircaloy-4 and irradiated in Reactor B are shown in Fig. 3. The databands for the high-tin and low-tin cladding rods represent individual measurements on 29 and 35 fuel rods, respectively. At a burnup of about 45 GWd/MTU in Reactor B, low-tin cladding shows about 30% lower oxide thickness than the high-tin cladding.

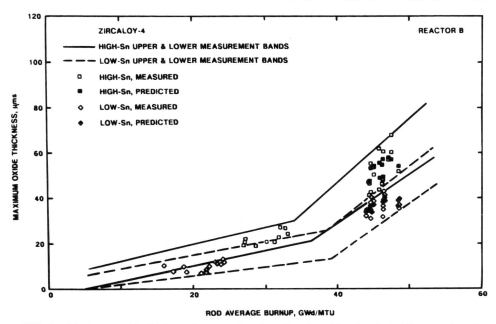

FIG. 3—*Maximum oxide thickness (measured and predicted) on Zircaloy-4 cladding as a function of rod average burnup for fuel rods irradiated in PWR-B.*

The cladding variants listed in Table 1 were irradiated side by side in the same assemblies in Reactor A to minimize the effect on corrosion of differences in power history and thermal-hydraulic environment. The cladding oxide thicknesses measured on these variants are presented in Fig. 4. Different fuel rods in the same fuel assembly had minor variations in the rod average burnup. For this reason, the oxide thickness for different rods of each cladding variant was averaged and this average value was associated with the average rod burnup for the specific cladding variant. The best-estimate correlation for oxide thickness versus rod average burnup for the high-tin cladding in Reactor A is presented in Fig. 4 for comparison. The best-estimate correlation was obtained by curve fitting the high-tin data shown in Fig. 2. As expected, the data points associated with Variants 1, 2, and 5 fall close to the high-tin curve in Fig. 4. The data points for the low-tin Variants 3 and 4 fall below the curve, confirming the in-PWR benefit of low-tin material indicated by the 633 K water autoclave results. At a burnup of about 35 GWd/MTU, the low-tin Zircaloy-4 has oxide thickness values 30 to 40% lower than those for high-tin cladding. Consistent with the 633 K water autoclave results, no significant effect of the annealing parameter variation (over the range studied) is seen in the in-reactor oxide thickness measurements for either group of cladding.

Discussion

Effect of Tin Level on the Oxide Thickness

The long-term water autoclave weight gain data and in-PWR oxide thickness measurement data presented in this paper demonstrate the improvement in the uniform corrosion resistance of Zircaloy-4 with a decrease in the tin content of the material. The extent of improvement appears to be greater in-reactor than what is indicated by the 633 K water autoclave test

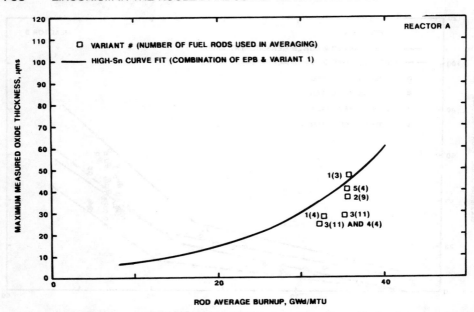

FIG. 4—*Maximum measured cladding oxide thickness as a function of fuel rod average burnup for Zircaloy-4 variants compared to the high-tin curve.*

results. This may be related to the effect of tin level on the microstructure of irradiated Zircaloy or the hydrogen uptake of the irradiated Zircaloy.

The uniform corrosion of zirconium alloys in high temperature water is postulated to be controlled by oxygen ion migration (via a vacancy exchange mechanism) through the oxide layer from the coolant/oxide interface to the oxide/metal interface [27]. Additions of Sn^{3+} ions (which replace a Zr^{4+} ion) to the oxygen deficient ZrO_{2-x} lattice increase the O^{2-} vacancy concentration in the oxide lattice and, thereby, increase the oxygen diffusion in the oxide lattice. The corrosion rate of the binary alloy is therefore increased [27,39]. Conversely, a reduction in tin content is expected to reduce the O^{2-} vacancy concentration in the oxide lattice and, thereby, reduce the oxygen transport through the oxide. Thus, the beneficial effect of lower tin level on the corrosion resistance of Zircaloy-4 may be related to lowering of the O^{2-} ion vacancy concentration in the oxide layer.

However, Zircaloy-4 is an alloy more complex than a zirconium-tin binary. A more applicable mechanism may also involve interaction with nitrogen. It has been known for a long time that the effect of tin on the uniform corrosion resistance of Zircaloy-4 is related to the nitrogen level in the Zircaloy-4 [27,31–34,39,40]. Corrosion rate enhancement due to nitrogen is believed to result from the creation of an O^{2-} ion vacancy by the presence of an N^{3-} ion instead of an O^{2-} ion. Although in binary zirconium alloys additions of tin [27–29] and nitrogen [28,41] individually degrade the corrosion resistance, the addition of both elements together in a ternary zirconium alloy appears to negate the detrimental effect of the other alloying element on the corrosion resistance [27,42]. This combined effect of tin-nitrogen (Sn-N) additions to Zircaloy is postulated to be similar to the observed, combined effect of tin and bismuth additions on the uniform corrosion resistance of zirconium-bismuth-tin (Zr-Bi-Sn) ternary alloys [43].

When both Sn^{3+} and N^{3-} ions are present, these ions are expected to form a complex with the oxygen ion vacancy and, thereby, reduce the mobility of the vacancy. Such a reduction in the vacancy mobility decreases the oxygen migration and improves the corrosion resistance of the ternary alloy. The possibility of a Sn^{3+}-N^{3-}-oxygen ion vacancy complex formation is consistent with the size of the ions. The ionic radii are [44]: Zr^{4+}, 0.079 nm; Sn^{3+} >0.071 nm, <0.102 nm, and probably about 0.080 nm; N^{3-}, 0.171 nm; and O^{2-}, 0.140 nm. An association of an oxygen ion vacancy with a Sn^{3+} ion in place of a comparable size Zr^{4+} ion and a neighboring, slightly larger N^{3-} ion at an O^{2-} ion lattice site will reduce the overall volumetric strain associated with individual O^{2-} vacancies and N^{3-} and Sn^{3+} ions.

The observed variability of the effect of tin on the uniform corrosion resistance of Zircaloy-4 (for example, only 15% improvement in corrosion resistance with respect to normal tin materials reported in Ref 20 as opposed to larger improvements observed in this investigation) is likely to be caused, at least partly, by differences in the nitrogen concentration present in these materials. Unfortunately, the nitrogen levels in different Zircaloy-4 materials investigated are rarely reported. The nitrogen concentration for the variants used in the current investigation are listed in Table 1 and are fairly constant at 14 to 27 ppm. It is suggested that an optimization of the tin and nitrogen levels in Zircaloy may be beneficial to improve the uniform corrosion resistance of the Zircaloy. The optimum tin level can be selected on the basis of the strength and creep resistance requirements of the Zircaloy. The nitrogen level then can be adjusted to negate the deleterious effect of the selected tin level on the corrosion properties of the Zircaloy.

Effect of Tin Level on Hydrogen Uptake

Hydrogen uptake data presented in Table 2 indicate that the variation of tin level within the Zircaloy-4 composition range does not alter the hydrogen uptake significantly. The small variation in hydrogen uptake fraction values listed in Table 2 is considered to be within the experimental measurement accuracy. The hydrogen uptake value for autoclaved specimens listed in Table 2 is higher than the upper limit on hydrogen uptake (18%) observed in PWRs [45]. The hydrogen generated by the corrosion process builds up in a static autoclave and, thereby, the hydrogen uptake by the metal in a static autoclave is higher. Under the flow conditions of a refreshed autoclave [46] or a PWR, a part of the generated hydrogen is swept away by the coolant flow and the hydrogen pickup fraction for the metal is lower than that in a static autoclave.

The hydrogen migration from the water/oxide interface to the metal/oxide interface is probably by the migration of H^+ ions (protons) along the oxide grain boundaries [47]. As the oxide porosity increases with the increase in the oxide thickness, the mode of hydrogen transport through the oxide layer is expected to change. The location within the oxide where H^+ reacts with the electron (moving in the opposite direction) influences whether hydrogen will be lost to the coolant flow or gets charged into the metal. The electron migration may occur through the oxide matrix or via the second phase particles, which are better conductors than the oxide for the electron transport. Consider the transport through the oxide (primarily via grain boundaries) first. If the proton combines with an electron (moving in the opposite direction) close to the oxide/water interface, there is a greater chance for the charge-neutralized hydrogen atom to be carried away by the coolant, and the hydrogen uptake of the metal would be lower [48,49]. This is likely to occur if the *n*-type semiconductor zirconia [50] has excess electrons available due to the replacement of a Zr^{4+} ion by a higher valency ion such as Nb^{5+}. This argument suggests that a replacement of a Zr^{4+} ion by a trivalent tin ion would increase the chance of hydrogen migration towards the metal. This will lead to a higher hydrogen uptake fraction with increasing tin content. The tin range investigated in the current work is probably not large enough to observe this change in the hydrogen uptake fraction. It is also possible that the tin level may influence the solubility of iron and chromium in Zircaloy-4 [28]. The variation of tin level in Zircaloy-4, therefore, may affect the extent of the second-phase particle precipitation. A change in the size and volume fraction of second-phase particles may change the short-circuit transport paths for electrons and, thereby, affect the hydrogen transport. Additional experimental work on irradiated Zircaloy-4 fuel cladding with variations in tin content is needed to resolve the mechanisms of in-PWR hydrogen uptake. The limited available data on irradiated Zircaloy-2 [51] are not directly applicable to the modern Zircaloy-4 used in PWRs due to irradiation under oxygenated coolant conditions and the absence of a beta-quenching step in the old Zircaloy-2 included in the hot-cell examination [51].

Correlation Between Autoclave Corrosion Results and In-PWR Cladding Oxide Thickness Measurements

The autoclave corrosion weight gain data presented in Fig. 1 show the following trend in descending order of weight gain: Variant 5 (highest weight gain), Variants 1 and 2 together, and Variants 4 and 3 (with the lowest weight gain). It is to be noted that the difference in weight gain between Variants 5, 1, and 2 is small. The in-PWR oxide thickness measurements (Fig. 4) for the same cladding variants rank essentially in the same order as the autoclave measurements. In-PWR measurements seem to segregate into two groups. One group is the high-tin data for Variants 1, 2, and 5, and the other group is the low-tin data of Variants 3 and 4. Small variations within each group are difficult to interpret, but could be associated with minor differences in power history and local thermal hydraulic conditions. This data comparison supports the suggestion [36] that a long-term 633 K water autoclave corrosion test is a good screening test for ranking the in-PWR corrosion resistance of Zircaloy-4 variants.

Effect of Annealing Process Variation

A comparison of Variants 1 and 2 is useful to evaluate the effect of annealing process variation in high-tin Zircaloy-4. The autoclave data in Fig. 1 shows no difference in weight gain for these two variants, while the in-PWR oxide thickness data (Fig. 4) show that a higher annealing parameter may improve the corrosion resistance slightly. A similar comparison for low-tin cladding variants shows that the lower annealing parameter, Variant 3, has slightly better corrosion resistance in the autoclave tests, while the in-PWR oxide thickness data (Fig. 4) show no effect of annealing process variation. These comparisons show that any effect of annealing parameter variation (within the investigated range) on the corrosion rate, if it exists, is minor. Relative insensitivity of cladding corrosion to annealing parameter variation in the range of approximately 1 to 4×10^{-17} h (with $Q/R = 40\,000°K$) is consistent with the results reported by Garzarolli et al. [5].

Corrosion Modeling

Representative subsets of fuel rods measured for oxide thicknesses from two relatively high coolant inlet temperature reactors were used to perform data comparisons with an analytical corrosion model. The purpose was to analytically describe the differences in corrosion due to cladding tin content. Table 3 delineates the characteristics of the subsets addressed. Eight of the ten rods in Subset A were measured for oxide after two cycles of irradiation as well as after three cycles of irradiation. Subset AA is composed of those eight rods.

Reactor A has a nominal coolant inlet temperature of about 569 K and operated with reactor cycle lengths of from about a year to 18 months. Reactor B's coolant inlet temperature was slightly lower, at about 563 K. Reactor B also operated with reactor cycles of slightly larger than 18 months. Reactor A and B both had an average subchannel coolant mass velocity of

TABLE 3—*Cladding fabrication parameters and irradiation conditions for fuel rods of different subsets included in the corrosion modeling.*

Subset	Reactor	Reactor Cycles Irradiated	Number of Rods for Data Comparison	Range of Rod Average Burnup, GWd/MTU	Cladding,[a] tin content	Fabrication Corresponds to Variant Number from Table 1	Pre-Exponential Multiplier,[b] Normalized to ESCORE
A	A	3	10	37.3 to 48.1	high	EPB[c]	1.31
AA	A	2	8	22.5 to 31.3	high	EPB	1.03
B	A	3	5	45.4 to 48.1	intermediate	EPB	1.08
C	A	2	10	32.6 to 36.1	high	Variant 1	0.84
D	A	2	10	32.0 to 36.0	low	Variant 3	0.60
E	B	2	19	44.4 to 48.4	high	Variant 1	0.92
F	B	2	21	44.0 to 48.6	low	Variant 3	0.68

[a] High tin content range, 1.51 to 1.70% by weight; intermediate tin content range, 1.45 to 1.50% by weight; and low tin content range, 1.20 to 1.44% by weight.
[b] The indicated pre-exponential multipliers provided best-fit predictions of the data shown in Figs. 2 and 3.
[c] EPB = early production batches fabricated in 1981 and 1982.

about 3730 Kg/m² s. Since prediction of oxide thickness at high burnup was of primary interest, high burnup rods from both Reactor A and B were selected. It should be noted, however, that the cladding for high burnup rods in Reactor A was fabricated about five years earlier than that for the high burnup fuel rods irradiated in Reactor B. Condensed power histories for the data comparison rods are provided in Table 4.

The ESCORE PWR corrosion model [35] was incorporated in a version of the ABB-CENF FATES fuel performance code [52]. The ESCORE PWR corrosion model describes corrosion as a two-stage process. An initial period of nonlinear time dependence (pre-transition) is followed by a period of linear time dependence (post-transition). The thickness at transition is dependent on temperature at the metal oxide interface and generally lies in the range of 2 to 3 μm for the oxide thicknesses of interest. The corrosion code accounts for the detailed fuel rod power history and thermal hydraulic environment.

A best fit was first independently obtained for each subset of rods with high-tin cladding. The best fit for each subset was derived by adjusting the pre-exponential coefficient (C) for the post-transition oxidation rate expression of the form

$$\frac{ds}{dt} = C \exp(-Q/RT)$$

The coefficients that yielded the best fit for the oxide thickness values associated with representative fuel rods irradiated to the indicated burnup interval of the subsets analyzed are listed in Table 3. Addressing first the subsets of highest burnup high-tin clad fuel rods (A and E), different coefficients were needed to obtain the best fits. It is worth noting that the difference in the metal-oxide interface temperature that may be present between Subsets A and E is accounted for by the exponential term in the oxidation rate equation. Therefore, Subsets A and E are expected to be predicted by essentially the same pre-exponential term if there was no difference in the material characteristics between these two groups of fuel cladding. The observed difference suggests some other difference in material beyond the nominal tin content reported. It is believed that at least part of this difference is related to the time period of cladding tube fabrication for Subsets A and E. The cladding for the rods in Subsets A (and

TABLE 4—*Power history for fuel rods included in the corrosion modeling.*

Reactor	Subset	Reactor Cycle of Irradiation	Cycle Length, EFPDs[a]	Range of Average LHGR for Cycle for Data Comparison Rods, W/cm
A	A, B	1	445	83 to 186
		2	285	186 to 220
		3	500	150 to 169
A	AA	1	445	83 to 182
		2	285	186 to 220
A	C, D	1	285	135 to 217
		2	500	216 to 230
B	E, F	1	535	212 to 245
		2	515	181 to 200

[a]EFPD = effective full power days.

AA) and B were fabricated from 1981 to 1982, well before the impact of annealing conditions on the corrosion resistance was recognized. On the other hand, cladding for the rods in Subsets C, D, E, and F were fabricated in late 1985 to early 1986, when the importance of annealing history was beginning to receive significant attention by the cladding fabricators. In addition, in the earlier period, the ingot chemical homogeneity probably was not at the level associated with later production. Because Subset A was fabricated much earlier than Subset E, the cladding for Subset A is believed to offer less corrosion resistance than Subset E. Because it is corroding more rapidly, Subset A rods have reached an oxide thickness beyond which corrosion is even further accelerated. The fuel cladding corrosion rate acceleration at high burnups shown in Figs. 2 and 3 under heat flux conditions is similar to the corrosion rate acceleration observed in pressure tube materials [51] operating without heat flux that has been attributed to the "thick film" hypothesis [51]. Subset E is corroding less rapidly than Subset A because of its improved fabrication, and Subset E has not reached or has not significantly exceeded that oxide thickness at which further corrosion acceleration occurs. The difference in the multipliers between Subsets AA and C is also attributed primarily to the fabrication time period differences between Subsets AA and C.

A comparison of the best-fit multipliers for Subset A and AA suggests that corrosion has accelerated during the third cycle of irradiation for these rods. The reason for the corrosion rate acceleration may be hydride precipitation at the metal-oxide interface [36]. In such a case, a feasible approach for modeling the enhancement would be to include a second transition point that is a function of oxide thickness to cause significant hydride precipitation, followed by corrosion associated with a higher rate compared to that of the rate after the first transition. Another possibility is degradation of oxide thermal conductivity due to radiation damage and growth stresses that eventually lead to oxide spalling. The current corrosion model in ESCORE does not include modeling of such high-burnup phenomena to cause acceleration.

To describe the effect of cladding tin content, the subsets of rods with different tin contents were compared only with subsets of rods with comparable irradiation histories from the same reactor and with similar cladding fabrication periods. Subset B was compared with Subset A; Subset D with C, and Subset F with E. For each pair of high-tin and lower tin content cladding subset, the post-transition corrosion rate derived for the subset of high-tin cladding was reduced to best predict the lower tin content cladding subset. Subset A's post-transition corrosion rate was reduced to predict Subset B's data, etc. The post-transition corrosion rate was fit independently for each subset of rods with low-tin cladding.

To predict lower oxide thickness for the low-tin Zircaloy, two adjustments were considered to the post-transition corrosion equation: lowering the pre-exponential factor (C) or an increase of the activation energy (Q). Lowering of the pre-exponential term is consistent with fewer anion vacancies introduced in the oxide layer in a low-tin cladding. Increasing the activation energy for lower tin content also can be justified on the basis of the high temperature (823 to 1173 K) oxidation data of Mallet and Albrecht [50]. However, changing the activation energy may make the model predictions more sensitive to change in temperature. Since the predicted metal/oxide interface temperature of the fuel rods investigated in this work did not vary significantly, it was decided to incorporate the tin content effect in the pre-exponential term (C) of the post-transition corrosion rate equation.

The pre-exponential multiplier for low-tin cladding was found to vary within the range of 71 to 74% of the high-tin cladding. For intermediate-tin cladding, the multiplier was about 82% of that for the high-tin cladding. Table 3 contains the pre-exponential multipliers that gave the best fit for the highest burnup rod in each subset of rods. The values listed are normalized with respect to the pre-exponential multiplier of ESCORE [35].

Predicted oxide thicknesses for the selected high burnup rods for Reactors A and B are shown in Figs. 2 and 3 along with the measured values. Generally, the agreement between measured oxide thickness and model predictions is good.

Conclusions

1. The in-PWR corrosion resistance of Zircaloy-4 at high burnups improves by 30 to 40% when the tin content is reduced to the lower limit of the ASTM composition range. Additional benefit may be possible by optimizing the tin and nitrogen concentration levels in Zircaloy-4.
2. Based on the autoclave test results, tin level within the examined range does not appear to affect the hydrogen uptake in Zircaloy-4.
3. The corrosion resistance improvement with lower tin levels can be modeled with an ESCORE-type corrosion model by adjusting the pre-exponential term in the post-transition corrosion rate equation. Reduction of the pre-exponential term with decreasing tin content, needed to provide a good prediction of the data, is consistent with the mechanism of a lower O^{2-} ion vacancy concentration in the zirconia lattice with a lower tin level in Zircaloy-4.
4. Results of 633 K long-term water autoclave tests correlate well with the in-PWR cladding corrosion resistance of the investigated cladding variants.

Acknowledgments

The authors are grateful to Sandvik Special Metals (SSM) for supplying the tubes with different fabrication histories. Stimulating discussions with E. R. Bradley of SSM and A. B. Johnson, Jr., of Battelle Pacific Northwest Laboratories are acknowledged. The authors are thankful to the Electric Power Research Institute for providing standard Zircaloy-4 material for corrosion testing. The authors thank the following colleagues at ABB Combustion Engineering Nuclear Fuel for different tasks: J. F. Kraus (data reduction) and P. A. Nelson (hydrogen analyses). The efforts of N. E. Adams and F. V. Rossano of ABB Combustion Engineering Nuclear Services, regarding autoclave testing are acknowledged. The authors acknowledge with thanks the efforts of P. Rudling, ABB ATOM, Sweden, for independent hydrogen analysis of autoclaved specimens.

References

[1] Garzarolli, F. and Stehle, H., "Behavior of Structural Materials for Fuel and Control Elements in Light Water Cooled Power Reactors," *Improvements in Water Reactor Fuel Technology and Utilization, Proceedings,* IAEA Symposium, Stockholm, 15–19 Sept. 1986, IAEA-SM-288/24, International Atomic Energy Agency, Vienna, 1987, pp. 387–407.
[2] Weidinger, H. G. and Lettau, H., "Advanced Material and Fabrication Technology for LWR Fuel," *Improvements in Water Reactor Fuel Technology and Utilization, Proceedings,* IAEA Symposium, Stockholm, 15–19 Sept. 1986, International Atomic Energy Agency, Vienna, 1987, pp. 451–467.
[3] Eucken, C. M., Finden, P. T., Trapp-Pritsching, S., and Weidinger, H. G., "Influence of Chemical Composition on Uniform Corrosion of Zirconium-Base Alloys in Autoclave Tests," *Zirconium in the Nuclear Industry: Eighth International Symposium, ASTM STP 1023,* L. F. P. Van Swam and C. M. Eucken, Eds., American Society for Testing and Materials, Philadelphia, 1989, pp. 113–127.
[4] Thorvaldsson, T., Andersson, T., Wilson, A., and Wardle, A., "Correlation between 400°C Steam Corrosion Behavior, Heat Treatment, and Microstructure of Zircaloy-4 Tubing," *Zirconium in the Nuclear Industry Eighth International Symposium, ASTM STP 1023,* L. F. P. Van Swam and C. M. Eucken, Eds., American Society for Testing and Materials, Philadelphia, 1989, pp. 128–140.
[5] Garzarolli, G., Steinberg, E., and Weidinger, H. G., "Microstructure and Corrosion Studies for Optimized PWR and BWR Zircaloy Cladding," *Zirconium in the Nuclear Industry: Eighth International Symposium; ASTM STP 1023,* L. F. P. Van Swam and C. M. Eucken, Eds., American Society for Testing and Materials, Philadelphia, 1989, pp. 202–212.
[6] Rudling, P., Pettersson, H., Andersson, T., and Thorvaldsson, T., "Corrosion Performance of Zircaloy-2 and Zircaloy-4 PWR Fuel Cladding," *Zirconium in the Nuclear Industry: Eighth Interna-*

tional Symposium, ASTM STP 1023, L. F. P. Van Swam and C. M. Eucken, Eds., American Society for Testing and Materials, Philadelphia, 1989, pp. 213–226.

[7] Sabol, G. P., Kilp, G. R., Balfour, M. G., and Roberts, E., "Development of a Cladding Alloy for High Burnup," *Zirconium in the Nuclear Industry: Eighth International Symposium, ASTM STP 1023,* L. F. P. Van Swam and C. M. Eucken, Eds., American Society for Testing and Materials, Philadelphia, 1989, pp. 227–244.

[8] Kilp, G. R., Thornburg, D. R., and Comstock, R. J., "Improvements in Zirconium Alloy Corrosion Resistance," *Proceedings,* IAEA Technical Committee Meeting on Fundamental Aspects of Corrosion of Zirconium-Base Alloys in Water Reactor Environments, International Atomic Energy Agency, Vienna, 1990, IWGFPT/34, ISSN 1011-2766, Paper 13.

[9] Glazkov, A. G., Grigor'ev, V. M., Kon'kov, V. F., Moinov, A. S., Nikulina, A. V., and Sidorenko, V. I., "Corrosion Behavior of Zr-Nb-Sn-Fe Alloy," *Proceedings,* IAEA Technical Committee Meeting on Fundamental Aspects of Corrosion of Zirconium-Base Alloys in Water Reactor Environments, International Atomic Energy Agency, Vienna, 1990, IWGFPT/34, ISSN 1011-2766, Paper 14.

[10] Thomazet, J., Mardon, J. P., Charquet, D., Senevat, J., and Billot, P., "Fragema Zirconium Alloy Corrosion Behavior and Development," *Proceedings,* IAEA Technical Committee Meeting on Fundamental Aspects of Corrosion of Zirconium-Base Alloys in Water Reactor Environments, International Atomic Energy Agency, Vienna, 1990, IWGFPT/34, ISSN 1011-2766, Paper 24.

[11] Grigoriev, V. M., Nikulina, A. V., and Peregud, M. M., "Evolution of Zr-Nb Base Alloys for LWR Fuel Clads," *Proceedings,* IAEA Technical Committee Meeting on Fundamental Aspects of Corrosion of Zirconium-Base Alloys in Water Reactor Environments, International Atomic Energy Agency, Vienna, 1990, IWGFPT/34, ISSN 1011-2766, Paper 26.

[12] Graham, R. A. and Eucken, C. M., "Controlled Composition Zircaloy-2 Uniform Corrosion Resistance," *Zirconium in the Nuclear Industry: Ninth International Symposium, ASTM STP 1132,* C. M. Eucken and A. M. Garde, Eds., American Society for Testing and Materials, Philadelphia, 1991, pp. 279–303.

[13] Isobe, T. and Matsuo, Y., "Development of Highly Corrosion Resistant Zirconium-Base Alloys," *Zirconium in the Nuclear Industry: Ninth International Symposium, ASTM STP 1132,* C. M. Eucken and A. M. Garde, Eds., American Society for Testing and Materials, Philadelphia, 1991, pp. 346–367.

[14] Harada, M., Kimpara, M., and Abe, K., "Effect of Alloying Elements on Uniform Corrosion Resistance of Zirconium-Based Alloys in 360°C Water and 400°C Steam," *Zirconium in the Nuclear Industry: Ninth International Symposium, ASTM STP 1132,* C. M. Eucken and A. M. Garde, Eds., American Society for Testing and Materials, Philadelphia, 1991, pp. 368–391.

[15] Riter, G. L., Van Swam, L. F. P., Kilian, D. C., and Yates, J., "Summary Overview of ANF's Fuel Performance," *Proceedings,* ANS-ENS International Topical Meeting on LWR Fuel Performance, Avignon, France, 21–24 April 1991, pp. 84–93.

[16] Melin, P., Gautier, B., and Combette, P., "Behavior of Fragema Fuel in Power Reactors," *Proceedings,* ANS-ENS International Topical Meeting on LWR Fuel Performance, Avignon, France, 21–24 April 1991, pp. 122–133.

[17] Charquet D., Gros, J. P., and Wadier, J. F., "The Development of Corrosion Resistant Zirconium Alloys," *Proceedings,* ANS-ENS International Topical Meeting on LWR Fuel Performance, Avignon, France, 21–24 April 1991, pp. 143–152.

[18] Fuchs, H. P., Garzarolli, F., and Weidinger, H. G., "Cladding and Structural Material Development for the Advanced Siemens PWR Fuel "FOCUS," *Proceedings,* ANS-ENS International Topical Meeting on LWR Fuel Performance, Avignon, France, 21–24 April 1991, pp. 682–690.

[19] Massih, A. R. and Rudling, P., "Corrosion Behavior of Zircaloy-2 and Zircaloy-4 Claddings in Pressurized Water Reactors," *Proceedings,* ANS-ENS International Topical Meeting on LWR Fuel Performance, Avignon, France, 21–24 April 1991, pp. 716–729.

[20] Kilp, G. R., Balfour, M. G., Stanutz, R. N., McAtee, K. R., Miller, R. S., Bowman, L. H., Wolfhope, N. P., Ozer, O., and Yang, R. L., "Corrosion Experience with Zircaloy and ZIRLO in Operating PWRs," *Proceedings,* ANS-ENS International Topical Meeting on LWR Fuel Performance, Avignon, France, 21–24 April 1991, pp. 730–741.

[21] Anada, H., Kuroda, T., and Shida, Y., "Corrosion Behavior of Zircaloy-4 with Several Hot Rolling and Annealing Conditions," in this volume.

[22] Mardon, J. P., Charquet, D., and Senevat, J., "Optimization of PWR Behavior of Stress-Relieved Zircaloy-4 Cladding Tubes by Improving the Manufacturing and Inspection Process," in this volume.

[23] Isobe, T., Matsuo, Y., and Mae, Y., "Micro-Characterization of Corrosion Resistant Zirconium-Based Alloys," in this volume.

[24] Nystrom, A-L., and Bradley, E. R., "Microstructure and Properties of Corrosion Resistant Zr-Sn-Fe-Cr-Ni Alloys," in this volume.

[25] Rudling, P., Mikes-Lindback, M., Lethinen, B., Andréa, H.-O., and Stiller, K., "Corrosion Performance of New Zircaloy-2 Base Alloys," in this volume.
[26] Sabol, G. P., Comstock, R. J., Weiner, R. A., Larouere, P., and Stanutz, R. N., "In-Reactor Corrosion Performance of ZIRLO and Zircaloy-4," in this volume.
[27] Thomas, D. E., "Corrosion in Water and Steam," *Metallurgy of Zirconium*, B. Lustman and F. Kerze, Jr., Eds., McGraw Hill, New York, 1955, pp. 608–640.
[28] Kass, S., "The Development of Zircaloys," *Corrosion of Zirconium Alloys, ASTM STP 368*, American Society for Testing and Materials, Philadelphia, 1964, pp. 3–27.
[29] Schleicher, H. W., *Werkstoffe and Korrosion*, Vol. 11, No.11, 1960, p. 691.
[30] Hillner, E., "Corrosion of Zirconium-Base Alloys—An Overview," *Zirconium in the Nuclear Industry, ASTM STP 633*, A. L. Lowe, Jr., and G. W. Parry, Eds., American Society for Testing and Materials, Philadelphia, 1977, pp. 211–235.
[31] Janitschek, F., *Atom Kernenergie*, Vol. 5, No. 6, 1960, p. 222.
[32] Ito Goro, et al., *Kagaku-kenkynjo Hokoku*, Abstracts, 1960–1963, pp. 35–38, 5–6.
[33] Kiselev, A. A. et al., *Proceedings, Corrosion of Reactor Materials*, Vol. 2, International Atomic Energy Agency, Vienna, 1962, p. 67.
[34] Ambartsumyan, R. S. et al., *Proceeding*, Second International Conference on the Peaceful Uses of Atomic Energy, Geneva, P/2044, Vol. 5, 1958.
[35] Fancher, R., Fiero, I., Freeburn, H., Garde, A., Kennard, M., Krammen, M., Smerd, P., and Yackle, N., "ESCORE—the EPRI Steady-State Core Reload Evaluator Code: General Description," EPRI Report EPRI NP-5100, Electric Power Research Institute, Palo Alto, Feb. 1987.
[36] Garde, A. M., "Enhancement of Aqueous Corrosion of Zircaloy-4 Due to Hydride Precipitation at the Metal-Oxide Interface," *Zirconium in the Nuclear Industry: Ninth International Symposium, ASTM STP 1132*, C. M. Eucken and A. M. Garde, Eds., American Society for Testing and Materials, Philadelphia, 1991, pp. 566–594.
[37] Goddard, H. D. and Weber, R. G., "Nondestructive Measurement of Zirconium Oxide Corrosion Films on Irradiated Zircaloy Clad Fuel Rods," presented at 1981 ANS Summer Meeting, 29th Conference on Remote Systems Technology, 7–12 June 1981, Bal Harbor, FL, CE Power Systems, TIS-6837, Windsor, CT, 1981.
[38] Schemel, J. H., Charquet, D., and Wadier, J. F., "Influence of the Manufacturing Process on the Corrosion Resistance of Zircaloy-4 Fuel Cladding," *Zirconium in the Nuclear Industry: Eighth International Symposium, ASTM STP 1023*, L. F. P. Van Swam and C. M. Eucken, Eds., American Society for Testing and Materials, Philadelphia, 1989, pp. 141–152.
[39] Chirigos, J. and Thomas, D. E., "The Mechanism of Oxidation and Corrosion of Zirconium," *Proceedings*, AEC Metallurgy Conference, March 1952, Report TID-5084 and Report WAPD-53, p. 337.
[40] Parfenov, B. G., Gerasimov, V. V., and Venediktova, G. I., *Corrosion of Zirconium and Zirconium Alloys*, AEC-tr-6978, VC-25, TT-69-55014, Israel Program for Scientific Translations, Jerusalem, 1969.
[41] Emel'vanov, V. S. and Barkov, N. V., "Metallurgiya; Metallovedenie Chistykh Metallov," *Atomizdat*, Issue 4, (MIFI), Moskva, 1963.
[42] Rubenstein, L. S. et al., *Corrosion*, Vol. 18, No. 2 45t, 1962.
[43] Taylor, D. F., "An Oxide-Semi Conductance Model of Nodular Corrosion and its Application to Zirconium Alloy Development," *Journal of Nuclear Materials*, Vol. 184, 1991, pp. 65–77.
[44] *Lange's Handbook of Chemistry*, Thirteenth ed., J. A. Dean, Ed., McGraw Hill, New York, 1985, p. 3–126.
[45] Garde, A. M., "Effects of Irradiation and Hydriding on the Mechanical Properties of Zircaloy-4 at High Fluence," *Zirconium in the Nuclear Industry: Eighth International Symposium, ASTM STP 1023*, L. F. P. Van Swam and C. M. Eucken, Eds., American Society for Testing and Materials, Philadelphia, 1989, pp. 548–569.
[46] Maffei, H. P. and Shannon, D. W., "Hydrogen Pickup of Zircaloy-2 and Zircaloy-4 in Static and Refreshed Autoclaving Systems," HW67437, General Electric Company, Hanford, WA, Nov. 1960.
[47] *Corrosion of Zirconium Alloys in Nuclear Power Plants*, IAEA-TECDOC-684, International Atomic Energy Agency, Vienna, Jan. 1993.
[48] Taylor, D. F., Cheng, B., and Adamson, R. B., "Nodular Corrosion Mechanisms and Their Application to Alloy Development," *Proceedings*, IAEA Technical Committee Meeting on Fundamental Aspects of Corrosion of Zirconium-Base Alloys in Water Reactor Environments, International Atomic Energy Agency, Vienna, IWGFPT/34, ISSN 10011-2766, Paper 1, 1990.
[49] Franklin, D. G. and Lang, P. M., "Zirconium-Alloy Corrosion: A Review Based on an International Atomic Energy Agency (IAEA) Meeting, *Zirconium in the Nuclear Industry: Ninth International*

Symposium, ASTM STP 1132, C. M. Eucken and A. M. Garde, Eds., American Society for Testing and Materials Philadelphia, 1991, pp. 3–32.
[50] Mallet, M. W. and Albrecht, W. M., "High Temperature Oxidation of Two Zirconium-Tin Alloys," *Journal, Electrochemical Society,* Vol. 102, 1955, p. 407.
[51] Johnson, A. B., Jr., "Zirconium Alloy Oxidation and Hydriding Under Irradiation: Review of Pacific Northwest Laboratories' Test Program Results," NP-5132, Research Project 1250-4, Final Report, Electric Power Research Institute, Palo Alto, CA, April 1987.
[52] "Improvements to Fuel Evaluation Model," Report CEN-161 (B) -NP, Supplement 1-NP-A, ABB Combustion Engineering Nuclear Power, Jan. 1992.

DISCUSSION

N. Ramasubramanian[1] (written discussion)—(1) Why should tin be in solid solution in a lower valence state in ZrO_2? Assuming it is so, and that it is in a 3+ state, each tin atom replacing a zirconium in the lattice would produce an additional 0.5 anion vacancy than the thermally produced vacancy concentration. A 1.3Sn or a 1.5Sn would add about 10^{20} additional vacancies, a high concentration. How can a difference in corrosion of about two times be ascribed to a small difference in the increase in concentration?

(2) If we propose a solid-state model based on Sn^{2+} or Sn^{3+} substitution, then how do we reconcile with oxygen transport via vacancies at the oxide grain boundary?

A. M. Garde et al. (authors' closure)—

(1) The observed effect of tin level on the uniform corrosion resistance of Zircaloy-4 is consistent with an average valancy of Sn atoms dissolved in ZrO_2 to be lower than four. Tin can exist in a 2+ valancy[2] state and the possibility of 3+ state is based on the model proposed by Thomas (Ref 27 of the paper). A presence of fraction of Sn atoms at these lower valancy states will reduce the average valance of tin below four.

The number of vacancies in oxygen deficient zirconium oxide ($ZrO_{2-\delta}$) on 100 g of Zirconium at 300°C is estimated to be 6.86×10^{20}.[3,4] Assuming that the average valancy of Sn is 3+, for 100 g of 1.3% Sn, 3.29×10^{21} additional vacancies are introduced in the oxide phase. For 1.5% Sn, the corresponding number is 3.8×10^{21}. The actual number of vacancies introduced due to tin addition will depend on what fraction of tin atoms have a valancy state lower than four. In any case, the number of additional vacancies introduced by tin addition is of the same order of magnitude as the thermal vacancy concentration in the oxygen deficient oxide phase. Therefore it is possible to associate the lower corrosion resistance of high-tin Zircaloy-4 with the introduction of additional vacancies in the oxide due to the tin addition to zirconium.

(2) Information regarding the effect of tin level on the oxide grain boundary configuration is not available. It is possible that the tin addition to the oxide may increase the vacancy concentration at the grain boundary region also.

---

[1] Ontario Hydro, Toronto, Ontario, Canada.
[2] R. C. Weast, Ed., *CRC Handbook of Chemistry and Physics,* The Chemical Rubber Co., Cleveland, OH, 1970, pF-152.
[3] D. L. Douglas, *The Metallurgy of Zirconium,* International Atomic Energy Association (IAEA) Supplement 1971, p. 406.
[4] T. Smith, "Diffusion Coefficients and Anion Vacancy Concentrations for the Zirconium-Zirconium Dioxide System," *Journal of the Electrochemical Society,* Vol. 112, 1965, p. 560.

Torill Karlsen[1] and Carlo Vitanza[1]

Effects of Pressurized Water Reactor (PWR) Coolant Chemistry on Zircaloy Corrosion Behavior

REFERENCE: Karlsen, T. and Vitanza, C., "**Effects of Pressurized Water Reactor (PWR) Coolant Chemistry on Zircaloy Corrosion Behavior,**" *Zirconium in the Nuclear Industry: Tenth International Symposium, ASTM STP 1245,* A. M. Garde and E. R. Bradley, Eds., American Society for Testing and Materials, Philadelphia, 1994, pp. 779–789.

ABSTRACT: A number of the test programs currently in progress at the Halden Project are aimed at evaluating Zircaloy cladding performance in terms of corrosion behavior, particularly at high burnup.

A pressurized water reactor (PWR) facility has been used to determine the effects of high (4 to 4.5 ppm) lithium on the corrosion behavior of Zircaloy-4 cladding material exposed to nucleate boiling (1% void) and to one-phase cooling conditions. The test employed four fuel rod segments, base-irradiated to an average burnup of 28.5 MWd/kg UO_2 in a commercial power plant, and with average initial oxide layers of either 10, 20, or 40 μm, respectively. Four sets of oxide layer thickness measurements were performed on the four rods during the course of the investigation. Pre-test eddy current measurements were performed on the as-received segments and two interim sets of measurements were made after 80 and 245 full power days of irradiation. Final oxide measurements were made after 425 full power days, when maximum rod burnup was 45 MWd/kg UO_2.

Average oxide layer increases of 30 and 55 μm were observed for the rods exposed to one-phase cooling and nucleate boiling conditions, respectively. Although maximum oxide thickness exceeded 95 μm, no evidence of spalling was apparent. The measured oxide layer thicknesses compared favorably with predicted values based on model calculations derived from power plant data, with little evidence of enhanced corrosion rates in the presence of increased lithium concentrations.

KEY WORDS: zirconium alloys, lithium hydroxide, surface boiling, in-pile tests, model prediction, zirconium, nuclear materials, nuclear applications, radiation effects

A common practice in operating pressurized water reactors (PWRs) is to maintain the coolant pH at 300°C at 6.9 in order to reduce the formation of crud on the fuel rods and thereby limit activation of the primary circuit. An increasing number of utilities are adopting even higher pH levels as a consequence of solubility studies [1,2] that indicate that optimum pH_{300} lies in the 7.2 to 7.4 range. Also, the coolant inlet temperature in modern PWRs may be raised to increase the power generation efficiency. Lithium (Li), a commonly used alkalizing agent has been shown to enhance the corrosion rate of Zircaloy in out-of-pile studies performed with concentrated solutions. As a result, in PWRs operating with an elevated lithium regime (3.2 to 3.5 ppm Li), there is some concern that, for highly rated rods where subcooled (nucleate)

---

[1] Research scientist and program manager, respectively, OECD Halden Reactor Project, 1781 Halden, Norway.

boiling occurs, lithium may enhance corrosion by becoming entrained in the oxide layer, particularly in cases where thick oxide layers with pores or cracks or both are present.

In order to address these issues, a PWR facility installed in the Halden Reactor was used to determine the effects of a high (4 to 4.5 ppm) lithium concentration on the corrosion behavior of high burnup Zircaloy-4 fuel rods subjected either to nucleate boiling or to one-phase cooling conditions. The irradiation rig consisted of two parallel flow channels containing four test segments that had been pre-irradiated in a commercial reactor. During the irradiation, close control of the chemistry and thermal-hydraulics were maintained. Oxide layer measurements were performed after 0, 80, 245, and 425 full power days of irradiation. Conclusions on the lithium effect were drawn by comparing the measured oxide thicknesses with the EPRI/C-E/KWU model predictions [3,4].

Experimental Details

In-Pile Test Section

A schematic representation of the PWR test facility is presented in Fig. 1. Coolant water enters the rig via four downcomer tubes and flows upwards into the two parallel coolant channels that have flow cross sections representative of those found in PWR subchannels. Two or more axially displaced fuel rod segments may be accommodated in each of the channels

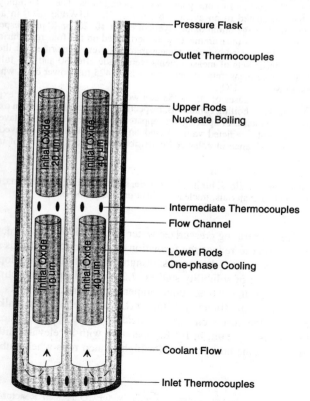

FIG. 1—*Schematic representation of in-pile test section of PWR facility.*

where coolant conditions are monitored by means of coolant thermocouples installed at the inlet, middle, and outlet of the test section. In the investigation described in this paper, each channel contained two pre-irradiated rod segments, referred to as the upper and lower rods, respectively.

The four fuel rods were taken from segmented rod assemblies that were base-irradiated in the Gösgen Nuclear Power Plant, Switzerland. The cladding, manufactured by NRG (Nuklearrohr-Gesellschaft), was standard stress-relief annealed Zircaloy-4 with a tin content of 1.5%. The rods were exposed to two irradiation cycles of 327 and 288 full power days (FPD) at an average linear heat rating of 22 kW/m. The accumulated average rod burnup obtained during the two pre-irradiation cycles was 28.2 MWd/kg UO_2. A summary of the fuel rod and cladding characteristics after pre-irradiation are presented in Table 1.

Following base irradiation, two of the segments had average oxide thicknesses of 10 and 20 μm, respectively, and two had an average thickness of about 40 μm, thereby enabling the effects of both lithium content and initial oxide thickness to be assessed in the study.

By adjusting the coolant inlet temperature (310°C), pressure (15 MPa), and the flow rate (2 m/s) in the test facility, nucleate boiling (with a calculated 1% void fraction) was created for the upper rods, located at the peak of the axial neutron flux. The two lower rods, experiencing lower coolant temperature and lower neutron flux, were subjected to one-phase cooling conditions.

For the upper rods, the time-averaged coolant temperature lay in the 315 to 325°C range, simulating conditions found in high-temperature PWRs, while for the lower rods, coolant temperature varied from 307 to 314°C, representative of typical PWRs. Based on results from an earlier PWR cladding corrosion study [5], in which thermocouples were attached to the cladding surface, cladding wall (that is, oxide-to-coolant) temperatures for the upper and lower rods could be estimated. The upper rods, operating under a nucleate boiling regime, have a fairly uniform cladding wall temperature of 346°C along the entire rod lengths. The cladding temperature for the lower rods ranged from 335 to 343°C. A schematic representation of the axial cladding wall and coolant temperatures is presented in Fig. 2.

TABLE 1—*Summary of fuel rod and cladding characteristics after pre-irradiation.*

	Lower Rods		Upper Rods	
	Channel 1	Channel 2	Channel 1	Channel 2
Alloy additions, % by weight				
Tin	...	1.49	...	...
Iron	...	0.22	...	...
Chromium	...	0.11	...	...
Oxygen	...	0.12	...	...
Impurities, ppm				
Carbon	...	132	...	...
Nitrogen	...	24	...	...
Silicon	...	<30	...	...
High cold work	...	74%	...	...
Final heat treatment	...	520°C/5 h electropolished	...	...
Outer surface finish	...		...	
Initial burn-up, MWd/kg UO_2	28.8	27.5	29.7	27.8
Initial oxide thickness, μm				
Minimum	9	38	18	39
Average	13	41	24	43
Maximum	14	43	28	44

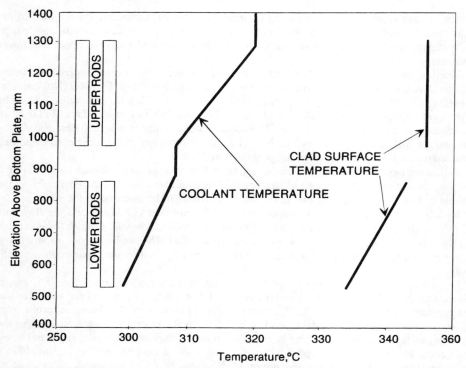

FIG. 2—*Schematic representation of coolant and cladding temperature profiles in the test section.*

The full time averaged linear heat rating of the upper and lower rods were 233 and 140 W/cm, respectively, with no discernible difference in power between the parallel channels (Table 2.).

Water Chemistry

The main water chemistry parameters of the facility are presented in Table 3. The 4 to 4.5 ppm lithium, added as LiOH and the 1000 ppm boron, added as H_3BO_3, resulted in a pH_{300} of 7.2 to 7.3. After 250 FPD, the boron concentration was reduced to 700 ppm (pH_{300} 7.4).

The concentrations of lithium and boron, monitored on a regular basis, as well as those of other chemical species measured less frequently, were maintained well within specification, and the water chemistry of the loop, in general, can be described as typical of operating PWRs.

TABLE 2—*Main operating parameters of PWR test facility.*

	Upper Rods	Lower Rods
Coolant velocity, m/s	1.9 ± 0.1	1.9 ± 0.1
Coolant pressure, MPa	14.9 ± 0.1	14.9 ± 0.1
Time-averaged coolant temperature, °C	315 to 325	307 to 314
Oxide-to-coolant temperature, °C	346	335 to 343
Time-averaged linear heat rating, W/cm	140	233

TABLE 3—*Water chemistry specifications for PWR loop system.*

Chemical Species	Concentration, ppm
Li	3.85 to 4.54
B	940 to 1060[a]
H_2	2.7 to 4.0
O_2	<0.005
Cl^-	<0.02
F^-	<0.02
SiO_2	<0.17
Fe^{2+}	5.3×10^{-3}
Ni^{2+}	2.7×10^{-4}

[a]Reduced to 700 ppm after 250 FPD.

Oxide Layer Thickness Measurements

An eddy current technique was used to measure the oxide layer thicknesses of the fuel rod segments. The measurements, performed in a handling compartment situated within the containment of the Halden Reactor, enabled the axial oxide layer profiles over the entire cladding surface to be mapped at 30° intervals around the rods. The accuracy of the eddy current measurements was empirically estimated to be ±2 μm.

The measured oxide layer thicknesses were compared with predicted values based on the semi-empirical EPRI/C-E/KWU model [3,4]. The post-transition correlation that is applicable to this study, where initial oxide thicknesses exceed 10 μm, is described by

$$dS/dt = (C + U(M\Phi)^{0.24}) \cdot \exp(-Q/RT) \qquad (1)$$

where

dS/dt = post transition corrosion rate (μm/day),
S = oxide layer thickness (μm),
$C = 8.04 \times 10^7$ (μm/day),
$U = 2.59 \times 10^8$ (μm/day),
$M = 7.46 \times 10^{-15}$ (cm²s/n),
Φ = fast neutron flux (n/cm²s),
$Q = 27\,354$ (cal/mole),
R = gas constant (cal/mole K), and
T = metal/oxide interface temperature (K).

The metal/oxide interface temperature was derived from local heat flux, cladding surface temperature, and an assumed thermal conductivity of 1.5 W/(mK) for the zirconium oxide (ZrO_2) film. The cladding surface temperatures we estimated from the measured coolant temperatures, flow velocity, and local heat rating by means of a relationship determined during commissioning tests.

Results and Discussion

A total of four sets of oxide layer measurements were performed on the rods. Initial oxide thicknesses were measured on the as-received segments and two interim oxide thickness measurements were made after 80 and 245 FPD of irradiation. A final set of measurements was performed after 425 FPD, when the maximum burnup of the rods was 45 MWd/kgUO$_2$.

Oxide layer increases of 55 μm were observed for both the upper rods despite the 20 μm difference in initial oxide layer thickness. The oxide layer traces recorded at an orientation of 180° for the upper rod with an initial oxide layer of 40 μm, are summarized in Fig. 3. Although maximum oxide thicknesses detected along some orientations exceeded 95 μm, no signs of oxide spalling were apparent.

A cross-sectional view of the oxide layer increases at mid-height for the rod are presented in Fig. 4. Circumferentially, increases in oxide thickness were uniform, indicating homogenous thermal hydraulic conditions at the fuel rod surface in the presence of a subcooled (nucleate boiling) regime.

For the lower rods, average oxide increases of 30 μm were observed. An example of the oxide profiles measured at 180° orientation for one of the lower rods is presented in Fig. 5.

The measured oxide thicknesses were compared with predicted values based on the model calculation. A comparison between the average measured and calculated oxide layer profiles for the upper rod with the highest initial oxide thickness is presented in Fig. 6. At the lower end of the rod, the measured oxide thickness was less than that predicted by the model. At the upper end of the rod, where the channel temperature is highest, the measured thickness was 20 to 30% higher than the model prediction. While the model was inclined to under and over-estimate the measured thickness to equal extents, the differences lie well within the accuracy predicted for the model on the basis of comparisons with in-pile and out-of-pile data [6].

The anomalously high oxide thicknesses recorded for the lower regions of the lower rods (Fig. 7) are attributed to preferential crud deposition in this region where the combined high pH and low coolant temperature may have resulted in a positive temperature coefficient of

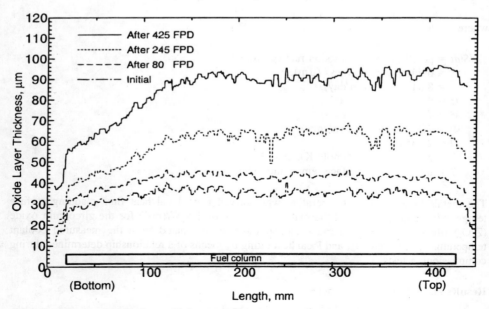

FIG. 3—*Measured oxide layer thickness profiles for upper rod with initial oxide layer thickness of 40 μm (180° orientation).*

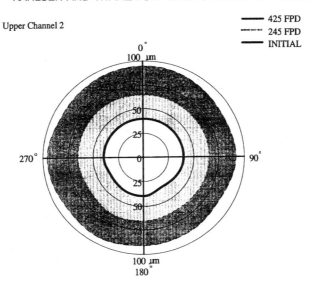

(FPD = Full Power Days)

FIG. 4—*Azimuthal oxide profiles at mid-height for upper rod exposed to surface boiling (FPD full power days.)*

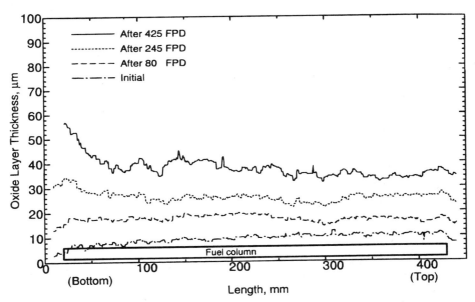

FIG. 5—*Measured oxide layer thickness profiles for one of the lower rods (180° orientation).*

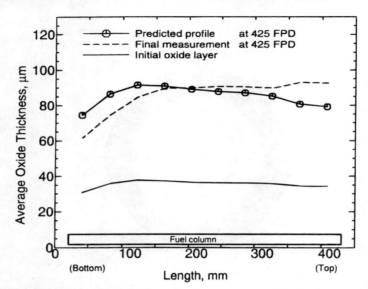

FIG. 6—*Comparison between predicted and final average measured oxide layer profiles for upper rod with surface boiling.*

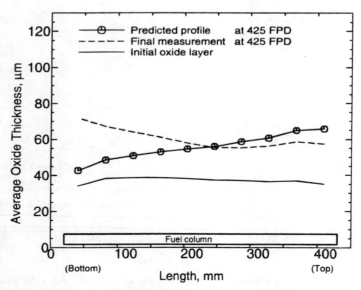

FIG. 7—*Comparison between predicted and final average measured oxide layer profiles for lower rod under one-phase cooling.*

solubility for nickel-ferrite-based crud [1]. Following the set of oxide thickness measurements performed after 245 FPD, the boron content in the coolant was reduced to 700 ppm (pH_{300} 7.4) in an attempt to increase crud solubility.

Post-irradiation examination of both the upper and the lower rods is to be undertaken in order to fully characterize both the oxide layers and the nature of the deposits at the lower ends of the lower rods.

The metal oxide interface temperature that rises as the oxide thickness increases is, according to model predictions, expected to result in accelerated rates of corrosion. Results from the present study, however, indicate that the effect is less pronounced than anticipated, as shown in Fig. 8 where the increases in oxide thickness for the two upper rods are compared. Despite initial differences in oxide thickness of the order of 20 µm, both rods display similar rates of oxide layer growth, with a comparatively minor dependence of corrosion rate on oxide layer thickness being observed in this study in which an oxide thermal conductivity of 1.5 W/(mK) was assumed. While this value for thermal conductivity resulted in good predictions for the upper rod with the thickest initial oxide layer, a lower thermal conductivity (1.1 to 1.2 W/(mK)) leads to better agreement between measured and predicted oxide growth rates in the other rods. Whether this is due to intrinsic differences in the properties of the oxide layers on the different segments, or to deficiencies in the model employed in the analyses, remains to be investigated. Further in-pile studies at Halden, aimed at measuring the oxide thermal conductivity directly, will help in clarifying this issue.

Summary and Conclusions

An in-pile PWR loop facility was used to determine the effects of 4 to 4.5 ppm Li (and a 700 to 1000 ppm boron level) on the corrosion behavior of Zircaloy-4 cladding material. The

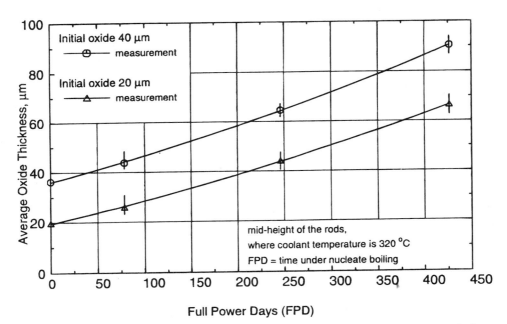

FIG. 8—*Oxide layer growth rates for the two upper rods that had average initial oxide layer thicknesses of 20 and 40 µm, respectively.*

fuel rod segments used in the study had been pre-irradiated to an average burnup of 28.5 MWd/kg UO_2 in a commercial reactor and had initial oxide thicknesses ranging from 10 to 40 μm. The rods, arranged as axially displaced pairs in two coolant flow channels, were exposed to one-phase cooling (lower rods) and surface boiling conditions (upper rods), respectively.

After 425 FPD of exposure (maximum burnup of 45 MWd/kg UO_2), the maximum detected oxide thickness for one of the upper rods exceeded 95 μm, with no signs of oxide spalling.

Measured oxide layer thickness values were compared with predictions based on the EPRI/C-E/KWU model. While the model appeared to substantially underestimate the oxide thicknesses at the lower ends of the lower rods, the discrepancy is attributed to crud deposits as opposed to a lithium enhancement effect.

In general, the predictions for the upper rods (where nucleate boiling is present) are in reasonable agreement with the measurements, with little evidence of substantial corrosion enhancement effects. The experimental data do, however, show that oxide growth is more pronounced towards the top of the upper rod, suggesting that a lithium effect may be present for hot conditions. At the upper end of the upper rod, where the channel temperature is highest, the measured oxidation rate was about 20 to 30% higher than predicted.

While the 20 to 30% higher corrosion rate lies within the scatterband of corrosion data used in developing the model, the possibility of a lithium effect is indicated, with the potential for more marked corrosion enhancement under hotter conditions.

The effects of initial oxide thickness and calculated metal/oxide interface temperature were studied by comparing of the rods located adjacent to one another in the flow channels. The results revealed that corrosion rates appear to be less sensitive to metal/oxide interface temperature than predicted by model calculations where an oxide thermal conductivity of 1.5 W/(mK) was assumed.

Post-irradiation examination of the test rods, scheduled for the near future, will include oxide layer characterization in terms of lithium entrainment and structural changes, as well as compositional analysis of the crud deposits at the lower end of the test section.

Studies of the effects of coolant chemistry on Zircaloy corrosion and crud deposition behavior are being continued. Currently, in the PWR facility, an investigation is in progress aimed at quantifying the effects of alloy composition and various lithium/boron ratios on the in-pile corrosion behavior of pre- and unirradiated Zircaloy fuel rod segments. Additional studies addressing the effects of applying chemical additives such as zinc to control radiation buildup in PWRs are also planned.

Acknowledgments

The authors would like to thank M. V. Polley (Nuclear Electric, UK) for useful discussions and are grateful for the experimental information provided by K. Fjellestad and T. Johnsen (Halden Project), T. Kido (NDC, Japan), and K. Ranta-Puska (VTT, Finland).

References

[1] Abe, K., Mizusaki, H., Ohta H., Hemmi, Y., Umehara R., Ooshima, S., Fukuda, F., and Kasahara, H., "Solubility Measurement of Crud and Evaluation of Optimum pH," *Proceedings*, 1991 JAIF International Conference on Water Chemistry in Nuclear Power Plants, JAIF, Inc., Fukui City, 1991.

[2] Lambert, I., Lecompte, J., Beslu, P., and Joyer, F., "Corrosion Product Solubility in the PWR Primary Coolant," *Water Chemistry of Nuclear Reactor Systems 4*, Vol. 1, British Nuclear Energy Society (BNES), London, 1986.

[3] Gazarolli, F., Jung, W., Garde, A. M., Parray, G. W., and Smerd, P. G., "Waterside Corrosion of Zircaloy Fuel Rods," EPRI-NP-2789, Electric Power Research Institute, Palo Alto, CA, 1982.

[4] Gazarolli, F., Bodmer, R. P., Stehle, H., and Trapp-Pritsching, S., "Progress in Understanding PWR Fuel Rod Waterside Corrosion," *Proceedings*, American Nuclear Society (ANS) Topical Meeting on Light Water Reactor Performance, Orlando, FL., 1985.

[5] Fjellestad, K., Gunnerud, P., Tsukada, T., and Vitanza, C., "Corrosion Research at the OECD Halden Reactor Project," *Proceedings,* 1988 JAIF International Conference on Water Chemistry in Nuclear Power Plants, JAIF, Inc., Tokyo, 1988.

[6] Billot, P. and Giordano, A., "Comparison of Zircaloy Corrosion Models from the Evaluation of In-Reactor and Out-of-Pile Loop Performance," *Zirconium in the Nuclear Industry: Ninth International Symposium, ASTM STP 1132,* C. M. Eucken and A. M. Garde, Eds., American Society for Testing and Materials, Philadelphia, 1991, pp. 539–565.

Author Index

A–D

Abe, H., 285
Adamson, R.B., 400, 499
Akiyama, T., 183
Aldridge, S.A., 221
Amanrich, H., 351
Anada, H., 307
Andrén, H.-O., 579, 599
Armijo, J.S., 3
Aubin, J.L., 245
Balakrishnan, P.V., 378
Barberis, P., 80
Beie, H.-J., 615, 644
Billot, P., 351
Bowden, J.W., 264
Bradley, E.R., 483
Bramwell, I.L., 450
Causey, A.R., 168, 202
Chan, P.K., 116
Charquet, D., 80, 328, 687
Cheng, B-C., 400
Choubey, R., 221
Chow, C.K.(P.), 135
Coffin, L.F., 3
Coleman, C.E., 35, 221, 264
Comstock, R.J., 724
Davies, P.H., 135
Davis, L., 221
Döbler, U., 644
Dorado, A.O., 559
Doubt, G.L., 264

E–G

Elder, J.E., 168, 202
Elmoselhi, M.B., 62
Endter, R.K., 760
Fleck, R.G., 168, 202
Fong, R.W.L., 264
Furugen, M., 285
Garde, A.M., 760
Garzarolli, F., 615, 709
Gilbon, D., 521
Giordano, A., 351
Girard, E., 245
Godlewski, J., 663
Graham, R.A., 221
Griffiths, M., 35, 135

H–K

Haddad, R.E., 559
Hama, T., 285
Haste, T.J., 450
Herb, B.J., 419
Holt, R.A., 168, 202
Hosbons, R.R., 135
Isobe, T., 437
Karlsen, T., 779
Khatamian, D.J., 116
Kim, Y.S., 745
Knop, A., 644
Koike, M.H., 183
Konishi, T., 285
Krammen, M.A., 760
Kruger, R.M., 400

L–N

Lai, Z., 579
Lansiart, S., 549
Larouere, P., 724
Lefebvre, F., 687
Lemaignan, C., 687
Lethinen, B., 599
Lindbäck, Mikes-, M., 80, 599
Mae, Y., 437
Mardon, J.-P., 328, 549
Matsuda, K., 285
Matsuo, Y., 437
McCarthy, J.M., 419
McIntyre, S., 62
Mikes-Lindbäck, M., 80, 599
Min, D.K., 745
Mitwalsky, A., 615
Moan, G.D., 221
Montmitonnet, P., 245
Motta, A.T., 687
Nagamatsu, K., 183
Nomoto, K.-I, 307
Nyström, A.-L., 483, 579

P–R

Parsons, P.D., 450
Pati, S.R., 760
Pearce, J.H., 98
Pêcheur, D., 687
Perkins, R.A., 469

Pettersson, H., 579
Peybernès, J., 351
Pickman, D.O., 19
Ramasubramanian, N., 378
Reynolds, M.B., 499
Rheem, K.S., 745
Robin, J.-C.C., 351
Root, J.H., 264
Rosenbaum, H.S., 3
Royer, J., 549
Rudling, P., 80, 579, 599
Ruhmann, H., 419, 615, 644

S–W

Sabol, G.P., 724
Sagat, S., 35, 264
Schrire, D.I., 98
Schumann, R., 709
Sell, H.-J., 615
Senevat, J., 328
Shann, S.-H, 469
Shibahara, I., 183
Shida, Y., 307
Smith, G.P., 760
Simonot, C., 521
Soniak, A., 549
Stanutz, R.N., 724
Steinberg, E., 709
Stiller, K., 599
Theaker, J.R., 221
Thomazet, J., 351
Urbanic, V.F., 116
Vitanza, C., 779
Wadman, B., 579
Waeckel, N., 549
Wang, C.T., 419
Warr, B.D., 62
Webster, R.T., 264
Weiner, R.A., 724
Wilkins, B.J.S., 35
Wisner, S.B., 499
Woo, O-T.T., 116
Worswick, D., 450

Subject Index

A

Accumulated annealing parameter, 437
Amorphization, 521, 687
Analytical scanning, 687
Anion vacancy migration, 760
Anisotropy, pressure tubes, 202, 469
Annealing parameter, 307, 328, 663, 760
Applied stress, 559

ASTM STANDARDS

B-531-92, 80

Autoclave testing, 307, 351, 483, 579, 724, 760
Axial fatigue, 499

B

Barrier fuel, 3
Barrier layer, 615
Beta quenched phase, 663
Boiling water reactors
    boiling conditions, 351
    corrosion behavior of zircaloy, 400, 599
    corrosion environment, 709
    creep anisotropy in zircaloy cladding, 469
    fatigue behavior of zircaloy and zirconium, 499
    heat treated pressure tube, 183
    zirconium barrier fuel, 3
Boric acid, 378
Boron, 378
Burnups, in pressurized water reactors, 438, 521, 760
Burst strength, welds in zirconium, 264

C

CANDU (Canada Deuterium Uranium) pressure tubes, 62, 116, 135, 168, 202
Calandria tubes in CANDU reactor, 264
Carbon, 221
Chemical composition, zircaloy oxidation, 80
Chlorine, 221
Chromium, in zirconium alloys, 438
Cladding, fuel, 3, 19
Cladding tubes, 549, 599
Cladding, zircaloy, 245, 285, 307, 328, 469, 599, 687, 745, 760
Cold pilgering, 245, 285
Compact tension specimen, fatigue behavior, 499
Contractile strain ratio, 469
Coolant chemistry in water reactors, effects on zircaloy corrosion behavior, 779
Corrosion, 19, 98, 116, 183
Corrosion acceleration, 378
Corrosion enhancement, 745
Corrosion inhibition, 378
Corrosion mechanism of zirconium alloys, 615, 644, 724, 760, 779
Corrosion modeling, 760
Corrosion resistance, zircaloy, 285, 307
Corrosion resistance, zirconium based alloys, 437, 483, 579, 709
Corrosion, zircaloy-2, 599
Corrosion, zircaloy-4, 351
Crack driving force, 35
Crack growth resistance, 135
Crack propagation of zircaloy-4, 559
Crack velocity, 35
Cracking, 35
Cracks (materials), zircaloy-2 tubing, 285
Creep (materials), 202, 469, 483, 521, 724
Crystallographic orientation, 559
Cumulative annealing parameter, 307, 328

D

Damage (materials), 245
Defected rods, 745
Deformation (materials), 419, 469
Delayed hydride cracking, 35, 264
Deuterium concentration, 62, 116
Diffusion, 116
Dislocations, 521
Dissolution, 521

E–F

Elongation, 168
Electron microscopy, 116

Examination methods, corrosion mechanism, 615, 644
Fabrication process, corrosion resistant zircaloy, 285, 307, 328
Failure mechanisms, nuclear fuel, 3
Fatigue behavior, zircaloy-4 tubes, 549
Fatigue crack growth, 499
Fatigue testing, 549
Fracture (materials), 221
Fracture resistance, 221
Fracture toughness, 135, 183
Fretting corrosion, 19
Fuel rods, cladding, 599, 709, 724, 745, 760
Fuel rod defects, 19
Fuel utilization in pressurized water reactors, 483
Fugen (165MWe), heavy water, boiling light water cooled reactor, 183

G–H

Granular oxides, 745
Growth, in-reactor corrosion performance of zircaloy, 724
Heat affected zone, 264
Heat treated pressure tube, 183
High pressure, 450
High temperature oxidation of zircaloy-4, 450
High temperature pressurized water reactors, 760
Hot rolling and annealing conditions, 307
Hydride cracking, 264
Hydride morphology, 98
Hydrides, 745
Hydriding, 19, 35, 80, 98, 599, 709
Hydrogen absorption, 579
Hydrogen analysis, 80, 116
Hydrogen concentration, 98
Hydrogen diffusion, 62, 116
Hydrogen in pressure tubes, 221
Hydrogen ingress, 62
Hydrogen mobility, 62
Hydrogen pickup, irradiation, 19, 183, 483
Hydrogen uptake in zirconium alloys, 62, 116, 760
Hydroxide, 724

I–K

In pile tests, 779
In reactor behavior, 328, 709
In reactor deformation, 469
In reactor fuel tests, 3
In reactor uniform corrosion, 745
Inspection process in manufacturing, 328
Intermetallic compounds, 307
Intermetallic precipitates
  corrosion mechanism of zirconium alloys, 615
  microstructure analysis, 419, 437, 483
  oxidation, 664, 687
Investigation methods, corrosion mechanism of zirconium alloys, 615, 644
Iodine stress corrosion cracking of zircaloy-4, 559
Ion mass spectroscopy, 116
Iron, 168
Iron, in zirconium alloys, 438
Irradiation, 135, 168, 183, 549, 687
Irradiation enhancement of corrosion and hydrogen pickup, 19
Irradiation fluence, 35
Irradiation growth, 521
Irradiation temperature, 35
J-R curves, 135
Kinetics, oxidation in zircaloy-4, 687

L–M

Large scale demonstration, 3
Light water reactors, 19, 400, 469
Lithium, 351, 378
Lithium hydroxide, 378, 779
Lithium hydroxide boric acid aqueous chemistry, 378
Loops, 351
Loss of coolant accidents, 19
Manufacturing process
  corrosion and stress behavior of zircaloy, 285, 307, 328
  pressure tubes, 221
  zircaloy, 285, 307, 328
Material chemistry effect, 709
Material samples, 709
Mechanical modeling, 245
Microchemistry, 400
Microscopy, 116

SUBJECT INDEX   795

Microstructure
  corrosion perameters, 351, 483, 615, 709
  effects of irradiation, 521
  in-reactor corrosion performance, 724
  oxide layers, 579
Model prediction, 779

N

Neutron fluence, 168, 183, 202, 499
Neutron irradiation, 521
Nickel, in zirconium alloys, 438
Niobium, zirconium- alloys, 116, 437, 579
Nodular corrosion, zircaloy, 285, 400, 419, 709
Nondestructive testing, 328
Nuclear applications
  anisotropy of pressure tubes, 202
  corrosion behavior, irradiated zircaloy, 400
  corrosion behavior, zircaloy, 285, 307, 599, 724, 760
  corrosion environment for boiling water reactors, 709, 745, 779
  corrosion mechanism of zirconium alloys, 615, 644
  corrosion of zircaloy-4, 351, 378, 400
  corrosion resistant zirconium based alloys, 437, 483
  crack propagation of zircaloy-4, 559
  creep anisotropy, 469
  damage in cold pilgering, 245
  fatigue behavior of zircaloy and zirconium, 499, 549
  fracture toughness, 135
  heat treated pressure tubes, 183
  hydride cracking, 35
  hydrogen absorption, 80
  hydrogen uptake, 62
  in-reactor corrosion performance of zircaloy, 724, 745
  iodine stress corrosion cracking of zircaloy-4, 559
  irradiation effect on zircaloy-4, 521
  irradiation growth in pressure tubes, 168
  light water reactors, 19
  microstructure analysis, 419
  nodular corrosion behavior, 419
  oxidation of zircaloy-4, 450, 687
  oxide layers, microstructure, 579
  pressure tubes, anisotropy, 202
  pressure tubes, fabrication, 221
  pressure tubes, fracture toughness, 135
  pressure tubes, heat treated, 183
  pressure tubes, irradiation growth, 168
  scanning electron microscope techniques, 98
  welds in zirconium, mitigation, 264
  zircaloy corrosion behavior, 98, 285, 307, 328, 400
  zirconium barrier fuel cladding, 3
  zirconium, oxide films, 116
Nuclear electricity, 549
Nuclear fuel
  hydride cracking, 35
  hydrogen absorption, 80
  hydrogen uptake, 62
  light water reactors, 19
  zirconium barrier cladding, 3, 19
Nuclear materials
  anisotropy of tubes, 202
  corrosion behavior of zircaloy, 285, 307, 328, 400, 760
  corrosion environment for boiling water reactors, 709
  corrosion mechanism of zirconium alloys, 615, 644, 724, 779
  corrosion of zircaloy-4, 351, 378
  corrosion resistant zirconium based alloys, 437, 483
  crack propagation of zircaloy-4, 559
  creep anisotropy, 469
  damage in cold pilgering, 245
  fatigue behavior of zircaloy and zirconium, 499, 549
  fracture toughness, 135
  fuel cladding, 3
  heat treated pressure tubes, 183
  hydride cracking, 35
  hydrogen absorption, 80
  in-reactor corrosion performance of ziraloy, 724
  irradiation effect on zircaloy-4, 521
  irradiation in pressure tubes, 168
  light water reactors, 19
  microstructure analysis, 419
  nodular corrosion behavior, 419
  oxidation of zircaloy-4, 450, 687

oxide layers, microstructure, 579
pressure tubes, anisotropy, 202
pressure tubes, fabrication, 221
pressure tubes, fracture toughness, 135
pressure tubes, irradiation growth, 168
pressure tubes, heat treated, 183
scanning electron microscopy techniques, 98
stress, zircaloy tubes, 328
welds in zirconium, mitigation, 264
zircaloy corrosion and hydriding, 98, 307, 328
zirconium, oxide films, 116
Nuclear reaction analysis, 116
Nuclear reactors, 202
Nuclear submarines, fuel elements, 19
Neutron irradiation, 19, 35

O–P

Oxidation behavior of zircaloy-4, 351, 378, 450, 687
Oxide barrier, hydrogen uptake in zirconium, 62
Oxide growth, 579
Oxide layers, corrosion mechanism of zirconium alloys, 615, 663, 709, 779
Oxide layers, microstructure, 579
Oxide lithium hydroxide, 724
Oxide metal interface, 579
Oxide scale, 307
Oxides, 116
Oxygen, 221
Oxygen reactivity, 644
Pellet cladding interaction, 3, 19
Phosphorous, 221
Photoelectron spectroscopy, 116
Pilgering, cold, 245, 285
Porosity, 62, 116
Post irradiation examination, 183
Post weld heat treatment, 264
Power cycling effects, 549
Power reactors, 351
Precipitates, 328, 400, 521
Pressure tubes
  anisotropy of reactor creep, 202
  fracture toughness, 135
  hydrogen uptake in zircaloy, 62
  irradiation growth, 168, 183

Pressure tubes, cold pilgering, 245
Pressure tubes, corrosion resistant zircaloy, 285, 328
Pressure tubes, manufacturing and inspection process, 328
Pressure tubes, trace elements, 221
Pressure tubes, zirconium alloy welds, 264
Pressurization, repeated, 549
Pressurized water reactors
  corrosion performance of zirconium alloys, 351, 483, 599, 709, 760, 779
  fatigue behavior, 499, 521
  fuel rods, cladding, 579, 760
  in reactor uniform corrosion, 745
  microstructure of oxide layers, 579
  oxidation of intermetallic precipitates, 687
  stress corrosion, 549
Processing creep, 724
Processing route, hydrogen absorption, 80

Q–R

Quenching, 285, 328
Q-value, 285
Radiation effects
  anisotropy of tubes, 202
  corrosion behavior of zircaloy, 285, 307, 328, 400, 779
  corosion mechanism of zirconium alloys, 615, 644, 709, 760
  corrosion of zircaloy-3, 599
  corrosion of zircaloy-4, 351, 724
  crack propagation of zircaloy-4, 559
  creep anisotropy, 469
  damage in cold pilgering, 245
  effect on microstructure of zircaloy-4, 521
  fatigue behavior of zircaloy and zirconium, 499, 549
  fracture toughness, pressure tubes, 135
  fuel elements, boiling water reactors, 709
  heat treated pressure tubes, 183
  hydride cracking, 35
  hydrogen absorption, 80
  hydrogen uptake, 62
  in-reactor corrosion of zircaloy, 724
  irradiation growth in pressure tubes, 168

SUBJECT INDEX 797

light water reactors, 19
lithium hydroxide and boric acid,
  corrosion of zircaloy-4, 378, 724
microstructure and properties of
  zirconium based alloys, 483
oxidation of zircaloy-4, 450, 687
oxide films on zirconium-niobium
  alloys, 116
oxide layers, microstructure, 579
pressure tubes, anisotropy, 202
pressue tubes, fabrication, 221
pressure tubes, fracture toughness, 135
pressure tubes, heat treated, 183
pressure tubes, irradiation growth, 168
scanning electron microscope
  techniques, 98
welds in zirconium, 264
zircaloy corrosion and hydriding, 98
zirconium barrier cladding, 3, 19
Reactor fuel, 3, 19
Recrystallized state, 663
Reduction in area, zircaloy tubing, 285
Repeated pressurization, 549
Residual stress, 264, 663

S-T

Scanning electron microscope techniques,
  98, 116
Second phase particles, 599s
Secondary ion mass spectrometry, 62
Single crystals, 559
Solutes, 400
Spectroscopy, 116
Steam corrosion, zircaloy oxidation, 80,
  450
Stress corrosion, 3, 19, 328, 559
Stress gradients, 663
Surface boiling, 779
Surface defects, 245
Surveillance specimens, 183
Temperature, 168
Tensile properties, corrosion resistant
  zirconium based alloys, 437, 483
Tensile strength, 183
Tensile stress, 663
Tensile stress, welds in zirconium
  alloys, 264
Tetragonal zirconia, 663
Texture (materials), 202, 419

Thermal reactor, 183
Threshold stress intensity factor, 35
Tin content, zircaloy-4 cladding, 760
Tin, effect on corrosion resistance under
  irradiation, 328
Tin, in zirconium alloys, 438
Tool design, 245
Trace elements, 221
Transmission electron microscopy, 307,
  419, 579, 615, 687
Tube shells, 285
Tubes, 202
Twin boundaries, 559

U-X

Uniform corrosion
  in reactor, 745, 760
  irradiated zircaloy, 400
  manufacturing and inspection process,
    328
  oxide layers, 579
  under hot rolling and annealing
    conditions, 307
  zirconium based alloys, 437, 483, 615
Uniform protective oxide, 400
Water cooled nuclear reactors, 521
Water, in-reactor corrosion performance of
  zircaloy, 724
Water rods, 285
Welds, in zirconium alloy components,
  264, 400
X-ray absorption spectroscopy, 644
X-ray diffraction, 663
X-ray microanalysis, 98, 116
X-ray photoelectron spectroscopy, 644

Z

Zr-2.5Nb pressure tubes
  anisotropy of in reactor creep, 202
  corrosion, 378
  fabrication, 221
  fracture toughness, 135
  heat treated, 183
  hydrogen uptake, 62
  irradiation growth, 168
  oxide films, 116
Zircaloy corrosion and hydriding, 98
Zircaloy-4 cladding, 285, 307, 328, 378,
  760

Zircaloy-4, corrosion performance, 724, 745, 779
Zircaloy-4, effect of irradiation on microstructure, 521, 687
Zirconia, 687, 760
Zirconium
  anisotropy of pressure tubes, 202
  cladding tubes, 307, 328, 351, 599
  coolant chemistry in water reactor, effect on zircaloy corrosion, 779
  corrosion behavior, 307, 351, 378, 400, 599, 724, 760
  corrosion environment for boiling water reactors, 709, 724
  corrosion mechanism of alloys, 615, 644, 663
  corrosion resistant alloys, 437, 483
  corrosion resistant tubing, 285
  crack propagation, 559
  creep anisotropy in zircaloy cladding, 469
  damage in cold pilgering, 245
  fatigue behavior, 499, 549
  fracture toughness, 135
  heat treated pressure tube, 183
  hydride cracking, 35
  hydrogen absorption kinetics, 80
  hydrogen uptake, 62
  in reactor corrosion, 745
  intermetallic precipitates, 687
  iodine stress corrosion cracking, 559
  irradiation effect on microstructure, 521, 687
  irradiation growth in pressure tubes, 168
  light water reactors, 19
  manufacturing and inspection process, 328
  microstructure analysis, 419
  nodular corrosion behavior, 419
  oxidation in steam, 80
  oxidation of zircaloy-4, 450, 687
  oxide films, 116
  oxide formation, 644, 663
  oxide layers, 579
  pressure tubes, anisotropy, 202
  pressure tubes, fracture toughness, 135
  pressure tubes, heat treated, 183
  scanning electron miocroscope techniques, 98

trace elements in pressure tubes, 221
welds, mitigation of harmful effects, 264
zirconium barrier cladding, 3, 19
Zirconium alloys
  anisotropy of pressure tubes, 202
  cladding tubes, 307, 328, 351, 599
  composition, 499
  coolant chemistry in water reactors, effect on zircaloy corrosion, 779
  corrosion behavior, 307, 351, 378, 400, 599, 724, 760
  corrosion environment for boiling water reactors, 709, 724, 779
  corrosion mechanism, 615, 644, 663
  corrosion resistance, 437, 483
  corrosion resistant tubing, 285
  crack propagation mechanisms, 559
  creep anisotropy in cladding, 469
  damage in cold pilgering, 245
  fatigue behavior, 499, 549
  fracture toughness of pressure tubes, 135,
  fuel cladding, 3
  heat treated pressure tube, 183
  hydride cracking, 35
  hydrogen absorption kinetics, 80
  hydrogen uptake, 62
  in reactor corrosion of zircaloy-4, 745
  intermetallic precipitates, 687
  iodine stress corrosion cracking, 559
  irradiation effect on microstructure, 521, 687
  irradiation growth in pressure tubes, 168
  microstructure analysis, 419
  nodular corrosion behavior, 419
  oxidation in steam, 80
  oxidation of zircaloy-4, 450, 687
  oxide films on zirconium-niobium, 116
  oxide formation, 644, 663
  oxide layers, 579
  oxides, 116
  performance in light water reactors, 19
  pressure tubes, 135, 168, 183, 202
  scanning electron microscope techniques, 98
  stress corrosion cracking, 559
  trace elements in Zr-2.5Nb pressure tubes, 221

welds, mitigation of harmful effects, 264
Zirconium barrier fuel cladding, 3
Zirconium niobium alloys, 116
Zirconium oxide surface chemistry, 378
Zirconium oxides, 62
Zirconium X-bar, 437
ZIRLO. *See* Zircaloy-4.